Student's t (single mean)

$$t = \frac{\bar{x} - \mu_0}{s/\sqrt{n}}$$

Student's t (comparing two means)

$$t = \frac{(\bar{x}_1 - \bar{x}_2) - D_0}{s\sqrt{\dfrac{1}{n_1} + \dfrac{1}{n_2}}}$$

Pooled Estimate of σ^2

$$s^2 = \frac{\sum_{i=1}^{n_1} (x_{1i} - \bar{x}_1)^2 + \sum_{i=1}^{n_2} (x_{2i} - \bar{x}_2)^2}{(n_1 - 1) + (n_2 - 1)}$$

Correlation Coefficient

$$r = \frac{S_{xy}}{\sqrt{S_{xx}S_{yy}}}$$

where

$$S_{yy} = \sum_{i=1}^{n} (y_i - \bar{y})^2 = \sum_{i=1}^{n} y_i^2 - \frac{\left(\sum_{i=1}^{n} y_i\right)^2}{n}$$

$$S_{xx} = \sum_{i=1}^{n} (x_i - \bar{x})^2 = \sum_{i=1}^{n} x_i^2 - \frac{\left(\sum_{i=1}^{n} x_i\right)^2}{n}$$

$$S_{xy} = \sum_{i=1}^{n} (x_i - \bar{x})(y_i - \bar{y}) = \sum_{i=1}^{n} x_i y_i - \frac{\left(\sum_{i=1}^{n} x_i\right)\left(\sum_{i=1}^{n} y_i\right)}{n}$$

Least Squares Estimators of β_0 and β_1

$$\hat{\beta}_1 = \frac{S_{xy}}{S_{xx}} \quad \text{and} \quad \hat{\beta}_0 = \bar{y} - \hat{\beta}_1 \bar{x}$$

Least Squares Line (single independent variable)

$$\hat{y} = \hat{\beta}_0 + \hat{\beta}_1 x$$

Mann-Whitney U

$$U_1 = n_1 n_2 + \frac{n_1(n_1 + 1)}{2} - T_1 \quad \text{and} \quad U_2 = n_1 n_2 + \frac{n_2(n_2 + 1)}{2} - T_2$$

Chi-Square Statistic for Contingency Tables

$$X^2 = \sum_{i=1}^{k} \frac{(n_i - np_i)^2}{np_i}$$

Spearman's Rank Correlation Coefficient

$$r_s = 1 - \frac{6 \sum_{i=1}^{n} d_i^2}{n(n^2 - 1)}$$

Take 2 sets of measurements
Look at the difference as means. Are locally
[illegible] in samples, and we need to be sure.
Proceed to the same way as last time. [illegible] we
[illegible] [illegible] by [illegible] at the distribution of means
[illegible] [illegible] [illegible]

Now, we use the properties of the
distribution of difference is normal. The mean we have is given
[illegible] [illegible] [illegible] [illegible]
the properties of the sampling distribution of
the difference $\bar{x}_1 - \bar{x}_2$ has a mean of $\mu_1 - \mu_2$,
and a s.d. of

~~[illegible] $= 3.7$, and σ~~
Control.
3.7 4.3 0.8

NINTH EDITION

Introduction to Probability and Statistics

WILLIAM MENDENHALL
University of Florida, Emeritus

ROBERT J. BEAVER
University of California, Riverside

Duxbury Press
An Imprint of Wadsworth Publishing Company
Belmont, California

Duxbury Press
An Imprint of Wadsworth Publishing Company
A division of Wadsworth, Inc.

Assistant Editor *Jennifer Burger*
Editorial Assistant *Michelle O'Donnell*
Production *Cecile Joyner/The Cooper Company*
Designer *Cloyce Wall*
Print Buyer *Barbara Britton*
Copy Editors *Betty Duncan and Micky Lawler*
Cover Designer *Harry Voigt*
Cover Photograph *Grant V. Faint/The Image Bank*
Compositor *Interactive Composition Corporation*
Printer *R. R. Donnelley & Sons*

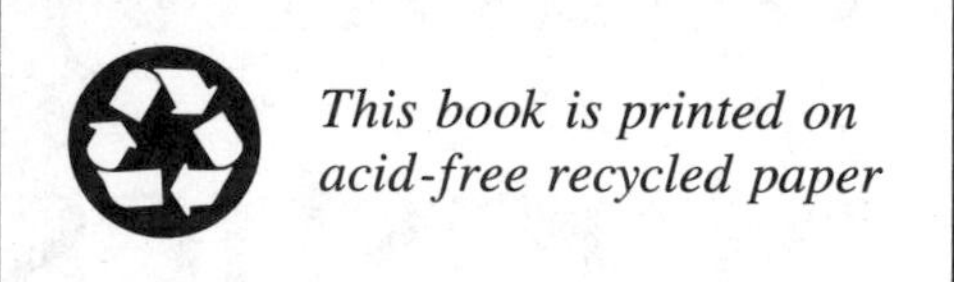

International Thomson Publishing
The trademark ITP is used under license

Printed in the United States of America
1 2 3 4 5 6 7 8 9 10—98 97 96 95 94

Library of Congress Cataloging-in-Publication Data

Mendenhall, William
Introduction to probability and statistics / William Mendenhall, Robert J. Beaver—9th ed.
p. cm.
"An imprint of Wadsworth Publishing Company."
Includes index.
ISBN 0-534-20886-X
1. Mathematical statistics. 2. Probabilities. I. Beaver, Robert J. II. Title.
QA276.M425 1994
519.5—dc20 93-24217

CONTENTS

PREFACE

In this ninth edition of *Introduction to Probability and Statistics,* we have attempted to revise the text to improve the students' understanding of statistics and its use in everyday life, as well as its use in scientific research and experimental situations. This edition, like its predecessors, presents statistical inference, the **objective of statistics**, as its theme. In the preparation of the ninth edition, our primary thrust has been to make the material more relevant and to increase its pedagogical effectiveness. To this end, we have integrated the discussion of statistical software programs for personal computers throughout the text, using and relying on computer printouts of statistical analyses when appropriate. We have revised exercise sets to provide a broad range of practical, relevant exercises, many of which are drawn from current periodicals and journals. These exercises will be meaningful both for those whose careers will utilize the results of statistical analysis of data and for others who merely need to make sense of the reports and claims that appear daily in newspapers, magazines, and television presentations.

The following are some specific changes in this edition:

1 A rewritten unnumbered introductory chapter opens with an invitation for the reader to experience the language of statistics as it is encountered in our everyday lives.

2 A new Chapter 1 focuses on graphical techniques in describing sets of measurements. Statistical tables and various kinds of graphs are used in describing and comparing sets of univariate and bivariate data. The latter half of the chapter deals with relative frequency histograms and stem and leaf plots.

3 In Chapter 2, dealing with numerical descriptive measures, the shortcut calculational formula for s^2 is now integrated into the section in which the sample variance is defined. The new emphasis is presently on the use of summary measures in description rather than on their calculation.

4 The intersection of two events is now denoted as $A \cap B$ rather than AB throughout Chapter 3. Bayes' Rule now appears along with counting rules in a supplement at the end of the chapter. Tree diagrams and probability tables are new to this edition.

5 Discussion of the discrete uniform and geometric distributions in Chapter 4 and the continuous uniform and exponential distributions in Chapter 5 has been eliminated.

6 Random sampling has been moved to the beginning of Chapter 6, "Sampling Distributions." An introduction to statistical process control in the form of $\bar{x}$ and p control charts is a new addition to this chapter. The sampling distributions for the difference in sample means and sample proportions have been deleted from Chapter 6. This information, however, is summarized and inserted in appropriate sections of Chapter 7.

7 The term *bound on error* has been replaced by the more commonly used term *margin of error* in Chapter 7. The introductory discussions in Chapters 7 and 8 have been streamlined to avoid unnecessary notation.

8 Computer printouts are now integrated throughout Chapters 10–13 rather than appearing in the last section of the chapter.

9 Chapters 10 and 11, "Linear Regression and Correlation" and "Multiple Regression Analysis," have been rewritten with an emphasis on understanding and interpreting computer printouts for regression analyses. Calculational formulas for linear regression are given in a separate section at the end of Chapter 10. Revised examples in Chapter 11 begin with a multiple regression of the selling price of a home as a function of four independent variables, continue with a quadratic polynomial model, and conclude with the use of dummy (indicator) variables in comparing two simple regression lines.

10 The thrust of the major changes in Chapter 13, "The Analysis of Variance," consists of using computer printouts throughout the chapter, rather than step-by-step calculation of various sums of squares. Calculational formulas are given in a separate optional section following each design for those who wish to verify the computer results. A section concerning the use of Tukey's procedure in comparing pairs of means is new to Chapter 13.

11 The F tables have been replaced by an alternative table in which right-tail critical values corresponding to tail areas of .10, .05, .025, .01 and .005 are given as a single cell entry for various combinations of υ_1 and υ_2 degrees of freedom.

12 Case studies in Chapters 2, 7, 9–12, and 14 are new to this edition.

13 Exercises throughout the book have been revised through elimination, revision, and replacement of dated exercises. However, there are still a large number and variety of exercises. A new kind of exercise, "Exercises Using the Data Disk," which utilize the three data sets on the accompanying data disk, follow the supplementary exercises at the end of most chapters.

Exercises are graduated in level of difficulty so that all students can solve some of them, a substantial number can solve most, and a few of the best students are challenged to solve all without error. Symbols are used to identify areas of application such as business, engineering, education, the social and biological sciences, and so on. The coding used to indicate the area of application appears in the inside front cover.

Color-coded exercise numbers indicate that the solutions to those exercises appear in the *Student Study Guide and Partial Solutions Manual*, which is available to students. Also, starred exercises (*) signal those problems that are optional.

The authors are grateful to Alex Kugushev and the editorial staff of Duxbury for their patience, assistance, and cooperation in the preparation of this edition. Special thanks go to Barbara Beaver for reading, editing, and typing manuscript material and for her diligence in preparing the solutions manual for this and previous editions. Thanks are also due to Dennis Kimzey, Rogue Community College; Yukiko K. Niiro, Gettysburg College; and Mary Sue Younger, The University of Tennessee, for their helpful reviews of the manuscript. We wish to thank authors and organizations for allowing us to reprint selected material; acknowledgments are made wherever such material appears in the text.

Robert J. Beaver
William Mendenhall

CREDITS

Table 1.5 (p. 18), Exercises 1.15 (p. 35), 1.16 (p. 36), 1.45 (p. 56), 1.48 (p. 58), 1.49 (p. 58), and 2.58 (p. 100), and data sets B and C on the data disk from *The Universal Almanac, 1993* (pp. 2, 123, 209, 295, 642, 674), by John W. Wright, 1992, Universal Press Syndicate. *The Universal Almanac* copyright by John W. Wright. Reprinted with permission of Andrews and McMeel. All rights reserved.

Exercise 1.13 (p. 35) from *The Sporting News,* December 14, 1992.

Exercises 1.26 (p. 48), 1.38 (p. 53), and 14.52 (p. 635) are reprinted with permission of *PC World,* November 1992, p. 9.

Exercises 1.28 (p. 49), 7.15 (p. 277), 10.35 (p. 453), and 14.53 (p. 635): Data from *Psychology in the Schools,* Vol. 29, 1992. Used with permission.

Exercise 1.35 (Figure 1.26, pp. 51–52): Data from *Youth Policy,* Vol. 14, 7/8, Dec., 1992, p. 50. (Original data from U.S. Dept. of Labor, and Bureau of Labor Statistics).

Exercises 1.36 (p. 52) and 1.40 (p. 54): Data from *American Demographics Desk Reference,* July, 1992. Reprinted with permission © American Demographics, July, 1992. For subscription information, please call (800) 828-1133.

Exercises 1.42 (p. 55), 4.22 (p. 176), 5.67 (p. 216), and 7.39 (p. 288) are reprinted by permission of the *New York Times.* Copyright 1988/89/93 by The New York Times Company.

Exercises 3S.21 (p. 163) from article entitled "More Companies Considering Flextime Option for Employees," by Sharilyn Bankole, *The Press-Enterprise,* April 7, 1992. Exercise 12.50 (p. 541) from "Motorists Have Little Respect for Others' Skills", *The Press-Enterprise,* Riverside, CA, May 15, 1991.

Exercises 4.58 (p. 189), 4.67 (pp. 190–191), 6.25 (p. 245), 8.26 (p. 332), 10.9 (p. 430), and Supplemental Exercise 6.41 (p. 253) are reprinted by permission of the *Wall Street Journal,* © Dow Jones & Company, Inc., 1988/89. All Rights Reserved Worldwide.

Exercises 5.29 (p. 206), 7.11 (p. 276), 14.7 (p. 599), 14.70, page 642, and Supplementary Exercise 6.50 (p. 254) are reprinted with permission from *Science News,* the weekly newsmagazine of science, copyright 1989 by Science Service, Inc.

Supplementary Exercise 6.65 (p. 256) is adapted from J. Hackl, *Journal of Quality Technology,* April, 1991.

Exercise 7.13 (p. 276) from Malvy, J. M. D. and Colleagues, "Retinal b-Carotene and a-Tocopherol Status in a French Population of Healthy Children" *International Journal of Vitamin and Nutrition Research,* Vol. 59, 1989, p. 29. Hans Huber Publishers, Toronto, Bern.

Exercise 7.26 (p. 283): Data from *Behavioral Research Therapy,* 1993, Alice F. Chang et al, 1993.

Exercises 9.24 (p. 375), 9.35 (p. 382), and 9.64 (p. 402): Data from "A Comparison of Anterior Compartment Pressures in Competitive Runners and Cyclists," by S. J. Beckham et al., *American Journal of Sports Medicine,* 1992, *21*(1), page 36.

Exercises 9.44 (pp. 385–386) and 9.118 (pp. 416–417) from "Well-Being of Cancer Survivors." *Psychosomatic Medicine, 45*(21) 1983. Reprinted by permission of Elsevier Publishing Co., Inc. Copyright 1983 by the American Psychosomatic Society, Inc.

Exercise 10.41 (p. 457) reprinted from *The World Almanac & Book of Fact 1993.* Copyright 1992. All rights reserved. The World Almanac is an imprint of Funk & Wagnalls.

Exercises 12.20 (p. 527) and 12.51 (p. 542): Data from the *Research Quarterly for Exercise and Sport,* June, 1983.

Exercise 12.35 (p. 537) reprinted with permission from "Changing Evaluations of Flood Plain Hazard: The Hunter Valley, Australia," *Environment and Behavior 13,* no. 4, 1981. © 1981 Sage Publications, Inc.

Exercises 12.49 (p. 541): Data from Costa, M. Martin and M. Gatz, "Determination of Authorship Credit in Published Dissertation," *Psychological Science, 3*(6), 1992, p. 54.

Table in Case Study, Chapter 12 (p. 543): Data from *British Journal of Educational Psychology, 62,* 1992, pp. 324–340.

Exercise 14.46 (p. 633): Data from "Ventilatory Threshold, Running Economy and Distance Running Performance of Trained Athletes," *The Research Quarterly for Exercise and Sport, 54*(3), September, 1983.

Case Study, Chapter 14 (p. 642): Data from the *San Francisco Chronicle,* February 10, 1993. © San Francisco Chronicle. Reprinted by permission.

Introduction: An Invitation to Statistics

What is statistics? Have you ever met a statistician? Do you know what a statistician does? Perhaps you are thinking of the person who sits in the broadcast booth at the Rose Bowl, recording the number of pass completions, yards rushing, or interceptions thrown on Super Bowl Sunday. Or perhaps the mere mention of the word *statistics* sends a shiver of fear through you. You may think you know nothing about statistics. However, it is almost inevitable that you encounter statistics in one form or another every time you pick up a daily newspaper. For example:

> **Poll shows candidates in statistical dead heat**
>
> The three major candidates for president are in a statistical dead heat, a new NBC News–Wall Street Journal poll showed yesterday.
>
> The poll shows undeclared candidate Ross Perot with 33 percent of the vote, President Bush with 31 percent and likely Democratic nominee Bill Clinton with 28 percent. The poll also found that more than 60 percent of the voters questioned have reservations about their choice.
>
> The poll was conducted July 5–7 among 1100 registered voters. There is a 3 percent margin of error.
>
> —*Press-Enterprise* (1992)

Articles similar to this one have become commonplace in our newspapers and magazines. In fact, in the period just prior to a presidential election, a new poll is reported almost every day. The language of this article is very familiar to us. However, it presents some unanswered questions to the inquisitive reader. How were the people in the poll selected? Will these people give the same response tomorrow? Will they give the same response on election day? Will they even vote? Are these people representative of all those who will vote on election day? It is the job of a statistician to ask these questions and to find answers for them in the language of the poll.

> **More graduate from high school**
>
> Americans aged 25–64 are more likely than those in other nations to have finished high school, but Japan and Germany have been educating a higher percentage of younger adults in recent years and may ultimately outpace the United States, a government study released yesterday shows.

The study showed that 82 percent of Americans aged 25–64 had completed high school by 1989, compared with 70 percent in Japan, 78 percent in Germany, 71 percent in Canada and 78 percent in Britain.

But among younger adults aged 25 to 34, 87 percent of Americans have graduated from high school, compared with nearly 91 percent in Japan, 83 percent in Canada and 77 percent in Britain. About 92 percent of Germans in that age group completed high school.

—Ross (1992)

When you see an article like this one in the newspaper, do you simply read the title and the first paragraph, or do you read further and try to understand the language of the second and third paragraphs? The percentage of American high school graduates jumps 5 points as the adults change from the 25–64-year age group to the 25–34-year age group, but the percentage of Japanese and German high school graduates jumps 21 points as the age group changes. What does this mean? How did they get these numbers? Did they really interview each American in a particular age category? It is the job of the statistician to interpret the language of this study.

Hot news: 98.6 not normal

After believing for more than a century that 98.6 was the normal body temperature for humans, researchers now say normal is not normal anymore.

For some people at some hours of the day, 99.9 degrees could be fine. And readings as low as 96 turn out to be highly human.

The 98.6 standard was derived by a German doctor in 1868. Some physicians have always been suspicious of the good doctor's research. His claim: 1 million readings—in an epoch without computers.

So Mackowiak & Co. took temperature readings from 148 healthy people over a three-day period and found that the mean temperature was 98.2 degrees. Only 8 percent of the readings were 98.6.

—Knight-Ridder Newspapers (1992)

What questions come to your mind when you read this article? How did the researcher select the 148 people, and how can we be sure that the results based on these 148 people are accurate when applied to the general population? How did the researcher arrive at the normal "high" and "low" temperatures given in the article? How did the German doctor record 1 million temperatures in 1868? Again, we encounter a statistical problem with an application to everyday life.

Statistics is a branch of mathematics that has applications in almost every facet of our daily life. However, it is a new and unfamiliar language for most people, and, like any new language, statistics can seem overwhelming at first glance. We invite you to learn this new language *one step at a time*. Once the language of statistics is learned and understood, it provides a powerful data analytic tool in many different fields of application.

0.1 The Population and the Sample

In the language of statistics, one of the most basic concepts is that of **sampling**. In most statistical problems, a specified number of measurements (or bits of information)—a **sample**—is drawn from a much larger body of measurements, called the **population**.

DEFINITION ■ A **population** is the set representing all measurements of interest to the investigator. ■

DEFINITION ■ A **sample** is a subset of measurements selected from the population of interest. ■

For the body-temperature experiment described above, the sample consists of body-temperature measurements for the 148 healthy people chosen by the experimenter. We would hope that the sample is representative of a much larger body of measurements—the population—the body temperatures of all healthy people in the world!

Which is of primary interest, the sample or the population? In most cases we are interested primarily in the population, but the population may be difficult or impossible to enumerate. Imagine trying to record the body temperature of every healthy person on earth or the presidential preference of every registered voter in the United States! Instead, **we try to describe or predict the behavior of the population on the basis of information obtained from a representative sample from that population.**

The words *sample* and *population* have two meanings for most people. For example, you read in the newspapers that a Gallup poll conducted in the United States was based on a sample of 1823 people. Presumably, each person interviewed is asked a particular question, and that person's response represents a single measurement in the sample. Is the sample the set of 1823 people, or is it the 1823 responses that they give?

When we use statistical language, we distinguish between the set of objects on which the measurements are taken and the measurements themselves. To experimenters, the objects on which measurements are taken are called **experimental units**. The sample survey statistician calls them **elements of the sample**.

0.2 Descriptive and Inferential Statistics

The first problem that confronts us when we are presented with a set of measurements is to find a way to organize and summarize them. Whether the set of measurements constitutes a population or is simply a sample drawn from a population, techniques are available that allow us to describe the measurements using graphical and numerical descriptive methods. This branch of statistics is called **descriptive statistics**. You have encountered descriptive statistics in many forms—graphical presentations such as bar charts, pie charts, and line charts presented by a political candidate; numerical tabulations in the newspaper; or the average rainfall amounts reported by the meteorologist on your local TV station. With the advent of computer software programs that generate spreadsheets as well as various types of computer graphics, descriptive procedures are becoming more and more common in everyday communication.

DEFINITION ■ **Descriptive statistics** consists of procedures used to summarize and describe the important characteristics of a set of measurements. ■

Descriptive statistics and its use in data analysis are the subject of the first two chapters of the text.

The second problem that arises when we are presented with a set of measurements involves the concept of **sampling**. It may be that we cannot enumerate the entire population within a reasonable time or that it is too costly to do so. Perhaps enumerating the population would be destructive, as in the case of "time to failure" testing. For these or other reasons, we may have only a sample from the population. By looking at the sample, we would like to be able to answer questions about the population as a whole. The branch of statistics that deals with this problem is called **inferential statistics**.

DEFINITION ■ **Inferential statistics** consists of procedures used to make inferences about population characteristics from information contained in a sample drawn from this population. ■

The objective of inferential statistics is to make inferences (that is, draw conclusions, make predictions, make decisions) about the characteristics of a population from information contained in a sample.

0.3 Achieving the Objective of Inferential Statistics: The Necessary Steps

How can we make inferences about a population using information contained in a sample? The task becomes simpler if we organize the problem into a series of logical steps.

1 **Specify the questions to be answered and identify the population of interest.** In the presidential election poll, the objective is to determine who will win the most votes on election day. Hence, the population of interest is the collection of all votes in the presidential election. In selecting a sample, it is important that the sample be representative of *this* population, not the population of voter preferences on July 5 or some other day prior to the election.

2 **Decide how to select the sample.** This is called the design of the experiment or the sampling procedure. Is the sample representative of the population of interest?

For example, if a sample of registered voters is selected from the state of Arkansas, will this sample be representative of the United States as a whole? Will it be the same as a sample of "likely voters"—those who are likely to actually vote in the election? Is the sample large enough to answer the questions posed in step 1 without wasting time and money on additional information? A good sampling design will answer the questions posed with minimal cost to the experimenter.

3 **Select the sample and analyze the sample information.** No matter how much information the sample contains, you must use an appropriate method of analysis to extract it. Many of these methods, which depend on the sampling procedure in step 2, are explained in the text.

4 **Use the information from step 3 to make an inference about the population.** Many different procedures can be used to make this inference, some of which are better than others. For example, ten different methods might be available to estimate human response to an experimental drug, but one procedure might be more accurate than others. You should use the best inference-making procedure available (many of these are explained in the text).

5 **Determine the goodness of the inference.** Since we are using only a fraction of the population in drawing the conclusions described in step 4, we might be wrong! How can this be? If an agency conducts a statistical survey for you and estimates that your company's product will gain 34% of the market this year, how much confidence can you place in this estimate? Is this estimate accurate to within 1, 5, or 20 percentage points? Is it reliable enough to be used in setting production goals? Every statistical inference should be accompanied by a measure of reliability that tells you how much confidence you can place in the inference.

Now that you have learned some of the basic terms and concepts in the language of statistics, we again pose the question asked at the beginning of this discussion: Do you know what a statistician does? It is the job of the statistician to implement all of the steps necessary to achieve the objective of inferential statistics. This may involve questioning the experimenter to make sure that the population of interest is clearly defined, developing an appropriate sampling plan or experimental design that will provide the best inference at a minimum cost to the researcher, providing the correct methods of analysis and drawing the proper conclusions using the sample information, and, finally, providing an accurate measure of the goodness or reliability of the inferences made on the basis of experimental results.

As you proceed through the text, you will learn more and more words, phrases, and concepts from this new language of statistics. Statistical procedures, for the most part, consist of commonsense steps that, given enough time, you would most likely have discovered for yourself. However, since statistics is an applied branch of mathematics, many of these basic concepts are mathematical—developed and based on results from calculus or higher mathematics. However, you do not have to be able to derive results in order to apply them in a logical way. This text will focus on statistical procedures that use numerical examples and intuitive arguments rather than more complicated mathematical arguments.

More important, the application of statistical concepts and techniques requires common sense and logical thinking in order to be successful. For example, if we proposed a study to find the average height of all students at a particular university, would we select our entire sample from the members of the basketball team? In the body-temperature example, the logical thinker would question an 1868 average based on 1 million measurements—when computers had not yet been invented. As you learn all the new terms, concepts, and techniques necessary for statistical analysis, remember to view all statistical problems with a critical eye and to be sure that the rule of common sense applies. Throughout the text, we will remind you of the pitfalls and dangers in the use or misuse of statistics. Benjamin Disraeli once said that there are three kinds of lies: *lies, damn lies,* and *statistics*! Our purpose is to dispel this claim—to show you how to make statistics *work* for you and not *lie* for you!

We encourage you to refer to this "Invitation to Statistics" periodically as you continue through the text. Each chapter will increase your knowledge of the language of statistics and should, in some way, help you achieve one of the steps described here. Each of these steps is essential in attaining the overall objective of inferential statistics: to make inferences about a population using information contained in a sample drawn from that population.

Exercises

Understanding the Concepts

0.1 You are a candidate for your state legislature, and you want to survey voter attitudes regarding your chances of winning. Identify the population that is of interest to you and from which you would like to select your sample. How will this population be dependent on time?

0.2 A medical researcher wants to estimate the survival time of a patient after the onset of a particular type of cancer and after a particular regimen of radiotherapy. Identify the population of interest to the medical researcher. Can you see some problems in sampling this population?

0.3 An educational researcher wants to evaluate the effectiveness of a new method for teaching reading to deaf students. Achievement at the end of a period of teaching is to be measured by a student's score on a reading test. Discuss the population (or populations) that might be of interest to the researcher.

0.4 During the presidential election campaign of 1992, the news media presented opinion polls on a daily basis that tracked the fortunes of the three major candidates. One such poll, taken for TIME/CNN ("In the Eye," 1992), showed the following results:

If the election for president were held today, for whom would you vote?

Clinton—46%
Bush —40%

The answers were based on a sample taken August 25–27 of 836 "likely" voters—those most likely to cast ballots in November.

a If the pollsters were planning to use these results to predict the outcome of the 1992 election, describe the population of interest to them.

b Describe the actual population from which they have drawn a sample.

c What is the difference between "registered voters" and "likely voters"? Why is this important?

d Is the sample selected by the pollsters representative of the population described in part (a)? Explain.

0.5 As the personal computer becomes a common tool in many households, parents, students, teachers, and others are becoming more and more "computer literate." Personal computers purchased as recently as 5 years ago are now "dinosaurs" that are being sold or traded in for newer and more powerful PCs. An article in *PC Source* ("Used Computer Prices," 1992) claimed that the average seller's asking price for a 20MB IBM PS/2 Model 30/286 was $950.

a If you wanted to buy a used IBM PS/2 such as the one described, you might be interested in the average seller's asking price. Describe the population of interest to you.

b Describe the population of interest to the *PC Source* researchers.

c Do you think that the researchers for *PC Source* enumerated the entire population in part (b), or did they take a sample?

d Are the two populations described in parts (a) and (b) identical? If not, how can you use the researchers' results to approximate the information you need?

0.6 A study in the *New England Journal of Medicine* (Feb. 16, 1989) reports that long-term heavy alcohol consumption has a major toxic effect on the strength of both the heart and the skeletal muscles. The team of investigators studied 50 white males, aged 25 to 59, who had *voluntarily* entered an alcohol treatment facility, all of whom were from emotionally stable environments and had no signs of ill health. An age-matched group of 50 white male physicians who were not heavy drinkers was also studied.

a Describe the two populations of interest to the experimenters.

b Describe the samples.

c Are the two samples representative of the populations in which the experimenters are interested?

d Why did the experimenters choose to isolate heavy drinkers who were from emotionally stable environments and who had no signs of ill health?

0.7 W. B. Jeffries, H. K. Voris, and C. M. Yang (1984) give data on the numbers of two types of barnacles found on 10 *T. orientalis* lobsters caught in the seas near Singapore. The data shown in the following table give the carapace length in millimeters (mm) and the numbers of two types of barnacles on each of the lobsters. The 10 lobsters were selected from among a total of 43 lobsters acquired from fishermen and fish markets and collected from the seas in the vicinity of Singapore. Suppose that we were interested in the number of *O. tridens* barnacles that one would find on a *T. orientalis* lobster.

		Number of Barnacles		
Field Number	**Carapace Length (mm)**	***O. tridens***	***O. lowei***	**Total**
AO61	78	645	6	651
AO62	66	320	23	343
AO66	65	401	40	441
AO70	63	364	9	373
AO67	60	327	24	351
AO69	60	73	5	78
AO64	58	20	86	106
AO68	56	221	0	221
AO65	52	3	109	112
AO63	50	5	350	355

a Describe the population that characterizes the number of *O. tridens* barnacles on a *T. orientalis* lobster.

b What do the numbers in the table under the *O. tridens* heading represent?

0.8 The California bar examination, described as one of the hardest in the nation, is given twice each year. Of the 4555 students who took the exam in February 1989, 2283 passed (*Press-Enterprise*, Riverside, Calif., May 31, 1989). This pass rate of 50.2% was a 15-year high;

however, state bar officials insist that the difficulty level of the exam has remained the same during this decade.

a Do the 4555 students who took the exam in February 1989 represent a population or a sample?

b If the 4555 students are a sample, describe the population from which the sample was drawn.

c What assumptions must we make about the examination in order to be certain that the sample was representative of the population of interest?

0.9 Smog experts in California have recently discovered that their estimates of auto emissions levels are not very accurate. In fact, the 26 million cars and trucks in the state of California emit two to four times as much pollutants as experts had estimated (*Press-Enterprise*, Riverside, Calif., July 4, 1992). Experts say that one reason for the inaccurate estimates is the sampling method. Pollution estimates in previous research had been based on emissions recorded during voluntary tests, and it is unlikely that the roughly 20% of motorists who have tampered with their pollution devices would volunteer for the tests.

a What is the population of interest to the air pollution experts?

b Are the air pollution experts actually sampling from the population of interest?

c Will estimates of auto emissions that were made several years ago be useful in drawing conclusions about auto emissions levels today? Explain.

1

Describing Sets of Measurements: Graphical Techniques

Case Study

Is your blood pressure normal, or is it too high or too low? The case study at the end of this chapter examines a large set of blood pressure data. You will use the methods of Chapter 1 to graphically describe this set of data and compare your blood pressure with that of others of your sex and age group.

General Objectives

Many of the sets of measurements that we collect represent a sample selected from a population. Other sets may represent the entire population, as in a national census. The objectives of this chapter are to explain what a "variable" is, to categorize variables into several types, and to explain how measurements or data are generated. We then present graphical descriptive methods used to describe different types of data sets.

Specific Topics

1 Variables, experimental units, samples and populations, data (1.1)
2 Univariate and bivariate data (1.1)
3 Qualitative and quantitative variables—discrete and continuous (1.2)
4 Pie graphs, bar graphs, line graphs (1.3)
5 Scatterplots for univariate data (1.4)
6 Relative frequency histograms for univariate data (1.4)
7 Stem and leaf displays for univariate data (1.5)
8 Scatterplots and histograms for bivariate data (1.7)

1.1 Variables and Data

Our primary objective in Chapters 1 and 2 will be to present some basic techniques in *descriptive statistics*—the branch of statistics concerned with describing sets of measurements, both *samples* and *populations*. Once we have collected a set of measurements, how can we display this set in a clear, understandable, and readable form? First, we must be able to define what is meant by measurements or "data" and to categorize the types of data that we are likely to encounter in real life. We begin by introducing some definitions—new terms in the statistical language that you need to know.

DEFINITION ■ A **variable** is a characteristic that changes or varies over time and/or for different individuals or objects under consideration. ■

For example, body temperature is a variable that changes over time within a single individual; it also varies from person to person. Religious affiliation, ethnic origin, income, height, age, and number of offspring are all variables—characteristics that vary depending on the individual chosen.

In the Introduction, we defined an *experimental unit* as the object on which a measurement is taken. Equivalently, we could define an experimental unit as the object on which a variable is measured. When a variable is actually measured on a set of experimental units, a set of measurements or **data** results.

DEFINITION ■ An **experimental unit** is the individual or object on which a variable is measured. A single **measurement** or data value results when a variable is actually measured on an experimental unit. ■

If a measurement is generated for every experimental unit in the entire collection, the resulting data set constitutes the *population* of interest. Any smaller subset of measurements is a *sample*.

EXAMPLE 1.1 A set of five students is selected from all undergraduate students at a large university, and the following measurements are recorded. Identify the various elements involved in generating this set of measurements.

Student	GPA	Sex	Year	Major	Current No. of Units Enrolled
1	2.0	F	Fr	Psychology	16
2	2.3	F	So	Mathematics	15
3	2.9	M	So	English	17
4	2.7	M	Fr	English	15
5	2.6	F	Jr	Business	14

Solution There are several *variables* in this example. The *experimental unit* on which the variables are measured is a particular undergraduate student on the campus. Five variables are measured for each student: grade point average (GPA), sex, year in college, major, and current number of units enrolled. Each of these characteristics varies from student to student. If we consider the GPAs of all students at this university to be the population of interest, the five GPAs represent a *sample* from this population.

If the GPA of each undergraduate student at the university had been measured, we would have generated the entire *population* of measurements for this variable.

The second variable measured on the students is sex, which can fall into one of two categories—male or female. It is not a numerically valued variable and hence is somewhat different from GPA. The population, if it could be enumerated, would consist of a set of Ms and Fs, one for each student at the university. Similarly, the third and fourth variables, year and major, generate nonnumerical data, the first generating four categories (Fr, So, Jr, Sr) and the second generating several categories, one for each undergraduate major on campus. The last variable, current number of units enrolled, is numerically valued, generating a set of numbers, rather than a set of qualities or characteristics.

Although we have discussed each variable individually, remember that we have measured each of these five variables on a single experimental unit: the student. Therefore, in this example, a "measurement" really consists of five observations, one for each of the five measured variables. For example, the measurement taken on student #2 produces the following observation:

(2.3, F, So, Mathematics, 15) ▪

You can see that there is a difference between a *single* variable measured on a single experimental unit and *multiple* variables measured on a single experimental unit, as in Example 1.1. If a single variable is measured, the resulting data are called **univariate data**. If two variables are measured on a single experimental unit (such as height and weight), the resulting data are called **bivariate data**. If more than two variables are measured, as in Example 1.1, the data are called **multivariate data**.

1.2 Types of Variables

Example 1.1 demonstrated that measuring variables produces data that might be either numerical or nonnumerical. Variables that give rise to nonnumerical data in which observations are categorized according to similarities or differences in kind are called **qualitative variables**. Political or religious affiliation, occupation, marital status, and high school attended are examples of qualitative variables.

The variables sex, year, and major in Example 1.1 are qualitative variables that give rise to qualitative data. However, if we had decided to assess class standing using "cumulative number of units completed" as the variable measured, the resulting data would have been numerical. When the variable used to measure an attribute produces numerical observations, the variable is said to be *quantitative*. The Dow Jones industrial average, the prime interest rate, the number of nonregistered taxicabs in a city, the number of students who pass a psychology midterm, and the daily power usage for an industrial plant are examples of **quantitative variables**, which give rise to quantitative data.

DEFINITION ▪ **Quantitative variables** give rise to numerical observations that represent an amount or quantity. **Qualitative variables** give rise to observations that represent a nonnumerical quality or characteristic. ▪

Quantitative variables can be further categorized according to the range of numerical values that a measurement can assume. Variables such as the number of family members, the number of new car sales, and the number of defective tires returned for replacement all take values corresponding to some subset of the counting integers 0, 1, 2, ... Specifically, these variables can take on a countable or *discrete* number of values and are called **discrete variables.** The name *discrete* arises because there are discrete gaps between the possible values that the data may assume. On the other hand, measurements on variables such as height, weight, time, distance, or volume can assume values corresponding to the many points on a line interval. Variables of this type are called **continuous**. For any two values of a continuous variable, a third value can always be found between them!

DEFINITION ▪ A **continuous variable** is one that can assume the infinitely many values corresponding to a line interval. **Discrete variables** can assume only a countable or discrete number of values. ▪

EXAMPLE 1.2 Identify each of the following variables as *qualitative* or *quantitative*:

1. The most frequent use of your microwave (reheating, defrosting, warming, other)
2. The number of consumers refusing to answer a telephone survey
3. The door chosen by a mouse in a maze experiment (A, B, or C)
4. The winning time for a horse running in the Kentucky Derby
5. The number of children in a fifth-grade class who are reading at or above grade level

Solution Variables 1 and 3 are both qualitative variables, since only an attribute is measured on each experimental unit. The categories for these two variables are shown in parentheses. The other three variables are quantitative. The number of consumers is a *discrete* variable, which could take on any of the values 0, 1, 2, . . . , with a maximum value depending on the number of consumers called. Similarly, the number of children reading at or above grade level could take any of the values 0, 1, 2, . . . , with a maximum value depending on the number of children in the class. The fourth variable, winning time for a Kentucky Derby horse, is the only *continuous* variable in the list. The winning time could be 121 seconds, 121.5 seconds, or 121.25 seconds, or it could take values in between any two values we have listed. ▪

Why should we be concerned about the different kinds of variables and the data that they generate? The techniques used for summarizing and describing data sets depend on the kind of data you have collected. Qualitative data are usually summarized by counting the number of observations in each of several categories, and the results are presented using tables and graphs. Graphical presentations differ somewhat for discrete and continuous quantitative variables but, in general, focus on graphs that are similar to bar charts. For each set of data that you encounter, the trick will be to determine what type of data is involved and how you can present it so that it is clear and understandable to your audience (see Figure 1.1).

FIGURE 1.1
Types of data

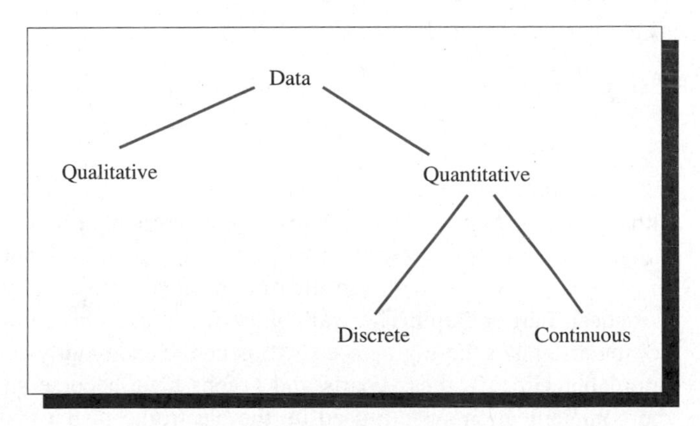

1.3 Statistical Tables and Graphs

When data are collected, the measurements on a variable need to be organized before any summary graphical presentations can be made. An important tool that you can use for data summary is a *statistical table*. Such tables will vary for each situation, but their primary purpose is to consolidate and summarize the data set so that it is more understandable to the reader. The statistical table can then be used to create charts and graphs that display the data pictorially.

Statistical tables for *qualitative data* consist of a list of the categories being measured along with either the frequency of occurrence of each category or a quantity or amount measured for each category. These categories must be chosen so that an observation will belong to one and only one category and so that each observation has a category to which it can be assigned. If categories are defined using these guidelines, there will be no ambiguity as to the category to which an observation belongs.

If we were interested in categorizing meat products according to the type of meat used, we might wish to use the categories in Table 1.1. Categories describing the ranks of college faculty are given in Table 1.2. The "other" category is included in both cases to allow for the possibility that an observation could not be assigned to one of the first five categories.

TABLE **1.1**
Source of meat

Categories	
Beef	Pork
Chicken	Turkey
Seafood	Other

TABLE **1.2**
Rank of college teachers

Categories	
Professor	Instructor
Associate Professor	Lecturer
Assistant Professor	Other

Once the observations have been categorized, the data summary is presented either as a *statistical table*, which displays an amount or quantity in each category, or as a *frequency distribution*, which displays the frequency or number of measurements in each category. The data can then be displayed graphically using an appropriate statistical chart or graph. The availability of computer graphics software for personal computers makes these graphs easy to generate and readily accessible to the general population. In fact, these charts and graphs have become such an integral part of the communication system used by the electronic media that you encounter them almost every day in newspapers and magazines and on television. In this section, you will learn how to use these tools to generate neat, understandable, and even colorful graphics for your own reports and presentations.

Pie Graphs and Bar Graphs

A *pie graph* is a circular graph that is useful in showing how a total quantity is distributed among a group of categories. The "pieces of the pie" represent the proportions of the total that fall into each category. A *bar graph* can also be used to distribute amounts or frequencies into categories, with the height of the bar representing the quantity or frequency for each category.

EXAMPLE **1.3** In a survey concerning public education, 400 school administrators were asked to rate the quality of education in the United States. Their responses are summarized in Table 1.3. Construct a pie graph and a bar graph for this set of data.

TABLE 1.3
U.S. education rating by 400 educators

Rating	Frequency	Percentage
A	35	9
B	260	65
C	93	23
D	12	3
Total	400	100

Solution In constructing a pie graph, we assign the sector of a circle to each category, so that the angle of the sector is in proportion to the percentage of the measurements in that category. Since a circle contains 360°, the proportion of the circle corresponding to each category is found using (percentage/100) × 360°.

Category	Angle
A	$.09 \times 360° = 32.4°$
B	$.65 \times 360° = 234.0°$
C	$.23 \times 360° = 82.8°$
D	$.03 \times 360° = 10.8°$
Total	360°

Figure 1.2 shows the pie graph for Example 1.3, which has been constructed using the table shown above. Although pie graphs use percentages to determine the relative sizes of the "pie slices," bar graphs in general plot frequency against the categories. A bar graph for these data is shown in Figure 1.3.

The visual impact of these two graphs is somewhat different. The pie graph is used to display the relationship of the parts to the whole, but the bar graph is used to emphasize the actual quantity or frequency for each category. Since the categories in this example are ordered "grades" (A, B, C, D), we would not want to rearrange

FIGURE 1.2
Pie graph for Example 1.3

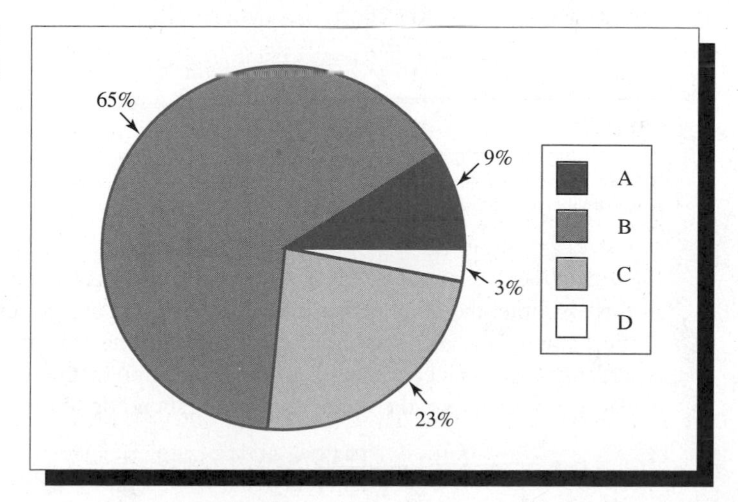

FIGURE 1.3
Bar graph for Example 1.3

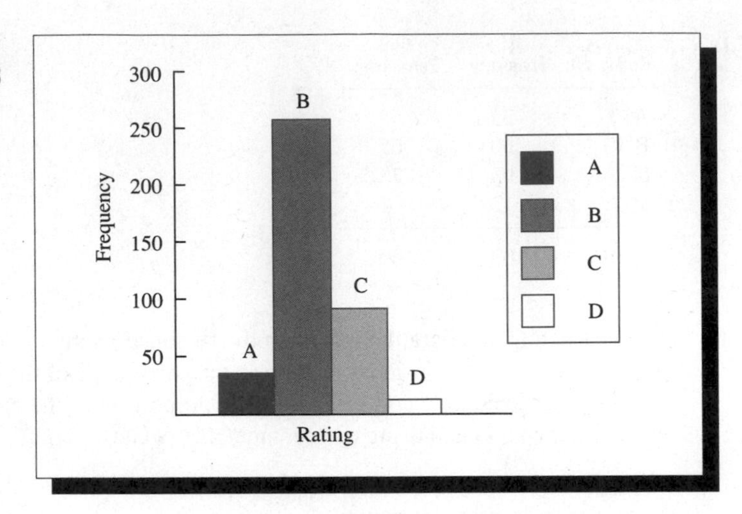

the bars in the graph to change its *shape*. In a pie graph, the order of presentation is irrelevant. ▪

EXAMPLE 1.4 The benefits paid for social insurance programs in the United States for fiscal year 1991 are summarized in Table 1.4. Construct a pie graph and a bar graph to graphically describe the data. Compare the two methods of presentation.

TABLE 1.4
Benefit data for Example 1.4

Program	Benefits ($ in billions)
OASDI†	$247.8
Medicare	113.9
Medicaid	64.9
Food stamps	18.2
AFDC‡	20.3
SSI	19.2
Total	484.3

†Old age, survivors, and disability insurance.
‡Aid to families with dependent children.
Source: Wright (1992), p. 123.

Solution In this example, the amounts paid for each of these categories are displayed on the vertical axis of the bar graph. For the pie graph, the "pie slice" or sector angle for a particular category corresponds to the proportion of the total benefits paid to that category. For example, the sector angle corresponding to the Medicare category is

$$(113.9/484.3) \times 360° = 84.7°$$

The pie graph and the bar graph are shown in Figures 1.4 and 1.5. In this case, the "amount paid," which determines the height of the bar or the size of the "pie slice," is a quantitative variable rather than a frequency of occurrence for each of the categories. Both figures show that the largest amounts of benefits are paid for OASDI and Medicare; however, Figure 1.5 allows us to assess the actual amounts paid, whereas Figure 1.4 does not. Since the categories in this example have no inherent order, we are free to rearrange the bars in the graph in any way we like. The *shape* of the bar graph has no bearing on its interpretation. ▪

FIGURE 1.4
Pie graph for Example 1.4

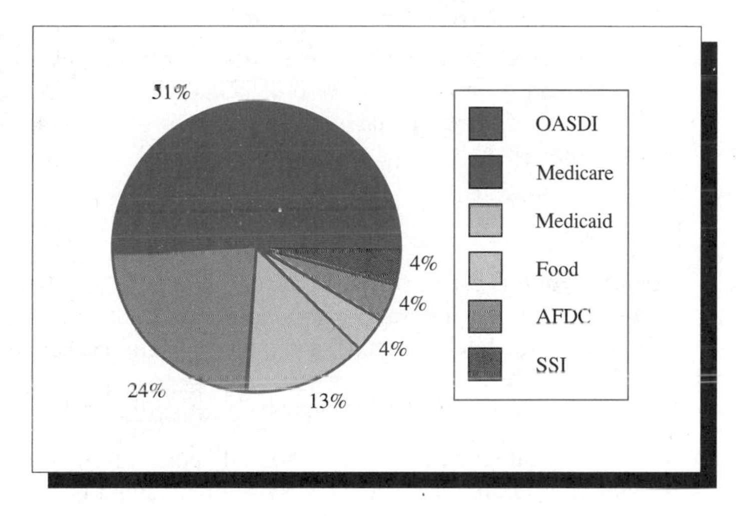

FIGURE 1.5
Bar graph for Example 1.4

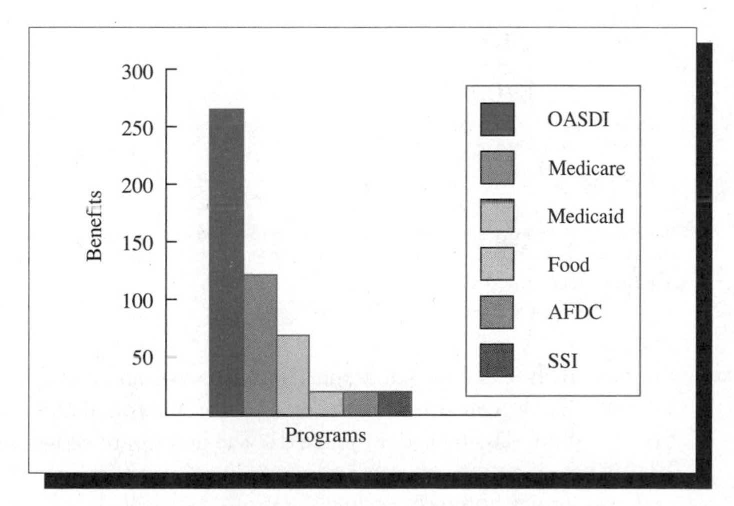

Guidelines for Constructing Pie Graphs and Bar Graphs

- Use a small number of categories if possible. Too many categories will produce a visual dilution of important information.
- For bar graphs, label frequencies or amounts along one axis and categories along the other. *You may leave a space between each bar* to emphasize the distinctiveness of each category.
- If the categories exhibit some inherent order or rank, the bars of the graph should be arranged in that order. Otherwise, you are free to change the shape of the bar graph by rearranging the bars as you wish.

Line Graphs

The *line graph* is used to display the change in a variable over time. In a line graph of this sort, the time intervals or categories (day, month, year, etc.) are located on one axis (usually the horizontal axis) and the variable to be charted on the other.

EXAMPLE **1.5** The average sales prices of new one-family houses from 1965 to 1990 are reported in Table 1.5. Graph the data using a line graph and a bar graph. Compare the two presentations.

TABLE **1.5** Average sales prices for Example 1.5

Year	Price
1965	$21,500
1970	26,600
1975	42,600
1980	76,400
1985	100,800
1990	149,800

Source: Wright (1992), p. 295.

Solution For both the line and the bar graphs, the categories are time intervals—years in 5-year intervals. These categories are located on the horizontal axis, and the "average sales price" is plotted using the vertical axis. The graphs are shown in Figures 1.6 and 1.7. Both types of graphs are very effective in showing the steadily increasing price of new one-family homes over time. ■

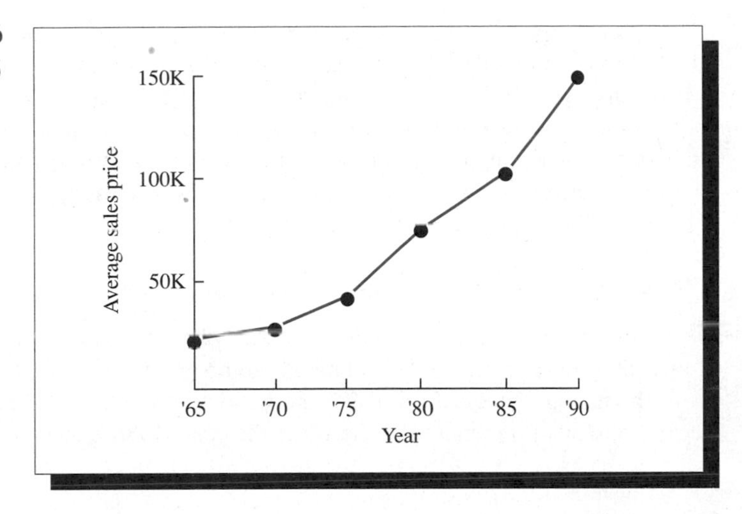

FIGURE 1.6
Line graph for Example 1.5

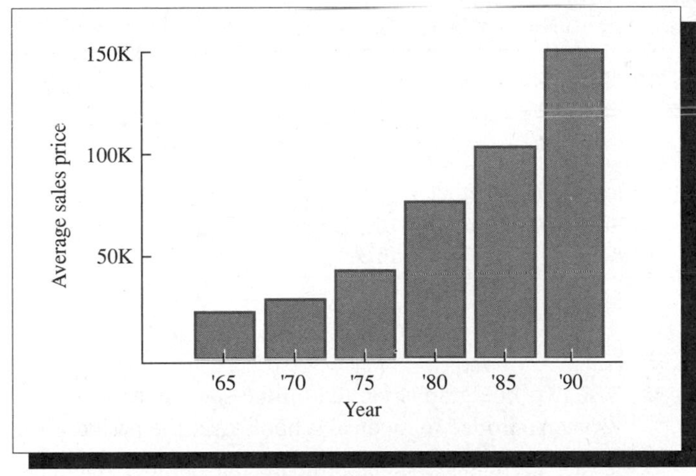

FIGURE 1.7
Bar graph for Example 1.5

Guidelines for Constructing Line Graphs

- The categories displayed along the horizontal axis are units of time.
- Label quantities as the measured variable along the vertical axis.
- The data are plotted on the graph, and the points are connected by line segments.

Comparative Graphs

When presented with two sets of data, one of the first things you might consider doing is to compare the sets using graphical techniques. Of course, the two sets of data must have a basis for comparison. It may be that the same categories have been measured at two or more points in time, in which case you may consider plotting two line graphs on the same set of axes or displaying two pie graphs side by side. Or it may be that the same categories are measured for two or more populations. In this case, you may consider using a bar graph in which, for each category, the bars for both populations are placed side by side. Another option might be to stack the bars for each category on top of each other. In any case, our objective is to display the similarities or differences in the frequencies or amounts for each category. These ideas will become clearer with the next examples.

EXAMPLE 1.6 According to a report by Kenneth Eskey (1992) of the Howard News Service, "Women are choosing medical careers in record numbers and could fill one of every two seats in the nation's medical schools if the trend continues." Medical school enrollments in thousands for men and women over the years 1986–1992 are given in Table 1.6.

TABLE 1.6
Medical school enrollments, in thousands

Year	Men	Women	Total
1986	10.5	5.6	16.1
1987	10.2	5.8	16.0
1988	10.1	5.9	16.0
1989	9.8	6.0	15.8
1990	9.8	6.2	16.0
1991	9.8	6.4	16.2
1992	9.5	6.8	16.3

Source: Eskey (1992), p. A13.

a Use two line graphs plotted on the same axes to show the number of men and women enrolled in medical schools over the period given.

b Using these same data, compare the number of men and women enrolled in medical schools using a stacked bar graph.

Solution **a** The line graphs must both use the same horizontal and vertical axes. With this in mind, the two lines are plotted in Figure 1.8. Although the two lines are not identical, it does appear that the number of women enrolled in medical schools in approaching the 50% ratio that more nearly reflects their numbers in the population at large.

b In preparing a stacked bar graph, for each time category we draw a rectangle equal to the total number enrolled and indicate with color or some other marking the proportion of each bar that represents the number of women enrolled in medical schools (see Figure 1.9). This presentation may be more effective than the first, in that the number of persons enrolled in medical schools has remained fairly constant, and the proportion of each bar representing women enrolled is approaching 50%. ▪

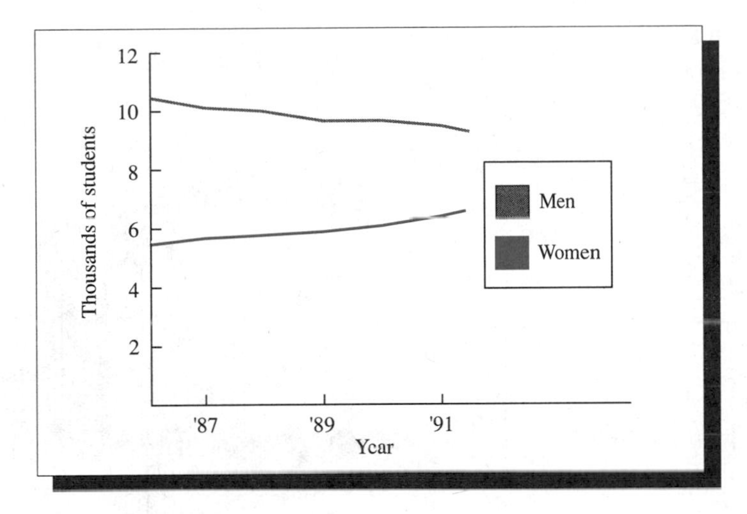

FIGURE 1.8
Comparative line graph for Example 1.6

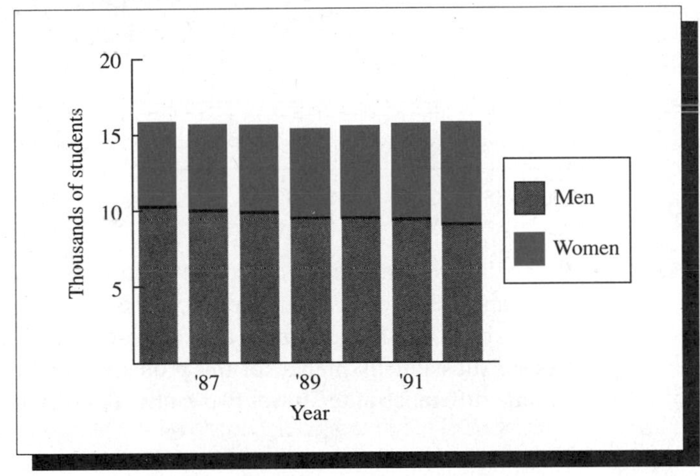

FIGURE 1.9
Stacked bar graph for Example 1.6

In Example 1.6, the medical school enrollments were classified both by gender and by year of enrollment. Thus, the data in Table 1.6 are *bivariate data.* A more elaborate graphical technique for bivariate data, which can be generated by computer software programs is a three-dimensional bar chart, as shown in Figure 1.10. The presentation is similar to one using a line graph or a side-by-side bar graph.

EXAMPLE 1.7

Are professors in private colleges paid more than professors at public colleges? If so, do the private colleges employ as many high-ranking professors as public colleges? The following data were collected using a sample of 400 college professors whose rank, type of college, and salary were recorded. The quantity in each cell is the average salary (in thousands of dollars) for all professors falling into that category, with the number of professors given in parentheses. Use graphical techniques to answer the questions posed for this sample.

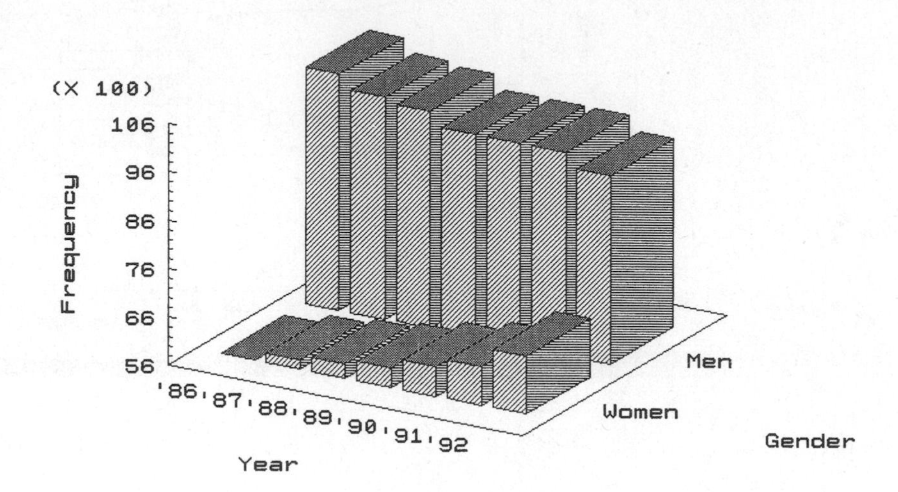

FIGURE 1.10 EXECUSTAT 3-D bar graph for Example 1.6

	Full Professor	Associate Professor	Assistant Professor
Public	55.8 (32)	42.2 (75)	35.2 (93)
Private	61.6 (48)	43.3 (62)	35.5 (90)

Solution To display the average salaries of these 400 professors, we choose to use a side-by-side bar graph, (Figure 1.11). The height of the bars is the average salary, with each group of bars along the horizontal axis representing a different professorial rank. Salaries are substantially higher for full professors in the private colleges, but there is very little difference at the lower two ranks. To compare the number of professors

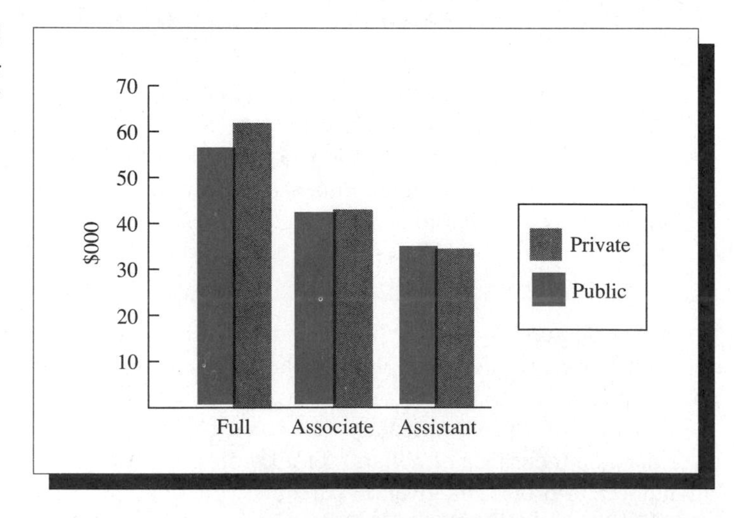

FIGURE 1.11 Comparative bar graphs for Example 1.7

FIGURE 1.12 Comparative pie graphs for Example 1.7

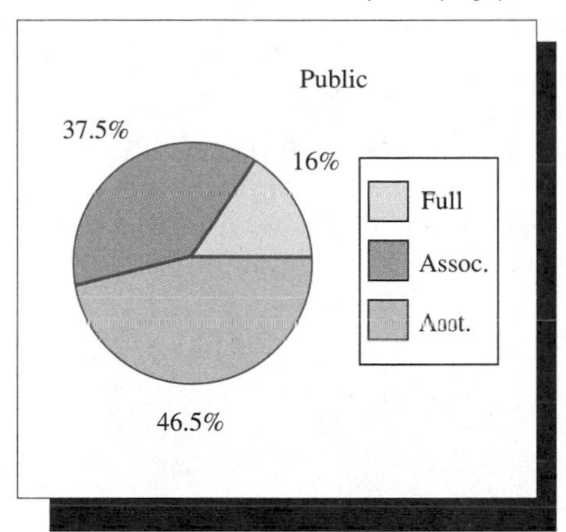

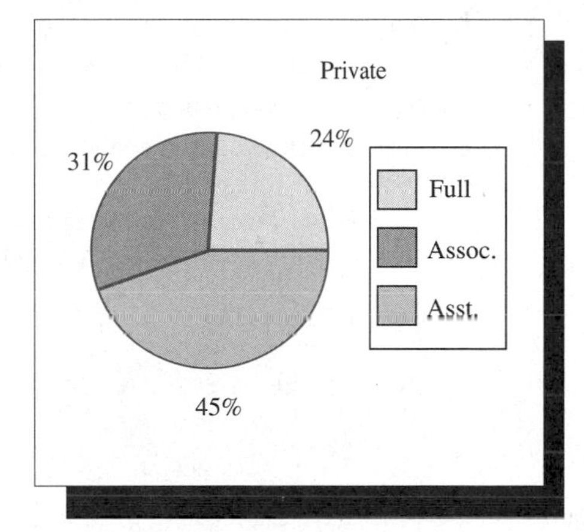

employed at each of the three ranks for public and private colleges, we choose to draw two pie graphs and display them side by side (Figure 1.12).

Although the two graphs are not strikingly different, you can see that public colleges have fewer full professors and more associate professors than private colleges. The reason for these differences is not clear, but we might speculate that private colleges, with their higher salaries, are able to attract more full professors. Or perhaps public colleges are not as willing to promote professors to the higher-paying ranks. In any case, the comparative graphs provide a means for comparing the two sets of data. ▪

All the graphs and charts we have discussed are important tools to provide a quick visual summary of data. The type of graph used is not important, as long as the graph is factual and easy to interpret. You must be careful when interpreting graphs that you are not fooled by changes in size and scale. We will discuss ways in which graphs can "lie" in Section 1.6.

Exercises

Basic Techniques

1.1 Identify the following variables as quantitative or qualitative:

a The amount of time it takes to assemble a simple puzzle

b The number of students in a first-grade classroom

c The rating of a newly elected politician (excellent, good, fair, poor)

d The state in which a person lives

1.2 Identify the following quantitative variables as discrete or continuous:

a The population per square mile for a particular area of the United States
b The weight of newspapers recovered for recycling on a single day
c The number of landfills in a particular state
d The time to completion of a sociology exam
e The number of consumers who consider nutritional labeling on food products to be important

1.3 Fifty people are grouped according to four categories—A, B, C, and D—and the number of people who fall into each category is shown below:

Category	Frequency
A	11
B	14
C	20
D	5

a Construct a pie graph to describe the data.
b Construct a bar graph to describe the data.
c Does the shape of the bar graph in part (b) change depending on the order of presentation of the four categories? Is the order of presentation important? Explain.

1.4 A group of items are categorized according to a certain attribute—X, Y, Z—and according to the state in which they were produced.

	X	Y	Z
New York	20	5	5
California	10	10	5

a Create a comparative (side-by-side) bar graph to compare the number of items of each type created in California and New York.
b Create a stacked bar graph to compare the number of items of each type made in the two states.
c Which of the two types of presentation is more easily understood? Explain.
d Are there any other graphical methods that you might use to describe the data?

1.5 An experimental psychologist measured the amount of time it took for a rat to successfully navigate a maze on each of 5 days. The results are shown below. Create a line graph to describe the data. Do you think that any learning is taking place?

Day	1	2	3	4	5
Time (sec.)	45	43	46	32	25

Applications

1.6 The respondents in a poll were asked to choose the most valued right enjoyed by democratic nations. The percentages of the 811 respondents choosing each of seven different rights are shown in the table:

Most Valued Right	Percentage
Right to own property	11
Right to vote	15
Freedom of the press	5
Right to pick career	5
Right to protest	4
Freedom of speech	31
Freedom of religion	24

Source: USA Today (July 4, 1990). Copyright 1990, *USA Today*. Reprinted with Permission

a Are all of the 811 opinions accounted for in the table? Add another category if necessary.

b Create a pie graph to describe the data.

c Create a bar graph to describe the data.

d Rearrange the bars in part (c) so that the categories are ranked from the largest percentage to the smallest.

e Which of the three methods of presentation—parts (b), (c), or (d)—is the most effective?

1.7 The price of health care in the United States has increased dramatically in the last 20 years and was a major campaign issue in the 1992 presidential race. The consumer price indexes for medical services in three different categories are shown below for the years 1970, 1980, and 1991:

Year	Total Services	Physician Fees	Hospital Room Charges
1970	30	37	24
1980	75	78	64
1991	180	172	192

Source: U.S. Department of Commerce (1992), p. 94.

a Use a comparative line graph to describe the three consumer price indexes over time.

b Use a side-by-side comparative bar chart to describe the indexes over time.

c What conclusions can you draw using these two graphs? Which is the most effective?

1.8 A Gallup poll surveyed 499 women and 502 men to assess the importance of family life in the United States. The results shown in the following table are responses to the question "How would you rate the importance of having a good family life?"

Sex	Very Important	Somewhat Important	Very Unimportant	No Opinion
Female	459	35	0	5
Male	432	70	0	0

Source: Adapted from Schmittroth (1991), p. 8.

a Define the sample and the population of interest to the researchers.

b Describe the variables that have been measured in this survey. Are the variables qualitative or quantitative? Are the data univariate or bivariate?

c What do the entries in the cells represent?

d Use comparative pie graphs to compare the responses for men and women.

e What other graphical techniques could be used to describe the data? Would any of these techniques be more informative than the pie graphs constructed in part (d)?

1.4 Relative Frequency Distributions

The statistical tables and graphs described in Section 1.3 are useful when data are categorized according to one or more *qualitative* variables; in this situation, pie graphs, bar graphs, and line graphs allow us to describe the frequencies or amounts falling into each category. These methods will not always be appropriate, however, when the data are *quantitative*.

Quantitative variables give rise to numerical observations that represent an amount or quantity. If the variable can take on only a finite or countable number of values, it is a *discrete* variable; a variable that can assume an infinite number of values corresponding to the points on a line interval is called *continuous*. A quantitative data set, either discrete or continuous, consists of a set of numerical values that cannot be easily separated into qualitative categories. How could we graphically describe this type of data?

The simplest graphical method available for quantitative data is the **scatterplot**. For a small set of measurements—for example, the set 2, 3, 5, 6, and 8—the measurements can simply be plotted as points on a horizontal line. Such a scatterplot is shown in Figure 1.13. However, as the number of measurements increases, the scatterplot becomes very uninformative, as shown in Figure 1.14.

FIGURE **1.13**
Scatterplot with a small set of measurements

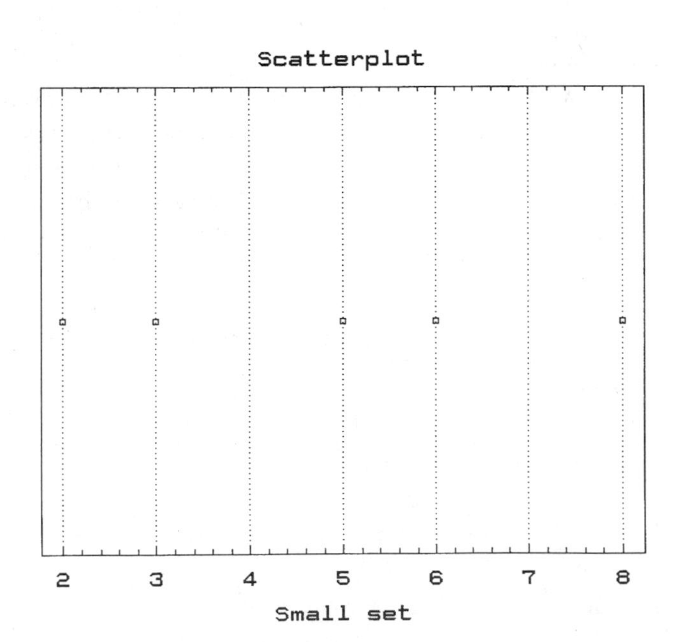

FIGURE 1.14
Scatterplot with a large set of measurements

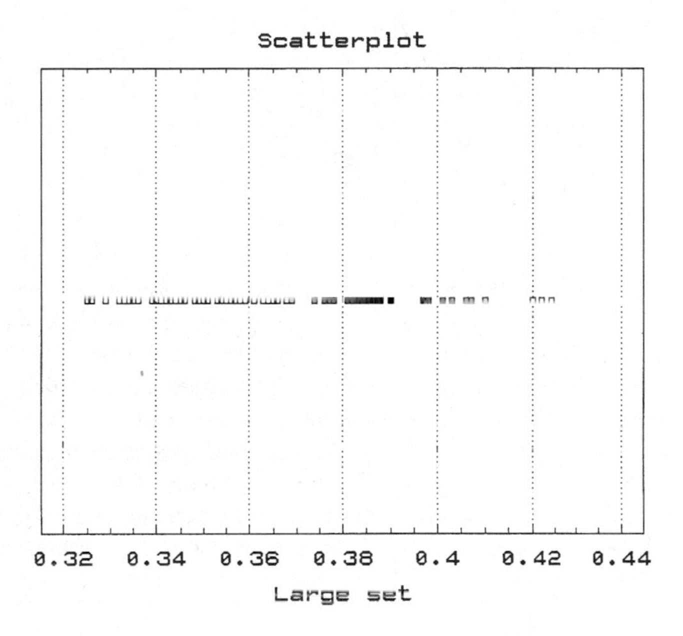

A better way to make sense out of quantitative data is to use a type of graph called a *relative frequency histogram*. The data in Table 1.7 present the GPAs of 30 Bucknell freshmen recorded at the end of the freshman year. The data, although recorded to only one decimal place accuracy, represent the continuous variable GPA, which can take values in the interval from 0 to 4. By examining the table, you can quickly see that the largest and smallest GPAs are 3.4 and 1.9, respectively. But how are the remaining GPAs distributed? To answer this question, we divide the interval into an arbitrary number of subintervals, or **classes**, of equal length. As a rule of thumb, the number of classes should range from 5 to 20; the more data available, the more classes you need. The classes must be chosen so that each measurement can fall into *one and only one* class. *These classes are similar to the categories used for qualitative data.* Once the classes are formed, the measurements are categorized according to the class into which they fall, and a graph resembling a bar graph is drawn. This graph is the **relative frequency histogram**.

DEFINITION ■

The **relative frequency histogram** for a quantitative data set is a bar graph in which the height of the bar represents the proportion or relative frequency of occurrence for a particular class or subinterval of the variable being measured. The classes or subintervals are plotted along the horizontal axis. ■

For the GPAs in Table 1.7, we choose to use eight intervals of equal length. Since the length of the total GPA span is $(3.4 - 1.9) = 1.5$, a convenient choice of interval length is $(1.5 \div 8) = .1875$, rounded off to .2. Rather than begin the first interval at the lowest value, 1.9, we choose a starting value of 1.85 and form subintervals from 1.85 to 2.05, 2.05 to 2.25, 2.25 to 2.45, and so on. By choosing 1.85 as the starting

TABLE 1.7
Grade point averages of 30 Bucknell University freshmen

2.0	3.1	1.9	2.5	1.9
2.3	2.6	3.1	2.5	2.1
2.9	3.0	2.7	2.5	2.4
2.7	2.5	2.4	3.0	3.4
2.6	2.8	2.5	2.7	2.9
2.7	2.8	2.2	2.7	2.1

value, we make it impossible for any measurement to fall on the class boundaries and eliminate any ambiguity regarding the disposition of a particular measurement.

The 30 measurements are now categorized according to the class into which they fall, as shown in Table 1.8. The eight classes are labeled from 1 to 8 for identification purposes. The boundaries for the eight classes, along with a tally of the number of measurements falling in each class, are given in the second and third columns of the table. The fourth column gives the *class frequency*, the number of measurements falling into a particular class. The last column of the table presents the fraction or proportion of the total number of measurements falling into each class. We call this proportion the **class relative frequency**. If we let n represent the total number of measurements—for instance, in our example, $n = 30$—then the relative frequency for a particular class is calculated as

$$\text{Relative frequency} = \frac{\text{Frequency}}{n}$$

TABLE 1.8
Relative frequencies for data of Table 1.7

Class	Class Boundaries	Tally	Class Frequency	Class Relative Frequency
1	1.85–2.05	|||	3	3/30
2	2.05–2.25	|||	3	3/30
3	2.25–2.45	|||	3	3/30
4	2.45–2.65	~~||||~~ ||	7	7/30
5	2.65–2.85	~~||||~~ ||	7	7/30
6	2.85–3.05	||||	4	4/30
7	3.05–3.25	||	2	2/30
8	3.25–3.45			1	1/30

To construct the relative frequency histogram, plot the class boundaries along the horizontal axis. Draw a bar over each class interval, with height equal to the relative frequency for that class. The relative frequency histogram for the GPA data, Figure 1.15, shows at a glance how GPAs are distributed over the interval 1.9 to 3.4.

FIGURE 1.15
Relative frequency histogram

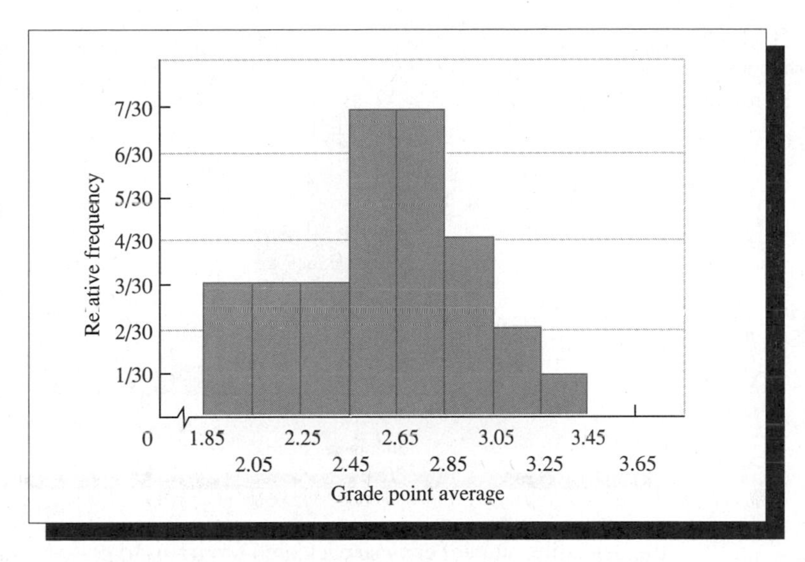

EXAMPLE 1.8

Twenty-five households are polled in a marketing survey, and the number of quarts of milk purchased during a particular week is recorded.

0	3	5	4	3
2	1	3	1	2
1	1	2	0	1
4	3	2	2	2
2	2	2	3	4

Construct a relative frequency histogram to describe the data.

Solution The variable being measured is "number of quarts of milk," which is a discrete variable taking only integer values. In this case, it is simplest to choose the classes or subintervals as the integer values over the range of observed values, 0, 1, 2, 3, 4, and 5. Table 1.9 shows the classes and their corresponding frequencies and relative frequencies. The relative frequency histogram is shown in Figure 1.16. ▪

TABLE 1.9
Frequency table for Example 1.8

Number of Quarts	Frequency	Relative Frequency
0	2	0.08
1	5	0.20
2	9	0.36
3	5	0.20
4	3	0.12
5	1	0.04

In recent years, computers and microcomputers have become readily available to many students, providing them with an invaluable tool. In the study of statistics, even

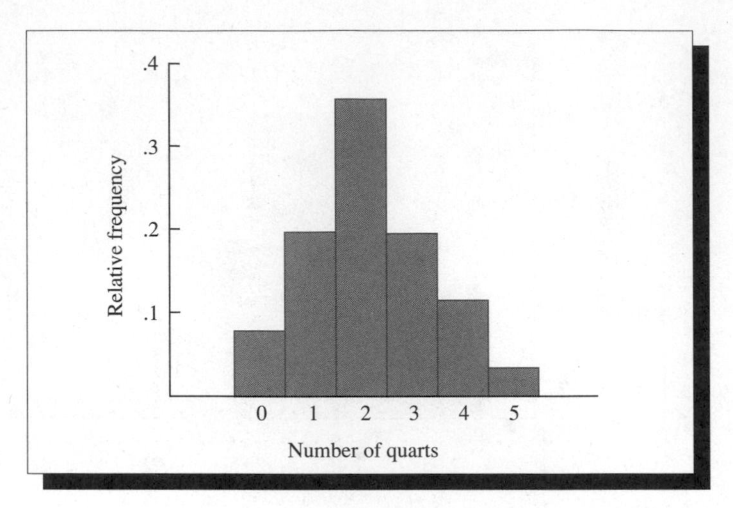

FIGURE 1.16
Relative frequency histogram for Example 1.8

the beginning student can use packaged programs to perform statistical analyses with a high degree of speed and accuracy. Some of the more common statistical packages available at computer facilities are MINITAB™†, SAS (Statistical Analysis System), SPSS (Statistical Package for the Social Sciences), and BMPD (Biomedical Package); personal computers will support packages such as MINITAB, EXECUSTAT, SYSTAT, and others.

These programs, called **statistical software**, differ in terms of the types of analyses available, the options within the programs, and the forms of printed results (called **output**). However, they are all similar. In this text, we will primarily use MINITAB as a statistical tool; understanding the basic output of this package will help you interpret the output from other software systems. However, a word of caution is necessary regarding the use of the computer for statistical analysis. Inaccurate entry of data, improper procedure commands, or incorrect interpretation of output will result in improper conclusions, even though you did use a computer!

The MINITAB software package was used to create a frequency histogram for the data in Example 1.8, as shown in Figure 1.17. The program automatically chooses classes with conveniently rounded midpoints. If a measurement falls on a class boundary, it is assigned to the class with the larger midpoint. The printout gives the midpoints of the class intervals and the number of measurements per interval; it uses a star (*) to represent a measurement in the histogram display. A histogram for the GPA data was generated by the EXECUSTAT program and is shown in Figure 1.18. This graph is almost identical to the one shown in Figure 1.15, except for the scale of the vertical axis.

†MINITAB is the trademark of MINITAB, Inc., 215 Pond Lab., University Park, PA 16802.

FIGURE 1.17
MINITAB histogram for the data of Example 1.8

```
Histogram of C1   N = 25

Midpoint   Count
       0       2  **
       1       5  *****
       2       9  *********
       3       5  *****
       4       3  ***
       5       1  *
```

FIGURE 1.18
EXECUSTAT histogram for the 30 grade point averages

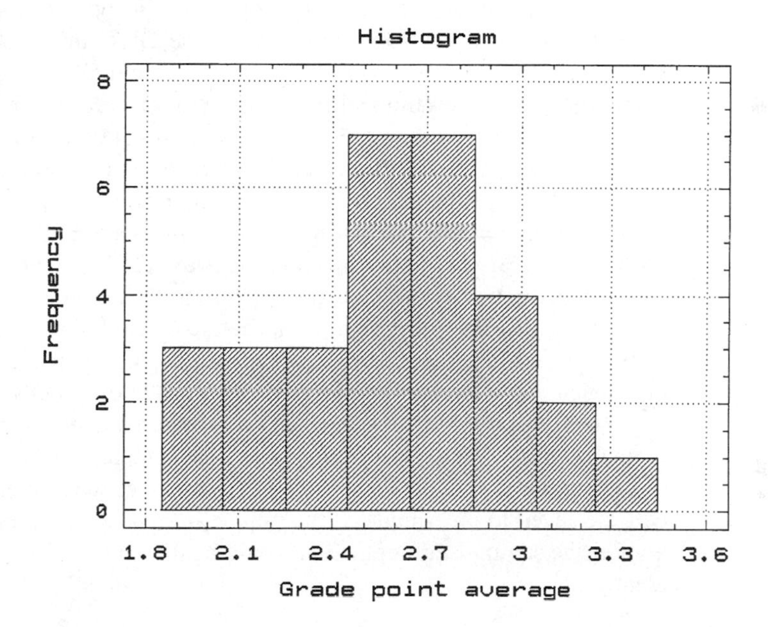

Consider the relative frequency histograms in Figures 1.15 and 1.18 in greater detail. What proportion of the students had GPAs equal to 2.7 or greater? Checking the relative frequency histograms, we see that the proportion involves all classes to the right of 2.65. Using Table 1.8, we see that 14 students had GPAs greater than or equal to 2.7. Hence, the proportion is 14/30, or approximately 47%. We note that this value is also the percentage of the total area of the histogram in Figure 1.15 lying to the right of 2.65.

Suppose we write each of the 30 GPAs on a piece of paper, place them in a hat, and then draw one piece of paper from the hat. What is the chance that this paper contains a GPA greater than or equal to 2.7? Since 14 of the 30 pieces of paper are marked with numbers greater than or equal to 2.7, we say that we have 14 chances out of 30 or that the probability is 14/30. You have undoubtedly encountered the word *probability* in ordinary conversation; we will defer a definition of it and a discussion of its significance until Chapter 3.

Although we are interested in describing the set of $n = 30$ measurements, we might also be interested in the population from which the sample was drawn, which consists of GPAs of all freshmen currently in attendance at Bucknell University. Or, if we are interested in the academic achievement of college freshmen in general, we might consider our sample as representative of the population of freshmen attending Bucknell or colleges *similar* to Bucknell. What proportion of the GPAs was greater than or equal to 2.7? If we possessed the relative frequency histogram for the population, we could give the exact answer to this question. Since we do not have this information, we are forced to make an **inference** using our sample information. Our estimate for the true population proportion, based on the sample information, would likely be 14/30, or 47%. Without knowledge of the population relative frequency histogram, we would infer that the population histogram is similar to the sample histogram and that approximately 47% of the GPAs in the population are greater than or equal to 2.7. Most likely this estimate would differ from the true population percentage. We will examine the magnitude of this error in Chapter 7.

The relative frequency histogram is often called a *relative frequency distribution* because it shows the manner in which the data are distributed along the horizontal axis of the graph. The bars constructed above each class are subject to two interpretations. They represent the proportion of observations falling in a given class. Also, if a measurement is drawn from the data, a particular class relative frequency is the chance or probability that the measurement will fall into that class. The most significant feature of the sample frequency histogram is that it provides information on the population frequency histogram that describes the population. An important point to note here is that different samples from the same population will result in different sample histograms, even when the class boundaries remain fixed. However, we would expect the sample and the population frequency histograms to be similar. The degree of resemblance will increase as more and more data are added to the sample. If the sample were enlarged to include the entire population, the sample and population would be the same and the histograms would be identical.

Guidelines for Constructing a Frequency Distribution

- *Determine the number of classes.* It is usually best to have from 5 to 20. The larger the amount of data, the more classes can be used.
- *Determine the class width.* As a rule of thumb, divide the difference between the largest and smallest measurements by the number of classes desired and round up to a convenient figure for class width. All the classes should be of equal width. This allows you to make uniform comparisons of the class frequencies.
- *If the data are discrete,* classes can be chosen to correspond to the integer values taken by the data. In this case, and if the number of integer values is small, the class width is usually 1. If there are a large number of integer values, you may need to group them into classes.

- *Locate the class boundaries.* The lowest class must include the smallest measurement. Then add the remaining classes. Class boundaries should be chosen so that it will be impossible for a measurement to fall on a boundary.
- *Construct a table* containing the classes along with their frequencies and relative frequencies.
- *Construct the relative frequency histogram* plotting the class intervals on the horizontal axis and drawing bars with heights corresponding to the appropriate relative frequencies.

Exercises

Basic Techniques

1.9 Construct a relative frequency histogram for the following set of data:

3.1	4.9	2.8	3.6	2.5
4.5	3.5	3.7	4.1	4.9
2.9	2.1	3.5	4.0	3.7
2.7	4.0	4.4	3.7	4.2
3.8	6.2	2.5	2.9	2.8
5.1	1.8	5.6	2.2	3.4
2.5	3.6	5.1	4.8	1.6
3.6	6.1	4.7	3.9	3.9
4.3	5.7	3.7	4.6	4.0
5.6	4.9	4.2	3.1	3.9

a Approximately how many class intervals should you use?

b Suppose you decide to use classes starting at 1.55 with a class width of .5 (i.e., 1.55 to 2.05, 2.05 to 2.55). Construct the relative frequency histogram for the data.

c What fraction of the measurements are less than 5.05?

d What fraction of the measurements are larger than 3.55?

1.10 The length of time (in months) between the onset of a particular illness and its recurrence was recorded for $n = 50$ patients. The times of recurrence are as follows:

2.1	4.4	2.7	32.3	9.9
9.0	2.0	6.6	3.9	1.6
14.7	9.6	16.7	7.4	8.2
19.2	6.9	4.3	3.3	1.2
4.1	18.4	.2	6.1	13.5
7.4	.2	8.3	.3	1.3
14.1	1.0	2.4	2.4	18.0
8.7	24.0	1.4	8.2	5.8
1.6	3.5	11.4	18.0	26.7
3.7	12.6	23.1	5.6	.4

a Construct a relative frequency histogram for the data.

b Give the fraction of recurrence times less than or equal to 10.

1.11 A discrete variable can take only the values 0, 1, or 2. A set of 20 measurements on this variable is shown below:

1	2	1	0	2
2	1	1	0	0
2	2	1	1	0
0	1	2	1	1

a Construct a relative frequency histogram for the data.

b What proportion of the measurements is greater than 1?

c What proportion of the measurements is less than 2?

d If a measurement is selected at random from the 20 measurements shown, what is the probability that it is a 2?

Applications

1.12 In the following table, the U.S. Bureau of the Census details the federal government's income from federal income taxes (on a per capita basis) as well as the amount spent per capita in federal aid for each of the 50 states in fiscal year 1991.

State	Taxes	U.S. Aid	State	Taxes	U.S. Aid
AL	964	566	MT	1007	755
AK	3168	1239	NE	1103	467
AZ	1256	396	NV	1312	330
AR	997	502	NH	565	399
CA	1477	578	NJ	1500	543
CO	951	466	NM	1347	595
CT	1514	646	NY	1567	816
DE	1713	532	NC	1165	454
FL	1040	345	ND	1189	772
GA	1080	510	OH	1056	507
HI	2325	644	OK	1216	475
ID	1159	516	OR	1037	623
IL	1151	416	PA	1088	465
IN	1102	460	RI	1251	737
IA	1233	518	SC	1104	575
KS	1120	453	SD	796	713
KY	1358	597	TN	870	551
LA	1013	668	TX	923	438
ME	1261	610	UT	1051	551
MD	1317	457	VT	1207	795
MA	1615	673	VA	1090	374
MI	1185	555	WA	1592	551
MN	1588	560	WV	1292	596
MS	948	704	WI	1416	532
MO	968	432	WY	1385	1218

Source: U.S. Department of Commerce (1992).

a Construct a relative frequency histogram to describe the data on per capita federal taxes for the 50 states.

b Construct a relative frequency histogram to describe the data on aid to the 50 states.

c Compare the shapes of the relative frequency histograms in parts (a) and (b). Are they similar or different?

1.13 The following data provide the rushing statistics for the top rushers in the National Football League (NFL) at the end of the 1992 season. Statistics include the number of attempts, the total number of rushing yards, and the average number of yard per carry.

American Football Conference				National Football Conference			
Name/Team	**Att.**	**Yds.**	**Avg.**	**Name/Team**	**Att.**	**Yds.**	**Avg.**
Foster, Pit.	328	1444	4.4	E. Smith, Dal.	304	1309	4.3
T. Thomas, Buf.	244	1155	4.7	B. Sanders, Det.	243	1048	4.3
White, Hou.	196	920	4.7	Cobb, T B	263	1016	3.9
Green, Cin.	209	868	4.2	Watters, S. F.	196	985	5.0
Warren, Sea.	183	805	4.4	Hampton, NY-G	215	971	4.5
Higgs, Mia.	204	718	3.5	Gary, Rams	230	923	4.0
Dickerson, Rai.	170	693	4.1	Allen, Min.	204	890	4.4
Butts, S. D.	183	616	3.5	Walker, Phi	215	820	3.8
Baxter, NY-J	135	623	4.6	Byner, Was.	213	801	3.8
Word, K. C.	158	603	3.8	Workman, G. B.	159	631	4.0
Green, Den.	136	568	4.2	Young, S. F.	62	484	7.8
K. Davis, Buf.	115	536	4.7	Anderson, Chi.	132	478	3.6
Johnson, Ind.	147	508	3.5	Cunningham, Phi.	72	448	6.2
Thomas, NY-J	97	440	4.5	Ervins, Was.	138	448	3.2
Bernstine, S. D.	83	439	5.3	Dunbar, N. O.	126	440	3.5
Fenner, Cin.	89	418	4.7	Hilliard, N. O.	97	391	4.0
Mack, Cle.	132	389	2.9	Sherman, Phi.	74	391	5.3
Vaughn, N. E.	89	370	4.2	Bunch, NY-G	85	386	4.5
Vardell, Cle.	98	369	3.8	Johnson, Pho.	94	341	3.6
Russell, N. E.	113	355	3.1	Heyward, N. O.	78	313	4.0

Source: The Sporting News (Dec. 14, 1992).

a Identify each of the three quantitative variables as either discrete or continuous.

b Construct two relative frequency histograms to describe the total rushing yards for the top rushers in the NFC and the AFC. Use the same class intervals for both histograms.

c Compare the two sets of data by comparing their relative frequency histograms.

1.14 Refer to Exercise 1.13.

a Construct a relative frequency histogram describing the average number of yards per carry for the top rushers in the NFL.

b What percentage of the top rushers in the NFL average 5 yards or more per carry?

1.15 Acquired immunodeficiency syndrome (AIDS) has become one of the most devastating diseases in modern society. The number of cases of AIDS (in thousands) reported in 25 major cities in the United States during 1992 are shown below:

38.3	6.2	3.7	2.6	2.1
14.6	5.6	3.7	2.3	2.0
11.9	5.5	3.4	2.2	2.0
6.6	4.6	3.1	2.2	1.9
6.3	4.5	2.7	2.1	1.8

Source: Wright (1992), p. 209.

a Construct a relative frequency histogram to describe the data.

b What proportion of the cities reported more than 10,000 cases of AIDS in 1992?

c What proportion of the cities reported fewer than 3000 cases of AIDS in 1992?

1.16 The officials of major league baseball have crowned a batting champion in the National League each year since 1876. A sample of winning batting averages is shown below.

Year	Name	Average	Year	Name	Average
1876	Roscoe Barnes	.403	1954	Willie Mays	.345
1893	Hugh Duffy	.378	1975	Bill Madlock	.354
1915	Larry Doyle	.320	1958	Richie Ashburn	.350
1917	Edd Roush	.341	1942	Ernie Lombardi	.330
1934	Paul Waner	.362	1948	Stan Musial	.376
1911	Honus Wagner	.334	1971	Joe Torre	.363
1898	Willie Keeler	.379	1913	Jake Daubert	.350
1924	Roger Hornsby	.424	1961	Roberto Clemente	.351
1963	Tommy Davis	.326	1968	Pete Rose	.335
1992	Gary Sheffield	.330	1885	Roger Connor	.371

Source: Wright (1992), p. 652.

a Construct a relative frequency histogram to describe the batting averages for these 20 champions.

b If you were to randomly choose one of the 20 names, what is the chance that you would choose a player whose average was above .400 for his championship year?

1.17 In order to decide on the number of service counters needed for stores to be built in the future, a supermarket chain wanted to obtain information on the length of time (in minutes) required to service customers. To obtain information on the distribution of customer service times, a sample of 1000 customers' service times was recorded. Sixty of these are shown in the following table:

3.6	1.9	2.1	.3	.8	.2
1.0	1.4	1.8	1.6	1.1	1.8
.3	1.1	.5	1.2	.6	1.1
.8	1.7	1.4	.2	1.3	3.1
.4	2.3	1.8	4.5	.9	.7
.6	2.8	2.5	1.1	.4	1.2
.4	1.3	.8	1.3	1.1	1.2
.8	1.0	.9	.7	3.1	1.7
1.1	2.2	1.6	1.9	5.2	.5
1.8	.3	1.1	.6	.7	.6

a Construct a relative frequency histogram for the data.

b What fraction of the service times are less than or equal to 1 minute?

1.5 Stem and Leaf Displays

The **stem and leaf display** is an alternative method for describing a set of data. This method was proposed by John Tukey as part of a newly emerging area of statistics called **exploratory data analysis** (EDA). The objective of EDA is to provide the experimenter with simple techniques that allow him or her to look more effectively at data. The stem and leaf display presents a histogram-like picture of the data, while allowing the experimenter to retain the actual observed values of each data point. Hence, the stem and leaf display is partly tabular and partly graphical in nature.

Table 1.10 gives the top 40 stocks on the over-the-counter (OTC) market ranked by percentage of outstanding shares traded on a particular day. In creating a stem and leaf display for these data, we divide each observation into two parts: the stem and the leaf. For example, we could divide each observation at the decimal point. The portion to the left of the point of division is the stem; the portion to the right is the leaf. Thus, the stem and leaf for the observation 7.15 are

Stem	Leaf
7	15

TABLE 1.10
Top 40 stocks on the OTC market

22.88	5.49	4.40	3.44	2.88
7.99	5.26	4.05	3.36	2.74
7.15	5.07	3.94	3.26	2.74
7.13	4.94	3.93	3.20	2.69
6.27	4.81	3.78	3.11	2.68
6.07	4.79	3.69	3.03	2.63
5.98	4.55	3.62	2.99	2.62
5.91	4.43	3.48	2.89	2.61

Alternatively, we could choose the point of division between the tenths and hundredths decimal places, whereby

Stem	Leaf
71	5

The choice of the stem and leaf coding depends on the nature of the data set. The stem and leaf display is constructed by using the following steps.

Constructing a Stem and Leaf Display

1. List the stem values, in order, in a vertical column.
2. Draw a vertical line to the right of the stem values.
3. For each observation, record the leaf portion of that observation in the row corresponding to the appropriate stem.
4. Reorder the leaves from lowest to highest within each stem row.
5. If the number of leaves appearing in each stem row is too large, divide the stems into two groups, the first corresponding to leaves beginning with digits 0 through 4 and the second corresponding to leaves beginning with digits 5 through 9. (This subdivision can be increased to five groups if necessary.)
6. Provide a key to your stem and leaf coding, so that the reader can recreate the actual measurements from your display.

The stem and leaf display for the data in Table 1.10 is shown in Figure 1.19. If we were to choose the point of division between the tenths and hundredths decimal places, there would be a large number of stems, from 26 to 228, and the display would not provide a good visual description of the data. Hence, we choose to use the decimal point as the point of division. As an aid to decoding, we write

Leaf unit = 0.10

5 00 represents 5.00

FIGURE 1.19
Stem and leaf display for the data in Table 1.10

2	61	62	63	68	69	74	74	88	89	99		
3	03	11	20	26	36	44	48	62	69	78	93	94
4	05	40	43	55	79	81	94					
5	07	26	49	91	98							
6	07	27										
7	13	15	99									
HI	22.88											

How does the stem and leaf display in Figure 1.19 describe the 40 OTC stocks? If you turn the display sideways, you can see that the stocks are displayed in a histogram-like picture. *The data are not symmetric,* with most of the observations falling between 2.00% and 5.00%. Notice that one extremely large observation (22.88) is listed as HI rather than extending the stem and leaf display to accommodate the one large stem.

The stem and leaf display has an advantage over the relative frequency histogram in that it allows you to reconstruct the actual data set; also, it lists the observations in order of magnitude. Hence, we can tell from the display that the lowest observation is 2.61 and the highest is 22.88. If we wish to find the fifth smallest measurement, we can count up from the smallest measurement and identify the fifth smallest as 2.69.

Sometimes the available stem choices result in a display containing too many stems (and very few leaves within a stem) or too few stems (and many leaves within a stem). This is the case with the Bucknell University grade point data (Table 1.7). In this situation, we may divide the too few stems by stretching them into two or more lines, depending on the leaf values with which they will be associated. Two options are available:

1 The stem is divided into two parts. The first is associated with leaves having 0, 1, 2, 3, or 4 as their first digit.

2 The stem is divided into five parts. The first is associated with leaves having 0 or 1 as their first digit; the second is associated with leaves having 2 or 3 as their first digit; and so on. The last stem part is associated with leaves having 8 or 9 as their first digit.

Since the data in Table 1.7 vary from 1.9 to 3.4, using the decimal point as the point of division would produce only three stem rows. Such a stem and leaf display would not produce a good descriptive picture of the data. We could divide each stem into two parts; this division would result in four stems (since the lower portion of the first stem and the upper portion of the last stem are unnecessary). Dividing each stem into five parts would result in nine stems and would produce a more visually descriptive display. The stem and leaf display for this option is shown in Figure 1.20. Notice the similarity between the stem and leaf display in Figure 1.20 and the relative frequency histogram in Figure 1.15. However, the stem and leaf display allows the reproduction of the actual data set if necessary.

FIGURE 1.20
Stem and leaf display for the data in Table 1.7

```
1 | 9 9
2 | 0 1 1
2 | 2 3
2 | 4 4 5 5 5 5 5
2 | 6 6 7 7 7 7 7
2 | 8 8 9 9
3 | 0 0 1 1
3 |
3 | 4
Leaf unit = 0.1
1 2 represents 1.2
```

A stem and leaf display for the OTC stock data can be generated by using the STEM AND LEAF command in the MINITAB package, as shown in Table 1.11. The MINITAB program chooses the decimal point as the point of division and divides each stem into two parts. The first column in the display is the number of leaves on that stem or on a stem closer to the nearer end of the display. For example, there are 17 data values less than or equal to 3.4, and there are 6 data values greater than or equal to 6.0. For the stem class containing the data value midway from each end, the number in parentheses gives the number of leaves for that stem. Notice that the computer truncates the second decimal place in the data, sacrificing some accuracy in the display in order to retain its simplicity and clarity.

TABLE 1.11
MINITAB stem and leaf display for the OTC stock data

```
MTB  > STEM AND LEAF C1;
SUBC > TRIM.

Stem-and-leaf of C1          N = 40
Leaf Unit = 0.10

     10    2   6666677889
     17    3   0122344
    (5)    3   66799
     18    4   044
     15    4   5789
     11    5   024
      8    5   99
      6    6   02
      4    6
      4    7   11
      2    7   9

          HI   228,
```

To summarize: a stem and leaf display is easy to construct, and the display creates the same sort of figure produced by a relative frequency histogram. In addition, it permits the user to reconstruct the data set and to identify observations ordered by their relative magnitude.

Exercises

Basic Techniques

1.18 Construct a stem and leaf display for the following data (which are the same as those in Exercise 1.9):

3.1	4.9	2.8	3.6	2.5
4.5	3.5	3.7	4.1	4.9
2.9	2.1	3.5	4.0	3.7
2.7	4.0	4.4	3.7	4.2
3.8	6.2	2.5	2.9	2.8
5.1	1.8	5.6	2.2	3.4
2.5	3.6	5.1	4.8	1.6
3.6	6.1	4.7	3.9	3.9
4.3	5.7	3.7	4.6	4.0
5.6	4.9	4.2	3.1	3.9

a Compare the stem and leaf display with the relative frequency histogram constructed in Exercise 1.9.

b Use the stem and leaf display to find the smallest observation and to find the eighth and ninth largest observations.

1.19 Use the following set of data to answer questions (a) and (b):

4.5	3.2	3.5	3.9	3.5	3.9
4.3	4.8	3.6	3.3	4.3	4.2
3.9	3.7	4.3	4.4	3.4	4.2
4.4	4.0	3.6	3.5	3.9	4.0

a Construct a stem and leaf display by using the leading digit as the stem.

b Construct a stem and leaf display by using each leading digit twice. Has this technique improved the presentation of the data?

Applications

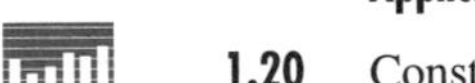

1.20 Construct a stem and leaf display for the supermarket service times given in Exercise 1.17. Compare the stem and leaf display with the relative frequency histogram constructed in that exercise. Do the two graphical descriptions of the data seem to convey the same information?

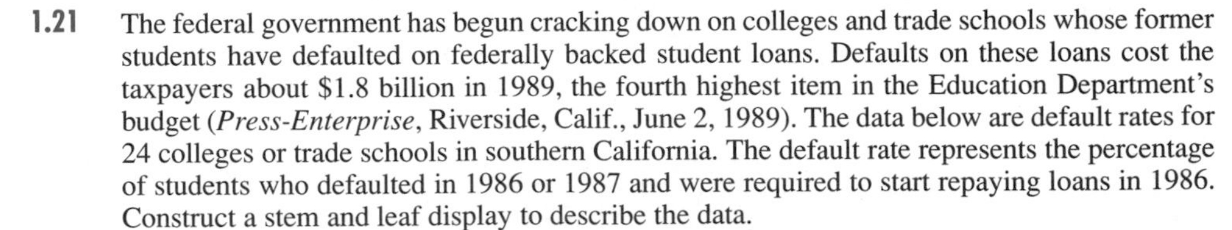

1.21 The federal government has begun cracking down on colleges and trade schools whose former students have defaulted on federally backed student loans. Defaults on these loans cost the taxpayers about $1.8 billion in 1989, the fourth highest item in the Education Department's budget (*Press-Enterprise*, Riverside, Calif., June 2, 1989). The data below are default rates for 24 colleges or trade schools in southern California. The default rate represents the percentage of students who defaulted in 1986 or 1987 and were required to start repaying loans in 1986. Construct a stem and leaf display to describe the data.

41.9	31.0	22.0	11.0	5.8
41.1	27.1	21.1	9.8	5.7
34.6	25.4	17.8	8.7	4.0
32.6	24.3	17.8	8.1	3.6
32.0	24.2	14.5	8.0	

Source: Data from U.S. Education Department.

1.22 Refer to Exercise 1.21. The MINITAB command STEM was used to produce the following stem and leaf display for the default rates in federally backed student loans for 24 southern California schools. Explain the choice of stem and leaf used in the MINITAB analysis. Does this display differ from the display you constructed in Exercise 1.21?

```
Stem-and-leaf of C1          N = 24
Leaf Unit = 1.0

    2     0   34
    8     0   558889
   10     1   14
   12     1   77
   12     2   1244
    8     2   57
    6     3   1224
    2     3
    2     4   11
```

1.6 Interpreting Graphs with a Critical Eye

Although graphical descriptive techniques are very useful for describing data, graphs must be interpreted with a critical eye. It is easy to construct graphs that may lead the unsuspecting reader to conclusions that are incorrect. One of the simplest ways to lead a reader astray is to shrink or stretch the scale of the axis of a graph.

For example, suppose that the number of accidents along a school bus route during the period September through January is recorded as 8, 9, 12, 13, and 12. If you want the number of accidents to appear large and volatile, you might present these data as in Figure 1.21(a). On the other hand, if you want the number of accidents to appear small and relatively constant, you might present these data as in Figure 1.21(b).

If you examine these two figures, you notice immediately that there is a marked difference in both the horizontal and vertical scales. *Stretching* the vertical scale and *shrinking* the horizontal scale cause relatively small changes to appear as large changes, whereas *shrinking* the vertical scale and *stretching* the horizontal scale cause both small and large changes to appear small. So, one must assess both the horizontal and vertical scales on any graph with a critical eye.

What *should* we look for when we are presented with a graph or graphs for a set of data? When examining a relative frequency histogram, there are several things to consider.

- First check the horizontal and vertical scales so that you are clear about what is being measured.

FIGURE 1.21 Two line graphs representing the same data set

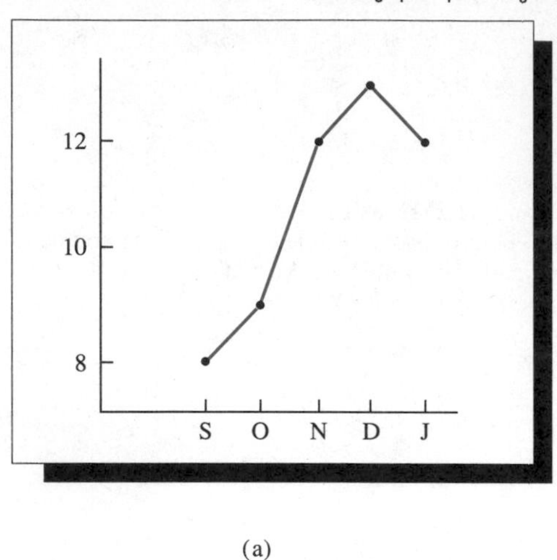

(a)

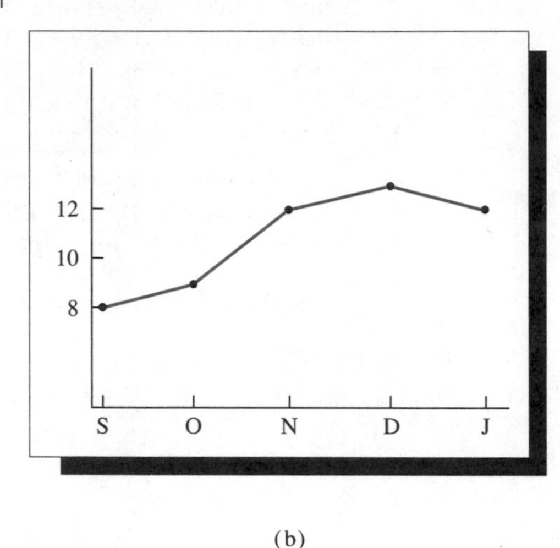

(b)

- Examine the *location* of the histogram. Where on the horizontal axis is the center of the histogram located? If you are comparing two histograms, are they both centered in the same place?
- Examine the *shape* of the histogram. Is there one relative frequency bar that is higher than any other, thereby identifying the most frequent value or class of values in the set? Are the relative frequency bars in the left and right halves of the graph equal?

Refer to the relative frequency histogram in Figure 1.22(a). The most frequent value is marked 4, which also marks the center of the histogram. If you were to fold the paper at this center, would the left and right halves be mirror images? If so, we would say that the relative frequency histogram is **symmetric**. The histogram in Figure 1.22(a) is symmetric about the value 4. However, the relative frequency histogram in Figure 1.22(b) is *not* symmetric, since the frequencies of the bars to the right of the highest bar marked 2 exceed the frequency of the bars to the left of 2.

How do we describe this lack of symmetry? A distribution is said to be **skewed to the right** if a greater proportion of the observations lies to the right of the highest relative frequency bar. We can also say that a distribution of measurements is *skewed to the right* when the "right tail" is longer than the "left tail." The distribution of a set of data may be **skewed left**, but, in general, nonsymmetric distributions are usually skewed to the right.

- Determine whether any of the measurements seem unusual; that is, are they bigger or smaller than all of the other measurements?

It is important to notice whether one or more observations lie *far* from the center. These measurements may be unusual observations that are not representative of the others in the set and are called **outliers.** The topic of outliers will be explained in Chapter 2.

FIGURE 1.22 Symmetric(a) and skewed(b) relative frequency histograms

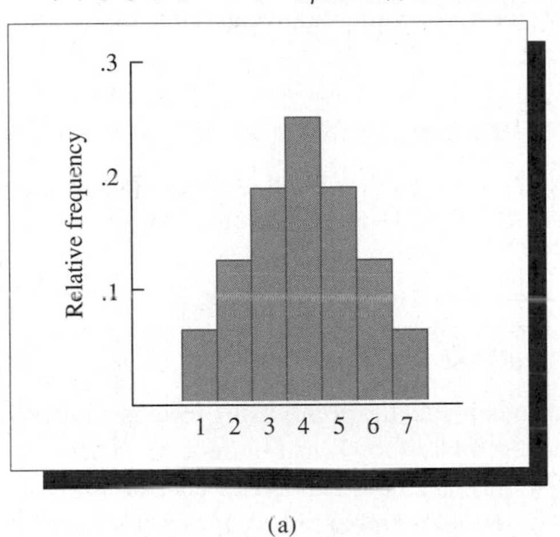

(a)

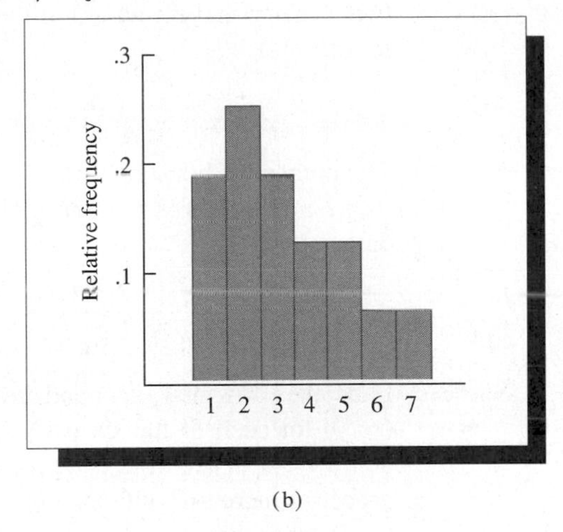

(b)

When comparing the graphical representations of two data sets, consider all the above characteristics. Of course, be especially careful to make sure that the scales on both graphs are the same. If the scales differ strongly, no valid comparisons or conclusions concerning the data sets can be made.

1.7 Graphical Techniques for Bivariate Data

When experiments or surveys are conducted, the researcher is often interested in two or more variables that can be measured during the investigation. For example, in a survey of policyholders conducted by an auto insurance company, the number of vehicles owned and the number of drivers per household are both important pieces of information for the company. In a survey of households in a certain city, the amount spent on groceries per week and the number of people in the household would be two quantitative variables of interest to an economist. In the same survey, the selling price of a residential property and the size of the living area in square feet are quantitative variables of interest to both a potential seller and a potential buyer in the city. It would not be difficult for us to think of other variables that would also be of interest to the investigator in these situations.

When two variables are measured on a single experimental unit, the resulting data are called *bivariate data*. How should we display the data? If we want to study each variable separately, the techniques described in earlier sections will be sufficient. However, if we wish to determine any inherent relationships between the two variables, other techniques must be used.

The comparative graphs illustrated in Section 1.3 were graphical techniques used for bivariate data when at least one of the two variables measured was *qualitative*. When both variables are *quantitative*, a bivariate plot is used, with one variable plotted along the horizontal axis and the second along the vertical axis. The first variable is often denoted by x and the second variable by y, so that the graph takes the form of a plot on the (x, y) axes, which are familiar to most of you. Each pair of data values is plotted as a point in this two-dimensional display, called a **scatterplot**. It is the two-dimensional equivalent of the scatterplot for one quantitative variable discussed in Section 1.4.

EXAMPLE 1.9 The number of household members, x, and the amount spent on groceries per week, y, are measured for six households in a local area. Draw a scatterplot of these six data points.

x	2	2	3	4	1	5
y	\$45.75	\$60.19	\$68.33	\$100.92	\$35.86	\$130.62

Solution Label the horizontal axis x and the vertical axis y. Plot the points using the coordinates (x, y) for each of the six pairs. The scatterplot is shown in Figure 1.23. There is definitely a discernible pattern even with only six data pairs. The cost of weekly groceries increases with the number of household members in an apparent straight-line relationship. ■

FIGURE 1.23 Scatterplot for the data of Example 1.9

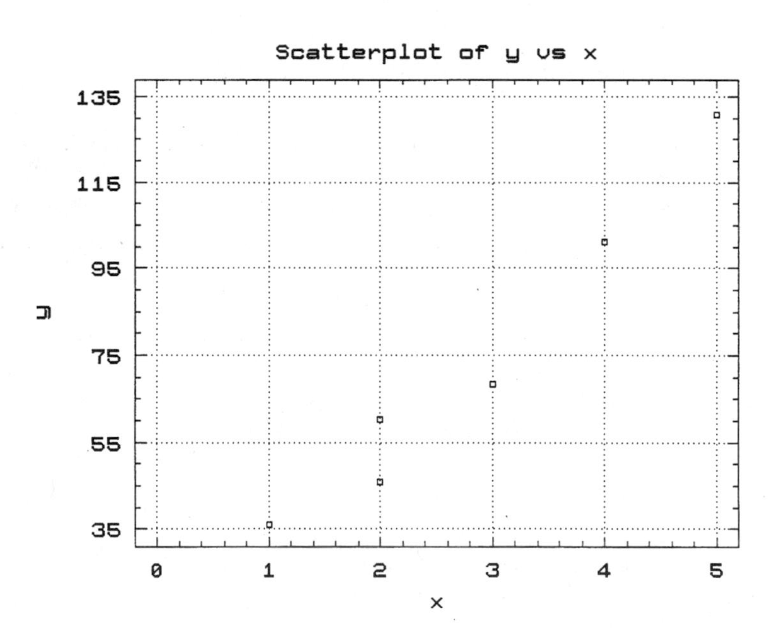

Although we expect demand for an item to decrease as the price increases when competing products are available at a lower price, this is not always the case. In fact, slight price increases may result in *reduced* demand for the item, and significant price increases result in the perception of increased quality and therefore *increased* demand.

EXAMPLE 1.10 In an attempt to develop an appropriate pricing policy, a distributor of table wines has conducted a study of the relationship between price and demand using a type of wine that ordinarily sells for $5.00 per 750 milliliter (ml) bottle. The distributor conducted the pricing study in ten different marketing areas over a 12-month period, using five different price levels. The data are given in Table 1.12. Construct a scatterplot for these data and comment on any apparent relationship that exists.

TABLE 1.12
Price data for Example 1.10

No. of Cases Sold/10,000 Population	Price per Bottle
23, 21	$5.00
19, 18	$5.50
15, 17	$6.00
19, 20	$6.50
25, 24	$7.00

Solution The ten data points are plotted in Figure 1.24. The plotted data decrease with an increase in price, reaching a low at $6.00 per bottle. The trend, then, is for sales to increase as the price increases, with demand for $7.00-per-bottle wine exceeding that for $5.00. Although we cannot say with certainty that this trend results because the increased price is an index of increased quality, the data seem to lend credence to this viewpoint. ▪

FIGURE 1.24
Scatterplot for the data of Example 1.10

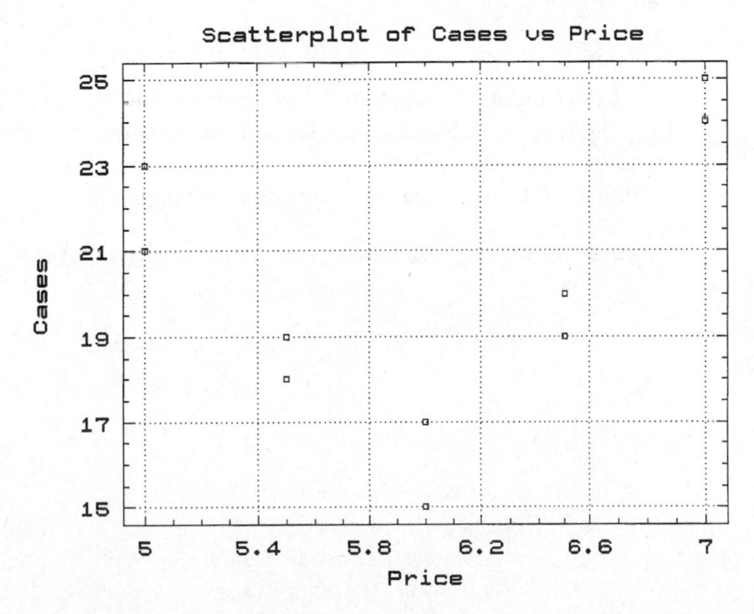

When data are collected on two variables, one of which is time, we may simply have a series of time-ordered observations. When this turns out to be the case and the observations are recorded at equally spaced units of time (such as daily, weekly, monthly, quarterly, or yearly), we refer to these data as a **time series.** These sets of data are most effectively presented using a line graph, with time as the horizontal axis. Again, the idea is to try to discern patterns that will likely continue into the future, thereby providing us with the basis for making predictions for the immediate future.

In Figure 1.10, a *3-D bar graph* was used to display the number of men and women enrolled in medical schools for each of the years 1986 through 1992. Notice that one variable used in constructing the graph was qualitative (gender: men or women) and the other (time) was quantitative. The resulting data were presented in $2 \times 7 = 14$ cross-classified categories.

If both variables are quantitative, a **3-D frequency histogram** can be created, using a procedure similar to the one described in Section 1.4. For each variable, we divide the range of observed values into several subintervals or classes. A two-dimensional system of classes is formed, and each pair of observations is classified according to the appropriate class cell. The histogram can be graphed by plotting the classes along two axes and then recording frequency (or relative frequency) of occurrence along the third axis.

EXAMPLE **1.11** In an experiment to determine the effect of ozone on the yield characteristics of broccoli, an experimenter recorded the fresh weight and the market weight for two varieties of broccoli grown under controlled ozone conditions. The resulting data consisted of pairs of observations, the first measurement being "fresh weight" and the second being "market weight." Fifty observations were taken, with typical observations such as the following:

$$(776, 171) \quad (449, 130)$$

The fresh weights varied from 280 to 780, and the market weights varied from 60 to 240. An EXECUSTAT computer software program was used to generate a 3-D frequency histogram for the data set. The histogram (Figure 1.25) shows a distinct "piling up" toward the center of the graph, although the pattern is quite irregular. ▪

FIGURE 1.25
3-D frequency histogram for two quantitative variables

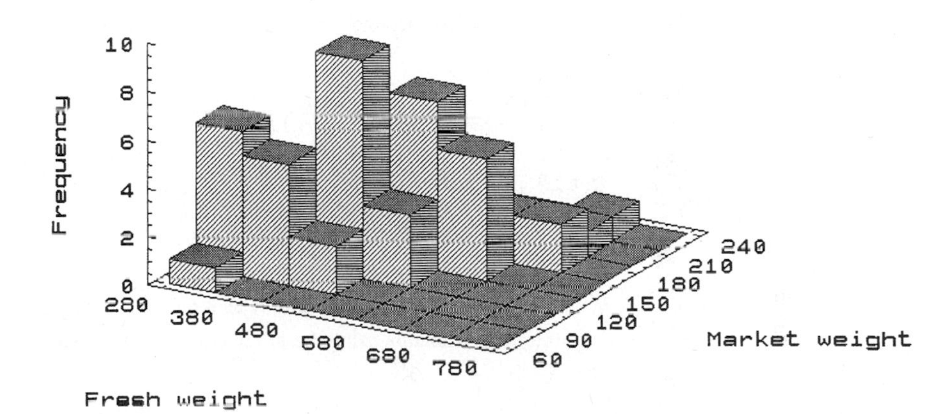

As you continue to work through the exercises in this chapter, you will become more experienced in recognizing different types of data and in determining the most appropriate type of graphical method to use. Remember that the type of graphic you use is not as important as the interpretation that accompanies the picture. The important characteristics to look for are the following:

- Location of center of the data
- Shape of the distribution of data
- Unusual observations in the data set
- Relationships between variables

Using these characteristics as a guide, you can interpret and compare sets of data using graphical methods, which are only the first of many statistical tools that you will soon have at your disposal.

Exercises

Basic Techniques

1.23 A set of bivariate data consists of measurements on two variables, x and y, as follows:

$$(3, 6) \quad (5, 8) \quad (2, 6) \quad (1, 4) \quad (4, 7) \quad (4, 6)$$

a Draw a scatterplot to describe the data.

b Does there appear to be a relationship between x and y? If so, how would you describe it?

1.24 Consider the set of bivariate data shown below:

x	1	2	3	4	5	6
y	5.6	4.6	4.5	3.7	3.2	2.7

a Draw a scatterplot to describe the data.

b Does there appear to be a relationship between x and y? If so, how would you describe it?

1.25 The value of a quantitative variable is measured once a year for a 10-year period. The data are shown below:

Year	Measurement	Year	Measurement
1	61.5	6	58.2
2	62.3	7	57.5
3	60.7	8	57.5
4	59.8	9	56.1
5	58.0	10	56.0

a Draw a line graph to describe the variable as it changes over time.

b Describe the measurements using the graph constructed in part (a).

Applications

1.26 The following data represent the average seller's asking price, the average buyer's bid, and the average closing bid for each of ten types of used computer equipment:

Machine	Average Seller's Asking Price	Average Buyer's Bid	Average Closing Bid
20MB PC XT	$ 400	$ 200	$ 300
20MB PC AT	700	400	575
IBM XT 089	450	200	325
IBM AT 339	700	350	600
20MB IBM PS/2 30	950	500	725
20MB IBM PS/2 50	1050	700	875
60MB IBM PS/2 70	2000	1600	1725
20MB Compaq SLT	1200	700	875
Toshiba 1600	1000	700	900
Toshiba 1200HB	1150	800	975

Source: "Used Computer Prices" (1992), p. 96.

a Draw a scatterplot relating average seller's asking price to average closing bid.

b Draw a scatterplot relating average buyer's bid to average closing bid.

c Compare the two scatterplots. Describe the data using the graphs from parts (a) and (b). What relationships appear to exist among the three variables?

1.27 Professor Isaac Asimov was one of the most prolific writers of all time. He wrote nearly 500 books during a 40-year career prior to his death in 1992. In fact, as his career progressed, he became even more productive in terms of the number of books written within a given period of time (Ohlsson, 1992). The data below give the time in months required to write his books in increments of 100:

Number of books	100	200	300	400	490
Time in months	237	350	419	465	507

a Plot the accumulated number of books as a function of time using a line graph.

b Describe the productivity of Professor Asimov in light of the data set graphed in part (a).

1.28 A social skills training program was implemented with seven mildly handicapped students in a study to determine whether the program caused improvement in pre/post measures and behavior ratings. For one such test, the pre- and posttest scores for the seven students are given below:

SSRS-T (Standard Score)

Subject	Pretest	Posttest
Earl	101	113
Ned	89	89
Jasper	112	121
Charlie	105	99
Tom	90	104
Susie	91	94
Lori	89	99

Source: Torrey, Vasa, Maag, and Kramer (1992), p. 248.

a Draw a scatterplot relating posttest score to pretest score.

b Describe the relationship between pre- and posttest scores using the graph in part (a). Can you see any trend?

1.29 As the United States becomes more aware of environmental problems, many cities are instituting curbside recycling programs in an attempt to conserve space in local landfills. These investments in recycling programs are more prevalent in some regions of the country than others. The following table shows the number of curbside recycling programs and the number of landfills in each of seven regions of the United States:

Region	Curbside Recycling Programs	Landfills
West	569	1374
Rocky Mountain	44	661
Midwest	108	1402
Great Lakes	1148	531
South	402	1007
Mid-Atlantic	1379	334
New England	305	503

Source: EPA Journal (July/August 1992).

a Draw a scatterplot relating number of curbside recycling programs to number of landfills for these seven regions.

b Describe the relationship between these two variables using the scatterplot from part (a).

c Can you explain why there is such a large difference in the pattern of these two variables from region to region?

1.8 Summary

A variable is a trait or characteristic that changes over time or from individual to individual. Data result when variables are measured for individuals or experimental units in the population. Our immediate goal is to present a rapid, visual presentation of a set of data, whether it be a population or a sample from a population. Variables are classified as being qualitative, denoting a category or group of items similar in kind, or quantitative, denoting an amount or quantity. Quantitative variables are further subdivided into two categories. Discrete variables are those whose values can be counted, such as the number of individuals who have carpooled for the last 7 days or the number of tosses of a penny until the first head appears. Continuous variables take values that correspond to all possible points on a given line interval, such as heights and weights. Bivariate data, in which two variables are measured on a single experimental unit, can be either qualitative or quantitative.

Pie graphs, bar graphs, and line graphs are commonly used in visual summaries or presentations of data. Qualitative data are usually presented using pie graphs, which emphasize the relation of the parts to the whole, and bar graphs, which highlight actual amounts for each of the categories considered. If the categories are ordered, a line graph can be used to indicate trends or changes when moving from one category to another.

Graphical summaries of quantitative variables are presented as frequency or relative frequency histograms or as stem and leaf displays. In constructing histograms, the number of classes depends on the size of the data set, and, in general, should vary from 5 or 6 classes for small data sets to as many as 20 classes for large data sets. The classes in a histogram correspond to the stems in a stem and leaf display. Both summaries display how the measurements are distributed across the classes or stems.

In general, graphs should be straightforward in interpretation, with axes clearly labeled. We should always examine graphs with a critical eye, checking for stretching or shrinking of axes that could distort the graph.

Supplementary Exercises

1.30 Identify the following variables as quantitative or qualitative:

a The ethnic origin of a candidate for public office

b The score (0–100) on a placement examination

c The fast-food establishment preferred by a student (McDonald's, Burger King, or Carl's Jr.)

d The mercury concentration in a sample of tuna

1.31 Would you expect the distributions of the following variables to be symmetric or skewed? Explain.

a The size in dollars of nonsecured loans

b The size in dollars of secured loans

c The price of an 8-ounce can of peas

d The height in inches of freshman women at your university

- **e** The number of broken taco shells in a package of 100 shells
- **f** The number of ticks found on each of 50 trapped cottontail rabbits

1.32 Identify the following as continuous or discrete variables:

- **a** The number of homicides in Detroit during a 1-month period
- **b** The length of time between arrivals at an outpatient clinic
- **c** The number of typing errors on a page of manuscript
- **d** The number of defective light bulbs in a packet containing four bulbs
- **e** The time required to finish an examination

1.33 Identify the following as continuous or discrete variables:

- **a** The weight of two dozen shrimp
- **b** A person's body temperature
- **c** The number of people waiting for treatment at a hospital emergency room
- **d** The number of properties for sale by a real estate agency
- **e** The number of claims received by an insurance company during 1 day

1.34 Identify the following as continuous or discrete variables:

- **a** The number of people in line at a supermarket checkout counter
- **b** The depth of a snowfall
- **c** The length of time for a driver to respond when faced with an impending collision
- **d** The number of aircraft arriving at the Atlanta airport in a given hour

1.35 As more and more mothers have entered the work force, problems have arisen concerning adequate child care, paid family leave, latchkey children, equal pay, and so on. The graph in Figure 1.26 shows the percentage of mothers in the paid work force for the years 1970–1990.

FIGURE **1.26** Graph for Exercise 1.35

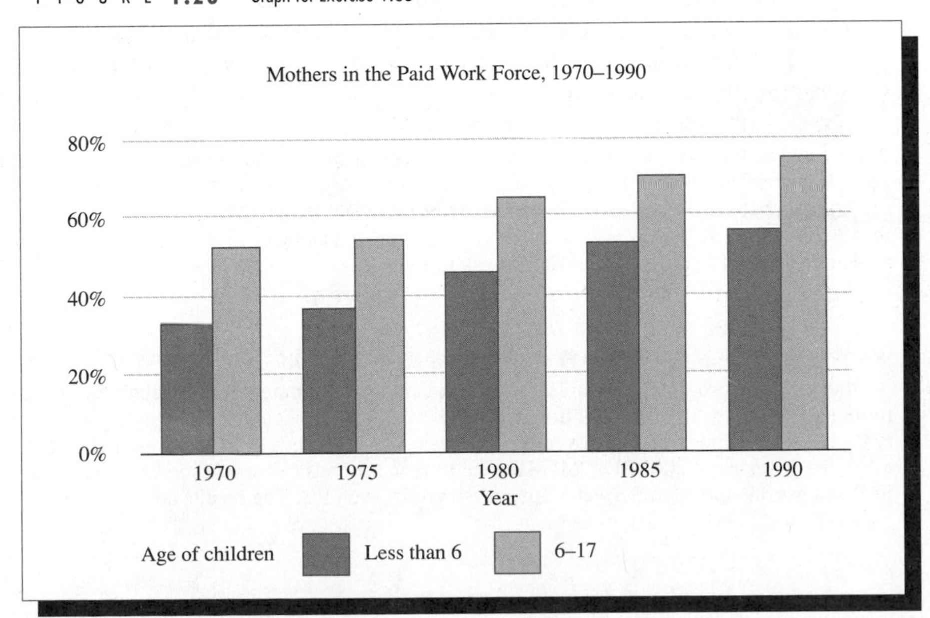

Source: U.S. Department of Labor (1992), p. 50

a What variables have been measured in this study? Are the variables qualitative or quantitative?

b Describe the population of interest. Do these data represent a population or a sample drawn from the population?

c What type of graphical presentation has been used? What other type could have been used?

d If you wanted to make the increase in the percentage of mothers in the paid work force look as dramatic as possible, what changes would you make in the graphical presentation?

1.36 An American household is defined by the census bureau as "an occupied dwelling that has its own entrance and kitchen," a definition that allows for great diversity in the makeup of a household. Over the last two decades, household types have changed considerably, as shown in the following table:

Household Type	1970	1980	1990
Married couples with children	38.7%	30.2%	26.7%
Married couples without children	30.8	30.0	28.4
Single-parent households	5.4	7.3	9.3
Other family households	5.5	5.8	5.7
Single-person households	17.6	22.7	24.6
Other nonfamily households	2.1	4.1	5.3

Source: "Changing American Household" (1992), p. 2.

a What variables have been measured in this study? Are they qualitative or quantitative?

b Construct segmented bar graphs for the 3 years to describe the percentages of the total households belonging to each of the six types. Do you see any trends over time?

c What other types of comparative graphs might you use to describe these data? Which method of presentation would be most effective?

1.37 Aqua running has been suggested as a method for cardiovascular conditioning for injured athletes and others who want a low-impact aerobics program. In a study to investigate the relationship between exercise cadence and heart rate (Wilder, Brennan, and Schotte, 1993), the heart rates of 20 healthy volunteers were measured at a cadence of 48 cycles per minute (a cycle consisted of two steps). The data are as follows:

87	109	79	80	96	95	90	92	96	98
101	91	78	112	94	98	94	107	81	96

Construct a stem and leaf display to describe the data. Discuss the characteristics of the display.

1.38 As the software programs available for use on personal computers (PCs) become more and more sophisticated, the speed with which a PC executes a particular task becomes important to users. In a comparison of the overall performance of 25 brands of PCs (all with 8MB RAM and hard drives from 200MB to 244MB), the time in seconds to completion of an application in Windows 3.0 and WordPerfect 5.1 is measured for each PC. The results are shown below:

Windows 3.0					WordPerfect 5.1				
46	46	47	51	55	32	33	35	35	36
54	56	59	56	61	36	37	38	38	37
59	63	58	59	55	38	39	37	39	40
63	58	60	54	60	40	40	40	42	41
69	74	69	70	69	41	42	42	42	46

Source: "Overdrive Boosts 486SX-25s" (1992), p. 192.

a Construct a stem and leaf display to describe the completion times for the Windows 3.0 application.

b Construct a stem and leaf display to describe the completion times for the WordPerfect 5.1 application.

c Compare the displays in parts (a) and (b). Are the locations of the two displays different? Are the shapes different? Are there any unusual observations?

1.39 Refer to Exercise 1.38. The MINITAB command STEM produced the output shown below for the Windows 3.0 and WordPerfect 5.1 completion times. Compare the displays with the stem and leaf displays constructed in Exercise 1.38. Explain the similarities and differences in the computer and hand-generated displays.

```
Stem-and-leaf of WIN3-0    N  =  25
Leaf Unit = 1.0

    3     4 667
    6     5 144
   (9)    5 556688999
   10     6 00133
    5     6 999
    2     7 04


Stem-and-leaf of WP5-1     N  =  25
Leaf Unit = 1.0

    2     3 23
    4     3 55
    9     3 66777
   (5)    3 88899
   11     4 000011
    5     4 2222
    1     4
    1     4 6
```

1.40 In the study of demographics, there is a difference in terminology among cities, urbanized areas, and metropolitan areas, with each of the three representing a subset of the next. That is, cities are included within urbanized areas, and urbanized areas are included within metropolitan areas. Metro areas are typically less dense and cover a larger area than do urbanized areas. Similarly, cities are more dense and cover less area than urbanized areas. The following data give the 50 densest urbanized areas in the United States, classified by state, population per square mile, and total areas in square miles:

Rank	Name of Urbanized Area	Population per Square Mile	Total Area in Square Miles
1	Los Angeles, CA	5801	1966
2	Miami–Hialeah, FL	5429	353
3	New York, NY–Northeastern New Jersey	5409	2967
4	Davis, CA	4956	11
5	Honolulu, HI	4561	139
6	Modesto, CA	4424	52
7	San Luis Obispo, CA	4357	12
8	Chicago, IL–Northwestern Indiana	4287	1585
9	San Jose, CA	4240	338
10	San Francisco–Oakland, CA	4152	874
11	New Orleans, LA	3852	270
12	Champaign–Urbana, IL	3815	30
13	Longmont, CO	3806	14
14	Fort Lauderdale–Hollywood–Pompano Beach, FL	3785	327
15	Laredo, TX	3757	33
16	Santa Barbara, CA	3733	49
17	Philadelphia, PA-NJ	3627	1164
18	Lodi, CA	3616	15
19	Washington, DC–MD–VA	3560	945
20	Stockton, CA	3552	74
21	Salinas, CA	3524	35
22	Santa Maria, CA	3520	25
23	Fresno, CA	3417	133
24	San Diego, CA	3403	690
25	Bloomington, IN	3383	21
26	Buffalo–Niagara Falls, NY	3343	286
27	New Bedford, MA	3313	42
28	Denver, CO	3308	459
29	Detroit, MI	3303	1119
30	Sacramento, CA	3286	334
31	Kailua, HI	3268	35
32	Merced, CA	3265	20
33	Napa, CA	3233	21
34	Baltimore, MD	3190	593
35	Lafayette–West Lafayette, IN	3155	32
36	Vacaville, CA	3142	23
37	Racine, WI	3135	39
38	Fall River, MA–RI	3135	46
39	Trenton, NJ–PA	3116	96
40	Reading, PA	3115	60
41	Great Falls, MT	3114	20
42	Boston, MA	3114	891
43	Salt Lake City, UT	3107	254
44	Brownsville, TX	3088	38
45	Erie, PA	3080	58
46	Bakersfield, CA	3079	98
47	Grand Forks, ND–MN	3074	19
48	Boulder, CO	3071	32
49	Oxnard–Ventura, CA	3059	157
50	Bloomington–Normal, IL	3055	31

Source: Larson (1993), p. 38.

a Tabulate the 50 areas according to state. (Use the first state listed.) Draw a pie graph and a bar graph to describe the data. Which is the most effective? What conclusions can you draw from the graphical presentation?

b Use a relative frequency histogram to describe the distribution of total areas for the 50 urbanized areas. Describe the shape of the histogram. Are there any unusual observations? Explain.

c Use a relative frequency histogram to describe the distribution of population per square mile.

d Are these 50 measurements a representative sample of the same variables measured on all urbanized areas in the United States? Explain.

1.41 An advertising flyer for the *Princeton Review* (Irvine, Calif., 1993), a review course designed for high school students taking the SAT tests, presents a bar graph showing the average score improvements for students using various study methods. The graph is shown in Figure 1.27.

a What graphical techniques did the *Princeton Review* use to make their average improvement figures look as dramatic as possible?

b If you were in charge of promoting a review course at your high school, how would you modify the graph to make the average improvement for students using a school review course look more impressive?

FIGURE 1.27
Graph for Exercise 1.41

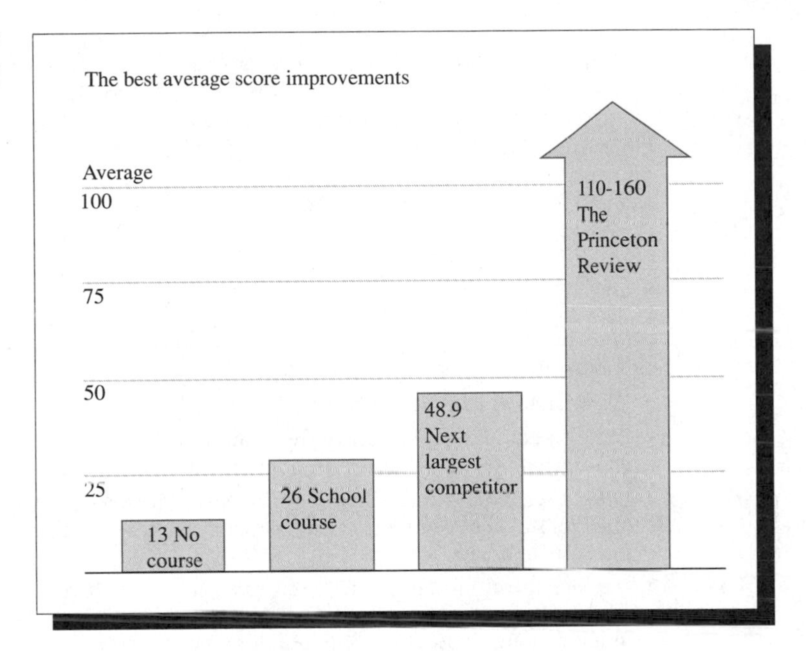

1.42 In anticipation of a family-leave bill that was passed shortly after the election of Bill Clinton, 163 mid- and large-sized companies were asked how costly and/or disruptive it would be to add or expand family-leave benefits. The survey found the following results:

	Costly	**Disruptive**
Very	17%	21%
Somewhat	51	49
Slightly	31	25
Not at All	0	7

Source: Rowland (1993), p. F16.

a Identify the population of interest, and describe the sample. What variables have been measured? Are they quantitative or qualitative?

b The survey notes that the figures do not add up to 100 due to rounding. What does this mean?

c Describe the data using comparative pie graphs for the two questions.

d What other types of graphical techniques could you use in this instance?

1.43 A study in *Psychological Reports* explored the relationship of mental health services (as reflected in per capita expenditures on psychiatric hospitals, the number of beds per 1000 population, and length of stay in days) to the gross domestic product (GDP) for selected countries. A portion of the results are shown in the following table:

Country	GDP per capita	Expenditures per capita	Beds/1000 Population	Length of stay
Finland	18,045.00	83.53	2.6	27.2
The Netherlands	14,449.80	71.38	1.7	35.6
Denmark	19,842.15	64.79	0.9	11.3
USA	17,670.15	41.21	0.7	12.3
Spain	7563.70	9.17	0.8	156.8
Japan	19,763.63	54.98	2.8	310.8

Source: Pillay (1992), pp. 723–726.

a What variables have been measured in this study?

b Draw a scatterplot to examine the relationship between per capita gross domestic product and expenditures on psychiatric hospitals for these six countries. Describe the relationship.

c Draw a scatterplot to examine the relationship between beds per 1000 population and length of stay for these six countries. Describe the relationship. Are there any observations that seem unusual?

1.44 Many nonprofit organizations have used newspaper recycling as a means of fund-raising. However, in the last few years the price per ton for old newspapers has dropped to the point at which it is no longer cost effective to use this type of fund-raising. The prices per ton for old newspapers for the years 1988–1992 are given below for four cities in the United States:

City	1988	1989	1990	1991	1992
SF/LA	61	32	30	40	31
Atlanta	50	22	16	18	16
Chicago	43	14	10	9	8
NYC	48	3	−3	−1	−3

Source: Adapted from *EPA Journal* (July/August 1992), p. 31.

a On the same set of axes, draw a line graph for each of the cities showing the changes in price over time. Label each line graph.

b Describe the similarities and differences in the four graphs in part (a).

c What other types of graphical presentations could be used in this situation? Would any of them be more effective?

1.45 The following data set represents the winning times (in seconds) for the Kentucky Derby races from 1950 to 1991:

122.4	121.3	122.3	121.3	122.0	123.0	122.0
123.2	121.4	123.2	122.1	125.0	122.1	122.0
122.2	122.2	124.0	120.2	121.4	120.0	122.2
125.0	121.1	122.0	120.3	122.1	121.4	122.1
122.0	123.2	123.1	121.4	119.2[†]	124.0	122.2
123.0	122.0	121.3	122.1	121.1	122.2	120.1

[†]Record time set by Secretariat in 1973.

Source: Wright (1992), p. 674.

a Do you think there will be a relationship between the winning time and the year in which the horse raced?

b Construct a relative frequency histogram to describe the distribution of winning times. Is the distribution symmetric or skewed? Are there any unusual observations?

1.46 The following table gives the percentages of Ph.D. degree holders in the field of chemistry from 1980 to 1989 who were hired by various types of employers:

Employer	Percentage
Domestic academe	26.64
Foreign academe	4.93
Industry	34.85
Government	7.30
Medical facility	2.55
Other	3.28
Unknown	20.44

Source: Miller (1993), p. 5.

a Draw a pie graph to describe the data.

b The percentage falling into the category "unknown" is quite large. What effect does this have on the interpretation of the data? Why would this percentage be so high?

1.47 In an article entitled "If Fido Takes a Piece of Postman, Feds May Take a Chunk Out of You" (1981), the *Wall Street Journal* noted that the Postal Service was helping letter carriers assemble evidence needed to seek damages from dog owners for the harmful behavior of their pets. The article noted that the mean cost in medical bills and lost time per dog bite was $300; but, of course, the actual cost could run much higher. Suppose that the distribution of cost per bite has a mean equal to $300. Why might the relative frequency distribution of costs per dog bite be skewed to the right as shown in Figure 1.28?

FIGURE 1.28
Graph for Example 1.47

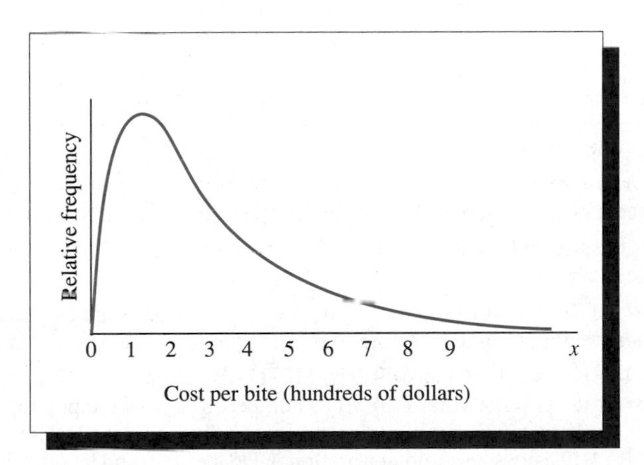

1.48 The 1992 election was a three-way race, in which Bill Clinton defeated incumbent president George Bush and independent Ross Perot by a substantial margin. The popular vote (in thousands) for President Clinton in each of the 50 states is shown in the following table:

AL	686.1	HI	178.9	MA	1315.0	NM	258.4	SD	124.9
AK	63.5	ID	136.2	MI	1858.3	NY	3246.8	TN	933.5
AZ	525.0	IL	2379.5	MN	997.0	NC	1103.7	TX	2278.7
AR	497.9	IN	829.2	MS	392.9	ND	98.9	UT	182.9
CA	4812.3	IA	583.5	MO	1053.0	OH	1965.2	VT	125.8
CO	626.2	KS	386.8	MT	153.9	OK	473.1	VA	1034.8
CT	681.1	KY	664.2	NE	214.1	OR	525.1	WA	855.7
DE	126.0	LA	815.3	NV	185.4	PA	2226.0	WV	326.9
FL	2051.8	ME	261.9	NH	207.3	RI	198.9	WI	1035.9
GA	1005.6	MD	942.0	NJ	1366.6	SC	476.3	WY	67.9

Source: Wright (1992), p. 2.

a Draw a relative frequency histogram to describe the distribution of the popular vote for Clinton in the 50 states for the 1992 election.

b Is the distribution of popular vote symmetric or skewed? Are there any unusually small or large observations? How would you account for this?

1.49 Refer to Exercise 1.48. In the same article, the *percentage* of popular vote gleaned by President Clinton in each of the 50 states was listed. The data are shown below:

AL	41	HI	49	MA	48	NM	46	SD	37
AK	32	ID	29	MI	44	NY	50	TN	47
AZ	37	IL	48	MN	44	NC	43	TX	37
AR	54	IN	37	MS	41	ND	32	UT	26
CA	47	IA	44	MO	44	OH	40	VT	46
CO	40	KS	34	MT	38	OK	34	VA	41
CT	42	KY	45	NE	30	OR	43	WA	44
DE	44	LA	46	NV	38	PA	45	WV	49
FL	39	ME	39	NH	39	RI	48	WI	41
GA	44	MD	50	NJ	43	SC	40	WY	34

Source: Wright (1992), p. 2.

a Draw a relative frequency histogram to describe the distribution of President Clinton's percentage of the popular vote in the 50 states for the 1992 election.

b Is the distribution of popular vote symmetric or skewed? Are there any unusually small or large observations?

c In the District of Columbia, Clinton received 86% of the popular vote. Is this an unusual observation in light of the distribution in part (a)? Can you explain why?

1.50 Refer to Exercises 1.48 and 1.49. The MINITAB command STEM produced the output shown below for the popular vote (POPVTE) and percentage of the popular vote (PCTVTE) for the 50 states. Answer the questions posed in part (b) of Exercise 1.48 and parts (b) and (c) of Exercise 1.49 using the computer printouts. Do the stem and leaf displays resemble the relative frequency histograms constructed in Exercises 1.48 and 1.49?

```
Stem-and-leaf of POPVTE        N  = 50
Leaf Unit = 100

   22     0  0001111111112222333444
  (13)    0  5556666888999
   15     1  0000133
    8     1  89
    6     2  0223
    2     2
    2     3  2
    1     3
    1     4
    1     4  8
```

```
Stem-and-leaf of PCTVTE      N  = 50
Leaf Unit = 1.0

    1    2  6
    2    2  9
    3    3  0
    5    3  22
    8    3  444
   12    3  7777
   17    3  88999
   24    4  0001111
  (4)    4  2333
   22    4  444444455
   13    4  66677
    8    4  88899
    3    5  00
    1    5
    1    5  4
```

Exercises Using the Data Disk

1.51 Refer to the *Fortune 500* largest U.S. Industrial Corporations on the data disk.

- **a** Scan the data set. Do you think that the distribution of sales (in millions) is symmetric or skewed?
- **b** Draw a sample of 25 companies from the list of 500 companies. Construct a relative frequency histogram to describe the sales for this data set.
- **c** Draw a stem and leaf display for the sales of the 25 companies in part (b).
- **d** Which of the two methods of graphical presentation was more effective? Was your answer to part (a) confirmed by the graphs?

1.52 Refer to the batting averages for the batting champions of both the National and American Leagues given on the data disk.

- **a** Choose a period of 20 sequential years and record the batting averages for both the National and American League champions. Draw comparative line charts for the two leagues on the same graph.
- **b** Using the results of part (a), does it appear that there is any trend in the averages over time?

1.53 Draw a sample of 50 systolic blood pressures from the data set for males on the data disk. Construct a relative frequency histogram for the data. Suppose we were to regard the set of 965 male blood pressures as a population. Would your sample relative frequency histogram be similar to the population relative frequency histogram?

1.54 Construct a stem and leaf display for your sample of 50 male systolic blood pressures given in Exercise 1.53. Compare it with the relative frequency histogram you constructed in Exercise 1.53.

1.55 Draw a sample of 50 systolic blood pressures from the data set for females on the data disk. Construct a relative frequency histogram for the data using the same class intervals as used for the data set in Exercise 1.53. Does the histogram for women resemble the histogram for men constructed in Exercise 1.53?

1.56 Construct a stem and leaf display for the sample of 50 female systolic blood pressures drawn in Exercise 1.55. Compare it with the relative frequency histogram you constructed in Exercise 1.55.

ASE STUDY

How Is Your Blood Pressure?

Blood pressure is the pressure that the blood exerts against the walls of the arteries. When physicians or nurses measure your blood pressure, they take two readings. The **systolic** pressure is the pressure when the heart is contracting and therefore pumping. The **diastolic** blood pressure is the pressure in the arteries when the heart is relaxing. The diastolic blood pressure is always the lower of the two readings. Blood pressure varies from one person to another. It will also vary for a single individual from day to day and even within a given day.

If your blood pressure is too high, it can lead to a stroke or a heart attack. If it is too low, blood will not get to your extremities and you may feel dizzy. Low blood pressure is usually not serious.

So, what should *your* blood pressure be? A systolic blood pressure of 120 would be considered normal. One of 150 would be high. But since blood pressure varies with sex and increases with age, a better gauge of the relative standing of your blood pressure would be obtained by comparing it with the population of blood pressures of all persons of your sex and age in the United States. Of course, we cannot supply you with that data set, but we can show you a very large sample selected from it. The data disk provides blood pressure data on 1910 persons, 965 males and 945 females between the ages of 15 and 20. The data are part of a health survey conducted by the National Institutes of Health (NIH). Entries for each person include that person's age and systolic and diastolic blood pressures at the time the blood pressure was recorded.

1 Describe the variables that have been measured in this survey. Are the variables quantitative or qualitative? Discrete or continuous? Are the data univariate, bivariate, or multivariate?

2 What types of graphical methods are available for describing this data set? What types of questions could be answered using various types of graphical techniques?

3 Using the systolic blood pressure data set, construct a relative frequency histogram for the 965 men and another for the 945 women. Use a statistical software package if you have access to one. Compare the two histograms.

4 One question that might be of interest is the relationship between systolic blood pressure and age. Choose a sample of $n = 50$ men and $n = 50$ women, recording their systolic blood pressures and their ages. Draw two scatterplots to graphically display the relationship for men and women. What conclusions can you draw?

5 How does your blood pressure compare to that of others in a comparable sex and age group? Check your blood pressure against the scatterplot in part (4) to determine whether your blood pressure is "normal" or whether it is too high or too low.

2

Describing Sets of Measurements: Numerical Techniques

Case Study

Are the baseball champions of today better than those of "yesteryear"? Do players in the National League hit better than players in the American League? The case study at the end of this chapter involves the batting averages of major league batting champions. Numerical descriptive measures can be used to answer these and similar questions.

General Objectives

Graphical techniques are extremely useful for visual description of a data set. However, they are not easy to use when we want to make inferences about a population using the information contained in a sample. For this latter purpose, we turn instead to numerical descriptive measures, in which we use numbers to construct a mental picture of the data. These numerical measures will be useful in achieving our objective: making inferences about the population based on sample information.

Specific Topics

1. Measures of central tendency: mean, median, mode (2.2)
2. Measures of variability: range, variance, and standard deviation (2.3)
3. Tchebysheff's Theorem and the Empirical Rule (2.4)
4. Measures of relative standing: z-scores, percentiles, and quartiles (2.6)
5. Box plots (2.7)

2.1 Numerical Methods for Describing a Set of Data

Graphical methods are extremely useful for conveying an immediate pictorial description of data—"a picture is worth a thousand words." There are, however, limitations to the use of graphical techniques. For instance, suppose we want to describe our data verbally before a group of people, but there is no overhead projector available! We would thus be forced to find other ways to convey to the listeners a mental picture of the data.

A second less obvious limitation of graphical techniques is that they are difficult to use for statistical inference. For example, suppose we want to use a sample histogram to make inferences about a population histogram, which graphically describes the population (which is unknown to us) from which the sample was drawn. How can we measure the similarities and differences between the two histograms in some sort of concrete way? If they were identical, we could say "They are the same!" But, if they are different, it is difficult to describe the "degree of difference."

The limitations of graphical methods for describing data can be overcome by the use of **numerical descriptive measures**, which can be calculated for either a sample or a population of measurements. We use the data to calculate a set of *numbers* that will convey a good mental picture of the frequency distribution. These measures are called **parameters** when calculated from population measurements and are called **statistics** when calculated from sample measurements. Statistics, calculated from sample information, are useful in making inferences concerning their corresponding population parameters.

DEFINITION ▪ Numerical descriptive measures computed from population measurements are called **parameters**; those computed from sample measurements are called **statistics**. ▪

2.2 Measures of Central Tendency

In constructing a mental picture of the frequency distribution for a set of measurements, we would likely envision a histogram similar to that shown in Figure 1.15 for the data on grade point averages (GPAs). One of the first descriptive measures of interest would be a **measure of central tendency**—that is, a measure that locates the center of the distribution. The GPA data ranged from a low of 1.9 to a high of 3.4, with the center of the histogram being located in the vicinity of 2.6. Let's consider some definite rules for locating the center of a distribution of measurements.

The arithmetic average of a set of measurements is the most common and useful measure of central tendency. This measure is also often referred to as the **arithmetic mean**, or simply the **mean**, of a set of measurements. To distinguish between the mean for the sample and the mean for the population, we will use the symbol $\bar{x}$ (x-bar) to represent the sample mean and μ (Greek lowercase letter mu) to represent the mean of the population.

DEFINITION ▪ The **arithmetic mean** of the set of measurements $x_1, x_2, \ldots, x_n$ is equal to the sum of the measurements divided by n. ▪

The procedures for calculating a sample mean and many other statistics are usually expressed as formulas. Since many of these formulas involve adding or summing numbers, we will need a symbol to represent the process of summation. If we denote the n quantities that are to be summed as $x_1, x_2, x_3, \ldots, x_n$, then their sum is denoted by the symbol

$$\sum_{i=1}^{n} x_i$$

The symbol $\sum_{i=1}^{n}$ ($\sum$ is the Greek capital sigma) tells us to sum the elements that appear to the right of the $\sum$ sign, beginning with x_1 (i.e., $i = 1$) and continuing in order to x_n. Thus,

$$\sum_{i=1}^{3} x_i = x_1 + x_2 + x_3$$

and

$$\sum_{i=1}^{n} x_i = x_1 + x_2 + \cdots + x_n$$

Using this notation, we can express the formula for the sample mean as follows:

Notation

Sample mean: $\bar{x} = \dfrac{\sum_{i=1}^{n} x_i}{n}$

Population mean: μ

EXAMPLE 2.1 Use a scatterplot to display the $n = 5$ measurements 2, 9, 11, 5, 6. Find the sample mean of these observations, and compare its value with what you might consider the "center" of these observations on the scatterplot.

Solution The following scatterplot is a simple way to display a small set of data:

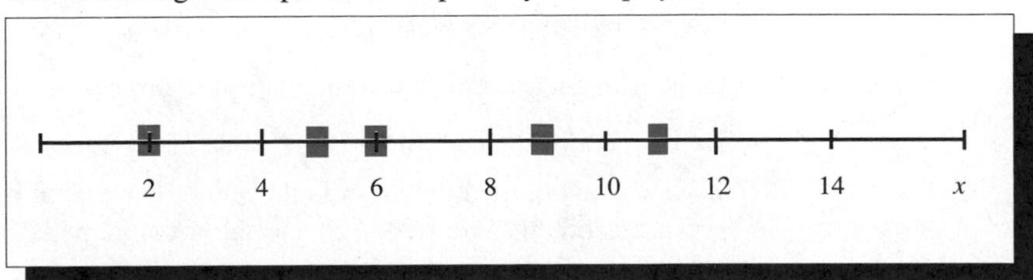

The center of the scatterplot seems to be a value between 6 and 8. Calculation of the sample mean yields

$$\bar{x} = \frac{\sum_{i=1}^{n} x_i}{n} = \frac{2 + 9 + 11 + 5 + 6}{5} = 6.6$$

The value of $\bar{x} = 6.6$ falls near the center of this small data set and quantifies our concept of the center of this distribution. ▪

A more important use of $\bar{x}$ is as an estimator of the value of the unknown population mean μ. For example, the mean of the sample of 30 observations given in Table 1.7 is

$$\bar{x} = \frac{\sum_{i=1}^{30} x_i}{30} = \frac{77.5}{30} = 2.58$$

When we visually assess the location of the center of this data from Figure 1.15, it would appear to be between 2.45 and 2.65. The value of $\bar{x} = 2.58$ falls in the interval we identified as the approximate center of the distribution. We do not know the mean of the entire population of GPAs, but, if we were to estimate its value, our estimate would be 2.58. Although the sample mean varies from sample to sample, the population mean stays the same.

A second measure of central tendency is the **median**, which is the value in the middle position in the set of measurements ordered from smallest to largest.

DEFINITION ▪

The **median** m of a set of measurements $x_1, x_2, x_3, \ldots, x_n$ is the value of x that falls in the middle position when the measurements are ordered from smallest to largest. ▪

The median divides a set of measurements into two equal parts. If the number n of measurements is odd, the median will be the measurement with rank equal to $(n+1)/2$. If the number of observations is even, the median is chosen as the value of x halfway between the two middle measurements—that is, halfway between the measurements ranked $n/2$ and $(n/2)+1$.

Rule for Calculating the Median

Rank the n measurements from smallest to largest.

1 If n is odd, the median m is the value of x with rank $(n+1)/2$.

2 If n is even, the median m is the value of x that is halfway between the measurement with rank $n/2$ and the measurement with rank $(n/2)+1$. In this case the median is the average of the two middle values.

EXAMPLE 2.2 Find the median for the set of measurements 2, 9, 11, 5, 6.

Solution We first rank the $n = 5$ measurements from the smallest to the largest: 2, 5, 6, 9, 11. Since n is odd, the median has rank $(n + 1)/2 = 6/2 = 3$ and is equal to 6. Notice that the value of the median does not differ strongly from the mean of 6.6 for this data set. ▪

EXAMPLE 2.3 Find the median for the set of measurements 2, 9, 11, 5, 6, 27.

Solution Since n is even, we rank the measurements 2, 5, 6, 9, 11, 27 and choose the measurement halfway between the two middle measurements, 6 and 9. Therefore the median is $m = 7.5$. However, the sample mean for these six observations is 10, which shows that the inclusion of the new measurement of 27 strongly affected the value of the mean but only slightly changed the value of the median. ▪

Although both the mean and the median are good measures of location of the center of a distribution of measurements, the median is less sensitive to extreme values. For example, if the distribution is symmetric about its mean as in Figure 2.1(a), the mean and median are equal. In contrast, if a distribution is skewed to the left or to the right, the mean shifts in the direction of skewness. Figure 2.1(b) shows a distribution skewed to the right. Since the large extreme values in the upper tail of the distribution increase the sum of the measurements, the mean shifts in the direction of skewness. The median is not affected by these extreme values because the numerical values of the measurements are not used in its calculation.

FIGURE 2.1 Relative frequency distributions showing the effect of extreme values on the mean and median

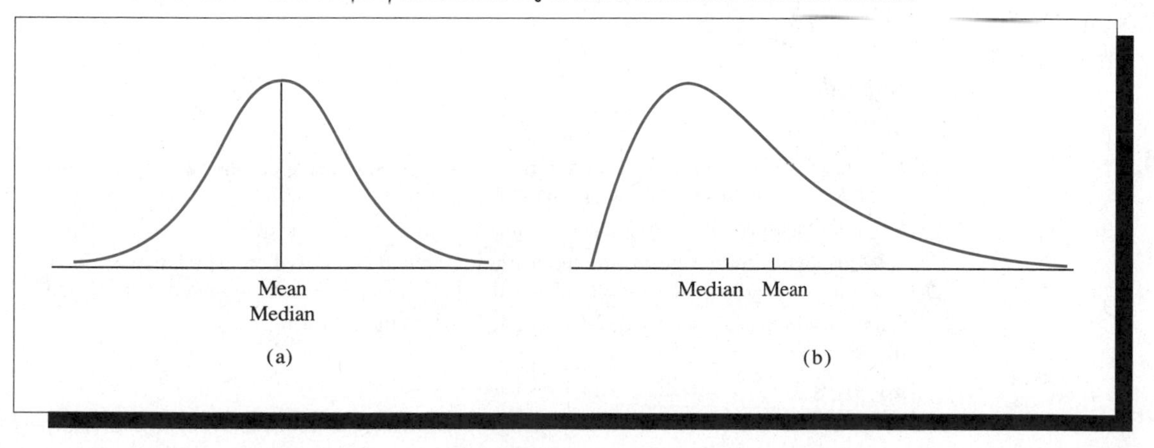

Another way to locate the center of a distribution is to look for the value of x that occurs with the highest frequency. This measure of the center of a distribution is called the **mode**.

DEFINITION ■ The **mode** is the category that occurs most frequently or the most frequently occurring value of x. When measurements on a continuous variable have been grouped as a frequency or relative frequency histogram, the class with the highest frequency is called the **modal class**, and the midpoint of that class is taken to be the mode. ■

The mode is generally used to describe large data sets, whereas the mean and median are used for both large and small data sets. In Example 1.8, the mode of the distribution of the number of quarts of milk purchased during one particular week is 2. Here the modal class and the value of x occurring with the highest frequency are one and the same. For the data in Table 1.7, a GPA of 2.5 occurs five times, and therefore the mode for the distribution of GPAs is 2.5. However, when we examine the frequency tabulation for these data given in Table 1.7, we find that there are two classes that occur with equal frequency. Fortunately, these classes are side by side in the tabulation, and our choice for the value of the mode would be 2.65, the value centered between the fourth and fifth classes.

It is possible for a distribution of measurements to have more than one mode. For example, if we were to tabulate the length of fish taken from a lake during one season, we might get a *bimodal distribution,* possibly reflecting a mixture of young and old fish from the population in the lake. Sometimes bimodal distributions of sizes or weights reflect a mixture of measurements taken on males and females. In any case, a set or distribution of measurements may have more than one mode.

Exercises

Basic Techniques

2.1 You are given $n = 5$ measurements, 0, 5, 1, 1, 3.

a Draw a scatterplot for the data. (Hint: If two measurements are the same, place one dot above the other.) Guess the approximate "center."

b Find the mean, the median, and the mode.

c Locate the three measures on the scatterplot in part (a). Based on the relative positions of the mean and median, would you say that the measurements are symmetric or skewed?

2.2 Given $n = 8$ measurements, 3, 2, 5, 6, 4, 4, 3, 5, find the following:

a $\bar{x}$

b m

c Based on the results of parts (a) and (b), are the measurements symmetric or skewed? Draw a scatterplot to confirm your answer.

2.3 Given $n = 10$ measurements, 3, 5, 4, 6, 10, 5, 6, 9, 2, 8, find the following:

a $\bar{x}$

b m

c The mode

Applications

2.4 Many computer buyers have discovered that they can save a considerable amount by purchasing a personal computer from a mail-order company—an average of $900 by their estimates ("Who's Tops," 1992). The satisfaction ratings (on a scale of 1 to 9) for seven such companies, based on a survey of 4000 buyers, are shown below:

CompuAdd	7.5	Insight	7.8
Del	7.9	Northgate	7.7
FastMicro	7.4	Zeos	8.0
Gateway	8.2		

a What is the average satisfaction rating for these seven companies?

b What is the median of the satisfaction ratings?

c If you were a computer buyer, would you be interested in the average satisfaction rating? If not, what measure would you be interested in? Explain.

2.5 The VCR (videocassette recorder) has become a very common piece of electronic equipment in most American households. In fact, 75% of all Americans own VCRs, and one owner in five has more than one ("Three-VCR Family," 1992). A sample of 25 households produced the following measurements on x, the number of VCRs in the household:

1	0	2	1	1	1	0	2	1
0	0	1	2	3	2	1	1	
1	0	1	3	1	0	1	1	

a Do you think that the distribution of x, the number of VCRs in a household, is symmetric or skewed? Explain.

b Guess the value of the mode, the value of x that occurs most frequently.

c Calculate the mean, the median, and the mode for these measurements.

d Draw a relative frequency histogram for the data set. Locate the mean, the median, and the mode along the horizontal axis. Are your answers to parts (a) and (b) correct?

2.6 Refer to Exercise 1.13. The average numbers of yards per carry for the top rushers in the American Football Conference are shown here in a different format:

4.4	4.7	4.7	4.2	4.4	3.5	4.1	3.5	4.6	3.8
4.2	4.7	3.5	4.5	5.3	4.7	2.9	4.2	3.8	3.1

a Find the average number of yards per carry for these 20 top rushers.

b Find the median for the 20 top rushers.

c Based on the results of parts (a) and (b), do you think that the distribution of average number of yards per carry for these 20 football players is symmetric or skewed?

2.7 In an article entitled "You Aren't Paranoid if You Think Someone Eyes Your Every Move" (1985), the *Wall Street Journal* noted that big business collects detailed statistics on your behavior. Jockey International knows how many undershorts you own; Frito-Lay, Inc., knows which you eat first—the broken pretzels in a pack or the whole ones; and, to get even more specific, Coca-Cola knows that you put 3.2 ice cubes in a glass. Have you ever put 3.2 ice cubes in a glass? What did the *Wall Street Journal* article mean by that statement?

2.8 The table in Exercise 1.12 gives the per capita federal tax for each of the 50 states in fiscal year 1991.

a Find the average per capita federal tax for the entire United States.

b Find the median per capita federal tax for the 50 states and compare it with the mean calculated in part (a).

c Based on your comparison in part (b), would you conclude that the distribution of per capita federal taxes is skewed? Explain.

2.9 The typical American household is shrinking in size, according to a recent report by the census bureau (Wright, 1992). This decline reflects the fact that families are having fewer children, young people are setting up housekeeping on their own rather than living with their parents, and the growing number of elderly persons in the population are now maintaining their own households. As of March 1991, the average population in the nation's 94.3 million households was 2.63 people, down from 2.76 people in 1980 and 5.79 people in 1790!

a Based on these statistics and on your general knowledge about the number of people in a household, describe the shape of the distribution of American household size.

b Based on your answer to part (a), would you expect the mean household size to be greater than, less than, or equal to the median household size? Explain.

2.10 In a psychological experiment, the time on task was recorded for ten subjects under a 5-minute time constraint. The measurements are in seconds.

175	190	250	230	240
200	185	190	225	265

a Find the average time on task.

b Find the median time on task.

c If you were writing a report to describe these data, which measure of central tendency would you use? Explain.

2.3 Measures of Variability

Once we have located the center of a distribution of data, the next step is to provide a measure of the **variability**, or **dispersion**, of the data. Consider the two distributions shown in Figure 2.2. Both distributions are located with a center at $x = 4$, but there is a vast difference in the variability of the measurements about the mean for the two distributions. The measurements in Figure 2.2(a) vary from 3 to 5; in Figure 2.2(b) the measurements vary from 0 to 8.

FIGURE 2.2 Variability or dispersion of data

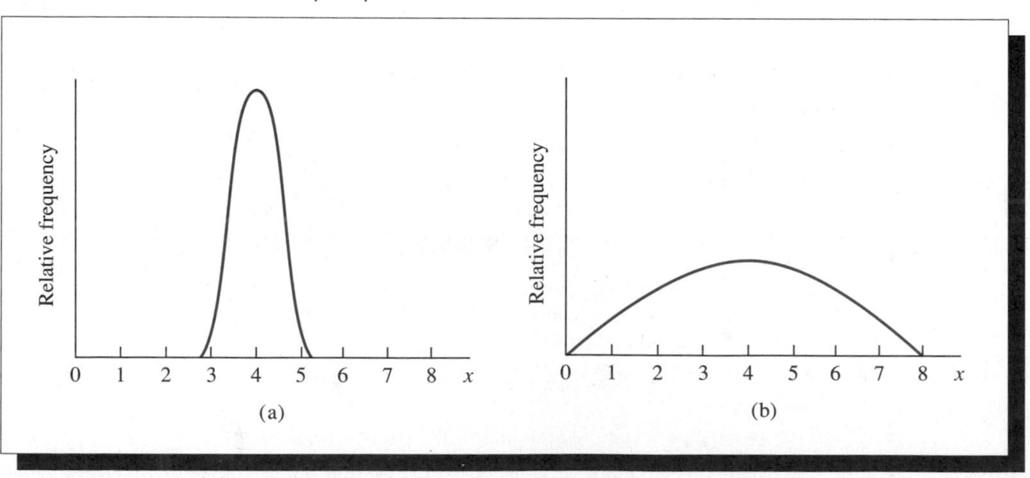

Variation is a very important characteristic of data. For example, if we are manufacturing bolts, excessive variation in the bolt diameter would imply a high percentage of defective product. On the other hand, if we are using an examination to discriminate between good and poor accountants, we would be most unhappy if the examination always produced test grades with little variation, since this would make discrimination very difficult.

In addition to the practical importance of variation in data, a measure of this characteristic is necessary to the construction of a mental image of the frequency distribution. We will discuss only a few of the many measures of variation.

The simplest measure of variation is the **range**.

DEFINITION ▪ The **range** of a set of n measurements $x_1, x_2, x_3, \ldots, x_n$ is defined as the difference between the largest and smallest measurements. ▪

For the GPA data in Table 1.7, the measurements vary from 1.9 to 3.4. Hence, the range is $(3.4 - 1.9) = 1.5$. The range is easy to calculate, easy to interpret, and quite adequate as a measure of variation for small sets of data. But, for large data sets, the range is not an adequate measure of variability. For example, the two relative frequency distributions in Figure 2.3 have the same range but have very different shapes and variability.

FIGURE 2.3 Distributions with equal range and unequal variability

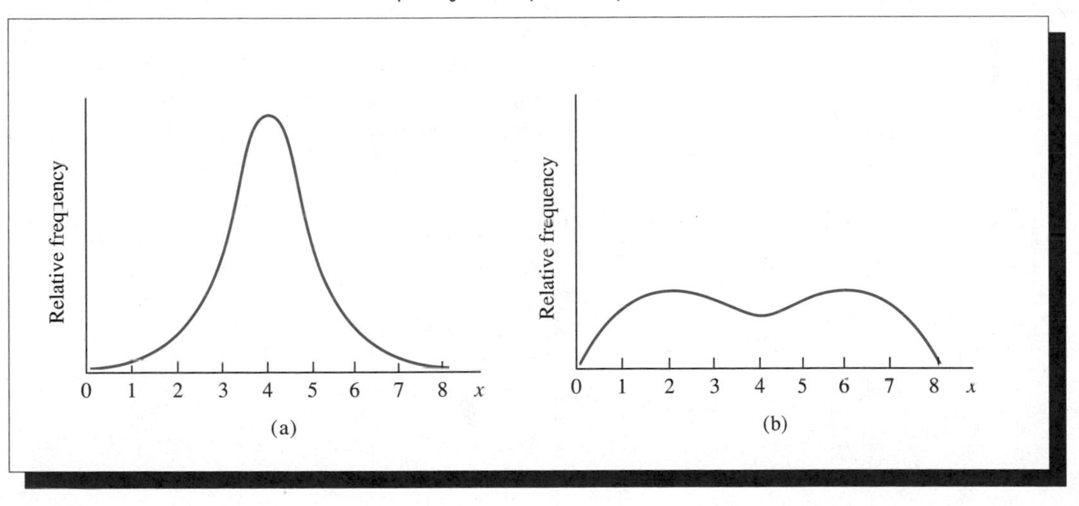

FIGURE 2.4
Scatterplot

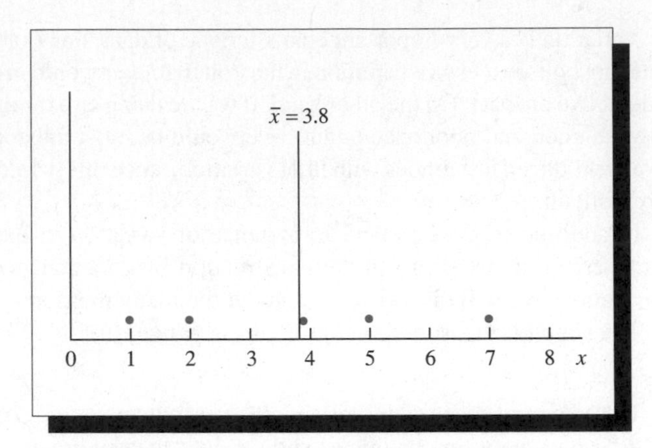

Can we find a measure of variability that is more sensitive than the range? Consider, as an example, the sample measurements 5, 7, 1, 2, 4, displayed as a scatterplot in Figure 2.4.

The mean of these five measurements is

$$\bar{x} = \frac{\sum_{i=1}^{n} x_i}{n} = \frac{19}{5} = 3.8$$

as indicated on the scatterplot.

We can now view variability in terms of distance between each dot (measurement) and the mean $\bar{x}$. If the distances are large, we can say that the data are more variable than if the distances are small. More explicitly, we define the **deviation** of a measurement from its mean to be the quantity $(x_i - \bar{x})$. Measurements to the right of the mean produce positive deviations, and those to the left produce negative deviations. The values of x and the deviations for our example are shown in the first and second columns of Table 2.1.

TABLE 2.1
Computation of $\sum_{i=1}^{n}(x_i - \bar{x})^2$

x_i	$(x_i - \bar{x})$	$(x_i - \bar{x})^2$
5	1.2	1.44
7	3.2	10.24
1	−2.8	7.84
2	−1.8	3.24
4	.2	.04
19	0.0	22.80

If we agree that deviations contain information on variation, our next step is to construct a measure of variation based on the deviations about the mean. As a first possibility, we might choose the average of the deviations. Unfortunately, the average will not work, because some of the deviations are positive, some are negative, and the sum is always zero (unless round-off errors have been introduced into the calculations). Note that the deviations in the second column of Table 2.1 sum to zero.

You may have observed an easy solution to this problem. Why not disregard the signs of the deviations and calculate the average of their absolute values?† This method has been used as a measure of variability in exploratory data analysis and in the analysis of time series data. We prefer, however, to overcome the difficulty caused by the signs of the deviations by working with their sum of squares. Using the sum of squared deviations, we calculate a single measure called the **variance** of a set of measurements. To distinguish between the variance of a *sample* and the variance of a *population*, we use the symbol s^2 to represent the sample variance and σ^2 (Greek lowercase sigma) to represent the population variance. *This measure will be relatively large for highly variable data and relatively small for less variable data.*

DEFINITION ▪ The **variance of a population** of N measurements $x_1, x_2, \ldots, x_N$ is defined to be the average of the squares of the deviations of the measurements about their mean μ. The population variance is given by the formula

$$\sigma^2 = \frac{\sum_{i=1}^{N}(x_i - \mu)^2}{N} \quad ▪$$

Most often, you will not have all the population measurements available but will need to calculate the *variance of a sample* of n measurements.

DEFINITION ▪ The **variance of a sample** of n measurements $x_1, x_2, \ldots, x_n$ is defined to be the sum of the squared deviations of the measurements about their mean $\bar{x}$ divided by $(n - 1)$. The sample variance is denoted by s^2 and is given by the formula

$$s^2 = \frac{\sum_{i=1}^{n}(x_i - \bar{x})^2}{n - 1} \quad ▪$$

For example, we may calculate the variance for the set of $n = 5$ sample measurements presented in Table 2.1. The square of the deviation of each measurement is recorded in the third column of Table 2.1. Adding, we obtain

$$\sum_{i=1}^{5}(x_i - \bar{x})^2 = 22.80$$

†The absolute value of a number is its magnitude, ignoring its sign. For example, the absolute value of −2, represented by the symbol $|-2|$, is 2. The absolute value of 2—that is, $|2|$—is 2.

The sample variance is

$$s^2 = \frac{\sum_{i=1}^{n}(x_i - \bar{x})^2}{n-1} = \frac{22.80}{4} = 5.70$$

The variance is measured in terms of the square of the original units of measurement. If the original measurements are in inches, the variance is expressed in square inches. Taking the square root of the variance, we obtain the **standard deviation**, which returns the measure of variability to the original units of measurement.

DEFINITION ■

The **standard deviation** of a set of measurements is equal to the positive square root of the variance. ■

Notation

n: number of measurements in the sample
s^2: sample variance
$s = \sqrt{s^2}$: sample standard deviation

N: number of measurements in the population
σ^2: population variance
$\sigma = \sqrt{\sigma^2}$: population standard deviation

For the set of $n = 5$ sample measurements in Table 2.1, the sample variance was $s^2 = 5.70$, so that the sample standard deviation is $s = \sqrt{s^2} = \sqrt{5.70} = 2.39$. The more variable the data set is, the larger the value of s will be.

For the small set of measurements we used, the calculation of the variance was not too difficult. However, for a larger set, the calculations can become very tedious. Most calculators with statistical capabilities have built-in programs that will calculate $\bar{x}$ and s or μ and σ, so that your computational work will be minimized. The sample or population mean key is usually marked with $\bar{x}$. The sample standard deviation key is usually marked with s or σ_{n-1} and the population standard deviation key with σ or σ_N. In using any calculator with these built-in function keys, be sure you know which calculation is being carried out by each key!

If you need to calculate s^2 and s by hand, it is much easier to use the alternative computing formula given below. This computational form is sometimes called the shortcut method for calculating s^2.

Computing Formula for s^2

$$s^2 = \frac{\sum_{i=1}^{n} x_i^2 - \frac{\left(\sum_{i=1}^{n} x_i\right)^2}{n}}{n-1}$$

where

$$\sum_{i=1}^{n} x_i^2 = \text{the sum of the squares of the individual observations}$$

$$\left(\sum_{i=1}^{n} x_i\right)^2 = \text{the square of the sum of the individual observations}$$

The *sample standard deviation, s*, is the positive square root of s^2.

EXAMPLE 2.4 Calculate the variance and standard deviation for the five measurements in Table 2.1, which are given as 5, 7, 1, 2, and 4. Use the computing formula for s^2 and compare your results with those obtained using the original definition of s^2.

Solution The entries in Table 2.2 are the individual measurements, x_i, and their squares, x_i^2, together with their sums. Using the computing formula for s^2, we have

$$s^2 = \frac{\sum_{i=1}^{n} x_i^2 - \frac{\left(\sum_{i=1}^{n} x_i\right)^2}{n}}{n-1}$$

$$= \frac{95 - \frac{(19)^2}{5}}{4} = \frac{22.80}{4} = 5.70$$

and $s = \sqrt{s^2} = \sqrt{5.70} = 2.39$, as before. ■

TABLE 2.2 Table for simplified calculation of s^2 and s

x_i	x_i^2
5	25
7	49
1	1
2	4
4	16
19	95

EXAMPLE 2.5 Calculate the sample variance and standard deviation for the $n = 30$ GPAs in Table 1.7.

Solution Using a calculator with built-in statistical functions, you can verify the following:

$$\sum_{i=1}^{n} x_i = 77.5$$

$$\sum_{i=1}^{n} x_i^2 = 204.19$$

Using the computing formula

$$s^2 = \frac{\sum_{i=1}^{n}(x_i - \bar{x})^2}{n-1} = \frac{\sum_{i=1}^{n} x_i^2 - \frac{\left(\sum_{i=1}^{n} x_i\right)^2}{n}}{n-1}$$

$$= \frac{204.19 - \frac{(77.5)^2}{30}}{29} = \frac{3.98}{29} = .137$$

and $s = \sqrt{s^2} = \sqrt{.137} = .37$. ■

You may wonder why we divide by $n - 1$ rather than n when we compute the sample variance. The sample mean $\bar{x}$ is used as an estimator of the population mean because it provides a good estimate of μ. If we wish to use the sample variance as an estimator of the population variance σ^2, the sample variance s^2 with $n - 1$ in the denominator provides better estimates of σ^2 than would an estimator calculated with n in the denominator. **For this reason, we will always divide by $n - 1$ when computing the sample variance s^2 and the sample standard deviation s.**

At this point, you have learned how to compute the variance and standard deviation of a set of measurements. Remember the following points:

- The larger the value of s^2 or s, the greater the variability of the data set.
- If s^2 or s is equal to zero, all the measurements must have the same value.
- The standard deviation s is computed in order to have a measure of variability measured in the same units as the observations.

This information allows us to compare several sets of data, with respect to their locations and their variability. How can we use these measures to say something more specific about a single set of data? The theorem and rule presented in the next section will help us answer this question.

2.4 On the Practical Significance of the Standard Deviation

We now introduce a useful theorem developed by the Russian mathematician Tchebysheff. Proof of the theorem is not difficult, but we are more interested in its application than its proof.

> **Tchebysheff's Theorem** Given a number k greater than or equal to 1 and a set of n measurements $x_1, x_2, \ldots, x_n$, at least $[1 - (1/k^2)]$ of the measurements will lie within k standard deviations of their mean. ▪

Tchebysheff's Theorem applies to any set of measurements and can be used to describe either a sample or a population. We will use the notation appropriate for populations, but you should realize that we could just as easily use the mean and the standard deviation for the sample.

The idea involved in Tchebysheff's Theorem is illustrated in Figure 2.5. An interval is constructed by measuring a distance $k\sigma$ on either side of the mean μ. Note that the theorem is true for any number we choose for k as long as it is greater than or equal to 1. Then, computing the fraction $\left[1 - (1/k^2)\right]$, we see that Tchebysheff's Theorem states that at least that fraction of the total number n measurements lies in the constructed interval.

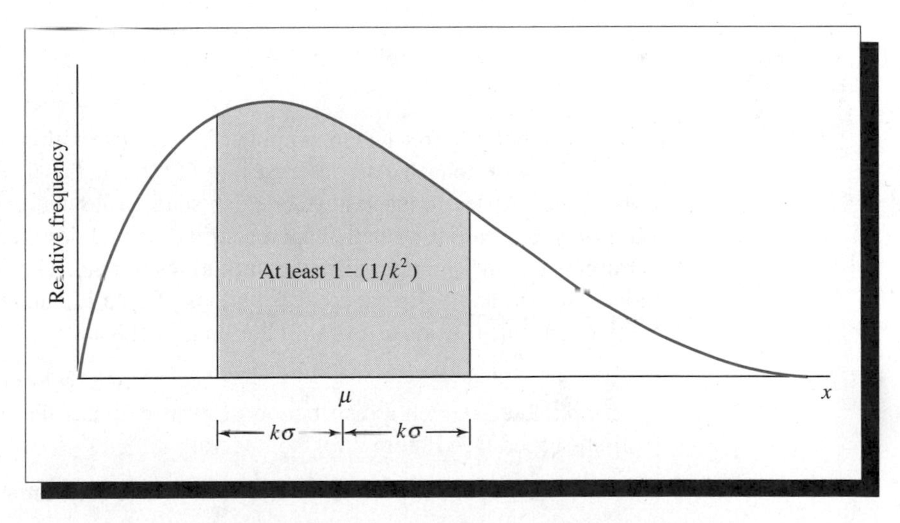

FIGURE 2.5 Illustrating Tchebysheff's Theorem

Let us choose a few numerical values for k and compute $[1 - (1/k^2)]$ (see Table 2.3). When $k = 1$, the theorem states that at least $1 - \left[1/(1)^2\right] = 0$ of the measurements lie in the interval from $(\mu - \sigma)$ to $(\mu + \sigma)$, a most unhelpful and uninformative result. However, when $k = 2$, we observe that at least $1 - \left[1/(2)^2\right] = 3/4$ of the measurements lie in the interval from $(\mu - 2\sigma)$ to $(\mu + 2\sigma)$. At least 8/9 of the measurements lie within three standard deviations of the mean—that is, in the interval from $(\mu - 3\sigma)$ to $(\mu + 3\sigma)$. Although $k = 2$ and $k = 3$ are very useful in practice, k need not be an integer. For example, the fraction of measurements falling within $k = 2.5$ standard deviations of the mean is at least $1 - [1/(2.5)^2] = .84$.

TABLE 2.3
Illustrative values of $[1 - (1/k^2)]$

k	$1 - (1/k^2)$
1	0
2	3/4
3	8/9

EXAMPLE 2.6 The mean and variance of a sample of $n = 25$ measurements are 75 and 100, respectively. Use Tchebysheff's Theorem to describe the distribution of measurements.

Solution We are given $\bar{x} = 75$ and $s^2 = 100$. The standard deviation is $s = \sqrt{100} = 10$. The distribution of measurements is centered about $\bar{x} = 75$, and Tchebysheff's Theorem states:

a *At least* 3/4 of the 25 measurements lie in the interval $\bar{x} \pm 2s = 75 \pm 2(10)$, that is, 55 to 95.

b *At least* 8/9 of the measurements lie in the interval $\bar{x} \pm 3s = 75 \pm 3(10)$, that is, 45 to 105. ■

Since Tchebysheff's Theorem applies to *any* distribution, it is very conservative. This is why we emphasize the "at least $1 - (1/k^2)$" in this theorem. If we are willing to be somewhat less conservative, we can state a rule that describes accurately the variability of a particular bell-shaped distribution and describes reasonably well the variability of other mound-shaped distributions of data. The frequent occurrence of mound-shaped and bell-shaped distributions of data in nature—hence, the applicability of our rule—leads us to call it the Empirical Rule.

Empirical Rule Given a distribution of measurements that is approximately bell-shaped (see Figure 2.6), the interval

$(\mu \pm \sigma)$ contains approximately 68% of the measurements.

$(\mu \pm 2\sigma)$ contains approximately 95% of the measurements.

$(\mu \pm 3\sigma)$ contains all or almost all of the measurements. ■

The bell-shaped distribution shown in Figure 2.6 is commonly known as the **normal distribution** and will be discussed in detail in Chapter 5. The Empirical Rule applies exactly to data that possess a normal distribution, but it also provides an excellent description of variation for many other types of data.

FIGURE 2.6
Normal (bell-shaped) distribution

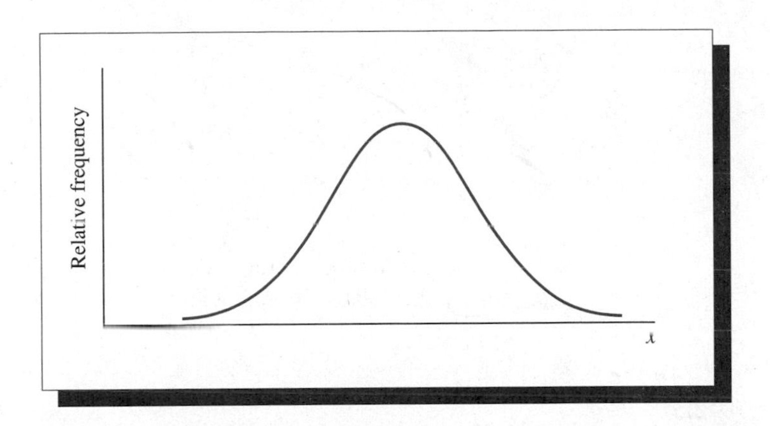

EXAMPLE 2.7 A time study is conducted to determine the length of time necessary to perform a specified operation in a manufacturing plant. The length of time necessary to complete the operation is measured for each of $n = 40$ workers. The mean and standard deviation are found to be 12.8 and 1.7, respectively. Describe the sample data by using the Empirical Rule.

Solution To describe the data, we calculate the intervals

$$(\bar{x} \pm s) = 12.8 \pm 1.7 \quad \text{or} \quad 11.1 \text{ to } 14.5$$
$$(\bar{x} \pm 2s) = 12.8 \pm 2(1.7) \quad \text{or} \quad 9.4 \text{ to } 16.2$$
$$(\bar{x} \pm 3s) = 12.8 \pm 3(1.7) \quad \text{or} \quad 7.7 \text{ to } 17.9$$

According to the Empirical Rule, we expect approximately 68% of the measurements to fall into the interval from 11.1 to 14.5, approximately 95% to fall into the interval from 9.4 to 16.2, and all or almost all to fall into the interval from 7.7 to 17.9.

If we doubt that the distribution of measurements is mound-shaped, or wish for some other reason to be conservative, we can apply Tchebysheff's Theorem and be absolutely certain of our statements. Tchebysheff's Theorem tells us that at least 3/4 of the measurements fall into the interval from 9.4 to 16.2 and at least 8/9 into the interval from 7.7 to 17.9. ▪

EXAMPLE 2.8 Student teachers are trained to develop lesson plans, on the assumption that the written plan will help them to perform successfully in the classroom. In a study to assess the relationship between written lesson plans and their implementation in the classroom, 25 lesson plans were scored on a scale of 0 to 34 according to a Lesson Plan Assessment Checklist. The 25 scores are shown in Table 2.4. Use Tchebysheff's Theorem and the Empirical Rule (if applicable) to describe the distribution of these assessment scores.

Solution By using the computing formulas for the sample mean and standard deviation, we can verify that $\bar{x} = 21.6$ and $s = 5.6$. The appropriate intervals are calculated and are shown in Table 2.5. We have also referred back to the original 25 measurements

TABLE **2.4**
Lesson plan assessment scores

26.1	26.0	14.5	29.3	19.7
22.1	21.2	26.6	31.9	25.0
15.9	20.8	20.2	17.8	13.3
25.6	26.5	15.7	22.1	13.8
29.0	21.3	23.5	22.1	10.2

and counted the actual number of measurements falling into each of these intervals. These frequencies and relative frequencies are shown in columns 3 and 4 of Table 2.5.

TABLE **2.5**
Intervals $\bar{x} \pm ks$ for the data of Table 2.4

k	Interval $\bar{x} \pm ks$	Frequency in Interval	Relative Frequency
1	16.0 – 27.2	16	0.64
2	10.4 – 32.8	24	0.96
3	4.8 – 38.4	25	1.00

Is Tchebysheff's Theorem applicable? Yes, since this theorem can be used for any set of data. According to Tchebysheff's Theorem,

- At least 3/4 of the measurements will fall between 10.4 and 32.8.
- At least 8/9 of the measurements will fall between 4.8 and 38.4.

If you consult Table 2.5, you can see that Tchebysheff's Theorem holds. In fact, the proportions of measurements falling into the specified intervals exceed the lower bound given by this theorem.

Is the Empirical Rule applicable? Yes, since the data are approximately mound-shaped (see Figure 2.7). The Empirical Rule states

- Approximately 68% of the measurements will fall between 16.0 and 27.2.
- Approximately 95% of the measurements will fall between 10.4 and 32.8.
- All or almost all of the measurements will fall between 4.8 and 38.4.

If you consult Table 2.5 again, you will see that the relative frequencies closely approximate those specified by the Empirical Rule. ▪

Guidelines for Using Tchebysheff's Theorem and the Empirical Rule

- Tchebysheff's Theorem can be proven mathematically. It applies to any set of measurements—sample or population, large or small, mound-shaped or skewed.

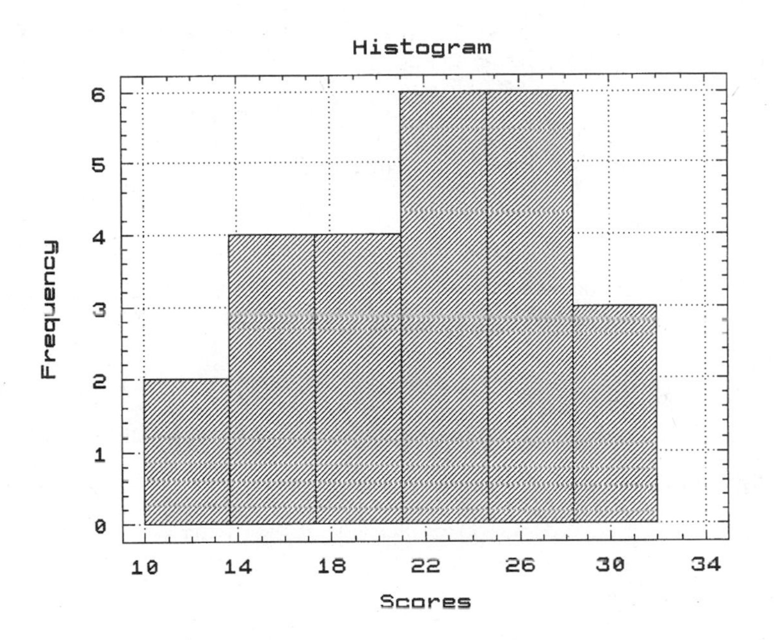

FIGURE 2.7
EXECUSTAT histogram for the data in Table 2.4

- Tchebysheff's Theorem gives a *lower bound* to the fraction of measurements to be found in an interval constructed as $\bar{x} \pm ks$. *At least* $1 - (1/k^2)$ of the measurements will fall into this interval, and probably more!
- The Empirical Rule is a "rule of thumb" that can be used as a descriptive tool only when the data tend to be roughly mound-shaped (the data tend to pile up near the center of the distribution).
- When you use these two tools for describing a set of measurements, Tchebysheff's Theorem will always be true, but it is a very conservative estimate of the fraction of measurements falling into a particular interval. If it is appropriate to use the Empirical Rule (mound-shaped data), this rule will give you a more accurate estimate of the fraction of measurements falling into the interval.

2.5 A Check on the Calculation of s

Tchebysheff's Theorem and the Empirical Rule can be used to detect gross errors in the calculation of s. For example, we know that at least three-fourths or, in the case of a mound-shaped distribution, nearly 95% of a set of measurements will lie within two standard deviations of their mean. Consequently, most of the sample measurements will lie in the interval $\bar{x} \pm 2s$, and the range will approximately equal $4s$. This is, of course, a very rough approximation, but it can serve as a useful check that will detect large errors in the calculation of s. If we let R equal the range,

$$R \approx 4s$$

then s is approximately equal to $R/4$; that is,

$$s \approx \frac{R}{4}$$

The computed value of s using the shortcut formula should be of roughly the same order as the approximation.

EXAMPLE 2.9 Use the approximation above to check the calculation of s for Table 2.2.

Solution The range of the five measurements is

$$R = 7 - 1 = 6$$

Then,

$$s \approx \frac{R}{4} = \frac{6}{4} = 1.5$$

This is the same order as the calculated value $s = \sqrt{\frac{22.8}{4}} = 2.4$. ■

Note that the range approximation is not intended to provide an accurate value for s. Rather, its purpose is to detect gross errors in calculating, such as the failure to divide the sum of squares of deviations by $(n - 1)$ or the failure to take the square root of s^2. Both errors yield solutions that are many times larger than the range approximation of s.

EXAMPLE 2.10 Use the range approximation to determine an approximate value for the standard deviation for the data in Table 2.4.

Solution The range is $R = 31.9 - 10.2 = 21.7$. Then

$$s \approx \frac{R}{4} = \frac{21.7}{4} = 5.4$$

We have shown that $s = 5.6$ for the data in Table 2.4. The approximation is very close to the actual value of s. ■

The range for a sample of n measurements will depend on the sample size of n. The larger the value of n, the more likely you will observe extremely large or small values of x. The range for large samples (say, $n = 50$ or more observations) may be as large as $6s$, whereas the range for small samples (say, $n = 5$ or less) may be as small as or smaller than $2.5s$.

The range approximation for s can be improved if it is known that the sample is drawn from a bell-shaped distribution of data. Thus, the calculated s should not differ substantially from the range divided by the appropriate ratio given in the following table:

Number of measurements	5	10	25
Expected ratio of range to s	2.5	3	4

Tips on Problem Solving

1 Be careful about rounding numbers. Use a calculator with built-in statistical functions if possible. If you calculate s^2 and s by hand, carry your calculations to at least six significant figures.

2 After you have calculated the standard deviation s for a set of data, compare its value with the range of the data. The Empirical Rule tells you that approximately 95% of the data should fall into the interval $\bar{x} \pm 2s$; that is, a very approximate value for the range will be $4s$. Consequently, a very rough rule of thumb is that

$$s \approx \frac{\text{range}}{4}$$

This crude check will help you to detect large errors—for example, failure to divide the sum of squares of deviations by $(n - 1)$ or failure to take the square root of s^2.

Exercises

Basic Techniques

2.11 Given $n = 8$ measurements, 4, 1, 3, 1, 3, 1, 2, 2, calculate

a $\bar{x}$

b s^2 using the formula given by the definition in Section 2.3.

c s^2 and s using the computing formula. Compare the results with those found in part (b).

2.12 Given $n = 5$ measurements, 2, 1, 1, 3, 5, calculate

a $\bar{x}$

b s^2 and s using the computing formula.

c A range estimate of s (Section 2.5) as a rough check on your calculations in part (b).

2.13 Given $n = 8$ measurements, 3, 1, 5, 6, 4, 4, 3, 5, find the following:

a The range

b The sample mean

c The sample variance and standard deviation

2.14 Suppose you want to create a mental picture of the relative frequency histogram for a large data set consisting of 1000 observations, and you know that the mean and standard deviation of the data set are equal to 36 and 3, respectively.

a If you are fairly certain that the relative frequency distribution of the data is mound-shaped, how might you picture the relative frequency distribution? (Hint: Use the Empirical Rule.)

b If you have no prior information concerning the shape of the relative frequency distribution, what can you say about the relative frequency histogram? (Hint: Construct intervals $\bar{x} \pm ks$ for several choices of k.)

Applications

2.15 The length of time required for an automobile driver to respond to a particular emergency situation was recorded for $n = 10$ drivers. The times, in seconds, were .5, .8, 1.1, .7, .6, .9, .7, .8, .7, .8.

a Scan the data and use the procedure in Section 2.5 to find an approximate value for s. Use this value to check your calculations in part (b).

b Calculate the sample mean $\bar{x}$ and the standard deviation s. Compare with part (a).

2.16 Refer to Exercise 1.13. The average numbers of yards per carry for the top rushers in the National Football League are shown here in a different format:

4.4	4.7	4.7	4.2	4.4	3.5	4.1
3.5	4.6	3.8	4.2	4.7	3.5	4.5
5.3	4.7	2.9	4.2	3.8	3.1	4.3
4.3	3.9	5.0	4.5	4.0	4.4	3.8
3.8	4.0	7.8	3.6	6.2	3.2	3.5
4.0	5.3	4.5	3.6	4.0		

a Find the mean and standard deviation of the data set.

b Find the percentage of measurements in the intervals $\bar{x} \pm s$, $\bar{x} \pm 2s$, and $\bar{x} \pm 3s$.

c How do the percentages obtained in part (b) compare with those given by the Empirical Rule? Explain.

2.17 The cost of educating our children has gone up again. Although graduation rates and college entrance exam scores have stayed roughly the same, the average expenditure per student has increased from \$9340 in 1990–1991 to \$10,017 in 1991–1992 (Wright, 1992, p. 223). Visualize the national distribution of expenditures per student in 1991–1992, and suppose that the standard deviation of these expenditures is \$1200.

a Within what limits would you expect at least 3/4 of the expenditures to lie?

b Within what limits would you expect at least 8/9 of the expenditures to lie?

2.18 Refer to Exercise 2.17.

a Do you think that the distribution of expenditures per student is symmetric about the mean of \$10,017 or skewed? Explain.

b If the distribution in part (a) is approximately mound-shaped, describe the distribution of expenditures per student using the Empirical Rule.

2.19 Is your breathing rate normal? Actually, there is no standard breathing rate for humans. It can vary from as low as 4 breaths per minute to as high as 70 or 75 for a person engaged in strenuous exercise. Suppose that the resting breathing rates for college-age students have a relative frequency distribution that is mound-shaped, with a mean equal to 12 and a standard deviation of 2.3 breaths per minute. What fraction of all students would have breathing rates in the following intervals?

a 9.7 to 14.3 breaths per minute

b 7.4 to 16.6 breaths per minute

c More than 18.9 or less than 5.1 breaths per minute

2.20 To estimate the amount of lumber in a tract of timber, an owner decided to count the number of trees with diameters exceeding 12 inches in randomly selected 50-by-50-foot squares. Seventy 50-by-50-foot squares were chosen, and the selected trees were counted in each tract. The data are as follows:

7	8	7	10	4	8	6
9	6	4	9	10	9	8
3	9	5	9	9	8	7
10	2	7	4	8	5	10
9	6	8	8	8	7	8
6	11	9	11	7	7	11
10	8	8	5	9	9	8
8	9	10	7	7	7	5
8	7	9	9	6	8	9
5	8	8	7	9	13	8

a Construct a relative frequency histogram to describe the data.

b Calculate the sample mean $\bar{x}$ as an estimate of μ, the mean number of timber trees for all 50-by-50-foot squares in the tract.

c Calculate s for the data. Construct the intervals $\bar{x} \pm s$, $\bar{x} \pm 2s$, and $\bar{x} \pm 3s$. Count the percentage of squares falling into each of the three intervals, and compare with the corresponding percentages given by the Empirical Rule and Tchebysheff's Theorem.

2.21 Suppose that some measurements occur more than once and that the data $x_1, x_2, \ldots, x_k$ are arranged in a frequency table as shown:

Observations	Frequency f_i
x_1	f_1
x_2	f_2
$\vdots$	$\vdots$
x_k	f_k
	n

Then,

$$\bar{x} = \frac{\sum_{i=1}^{k} x_i f_i}{n}, \quad \text{where} \quad n = \sum_{i=1}^{k} f_i$$

and

$$s^2 = \frac{\sum_{i=1}^{k} x_i^2 f_i - \frac{\left(\sum_{i=1}^{k} x_i f_i\right)^2}{n}}{n-1}$$

Although these formulas for grouped data are primarily of value when you have a large number of measurements, demonstrate their use for the sample 1, 0, 0, 1, 3, 1, 3, 2, 3, 0, 0, 1, 1, 3, 2.

a Calculate $\bar{x}$ and s^2 directly, using the formulas for ungrouped data.

b The frequency table for the $n = 15$ measurements is as follows:

x	f
0	4
1	5
2	2
3	4

$n = 15$

Calculate $\bar{x}$ and s^2 using the formulas for grouped data. Compare with your answers to part (a).

2.22 The International Baccalaureate (IB) program is an accelerated academic program offered at a growing number of high schools throughout the country. Students enrolled in this program are placed in accelerated or advanced courses and must take IB examinations in each of the six subject areas at the end of their junior or senior year. Students are scored on a scale of 1–7, with 1–2 being poor, 3 mediocre, 4 average, and 5–7 excellent. During its first year of operation at John W. North High School in Riverside, California, 17 juniors attempted the IB economics exam, with the following results:

Exam Grade	Number of Students
7	1
6	4
5	4
4	4
3	4

Calculate the mean and standard deviation for these scores.

2.23 As an illustration of the utility of the Empirical Rule, consider a distribution that is heavily skewed to the right, as shown in Figure 2.8.

a Calculate $\bar{x}$ and s for the data shown. (NOTE: There are 10 zeros, 5 ones, and so on.)

b Construct the intervals $\bar{x} \pm s$, $\bar{x} \pm 2s$, and $\bar{x} \pm 3s$ and locate them on the frequency distribution.

c Calculate the proportion of the $n = 25$ measurements falling into each of the three intervals. Compare with Tchebysheff's Theorem and the Empirical Rule. Note that, although the proportion falling into the interval $\bar{x} \pm s$ does not agree closely with the Empirical Rule, the proportion falling into the intervals $\bar{x} \pm 2s$ and $\bar{x} \pm 3s$ agree very well. Many times this is true, even for non-mound-shaped distributions of data.

FIGURE 2.8
Distribution for Exercise 2.23

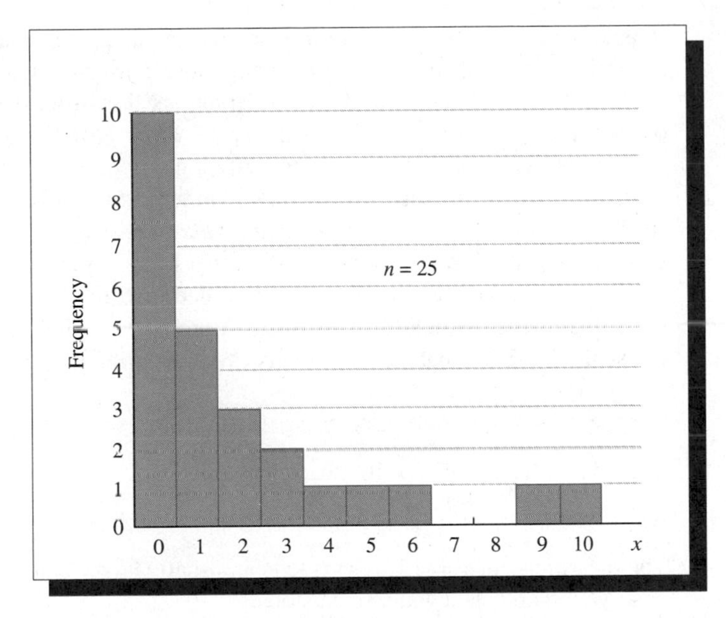

2.6 Measures of Relative Standing

Sometimes we want to know the position of one observation relative to others in a set of data. For example, if you took an examination based on a total of 35 points, you might want to know how your score of 30 compared to the scores of the other students in the class. The mean and standard deviation of the scores can be used to calculate a z-score, which will measure the relative standing of a measurement in a data set.

DEFINITION ■

The **sample z-score** is a measure of relative standing defined by

$$z\text{-score} = \frac{x - \bar{x}}{s} \quad \blacksquare$$

A z-score measures the distance between an observation and the mean, measured in units of standard deviation. For example, suppose we know that the mean and standard deviation of the test scores (based on a total of 35 points) are 25 and 4, respectively. The z-score for your score of 30 is calculated as follows:

$$z\text{-score} = \frac{x - \bar{x}}{s} = \frac{30 - 25}{4} = 1.25$$

Your score of 30 lies 1.25 standard deviations above the mean ($30 = \bar{x} + 1.25s$).

Sample z-scores by themselves may not convey a strong sense of relative position other than to indicate that a score is either above or below the mean of the data set. However, for any data set, the z-score used in conjunction with Tchebysheff's Theorem allows us to make some conservative statements about the relative standing of an observation. Moreover, for mound-shaped data, the Empirical Rule can be used in conjunction with the z-score to make more powerful statements. Since at least 75%, and more likely 95%, of the observations lie within two standard deviations of their mean, *z-scores between -2 and $+2$ are highly likely.* Therefore, the exam score of 30 with $z = 1.25$ is not unusually high or unusually low. At least 89%, and more likely all, of the observations lie within three standard deviations of their mean, so that *z-scores exceeding 3 in absolute value are very unlikely.* The experimenter should closely examine any observation that has a z-score exceeding 3 in absolute value. Perhaps the measurement was recorded incorrectly or does not belong to the population being sampled. Perhaps it is just a highly unlikely observation. Such an unusually large or small observation is called an **outlier**.

EXAMPLE 2.11 Consider a sample of $n = 10$ measurements:

$$1, 1, 0, 15, 2, 3, 4, 0, 1, 3$$

The measurement $x = 15$ appears to be unusually large. Calculate the z-score for this observation and state your conclusions.

Solution For the sample,

$$\sum_{i=1}^{10} x_i = 30 \qquad \text{and} \qquad \sum_{i=1}^{10} x_i^2 = 266$$

Then,

$$\bar{x} = \frac{\sum_{i=1}^{10} x_i}{n} = \frac{30}{10} = 3.0$$

$$s^2 = \frac{\sum_{i=1}^{n} x_i^2 - \frac{\left(\sum_{i=1}^{n} x_i\right)^2}{n}}{n-1}$$

$$= \frac{266 - \frac{(30)^2}{10}}{9} = 19.5556 \qquad \text{and} \qquad s = 4.42$$

The z-score for the suspected outlier, $x = 15$, is calculated as

$$z\text{-score} = \frac{x - \bar{x}}{s} = \frac{15 - 3}{4.42} = 2.71$$

Hence, the measurement $x = 15$ lies 2.71 standard deviations above the sample mean, $\bar{x} = 3.0$. Although the z-score does not exceed 3, it is close enough so that we suspect that $x = 15$ is an outlier. We should examine our sampling procedure to see whether $x = 15$ is a faulty observation. ▪

A **percentile** is another measure of relative standing and is most often used for large data sets. (Percentiles are not very useful for small data sets.)

DEFINITION ▪ Let $x_1, x_2, \ldots, x_n$ be a set of n measurements arranged in order of magnitude. The ***p*th percentile** is the value of x that exceeds $p\%$ of the measurements and is less than the remaining $(100 - p)\%$. ▪

EXAMPLE 2.12 Suppose you have been notified that your score of 610 on the Verbal Graduate Record Examination placed you at the 60th percentile in the distribution of scores. Where does your score of 610 stand in relation to the scores of others who took the examination?

Solution Scoring at the 60th percentile means that 60% of all the examination scores were lower than your score and 40% were higher. ▪

Viewed graphically, a particular percentile—say, the 60th percentile—is a point on the x-axis located so that 60% of the area under the relative frequency histogram for the data lies to the left of the 60th percentile (see Figure 2.9) and 40% of the area lies to the right. Thus, by our definition, the median of a set of data is the 50th percentile, because half of the measurements in a data set are smaller than the median and half are larger.

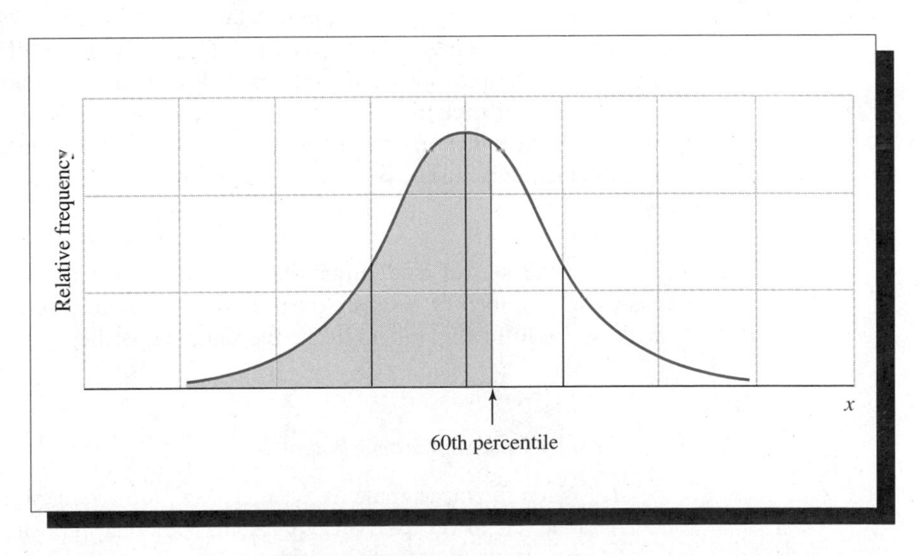

FIGURE 2.9 The 60th percentile shown on the relative frequency histogram for a data set

The 25th and 75th percentiles, called the **lower** and **upper quartiles**, along with the median (the 50th percentile), locate points that divide the data into four sets of equal number. Twenty-five percent of the measurements will be less than the lower (first) quartile, 50% will be less than the median (the second quartile), and 75% will be less than the upper (third) quartile. Thus, the median and the lower and upper quartiles are located at points on the x-axis so that the area under the relative frequency histogram for the data is partitioned into four equal areas, as shown in Figure 2.10.

FIGURE 2.10
Location of quartiles

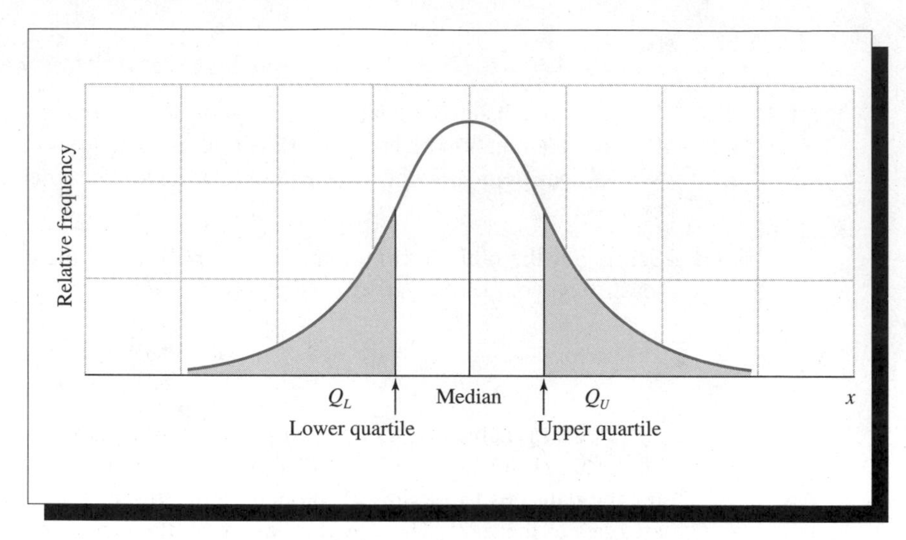

DEFINITION ■

Let $x_1, x_2, \ldots, x_n$ be a set of n measurements arranged in order of magnitude. The **lower quartile (first quartile)**, Q_L, is the value of x that exceeds one-fourth of the measurements and is less than the remaining three-fourths. The **second quartile** is the median. The **upper quartile (third quartile)**, Q_U, is the value of x that exceeds three-fourths of the measurements and is less than one-fourth. ■

For small sets of data, a quartile may fall between two observations, in which case many numbers may satisfy the preceding definition. To avoid this ambiguity, we will use the following rule to locate the sample quartiles.

Rules for Calculating Sample Quartiles

1 When the measurements $x_1, x_2, \ldots, x_n$ are arranged in order of magnitude, the **lower quartile**, Q_L, is the value of x in position $.25(n + 1)$, and the **upper quartile**, Q_U, is the value of x in position $.75(n + 1)$.

2 When $.25(n + 1)$ and $.75(n + 1)$ are not integers, the quartiles are found by interpolation, using the values in the two adjacent positions.[†]

We demonstrate the procedure with the next example.

EXAMPLE 2.13 Find the lower and upper quartiles for the following set of measurements:

$$16, 25, 4, 18, 11, 13, 20, 8, 11, 9$$

Solution Rank the $n = 10$ measurements from the smallest to largest:

$$4, 8, 9, 11, 11, 13, 16, 18, 20, 25$$

The lower quartile is the value in position $.25(n + 1) = .25(10 + 1) = 2.75$, and the upper quartile is the value in position $.75(10 + 1) = 8.25$. The lower quartile is taken to be the value 3/4 of the distance between the second and third ordered measurements, and the upper quartile is taken to be the value 1/4 of the distance between the eighth and ninth ordered measurements. Therefore,

$$Q_L = 8 + .75(9 - 8) = 8 + .75 = 8.75$$

and

$$Q_U = 18 + .25(20 - 18) = 18 + .5 = 18.5$$ ▪

Many of the numerical descriptive measures that we have discussed are easily found by using the command DESCRIBE in the MINITAB package. As part of its output, this program produces the values of the mean, the standard deviation, the median, and the lower and upper quartiles, as well as the values of some other statistics that we have not discussed. The **trimmed mean** (given as TRMEAN) is the mean of the middle 90% of the measurements after excluding the smallest 5% and the largest 5%. Unlike the ordinary arithmetic mean, the trimmed mean is not sensitive to extremely large or extremely small values in the data set. Using the command DESCRIBE to summarize the data in Example 2.13 produced the computer output shown in Table 2.6. Notice that the quartiles are identical to those calculated in that example.

[†]This definition of quartiles is consistent with the one used in the MINITAB package. However, some texts use ordinary rounding when finding quartile positions, whereas others compute a sample quartile as the average of the values in adjacent positions only when $.25(n + 1)$ and $.75(n + 1)$ end in the fraction .5.

TABLE 2.6
MINITAB output using the command DESCRIBE for the data in Example 2.13

```
MTB>DESCRIBE C1

              N      MEAN     MEDIAN     TRMEAN     STDEV     SEMEAN
C1           10     13.50      12.00      13.25      6.28       1.98

            MIN       MAX         Q1         Q3
C1         4.00     25.00       8.75      18.50
```

Exercises

Basic Techniques

2.24 Find the median and the lower and upper quartiles for the following data:

$$8, 7, 1, 4, 6, 6, 4, 5, 7, 6, 3, 0$$

2.25 Refer to the data in Exercise 2.24.

a Calculate $\bar{x}$ and s.

b Calculate the z-score for the smallest and largest observations. Are either of these observations unusually large or unusually small?

2.26 Find the median and the lower and upper quartiles for the following data:

$$19, 12, 16, 0, 14, 9, 6, 1, 12, 13, 10, 19, 7, 5, 8$$

Applications

2.27 The table in Exercise 1.12 gives the per capita federal tax for each of the 50 states in fiscal year 1991.

a Find the median (see Exercise 2.8) and the lower and upper quartiles for the data set.

b Are there any states that pay unusually high or unusually low per capita taxes? What are the z-scores associated with these states' taxes?

2.28 If you scored at the 69th percentile on a placement test, how would your score stand in relation to others?

2.29 According to researchers Edward Lazear and Robert Michael, the allocation of family income among all household members is far from uniform. In fact, about $2\frac{1}{2}$ times as much is spent on adult household members as on children (*Wall Street Journal*, Dec. 8, 1988). The article states, "About one in 10 households probably spends less than $20 a child for every $100 spent on each adult; another 10% spend more than $55 a child." Identify any percentiles for the distribution of amounts spent per child for every $100 spent per adult.

2.30 An article in *American Demographics* (Schreiner and Hamel, 1989) provides some interesting statistics on differences between residents of San Francisco and residents of Los Angeles. A survey of shoppers in the two areas showed that Los Angeles residents have a median income of $35,000 and that more than 10% have an income of $75,000 or more. The average income of San Franciscans is estimated at $32,000, and 7% have incomes in the $75,000+ range. Identify any percentiles that can be determined from this information.

2.31 Refer to Exercise 2.30. The survey of shoppers also revealed that 16% of San Franciscans are age 65 or older, compared with 13% in Los Angeles. In the distribution of ages for residents of San Francisco, at what percentile does age 65 lie? For the residents of Los Angeles, at what percentile does age 65 lie?

2.32 In recent years, increasing concern has begun to surface over the amount of toxic chemical emissions that are presently being pumped into the environment. All emissions data must be carefully examined, however, to determine whether the figures are biased by the inclusion or exclusion of certain information. The following table gives the Environmental Protection Agency's (EPA's) top ten counties for toxic waste in 1987, along with a *USA Today* computer analysis of the same data. The EPA reports all toxic emissions, including sodium sulfate, whereas *USA Today* reports only toxic emissions into the environment (not waste-treated or hauled away), excluding sodium sulfate, which is now considered nontoxic. Emissions are reported in millions of pounds.

County	EPA	*USA Today*
San Bernardino, CA	5200	22.5
Harris, TX	612	78.3
Calhoun, TX	581	579.2
Wayne, MI	512	17.7
Harrison, MS	423	53.0
Brazoria, TX	404	197.9
St. James, LA	349	312.5
Milam, TX	329	329.1
Jefferson, TX	309	198.0
St. Charles, LA	284	209.1

Source: Sanders (1989). Copyright 1989, *USA Today*. Excerpted with permission.

a Find the mean and standard deviation of the EPA emissions for the ten counties.

b Calculate the z-score for the emissions reported by the EPA for San Bernardino County. Is this an unusually high amount?

c If you were told that an industrial plant in San Bernardino County emitted a large amount of sodium sulfate, would this explain the difference in the emissions amounts for San Bernardino County as reported by the EPA and by *USA Today*?

2.7 The Box Plot

The **box plot** is another technique used in explanatory data analysis (EDA). It can be used to describe not only the behavior of the measurements in the middle of the distribution but also their behavior at the ends or tails of the distribution. Values that lie very far from the middle of the distribution in either direction are called **outliers**. An outlier may result from transposing digits when recording a measurement, from incorrectly reading an instrument dial, from a malfunctioning piece of equipment, and so on. Even when there are no recording or observational errors, a data set may contain one or more valid measurements that, for one reason or another, differ markedly from the others in the set. These outliers can cause a marked distortion in the values of commonly used numerical measures such as $\bar{x}$ and s. In fact, outliers may themselves contain important information not shared with the other measurements in the set. Therefore, isolating outliers, if they are present, is an important step in any preliminary analysis of a data set. The box plot is designed expressly for this purpose.

A box plot is constructed by using the median and two other measures, known as **hinges**. The median divides the ordered data set into two halves: the hinges are the values in the middle of each half of the data. Hinges are very similar to quartiles and, in effect, serve the same purpose. The actual difference in value between a hinge and a quartile is quite small and decreases as the number of measurements increases.

Nevertheless, we retain the distinction between these two measures and use hinges in constructing a box plot. *However, we can think of a hinge as playing the role of a quartile.* Hinges are calculated using the following procedure.

Calculating Hinges

1. Calculate the position of the median, $(n + 1)/2$, and drop the fraction 1/2 if there is one. This quantity is called $d(M)$ and measures the "depth" of the median from each end of the ordered measurements.
2. Calculate the position of the hinges as

$$\frac{d(M) + 1}{2}$$

The hinges are the values in position $[d(M) + 1]/2$ as measured from each end of the ordered data set.

For example, for a data set with $n = 10$, the position of the median is $(10 + 1)/2 = 5.5$, and $d(M) = 5$. The position of the hinges is then

$$\frac{d(M) + 1}{2} = \frac{5 + 1}{2} = 3$$

The hinges are those values in the third position, as measured from each end of the ten ordered measurements. If the position of the hinges ends in 1/2, a hinge will be the average of the two adjacent values in the ordered data set.

The dispersion of the measurements is now measured in terms of the difference between the hinges, called the **H-spread**, which is approximately equal to $(Q_U - Q_L)$, the **interquartile range**. A data value will be identified as an outlier depending on its relative position with respect to boundary points, called inner and outer fences. The **inner fences** are defined as follows:

Lower inner fence = lower hinge − 1.5(H-spread)

Upper inner fence = upper hinge + 1.5(H-spread)

The **outer fences** are defined as follows:

Lower outer fence = lower hinge − 3(H-spread)

Upper outer fence = upper hinge + 3(H-spread)

The data values in each tail closest to, but still inside, the inner fences are called the **adjacent values**. Values lying between an inner fence and its neighboring outer fence are termed "outside" and are considered to be mild outliers. Values outside the outer fences are termed "far outside" and are considered to be extreme outliers. A box plot combines all this information in a pictorial display as follows.

Constructing a Box Plot

1. Calculate the median and the upper and lower hinges, and locate them on a horizontal line representing the scale of measurement.
2. Draw a box whose ends are the upper and lower hinges. Draw a line through the box at the value of the median.
3. Calculate the **H-spread** = upper hinge − lower hinge. Then calculate the **inner and outer fences**. Determine mild and extreme outliers, and locate them on the box plot.
4. Locate the **adjacent values** on the box plot. Draw a dashed line from the box to the corresponding adjacent values.

EXAMPLE 2.14 Construct a box plot for the data in Example 2.11.

Solution The $n = 10$ measurements in the sample, ranked from smallest to largest are

$$0, 0, 1, 1, 1, 2, 3, 3, 4, 15$$

The position of the median is $(n + 1)/2 = 11/2 = 5.5$. Hence, the median is $(1 + 2)/2 = 1.5$ and the depth of the median is

$$d(M) = 5$$

The position of the hinges, then, is

$$\frac{d(M) + 1}{2} = \frac{5 + 1}{2} = 3$$

and the hinges are 1 and 3, the values in the third position as measured from each end of the ordered data set. The H-spread is $3 - 1 = 2$, and the outer and inner fences are calculated as

$$\text{Upper inner fence} = 3 + 1.5(2) = 6$$
$$\text{Lower inner fence} = 1 - 1.5(2) = -2$$
$$\text{Upper outer fence} = 3 + 3(2) = 9$$
$$\text{Lower outer fence} = 1 - 3(2) = -5$$

The value $x = 15$ lies outside the outer fences and hence is considered an extreme outlier.

To construct a box plot for the data, we draw a box whose ends are the upper and lower hinges, as shown in Figure 2.11. A line is drawn through the box at the value of the median. The outlier, $x = 15$, is plotted using a capital O. The adjacent values, lying just inside the inner fences, are 0 and 4 and are connected to the box with a dashed line.

The box plot emphasizes the fact that the outlier lies far from the central 50% of the measurements lying between the hinges. The box plot also indicates that these

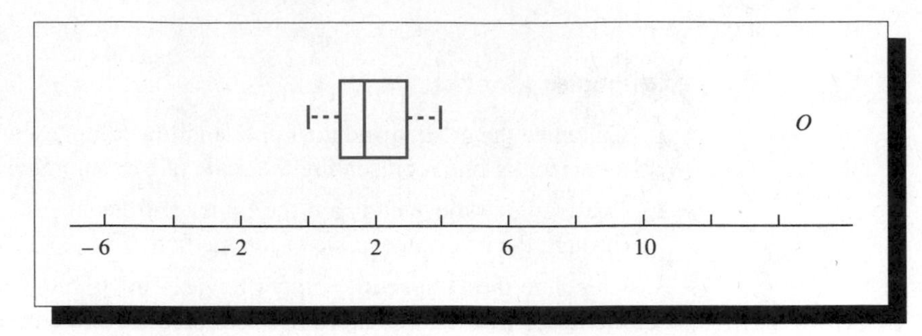

FIGURE 2.11
Box plot for the data in Example 2.14

data are positively skewed (skewed to the right), since the median is not equally spaced between the two hinges but, rather, lies closer to the lower hinge.

For these same data the command BOXPLOT in the MINITAB package produced the box plot in Table 2.7. Notice that, except for scale, the box plots are identical and lead to the same conclusions concerning the outlier, $x = 15$. ■

TABLE 2.7
MINITAB printout of the box plot for the data in Example 2.14

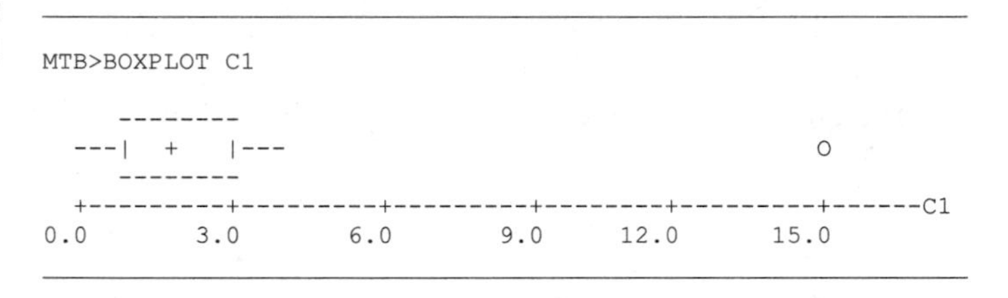

```
MTB>BOXPLOT C1

         --------
     ---I   +    I---                                    O
         --------

    +---------+---------+---------+--------+---------+------C1
   0.0       3.0       6.0       9.0     12.0      15.0
```

Exercises

Basic Techniques

2.33 Construct a box plot for the following data and identify any outliers:

25, 22, 26, 23, 27, 26, 28, 18, 25, 24, 12

2.34 Construct a box plot for the following data and identify any outliers:

3, 9, 10, 2, 6, 7, 5, 8, 6, 6, 4, 9, 22

Applications

2.35 In Exercises 2.8 and 2.27, we found the median and the lower and upper quartiles for the per capita federal tax data of Exercise 1.12.

a Find the upper and lower hinges for the data. Compare the hinges to the upper and lower quartiles found in Exercise 2.27.

b Construct a box plot for the data.

c Are there any outliers? If so, how would you explain them?

2.36 Environmental scientists are increasingly concerned with the accumulation of toxic elements in marine mammals and the transfer of such elements in the animals' offspring. The striped dolphin (*Stenella coeruleoalba*), considered to be the top predator in the marine food chain, was the subject of one such study. The mercury concentration (micrograms/gram) in the livers of 28 male striped dolphins are as follows:

1.70	183.00	221.00	286.00
1.72	168.00	406.00	315.00
8.80	218.00	252.00	241.00
5.90	180.00	329.00	397.00
101.00	264.00	316.00	209.00
85.40	481.00	445.00	314.00
118.00	485.00	278.00	318.00

a Construct a box plot for the data.

b Are there any outliers?

c If you knew that the first four dolphins were all less than three years old, while all the others were more than eight years old, would this information help explain the difference in the magnitude of those four observations? Explain.

2.37 The total toxic emissions reported by the EPA for ten counties in the United States (Exercise 2.32) are reproduced below. Data are reported in millions of pounds.

$$5200, 612, 581, 512, 423, 404, 349, 329, 309, 284$$

a Construct a box plot for the data and identify any outliers.

b Do the results of part (a) agree with the results obtained using the z-score in Exercise 2.32?

2.8 Summary

Numerical descriptive measures are numbers that attempt to create a mental image of the frequency histogram (or frequency distribution). We have restricted the discussion to measures of central tendency and variation, the most useful of which are the mean and standard deviation. Although the mean possesses intuitive descriptive significance, the standard deviation is significant only when used in conjunction with Tchebysheff's Theorem and the Empirical Rule. The objective of sampling is a description of the population from which the sample was obtained. This objective is accomplished by using the sample mean $\bar{x}$ and the quantity s^2 as estimators of the population mean μ and variance σ^2.

The many descriptive methods and numerical measures presented in this chapter constitute only a small percentage of those that might have been discussed. Many special computational techniques usually found in elementary texts have been omitted. Because calculators and computers have minimized the importance of special computational formulas, we have chosen to focus on the main objective of modern statistics and this text—statistical inference.

2.9 MINITAB Commands

At the end of this and other chapters, we will supply the MINITAB commands and subcommands used in implementing procedures and analyses introduced in the chapter.

In general, when one or more subcommands are to be used, the main command and subsequent subcommands are followed by a *semicolon*, and *the last subcommand is followed by a period.* In the presentation of a command and its admissible subcommands, the letter C represents a column number, K represents a constant, M represents a matrix, and E represents one or more of the three preceding letters. For example, the command DESCRIBE is used to calculate descriptive measures for data stored in a column, whereas the command PRINT can be used to print a column of data, one or more stored constants, a matrix, or some combination of all three.

```
BOX PLOT C

   INCREMENT = K
   START     = K [end = K]
   BY C

   LINES     = K
   NOTCH    [K%] (sign confidence interval)
   LEVELS    K . . . K
```

```
HISTOGRAM    C . . . C

   INCREMENT = K
   START     = K [end = K]
   BY C
   SAME scales for all columns
```

```
STEM-AND-LEAF display of C . . . C

   TRIM outliers
   INCREMENT = K
   BY C
```

```
DESCRIBE C . . . C

   BY C
```

Supplementary Exercises

2.38 Conduct the following experiment: toss ten coins and record x, the number of heads observed. Repeat this process $n = 50$ times, thus providing 50 values of x.

a Construct a relative frequency histogram for these measurements.

b Calculate $\bar{x}$, s^2, and s.

c Find the proportion of measurements lying in the interval $\bar{x} \pm s$.

d Find the proportion of measurements lying in the interval $\bar{x} \pm 2s$. Are these results consistent with Tchebysheff's Theorem?

e Is the frequency histogram relatively mound-shaped? Does the Empirical Rule adequately describe the variability of the data?

2.39 A report by the Department of Education (Wright, 1992) revealed that teacher salaries in the United States rose in 1991, to an average of \$34,413. The highest statewide average, \$47,300 per year, was reported in Connecticut, and the lowest average, \$24,145 per year, was reported in North Dakota.

a Use the range approximation to estimate the standard deviation of the average salaries by state.

b If the distribution of statewide average salaries is approximately mound-shaped, what proportion of the states have average salaries between \$28,624 and \$45,991?

2.40 The number of television viewing hours per household and the prime viewing times are two factors that affect television advertising income. A random sample of 25 households in a particular viewing area produced the following estimates of viewing hours per household:

3.0	6.0	7.5	15.0	12.0
6.5	8.0	4.0	5.5	6.0
5.0	12.0	1.0	3.5	3.0
7.5	5.0	10.0	8.0	3.5
9.0	2.0	6.5	1.0	5.0

a Scan the data and use the procedure in Section 2.5 to find an approximate value for s. Use this value to check your calculations in part (b).

b Calculate the sample mean $\bar{x}$ and the sample standard deviation s. Compare s with the approximate value obtained in part (a).

c Find the percentage of the viewing hours per household that falls into the interval $\bar{x} \pm 2s$. Compare with the corresponding percentage given by the Empirical Rule.

2.41 Refer to Exercise 1.10. The length of time (in months) between the onset of a particular illness and its recurrence was recorded, as follows:

2.1	4.4	2.7	32.3	9.9
9.0	2.0	6.6	3.9	1.6
14.7	9.6	16.7	7.4	8.2
19.2	6.9	4.3	3.3	1.2
4.1	18.4	.2	6.1	13.5
7.4	.2	8.3	.3	1.3
14.1	1.0	2.4	2.4	18.0
8.7	24.0	1.4	8.2	5.8
1.6	3.5	11.4	18.0	26.7
3.7	12.6	23.1	5.6	.4

a Find the range.

b Use the range approximation to find an approximate value for s.

c Compute s for the data and compare with your approximation from part (b).

2.42 Refer to Exercise 2.41.

a Examine the data, and count the number of observations falling into the interval $\bar{x} \pm s$, $\bar{x} \pm 2s$, and $\bar{x} + 3s$.

b Do the percentages falling into these intervals agree with Tchebysheff's Rule? With the Empirical Rule?

c Why might the Empirical Rule be unsuitable for describing these data?

2.43 Find the median and the lower and upper quartiles for the data on times until recurrence of an illness in Exercise 1.10. Use these descriptive measures to construct a mental image of the relative frequency histogram for the data. Compare your visualization with the relative frequency histogram that you constructed in Exercise 1.10.

2.44 An analytical chemist wanted to determine the number of moles of cupric ions in a given volume of solution by electrolysis. The solution was partitioned into $n = 30$ portions of .2 milliliter each. Each of the $n = 30$ unknown portions was tested. The average number of moles of cupric ions for the $n = 30$ portions was found to be .17 mole; the standard deviation was .01 mole.

a Describe the distribution of the measurements for the $n = 30$ portions of the solution, using Tchebysheff's Theorem.

b Describe the distribution of the measurements for the $n = 30$ portions of the solution, using the Empirical Rule. (Would you expect the Empirical Rule to be suitable for describing these data?)

c Suppose the chemist had employed only $n = 4$ portions of the solution for the experiment and obtained the readings .15, .19, .17, and .15. Would the Empirical Rule be suitable for describing the $n = 4$ measurements? Why?

2.45 According to the EPA, chloroform, which in its gaseous form is suspected of being a cancer-causing agent, is present in small quantities in all of the country's 240,000 public water sources. If the mean and standard deviation of the amounts of chloroform present in the water sources are 34 and 53 micrograms per liter, respectively, describe the distribution for the population of all public water sources.

2.46 In contrast to aptitude tests, which are predictive measures of what one can accomplish with training, achievement tests tell what an individual can do at the time of the test. Mathematics achievement test scores for 400 students were found to have a mean and a variance equal to 600 and 4900, respectively. If the distribution of test scores was mound-shaped, approximately how many of the scores would fall into the interval 530 to 670? Approximately how many scores would be expected to fall into the interval 460 to 740?

2.47 Petroleum pollution in seas and oceans stimulates the growth of some types of bacteria. A count of petroleumlytic microorganisms (bacteria per 100 milliliters) in ten portions of seawater gave the following readings:

$$49, 70, 54, 67, 59, 40, 61, 69, 71, 52$$

a Observe the data and guess the value for s by use of the range approximation.

b Calculate $\bar{x}$ and s and compare with the range approximation of part (a).

2.48 Why do statisticians generally prefer to divide the sum of squares of deviations of the sample measurements by $(n - 1)$ rather than n when estimating a population variance σ^2?

2.49 The percentage of city telephone subscribers who use unlisted numbers is on the increase. The distribution of the percentage of unlisted numbers in cities has a mean and standard deviation that are near 14% and 6%, respectively. If you were to pick a city at random, is it likely that the percentage of unlisted numbers would exceed 20%? Explain. Within what limits would you expect the percentage to fall?

2.50 An industrial concern uses an employee screening test with average score μ and standard deviation $\sigma = 10$. Assume that the test-score distribution is mound-shaped and that a score of 65 qualifies an applicant for further consideration. What is the value of μ such that approximately 2.5% of the applicants qualify for further consideration?

2.51 Attendances at a high school's basketball games were recorded and found to have a sample mean and variance of 420 and 25, respectively. Calculate $\bar{x} \pm s$, $\bar{x} \pm 2s$, and $\bar{x} \pm 3s$ and state the approximate fraction of measurements you would expect to fall into these intervals according to the Empirical Rule.

$\bar{x} \pm s$ __________ fraction__________

$\bar{x} \pm 2s$__________ fraction__________

$\bar{x} \pm 3s$__________ fraction__________

2.52 The College Board's verbal and mathematics scholastic aptitude tests are scored on a scale of 200 to 800. Although originally designed to produce mean scores approximately equal to 500, the mean verbal and math scores in recent years have been as low as 463 and 493, respectively, and have been trending downward. It seems reasonable to assume that a distribution of all GRE scores, either verbal or math, is mound-shaped. If σ is the standard deviation of one of these distributions, what is the largest value (approximately) that σ might assume? Explain.

2.53 The mean duration of television commercials on a given network is 75 seconds, with a standard deviation of 20 seconds. Assuming that duration times are approximately normally distributed:

a What is the approximate probability that a commercial will last less than 35 seconds?

b What is the approximate probability that a commercial will last longer than 55 seconds?

2.54 A random sample of 100 foxes was examined by a team of veterinarians to determine the prevalence of a particular type of parasite. Counting the number of parasites per fox, the veterinarians found that 69 foxes had no parasites, 17 had one parasite, and so on. The following is a frequency tabulation of the data:

Number of parasites, x	0	1	2	3	4	5	6	7	8
Number of foxes, f	69	17	6	3	1	2	1	0	1

a Construct a relative frequency histogram for x, the number of parasites per fox.

b Calculate $\bar{x}$ and s for the sample.

c What fraction of the parasite counts fall within two standard deviations of the mean? Within three standard deviations? Do these results agree with Tchebysheff's Theorem? With the Empirical Rule?

2.55 Consider a population consisting of the number of teachers per college at small 2-year colleges. Suppose that the number of teachers per college has an average $\mu = 175$ and a standard deviation $\sigma = 15$.

a Use Tchebysheff's Theorem to make a statement about the percentage of colleges that have between 145 and 205 teachers.

b Assume that the population is normally distributed. What fraction of colleges have more than 190 teachers?

2.56 From the following data, a student calculated s to be .263. On what grounds might we doubt his accuracy? What is the correct value (to the nearest hundredth)?

17.2	17.1	17.0	17.1	16.9
17.0	17.1	17.0	17.3	17.2
17.1	17.0	17.1	16.9	17.0
17.1	17.3	17.2	17.4	17.1

2.57 The state of California not only has some of the most polluted metropolitan areas in the country but also has some of the strictest pollution regulations in the country. As a result of an aggressive campaign by the South Coast Air Quality Management District (AQMD), the amount of pollution in southern California's atmosphere is beginning to decrease. One of its regulations requires employers of 100 or more workers to prepare a ride-sharing plan offering incentives for workers to share a ride, telecommute, ride a bike, or take public transportation to work. The goal is to achieve 1.5 workers per car during morning rush hours (Polakovic, 1993). What does the AQMD mean by this statement?

2.58 The following data represent median sales prices (in thousands of dollars) for existing single-family homes during the first quarters of 1989 and 1992 for 25 selected metropolitan areas in the United States:

Area	1989	1992	Area	1989	1992
Atlanta, GA	\$84.0	\$85.8	Milwaukee, WI	79.6	96.1
Baltimore, MD	96.3	111.5	Minneapolis, MN	87.2	94.8
Boston, MA	181.9	168.2	NY–NJ	183.2	168.6
Chicago, IL	107.0	132.9	Philadelphia, PA	103.9	120.7
Cincinnati, OH	75.8	87.5	Phoenix, AZ	78.8	84.7
Cleveland, OH	75.2	88.1	Pittsburgh, PA	65.8	74.8
Dallas, TX	92.4	90.8	San Diego, CA	181.8	182.7
Denver, CO	85.5	91.5	San Francisco, CA	260.2	243.9
Detroit, MI	73.7	77.5	Seattle, WA	115.0	141.3
Houston, TX	66.7	78.2	St. Louis, MO	76.9	83.0
Kansas City, MO	71.6	76.1	Tampa, FL	71.9	70.1
Los Angeles, CA	214.1	218.0	Washington, D.C.	144.4	152.5
Miami, FL	86.9	96.6			

Source: Wright (1992).

a Construct a relative frequency histogram for the median home prices during the first quarter of 1989.

b Calculate the range for the 1989 prices and the range approximation for s.

c Calculate $\bar{x}$, s^2, and s. Compare s to the approximation that was obtained in part (b).

d Find the proportion of the 1989 prices falling into the intervals $\bar{x} \pm s$, $\bar{x} \pm 2s$, and $\bar{x} \pm 3s$. Do these results agree with Tchebysheff's Theorem and the Empirical Rule?

2.59 Refer to Exercise 2.58. The MINITAB commands DESCRIBE and BOXPLOT produced the output shown below for the 1989 median prices:

```
MTB > DESCRIBE C1

                N      MEAN     MEDIAN     TRMEAN     STDEV     SEMEAN
1989           25     110.4       86.9      105.8      52.7       10.5

              MIN       MAX         Q1         Q3
1989         65.8     260.2       75.5      129.7

MTB > BOXPLOT C1

                -----------
            ---I  +        I-------            **          *           O
                -----------
       --------+---------+---------+---------+---------------1989
              80        120       160       200       240
```

a Compare the values for $\bar{x}$ and s with those obtained in Exercise 2.58.

b What are the upper and lower quartiles for the data?

c Are there any outliers in the data set? If so, how would you explain them?

2.60 Refer to Exercise 2.58. The MINITAB commands DESCRIBE and HISTOGRAM produced the output shown below for the 1992 median home prices:

```
MTB > DESCRIBE C2

                N     MEAN    MEDIAN    TRMEAN     STDEV    SEMEAN
1992           25   116.64     94.80    113.13     47.58      9.52

              MIN      MAX        Q1        Q3
1992        70.10   243.90     83.85    146.90

MTB > HISTOGRAM C2

Histogram of 1992      N = 25

Midpoint    Count
      80       10  **********
     100        5  *****
     120        2  **
     140        2  **
     160        3  ***
     180        1  *
     200        0
     220        1  *
     240        1  *
```

Compare the histogram for the 1992 prices generated by MINITAB with the relative frequency histogram that you constructed for the 1989 median home prices. Describe and compare the median home prices for these 2 years.

2.61 Refer to Exercise 1.15. The numbers of AIDS cases (in thousands) reported in 25 major cities in the United States during 1992 were as follows:

38.3	6.2	3.7	2.6	2.1
14.6	5.6	3.7	2.3	2.0
11.9	5.5	3.4	2.2	2.0
6.6	4.6	3.1	2.2	1.9
6.3	4.5	2.7	2.1	1.8

a Is the distribution of the measurements skewed or symmetric?

b Construct a box plot to determine whether there are any outliers.

c The three largest observations were obtained in the cities of New York, Los Angeles, and San Francisco, respectively. Does this help explain the magnitude of these three measurements?

Exercises Using the Data Disk

2.62 Refer to the sample of 50 male systolic blood pressures used in Exercise 1.53.

a Calculate a range estimate of s to use as a check against the calculation of s.

b Calculate $\bar{x}$ and s for the data. Check the value of s against the approximation found in part (a).

c Calculate the percentages of observations in the intervals $\bar{x} \pm s$, $\bar{x} \pm 2s$, and $\bar{x} \pm 3s$. Compare with the Empirical Rule.

2.63 Repeat the instructions in Exercise 2.62 using the 50 female systolic blood pressures from Exercise 1.55. Compare the two distributions by comparing their means and standard deviations.

2.64 The distribution of sales for the *Fortune 500* companies is skewed to the right. Take a sample of 25 observations, and scan the data.

a Are there any unusually large observations? Construct a box plot to determine whether or not there are any outliers in your sample.

b Calculate the mean and standard deviation for your sample.

c Would it be appropriate to use the Empirical Rule to describe this data set? Explain.

CASE STUDY

The Boys of Summer

Which baseball league has had the best hitters? Many of us have heard of baseball greats like Stan Musial, Hank Aaron, Roberto Clemente, and Pete Rose of the National League and Ty Cobb, Babe Ruth, Ted Williams, Rod Carew, and Wade Boggs of the American League. But have you ever heard of Willie Keeler, who batted .432 for the Baltimore Orioles, or Nap Lajoie, who batted .422 for the Philadelphia A's? The average number of hits for the batting champions of the National and American Leagues are given on the data disk. The batting averages for the National League begin in 1876 with Roscoe Barnes, whose batting average was .403 when he played with the Chicago Cubs. The last entry for the National League is for 1992, when Gary Sheffield of the San Diego Padres averaged .330. The American League records begin in 1901 with Nap Lajoie of the Philadelphia A's, who batted .422, and end in 1992, with Edgar Martinez of the Seattle Mariners, who batted .343. How will we summarize the information in this data set?

1 Use a statistical software package to describe the batting averages for the American and National League batting champions. Generate any graphics that may help you in interpreting these data sets.
2 Does one league appear to have a higher percentage of hits than the other? Does the batting average of one league appear to be more variable than the other?
3 Do there appear to be any outliers in either league?
4 How would you summarize your comparison of the two baseball leagues?

3

Probability and Probability Distributions

Case Study

In his exciting novel *Congo* (1980) author Michael Crichton describes an expedition racing to find boron-coated blue diamonds in the rain forests of eastern Zaire. Can probability help the heroine Karen Ross in her search for the Lost City of Zinj? The case study at the end of this chapter involved Ross's use of probability in decision-making situations.

General Objectives

Now that you have learned to describe a data set, how can you use sample data to draw conclusions about the sampled populations? The technique involves a statistical tool called *probability*. To use this tool correctly, you must first understand how it works. The first part of this chapter will teach you the new language of probability, presenting the basic concepts with simple examples.

The variables that we measured in Chapters 1 and 2 can now be redefined as random variables, whose values depend upon the chance selection of the elements in the sample. Using probability as a tool, we can develop probability distributions that serve as models for discrete random variables and can describe these random variables using a mean and standard deviation similar to those used in Chapter 2.

Specific Topics

1 Experiments and events (3.2)
2 Relative frequency definition of probability (3.3)
3 Intersections, unions, and complements (3.4)
4 Conditional probability and independence (3.5)
5 Additive and Multiplicative Laws of Probability (3.5)
6 Random variables (3.6)
7 Probability distributions for discrete random variables (3.7)
8 The mean and standard deviation for a discrete random variable (3.8)
9 Counting Rules and Bayes' Rule (3S.1, 3S.2)

3.1 The Role of Probability in Statistics

Probability and statistics are related in an important way. **Probability is the tool that allows the statistician to use sample information to make inferences about or to describe the population from which the sample was drawn.** We can illustrate this relationship with a simple example.

Consider a balanced die with its familiar six faces. When the die is tossed, any of the six sides has an equal chance of being the upper face. If we were to toss this die over and over again, we would generate a population of numbers, in which the upper face, x, would be 1, 2, 3, 4, 5, or 6. What would this population look like? Although it would be infinitely large, we can still say that it would consist of an equal number of 1s, 2s, . . . , 6s. Now let us toss the die once and observe the value of x. This is equivalent to taking a sample of size $n = 1$ from the population. What is the probability that $x = 2$? Knowing the structure of the population, we know that each of the six values of x has an equal chance of occurring, so that the probability that $x = 2$ is 1/6. This is a simple application of *probability*. When the population is known, we can calculate the probability of observing a particular sample.

Suppose now that the population is *not* known. That is, we do know whether the die is balanced or not. The population still consists of 1s, 2s, 3s, . . . , 6s, but we do not know in what proportion they occur! In an attempt to find out, we toss the die $n = 10$ times and record the upper face, x, after each toss. This is equivalent to taking a sample of size $n = 10$ from the population. Suppose that, on each toss, the upper face turns out to be $x = 1$. What would you infer about the population? Would you believe that the die is balanced? Probably not, because, if the die were balanced, the chance of observing a "1" ten times in a row is very small. Either we have observed this very unlikely event, or the die is unbalanced. We would likely be inclined toward the latter conclusion, because it is the more *probable* of the two choices.

This example shows the relationship between probability and statistics. When the population is known, probability is used to describe the likelihood of various sample outcomes. When the population is *unknown* and we have only a sample, we have the statistical problem of trying to make inferences about the unknown population. Probability is the tool we use to make these inferences. **Thus, probability reasons from the population to the sample, whereas statistics acts in reverse, moving from the sample to the population.**

As we explain the language of probability, we will assume that the population is known and will calculate the probability of drawing various samples. In doing so, we are really choosing a **model** for a physical situation, because the actual composition of a population is rarely known in practice. Thus the probabilist models a physical situation (the population) with probability much as the sculptor models with clay. In the sections that follow, we will use simple examples to help you grasp the concept of probability. Practical applications follow these simple examples.

3.2 Probability and the Sample Space

Data are obtained either by observing uncontrolled events in nature or by observing controlled situations in a laboratory. We will use the term **experiment** to describe either method of data collection.

DEFINITION ■ An **experiment** is the process by which an observation (or measurement) is obtained. ■

Note that the observation need not produce a numerical value. Here are some typical examples of experiments.

1 Recording a test grade.
2 Making a measurement of daily rainfall.
3 Interviewing a householder to obtain his or her opinion on a greenbelt zoning ordinance.
4 Testing a printed circuit board to determine whether it is a defective product or an acceptable product.
5 Tossing a coin and observing the face that appears.

Each experiment may result in one or more outcomes, which we will call **events** and denote by capital letters.

DEFINITION ■ An **event** is the outcome of an experiment. ■

EXAMPLE 3.1 Experiment: Toss a die and observe the number appearing on the upper face. Some events would be as follows:

Event A: Observe an odd number.
Event B: Observe a number less than 4.
Event E_1: Observe a 1.
Event E_2: Observe a 2.
Event E_3: Observe a 3.
Event E_4: Observe a 4.
Event E_5: Observe a 5.
Event E_6: Observe a 6. ■

There is a distinct difference between events A and B and events E_1, E_2, E_3, E_4, E_5, and E_6. Event A will occur if event E_1, E_3, or E_5 occurs—that is, if we observe a 1, 3, or 5. Thus, A could be decomposed into a collection of simpler events, namely, E_1, E_3, and E_5. Likewise, event B will occur if E_1, E_2, or E_3 occurs and could be viewed as a collection of smaller or simpler events. In contrast, it is impossible to decompose events $E_1, E_2, E_3, \ldots, E_6$. These events are called **simple events**.

DEFINITION ■ An event that cannot be decomposed is called a **simple event**. Simple events will be denoted by the symbol E with a subscript. ■

The events $E_1, E_2, \ldots, E_6$ represent a complete listing of all simple events associated with the experiment in Example 3.1. **An experiment will result in one and only one of the simple events.** For instance, if a die is tossed, we will observe a 1, 2, 3, 4, 5, or 6, but we cannot possibly observe more than one of the simple events at the same time. Hence, a list of simple events provides a breakdown of all possible indecomposable outcomes of the experiment.

DEFINITION ▪ The **sample space *S*** is the set of all possible outcomes of an experiment. ▪

EXAMPLE 3.2 Experiment: Toss a coin. The simple events are as follows:

E_1: Observe a head.

E_2: Observe a tail. ▪

EXAMPLE 3.3 Toss two coins and record the outcome. We can draw a picture for this experiment using a tree diagram. In a **tree diagram**, each successive branching of the tree corresponds to a step necessary to generate the possible outcomes of an experiment. A tree diagram for this example is given in Figure 3.1. The resulting simple events are shown in Table 3.1. ▪

FIGURE 3.1
Tree diagram for Example 3.3

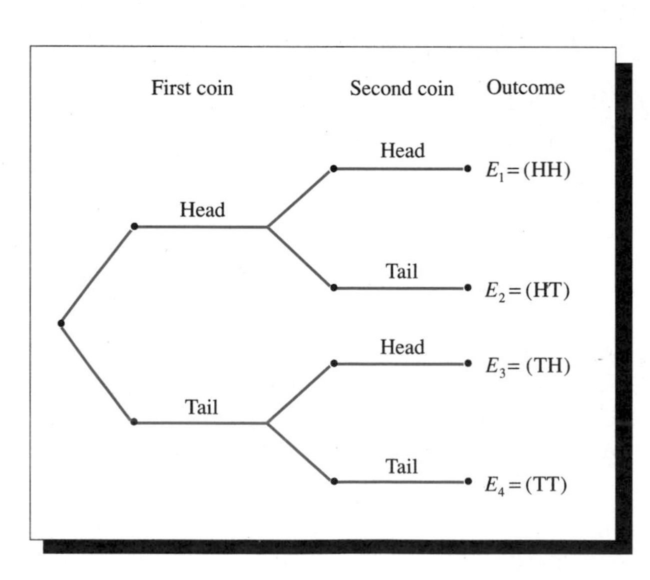

TABLE 3.1
Simple events for tossing two coins

Event	Coin 1	Coin 2
E_1	Head	Head
E_2	Head	Tail
E_3	Tail	Head
E_4	Tail	Tail

We can now define the outcomes of an experiment in terms of the associated simple events.

DEFINITION ■ An **event** is a collection of one or more simple events. ■

In order to visualize a problem, we can describe an experiment graphically using a **Venn diagram**, in which events are depicted as circular portions of the sample space. For example, the sample space for the toss of single die consists of the six simple events $E_1, E_2, \ldots, E_6$, and the event A of the simple events E_1, E_3, and E_5.

In Figure 3.2, the large outer box represents the sample space. The circle inside the box divides the sample space into two portions: the portion inside the circle, which represents the event A (an odd number is observed), and the portion outside the circle, which represents the event that A did not occur (an even number is observed). Keep in mind that a single repetition of this experiment will result in one and only one simple event. However, an arbitrary event such as A (an odd number is observed) will occur if any simple event within the circle (E_1, E_3, or E_5) occurs on that single repetition.

FIGURE 3.2
Venn diagram for die tossing

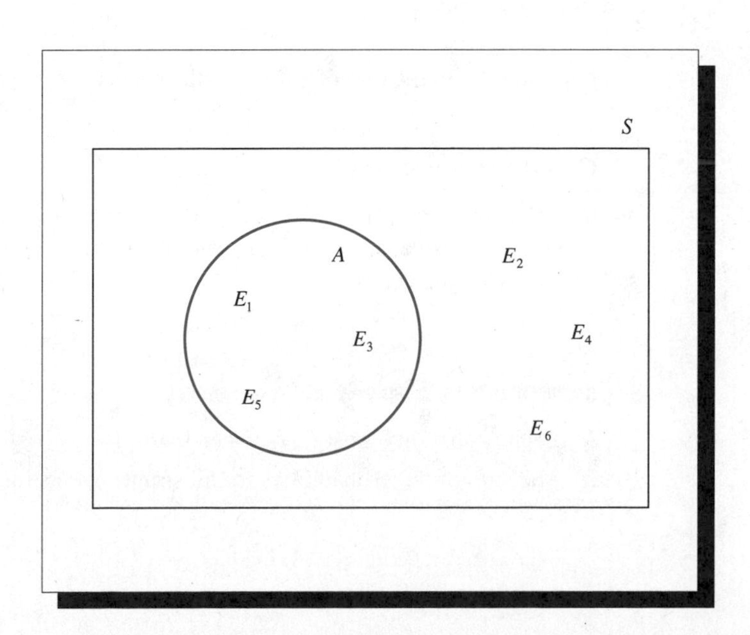

3.3 The Probability of an Event

The probability associated with an event is a measure of belief that the event will occur on the next repetition of the experiment. One way of assessing the probability of an event is to repeat the experiment a very large number of times N and to record n, the number of times that A occurs in these N repetitions. Then the **probability of the event A** is estimated by

$$P(A) = \frac{n}{N}$$

More properly, $P(A)$ is the limiting value of the fraction n/N as N becomes infinitely large. This interpretation of the meaning of probability, an interpretation held by many laypersons, is called the **relative frequency concept of probability**. In tossing a balanced die in a fair manner, we would expect that, in a larger number of tosses, the numbers 1, 2, 3, 4, 5, and 6 would appear with approximately the same frequency. Hence, it would be reasonable to take

$$P(E_1) = P(E_2) = \cdots = P(E_6) = \frac{1}{6}$$

From the relative frequency definition of probability, it is apparent that, for any event A, $P(A)$ is a fraction between 0 and 1 with $P(A) = 0$ if A never occurs and $P(A) = 1$ if A always occurs. Therefore, the closer its value to 1, the more likely A is to occur.

DEFINITION ▪ Two events, A and B, are **mutually exclusive** if, when one event occurs, the other cannot, and vice versa. ▪

For example, in the toss of a fair die, the events

A: Observe an odd number.

C: Observe an even number.

are mutually exclusive, since, if A occurs, C cannot, and vice versa. Simple events are mutually exclusive, and therefore the probabilities associated with simple events satisfy the following conditions.

Requirements for Simple-Event Probabilities

1 Each probability must lie between 0 and 1.

2 The sum of the probabilities for all simple events in S equals 1.

When it is possible to write down the simple events associated with an experiment and assess their respective probabilities, we can find the probability of an event A by summing the probabilities for the simple events contained in the event A.

DEFINITION ▪ The **probability of an event** A is equal to the sum of the probabilities of the simple events contained in A. ▪

EXAMPLE 3.4 Calculate $P(A)$ and $P(B)$ for the die-tossing experiment in Example 3.1.

Solution Event A occurs if E_1, E_3, or E_5 occurs. Therefore,

$$P(A) = P(E_1) + P(E_3) + P(E_5) = \frac{1}{6} + \frac{1}{6} + \frac{1}{6} = \frac{1}{2}$$

Event B will occur if E_1, E_2, or E_3 occurs. Therefore,

$$P(B) = P(E_1) + P(E_2) + P(E_3) = \frac{3}{6} = \frac{1}{2} \quad ▪$$

EXAMPLE 3.5 Calculate the probability of observing exactly one head in a toss of two coins.

Solution Construct the sample space letting H represent a head and T a tail.

Event	First Coin	Second Coin	$P(E_i)$
E_1	H	H	1/4
E_2	H	T	1/4
E_3	T	H	1/4
E_4	T	T	1/4

It would seem reasonable to assign a probability of 1/4 to each of the simple events. We are interested in

Event A: observe exactly one head.

Simple events E_2 and E_3 are in A. Hence,

$$P(A) = P(E_2) + P(E_3) = \frac{1}{4} + \frac{1}{4} = \frac{1}{2}$$

EXAMPLE 3.6 Consider the following experiment involving two urns. Urn 1 contains two white balls and one black ball. Urn 2 contains one white ball. A ball is drawn from urn 1 and placed in urn 2. Then a ball is drawn from urn 2. What is the probability that the ball drawn from urn 2 will be white? (See Figure 3.3.)

Solution The problem is easily solved once we have listed the simple events. For convenience, number the white balls 1, 2, and 3, with ball 3 residing in urn 2. The simple events are listed below using W_i to represent the ith white ball and B to represent the black ball. For example, E_1 is the event that the white ball 1 is drawn from urn 1, placed in urn 2, and then drawn from urn 2.

	Drawn from		
Event	**Urn 1**	**Urn 2**	$P(E_i)$
E_1	W_1	W_1	1/6
E_2	W_1	W_3	1/6
E_3	W_2	W_2	1/6
E_4	W_2	W_3	1/6
E_5	B	B	1/6
E_6	B	W_3	1/6

Event A, drawing a white ball from urn 2, occurs if E_1, E_2, E_3, E_4, or E_6 occurs. Once again, it would seem reasonable to assume that the simple events are equally likely and to assign a probability of 1/6 to each of the six simple events. Hence,

$$P(A) = P(E_1) + P(E_2) + P(E_3) + P(E_4) + P(E_6)$$

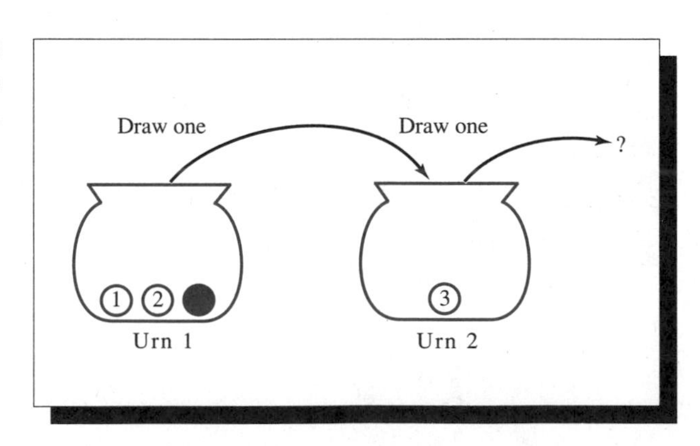

FIGURE 3.3 Representation of the experiment in Example 3.6

$$= \frac{1}{6} + \frac{1}{6} + \frac{1}{6} + \frac{1}{6} + \frac{1}{6}$$
$$= \frac{5}{6} \quad \blacksquare$$

To summarize, the probability of an event can be calculated by

- Listing all the simple events in the sample space.
- Assigning an appropriate probability to each simple event.
- Determining which simple events result in the event of interest.
- Summing the probabilities of the simple events that result in the event of interest.

However, there are some drawbacks to what looks like a simple procedure. Because the number of simple events can become very large, listing the simple events can become tedious if not impossible. It is essential that none be omitted, since this will result in incorrect calculations. Second, once the simple events are listed, the initial assignment of probabilities to these events must be accurate. Since this can be a very subjective task, you must take care that the probability assignments provide a realistic measure of the true likelihood of occurrence for the simple events.

To simplify the counting of simple events, we can use counting rules, which are included in a supplement at the end of this chapter. Although not necessary for understanding the basic concepts of probability, counting rules are required for solving more complex probability problems. If you wish to develop an ability to solve these more difficult problems using the simple-event approach, you should move directly to that section.

Tips on Problem Solving

Calculating the probability of an event: the simple-event approach

- Use the following steps for calculating the probability of an event by summing the probabilities of the simple events:
  1. Define the experiment.
  2. Identify a typical simple event. List the simple events associated with the experiment, and test each one to make certain that it cannot be decomposed. This defines the sample space S.
  3. Assign reasonable probabilities to the simple events in S, making certain that each probability is between 0 and 1 and that the sum of the simple-event probabilities equals 1.
  4. Define the event of interest A as a specific collection of simple events. (A simple event is in A if A occurs when the simple event occurs. Test *all* simple events in S to locate those in A.)

5 Find $P(A)$ by summing the probabilities of simple events in A.

- When the simple events are equiprobable, obtain the sum of the probabilities of the simple events in A (step 5) by counting the points in A and multiplying by the probability per simple event.
- Calculating the probability of an event by using the five-step procedure described above is systematic and will lead to the correct solution if all steps are followed correctly. Major sources of error include

1 Failure to define the experiment clearly (step 1).
2 Failure to specify simple events (step 2).
3 Failure to list all the simple events (step 3).
4 Failure to assign valid probabilities to the simple events.

Exercises

Basic Techniques

3.1 An experiment involves tossing a single die. Specify which of the following are simple events:

A: Observe a 2.
B: Observe an even number.
C: Observe a number greater than 2.
D: Observe both A and B.
E: Observe A or B or both.
F: Observe both A and C.

Calculate the probabilities of the events D, E, and F by summing the probabilities of the appropriate simple events.

3.2 A sample space contains five simple events, E_1, E_2, E_3, E_4, and E_5. If $P(E_3) = .4$, $P(E_4) = 2P(E_5)$, and $P(E_1) = P(E_2) = .15$, find the probabilities of E_4 and E_5.

3.3 A sample space contains ten simple events, $E_1, E_2, \ldots, E_{10}$. If $P(E_1) = 3P(E_2) = .45$ and the remaining simple events are equiprobable, find the probabilities of these remaining simple events.

3.4 A particular basketball player hits 70% of her free throws. When tossing a pair of free throws, the four possible simple events and three of their associated probabilities are as follows:

Simple Event	Outcome of First Free Throw	Outcome of Second Free Throw	Probability
1	Hit	Hit	.49
2	Hit	Miss	?
3	Miss	Hit	.21
4	Miss	Miss	.09

a Find the probability that the player will hit on the first throw and miss on the second.
b Find the probability that the player will hit on at least one of the two free throws.

3.5 The game of roulette uses a wheel containing 38 pockets. Thirty-six pockets are numbered $1, 2, \ldots, 36$, and the remaining two are marked 0 and 00. The wheel is spun, and a pocket is identified as the "winner." Assume that the observance of any one pocket is just as likely as any other.

a Identify the simple events in a single spin of the roulette wheel.

b Assign probabilities to the simple events.

c Let A be the event that you observe either a 0 or a 00. List the simple events in the event A and find $P(A)$.

d Suppose you were to place bets on the numbers 1 through 18. What is the probability that one of your numbers would be the winner?

3.6 A jar contains four coins: a nickel, a dime, a quarter, and a half-dollar. Three coins are randomly selected from the jar.

a List the simple events in S.

b What is the probability that the selection will contain the half-dollar?

c What is the probability that the total amount drawn will equal 60¢ or more?

3.7 Two dice are tossed.

a Construct a tree diagram for this experiment. How many simple events are there?

b What is the probability that the sum of the numbers shown on the upper faces is equal to 7? To 11?

Applications

3.8 A survey classified a large number of adults according to whether they were judged to need eyeglasses to correct their reading vision and whether they used eyeglasses when reading. The proportions falling into the four categories are shown below. (Note that a small proportion, .02, of adults used eyeglasses, when in fact, they were judged not to need them.)

Judged to Need Eyeglasses	Used Eyeglasses for Reading: Yes	No
Yes	.44	.14
No	.02	.40

If a single adult is selected from this large group, find the probability that:

a The adult is judged to need eyeglasses.

b The adult needs eyeglasses for reading but does not use them.

c The adult uses eyeglasses for reading whether he or she needs them or not.

3.9 According to *Webster's New Collegiate Dictionary*, a divining rod is "a forked rod believed to indicate [divine] the presence of water or minerals by dipping downward when held over a vein." To test the claims of success of a divining rod expert, skeptics bury four cans in the ground, two empty and two filled with water. The expert is to use the divining rod to test each of the four cans and decide which two contain water.

a Define the experiment.

b List the simple events in S.

c If the rod is completely useless in locating water, what is the probability that the expert will correctly identify (by guessing) the two cans containing water?

3.10 Refer to Exercise 3.9. Suppose the experiment were conducted using five cans: three empty and two containing water. Answer parts (a), (b), and (c) of the exercise.

3.11 A tea taster is required to taste and rank three varieties of tea, A, B, and C, according to the taster's preference.

a Define the experiment.

b List the simple events in S.

c If the taster has no ability to distinguish difference in taste among teas, what is the probability that the taster will rank tea type A as the most desirable? As the least desirable?

3.12 Four union men, two from a minority group, are assigned to four distinctly different one-man jobs, which can be ranked in order of desirability.

a Define the experiment.

b List the simple events in S.

c If the assignment to the jobs is unbiased—that is, if any one ordering of assignments is as probable as any other—what is the probability that the two men from the minority group are assigned to the least desirable jobs?

3.13 The odds are 2:1 that, when players A and B play racquetball, A wins. Suppose that A and B play three matches and that the winners of the matches are recorded. Using the letters A and B to denote the winner of each match, the eight simple events are listed in the following table. (As you will subsequently learn, under certain conditions it is reasonable to assume that the simple event probabilities are as listed in the table.)

Winner of Match			Simple Event	
1	2	3	i	$P(E_i)$
A	A	A	1	8/27
A	A	B	2	4/27
A	B	A	3	4/27
A	B	B	4	2/27
B	A	A	5	4/27
B	A	B	6	?
B	B	A	7	2/27
B	B	B	8	1/27

Using these probabilities,

a Find $P(E_6)$.

b Find the probability that A wins at least two of the three matches.

3.14 An investor has the option of investing in three of five recommended stocks. Unknown to her, only two will show a substantial profit within the next five years. If she selects the three stocks at random (giving every combination of three stocks an equal chance of selection), what is the probability that she selects the two profitable stocks? What is the probability that she will select only one of the two profitable stocks?

3.15 The Bureau of the Census reports that the median family income for all families in the United States during the year 1991 was \$35,353. That is, half of all American families had annual incomes exceeding this amount, and half had annual incomes equal to or below this amount (Wright, 1992, p. 242). Suppose that four families are surveyed and that each one reveals whether their income exceeded \$35,353 in 1991.

a How many simple events will be in the sample space?

b List the simple events.

c Identify the simple events in the following list of events:

A: At least two had income exceeding \$35,353.

B: Exactly two had income exceeding \$35,353.

C: Exactly one had income less than or equal to \$35,353.

d The definition of median income implies that each family is just as likely to exceed \$35,353 as to have income less than or equal to \$35,353. Using this information, assign reasonable probabilities to the simple events and find $P(A)$, $P(B)$, and $P(C)$.

3.16 Two city commissioners are to be selected from a total of five to form a subcommittee to study the city's traffic problems.

a Define the experiment.

b List the simple events in S.

c If all possible pairs of commissioners have an equal probability of selection, what is the probability that commissioners Jones and Smith will be selected?

3.4 Event Composition and Event Relations

Frequently we are interested in experimental outcomes that can be described as being formed by some composition of two or more events. **Compound events**, as the name suggests, can be formed in one of two ways or by some combination of the two, namely, a **union** or an **intersection**.

Let A and B be two events defined on the sample space S.

DEFINITION ■ The **intersection** of events A and B, denoted by $A \cap B$,[†] is the event that both A and B occur. ■

DEFINITION ■ The **union** of events A and B, denoted by the symbol $A \cup B$, is the event that A or B or both occur. ■

A Venn diagram representation of $A \cap B$, the intersection of events A and B, is given in Figure 3.4, and Figure 3.5 gives the Venn diagram representation of $A \cup B$. If we can enumerate the simple events and their probabilities, then $P(A \cap B)$ and $P(A \cup B)$ can be found by summing the probabilities associated with the simple events in $A \cap B$ and $A \cup B$.

EXAMPLE 3.7 Refer to the experiment in Example 3.3, in which two coins are tossed, and define

Event A: at least one head

Event B: at least one tail

Define the events A, B, $(A \cap B)$, and $(A \cup B)$ as collections of simple events.

[†]Some authors use the symbol AB.

FIGURE 3.4
Venn diagram of $A \cap B$

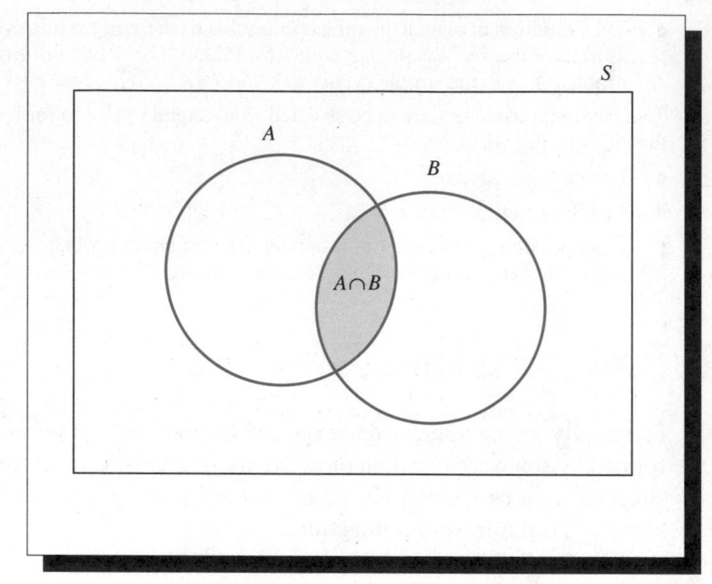

FIGURE 3.5
Venn diagram of $A \cup B$

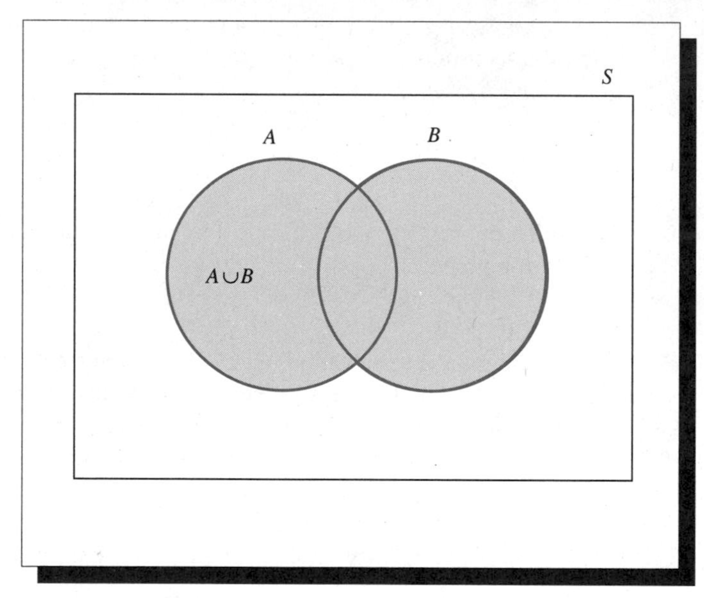

Solution Recall that the simple events for this experiment were:

E_1: HH (head on first coin, head on second)
E_2: HT
E_3: TH
E_4: TT

The occurrence of simple events E_1, E_2, and E_3 implies and hence defines event A. The other events could similarly be defined

Event B: E_2, E_3, E_4
Event $(A \cap B)$: E_2, E_3
Event $(A \cup B)$: E_1, E_2, E_3, E_4

Note that $(A \cup B) = S$, the sample space, and is thus certain to occur. ▪

When the two events A and B are mutually exclusive, it means that, when A occurs, B cannot, and vice versa. Figure 3.6 is a Venn diagram representation of two such events with no simple events in common. Mutually exclusive events are also referred to as **disjoint events**. When A and B are mutually exclusive:

1. $P(A \cap B) = 0$
2. $P(A \cup B) = P(A) + P(B)$

That is, if $P(A)$ and $P(B)$ are known, we do not need to enumerate the simple events in $A \cup B$ and sum their respective probabilities; rather, we simply sum $P(A)$ and $P(B)$.

FIGURE 3.6
Two disjoint events

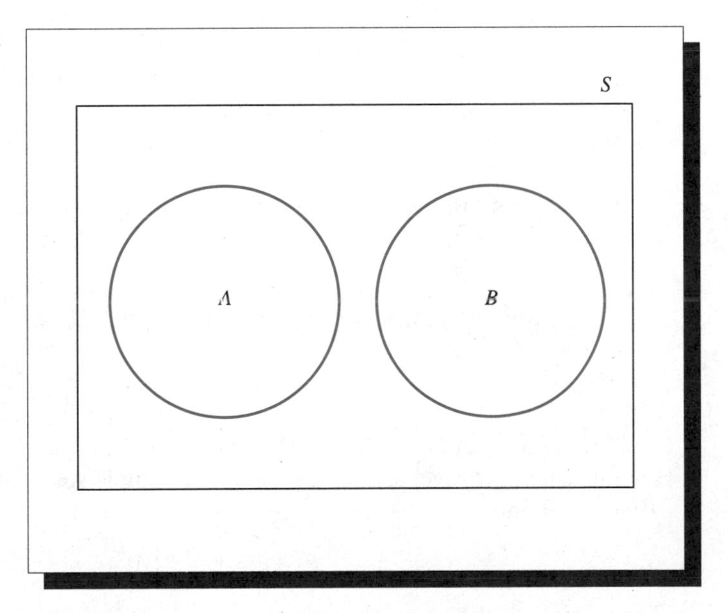

EXAMPLE 3.8 An oil-prospecting firm plans to drill two exploratory wells. Past evidence is used to assess the following possible outcomes:

Event	Description	Probability
A	Neither well produces oil or gas	.80
B	Exactly one well produces oil or gas	.18
C	Both wells produce oil or gas	.0

Find $P(A \cup B)$ and $P(B \cup C)$.

Solution By their definition, events A, B, and C are jointly mutually exclusive, since the occurrence of one event precludes the occurrence of either of the other two. Therefore,

$$P(A \cup B) = P(A) + P(B) = .80 + .18 = .98$$

and

$$P(B \cup C) = P(B) + P(C) = .18 + .02 = .20$$

The event $A \cup B$ can be described as the event that *at most* one well produces oil or gas, and $B \cup C$ describes the event that *at least* one well produces gas or oil. ■

The concept of unions and intersections can be extended to more than two events. For example, the union of three events A, B, and C, would be that A, B, C, or any combination of the three occurs. This event, which is denoted by the symbol $A \cup B \cup C$, would be the set of simple events that are in A or B or C or in any combination of those events. Similarly, the intersection of the three events A, B, and C, denoted as $A \cap B \cap C$, is the event that all three events A, B, and C occur. This event is the collection of simple events that are common to the three events A, B, and C.

Complementation is another event relationship that often simplifies probability calculations.

DEFINITION ■ The **complement** of an event A, denoted by A^C, consists of all the simple events in the sample space S that are not in A. ■

The complement of A is the event that A does not occur. Therefore, A and A^C are mutually exclusive, and $A \cup A^C = S$, the sample space. It follows that $P(A) + P(A^C) = 1$ and

$$P(A) = 1 - P(A^C)$$

For example, if the event A is the occurrence of at least one head in the toss of three fair coins, then the event A^C is the occurrence of no heads in the toss, and

$$P(A) = 1 - P(A^C) = 1 - (.5)^3 = 1 - .125 = .875$$

3.5 Conditional Probability and Independence

Two events are often related so that the probability of the occurrence of one depends on whether the second has or has not occurred. For instance, suppose one experiment consists in observing the weather on a specific day. Let A be the event "observe rain" and B be the event "observe an overcast sky." Events A and B are obviously related. The probability of rain, $P(A)$, is not the same as the probability of rain given prior information that the day is cloudy. The probability of A, $P(A)$, would be the fraction of the entire population of observations that result in rain. Now let us look only at the subpopulation of observations that result in B, a cloudy day, and the fraction of these that result in A. This fraction, called the **conditional probability of *A* given *B***, may equal $P(A)$, but we would expect the chance of rain, given that the day is cloudy, to be larger.

For example, consider a toss of a fair die. Let B be the outcome for which the number is less than 4 (i.e., a 1, 2, or 3), and let A be the outcome for which the number is odd. If the event B has occurred, and the outcome is a 1, 2, or 3, then the chance that the number is odd is 2/3. We have just described the conditional probability of A given B.

The conditional probability of *A*, given that *B* has occurred, is denoted as

$$P(A|B)$$

where the vertical bar in the parentheses is read "given" and events appearing to the right of the bar are the events that have occurred.

We will define the conditional probabilities of B given A and A given B as follows.

DEFINITION ▪

The **conditional probability of *B***, given that A has occurred, is

$$P(B|A) = \frac{P(A \cap B)}{P(A)} \qquad \text{if } P(A) \neq 0$$

The **conditional probability of *A***, given that B has occurred, is

$$P(A|B) = \frac{P(A \cap B)}{P(B)} \qquad \text{if } P(B) \neq 0 \quad ▪$$

By attaching some numbers to the probabilities in the weather example, we can see that this definition of conditional probability is consistent with the relative frequency concept of probability. Recall that A denotes rain on a given day; B denotes a day that is cloudy. Now suppose that 10% of all days are rainy and cloudy [that is, $P(A \cap B) = .10$] and 30% of all days are cloudy [$P(B) = .30$].

This situation is graphically portrayed in Figure 3.7. Each simple event in event B, denoted by the large circular area, is associated with a single cloudy day. Since 30% of all days will be cloudy, we can regard this area as .30. Ten percent of all days, or 1/3 of all cloudy days, will also be rainy. These days are included in the color-shaded event $A \cap B$. If a single day is selected from the set of all days representing

FIGURE 3.7
Events A and B

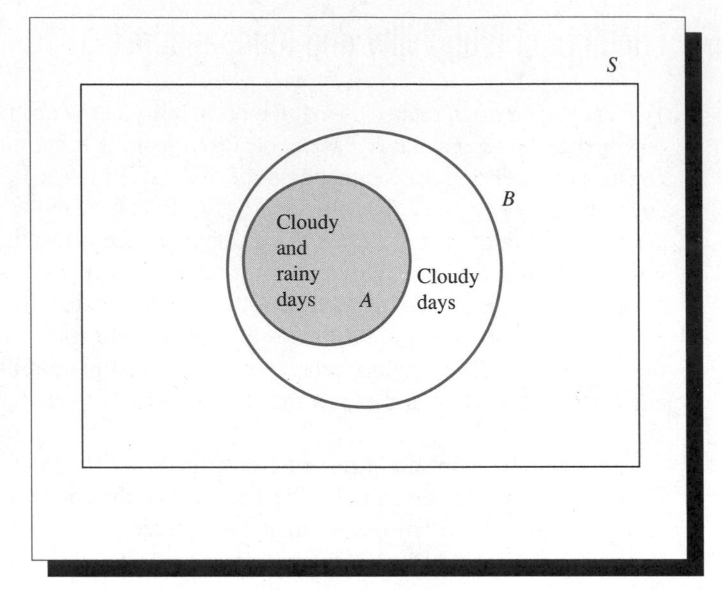

the population, what is the probability that we will select a rainy day, given that we know the day is cloudy? That is, what is $P(A|B)$?

Since we already know that the day is cloudy, we know that the simple event to be selected must fall in event B (Figure 3.7). One-third of these days will result in rain. Hence, the probability that we will select a rainy day is

$$P(A|B) = \frac{1}{3}$$

We can see that this result agrees with our definition for $P(A|B)$; that is,

$$P(A|B) = \frac{P(A \cap B)}{P(B)} = \frac{.10}{.30} = \frac{1}{3}$$

EXAMPLE 3.9 Calculate $P(A|B)$ for the die-tossing experiment described in Example 3.1, where

A: Observe an odd number.

B: Observe a number less than 4.

Solution Given that event B (a number less than 4) has occurred, we have observed a 1, 2, or 3, which occur with equal frequency. Of these three simple events, exactly two (1 and 3) result in event A (an odd number). Hence,

$$P(A|B) = \frac{2}{3}$$

Or, we could obtain $P(A|B)$ by substituting into the equation

$$P(A|B) = \frac{P(A \cap B)}{P(B)} = \frac{1/3}{1/2} = \frac{2}{3}$$

Note that $P(A|B) = 2/3$ and $P(A) = 1/2$, indicating that A and B are dependent on each other. ▪

DEFINITION ▪ Two events A and B are said to be **independent** if and only if either

$$P(A|B) = P(A)$$

or

$$P(B|A) = P(B)$$

Otherwise, the events are said to be **dependent**. ▪

Translating this definition into words: Two events are independent if the occurrence or nonoccurrence of one of the events does not change the probability of the occurrence of the other event. If $P(A|B) = P(A)$, then $P(B|A)$ will also equal $P(B)$. Similarly, if $P(A|B)$ and $P(A)$ are unequal, then $P(B|A)$ and $P(B)$ will also be unequal.

EXAMPLE **3.10** Refer to the die-tossing experiment in Example 3.1, where

A: Observe an odd number.

B: Observe a number less than 4.

Are events A and B mutually exclusive? Are they complementary? Are they independent?

Solution In the terms of simple events, we can write

Event A: E_1, E_3, E_5

Event B: E_1, E_2, E_3

Event $A \cap B$ is the set of simple events in both A and B. Since $A \cap B$ includes events E_1 and E_3, A and B are not mutually exclusive. They are not complementary because B is not the set of all outcomes in S that are not in A. The test for independence lies in the definition; that is, we will check to see if $P(A|B) = P(A)$. From Example 3.9, $P(A|B) = 2/3$. Then, since $P(A) = 1/2$, $P(A|B) \neq P(A)$ and by definition events A and B are dependent. ▪

A second approach to the solution of probability problems is based on the classification of compound events, event relations, and two probability laws, which we will now state and illustrate. The "laws" can be simply stated and taken as fact as long as they are consistent with our model and with reality. The first is called the Additive Law of Probability and applies to unions.

The Additive Law of Probability Given two events A and B, the probability of the union $(A \cup B)$ is equal to

$$P(A \cup B) = P(A) + P(B) - P(A \cap B)$$

If A and B are mutually exclusive, $P(A \cap B) = 0$ and

$$P(A \cup B) = P(A) + P(B)$$ ▪

The Additive Law conforms to reality and our model. Note in Figure 3.8 that the sum $P(A) + P(B)$ contains the sum of the probabilities of all simple events in $(A \cup B)$ but includes a double counting of the probabilities of all outcomes in the intersection $A \cap B$. Subtracting $P(A \cap B)$ gives the correct result.

The second law of probability is called the Multiplicative Law and applies to intersections.

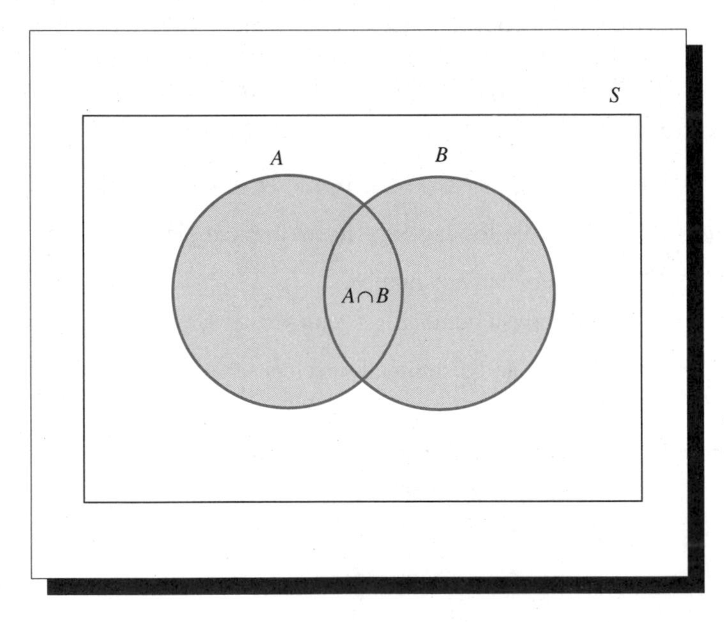

FIGURE 3.8
The union of two events, A and B ($A \cup B$ is shaded)

The Multiplicative Law of Probability Given two events A and B, the probability of the intersection $A \cap B$ is

$$\begin{aligned} P(A \cap B) &= P(A)P(B|A) \\ &= P(B)P(A|B) \end{aligned}$$

If A and B are independent, $P(A \cap B) = P(A)P(B)$. ▪

The Multiplicative Law follows from the definition of conditional probability.

Using probability laws to calculate the probability of a compound event is less direct than listing simple events and requires some experience and ingenuity. The approach involves expressing the event of interest as a union or intersection (or combination of both) of two or more events whose probabilities are known or easily calculated. This can often be done in many ways; the trick is to find the right combination—a task requiring no small amount of creativity in some cases. The usefulness of event relations is now apparent. If the event of interest is expressed as a union of mutually exclusive events, the probabilities of the intersections will equal zero. If the events are independent, we can use the unconditional probabilities to calculate the probability of an intersection. Examples 3.11 through 3.15 illustrate the use of the probability laws and the technique just described.

EXAMPLE **3.11** Calculate $P(A \cap B)$ and $P(A \cup B)$ for Example 3.1.

Solution Recall that $P(A) = P(B) = 1/2$ and $P(A|B) = 2/3$. Then

$$\begin{aligned} P(A \cap B) &= P(B)P(A|B) \\ &= \left(\frac{1}{2}\right)\left(\frac{2}{3}\right) \\ &= \frac{1}{3} \end{aligned}$$

and

$$\begin{aligned} P(A \cup B) &= P(A) + P(B) - P(A \cap B) \\ &= \frac{1}{2} + \frac{1}{2} - \frac{1}{3} \\ &= \frac{2}{3} \end{aligned}$$

These solutions will agree with those obtained using the simple-event approach. ▪

EXAMPLE 3.12 Consider the experiment in which two coins are tossed. Let A be the event that the toss results in at least one head. Find $P(A)$.

Solution A^C is the collection of simple events implying the event "two tails." Because A^C is the complement of A,

$$P(A) = 1 - P(A^C)$$

The event A^C will occur if both of two independent events occur: "tail on the first coin," T_1, and "tail on the second coin," T_2. Then A^C is the intersection of T_1 and T_2, or

$$A^C = T_1 \cap T_2$$

Applying the Multiplicative Law and noting that T_1 and T_2 are independent events,

$$P(A^C) = P(T_1)P(T_2) = \left(\frac{1}{2}\right)\left(\frac{1}{2}\right) = \frac{1}{4}$$

Then, $P(A) = 1 - P(A^C) = 1 - 1/4 = 3/4$. ▪

EXAMPLE 3.13 In a color preference experiment, eight toys are placed in a container. The toys are identical except for color—two are red, and six are green. A child is asked to choose two toys *at random*. What is the probability that the child chooses the two red toys?

Solution The experiment can be visualized using a tree diagram as shown in Figure 3.9.

Define the following events:

R: red toy is chosen.

G: green toy is chosen.

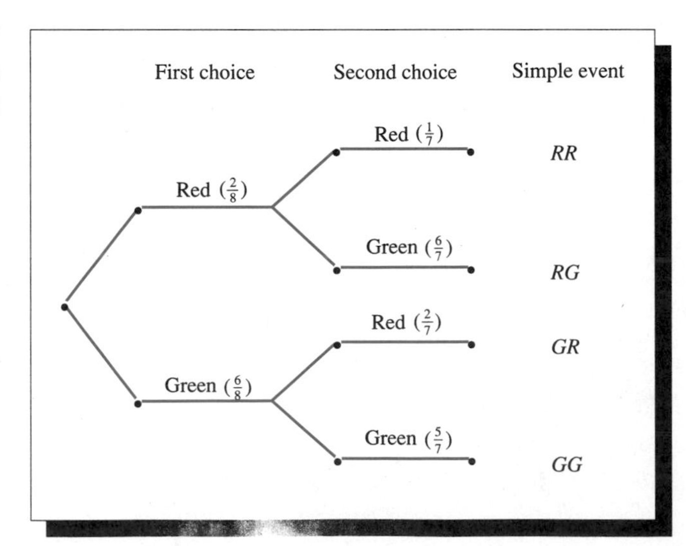

FIGURE 3.9 Tree diagram for color preference experiment

The event A (both toys are red) can be constructed as the intersection of two events:

$$A = (R \text{ on first choice }) \cap (R \text{ on second choice})$$

Since there are only two red toys in the container, the probability of choosing red on the first choice is 2/8. However, once this red toy has been chosen, the probability of red on the second choice is *dependent* on the outcome of the first choice (see Figure 3.9). If the first choice was red, the probability of choosing a second red toy is only 1/7, since there is only one red toy among the seven remaining. If the first choice was green, the probability of choosing red on the second choice is 2/7, since there are two red toys among the seven remaining. Using this information and the Multiplicative Law, we can find the probability of event A:

$$\begin{aligned} P(A) &= P(R \text{ on first choice } \cap R \text{ on second choice}) \\ &= P(R \text{ on first choice })P(R \text{ on second } | R \text{ on first}) \\ &= (2/8)(1/7) = 2/56 = 1/28 \quad \blacksquare \end{aligned}$$

Other probabilities can be found similarly, using the tree diagram. The probabilities on the second branching of the tree (the second choice) are the conditional probabilities, given that the first choice has occurred. Probabilities for individual simple events found at the ends of the branches can be calculated by multiplying probabilities along the branches leading to the event. Probabilities for more complex events are found as the sum of the appropriate simple event probabilities.

EXAMPLE 3.14 In a telephone survey of 1000 adults, respondents were asked about the expense of a college education and the relative necessity of some form of financial assistance. The respondents were classified according to whether they currently had a child in college and whether they thought the loan burden for most college students is too high, the right amount, or too little. The proportions responding in each category are shown in the following **probability table**:

	Too High (A)	Right Amount (B)	Too Little (C)
Child in college (D)	0.35	0.08	0.01
No child in college (E)	0.25	0.20	0.11

Source: Miko and Weilant (1991), p. 202.

Suppose one respondent is chosen at random from this group.

a What is the probability that the respondent has a child in college?

b Given that the respondent has a child in college, what is the probability that he or she ranks the loan burden as "too high"?

Solution The entries in the probability table represent the intersection probabilities for the events indicated in the margins. For example, the entry in the top-left corner of the table is the probability that a respondent has a child in college *and* thinks the loan burden is too high ($A \cap D$).

a The event that a respondent has a child in college will occur regardless of his or her response to the question about loan burden. That is, event D will occur if any of the events $A \cap D$, $B \cap D$, or $C \cap D$ occur. Hence, using the Additive Law,

$$\begin{aligned} P(D) &= P(A \cap D) + P(B \cap D) + P(C \cap D) \\ &= .35 + .08 + .01 = .44 \end{aligned}$$

In general, the probabilities of marginal events such as A, B, and C are found by summing the probabilities in the appropriate row or column.

b To find the probability of A given D, we use the definition of conditional probability:

$$P(A|D) = \frac{P(A \cap D)}{P(D)} = \frac{.35}{.44} = .80 \quad \blacksquare$$

EXAMPLE 3.15 Two cards are drawn from a deck of 52 cards. Calculate the probability that the draw will include an ace and a ten.

Solution Event A: Draw an ace and a ten.

Then $A = B \cup C$, where

B: Draw the ace on the first draw and the ten on the second.

C: Draw the ten on the first draw and the ace on the second.

B and C were chosen to be mutually exclusive and also to be intersections of events with known probabilities. Thus,

$$B = B_1 \cap B_2 \qquad \text{and} \qquad C = C_1 \cap C_2$$

where

B_1: Draw an ace on the first draw.

B_2: Draw a ten on the second draw.

C_1: Draw a ten on the first draw.

C_2: Draw an ace on the second draw.

Applying the Multiplicative Law,

$$\begin{aligned} P(B_1 \cap B_2) &= P(B_1)P(B_2|B_1) \\ &= \left(\frac{4}{52}\right)\left(\frac{4}{51}\right) \end{aligned}$$

and

$$P(C_1 \cap C_2) = \left(\frac{4}{52}\right)\left(\frac{4}{51}\right)$$

Then, applying the Additive Law,

$$\begin{aligned} P(A) &= P(B) + P(C) \\ &= \left(\frac{4}{52}\right)\left(\frac{4}{51}\right) + \left(\frac{4}{52}\right)\left(\frac{4}{51}\right) = \frac{8}{663} \end{aligned}$$

Check each composition carefully to be certain that it is actually equal to the event of interest. ▪

Tips on Problem Solving

Calculating the probability of an event: event-composition approach

- Use the following steps to calculate the probability of an event using the event-composition approach:
  1. Define the experiment.
  2. Clearly visualize the nature of the simple events. Identify a few to clarify your thinking.
  3. Write an equation expressing the event of interest (say, A) as a composition of two or more events using either or both of the two forms of composition (unions and intersections). Note that this equates point sets. Make certain that the event implied by the composition and event A represent the same set of simple events.
  4. Apply the Additive and Multiplicative Laws of Probability to step 3 and find $P(A)$.
- Be careful with step 3. You can often form many compositions that will be equivalent to event A. The trick is to form a composition in which all the probabilities appearing in step 4 will be known. Visualize the results of step 4 for any composition and select the one for which the component probabilities are known.
- Always write down letters to represent events described in an exercise. Then write down the probabilities that are given, and assign them to events. Identify the probability that is requested in the exercise. This can help you arrive at the appropriate event composition.

Exercises

Basic Techniques

3.17 An experiment can result in one of five equally likely simple events, $E_1, E_2, \ldots, E_5$. Events A, B, and C are defined as follows:

$A: E_1, E_3$
$B: E_1, E_2, E_4, E_5$
$C: E_3, E_4$

Find the probabilities associated with the following compound events by listing the simple events in each:

a A^C **b** B^C **c** $A \cap B$
d $A \cap C$ **e** $B \cap C$ **f** $A \cup B$
g $A \cup C$ **h** $B|C$ **i** $A|B$
j $A \cup B \cup C$ **k** $(A \cap B)^C$ **l** $A^C \cap B^C$

3.18 Refer to Exercise 3.17. Use the definition of a complementary event to find

a $P(A^C)$ **b** $P(A \cap B)^C$ **c** $P(A^C \cap B^C)$

Do the results agree with those obtained in Exercise 3.17?

3.19 Refer to Exercise 3.17. Use the definition of conditional probability to find

a $P(A|B)$ **b** $P(B|C)$

Do the results agree with those obtained in Exercise 3.17?

3.20 Refer to Exercise 3.17. Use the Additive and Multiplicative Laws of Probability to find

a $P(A \cup B)$ **b** $P(A \cap B)$ **c** $P(B \cap C)$

Do the results agree with those obtained in Exercise 3.17?

3.21 Refer to Exercise 3.17. Determine whether events A and B are

a Independent. **b** Mutually exclusive.

3.22 An experiment consists of tossing a single die and observing the number of dots shown on the upper face. Events A, B, and C are defined as follows:

A: Observe a number less than 4.

B: Observe a number less than or equal to 2.

C: Observe a number greater than 3.

Find the probabilities associated with the following compound events using either the simple event approach or the event composition approach.

a S **b** $A|B$ **c** B
d $A \cap B \cap C$ **e** $A \cap B$ **f** $A \cap C$
g $B \cap C$ **h** $A \cup C$ **i** $B \cup C$

3.23 Refer to Exercise 3.22.

a Are events A and B independent? Mutually exclusive?
b Are events A and C independent? Mutually exclusive?

Applications

3.24 In a survey designed to monitor the spending patterns of employed high school seniors, 70% of the males surveyed spent none or only a little of their earnings on long-term savings, while 21% spent all or almost all of their earnings on personal items (Schmittroth, 1991, p. 335). Suppose one male senior is chosen at random from this group.

a What is the probability that he spent none or only a little of his earnings on long-term savings?

b What is the probability that he did *not* spend all or almost all on personal items?

c If two male seniors are chosen at random from this group, what is the probability that *both* spent all or almost all of their earnings on personal items?

3.25 Many companies are testing prospective employees for drug use, with the intent of improving efficiency and reducing absenteeism, accidents, and theft. Opponents claim that this procedure is creating a class of unhirables and that some persons may be placed in this class because the tests themselves are not 100% reliable. Suppose that a company uses a test that is 98% accurate—that is, it correctly identifies a person as a drug user or nonuser with probability .98—and that, to reduce the chance of error, each job applicant is required to take two tests. If the outcomes of the two tests on the same person are independent events, what is the probability that

a A nondrug user will fail both tests?

b A drug user will be detected (i.e., he or she will fail at least one test)?

c A drug user will pass both tests?

3.26 A study of the behavior of a large number of drug offenders after treatment for drug abuse suggests that the likelihood of conviction within a two-year period after treatment may depend on the offender's education. The proportions of the total number of cases falling to four education/conviction categories are shown in the following table:

	Status Within 2 Years After Treatment		
Education	**Convicted**	**Not Convicted**	**Totals**
10 years or more	.10	.30	.40
9 years or less	.27	.33	.60
Totals	.37	.63	1.00

Suppose a single offender is selected from the treatment program. Define the events:

A: The offender has 10 or more years of education.

B: The offender is convicted within 2 years after completion of treatment.

Find the approximate probabilities for these events:

a A

b B

c $A \cap B$

d $A \cup B$

e A^C

f $(A \cup B)^C$

g $(A \cap B)^C$

h A given that B has occurred

i B given that A has occurred

3.27 Use the probabilities of Exercise 3.26 to show that

a $P(A \cap B) = P(A)P(B|A)$

b $P(A \cap B) = P(B)P(A|B)$

c $P(A \cup B) = P(A) + P(B) - P(A \cap B)$

3.28 When an American marriage ends in divorce, it is most likely that the woman is the partner who is dissatisfied. A study conducted by the National Center for Health Statistics (*Press-Enterprise*, Riverside, Calif., June 7, 1989) indicates that in 1986, 61.5% of all divorce petitions in the

United States were filed by wives, 32.6% by husbands, and the remaining petitions were filed jointly.

a What percentage of all divorce petitions were filed jointly?

b If a couple were divorced in 1986, what is the probability that the wife did not unilaterally file the divorce petition?

c If two couples were divorced in 1986, what is the probability that the wife unilaterally filed the divorce petition in both cases? In at least one case?

3.29 Refer to Exercise 3.28. The study reported that the filing rates for divorce petitions have changed within the past decade, as indicated in the following table:

	1975	1980	1986
Filed by wife	67.2%	63.4%	61.5%
Filed by husband	29.4%	30.2%	32.6%
Filed jointly	3.4%	6.4%	5.9%

Suppose two women are interviewed, one of whom was divorced in 1975 and the other in 1986.

a What is the probability that both women unilaterally filed their divorce petitions?

b What is the probability that one woman unilaterally filed the divorce petition, but the other petition was filed by the husband?

c What is the probability that at least one woman filed the divorce petition jointly with her husband?

3.30 Whether a grant proposal is funded quite often depends on the reviewers. Suppose a group of research proposals was evaluated by a group of experts as to whether the proposals were worthy of funding. When these same proposals were submitted to a second independent group of experts, the decision to fund was reversed in 30% of the cases. If the probability that a proposal is judged worthy of funding by the first peer review group is .2, what is the probability that a project worthy of funding will be:

a Approved by both groups?

b Disapproved by both groups?

c Approved by one group?

3.31 Moreno Valley, California, a bedroom community lying east of Los Angeles, has experienced rapid growth in the last decade, mainly because of the affordable housing and the 1-hour commute time to many parts of Los Angeles. The age distribution of its residents 18 years or older was reported as follows:

Age	Proportion
18–24 years	.15
25–34 years	.28
35–44 years	.21
45–54 years	.14
55–64 years	.11
65+ years	.12

Source: Press-Enterprise, Riverside, Calif. (Oct. 12, 1992).

If four persons 18 years or older are randomly selected from this population,

a What is the probability that all four will be under 25 years of age?

b What is the probability that all four will be 65 years or older?

c What is the probability that exactly one person will be under 25 and the remaining three 25 years or older?

d What is the probability that exactly two people will be in the 25–34 age group and the remaining two people will be in the 35–44 age group?

3.32 A certain article is visually inspected by two successive inspectors. When a defective article comes through, the probability that it gets by the first inspector is .1. Of those that get past the first inspector, the second inspector will "miss" five out of ten. What fraction of the defectives get by both inspectors?

3.33 A survey of people in a given region showed that 20% were smokers. The probability of death due to lung cancer, given that a person smoked, was roughly ten times the probability of death due to lung cancer, given that a person did not smoke. If the probability of death due to lung cancer in the region is .006, what is the probability of death due to lung cancer given that a person is a smoker?

3.34 A smoke-detector system uses two devices, A and B. If smoke is present, the probability that it will be detected by device A is .95; by device B, .98; and by both devices, .94.

a If smoke is present, find the probability that the smoke will be detected by device A or device B or both devices.

b Find the probability that the smoke will not be detected.

3.35 Many public schools are implementing a "no pass, no play" rule for athletes. Under this system, a student who fails a course is disqualified from participating in extracurricular activities during the next grading period. Suppose the probability that an athlete who has not previously been disqualified will be disqualified is .15 and that the probability that an athlete who has been disqualified will be disqualified again in the next time period is .5. If 30% of the athletes have been disqualified before, what is the probability that an athlete will be disqualified during the next grading period?

3.36 Tay-Sachs disease is a genetic disorder that is usually fatal in early childhood. If both parents are carriers of the disease, the probability that their offspring will develop the disease is approximately .25. Suppose a husband and wife are both carriers of the disease and the wife is pregnant on three different occasions. If the occurrence of Tay-Sachs in any one offspring is independent of the occurrence in any other, what are the following probabilities?

a All three children will develop Tay-Sachs disease.

b Only one child will develop Tay-Sachs disease.

c The third child will develop Tay-Sachs disease, given that the first two did not.

3.37 When Brian Bosworth announced that he would become a professional football player in the supplemental draft of 1987, National Football League officials explained the rules for determining the drafting order for NFL teams as follows (*New York Times*, May 12, 1987):

> The team with the worst record, Tampa Bay, puts 28 of its logos into a barrel, the [then] Indianapolis Colts put 27 in, and so on down to the Jets with eight and the [Super Bowl Champion] Giants with one. Twenty-eight logos are then picked to set the order. Once a team is picked, its remaining logos in the barrel are ignored.

a How many logos will be put in the barrel?

b What is the probability that the Giants will obtain the first pick?

c What is the probability that the Buccaneers will obtain the first pick?

d What is the probability that the Colts and Jets will have the first two picks?

3.38 Two people enter a room, and their birthdays (ignoring years) are recorded.

a Identify the nature of the simple events in S.

b What is the probability that the two people have a specific pair of birthdates?

c Identify the simple events in event A, "both persons have the same birthday."

d Find $P(A)$.

e Find $P(A^C)$.

3.39 If n people enter a room, find the probability that

A: None of the persons have the same birthday.

B: At least two of the persons have the same birthday.

Solve for

a $n = 3$ **b** $n = 4$

[NOTE: Surprisingly, $P(B)$ increases rapidly as n increases. For example, for $n = 20$, $P(B) = .411$; for $n = 40$, $P(B) = .891$.]

3.40 A worker-operated machine produces a defective item with probability .01 if the worker follows the machine's operating instructions exactly, and with probability .03 if he does not. If the worker follows the instructions 90% of the time, what proportion of all items produced by the machine will be defective?

3.6 Random Variables

In Chapter 1, we defined a *variable* as a characteristic that changes or varies over time and/or for different individuals or objects under consideration. Variables were classified as qualitative or quantitative, discrete or continuous, depending on the type of variable being measured.

In some cases, it is possible to convert qualitative data to quantitative data by assigning a numerical value to each category to form a scale. Industrial production is often scaled according to first grade, second grade, and so on. In other cases, the classification categories are simply assigned a number to simplify the coding of the observation. For example, for the toss of a coin, the observation is a "head" or a "tail." However, when we are interested in the number of heads, the recorded measurement is a "1" if a head or a "0" if a tail. In this way, observations on qualitative variables usually give rise to measurements reflecting the number of individuals or elements in each of the defined categories.

In general, most experiments give rise to numerical measurements. If we denote the variable being measured by x, then the value that x assumes changes or varies, depending on the particular outcome of the experiment. For example, suppose we toss a die and let x be the number observed on the upper face. The variable x can take any one of six values—1, 2, 3, 4, 5, or 6—depending on the random outcome of the experiment. Therefore, we refer to the variable x as a **random variable**.

DEFINITION ▪ A variable x is a **random variable** if the value that it assumes, corresponding to the outcome of an experiment, is a chance or random event. ▪

Observing the number of defects on a randomly selected piece of furniture, selecting a college applicant at random and observing the person's SAT score, or measuring the number of telephone calls received by a crisis intervention hotline during a randomly selected time period give rise to random numerical events.

Quantitative random variables are similarly classified as either *discrete* or *continuous*, according to the values that x may assume. The distinction between discrete and continuous random variables is important, since different probability models are required for each type of variable. We will focus our attention on discrete random variables in the remainder of this chapter. Continuous random variables are the subject of Chapter 5.

3.7 Probability Distributions for Discrete Random Variables

Since each value of the random variable x is a numerical event, we can apply the methods of this chapter to obtain $p(x)$, the probability associated with each of the values of x.

DEFINITION ■ The **probability distribution** for a discrete random variable is a formula, table, or graph that provides $p(x)$, the probability associated with each of the values of x. ■

The events associated with different values of x cannot overlap, because one and only one value of x is assigned to each simple event; hence, the values of x represent mutually exclusive numerical events. Summing $p(x)$ over all values of x equals the sum of the probabilities of all simple events and hence equals 1. We can therefore state two requirements for a discrete probability distribution.

Requirements for a Discrete Probability Distribution

1 $0 \le p(x) \le 1$

2 $\sum_{\text{all } x} p(x) = 1$

EXAMPLE 3.16 Consider an experiment that consists of tossing two coins and let x equal the number of heads observed. Find the probability distribution for x.

Solution The simple events for this experiment with their respective probabilities are as follows:

Simple Event	Coin 1	Coin 2	$P(E_i)$	x
E_1	H	H	1/4	2
E_2	H	T	1/4	1
E_3	T	H	1/4	1
E_4	T	T	1/4	0

Because E_1 is associated with the simple event "observe a head on coin 1 and a head on coin 2," we assign it the value $x = 2$. Similarly, we assign $x = 1$ to event E_2, and so on. The probability of each value of x can be calculated by adding the probabilities of the simple events in that numerical event. The numerical event $x = 0$ contains one simple event, E_4; $x = 1$ contains two simple events, E_2 and E_3; and

$x = 2$ contains one event, E_1. The values of x with respective probabilities are given in Table 3.2. Observe that

$$\sum_{x=0}^{2} p(x) = 1 \quad \blacksquare$$

TABLE 3.2 Probability distribution for x (x = number of heads)

x	Simple Events in x	$p(x)$
0	E_4	1/4
1	E_2, E_3	1/2
2	E_1	1/4
	$\sum_{x=0}^{2} p(x) = 1$	

The probability distribution in Table 3.2 can be presented graphically in the form of the relative frequency histogram (see Section 1.4).[†] The histogram for the random variable x would contain three classes, corresponding to $x = 0$, $x = 1$, and $x = 2$. Since $p(0) = 1/4$, the theoretical relative frequency for $x = 0$ is 1/4; $p(1) = 1/2$, and hence the theoretical frequency for $x = 1$ is 1/2. The histogram is given in Figure 3.10.

If you were to draw a sample from this population—that is, if you were to throw two balanced coins (say, $n = 100$ times), record the number of heads observed, x, and then construct a histogram using the 100 measurements on x—you would find that the histogram for the sample would appear very similar to that for $p(x)$ in Figure 3.10. If you were to repeat the experiment $n = 1000$ times, the similarity would be much more pronounced.

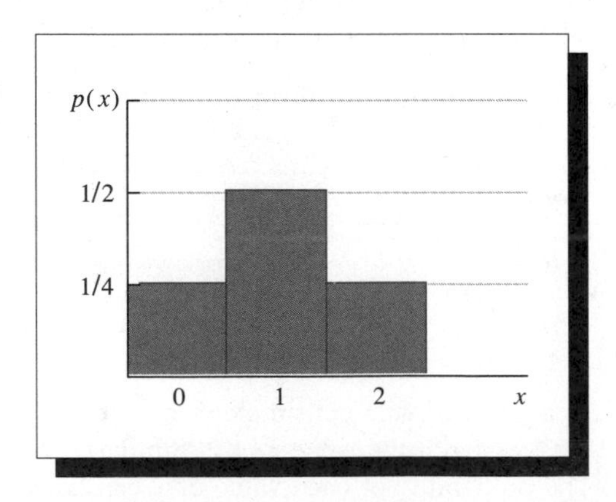

FIGURE 3.10 Probability histogram showing $p(x)$ for Example 3.16

[†]The probability distribution in Table 3.2 can also be presented using a formula, which is given in Section 4.2.

EXAMPLE 3.17 Let x equal the number observed on the throw of a single balanced die. Find $p(x)$.

Solution The simple events for this experiment are given in Table 3.3. Assign $x = 1$ to E_1, $x = 2$ to E_2, and so on. Since each value of x contains only one simple event, $p(x)$, the probability distribution for x would appear as shown in the fifth column of Table 3.3. Note that

$$p(x) = \frac{1}{6}, \qquad x = 1, 2, 3, 4, 5, 6$$

gives the probability distribution as a formula. The corresponding histogram is given in Figure 3.11.

TABLE 3.3 Tossing a die: probability distribution for x

Simple Event	Number on Upper Face	$P(E_i)$	x	$p(x)$
E_1	1	1/6	1	1/6
E_2	2	1/6	2	1/6
E_3	3	1/6	3	1/6
E_4	4	1/6	4	1/6
E_5	5	1/6	5	1/6
E_6	6	1/6	6	1/6

There are a number of useful discrete random variables—useful because their probability distributions, given as formulas, provide good models for the relative frequency distributions of many types of data observed in real life. Formulas for several useful discrete distributions are presented in Chapter 4. ▪

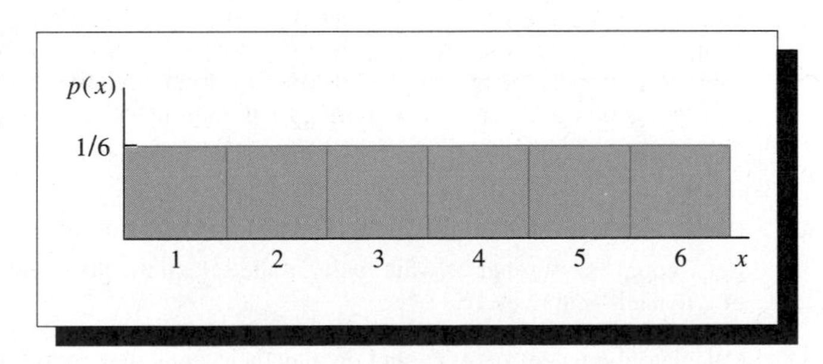

FIGURE 3.11 Probability histogram for $p(x) = 1/6$ in Example 3.17

Exercises

Basic Techniques

3.41 Identify the following as discrete or continuous random variables:

a The total number of points scored in a football game

b The shelf life of a particular drug

c The height of the ocean's tide at a given location

d The length of a two-year-old black bass

e The number of aircraft near-collisions in a year

3.42 Identify the following as discrete or continuous random variables:

a The increase in length of life achieved by a cancer patient as a result of surgery

b The tensile breaking strength, in pounds per square inch, of 1-inch-diameter steel cable

c The number of deer killed per year in a state wildlife preserve

d The number of overdue accounts in a department store at a particular time

e Your blood pressure

3.43 A random variable x has the following probability distribution:

x	0	1	2	3	4	5
$p(x)$	.1	.3	.4	.1	?	.05

a Find $p(4)$.

b Construct a probability histogram to describe $p(x)$.

3.44 A random variable x can assume five values, 0, 1, 2, 3, and 4. A portion of the probability distribution is as follows:

x	0	1	2	3	4
$p(x)$	.1	.3	.3	?	.1

a Find $p(3)$.

b Construct a probability histogram for $p(x)$.

c Simulate the experiment by marking ten poker chips (or coins)—one with a 0, three with a 1, three with a 2, and so on. Mix the chips thoroughly, draw one, and record the observed value of x. Repeat the process 100 times. Construct a relative frequency histogram for the 100 values of x and compare with the probability histogram in part (b).

3.45 A jar contains two black balls and two white balls. Suppose the balls have been thoroughly mixed and two are randomly selected from the jar.

a List all simple events for this experiment and assign appropriate probabilities to each.

b Let x equal the number of white balls in the selection. Then assign the appropriate value of x to each simple event.

c Calculate the values of $p(x)$, and display them in tabular form. Show that $\sum_{x=0}^{2} p(x) = 1$.

d Construct a probability histogram for $p(x)$.

e Simulate the experiment by actually drawing two balls from a jar that contains two black balls and two white balls. (This experiment could also be conducted using coins instead of balls.) Repeat the drawing process 100 times, each time recording the value of x that was observed. Construct a relative frequency histogram for the 100 observed values of x and compare with the probability histogram in part (d).

Applications

3.46 If you toss a pair of dice, the sum T of the number of dots appearing on the upper faces of the dice can assume the value of an integer in the interval $2 \leq T \leq 12$.

a Find the probability distribution for T, and display it in tabular form.

b Construct a probability histogram for $p(T)$.

3.47 A key ring contains four office keys that are identical in appearance. Only one will open your office door. Suppose you randomly select one key and try it. If it does not fit, you randomly select one of the three remaining keys. If it does not fit, you randomly select one of the last two. Each different sequence that could occur in selecting the keys represents one of a set of equiprobable simple events.

a List the simple events in S and assign probabilities to the simple events.

b Let x equal the number of keys that you try before you find the one that opens the door ($x = 1, 2, 3, 4$). Then assign the appropriate value of x to each simple event.

c Calculate the values of $p(x)$ and display them in tabular form.

d Construct a probability histogram for $p(x)$.

3.48 Are you concerned about the educational system in the United States? A survey conducted by Market Opinion Research, Inc., indicated than many parents and other adults support a number of educational reform initiatives. For example, 71% of those surveyed support the concept of merit pay for teachers, and a longer elementary school day was endorsed by 43% (*Washington Post*, May 25, 1990). Suppose we randomly select three people and ask them to indicate whether or not they support the concept of merit pay for teachers.

a Find the probability distribution for x, the number of people who support merit pay for teachers.

b Construct the probability histogram for $p(x)$.

c Find the probability that more than one person will support merit pay for teachers.

3.49 A company has five applicants for two positions: two women and three men. Suppose that the five applicants are equally qualified and that no preference is given for choosing either sex. Let x equal the number of women chosen to fill the two positions.

a Find $p(x)$.

b Construct a probability histogram for x.

3.50 A piece of electronic equipment contains six transistors, two of which are defective. Three transistors are selected at random, removed from the piece of equipment, and inspected. Let x equal the number of defectives observed, where $x = 0, 1,$ or 2. Find the probability distribution for x. Express the results graphically as a probability histogram.

3.51 Past experience has shown that, on the average, only one in ten wells drilled hits oil. Let x be the number of drillings until the first success (oil is struck). Assume that the drillings represent independent events.

a Find $p(1)$, $p(2)$, and $p(3)$.

b Give a formula for $p(x)$.

c Graph $p(x)$.

3.52 Two tennis professionals A and B are scheduled to play a match; the winner is determined by the first player to win three sets in a total that cannot exceed five sets. The event that A wins any one set is independent of the event that A wins any other, and the probability that A wins any one set is equal to .6. Let x equal the total number of sets in the match; that is $x = 3, 4,$ or 5. Find $p(x)$.

3.53 Shortly before the November 1992 presidential election, a Gallup poll indicated that a near record level of 73% of adults expressed dissatisfaction with the way things were going in the United States (*Gallup Monthly Poll*, Sept. 1992). Suppose you had conducted your own telephone survey at the same time. You randomly called people and asked them if they were satisfied with the state of the nation. Find the probability distribution for x, the number of calls until the first person is found who *is* satisfied with the state of the nation.

3.8 Numerical Descriptive Measures for Discrete Random Variables

The probability distribution for a discrete random variable provides a model that indicates the relative frequency (or probability) with which you should expect any given value of x to occur. You have probably noticed the similarity between the probability distribution for a discrete random variable and the relative frequency histograms discussed in Chapter 1. The difference is that the relative frequency histograms are constructed for a *sample* of n measurements drawn from the population, while the probability histogram is constructed as a model for the *entire population* of measurements. Just as we calculated the mean and standard deviation for a sample of n measurements to measure the location and the variability of the relative frequency distribution, we can also calculate a mean and standard deviation to describe the probability distribution for a random variable x. The population mean, which measures the average value of x in the population, is also called the **expected value** of the random variable x.

The method for calculating the population mean or expected value of a random variable can be more easily understood by considering an example. Let x equal the number of heads observed in the toss of two coins. For convenience, $p(x)$ is given as

x	0	1	2
$p(x)$	1/4	1/2	1/4

Suppose the experiment is repeated a large number of times—say, $n = 4{,}000{,}000$ times. Intuitively, we would expect to observe approximately 1 million zeros, 2 million ones, and 1 million twos. Then the average value of x would equal

$$\begin{aligned}
\frac{\text{sum of measurements}}{n} &= \frac{1{,}000{,}000(0) + 2{,}000{,}000(1) + 1{,}000{,}000(2)}{4{,}000{,}000} \\
&= \frac{1{,}000{,}000(0)}{4{,}000{,}000} + \frac{2{,}000{,}000(1)}{4{,}000{,}000} + \frac{1{,}000{,}000(2)}{4{,}000{,}000} \\
&= (1/4)(0) + (1/2)(1) + (1/4)(2)
\end{aligned}$$

Note that the first term in this sum is equal to $(0)p(0)$, the second is equal to $(1)p(1)$, and the third is equal to $(2)p(2)$. The average value of x, then, is

$$\sum_{x=0}^{2} xp(x) = 1$$

This result is not an accident, and it provides some intuitive justification for the definition of the expected value of a discrete random variable x.

DEFINITION ■ Let x be a discrete random variable with probability distribution $p(x)$. The mean or **expected value of x** is given as

$$\mu = E(x) = \sum_x xp(x)$$

where the elements are summed over all values of the random variable x. ■

EXAMPLE 3.18 Consider the random variable x representing the number observed on the toss of a single die. The probability distribution for x is given in Example 3.17. Then the expected value of x would be

$$\begin{aligned} E(x) = \sum_{x=1}^{6} xp(x) &= (1)p(1) + (2)p(2) + \cdots + (6)p(6) \\ &= (1)(1/6) + (2)(1/6) + \cdots + (6)(1/6) \\ &= \frac{1}{6}\sum_{x=1}^{6} x = \frac{21}{6} = 3.5 \end{aligned}$$

Note that this value $\mu = E(x) = 3.5$ locates the center or "balancing point" of the probability distribution exactly (see Figure 3.11). ■

EXAMPLE 3.19 In a lottery conducted to benefit the local fire company, 8000 tickets are to be sold at \$5 each. The prize is a \$12,000 automobile. If you purchase two tickets, what is your expected gain?

Solution Your gain x may take one of two values. You will either lose \$10 (i.e., your gain will be −\$10) or win \$11,990, with probabilities 7998/8000 and 2/8000, respectively. The probability distribution for the gain x is as follows:

x	$p(x)$
−\$10	7998/8000
\$11,990	2/8000

The expected gain will be

$$\begin{aligned} E(x) &= \sum_x xp(x) \\ &= (-\$10)\left(\frac{7998}{8000}\right) + (\$11{,}990)\left(\frac{2}{8000}\right) \\ &= -\$7 \end{aligned}$$

Recall that the expected value of x is the average of the theoretical population that would result if the lottery were repeated an infinitely large number of times. If this were done, your average or expected gain per lottery would be a loss of \$7. ■

EXAMPLE 3.20 Determine the yearly premium for a \$1000 insurance policy covering an event that, over a long period of time, has occurred at the rate of two times in 100. Let x equal the yearly financial gain to the insurance company resulting from the sale of the policy, and let C equal the unknown yearly premium. Calculate the value of C such that the expected gain $E(x)$ will equal zero. Then C is the premium required to break even. To this, the company would add administrative costs and profit.

Solution The first step in the solution is to determine the values that the gain x may take and then to determine $p(x)$. If the event does not occur during the year, the insurance company will gain the premium of $x = C$ dollars. If the event does occur, the gain will be negative; that is, the company will lose \$1000 less the premium of C dollars already collected. Then $x = -(1000 - C)$ dollars. The probabilities associated with these two values of x are 98/100 and 2/100, respectively. The probability distribution for the gain would be

x = gain	$p(x)$
C	98/100
$-(1000 - C)$	2/100

Since we want the insurance premium C such that, in the long run (for many similar policies), the mean gain will equal zero, we will set the expected value of x equal to zero and solve for C. Then

$$E(x) = \sum_x xp(x)$$

$$= C\left(\frac{98}{100}\right) + [-(1000 - C)]\left(\frac{2}{100}\right) = 0$$

or

$$\frac{98}{100}C + \frac{2}{100}C - 20 = 0$$

Solving this equation for C, we obtain $C = \$20$. Therefore, if the insurance company were to charge a yearly premium of \$20, the average gain calculated for a large number of similar policies would equal zero. The actual premium would equal \$20 plus administrative costs and profit. ■

Just as we used numerical descriptive measures to describe a relative frequency distribution (Chapter 2), we can use the mean and standard deviation of a random variable to describe its probability distribution. Knowing μ and σ, we could use Tchebysheff's Theorem or the Empirical Rule to describe $p(x)$.

In Chapter 2, we defined the population variance σ^2 to be the average of the squares of the deviations of the measurements from their mean. Because taking an expectation is equivalent to "averaging," we define the **variance** and the **standard deviation** as shown in the displays.

DEFINITION ■

Let x be a discrete random variable with probability distribution $p(x)$ and expected value $E(x) = \mu$. The **variance of x** is

$$\sigma^2 = E[(x-\mu)^2] = \sum_x (x-\mu)^2 p(x)$$

where the summation is over all values of the random variable x.† ■

DEFINITION ■

The **standard deviation σ of a random variable x** is equal to the square root of its variance. ■

EXAMPLE 3.21 Find the variance σ^2 for the population associated with Example 3.16, the coin-tossing problem. At the beginning of this section, the expected value of x was calculated to be $\mu = 1$.

Solution The variance is equal to the expected value of $(x-\mu)^2$, or

$$\begin{aligned}\sigma^2 &= E[(x-\mu)^2] = \sum_x (x-\mu)^2 p(x) \\ &= (0-1)^2 p(0) + (1-1)^2 p(1) + (2-1)^2 p(2) \\ &= (1)(1/4) + (0)(1/2) + (1)(1/4) = 1/2\end{aligned}$$

Then $\sigma = \sqrt{1/2} = .707$. The values $\mu = 1$ and $\sigma = .707$ can be used to describe the probability distribution shown in Figure 3.10. ■

EXAMPLE 3.22 Let x be a random variable with the probability distribution given in the following table:

x	−1	0	1	2	3	4	5
$p(x)$	.05	.10	.40	.20	.10	.10	.05

Find μ, σ^2, and σ. Graph $p(x)$, and locate the interval $\mu \pm 2\sigma$ on the graph. What is the probability that x will fall in the interval $\mu \pm 2\sigma$?

†It can be shown (proof omitted) that

$$\sigma^2 = \sum_x (x-\mu)^2 p(x) = \sum_x x^2 p(x) - \mu^2$$

This result is analogous to the computing formula for the sum of squares of deviations given in Chapter 2.

Solution

$$\begin{aligned}\mu = E(x) &= \sum_{x=-1}^{5} xp(x)\\ &= (-1)(.05) + (0)(.10) + (1)(.40) + \cdots + (4)(.10) + (5)(.05)\\ &= 1.70\end{aligned}$$

$$\begin{aligned}\sigma^2 = E[(x-\mu)^2] &= \sum_{x=-1}^{5} (x-\mu)^2 p(x)\\ &= (-1-1.70)^2(.05) + (0-1.70)^2(.10) + \cdots + (5-1.70)^2(.05)\\ &= 2.11\end{aligned}$$

$$\begin{aligned}\sigma &= \sqrt{\sigma^2} = \sqrt{2.11}\\ &= 1.45\end{aligned}$$

The interval $\mu \pm 2\sigma$ is $1.70 \pm (2)(1.45)$ or -1.20 to 4.60.

The graph of $p(x)$ and the interval $\mu \pm 2\sigma$ are shown in Figure 3.12. Note that $x = -1, 0, 1, 2, 3,$ and 4 fall in the interval. Therefore,

$$\begin{aligned}P[\mu - 2\sigma < x < \mu + 2\sigma] &= p(-1) + p(1) + p(2) + \cdots + p(4)\\ &= .05 + .10 + .40 + .20 + .10 + .10\\ &= .95\end{aligned}$$

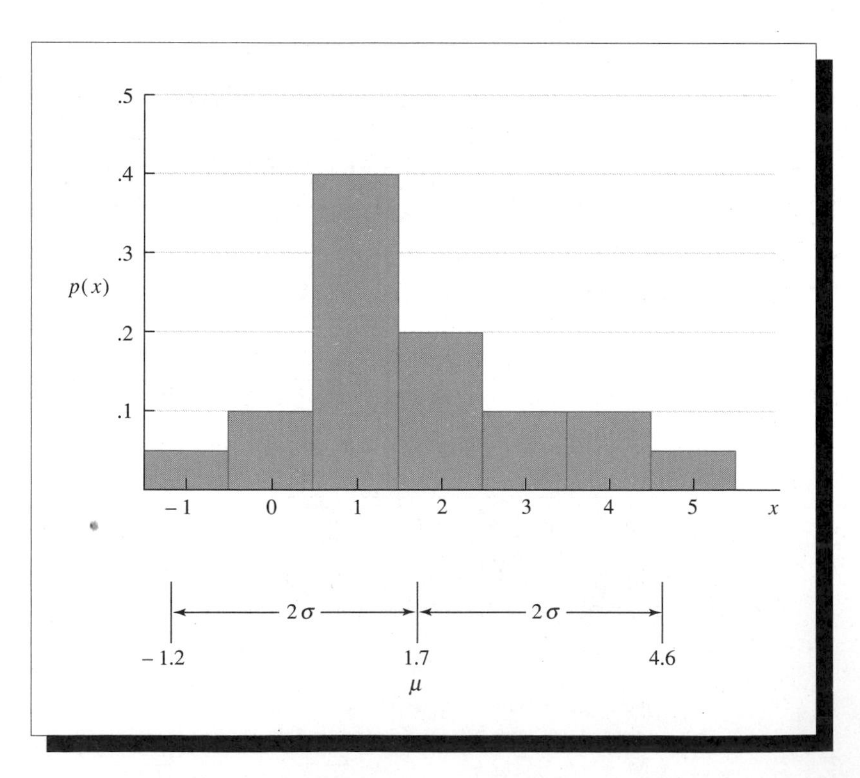

FIGURE 3.12
The probability histogram for $p(x)$ in Example 3.22

The method for calculating the expected value of x for a continuous random variable is similar to what we have done, but in practice it involves the use of calculus. However, the basic results concerning expectations are the same for continuous and discrete random variables. For example, regardless of whether x is continuous or discrete, $E(x) = \mu$ and $\sigma^2 = E(x - \mu)^2$.

Exercises

Basic Techniques

3.54 Let x be a discrete random variable with the probability distribution given in the following table:

x	0	1	2	3	4	5
$p(x)$	.05	.10	.20	.40	.20	.05

a Find μ, σ^2, and σ.

b Construct a probability histogram for $p(x)$.

c Locate the interval $\mu \pm 2\sigma$ on the x-axis of the histogram. What is the probability that x will fall in this interval?

d If you were to select a very large number of values of x from the population, would most fall in the interval $\mu \pm 2\sigma$? Explain.

3.55 Let x be a discrete random variable with the probability distribution given in the following table:

x	1	2	3	4	5	6	7
$p(x)$	.05	.20	.35	.20	.10	.05	.05

a Find μ, σ^2, and σ.

b Construct a probability histogram for $p(x)$.

c Locate the interval $\mu \pm 2\sigma$ on the x-axis of the histogram. What is the probability that x will fall in this interval?

d If you were to select a very large number of values of x from the population, would most fall in the interval $\mu \pm 2\sigma$? Explain.

3.56 Let x represent the number of times a customer visits a grocery store in a 1-week period. Assume that the following is the probability distribution of x:

x	0	1	2	3
$p(x)$	.1	.4	.4	.1

Find the expected value of x. This is the average number of times a customer visits the store.

Applications

3.57 Exercise 3.5 described the game of roulette. Suppose you were to bet \$5 on a single number—say, the number 18. The payoff on this type of bet is usually 35 to 1. What is your expected gain?

3.58 In Exercise 2.7, we learned that big business knows a lot about our habits. One of the statistics quoted by the *Wall Street Journal*, that the average number of ice cubes placed in a cold drink is 3.2, affects the amount of beverage consumed per drink. Since the number of ice cubes used per drink will vary, not only from person to person but from glass to glass, suppose that a study of a large number of drinks produced the following probability distribution for the number x of ice cubes used per glass.

x	1	2	3	4	5	6
$p(x)$	.05	.20	.40	.20	.10	.05

a Find the expected number of ice cubes used per glass.

b Find the variance and standard deviation of x.

c What did the Coca-Cola Company mean when it stated that we use 3.2 ice cubes per glass?

3.59 One professional golfer plays best on short-distance holes. Experience has shown that the number x of shots required for 3-, 4-, and 5-par holes has the probability distributions shown below:

Par-3 Holes		Par-4 Holes		Par-5 Holes	
x	$p(x)$	x	$p(x)$	x	$p(x)$
2	.12	3	.14	4	.04
3	.80	4	.80	5	.80
4	.06	5	.04	6	.12
5	.02	6	.02	7	.04

What is the golfer's expected score on the following?

a A par-3 hole

b A par-4 hole

c A par-5 hole

3.60 A \$50,000 diamond is insured for its total value by paying a premium of D dollars. If the probability of theft in a given year is estimated to be .01, what premium should the insurance company charge if it wants the expected gain to equal \$1000?

3.61 The maximum patent life for a new drug is 17 years. Subtracting the length of time required by the FDA for testing and approval of the drug provides the actual patent life of the drug—that is, the length of time that a company has to recover research and development costs and make a profit. Suppose the distribution of the length of patent life for new drugs is as shown:

Years x	3	4	5	6	7	8	9	10	11	12	13
$p(x)$	.03	.05	.07	.10	.14	.20	.18	.12	.07	.03	.01

a Find the expected number of years of patent life for a new drug.

b Find the standard deviation of x.

c Find the probability that x falls in the interval $\mu \pm 2\sigma$.

3.62 From experience, a shipping company knows that the cost of delivering a small package within 24 hours is \$14.80. The company charges \$15.50 for shipment but guarantees to refund the charge if delivery is not made within 24 hours. If the company fails to deliver only 2% of its packages within the 24-hour period, what is the expected gain per package?

3.63 A manufacturing representative is considering taking out an insurance policy to cover possible losses incurred by marketing a new product. If the product is a complete failure, the representative feels that a loss of \$80,000 would be incurred; if it is only moderately successful, a loss of \$25,000 would be incurred. Insurance actuaries have determined from market surveys and other available information that the probabilities that the product will be a failure or only moderately successful are .01 and .05, respectively. Assuming that the manufacturing representative is willing to ignore all other possible losses, what premium should the insurance company charge for policy in order to break even?

3.64 The probability that a tennis player A can win a set from tennis player B is one measure of the comparative abilities of the two players. In Exercise 3.52 you found the probability distribution for x, the number of sets required to play a best-of-five-sets match, given that the probability that A wins any one set—call this $P(A)$—is .6.

a Find the expected number of sets required to complete the match for $P(A) = .6$.

b Find the expected number of sets required to complete the match when the players are of equal ability, that is, $P(A) = .5$.

c Find the expected number of sets required to complete the match when the players differ greatly in ability, that is, say, $P(A) = .9$.

3.9 Summary

The theories of both probability and statistics are concerned with samples drawn from populations. Probability assumes that the population is known and calculates the probability of observing a particular sample. Statistics assumes the sample to be known and, with the aid of probability, attempts to describe the unknown population frequency distribution.

A random variable is a numerically valued outcome of an experiment. Random variables can be classified as discrete or continuous, depending on whether the number of simple events in the sample space is or is not countable. The theoretical frequency distribution for a discrete random variable is called a probability distribution and often can be derived using the techniques in this chapter.

The expected value of a random variable is the average of the random variable calculated for the theoretical population defined by its probability distribution. Mathematical expectation can be used to find the mean and variance, and hence the standard deviation, of a random variable. These quantities can be used to describe a probability distribution in the same way that the mean and standard deviation were used to describe distribution of data in Chapter 2.

Supplementary Exercises

3.65 The discrete random variable x and its probability distribution $p(x)$ are as follows:

x	0	2	3	4
$p(x)$	1/8	1/4	1/2	1/8

a Find $\mu = E(x)$. **b** Find $E(x^2)$. **c** Find σ^2 and σ.

3.66 Let x equal the number of dots observed when a die is tossed, and let $p(x)$ equal 1/6, $x = 1, 2, 3, \ldots, 6$. We found in Example 3.18 that $\mu = E(x) = 3.5$. Find σ^2 and show that $\sigma = 1.71$. Then find the probability that x will fall in the interval $\mu \pm 2\sigma$.

3.67 Draw a sample of $n = 50$ measurements from the die-throwing population of Example 3.17 by tossing a die 50 times and recording x after each toss. Calculate $\bar{x}$ and s^2 for the sample. Compare $\bar{x}$ with the expected value of x in Example 3.18 and s^2 with the variance of x obtained in Exercise 3.66. Do $\bar{x}$ and s^2 provide good estimates of μ and σ^2?

3.68 Whistle blowers is the name given to employees who report corporate fraud, theft, and other unethical and perhaps criminal activities by fellow employees or by their employer. Although there is legal protection for whistle blowers, it has been reported that approximately 23% of those who reported fraud suffered reprisals such as demotion, poor performance ratings, and so on. Suppose that the probability that an employee will fail to report a case of fraud is .69. Find the probability that a worker who observes a case of fraud will report it and will subsequently suffer some form of reprisal.

3.69 Two cold tablets are accidentally placed in a box containing two aspirin tablets. The four tablets are identical in appearance. One tablet is selected at random from the box and is swallowed by patient A. A tablet is then selected at random from the three remaining tablets and is swallowed by patient B. Define the following events as specific collections of simple events:

a The sample space S

b The event A that patient A obtained a cold tablet

c The event B that exactly one of the two patients obtained a cold tablet

d The event C that neither patient obtained a cold tablet

3.70 Refer to Exercise 3.69. By summing probabilities of simple events, find $P(A)$, $P(B)$, $P(A \cap B)$, $P(A \cup B)$, $P(C)$, $P(A \cap C)$, and $P(A \cup C)$.

3.71 A coin is tossed four times, and the outcome is recorded for each toss.

a List the simple events for the experiment.

b Let A be the event that the experiment yields exactly three heads. List the simple events in A.

c Assign reasonable probabilities to the simple events and find $P(A)$.

3.72 A retailer sells two styles of high-priced compact disc players that experience indicates are in equal demand. (Fifty percent of all potential customers prefer style 1, and 50% favor style 2.) If the retailer stocks four of each, what is the probability that the first four customers seeking a CD player all purchase the same style?

a Define the experiment.

b List the simple events.

c Define the event of interest A as a specific collection of simple events.

d Assign probabilities to the simple events and find $P(A)$.

3.73 A boxcar contains seven complex electronic systems. Unknown to the purchaser, three are defective. Two of the seven are selected for thorough testing and are then classified as defective or nondefective.

a List the simple events for this experiment.

b Let A be the event that the selection includes no defectives. List the simple events in A.

c Assign probabilities to the simple events and find $P(A)$.

3.74 A heavy-equipment salesman can contact either one or two customers per day with probability 1/3 and 2/3, respectively. Each contact will result in either no sale or a $50,000 sale with probability 9/10 and 1/10, respectively. What is the expected value of his daily sales?

3.75 A county containing a large number of rural homes is thought to have 60% of those homes insured against fire. Four rural homeowners are chosen at random from the entire population, and x are found to be insured against fire. Find the probability distribution for x. What is the probability that at least three of the four will be insured?

3.76 A fire-detection device uses three temperature-sensitive cells acting independently of one another in such a manner that any one or more can activate the alarm. Each cell has a probability of $p = .8$ of activating the alarm when the temperature reaches 100° or more. Let x equal the number of cells activating the alarm when the temperature reaches 100°. Find the probability distribution for x. Find the probability that the alarm will function when the temperature reaches 100°.

3.77 Find the expected value and variance for the random variable x defined in Exercise 3.76.

3.78 If you toss a pair of dice, the sum T of the numbers of dots appearing on the upper faces of the dice can assume the value of an integer in the interval $2 \le T \le 12$.

a Find $E(T)$.

b Find σ^2.

c Use your graph of $p(T)$, found in Exercise 3.46, and locate the interval $\mu \pm 2\sigma$.

d What is the probability that T will assume a value in the interval $\mu \pm 2\sigma$?

3.79 A die is tossed twice. What is the probability that the sum of the numbers observed will be greater than nine?

3.80 Toss a die and a coin. If event A is the occurrence of a head and an even number, and event B is the occurrence of a head and a 1, find $P(A)$, $P(B)$, $P(A \cap B)$, and $P(A \cup B)$. (Solve by listing the simple events.)

3.81 Refer to Exercise 3.80 and calculate $P(A \cap B)$ and $P(A \cup B)$. Use the laws of probability.

3.82 A salesperson figures that the probability of her consummating a sale during the first contact with a client is .4 but improves to .55 on the second contact if the client did not buy during the first contact. Suppose this salesperson makes one and only one callback to any client. If she contacts a client, calculate the following:

a The probability that the client will buy

b The probability that the client will not buy

3.83 A man takes either a bus or the subway to work with probabilities .3 and .7, respectively. When he takes the bus, he is late 30% of the days. When he takes the subway, he is late 20% of the days. If the man is late for work on a particular day, what is the probability that he took the bus?

3.84 Suppose independent events A and B have nonzero probabilities. Show that A and B cannot be mutually exclusive.

3.85 The failure rate for a guided missile control system is 1 in 1000. Suppose that a duplicate, but completely independent, control system is installed in each missile so that, if the first fails, the second can take over. The reliability of a missile is the probability that it does not fail. What is the reliability of the modified missile?

3.86 Suppose that at a particular supermarket the possibility of waiting five minutes or longer for checkout at the cashier's counter is .2. On a given day, a man and his wife decide to shop individually at the market, each checking out at different cashier counters. If they both reach cashier counters at the same time, answer the following questions:

a What is the probability that the man will wait less than 5 minutes for checkout?

b What is the probability that both the man and his wife will be checked out in less than 5 minutes? (Assume that the checkout times for the two are independent events.)

c What is the probability that one or the other or both will wait 5 minutes or more?

3.87 A quality-control plan calls for accepting a large lot of crankshaft bearings if a sample of seven is drawn and none are defective. What is the probability of accepting the lot if none in the lot are defective? If 1/10 are defective? If 1/2 are defective?

3.88 It is said that only 40% of all people in a community favor the development of a mass transit system. If four citizens are selected at random from the community, what is the probability that all four favor the mass transit system? That none favor the mass transit system?

3.89 A TV meteorologist forecasts rain with probability .6 today and .4 tomorrow. Experience has shown that, in this particular locale, it rains one day in four and the probability of rain on two successive days is .15. If it is raining outside when we hear the TV forecast (as so often happens), what is the probability of rain tomorrow?

3.90 A research physician compared the effectiveness of two blood pressure drugs A and B by administering the two drugs to each of four pairs of identical twins. Drug A was given to one member of a pair, drug B to the other. If, in fact, there is no difference in the effect of the drugs, what is the probability that the drop in the blood pressure reading for drug A would exceed the corresponding drop in the reading for drug B for all four pairs of twins? Suppose drug B created a greater drop in blood pressure than drug A for each of the four pairs of twins. Do you think this provides sufficient evidence to indicate that drug B is more effective in lowering blood pressure than drug A?

3.91 To reduce the cost of detecting a disease, blood tests are conducted on a pooled sample of blood collected from a group of n people. If no indication of the disease is present in the pooled blood sample (as is usually the case), none have the disease. If analysis of the pooled blood sample indicates that the disease is present, each individual must submit to a blood test. The individual tests are conducted in sequence. If, among a group of five people, one person has the disease, what is the probability that six blood tests (including the pooled test) are required to detect the single diseased person? If two people have the disease, what is the probability that six tests are required to locate both diseased people?

3.92 How many times should a coin be tossed to obtain a probability equal to or greater than .9 of observing at least one head?

3.93 An oil prospector will drill a succession of holes in a given area to find a productive well. The probability that she is successful on a given trial is .2.

a What is the probability that the third hole drilled is the first hole to locate a productive well?

b If her total resources allow the drilling of only three holes, what is the probability that she locates a productive well?

3.94 Suppose two defective refrigerators are included in a shipment of six refrigerators. The buyer begins to test the six refrigerators one at a time.

a What is the probability that the last defective refrigerator is found on the fourth test?

b What is the probability that no more than four refrigerators need be tested before both defective refrigerators are located?

c Given that one defective refrigerator has been located in the first two tests, what is the probability that the remaining defective refrigerator is found on the third or fourth test?

3.95 Two men each toss a coin. They obtain a "match" if either both coins are heads or both are tails. Suppose the tossing is repeated three times.

a What is the probability of three matches?

b What is the probability that all six tosses (three for each man) result in tails?

c Coin tossing provides a model for many practical experiments. Suppose that the coin tosses represented the answers given by two students for three specific true-false questions on an examination. If the two students gave three matches for answers, would the low probability found in (a) suggest collusion?

3.96 Experience has shown that, 50% of the time, a particular union–management contract negotiation led to a contract settlement within a 2-week period, 60% of the time the union strike fund was adequate to support a strike, and 30% of the time both conditions were satisfied. What is the probability of a contract settlement given that the union strike fund is adequate to support a strike? Is settlement of a contract within a 2-week period dependent on whether the union strike fund is adequate to support a strike?

3.97 Suppose the probability of remaining with a particular company 10 years or more is 1/6. A man and a woman start work at the company on the same day.

a What is the probability that the man will work there less than 10 years?

b What is the probability that both the man and the woman will work there less than 10 years? (Assume they are unrelated and their lengths of service are independent of each other.)

c What is the probability that one or the other or both will work 10 years or more?

3.98 Accident records collected by an automobile insurance company give the following information: The probability that an insured driver has an automobile accident is .15; if an accident has occurred, the damage to the vehicle amounts to 20% of its market value with probability .80, 60% of its market value with probability .12, and a total loss with probability .08. What premium should the company charge on a $12,000 car so that the expected gain by the company is zero?

3.99 If you own common stock in a corporation, you can sometimes increase the return on your investment by selling an option. For a stated price, the purchaser of the option gains the right to buy your stock at any time up to a specified expiration date. Suppose you purchased 200 shares of a stock at $25 per share and you sell an option for the option purchaser to buy the stock at any time within the next six months for $30 per share. If the stock reaches $30 per share within the next six months, your stock will be sold and you will gain $5 per share plus $2 per share (from the dividends and the sale of the option) less $109 commission to the broker for selling your stock. If you do not sell, you will gain $2 per share. If the probability of the stock reaching $30 per share within the next six months is .7, what is the expected return (in dollars) from your 200 shares of stock? What is the expected annual rate of return, in percentage, on your investment? (NOTE: The $15 commission on the sale of the option has been ignored.)

CASE STUDY

Probability and Decision Making in the Congo

In his exciting novel *Congo* (1980), Michael Crichton describes a search by Earth Resources Technology Service (ERTS), a geological survey company, for deposits of boron-coated blue diamonds, diamonds that ERTS believes to be the key to a new generation of optical computers. In the novel, ERTS is racing against an international consortium to find the Lost City of Zinj, a city that thrived on diamond mining and existed several thousand years ago (according to African fable), deep in the rain forests of eastern Zaire.

After the mysterious destruction of its first expedition, ERTS launches a second expedition under the leadership of Karen Ross, a 24-year-old computer genius who is accompanied by Professor Peter Elliot, an anthropologist; Amy, a talking gorilla; and the famed mercenary and expedition leader, "Captain" Charles Munro. Ross's efforts to find the city are blocked by the consortium's offensive actions, by the deadly rain forest, and by hordes of "talking" killer gorillas whose perceived mission is to defend the diamond mines. Ross overcomes these obstacles by using space-age computers to evaluate the probabilities of success for all possible circumstances and all possible actions that the expedition might take. At each stage of the expedition, she is able to quickly evaluate the chances of success.

At one stage in the expedition, Ross is informed by her Houston headquarters that their computers estimate that she is 18 hours, 20 minutes behind the competing Euro-Japanese team, instead of 40 hours ahead. She changes plans and decides to have the 12 members of her team—Ross, Elliot, Munro, Amy, and eight native porters—parachute into a volcanic region near the estimated location of Zinj. As Crichton relates, "Ross had double-checked outcome probabilities from the Houston computer, and the results were unequivocal. The probability of a successful jump was .7980, meaning that there was approximately one chance in five that someone would be badly hurt. However, given a successful jump, the probability of expedition success was .9943, making it virtually certain that they would beat the consortium to the site."

Keeping in mind that this is an excerpt from a novel, let us examine the probability, .7980, of a successful jump. If you were one of the 12-member team, what is the probability that you would successfully complete your jump? In other words, if the probability of a successful jump by all 12 team members is .7980, what is the probability that a single member could successfully complete the jump?

Supplement: Optional Topics

3S.1 Useful Counting Rules (Optional)

This section presents three rules that fall into the realm of combinatorial mathematics and that can be used to solve probability problems involving a large number of simple events. For example, suppose that you are interested in the probability of an event A and you know that the simple events in the sample space S are equiprobable. Then

$$P(A) = \frac{n_A}{N}$$

where

$$n_A = \text{number of simple events in } A$$
$$N = \text{number of simple events in } S$$

Often we can use counting rules to find the values of n_A and N and thereby eliminate the necessity of listing the simple events in S. The first rule, called the mn rule, is as follows:

> ***mn* Rule** With m elements $a_1, a_2, a_3, \ldots, a_m$ and n elements $b_1, b_2, b_3, \ldots, b_n$, it is possible to form mn pairs that contain one element from each group. ▪

To illustrate, suppose four companies have job openings in each of three areas: sales, manufacturing, and personnel. How many job opportunities are available? This situation contains two sets of "things": companies (four) and types of jobs (three). Therefore, as shown in Figure 3S.1, there are three jobs for each of the four companies, or (4)(3) = 12 possible pairings of companies and jobs. This example illustrates a use of the mn rule.

FIGURE 3S.1
Company–job combinations

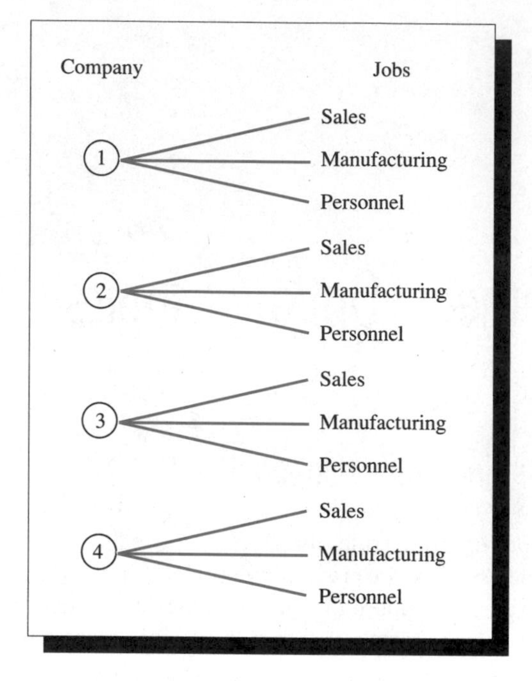

EXAMPLE 3S.1 Two dice are tossed. How many simple events are associated with the experiment?

Solution The first die can fall in one of six ways; that is, $m = 6$. Similarly, the second die can fall in $n = 6$ ways. Since an outcome of this experiment involves pairing of the numbers showing on the faces of the two dice, the total number N of simple events is

$$N = mn = 6(6) = 36 \quad \blacksquare$$

EXAMPLE 3S.2 Urn 1 contains two white balls and one black ball; urn 2 contains one white ball. A ball is drawn from urn 1 and placed in urn 2. A ball is then drawn from urn 2. How many simple events are associated with the experiment?

Solution A ball can be chosen from urn 1 in one of $m = 3$ ways. After one of these ways has been chosen, a ball may be drawn from urn 2 in $n = 2$ ways. The total number of simple events is

$$N = mn = 3(2) = 6 \quad \blacksquare$$

A Counting Rule for Forming Pairs, Triplets, and So On

Given k groups of elements, n_1 elements in the first group, n_2 in the second, ..., and n_k in the kth group, the number of ways of selecting one element from each of the k groups is

$$n_1 n_2 n_3 \ldots n_k$$

EXAMPLE 3S.3 How many simple events are in the sample space when three coins are tossed?

Solution Each coin can land in one of two ways. Hence,

$$N = 2(2)(2) = 8 \quad \blacksquare$$

EXAMPLE 3S.4 A truck driver can take three routes from city A to city B, four from city B to city C, and three from city C to city D. If, when traveling from A to D, the driver must proceed from A to B to C to D, how many possible A-to-D routes are available?

Solution Let

m = number of routes from A to B = 3

n = number of routes from B to C = 4

t = number of routes from C to D = 3

Then the total number of ways to construct a complete route, taking one subroute from each of the three groups (A to B), (B to C), and (C to D), is

$$mnt = (3)(4)(3) = 36 \quad \blacksquare$$

A second useful mathematical rule is associated with orderings or permutations. For instance, suppose we have three books, b_1, b_2, and b_3. In how many ways can the books be arranged on the shelf, taking them two at a time? Listing all combinations of two in the first column and a reordering of each in the second column, we get the following enumeration:

Combinations of Two	Reordering of Combinations
$b_1 b_2$	$b_2 b_1$
$b_1 b_3$	$b_3 b_1$
$b_2 b_3$	$b_3 b_2$

The number of permutations is six, a result easily obtained from the mn rule. The first book can be chosen in $m = 3$ ways; once it is selected, the second book can be chosen $n = 2$ ways. The result is $mn = 6$.

In how many ways can three books be arranged on a shelf, taking three at a time? Enumerating, we obtain

$$\begin{array}{ccc} b_1b_2b_3 & b_2b_1b_3 & b_3b_1b_2 \\ b_1b_3b_2 & b_2b_3b_1 & b_3b_2b_1 \end{array}$$

for a total of six. This, again, could be obtained easily by the extension of the mn rule. The first book can be chosen and placed in $m = 3$ ways. Then the second can be chosen in $n = 2$ ways and, finally, the third in $t = 1$ way. Hence, the total number of ways is

$$N = mnt = 3 \cdot 2 \cdot 1 = 6$$

A Counting Rule for Permutations

The number of ways we can arrange n distinct objects taking them r at a time is

$$P_r^n = n(n-1)(n-2)\cdots(n-r+1) = \frac{n!}{(n-r)!}$$

where $n! = n(n-1)(n-2)\cdots(3)(2)(1)$ and $0! = 1$.

We are concerned with the number of ways of filling r positions with n distinct objects. Applying the extension of the mn rule, the first object can be chosen in one of n ways. After the first is chosen, the second can be chosen in $(n - 1)$ ways, the third in $(n - 2)$ ways, and the rth in $(n - r + 1)$ ways. Hence, the total number of ways is

$$P_r^n = n(n-1)(n-2)\cdots(n-r+1)$$

EXAMPLE **3S.5** Three lottery tickets are drawn from a total of 50. Assume that order is of importance. How many simple events are associated with the experiment?

Solution The total number of simple events is

$$P_3^{50} = \frac{50!}{47!} = 50(49)(48) = 117{,}600 \quad ■$$

EXAMPLE 3S.6 A piece of equipment is composed of five parts that can be assembled in any order. A test is to be conducted to determine the time necessary for each order of assembly. If each order is to be tested once, how many tests must be conducted?

Solution The total number of tests would equal

$$P_5^5 = \frac{5!}{0!} = 5(4)(3)(2)(1) = 120 \quad \blacksquare$$

The enumeration of the permutations of books in the previous discussion was performed in a systematic manner, first writing the combinations of n books taken r at a time, and then writing the rearrangements of each combination. In many situations, ordering is unimportant and we are interested solely in the number of possible combinations. For instance, suppose an experiment involves the selection of five people, a committee, from a total of 20 candidates. Then the simple events associated with this experiment correspond to the different combinations of persons selected from the group of 20. How many simple events (different combinations) are associated with this experiment? Since order in a single selection is unimportant, permutations are irrelevant. Thus, we are interested in the number of combinations of $n = 20$ things taken $r = 5$ at a time.

A Counting Rule for Combinations

The number of distinct combinations of n distinct objects that can be formed, taking them r at a time, is

$$C_r^n = \frac{n!}{r!(n-r)!}$$

The relationship between the number of **combinations** and the number of **permutations** of n things taken r at a time is given by

$$C_r^n = \frac{P_r^n}{r!}$$

Expressed in this way, we see that C_r^n results from dividing the number of permutations by $r!$, the number of ways of rearranging each distinct selection of r items from n.

EXAMPLE 3S.7 A printed circuit board may be purchased from five suppliers. In how many ways can three suppliers be chosen from the five?

Solution

$$C_3^5 = \frac{5!}{3!2!} = \frac{(5)(4)}{2} = 10 \quad \blacksquare$$

The following example illustrates the use of the counting rules in the solution of a probability problem.

EXAMPLE 3S.8 Five manufacturers, of varying but unknown quality, produce a certain type of electronic device. If we were to select three manufacturers at random, what is the chance that the selection would contain exactly two of the best three?

Solution Without enumerating the simple events, we would likely agree that each point (i.e., any combination of three) would be assigned equal probability. If N points are in S, then each event receives probability

$$P(E_i) = \frac{1}{N}$$

Let n be the number of events in which two of the best three manufacturers are selected. Then the probability of including two of the best three manufacturers in a selection of three is

$$P = \frac{n}{N}$$

Our problem is to use the counting rules to find n and N. Since order within a selection is not important and is not recorded, each selection is a combination and, hence,

$$N = C_3^5 = \frac{5!}{3!2!} = 10$$

Determination of n is more difficult, but it can be obtained using the mn rule. Let a be the number of ways of selecting exactly two from the best three, or

$$C_2^3 = \frac{3!}{2!1!} = 3$$

and let b be the number of ways of choosing the remaining manufacturer from the two poorest, or

$$C_1^2 = \frac{2!}{1!1!} = 2$$

Then the total number of ways of choosing two of the best three in a selection of three is $n = ab = 6$.

Hence, the probability P is equal to

$$P = \frac{6}{10} \quad \blacksquare$$

Many other counting rules are available in addition to the three presented in this section. If you are interested in this topic, you should consult one of the many texts on combinatorial mathematics.

Exercises

Basic Techniques

3S.1 You have *two* sets of distinctly different elements, ten in the first group and eight in the second. If you select one element from each group, how many different pairs can you form?

3S.2 You have *three* sets of distinctly different elements, four in the first set, seven in the second, and three in the third. If you select one element from each set, how many triplets can you select?

3S.3 You have *one* set of ten distinctly different elements. If you select two elements from among the ten, how many different pairs can you select?

Applications

3S.4 A salesperson in New York is preparing an itinerary for a visit to six major cities. The distance traveled, and hence the cost of the trip, will depend on which city is visited first, second, . . . , sixth. How many different itineraries (and hence trip costs) are possible?

3S.5 Probability played a role in the rigging for the April 24, 1980, Pennsylvania state lottery (*Los Angeles Times*, Sept. 8, 1980). To determine each digit of the three-digit winning number, each of the numbers 0, 1, 2, . . . , 9 is written on a Ping-Pong ball; the ten balls are blown into a compartment; and the number selected for the digit is the one on the ball that floats to the top of the machine. To alter the odds, the conspirators injected a liquid into all balls used in the game except those numbered 4 and 6, making it almost certain that the lighter balls would be selected and determine the digits in the winning number. They then proceeded to buy lottery tickets bearing the potential winning numbers. How many potential winning numbers were there (666 was the eventual winner)?

3S.6 Refer to Exercise 3S.5. Hours after the rigging of the Pennsylvania state lottery was announced, Connecticut state lottery officials were stunned to learn that *their* winning number for the day was 666 (*Los Angeles Times*, Sept. 21, 1980).

a All evidence indicates that the Connecticut selection of 666 was due to pure chance. What is the probability that a 666 would be drawn in Connecticut, given that a 666 had been selected in the April 24, 1980, Pennsylvania lottery?

b What is the probability of drawing a 666 in the April 24, 1980, Pennsylvania lottery (remember, this drawing was rigged) *and* a 666 on the September 19, 1980, Connecticut lottery?

3S.7 A study is to be conducted in a hospital to determine the attitudes of nurses toward various administrative procedures that are currently employed. If a sample of ten nurses is to be selected from a total of 90, how many different samples could be selected? (NOTE: Order within a sample is unimportant.)

3S.8 If five cards are to be selected, one after the other, in sequence, from a 52-card deck, and each card is to be replaced in the deck before the next draw, how many different selections are possible?

3S.9 Refer to Exercise 3S.8. Suppose the five cards are drawn from the 52-card deck simultaneously and without replacement. How many different hands could be selected?

3S.10 The following case occurred in the city of Gainesville, Florida. The eight-member Human Relations Advisory Board considered the complaint of a woman who claimed discrimination, based on her sex, on the part of a local surveying company. The board, composed of five women and three men, voted 5–3 in favor of the plaintiff, the five women voting for the plaintiff and the three men against. The attorney representing the company appealed the board's decision by claiming sex bias on the part of the board members. If the vote in favor of the plaintiff was 5–3 and the board members were not biased by sex, what is the probability that the vote would split along gender lines (five women for, three men against)?

3S.11 A student prepares for an exam by studying a list of ten problems. She can solve six of them. For the exam, the instructor selects five questions at random from the list of ten. What is the probability that the student can solve all five problems on the exam?

3S.12 A monkey is given 12 blocks—three shaped like squares, three like rectangles, three like triangles, and three like circles. If it draws three of each kind in order—say, three triangles, then three squares, and so on—would you suspect that the monkey associates identically shaped figures? Calculate the probability of this event.

3S.2 Bayes' Rule (Optional)

Frequently, we wish to find the conditional probability of an event A, given that an event B has already occurred. For example, we might wish to know the probability of rain tomorrow, given that it has rained during the preceding seven days. Hence, we assume that some state of nature exists, and we wish to calculate the probability of some event that will occur in the future.

Equally interesting is the probability that a certain state of nature exists given that a certain sample is observed. One such problem occurs in screening tests, which used to be associated primarily with medical diagnostic tests but are now finding application in a variety of fields. Camera technology and automatic test equipment are routinely used in inspecting parts in high-volume production processes. Corporate drug testing of employees, steroid testing of athletes, home pregnancy tests, and AIDS (acquired immunodeficiency syndrome) testing are some other familiar applications.

Very few, if any, screening tests are perfect. There is always the risk that defective parts, diseased persons, or contaminated products will go undetected (a false negative). On the other hand, good parts, healthy people, or safe products may be classified as defective, sick, or unsafe (a false positive). The effectiveness of a screening test is evaluated by assessing the probability of a false negative or a false positive. In this section we present an application of the Multiplicative Law, together with conditional probabilities, in a form derived by the probabilist Thomas Bayes. It allows for the determination of the probability of false positives and false negatives, both of which are conditional probabilities.

Consider an experiment that involves selection of a sample from one of k mutually exclusive and only possible populations, referred to as **states of nature** and denoted by $S_1, S_2, \ldots, S_k$ with *prior* probabilities $P(S_1), P(S_2), \ldots, P(S_k)$, respectively. The sample is selected and results in the event A, but the population (or state of nature) giving rise to the sample is unknown. The problem is to determine the population from which the sample was selected using the conditional probabilities $P(S_i|A)$ for $i = 1, 2, \ldots, k$. These probabilities are known as **posterior probabilities**—that is, probabilities that result when prior probabilities are updated using sample information.

To find the conditional probability that the sample was selected from population i given that event A was observed, $P(S_i|A), i = 1, 2, \ldots, k$, note that A could have been observed if the sample was selected from population 1, population 2, or any one of the k populations $S_1, S_2, \ldots, S_k$. The probability that population i was selected *and* that event A occurred is the intersection of the events S_i and A, or $(A \cap S_i)$. These events, $(A \cap S_1), (A \cap S_2), \ldots, (A \cap S_k)$, are mutually exclusive and hence

$$P(A) = P(A \cap S_1) + P(A \cap S_2) + \cdots + P(A \cap S_k)$$

This relationship is shown in Figure 3S.2 for $k = 3$ subpopulations. Then the probability that the sample came from population i is

$$P(S_i|A) = \frac{P(A \cap S_i)}{P(A)} = \frac{P(S_i)P(A|S_i)}{\sum_{j=1}^{k} P(A \cap S_j)} = \frac{P(S_i)P(A|S_i)}{\sum_{j=1}^{k} P(S_j)P(A|S_j)}$$

The expression for $P(S_i|A)$ is known as Bayes' Rule for the probability of causes. As you can see, it follows easily from the definition of conditional probability.

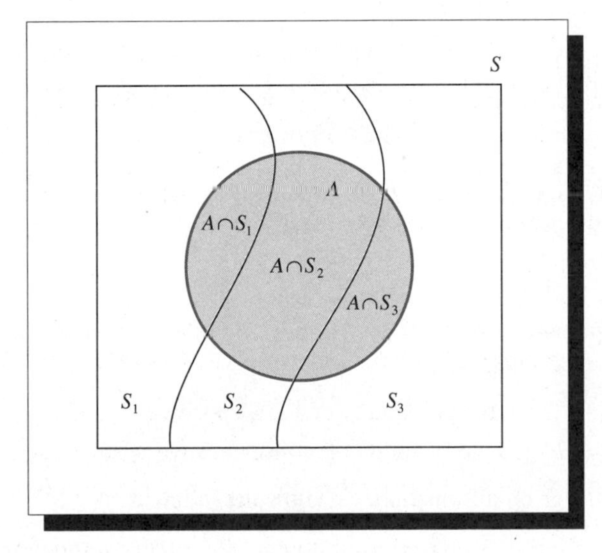

FIGURE 3S.2
Representation for the event A for $k = 3$ subpopulations

Bayes suggested that, if the prior probabilities, $P(S_i)$, were unknown, they be taken as equally probable—that is, $P(S_i) = 1/k$, for $i = 1, 2, \ldots, k$. In many instances, however, the experimenter does have at least some approximation to the prior probabilities for $S_1, S_2, \ldots, S_k$.

Bayes' Rule Let $S_1, S_2, \ldots, S_k$ represent the k mutually exclusive, only possible states of nature with prior probabilities $P(S_1), P(S_2), \ldots, P(S_k)$. If an event A occurs, the posterior probability of S_i given A is the conditional probability

$$P(S_i|A) = \frac{P(S_i)P(A|S_i)}{\sum_{j=1}^{k} P(S_j)P(A|S_j)}$$

for $i = 1, 2, \ldots, k$. ▪

EXAMPLE 3S.9 To evaluate the effectiveness of a screening procedure, we will evaluate the probabilities of a false negative or a false positive using the following notation:

T^+: The test is positive and indicates that the person has the disease.

T^-: The test is negative and indicates that the person does not have the disease.

D : The person tested has the disease.

D^C: The person tested does not have the disease.

In an assessment of a test for a particular disease, the "sensitivity" of the test, defined as the probability of a positive result given the person has the disease, is

$$P[T^+|D] = .98$$

and the "specificity" of the test, defined as the probability that the test is negative given the person does not have the disease, is

$$P[T^-|D^C] = .99$$

If the proportion of the general population infected with this disease is 2 per million, or .000002, find

a $P[D^C|T^+]$, the probability of a false positive

b $P[D|T^-]$, the probability of a false negative

Solution From the given information, we know the following:

$$P(D) = .000002 \qquad P(D^C) = .999998$$
$$P[T^+|D] = .98 \qquad P[T^-|D] = .02$$
$$P[T^+|D^C] = .01 \qquad P[T^-|D^C] = .99$$

a From Bayes' Rule,

$$P[D^C|T^+] = \frac{P(D^C \cap T^+)}{P(T^+)}$$

with

$$\begin{aligned} P(T^+) &= P(T^+ \cap D) + P(T^+ \cap D^C) \\ &= P(D)P(T^+|D) + P(D^C)P(T^+|D^C) \\ &= (.000002)(.98) + (.999998)(.01) \\ &= .00000196 + .00999998 \\ &= .01000194 \end{aligned}$$

Therefore, the probability of a false positive is

$$P[D^C|T^+] = \frac{.00999998}{.01000194} = .999804038$$

b Using a similar calculation,

$$P[D|T^-] = \frac{P(D \cap T^-)}{P(T^-)}$$

with

$$\begin{aligned} P(T^-) &= P(T^- \cap D) + P(T^- \cap D^C) \\ &= P(D)P(T^-|D) + P(D^C)P(T^-|D^C) \\ &= (.000002)(.02) + (.999998)(.99) \\ &= .00000004 + .98999802 \\ &= .98999806 \end{aligned}$$

Therefore, the probability of a false negative is

$$P[D|T^-] = \frac{.00000004}{.98999806} = .00000004$$

Hence, the probability of a false positive is near 1 and very likely, while the probability of a false negative is quite small and very unlikely. ▪

Another way to view Bayes' Rule is as a method of incorporating the information from sample observations to adjust the probability of some event. For example, if you had no information on the result of a person's diagnostic test, you would regard the probability that he or she is infected with the disease as .000002 (since .0002% of all people are infected with the disease). However, given the added information that the person's test was positive, the probability that he or she is infected with the disease is

$$P[D|T^+] = .0002$$

Thus, based on the test information, the probability that the person is infected is adjusted from a very small probability, .000002, to .0002—a 100-fold increase.

Exercises

Basic Techniques

3S.13 A sample is selected from one of two populations A_1 and A_2, with probabilities $P(A_1) = .7$ and $P(A_2) = .3$. If the sample has been selected from A_1, the probability of observing an event B is $P(B|A_1) = .2$. Similarly, if the sample has been selected from A_2, the probability of observing B is $P(B|A_2) = .3$. If a sample is selected and event B is observed, what is the probability that the sample was selected from population A_1? From A_2?

3S.14 If an experiment is conducted, one and only one of three mutually exclusive events A_1, A_2, and A_3 can occur, with these probabilities:

$$P(A_1) = .2 \qquad P(A_2) = .5 \qquad P(A_3) = .3$$

The probabilities of a fourth event B occurring, given that events A_1, A_2, or A_3 occur, are

$$P(B|A_1) = .2 \qquad P(B|A_2) = .1 \qquad P(B|A_3) = .3$$

If event B is observed, find $P(A_1|B)$, $P(A_2|B)$, and $P(A_3|B)$.

Applications

3S.15 City crime records show that 20% of all crimes are violent and 80% are nonviolent, involving theft, forgery, and so on. Ninety percent of violent crimes are reported versus 70% of nonviolent crimes. If a crime in progress is reported to the police, what is the probability that the crime is violent?

3S.16 Suppose that, in a particular city, airport A handles 50% of all airline traffic, while airports B and C handle 30% and 20%, respectively. The detection rates for weapons at the three airports are .9, .5, and .4, respectively. If a passenger at one of the airports is found to be carrying a weapon through the boarding gate, what is the probability that the passenger is using airport A? Airport C?

3S.17 A particular football team is known to run 30% of its plays to the left and 70% to the right. A linebacker on an opposing team notes that the right guard shifts his stance most the time (80%) when plays go to the right and that he uses a balanced stance the remainder of the time. When plays go to the left, the guard takes a balanced stance 90% of the time and the shift stance the remaining 10%. On a particular play, the linebacker notes that the guard takes a balanced stance. What is the probability that the play will go to the left?

3S.18 Do men and women have the same goals in life, or are their expectations quite different? In a poll conducted by Yankelovich Clancy Shulman for *Time* magazine involving 505 Americans aged 18–24, 32% of the men considered "a successful career" as their single most important goal, while only 27% of the women considered "a successful career" their single most important goal. On the other hand, 39% of the women chose "a happy marriage" as their single most important goal and only 30% of the men made this same choice. Suppose the proportion of men in the survey was .6 and the proportion of women was .4.

a If a person is chosen at random from the 505 surveyed, and that person chose "a successful career" as the single most important goal, what is the probability that the respondent was a man?

b If a person chosen at random from the 505 surveyed picked "a happy marriage" as the single most important goal, what is the probability that the respondent was a woman?

c What is the unconditional probability that a person chosen at random from this group would select "a successful career" as the single most important goal in life?

3S.19 Medical case histories indicate that different illnesses may produce identical symptoms. Suppose a particular set of symptoms, which we will denote as event H, occurs only when any one of three illnesses—A, B, or C—occurs (for the sake of simplicity, we will assume that illnesses A, B, and C are mutually exclusive). Studies show that the probabilities of getting the three illnesses are

$$P(A) = .01$$
$$P(B) = .005$$
$$P(C) = .02$$

The probabilities of developing the symptoms H, given a specific illness, are

$$P(H|A) = .90$$
$$P(H|B) = .95$$
$$P(H|C) = .75$$

Assuming that an ill person shows the symptoms H, what is the probability that the person has illness A?

3S.20 Suppose 5% of all people filing the long income tax form seek deductions that they know are illegal, and an additional 2% incorrectly list deductions because they are unfamiliar with income tax regulations. Of the 5% who are guilty of cheating, 80% will deny knowledge of the error if confronted by an investigator. If the filer of the long form is confronted with an unwarranted deduction and he or she denies the knowledge of the error, what is the probability that he or she is guilty?

3S.21 The number of companies offering flexible work schedules has increased as companies try to help employees cope with the demand of home and work. One flextime schedule is to work four 10-hour shifts. However, a big obstacle to flextime schedules for workers paid hourly is state legislation on overtime. A Thomas Temporaries survey that appeared in the *Press-Enterprise* (Riverside, Calif., Apr. 7, 1992) provided the following information for 220 firms located in the cities of Riverside and Temecula, both in California.

	Flextime Schedule		
City	**Available**	**Not Available**	**Total**
Riverside	39	75	114
Temecula	25	81	106
Totals	64	156	220

If a company is selected at random from this pool of 220 companies,

d What is the probability that the company will be located in Riverside?

e What is the probability that the company is located in Temecula and offers flextime work schedules?

f What is the probability that the company will not have flextime schedules?

g What is the probability that the company is located in Temecula, given that the company has flextime schedules available?

4

Several Useful Discrete Distributions

Case Study

Is the Pilgrim I nuclear reactor responsible for an increase in cancer cases in the surrounding area? A political controversy was set off when the Massachusetts Department of Public Health found an unusually large number of cases in a 4-mile-wide coastal strip just north of the nuclear reactor in Plymouth, Massachusetts. The case study at the end of this chapter examines how this question can be answered using one of the discrete probability distributions presented here.

General Objectives

In many applications, the variable of interest is discrete, and its probability distribution can be found by using the concepts of probability. The binomial, hypergeometric, and Poisson random variables presented in this chapter are often used when the data represent the number of occurrences of a specified event in a fixed number of trials or a fixed unit of time or space.

Specific Topics

1 The binomial probability distribution (4.2)
2 The mean and variance for the binomial random variable (4.2)
3 The Poisson probability distribution (4.3)
4 The hypergeometric probability distribution (4.4)

4.1 Introduction

In Chapter 3, we found that random variables defined over a finite or countably infinite number of simple events are called **discrete random variables**. Examples of discrete random variables abound in the physical and social sciences, as well as in business and economics, but three discrete probability distributions serve as **models** for a large number of these applications. These three distributions are **the binomial, the Poisson, and the hypergeometric probability distributions**. In this chapter, we will study these distributions and discuss their development as logical models for discrete processes observed in different physical settings.

4.2 The Binomial Probability Distribution

One of the most elementary, useful, and interesting discrete random variables, the binomial random variable, is associated with the coin-tossing experiment described in Examples 3.5 and 3.16. As an illustration, consider a sample survey conducted to predict voter preference in a political election. Interviewing a single voter bears a similarity, in many respects, to tossing a single coin, because the voter's response may be in favor of our candidate—a "head"—or may be against (or indicate indecision)—a "tail." In most cases, the fraction of voters favoring a particular candidate does not equal one-half, but in most national presidential elections, the fraction of the total vote favoring the winning presidential candidate is *very near* one-half.

Similar polls are conducted in the social sciences, in industry, and in education. The sociologist is interested in the fraction of army recruits who are white women; the marketer of soft drinks desires knowledge concerning the fraction of cola drinkers who prefer his or her brand; the teacher is interested in the fraction of high school seniors who pass basic proficiency tests in reading, language, and mathematics. Each person sampled is analogous to the toss of an unbalanced coin, since the probability of a "head" is usually not one-half. Although dissimilar in some respects, the surveys described above often exhibit, to a reasonable degree of approximation, the characteristics of a **binomial experiment.**

DEFINITION ▪

A **binomial experiment** is one that has the following properties:

1 The experiment consists of n identical trials.
2 Each trial results in one of two outcomes. For lack of a better nomenclature, the one outcome is called a success, S, and the other a failure, F.
3 The probability of success on a single trial is equal to p and remains the same from trial to trial. The probability of failure is equal to $(1 - p) = q$.
4 The trials are independent.
5 We are interested in x, the number of successes observed during the n trials. ▪

EXAMPLE 4.1

Suppose there are approximately 1,000,000 adults in a county and that an unknown proportion p favor the Equal Rights Amendment (ERA). A sample of 1000 adults will be chosen in such a way that every one of the 1,000,000 adults has an equal chance of being selected, and each adult is asked whether he or she favors the ERA. (The ultimate objective of this survey is to estimate the unknown proportion p, a problem that we will discuss in Chapter 7.) Is this a binomial experiment?

Solution To decide whether this is a binomial experiment, we must see whether the sampling satisfies the five characteristics described in the preceding definition.

1 The sampling consists of $n = 1000$ identical trials. One trial represents the selection of a single adult from the 1,000,000 adults in the county.

2 Each trial will result in one of two outcomes. A person will either favor the amendment or will not. These two outcomes could be associated with the "success" and "failure" of a binomial experiment.†

3 The probability of a success will equal the proportion of adults favoring the ERA. For example, if 500,000 of the 1,000,000 adults in the county favor the ERA, then the probability of selecting an adult favoring the ERA out of the 1,000,000 in the county is $p = .5$. For all practical purposes, this probability will remain the same from trial to trial, even though adults selected in the earlier trials are not replaced as sampling continues.

4 For all practical purposes, the probability of a success on any one trial will be unaffected by the outcome on any of the others (it will remain very close to p).

5 We are interested in the number x of adults in the sample of 1000 who favor the ERA.

Because the survey satisfies the five characteristics reasonably well, for all practical purposes it can be viewed as a binomial experiment. ▪

EXAMPLE 4.2 A purchaser who has received a shipment containing 20 personal computers (PCs) wants to sample 3 of the PCs to see if they are in working order before accepting the shipment. The nearest 3 PCs are selected for testing and, afterward, are declared either defective or nondefective. Unknown to the purchaser, 2 of the PCs in the shipment of 20 are defective. Is this a binomial experiment?

Solution Again, we check the sampling procedure against the characteristics of a binomial experiment.

1 The experiment consists of $n = 3$ identical trials. Each trial represents the selection and testing of 1 PC from the total of 20.

2 Each trial results in one of two outcomes. Either a PC is defective (call this a "success") or it is not (a "failure").

3 Suppose the PCs were randomly loaded into a boxcar so that any one of the 20 PCs could have been placed near the boxcar door. Then the unconditional probability of drawing a defective PC on a given trial would be 2/20.

4 The condition of independence between trials is *not* satisfied because the probability of drawing a defective PC on the second and third trials will be dependent on the outcome of the first trial. For example, if the first trial results in a defective PC, then there is only 1 defective left in the remaining 19 in the boxcar. Therefore, the conditional probability of success on trial 2, given a success on trial 1, is 1/19. This differs from the unconditional probability of a success on the second trial

† Although it is traditional to call the two possible outcomes of a trial "success" and "failure," they could have been called "head" and "tail," "red" and "white," or any other pair of words. Consequently, the outcome called a "success" does not need to be viewed as a success in the ordinary usage of the word.

(which is 2/20). Thus the trials are dependent and the sampling does not represent a binomial experiment. ■

Example 4.2 illustrates an important point. If the sample size n is large relative to the population size N, then the probability of success p will not remain constant from trial to trial. Hence, the trial outcomes will be dependent and the resulting experiment will not be a binomial experiment. **As a rule of thumb, if $n/N \geq .05$, the resulting experiment will not be binomial.**

The probability distribution for a simple binomial random variable (the number of heads in the tosses of two coins) was derived in Example 3.16. The probability distribution for a binomial experiment consisting of n tosses is derived in exactly the same way, but the procedure is much more complex when the number n of trials is large. We will omit this derivation and will simply present the **binomial probability distribution** and its mean, variance, and standard deviation, as shown in the following display.

Binomial Probability Distribution

$$p(x) = C_x^n p^x q^{n-x} = \frac{n!}{x!(n-x)!} p^x q^{n-x}$$

x, the number of successes in n trials, may take values $0, 1, 2, \ldots, n$;
p is the probability of success on a single trial;
C_x^n is defined as

$$\frac{n!}{x!(n-x)!}$$

where $n! = n(n-1)(n-2)\cdots(2)(1)$ and $0! \equiv 1$.

Mean:	$\mu = np$
Variance:	$\sigma^2 = npq$
Standard deviation:	$\sigma = \sqrt{npq}$

In the formula for $p(x)$ the quantity $p^x q^{n-x}$ represents the probability of observing a simple event with x successes and $(n-x)$ failures; the term C_x^n, defined as $n!/x!(n-x)!$, counts the number of such simple events. The probability $p(x) = C_x^n p^x q^{n-x}$ associated with a particular value of x in n independent trials is the term involving p to the power x in the series expansion of the binomial $(p+q)^n$—hence, the name *binomial probability.*

Graphs of three binomial probability distributions are shown in Figure 4.1—the first for $n = 10$, $p = .1$; the second for $n = 10$, $p = .5$; and the third for $n = 10$, $p = .9$.

FIGURE 4.1
Binomial probability distributions

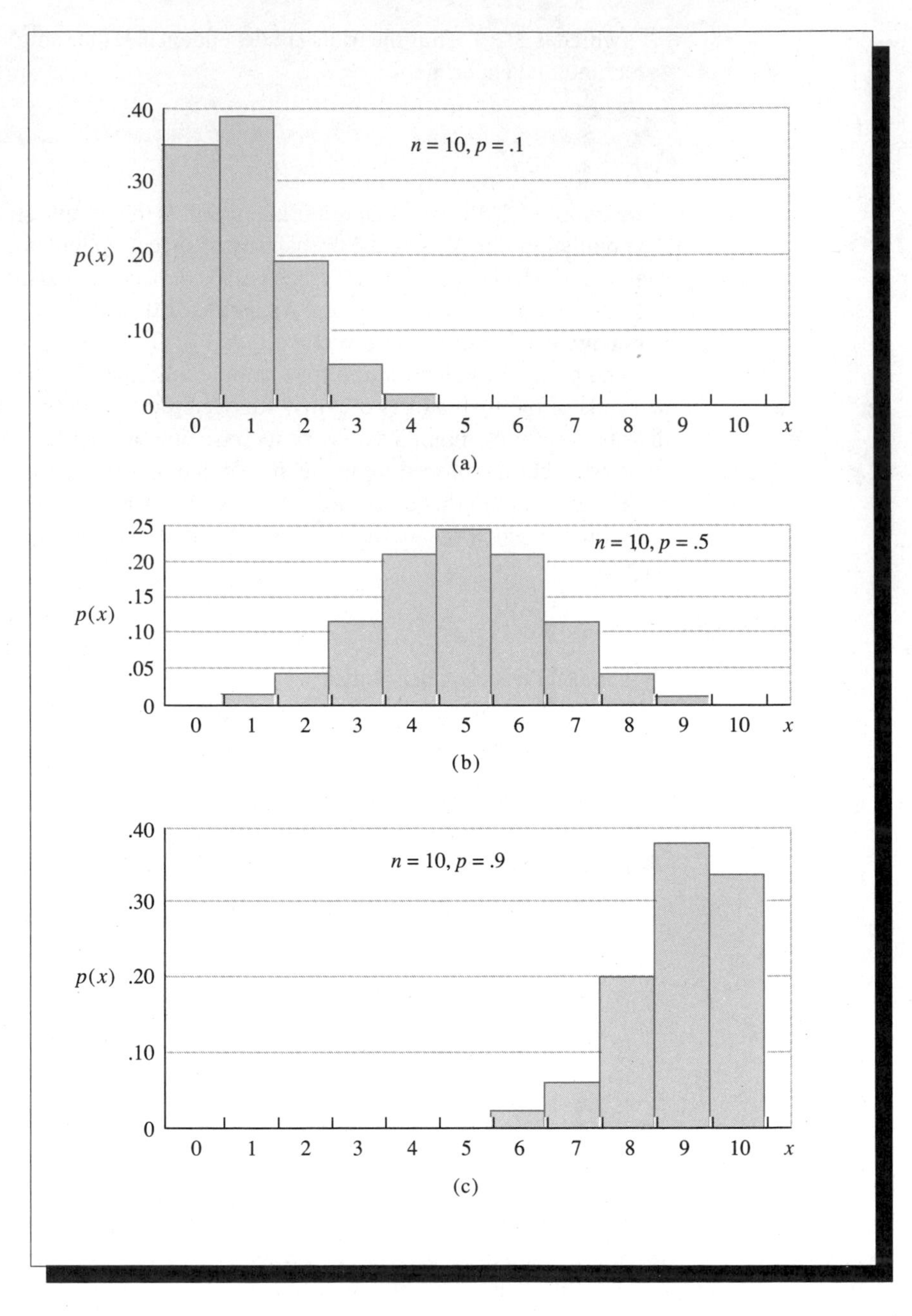

EXAMPLE **4.3** Over a long period of time it has been observed that a given rifleman can hit a target on a single trial with probability equal to .8. Suppose he fires four shots at the target.

1 What is the probability that he will hit the target exactly two times?

2 What is the probability that he will hit the target at least two times?

3 What is the probability that he will hit the target exactly four times?

Solution Assume that the trials are independent and that the probability p of hitting the target remains constant from trial to trial, $n = 4$ and $p = .8$. Let x denote the number of shots that hit the target. Then, for $x = 0, 1, 2, 3, 4$, we have

$$p(x) = C_x^4(.8)^x(.2)^{4-x}$$

1

$$\begin{aligned} p(2) &= C_2^4(.8)^2(.2)^{4-2} \\ &= \frac{4!}{2!2!}(.64)(.04) \\ &= \frac{4(3)(2)(1)}{2(1)(2)(1)}(.64)(.04) \\ &= .1536 \end{aligned}$$

The probability is .1536 that he will hit the target exactly two times.

2

$$\begin{aligned} P(\text{at least two}) &= p(2) + p(3) + p(4) \\ &= 1 - p(0) - p(1) \\ &= 1 - C_0^4(.8)^0(.2)^4 - C_1^4(.8)(.2)^3 \\ &= 1 - .0016 - .0256 \\ &= .9728 \end{aligned}$$

The probability is .9728 that he will hit the target at least two times.

3

$$\begin{aligned} p(4) &= C_4^4(.8)^4(.2)^0 \\ &= \frac{4!}{4!0!}(.8)^4(1) \\ &= .4096 \end{aligned}$$

The probability is .4096 that he will hit the target exactly four times.

Note that these probabilities would be incorrect if the rifleman could observe the location of each hit on the target and thereby adjust his aim. In that case, the trials would be dependent and p would likely increase from trial to trial. ▪

Calculating binomial probabilities is a tedious task when n is large. To simplify our calculations, the sum of the binomial probabilities from $x = 0$ to $x = a$ is presented in Table 1 of Appendix I for $n = 2, 3, \ldots, 12, 15, 20$, and 25.

To illustrate the use of Table 1, let's find the sum of the binomial probabilities from $x = 0$ to $x = 3$ for $n = 5$ trials and $p = .6$. That is, we wish to find

$$P(x \leq 3) = \sum_{x=0}^{3} p(x) = p(0) + p(1) + p(2) + p(3)$$

where

$$p(x) = C_x^5(.6)^x(.4)^{5-x}$$

Since the table values give

$$P(x \leq a) = \sum_{x=0}^{a} p(x)$$

we seek the table value in the row corresponding to $a = 3$ and the column for $p = .6$. The table value, .663, is shown in Table 4.1 as it appears in Table 1, Appendix I. Therefore, the sum of the binomial probabilities from $x = 0$ to $x = a = 3$ (for $n = 5$, $p = .6$) is .663.

TABLE **4.1** Portion of Table 1, Appendix I, for $n = 5$

							p							
a	**0.01**	**0.05**	**0.10**	**0.20**	**0.30**	**0.40**	**0.50**	**0.60**	**0.70**	**0.80**	**0.90**	**0.95**	**0.99**	a
0	—	—	—	—	—	—	—	—	—	—	—	—	—	0
1	—	—	—	—	—	—	—	—	—	—	—	—	—	1
2	—	—	—	—	—	—	—	—	—	—	—	—	—	2
3	—	—	—	—	—	—	—	.663	—	—	—	—	—	3
4	—	—	—	—	—	—	—	—	—	—	—	—	—	4

Table 1 can also be used to find individual binomial probabilities. For example, suppose we wish to find $p(3)$ when $n = 5$ and $p = .6$. Since $P(x = 3) = P(x \leq 3) - P(x \leq 2)$, we write

$$p(3) = \sum_{x=0}^{3} p(x) - \sum_{x=0}^{2} p(x) = .663 - .317 = .346$$

The values $\sum_{x=0}^{3} p(x)$ and $\sum_{x=0}^{2} p(x)$ are found directly from Table 1, indexing $n = 5$ with $p = .6$. In general, an individual binomial probability may be found by subtracting successive entries in the table for a given value of p.

EXAMPLE **4.4** A regimen consisting of a daily dose of vitamin C was tested to determine its effectiveness in preventing the common cold. Ten people who were following the prescribed regimen were observed for a period of 1 year. Eight survived the winter without a cold. Suppose the probability of surviving the winter without a cold is .5 when the vitamin-C regimen is not followed. What is the probability of observing eight or more survivors, given that the regimen is ineffective in increasing resistance to colds?

Solution Assuming that the vitamin-C regimen is ineffective, the probability p of surviving the winter without a cold is .5. The probability distribution for x, the number of survivors, is

$$p(x) = C_x^{10}(.5)^x(.5)^{10-x}$$

Then, by direct calculation,

$$\begin{aligned} P(8 \text{ or more}) &= p(8) + p(9) + p(10) \\ &= C_8^{10}(.5)^{10} + C_9^{10}(.5)^{10} + C_{10}^{10}(.5)^{10} \\ &= .055 \end{aligned}$$

From Table 1 in Appendix I for $n = 10$, we find

$$P(x \geq 8) = \sum_{x=8}^{10} p(x) = 1 - \sum_{x=0}^{7} p(x)$$

The quantity

$$\sum_{x=0}^{7} p(x)$$

can be found by moving across the top of Table 1 to the column for $p = .5$ and down that column to the row corresponding to $a = 7$. We read

$$\sum_{x=0}^{7} p(x) = .945$$

Then, as before,

$$P(x \geq 8) = \sum_{x=8}^{10} p(x) = 1 - \sum_{x=0}^{7} p(x) = 1 - .945 = .055 \quad \blacksquare$$

Both individual and cumulative binomial probabilities are available through the MINITAB package. Individual binomial probabilities associated with each value of x for any combination of n and p can be found by using the probability density function command PDF followed by a semicolon (;) and then the subcommand BINOMIAL N P followed by a period. The variable N is the given sample size, and P is the probability of success. Cumulative binomial probabilities can be found by using the cumulative distribution function command CDF followed by a semicolon (;) and then the subcommand BINOMIAL N P followed by a period.

The MINITAB output for both the PDF and CDF commands when $n = 10$ and $p = .5$ is given in Table 4.2. The PDF command gives rise to the individual probabilities $P(x = K)$; the CDF command gives rise to the cumulative probabilities $P(x \leq K)$. (The letter K plays the role of the letter a in Table 1 of Appendix I.) Notice that, in the MINITAB output, $P(x \leq 7) = .9453$, so that $P(x \geq 8) = (1 - .9453) = .0547$, which, to three-decimal accuracy, agrees with our earlier results using Table 1.

TABLE 4.2 MINITAB output of binomial probabilities when $n = 10$ and $p = .5$

```
MTB  > PDF;
SUBC > BINOMIAL  10  .5.

BINOMIAL  WITH N = 10  P = 0.500000
    K            P(X = K)
    0             0.0010
    1             0.0098
    2             0.0439
    3             0.1172
    4             0.2051
    5             0.2461
    6             0.2051
    7             0.1172
    8             0.0439
    9             0.0098
   10             0.0010
```

```
MTB  > CDF;
SUBC > BINOMIAL  10  .5.

BINOMIAL  WITH N = 10  P = 0.500000
    K   P(X  LESS  OR  =  K)
    0              0.0010
    1              0.0107
    2              0.0547
    3              0.1719
    4              0.3770
    5              0.6230
    6              0.8281
    7              0.9453
    8              0.9893
    9              0.9990
   10              1.0000
```

The MINITAB output may not list the probabilities for all values of $x = 0, 1, 2, \ldots, n$ for various combinations of n and p, since the MINITAB package has an internal check that stops the calculations when $P(x = K) = 0$ [or, equivalently, when $P(x \leq K) = 1$] to within a preassigned level of accuracy.

EXAMPLE 4.5 Find the mean and standard deviation for a binomial probability distribution with $n = 10$ and $p = .5$. Find the probability that x falls in the interval $\mu \pm 2\sigma$.

Solution The mean and standard deviation are

$$\mu = np = 10(.5) = 5$$
$$\sigma = \sqrt{npq} = \sqrt{10(.5)(.5)} = 1.58$$

Therefore, the interval $\mu \pm 2\sigma$ is given by

$$5 \pm 2(1.58) = 5 \pm 3.2$$

or from 1.8 to 8.2. This interval includes the values $x = 2, 3, \ldots, 8$. Therefore,

$$P[2 \leq x \leq 8] = P[x \leq 8] - P[x \leq 1]$$
$$= .989 - .011 = .978$$

a result that agrees with Tchebysheff's Theorem and agrees approximately with the Empirical Rule. ■

EXAMPLE **4.6** How are scores on a multiple-choice test evaluated? A score of 0 on an objective test (questions requiring complete recall of the material) indicates that the person was unable to recall the test material at the time the test was given. In contrast, a person with little or no recall knowledge of the test material can achieve a higher score on a multiple-choice test because the person needs only to recognize (in contrast to recall) the correct answer and because some questions will be answered correctly just by chance, even if the person does not know the correct answers. Consequently, the no-knowledge score for a multiple-choice test may be well above 0. If a multiple-choice test contains 100 questions, each with six possible answers, what is the expected score for a person who has no knowledge of the test material? Within what limits would a no-knowledge score fall?

Solution Let p equal the probability of a correct choice on a single question, and let x equal the number of correct responses out of the $n = 100$ questions. Assume that "no knowledge" means that a student will randomly select one of the six possible answers for each question and hence that $p = 1/6$. Then, for $n = 100$ questions, the expected score for a student with no knowledge would be $E(x)$, where

$$E(x) = np = 100\left(\frac{1}{6}\right) = 16.7 \text{ correct questions}$$

To evaluate the variation of no-knowledge scores, we need to know σ, where

$$\sigma = \sqrt{npq} = \sqrt{(100)(1/6)(5/6)} = 3.7$$

Based on Tchebysheff's Theorem and the Empirical Rule, we would expect x to fall within the interval $(\mu \pm 2\sigma)$ with a high probability and almost certainly within the interval $(\mu \pm 3\sigma)$. The intervals are

$$(\mu \pm 2\sigma) = (16.7 \pm 7.4) \quad \text{or} \quad 9.3 \text{ to } 24.1$$
$$(\mu \pm 3\sigma) = (16.7 \pm 11.1) \quad \text{or} \quad 5.6 \text{ to } 27.8$$

This compares with a score of 0 for a no-knowledge student taking an objective recall test. (COMMENT: A histogram of the binomial probability distribution will be very mound-shaped for $n = 100$ and $p = 1/6$. Hence, we would expect the Empirical Rule to work very well. The justification for this statement will be given in Chapter 6.) ▪

Exercises

Basic Techniques

4.1 A jar contains five balls: three red and two white. Two balls are randomly selected without replacement from the jar, and the number x of red balls is recorded. Explain why x is or is not a binomial random variable. (HINT: Compare the characteristics of this experiment with the characteristics of a binomial experiment given in Section 4.1.) If the experiment is binomial, give the values of n and p.

4.2 Refer to Exercise 4.1. Assume that the sampling was conducted with replacement. That is, assume that the first ball was selected from the jar, observed, and then replaced, and that the balls were then mixed before the second ball was selected. Explain why x, the number of red balls observed, is or is not a binomial random variable. If the experiment is binomial, give the values of n and p.

4.3 Evaluate the following binomial probabilities:

a $C_2^8(.3)^2(.7)^6$ **b** $C_0^4(.05)^0(.95)^4$

c $C_3^{10}(.5)^3(.5)^7$ **d** $C_1^7(.2)^1(.8)^6$

4.4 Use the formula for the binomial probability distribution to calculate the values of $p(x)$ and construct the probability histogram for the following:

a $n = 7,\ p = .2$ **b** $n = 7,\ p = .5$ **c** $n = 7,\ p = .8$

4.5 Refer to Exercise 4.4. For each of the binomial random variables given in that exercise, calculate the following:

a $P(x = 1)$ **b** $P(x \geq 1)$ **c** $P(x > 1)$

d $P(x \leq 1)$ **e** $\mu = np$ **f** $\sigma = \sqrt{npq}$

4.6 Use Table 1 in Appendix I to find the sum of the binomial probabilities from $x = 0$ to $x = a$ for the following:

a $n = 10,\ p = .1,\ a = 3$ **b** $n = 15,\ p = .6,\ a = 7$ **c** $n = 25,\ p = .5,\ a = 14$

4.7 Use Table 1 in Appendix I to evaluate the following probabilities for $n = 7$ and $p = .8$:

a $P(x \geq 4)$ **b** $P(x = 2)$ **c** $P(x < 2)$ **d** $P(x > 1)$

Verify these answers using the values of $p(x)$ calculated in part (c), Exercise 4.4.

4.8 Find $\sum_{x=0}^{a} p(x)$ for the following:

a $n = 20,\ p = .05,\ a = 2$ **b** $n = 15,\ p = .7,\ a = 8$ **c** $n = 10,\ p = .9,\ a = 9$

4.9 Use Table 1, Appendix I, to find the following:

a $P\{x < 12\}$ for $n = 20,\ p = .5$ **b** $P\{x \leq 6\}$ for $n = 15,\ p = .4$

c $P\{x > 4\}$ for $n = 10,\ p = .4$ **d** $P\{x \geq 6\}$ for $n = 15,\ p = .6$

e $P\{3 < x < 7\}$ for $n = 10,\ p = .5$

4.10 Find the mean and standard deviation for a binomial distribution with the following:

a $n = 1000,\ p = .3$ **b** $n = 400,\ p = .01$

c $n = 500,\ p = .5$ **d** $n = 1600,\ p = .8$

4.11 Find the mean and standard deviation for a binomial distribution with $n = 100$ and

a $p = .01$ **b** $p = .9$ **c** $p = .3$

d $p = .7$ **e** $p = .5$

4.12 In Exercise 4.11 the mean and standard deviation for a binomial random variable were calculated for a fixed sample size, $n = 100$, and for different values of p. Graph the values of the standard deviation for the five values of p given in Exercise 4.11. For what value of p does the standard deviation seem to be a maximum?

4.13 Let x be a binomial random variable with $n = 20$ and $p = .1$.

a Calculate $P(x \leq 4)$ using the binomial formula.

b Calculate $P(x \leq 4)$ using Table 1 in Appendix I.

c Use the MINITAB output given below to calculate $P(x \leq 4)$. Compare the results of parts (a), (b), and (c).

```
MTB  > PDF;
SUBC > BINOMIAL  20  .1.

     BINOMIAL  WITH N = 20   P = 0.100000
        K            P(X = K)
        0              0.1216
        1              0.2702
        2              0.2852
        3              0.1901
        4              0.0898
        5              0.0319
        6              0.0089
        7              0.0020
        8              0.0004
        9              0.0001
       10              0.0000
```

d Calculate the mean and standard deviation of the random variable x.

e Use the results of part (d) to calculate the intervals $\mu \pm \sigma$, $\mu \pm 2\sigma$, and $\mu \pm 3\sigma$. Find the probability that an observation will fall in each of these intervals.

f Are the results of part (e) consistent with Tchebysheff's Theorem? With the Empirical Rule? Why or why not?

Applications

4.14 In a national survey of 1387 adults conducted for CNN/Knight-Ridder prior to the 1992 elections (Hugick and McAneny, 1992), 62% of the 351 young adults (aged 18–31 years) surveyed indicated that the standard of living for people in their age group was not going up. If we let x be the number of young adults in the survey who expressed this opinion, would x have a binomial distribution? Explain.

4.15 The fear of air travel is becoming more and more prevalent among American adults. In a survey of 1162 American adults conducted by Media General–Associated Press (*New York Times*, Feb. 3, 1989), 56% of those surveyed said that airline security on international flights was inadequate. Does this sampling represent a binomial experiment? Explain.

4.16 According to an article in the *Press-Enterprise* (Shapiro, 1992), caffeinated colas had 48% of the market share among all carbonated soft drinks. Suppose that $n = 100$ randomly chosen people who drink carbonated soft drinks are asked which soft drink they prefer. If x is the number in the sample who prefer a caffeinated cola, does x have a binomial distribution? If so, what are n and p?

4.17 The experiment described in Exercise 0.10 involves measuring the levels of essential vitamins in violent inmates from the California Youth Authority. Does the experiment satisfy the requirements of a binomial experiment? Why or why not?

4.18 A new surgical procedure is said to be successful 80% of the time. Suppose the operation is performed five times and the results are assumed to be independent of one another. What is the probability that:

a All five operations are successful?

b Exactly four are successful?

c Less than two are successful?

4.19 Refer to Exercise 4.18. If less than two operations were successful, how would you feel about the performance of the surgical team?

4.20 Records show that 30% of all patients admitted to a medical clinic fail to pay their bills and that eventually the bills are forgiven. Suppose $n = 4$ new patients represent a random selection from the large set of prospective patients served by the clinic. Find the probability that

a All the patients' bills will eventually have to be forgiven.

b One will have to be forgiven.

c None will have to be forgiven.

4.21 Consider the medical payment problem in Exercise 4.20 in a more realistic setting. Of all patients admitted to a medical clinic, 30% fail to pay their bills, and the bills are eventually forgiven. If the clinic treats 2000 different patients over a period of 1 year, what is the mean (expected) number of bills that would have to be forgiven? If x is the number of forgiven bills in the group of 2000 patients, find the variance and standard deviation of x. What can you say about the probability that x will exceed 700? (HINT: Use the values of μ and σ, along with Tchebysheff's Theorem, to answer this question.)

4.22 High serum cholesterol levels are becoming increasingly prevalent in the American adult population. In fact, approximately one-third of Americans 20 years of age or older are at high risk for coronary disease and are considered "candidates for medical advice and intervention" (*New York Times*, July 7, 1989). If six Americans 20 years of age or older are randomly chosen to undergo cholesterol testing, what is the probability that

a Exactly three are at high risk for coronary disease?

b At least one is at high risk for coronary disease?

c At most two are at high risk for coronary disease?

4.23 Consider a metabolic defect that occurs in approximately 1 of every 100 births. If four infants are born in a particular hospital on a given day, what is the probability that

a None has the defect?

b No more than one has the defect?

4.24 Early in the U.S. missile development program, government and industry defense officials were proclaiming that our missiles were highly reliable, and probabilities of successful firings in the neighborhood of .999 . . . were quoted. Such statements were made even though many missile firings resulted in failure, including the Navy Vanguard missile in the 1950s. If the reliability (probability of a successful launch) is even as high as .9, what is the probability of observing three or more failures in a total of four firings? One failure or more? If you observed two or more failures out of four, what would you think about the high claims of reliability of the missiles produced in the 1950s?

4.25 New excitement about the progress in the treatment of AIDS was generated at the Fifth International Conference on AIDS (*Press-Enterprise*, June 8, 1989). Dr. Samuel Broder announced that the chances of living 18 months or longer after having been diagnosed with AIDS have increased to more than 60%, compared to 30% in 1982. Suppose a hospital admitted 480 patients with AIDS in 1989.

a What is the minimum number of these patients you would expect to survive for at least 18 months? Justify your answer.

b What is the maximum number? Justify your answer.

4.26 Scientific jury research is concerned with assessing how jurors vote based upon characteristics of the jurors, together with the type and location of the case under consideration. For example, in a survey of 750 jury-eligible adults conducted by Behavioral Science Research of Coral Gables, Florida, it was found that older white homeowners are most likely to award large settlements in eminent-domain cases. In fact, 26% of the jury-eligible homeowners indicated that they would vote to award the homeowner over $500,000 as a settlement. Assume that the 26% figure is accurate for jury-eligible homeowners, and suppose we randomly select 200 such homeowners.

a What is the minimum number of jury-eligible homeowners you would expect to vote to award more than $500,000 in such cases?

b What is the maximum number of jury-eligible homeowners you would expect to vote to award more than $500,000 in such cases?

4.27 Mothers worry more about events over which they have no control than about events over which parents have influence (Schmittroth, 1991). Of 400 mothers aged 16–52, 72% said they feared child abduction, although the actual incidence of kidnapping by a nonfamily member is less than 2 in 10 million. If the 72% figure is accurate and if a sample of 500 mothers in this same age range were polled, find the mean and standard deviation of x, the number of mothers in the sample who feared child abduction. Within what range would you expect x to lie?

4.28 A report by the American Medical Association's Special Task Force on Professional Liability and Insurance notes that, every year, 16 of every 100 doctors are subject to malpractice claims. Suppose a hospital staff consists of 200 physicians and the occurrence of malpractice claims for one physician is independent of the occurrence of claims for any others.

a What is the expected number of the hospital's staff physicians who will be sued for malpractice in a given year? (Assume that the probability that a physician is sued for malpractice is .16)

b Is it likely that the number sued for malpractice could be as large as 50? Explain.

4.3 The Poisson Probability Distribution

Another discrete random variable that has numerous applications in business and economics is the **Poisson random variable**. Its probability distribution provides a good model for data that represent the number of occurrences of a specified event in a given unit of time or space. Here are some examples of experiments for which the random variable x can be modeled by the Poission random variable:

1 The number of calls received by a switchboard during a given period of time
2 The number of bacteria per small volume of fluid
3 The number of arrivals at a checkout counter during a given minute
4 The number of machine breakdowns during a given day
5 The number of traffic accidents at a given intersection during a given time period

In each example, **x represents the number of events occurring in a period of time or space during which an average of μ such events can be expected to occur**. The only assumptions needed when one uses the Poisson distribution to model experiments such as those described above are that the counts or events occur **randomly and independently** of one another. The formula for the Poisson probability distribution, as well as its mean and variance, are shown in the display.

The Poisson Probability Distribution

$$p(x) = \frac{\mu^x e^{-\mu}}{x!} \qquad x = 0, 1, 2, 3, \ldots$$

where

$$\mu = E(x) = \text{mean of random variable } x$$
$$\sigma^2 = \mu = \text{variance of random variable } x$$
$$e = 2.71828\ldots \ (e \text{ is the base of natural logarithms})$$

The value of $e^{-\mu}$ can be found by using a calculator or by using Table 2(b) in Appendix I, which provides the values of $e^{-\mu}$ for values of μ between 0 and

10 in increments of .05. Alternatively, the Poisson probabilities can be obtained from Table 2(a) in Appendix I, in which the cumulative probabilities, $P(x \leq a) = p(0) + p(1) + \cdots + p(a)$, are given for various values of μ. This table can be used to find Poisson probabilities just as the binomial tables were used in Section 4.2. Graphs of the Poisson probability distribution for $\mu = .5$, 1, and 4 are shown in Figure 4.2.

FIGURE **4.2** Poisson probability distributions for $\mu = .5$, 1, and 4

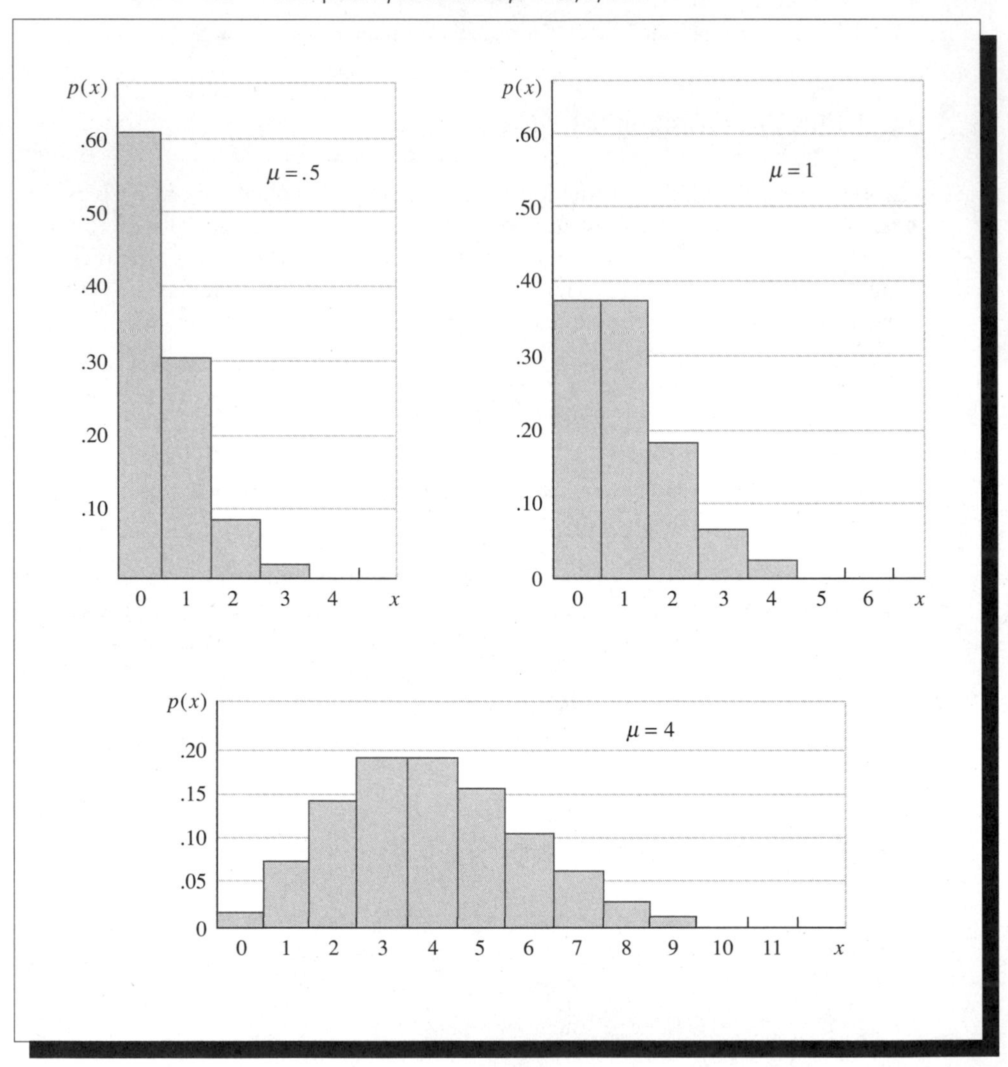

EXAMPLE **4.7** The average number of traffic accidents on a certain section of highway is two per week. Assume that the number of accidents follows a Poisson distribution with $\mu = 2$.

1 Find the probability of no accidents on this section of highway during a 1-week period.

2 Find the probability of at most three accidents on this section of highway during a 2-week period.

Solution 1 The average number of accidents per week is $\mu = 2$. Therefore, the probability of no accidents on this section of highway during a given week is

$$p(0) = \frac{2^0 e^{-2}}{0!} = e^{-2} = .135335$$

2 During a 2-week period the average number of accidents on this section of highway would be $2(2) = 4$. The probability of at most three accidents during a 2-week period is

$$P(x \leq 3) = p(0) + p(1) + p(2) + p(3)$$

where

$$p(0) = \frac{4^0 e^{-4}}{0!} = .018316 \qquad p(2) = \frac{4^2 e^{-4}}{2!} = .146528$$

$$p(1) = \frac{4^1 e^{-4}}{1!} = .073264 \qquad p(3) = \frac{4^3 e^{-4}}{3!} = .195367$$

Therefore,

$$P(x \leq 3) = .018316 + .073264 + .146528 + .195367 = .433475$$

This value could be read directly from Table 2(a) in Appendix I, indexing $\mu = 4$ and $a = 3$, as $P(x \leq 3) = .433$. ▪

Remember that we were able to simplify the calculation of binomial probabilities by using Table 1 in Appendix I. However, binomial tables are seldom available for n greater than 100, and many applications of the binomial experiment with $n = 100$ or more arise in practical situations. Consequently, we need simple, easy-to-compute approximation procedures for calculating binomial probabilities. **The Poisson probability distribution provides good approximations to binomial probabilities when n is large and $\mu = np$ is small, preferably with $np < 7$.** An approximation procedure suitable for larger values of $\mu = np$ will be presented in Chapter 5.

As an illustration of the Poisson approximation procedure, consider the following application. Suppose a life insurance company insures the lives of 5000 men of age 42. If actuarial studies show the probability of any 42-year-old man dying in a given year to be .001, the exact probability that the company will have to pay $x = 4$ claims during a given year is given by the binomial distribution as

$$P(x = 4) = p(4) = \frac{5000!}{4!4996!}(.001)^4(.999)^{4996}$$

for which binomial tables are not available. To compute $P(x = 4)$ without the aid of a computer would be very time consuming, but the Poisson distribution can be used to provide a good approximation to $P(x = 4)$. Computing $\mu = np = (5000)(.001) = 5$ and substituting into the formula for the Poisson probability distribution, we have

$$p(4) \approx \frac{\mu^4 e^{-\mu}}{4!} = \frac{5^4 e^{-5}}{4!} = \frac{(625)(.006738)}{24} = .175$$

The value of $p(4)$ could also be obtained using Table 2(a) in Appendix I with $\mu = 5$ as

$$p(4) = P(x \leq 4) - P(x \leq 3) = .440 - .265 = .175$$

EXAMPLE 4.8 A manufacturer of power lawn mowers buys one-horsepower, two-cycle engines in lots of 1000 from a supplier. She then equips each of the mowers produced by her plant with one of the engines. Past history shows that the probability of any one engine from that supplier proving unsatisfactory is .001. In a shipment of 1000 engines, what is the probability that none are defective? One is defective? Two are? Three are? Four are?

Solution This is a binomial experiment with $n = 1000$ and $p = .001$. The expected number of defectives in a shipment of $n = 1000$ engines is $\mu = np = (1000)(.001) = 1$. Since this is a binomial experiment with $np < 7$, the probability of x defective engines in the shipment may be approximated by

$$p(x) = \frac{\mu^x e^{-\mu}}{x!} = \frac{1^x e^{-1}}{x!} = \frac{e^{-1}}{x!}$$

(since $1^x = 1$ for any value of x). Therefore, we have

$$p(0) \approx \frac{e^{-1}}{0!} = \frac{.368}{1} = .368 \qquad p(3) \approx \frac{e^{-1}}{3!} = \frac{.368}{6} = .061$$

$$p(1) \approx \frac{e^{-1}}{1!} = \frac{.368}{1} = .368 \qquad p(4) \approx \frac{e^{-1}}{4!} = \frac{.368}{24} = .015$$

$$p(2) \approx \frac{e^{-1}}{2!} = \frac{.368}{2} = .184$$ ■

The individual and cumulative probabilities for a Poisson distribution with mean μ can be found by using the PDF and the CDF MINITAB commands, followed by the subcommand POISSON μ. The binomial probabilities for $n = 1000$ and $p = .001$, together with the Poisson probabilities for $\mu = 1$, are given in Table 4.3.

Comparing the actual binomial probabilities with the corresponding probabilities found by using the Poisson approximation to binomial probabilities, we see that they are quite accurate in this case. Furthermore, we see that the POISSON command also terminates when an individual probability equals 0 within a preassigned level of accuracy.

TABLE 4.3 MINITAB output of binomial and Poisson probabilities

```
MTB  > PDF;                                  MTB  > PDF;
SUBC > BINOMIAL 1000 .001.                   SUBC > POISSON 1.

 BINOMIAL   WITH N = 1000 P = 0.001000       POISSON   WITH MEAN = 1.000
     K              P(X = K)                     K              P(X = K)
     0               0.3677                      0               0.3679
     1               0.3681                      1               0.3679
     2               0.1840                      2               0.1839
     3               0.0613                      3               0.0613
     4               0.0153                      4               0.0153
     5               0.0030                      5               0.0031
     6               0.0005                      6               0.0005
     7               0.0001                      7               0.0001
     8               0.0000                      8               0.0000
```

Exercises

Basic Techniques

4.29 Let x be a Poisson random variable with mean $\mu = 2$. Calculate the following probabilities:

a $P(x = 0)$ **b** $P(x = 1)$ **c** $P(x > 1)$ **d** $P(x = 5)$

4.30 Let x be a Poisson random variable with mean $\mu = 2.5$. Use Table 2(a) in Appendix I to calculate the following probabilities:

a $P(x \geq 5)$ **b** $P(x < 6)$ **c** $P(x = 2)$ **d** $P(1 \leq x \leq 4)$

4.31 Let x be a binomial random variable with $n = 20$ and $p = .1$.

a Calculate $P(x \leq 2)$ using Table 1 in Appendix I to obtain the exact binomial probability.

b Use the Poisson approximation to calculate $P(x \leq 2)$.

c Compare the results of parts (a) and (b). Is the approximation accurate?

4.32 To illustrate how well the Poisson probability distribution approximates the binomial probability distribution, calculate the Poisson approximate values for $p(0)$ and $p(1)$ for a binomial probability distribution with $n = 25$ and $p = .05$. Compare the answers with the exact values obtained from Table 1, Appendix I.

Applications

4.33 The increased number of small commuter planes in major airports has heightened concern over air safety. An eastern airport has recorded a monthly average of five near-misses on landings and takeoffs in the past 5 years.

a Find the probability that during a given month there are no near-misses on landings and takeoffs at the airport.

b Find the probability that during a given month there are five near-misses.

c Find the probability that there are at least five near-misses during a particular month.

4.34 The number x of people entering the intensive-care unit at a particular hospital on any one day has a Poisson probability distribution with mean equal to five persons per day.

a What is the probability that the number of people entering the intensive-care unit on a particular day is two? Less than or equal to two?

b Is it likely that x will exceed ten? Explain.

4.35 Parents who are concerned that their children are "accident prone" can be reassured, according to a study conducted by the Department of Pediatrics at the University of California, San Francisco (*Physician and Sportsmedicine*, 1989). Children who are injured two or more times tend to sustain these injuries during a relatively limited time, usually 1 year or less. This suggests that the children are experiencing only a "temporary period of heightened vulnerability to injury," perhaps triggered by the biological and psychological changes of adolescence or by stress in the child's environment. If the average number of injuries per year for school-age children is two, what is the probability that

a A child will sustain two injuries during the year?

b A child will sustain two or more injuries during the year?

c A child will sustain at most one injury during the year?

4.36 Refer to Exercise 4.35.

a Calculate the mean and standard deviation for x, the number of injuries per year sustained by a school-age child.

b Within what limits would you expect the number of injuries per year to fall?

4.37 If a drop of water is placed on a slide and examined under a microscope, the number x of a particular type of bacteria present has been found to have a Poisson probability distribution. Suppose the maximum permissible count per water specimen for this type of bacteria is five. If the mean count for your water supply is two and you test a single specimen, is it likely that the count will exceed the maximum permissible count? Explain.

4.38 Corporate drug testing is becoming a more established practice in the U.S. business community. As an example, MetPath, a commercial drug-testing company, claims that it is performing about 22,000 tests a month for its corporate clients and that between 5% and 15% of the samples test positive for illegal substances (*Wall Street Journal*, Nov. 29, 1988). On a particular day, suppose that MetPath performs 100 independent drug tests between the hours of 9:00 and 10:00 A.M. and that the probability of a positive test is approximately .05.

a Let x be the number of positive tests observed. What is the probability distribution for x? What is the mean of this distribution?

b Use the Poisson distribution to approximate the probability of at most four positive drug tests.

4.4 The Hypergeometric Probability Distribution

Suppose you are selecting a sample of elements from a population and you record whether each element does or does not possess a certain characteristic. Consequently, you are dealing with the "success" or "failure" type of data encountered in the binomial experiment. The ERA survey of Example 4.1 and the sampling for defectives of Example 4.2 are practical illustrations of these sampling situations.

If the number of elements in the population is large relative to the number in the sample (as in Example 4.1), the probability of selecting a success on a single trial is equal to the proportion p of successes in the population. Because the population is large in relation to the sample size, this probability will remain constant (for all practical purposes) from trial to trial, and the number of x successes in the sample will follow a binomial probability distribution. However, **if the number of elements in the population is small in relation to the sample size ($n/N \geq .05$), the probability of a success for a given trial is dependent on the outcomes of preceding trials. Then the number x of successes follows what is known as a *hypergeometric probability distribution.***

We define the following notation, which is necessary in order to present the formula for the hypergeometric probability distribution.

N = number of elements in population

k = number of elements in population that are successes (i.e., number possessing one of two characteristics)

$N - k$ = number of elements in population that are not successes

n = number of elements in sample, selected from N elements in population

x = number of successes in sample

The hypergeometric probability distribution for the random variable x is then as given in the following display.

Hypergeometric Probability Distribution

$$p(x) = \frac{C_x^k C_{n-x}^{N-k}}{C_n^N}$$

where x can assume integer values $0, 1, 2, \ldots, n$ subject to the restrictions $x \leq k$ and $x \geq k + n - N$, and where

$$C_r^n = \frac{n!}{r!(n-r)!}$$

The mean and variance of a hypergeometric random variable are very similar to those of a binomial random variable with a correction for the finite population size.

$$\mu = n\left(\frac{k}{N}\right)$$

$$\sigma^2 = n\left(\frac{k}{N}\right)\left(\frac{N-k}{N}\right)\left(\frac{N-n}{N-1}\right)$$

EXAMPLE 4.9 A case of wine contains 12 bottles, 3 of which contain spoiled wine. A sample of 4 bottles is randomly selected from the case.

1. Find the probability distribution for x, the number of bottles of spoiled wine in the sample.
2. What are the mean and variance of x?

Solution For this example, $N = 12, n = 4, k = 3$, and $(N - k) = 9$. Then

$$p(x) = \frac{C_x^3 C_{4-x}^9}{C_4^{12}}$$

1 The possible values for x are 0, 1, 2, and 3. Therefore,

$$p(0) = \frac{C_0^3 C_4^9}{C_4^{12}} = \frac{1(126)}{495} = \frac{126}{495} = \frac{14}{55}$$

$$p(1) = \frac{C_1^3 C_3^9}{C_4^{12}} = \frac{3(84)}{495} = \frac{252}{495} = \frac{28}{55}$$

$$p(2) = \frac{C_2^3 C_2^9}{C_4^{12}} = \frac{3(36)}{495} = \frac{108}{495} = \frac{12}{55}$$

$$p(3) = \frac{C_3^3 C_1^9}{C_4^{12}} = \frac{1(9)}{495} = \frac{9}{495} = \frac{1}{55}$$

2 The mean is given by

$$\mu = 4\left(\frac{3}{12}\right) = 1$$

and the variance is given by

$$\sigma^2 = 4\left(\frac{3}{12}\right)\left(\frac{9}{12}\right)\left(\frac{12-4}{11}\right) = .5455$$ ■

EXAMPLE 4.10 A particular industrial product is shipped in lots of 20. Testing to determine whether an item is defective is costly; hence, the manufacturer samples production rather than using a 100% inspection plan. A sampling plan constructed to minimize the number of defectives shipped to customers calls for sampling five items from each lot and rejecting the lot if more than one defective is observed. (If rejected, each item in the lot is then tested.) If a lot contains four defectives, what is the probability that it will be accepted?

Solution Let x be the number of defectives in the sample. Then $N = 20, k = 4, (N - k) = 16$, and $n = 5$. The lot will be rejected if $x = 2, 3$, or 4. Then

$$P(\text{accept the lot}) = P(x \leq 1) = p(0) + p(1) = \frac{C_0^4 C_5^{16}}{C_5^{20}} + \frac{C_1^4 C_4^{16}}{C_5^{20}}$$

$$= \frac{\left(\frac{4!}{0!4!}\right)\left(\frac{16!}{5!11!}\right)}{\frac{20!}{5!15!}} + \frac{\left(\frac{4!}{1!3!}\right)\left(\frac{16!}{4!12!}\right)}{\frac{20!}{5!15!}}$$

$$= \frac{91}{323} + \frac{455}{969} = .2817 + .4696 = .7513$$ ■

Familiarity with discrete probability distributions and the properties of the experiments that generate them is extremely helpful. Rather than solve the same probability problem over and over again from first principles (as was done in Chapter 3), you need only recognize the type of random variable involved and then substitute into the formula for its probability distribution.

Exercises

Basic Techniques

4.39 Evaluate the following probabilities:

a $\dfrac{C_1^3 C_1^2}{C_2^5}$ **b** $\dfrac{C_2^4 C_1^3}{C_3^7}$ **c** $\dfrac{C_4^5 C_0^3}{C_4^8}$

4.40 Let x be the number of successes observed in a sample of $n = 5$ items selected from $N = 10$. Suppose that, of the $N = 10$ items, 6 are considered "successes."

a Find the probability of observing no successes.

b Find the probability of observing at least two successes.

c Find the probability of observing exactly two successes.

4.41 Let x be a hypergeometric random variable with $N = 15$, $n = 3$, and $k = 4$.

a Calculate $p(0)$, $p(1)$, $p(2)$, and $p(3)$.

b Construct the probability histogram for x.

c Use the formulas given in Section 4.4 to calculate $\mu = E(x)$ and σ^2.

d What proportion of the population of measurements fall within the interval $(\mu \pm 2\sigma)$? Within the interval $(\mu \pm 3\sigma)$? Do these results agree with those given by Tchebysheff's Theorem?

4.42 A jar contains two black balls and two white balls. Suppose the balls have been thoroughly mixed and that two are randomly selected from the jar. Let x be the number of white balls in the selection.

a Use the formulas given in this section to calculate the values of $p(x)$, $x = 0, 1, 2$. Compare with the results of Exercise 3.45.

b Using the probability distribution from part (a) and the methods of Section 3.8, calculate $\mu = E(x)$ and $\sigma^2 = E(x - \mu)^2$

c Calculate μ and σ^2 using the formulas given in this section. Compare the results of parts (b) and (c).

Applications

4.43 Refer to Exercise 3.50. Use the hypergeometric probability distribution to calculate $p(x)$ for each value of x. Compare your answers with those obtained in Exercise 3.50.

4.44 A company has five applicants for two positions: two women and three men. Suppose that the five applicants are equally qualified and that no preference is given for choosing either sex. Let x equal the number of women chosen to fill the two positions.

a Write the formula for $p(x)$, the probability distribution of x.

b What are the mean and variance of this distribution?

c Construct a probability histogram for x.

4.45 According to the census bureau, about one in four Americans lack continuous health-insurance coverage, with strong inequalities in coverage depending on ethnic, racial, and age classification ("Quarter of Americans," June 25, 1992). Suppose 10 in a group of 40 workers chosen for the study have no health-insurance coverage. If 5 workers are chosen for a new type of work, what is the probability that exactly 1 will have health-insurance coverage? At least 1?

4.46 When parents stop reading to their children, the TV takes over, according to a recent study ("When Parents Stop," November 17, 1992). Researchers have found that 52% of parents read daily to their children under the age of 8 but that only 13% of those with children aged 9–14 read to their children every day. Twenty parents of 9–14-year-olds are interviewed, four of whom read to their children daily. If 3 of the 20 parents are selected at random, construct the probability distribution for x, the number of parents who read to their 9–14-year-old children daily. Calculate the mean and standard deviation for x. Within what limits would you expect x to fall?

4.47 Refer to Exercise 4.46. If 3 of the 20 parents are interviewed, what is the probability that all three of the parents read daily to their 9–14-year-olds? If this event were to occur, what conclusions might you draw?

Tips on Problem Solving

1. If the random variable is the number of occurrences of a specified event in a given unit of *time or space* for which the average number of occurrences per unit time or space is μ, then the random variable has a Poisson distribution.
2. Suppose that a random sample of n items is selected without replacement from a population of N items in which a proportion p of the items possess a specified property and $q = 1 - p$ do not. The following flowchart is provided to help determine which distribution is appropriate for the situations presented.

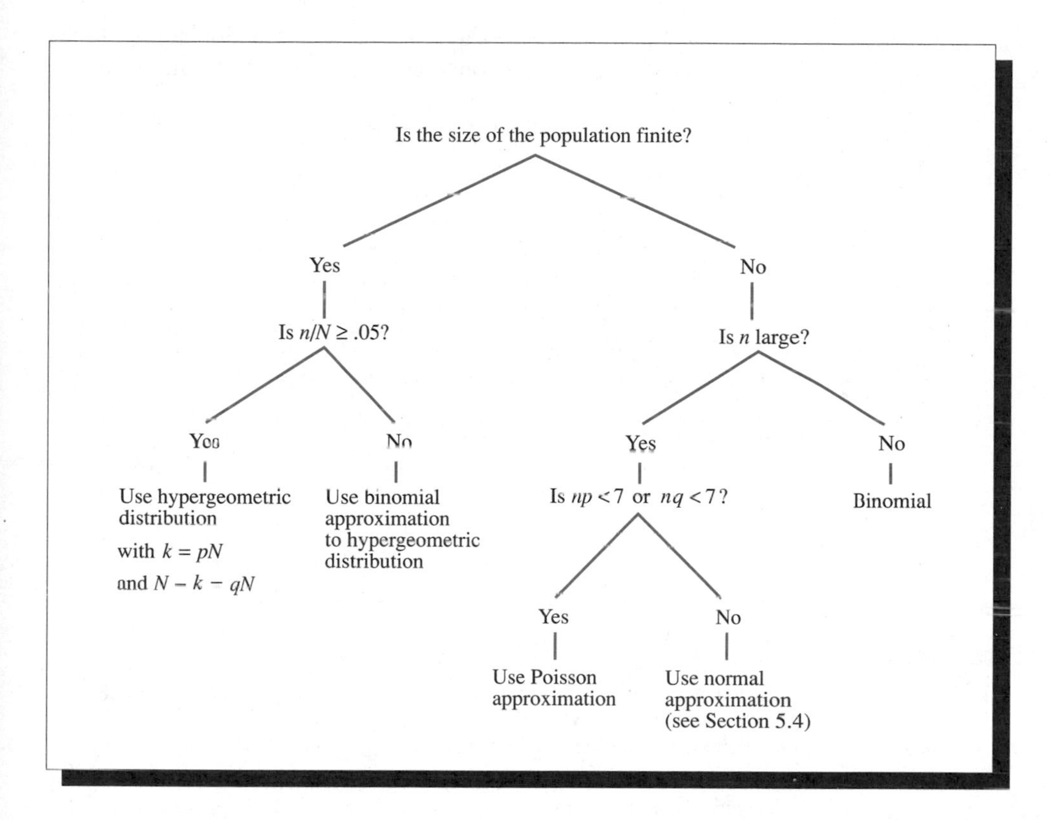

4.5 Summary

Several useful discrete probability distributions were presented in this chapter: the binomial, the Poisson, and the hypergeometric distributions. These probability distributions enabled us to calculate the probabilities associated with events that are of interest in the sciences, in business, and in marketing.

The binomial probability distribution allows us to calculate the probability of x successes in a series of n identical independent trials, where the probability of a success in a single trial is equal to p. The binomial experiment is an excellent model for many sampling situations, particularly surveys that result in "yes" or "no" types of data.

The Poisson probability distribution is important because it can be used to approximate certain binomial probabilities when n is large and p is small. Consequently, it can greatly reduce the computations involved in calculating binomial probabilities. In addition, the Poisson probability distribution is important in its own right. It provides an excellent probabilistic model for the number of occurrences of rare events in time or space.

The hypergeometric probability distribution is also related to the binomial probability distribution. It gives the probability of drawing x elements of a particular type from a population where the number N of elements in the population is small in relation to the sample size n. The binomial probability distribution applies to the same situation except that it is appropriate only when N is large in relation to n.

MINITAB Commands

```
PDF for values in E [put into E]

     BINOMIAL n = K   p = K
     POISSON  μ = k
```

```
CDF for values in E...E [put into E...E]

     BINOMIAL n = K   p = K
     POISSON  μ = K
```

Supplementary Exercises

4.48 List the five identifying characteristics of the binomial experiment.

4.49 Under what conditions can the Poisson random variable be used to approximate the probabilities associated with the binomial random variable? What application does the Poisson distribution have other than to estimate certain binomial probabilities?

4.50 Under what conditions would one use the hypergeometric probability distribution in evaluating the probability of x successes in n trials?

4.51 A balanced coin is tossed three times. Let x equal the number of heads observed.

a Use the formula for the binomial probability distribution to calculate the probabilities associated with $x = 0, 1, 2,$ and 3.

b Construct a probability distribution.

c Find the expected value and standard deviation of x, using the formulas

$$E(x) = np$$
$$\sigma = \sqrt{npq}$$

d Using the probability distribution in (b), find the fraction of the population measurements lying within one standard deviation of the mean. Repeat for two standard deviations. How do your results agree with Tchebysheff's Theorem and the Empirical Rule?

4.52 Refer to Exercise 4.51. Suppose the coin is definitely unbalanced and the probability of a head is equal to $p = .1$. Follow instructions (a), (b), (c), and (d). Note that the probability distribution loses its symmetry and becomes skewed when p is not equal to 1/2.

4.53 Suppose the four engines of a commercial aircraft are arranged to operate independently and that the probability of in-flight failure of a single engine is .01. What is the probability that, on a given flight,

a No failures are observed?

b No more than one failure is observed?

4.54 The 10-year survival rate for bladder cancer is approximately 50%. If 20 bladder cancer patients are properly treated for the disease, what is the probability that

a At least 1 will survive for 10 years?

b At least 10 will survive for 10 years?

c At least 15 will survive for 10 years?

4.55 A city commissioner claims that 80% of all people in the city favor garbage collection by contract to a private concern (in contrast to collection by city employees). To check the theory that the proportion of people in the city favoring private collection is .8, you randomly sample 25 people and find that x, the number of people who support the commissioner's claim, is 22.

a What is the probability of observing at least 22 who support the commissioner's claim if, in fact, $p = .8$?

b What is the probability that x is exactly equal to 22?

c Based on the results of part (a), what would you conclude about the claim that 80% of all people in the city favor private collection? Explain.

4.56 If a person is given the choice of an integer from 0 to 9, is it more likely that he or she will choose an integer near the middle of the sequence than one at either end?

a If the integers are equally to be chosen, find the probability distribution for x, the number chosen.

b What is the probability that a person will choose a 4, 5, or 6?

c What is the probability that a person will not choose a 4, 5, or 6?

4.57 Refer to Exercise 4.56. Twenty persons are asked to select a number from 0 to 9. Eight of them choose a 4, 5, or 6.

a If the choice of any one number is as likely as any other, what is the probability of observing eight or more choices of the interior numbers 4, 5, or 6?

b What conclusions would you draw from the results of part (a)?

4.58 "What parents don't know may hurt them—and their children," according to Alan Otten (1988). In his column "People Patterns," he reported the results of a survey of primary and secondary students and parents across the United States, in which 41% of the children say they have smoked cigarettes, whereas only 14% of the parents *think* that their children have smoked cigarettes. If a random sample of 25 parents is taken, what are the mean and standard deviation of x, the number of parents who think their children have smoked cigarettes? Within what limits would you expect x to lie?

4.59 Refer to Exercise 4.58. A similar survey is conducted, based on a group of 50 students. Of the 50, 15 say they have smoked cigarettes. If ten of these students are randomly selected to appear on a television talk show involving student drug use, what is the probability that at least three have smoked cigarettes? How many of the ten students would you expect to have smoked cigarettes?

4.60 Refer to Exercise 4.27. If the actual number of kidnappings by nonfamily members is 1.5 per 10 million children, as reported by Schmittroth (1991), then the number of kidnappings per million should have an approximate Poisson distribution with mean $\mu = .15$ per million. The records for a particular region of the country, which has a population of approximately 1 million children, are checked.

a What is the probability that there are no kidnappings?

b What is the probability that two or fewer kidnappings have occurred?

c Would you expect to see more than three cases of kidnapping?

4.61 Suppose that early statewide election returns indicate totals of 33,000 votes for candidate A versus 27,000 for candidate B and these early returns can be regarded as a random sample selected from the population of all 10,000,000 eligible voters in the state.

a If the statewide vote will be split 50–50 (i.e., the probability that A will win is .5), find the expected number of x votes for A in the sample of 60,000 early returns.

b Find the standard deviation of x.

c Is the observed value of $x = 33,000$ consistent with the theory in part (a) that the vote will split (i.e., $p = .5$), or is x a highly unlikely value?

d Do you think that the value $x = 33{,}000$ is sufficient evidence to indicate that A will win?

4.62 A psychiatrist believes that 80% of all people who visit doctors have problems of a psychosomatic nature. She decides to select 25 patients at random to test her theory.

a Assuming that the psychiatrist's theory is true, what is the expected value of x, the number of the 25 patients who have psychosomatic problems?

b What is the variance of x, assuming that the theory is true?

c Find $P(x \leq 14)$. (Use tables, and assume that the theory is true.)

d Based on the probability in part (c), if only 14 of the 25 sampled had psychosomatic problems, what conclusions would you make about the psychiatrist's theory? Explain.

4.63 A student government states that 80% of all students favor an increase in student fees to subsidize a new recreational area. A random sample of $n = 25$ students produced 15 in favor of increased fees. What is the probability that 15 or fewer in the sample would favor the issue if student government is correct? Do the data support the student government's assertion, or does it appear that the percentage favoring an increase in fees is less than 80%?

4.64 College campuses are graying! According to an Associated Press article (Bovee, 1991), one in four college students is age 30 or older. Many of these students are women updating their job skills. Assume that the 25% figure is accurate, that your college is representative of colleges at large, and that you sample $n = 200$ students, recording x, the number of students age 30 or older.

a What are the mean and standard deviation of x?

b If there are 35 students in your sample who are age 30 years or older, would you be willing to assume that the 25% figure is representative of your campus? Explain.

4.65 Most weather forecasters seem to protect themselves very well by attaching probabilities to their forecasts, such as "The probability of rain today is 40%." Then, if a particular forecast is incorrect, you are expected to attribute the error to the random behavior of the weather rather than the inaccuracy of the forecaster. To check the accuracy of a particular forecaster, records were checked only for those days when the forecaster predicted rain "with 30% probability." A check of 25 of those days indicated that it rained on 10 of the 25.

a If the forecaster is accurate, what is the appropriate value of p, the probability of rain on one of the 25 days?

b What are the mean and standard deviation of x, the number of days on which it rained, assuming that the forecaster is accurate?

c Calculate the z-score for the observed value, $x = 10$. [HINT: Recall from Section 2.6 that $z\text{-score} = (x - \mu)/\sigma$.]

d Do these data disagree with the forecast of a "32% probability of rain"? Explain.

4.66 A packaging experiment is conducted by placing two different package designs for a breakfast food side by side on a supermarket shelf. The objective of the experiment is to see whether buyers indicate a preference for one of the two package designs. On a given day, 25 customers purchased a package from the supermarket. Let x equal the number of buyers who choose the second package design.

a If there is no preference for either of the two designs, what is the value of p, the probability that a buyer chooses the second package design?

b If there is no preference, use the results of part (a) to calculate the mean and standard deviation of x.

c If 5 of the 25 customers choose the first package design and 20 choose the second design, what would you conclude about the customer's preference for the second package design?

4.67 Although employers are becoming increasingly concerned about the fitness of their employees, few employers encourage their workers by providing personal exercise or fitness programs. In a survey of 200 companies in the high-tech Silicon Valley of California, only one-third of the companies said they offered preventive health-care programs such as weight control or smoking cessation (*Wall Street Journal*, Dec. 6, 1988). If we assume that 133 of the 200 companies offer such programs, and we randomly select 5 of these companies for a follow-up investigation,

a What is the probability that none of the 5 offer preventive health-care programs?

b What is the probability that at least two offer preventive health-care programs?

4.68 Refer to Exercise 4.67. Let x be the number of companies offering preventive health-care programs.

a Calculate the mean and standard deviation of x.

b Within what limits would you expect x to lie?

4.69 The safety record of the Ford Bronco II was called into question when *Consumer Reports* magazine quoted federal statistics concerning Bronco II's fatal accident rate. According to an article in the *Press-Enterprise* (Riverside, Calif., May 18, 1989), there were 19 fatal rollovers for every 100,000 Bronco IIs in 1987. Suppose that, in an attempt to prove or disprove this claim, a consumer activist locates the records of 1000 randomly chosen Bronco II owners during that year. Of the 1000 cars, 1 had been involved in a fatal rollover.

a What is the distribution of x, the number of fatal rollovers in a sample of 1000 Bronco IIs?

b If the *Consumer Reports* statistics are correct, what is the probability of having one or more fatal rollovers in a sample of 1000 Bronco IIs?

c What are the mean and standard deviation of x?

d Is the observed value of $x = 1$ consistent with the *Consumer Reports* statistics on fatal rollovers? Explain.

4.70 Refer to Exercise 4.69. Use the Poisson approximation to the binomial to approximate the probability found in part (b). Compare the results.

4.71 A manufacturer of videotapes ships them in lots of 1200 tapes per lot. Before shipment, 20 tapes are randomly selected from each lot and tested. If none are defective, the lot is shipped. If 1 or more are defective, every tape in the lot is tested.

a What is the probability distribution for x, the number of defective tapes in the sample of 20 tapes?

b What distribution can be used to approximate probabilities for the random variable x in part (a)?

c What is the probability that a lot will be shipped if it contains 10 defectives? 20 defectives? 30 defectives?

4.72 The *Orlando Sentinel*, (Orlando, Fla., Jan. 29, 1985) reports on a study conducted by the Massachusetts Department of Health on the death rate from cancer for Vietnam veterans. The researchers examined the cause of death for 804 Vietnam veterans and found that 9 had died of tumors in muscle or other soft tissue. The expected number of deaths in an equal-size group of non-Vietnam veterans is 1.9.

a What is the probability distribution for x, the number of veterans in the sample of 804 who died of tumors?

b What distribution would provide a good approximation to the number x of veterans who died of tumors if the mean number was 1.9 in a sample of 804? Explain.

c If you observed $x = 9$ deaths in the sample of 804, would you believe that the observation was selected from a binomial population with mean equal to 1.9? Explain.

4.73 During the 1992 football season, the Los Angeles Rams had a bizarre streak of coin-toss losses. In fact, they lost the call 11 weeks in a row ("Call It," 1992).

a The Rams' computer system manager said that the odds against losing 11 straight tosses are 2047 to 1. Is he correct?

b After this newspaper article was published, the Rams lost the call for the next two games, for a total of 13 straight losses. What is the probability of this happening if, in fact, the coin was fair?

CASE STUDY

A Mystery: Cancers Near a Reactor

How safe is it to live near a nuclear reactor? Men who lived in a coastal strip that extends 20 miles north from a nuclear reactor in Plymouth, Massachusetts, developed some forms of cancer at a rate 50% higher than the statewide rate, according to a study endorsed by the Massachusetts Department of Public Health and reported in the May 21, 1987, edition of the *New York Times*.

The cause of the excess cancers is a mystery, but it was suggested that the cause was linked to the Pilgrim I reactor, which had been shut down for 13 months because of management problems. Boston Edison, the owner of the reactor, acknowledged radiation releases in the mid-1970s that were just above permissible levels. If the reactor was in fact responsible for the excess cancers, the currently acknowledged level of radiation required to cause cancer would have to change. However, confounding the mystery was the fact that women in this same area were seemingly unaffected.

In his report, Dr. Sidney Cobb, an epidemiologist, noted the connection between the radiation releases at the Pilgrim I reactor and 52 cases of hematopoietic cancers. The report indicated that this unexpectedly large number might be attributable to airborne radioactive effluents from Pilgrim I, concentrated along the coast by wind patterns and not dissipated, as assumed by government regulators. How unusual was this number of cancer cases? That is, statistically speaking, is 52 a highly improbable number of cases? If the answer is yes, then either some external factor (possibly radiation) caused this unusually large number, or we have observed a very rare event!

The Poisson probability distribution provides a good approximation to the distributions of variables such as the number of deaths in a region due to a rare disease, the number of accidents in a manufacturing plant per month, or the number of airline crashes per month. Therefore, it is reasonable to assume that the Poisson distribution will provide an appropriate model for the number of cancer cases in this instance.

1 If the 52 reported cases represented a rate 50% higher than the statewide rate, what is a reasonable estimate of μ, the average number of such cancer cases statewide?

2 Based on your estimate of μ, what is the estimated standard deviation of the number of cancer cases statewide?

3 What is the z-score for the $x = 52$ observed cases of cancer? How would you interpret this z-score in light of the concern about an elevated rate of hematopoietic cancers in this area?

5

The Normal Probability Distribution

Case Study

If you were the boss, would height play a role in your selection of a successor for your job? Would you purposely choose a successor who was shorter than you? The case study at the end of this chapter examines how the normal curve can be used in a situation involving the height distribution of Chinese males eligible for a very prestigious job.

General Objectives

In Chapters 3 and 4, you learned about discrete random variables and their probability distributions. In this chapter, you will learn about continuous random variables and their probability distributions and about one very important continuous random variable—the normal. You will learn how to calculate normal probabilities and, under certain conditions, how to use the normal probability distribution to approximate the binomial probability distribution. Then, in Chapter 6 and in the chapters that follow, you will see how the normal probability distribution plays a central role in statistical inference.

Specific Topics

1 Probability distributions for continuous random variables (5.1)

2 The normal probability distribution (5.2)

3 Calculation of areas associated with the normal probability distribution (5.3)

4 The normal approximation to the binomial probability distribution (5.4)

5.1 Probability Distributions for Continuous Random Variables

When a random variable x is discrete, we can assign a positive probability to each value that x can take and obtain the probability distribution for x. The sum of all of the probabilities associated with the different values of x is 1. However, not all experiments result in random variables that are discrete. **Continuous random variables**, such as heights and weights, length of life of a particular product, or experimental laboratory error, can assume the infinitely many values corresponding to points on a line interval. If we try to assign a positive probability to each of these uncountable values, the probabilities will no longer sum to 1, as with discrete random variables. Therefore, we must use a different approach to generate the probability distribution for a continuous random variable.

Suppose we have a set of measurements, such as the grade point averages in Chapter 2, for which we have created a relative frequency histogram. For a small number of measurements, we use a small number of classes; but, if more and more measurements are collected, we can use more classes and reduce the class width. The outline of the histogram would change slightly, for the most part becoming less and less irregular, as shown in Figure 5.1. As the number of measurements becomes very large and the class widths become very narrow, the relative frequency histogram appears more and more like the smooth curve shown in Figure 5.1(d).

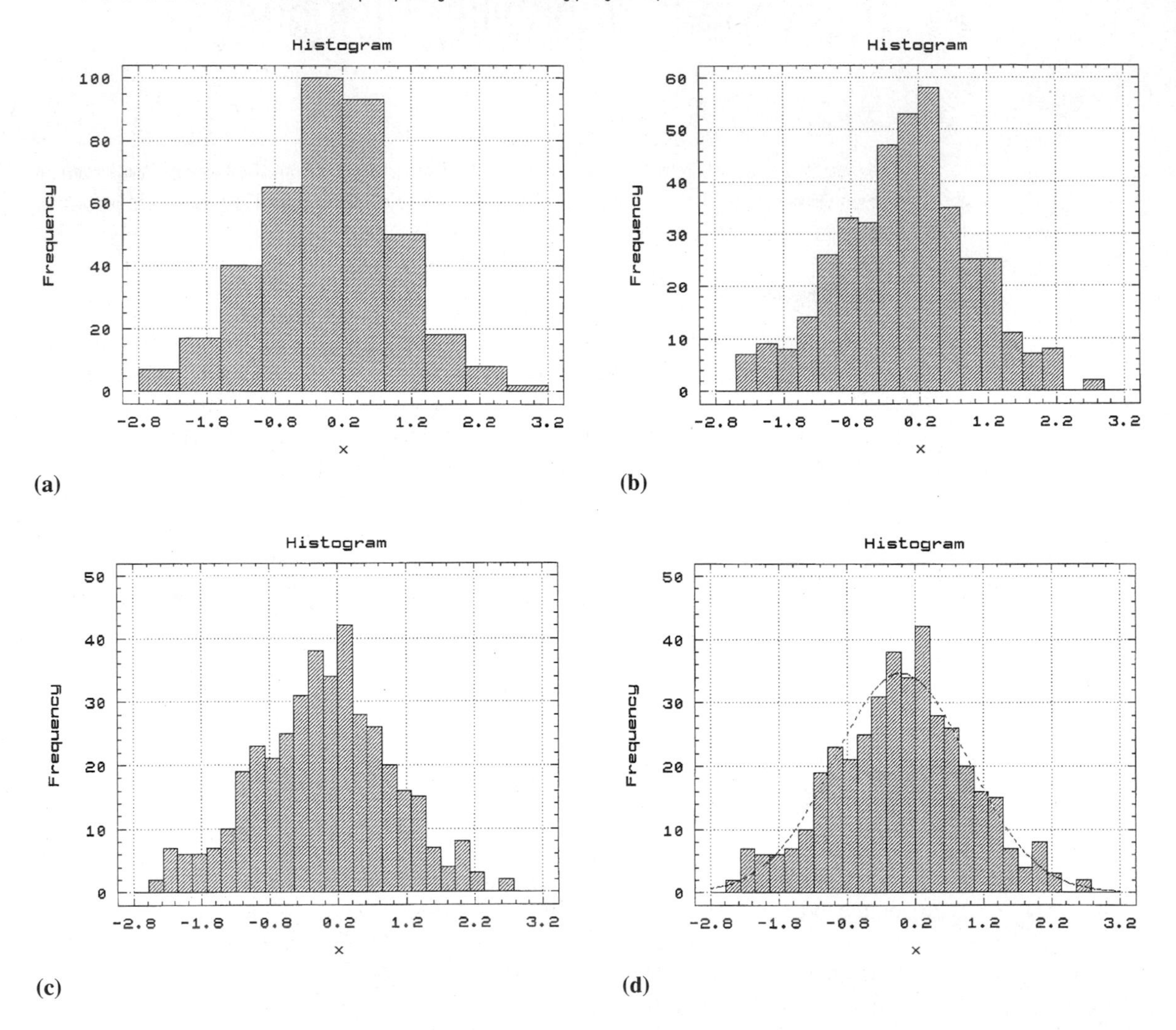

FIGURE 5.1 Relative frequency histograms for increasingly larger sample sizes

The relative frequency associated with a particular class in the population is the fraction of measurements in the population falling in that interval and also is the probability of drawing a measurement from that class. If the total area under the relative frequency histogram is adjusted to equal 1, areas under the frequency curve correspond to probabilities. In fact, this correspondence was the basis for the application of the Empirical Rule in Chapter 2, which applies when the data are approximately bell-shaped.

Let us construct a model for the probability distribution for a continuous random variable. Assume that the random variable x may take on any value on a real line, as in Figure 5.1(d). We then distribute 1 unit of probability along the line, much as a person might distribute a handful of sand, each measurement in the population corresponding to a single grain. The probability—grains of sand or measurements—will pile up in certain places, and the result will be the probability distribution shown if Figure 5.2. The depth or density of the probability, which varies with x, may be represented by a mathematical formula $f(x)$, called the *probability distribution*, or the *probability density function*, for the random variable x. The density function $f(x)$, represented graphically in Figure 5.2, provides a mathematical model for the population relative frequency histogram that exists in reality. *Since the total area under the curve* f(x) *is equal to 1, the area lying above a given interval equals the probability that* x *will fall in that interval.* Thus, the probability that $x_1 < x < x_2$ (x is greater than x_1 and x is less than x_2) is equal to the area under the density function between the two points x_1 and x_2. This is the shaded area in Figure 5.2.

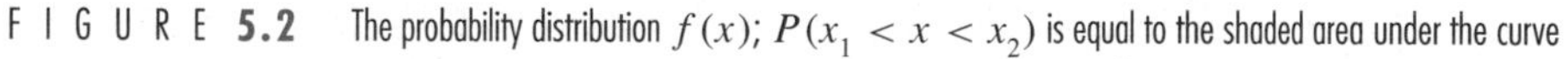
FIGURE 5.2 The probability distribution $f(x)$; $P(x_1 < x < x_2)$ is equal to the shaded area under the curve

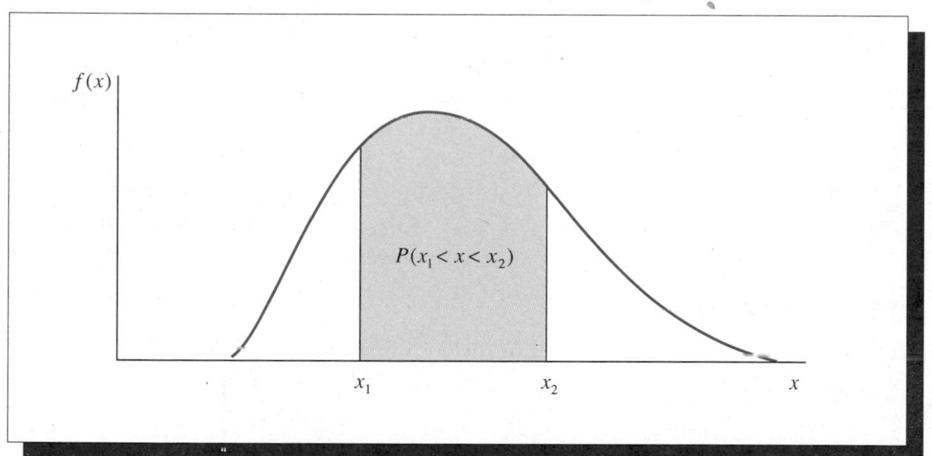

How do we choose the model—that is, the probability distribution $f(x)$— appropriate for a given experiment? Many types of continuous curves are available for modeling. Some are mound-shaped, like the one in Figure 5.1(d), but others are not. In general, we try to pick a model that

- Fits the accumulated body of data.
- Allows us to make the best possible inferences using the data.

Our model may not always fit the experimental situtation perfectly, but we try to choose a model that *best fits* the population relative frequency histogram. The better our model approximates reality, the better our inferences will be. Fortunately, we will find that many continuous random variables have mound-shaped frequency distributions, such as the data in Figure 5.1(d). A probability model that provides a good approximation to such a distribution is the **normal probability distribution**, the subject of Section 5.2.

5.2 The Normal Probability Distribution

In Section 5.1, we saw that the probabilistic model for the frequency distribution of a continuous random variable involves the selection of a curve, usually smooth, called the **probability distribution**. Although these distributions may assume a variety of shapes, a large number of random variables observed in nature possess a frequency distribution that is approximately bell-shaped or, as the statistician would say, is approximately a normal probability distribution. The formula that generates this distribution is shown below.

Normal Probability Distribution

$$f(x) = \frac{1}{\sigma\sqrt{2\pi}} e^{-(x-\mu)^2/2\sigma^2} \qquad -\infty \leq x \leq \infty$$

The symbols e and π are mathematical constants given approximately by 2.7183 and 3.1416, respectively; μ and σ $(\sigma > 0)$ are parameters representing the population mean and standard deviation.

The graph of a normal probability distribution with mean μ and standard deviation σ is shown in Figure 5.3. The mean μ locates the *center* of the distribution, and the

FIGURE 5.3 Normal probability distribution

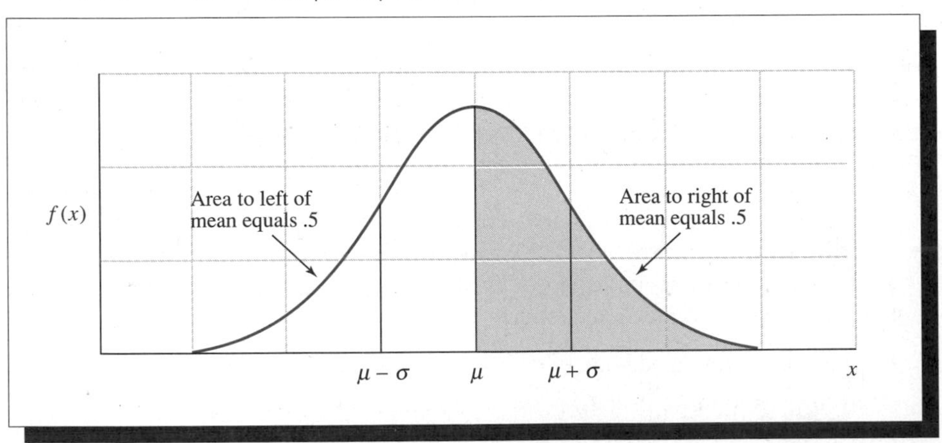

distribution is *symmetric* about its mean μ. Since the total area under the normal probability distribution is equal to 1, this implies that the area to the right of μ is .5

and the area to the left of μ is also .5. The *shape* of the distribution is determined by σ, the population standard deviation. Large values of σ reduce the height of the curve and increase the spread; small values of σ increase the height of the curve and reduce the spread. Figure 5.4 is an EXECUSTAT printout showing three normal probability distributions with different means and standard deviations. Notice the differences in shape and location.

FIGURE 5.4 EXECUSTAT printout showing normal probability distributions with differing values of μ and σ.

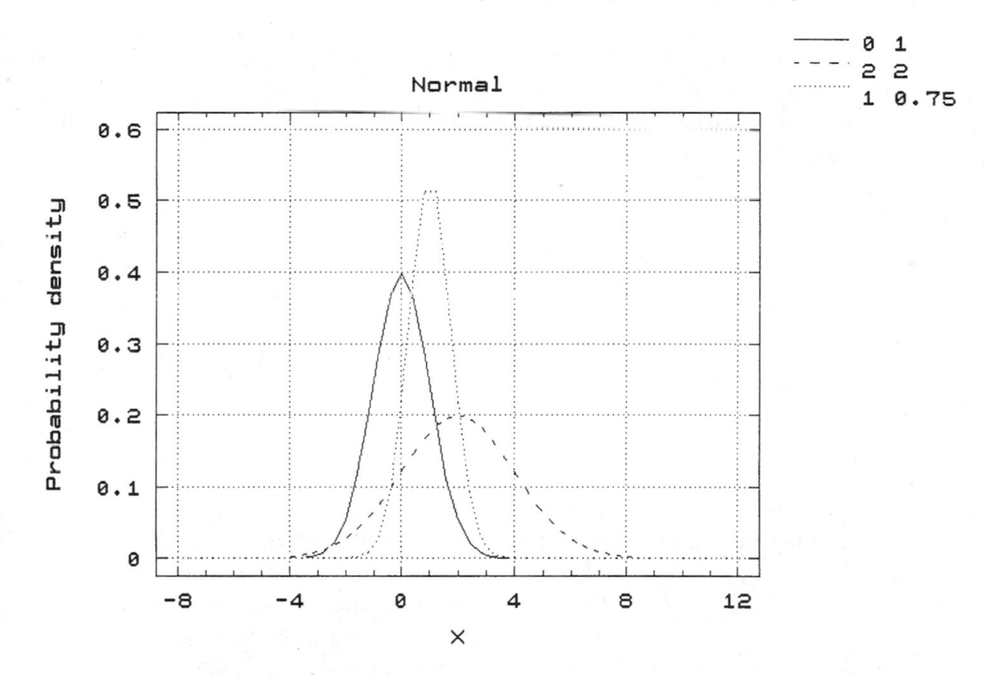

In practice, we seldom encounter variables that range from infinitely small negative values to infinitely large positive values. Nevertheless, many positive random variables (such as heights, weights, and times) generate a frequency histogram that is well approximated by a normal distribution. The approximation applies because almost all of the values of a normal random variable lie within three standard deviations of the mean, and in these cases $(\mu \pm 3\sigma)$ almost always encompasses positive values.

5.3 Tabulated Areas of the Normal Probability Distribution

In Section 5.1, we explained that the probability that a continuous random variable assumes a value in the interval x_1 to x_2 is the area under the probability density function between the points x_1 and x_2 (see Figure 5.2). The probability model for a continuous random variable differs greatly from the model for a discrete random variable when we consider the probability that x equals some particular value, say a. Since **the area**

lying over any particular point—say, $x = a$—is 0, it follows from our probability model that the probability that $x = a$ is 0. Thus, the expression $P(x \leq a)$ is the same as $P(x < a)$ because $P(x = a) = 0$. Similarly, $P(x \geq a) = P(x > a)$. This statement is, of course, not true for a discrete random variable because $P(x = a)$ may not equal 0.

Since normal curves have different means and standard deviations (see Figure 5.4), we could generate an infinitely large number of normal distributions. A separate table of areas for each of these curves is obviously impractical. Instead, we would like to devise a standardization procedure that will allow us to use the same normal curve areas for all normal distributions.

Standardization is most easily accomplished by expressing the value of a normal random variable as the number of standard deviations to the left or right of the mean. In other words, the value of a normal random variable x with mean μ and standard deviation σ can be expressed as

$$z = \frac{x - \mu}{\sigma}$$

or, equivalently,

$$x = \mu + z\sigma$$

- When z is negative, x lies to the left of the mean μ.
- When $z = 0$, $x = \mu$.
- When z is positive, x lies to the right of the mean μ.

We will learn to calculate probabilities for x using $z = \frac{x-\mu}{\sigma}$, which is called the **standard normal random variable**. The probability distribution for z is called the **standardized normal distribution** because its mean is 0 and its standard deviation is 1. It is shown in Figure 5.5. The area under the standard normal curve between the mean $z = 0$ and a specified value of z—say, z_0—is the probability $P(0 \leq z \leq z_0)$. This area is recorded in Table 3 of Appendix I and is shown as the shaded area in Figure 5.5. An abbreviated version of Table 3 in Appendix I is shown here in Table 5.1.

How can we find areas to the left of the mean? Since the standard normal curve is symmetric about $z = 0$ (see Figure 5.5), any area to the left of the mean can be found by using the equivalent area to the right of the mean.

Note that z, correct to the nearest tenth, is recorded in the left-hand column of the table. The second decimal place for z, corresponding to hundredths, is given across the top row. Thus, the area between the mean and $z = .7$ standard deviation to the right, read in the second column of the table opposite $z = .7$, is found to be .2580. Similarly, the area between the mean and $z = 1.0$ is .3413. The area between $z = -1.0$ and the mean is also .3413. Thus, the area lying within one standard deviation on either side of the mean would be two times .3413, or .6826. The area lying within two standard deviations of the mean, correct to four decimal places, is $2(.4772) = .9544$. These numbers agree with the approximate values, 68% and 95%, used in the Empirical Rule in Chapter 2.

To find the area between the mean and a point $z = .57$ standard deviation to the right of the mean, proceed down the left-hand column to the 0.5 row. Then move

FIGURE 5.5 Standardized normal distribution

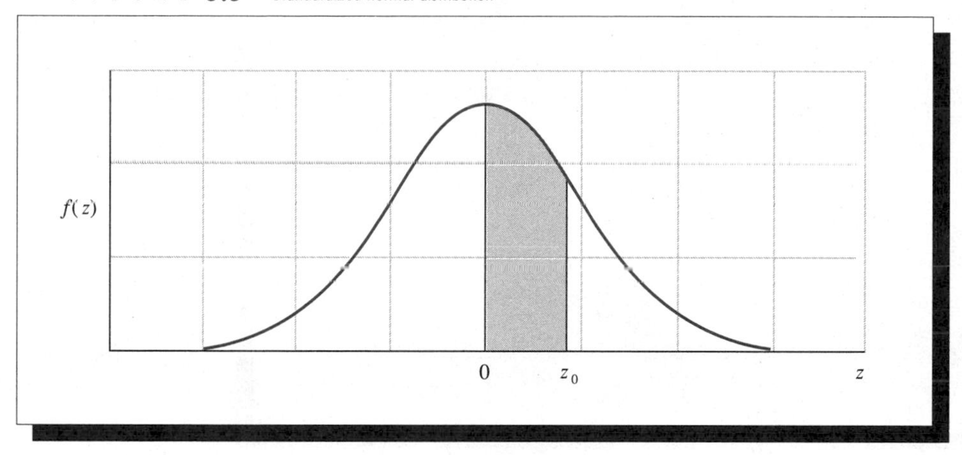

TABLE 5.1 Abbreviated version of Table 3 in Appendix I

z_0	.00	.01	.02	.03	.04	.05	.06	.07	.08	.09
0.0	.0000	.0040	.0080	.0120	.0160	.0199	.0239	.0279	.0319	.0359
0.1	.0398	.0438	.0478	.0517	.0557	.0596	.0636	.0675	.0714	.0753
0.2	.0793	.0832	.0871	.0910	.0948	.0987	.1026	.1064	.1103	.1141
0.3	.1179	.1217	.1255	.1293	.1331	.1368	.1406	.1443	.1480	.1517
0.4	.1554	.1591	.1628	.1664	.1700	.1736	.1772	.1808	.1844	.1879
0.5	.1915	.1950	.1985	.2019	.2054	.2088	.2123	**.2157**	.2190	.2224
0.6	.2257	⋮	⋮	⋮	⋮	⋮	⋮	⋮	⋮	⋮
0.7	**.2580**									
⋮	⋮									
1.0	**.3413**									
⋮	⋮									
2.0	**.4772**									

across the top row of the table to the .07 column. The intersection of this row–column combination gives the appropriate area, .2157.

Since the normal distribution is continuous, the area under the curve associated with a single point is equal to 0. Keep in mind that this result applies only to continuous random variables. Later in this chapter, we will use the normal probability distribution to approximate the binomial probability distribution. The binomial random variable x is a discrete random variable. Hence, as you know, the probability that x takes some specific value, say $x = 10$, will not necessarily equal 0.

EXAMPLE 5.1 Find $P(0 \leq z \leq 1.63)$. This probability corresponds to the area between the mean ($z = 0$) and a point $z = 1.63$ standard deviations to the right of the mean (see Figure 5.6).

FIGURE 5.6 Probability required for Example 5.1

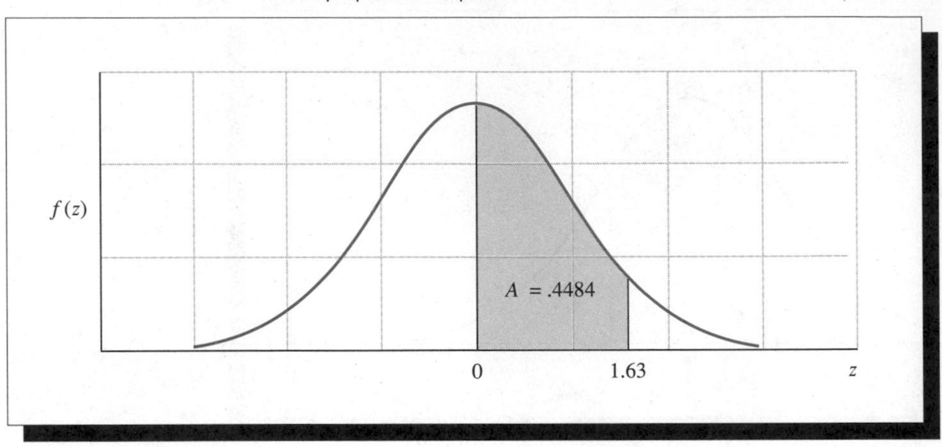

Solution The area is shaded and indicated by the symbol A in Figure 5.6. Since Table 3 in Appendix I gives areas under the normal curve to the right of the mean, we need only find the tabulated value corresponding to $z = 1.63$. Proceed down the left-hand column of the table to $z = 1.6$ and across the top of the table to the column marked .03. The intersection of this row and column combination gives the area $A = .4484$. Therefore, $P(0 \leq z \leq 1.63) = .4484$. ▪

EXAMPLE 5.2 Find $P(-.5 \leq z \leq 1.0)$. This probability corresponds to the area between $z = -.5$ and $z = 1.0$, as shown in Figure 5.7.

FIGURE 5.7 Area under the normal curve in Example 5.2

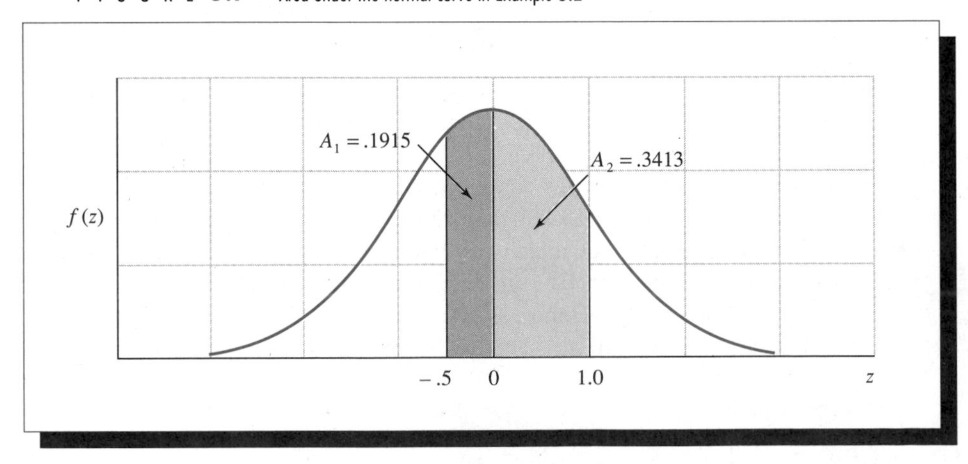

Solution The area required is equal to the sum of A_1 and A_2 shown in Figure 5.7. From Table 3 in Appendix I, we read $A_2 = .3413$. The area A_1 equals the area between $z = 0$ and $z = .5$, or $A_1 = .1915$. Thus, the total area is

$$A = A_1 + A_2 = .1915 + .3413 = .5328$$

That is, $P(-.5 \leq z \leq 1.0) = .5328$. ▪

EXAMPLE 5.3 Find the value of z, say z_0, such that (to four decimal places) .95 of the area is within $\pm z_0$ standard deviations of the mean.

Solution Half of the area, $(1/2)(.95) = .475$, will lie to the left of the mean and half to the right, because the normal distribution is symmetrical. Thus, we seek the value z_0 corresponding to an area equal to .475. The area .475 falls in the row corresponding to $z = 1.9$ and the .06 column. Hence, $z_0 = 1.96$. Note that this result is very close to the approximate value, $z = 2$, used in the Empirical Rule. ▪

EXAMPLE 5.4 Let x be a normally distributed random variable, with a mean of 10 and a standard deviation of 2. Find the probability that x lies between 11 and 13.6.

Solution As a first step, we must calculate the values of z corresponding to $x_1 = 11$ and $x_2 = 13.6$. Thus, we have

$$z_1 = \frac{x_1 - \mu}{\sigma} = \frac{11 - 10}{2} = .5 \qquad z_2 = \frac{x_2 - \mu}{\sigma} = \frac{13.6 - 10}{2} = 1.8$$

The desired probability is therefore $P(.5 \leq z \leq 1.8)$ and is the area lying between z_1 and z_2, as shown in Figure 5.8. The area between $z = 0$ and z_1 is $A_1 = .1915$, and the area between $z = 0$ and z_2 is $A_2 = .4641$; these areas are obtained from Table 3. The desired probability is equal to the difference between A_2 and A_1; that is,

$$P(.5 \leq z \leq 1.8) = A_2 - A_1 = .4641 - .1915 = .2726$$ ▪

FIGURE 5.8 Area under the normal curve in Example 5.4

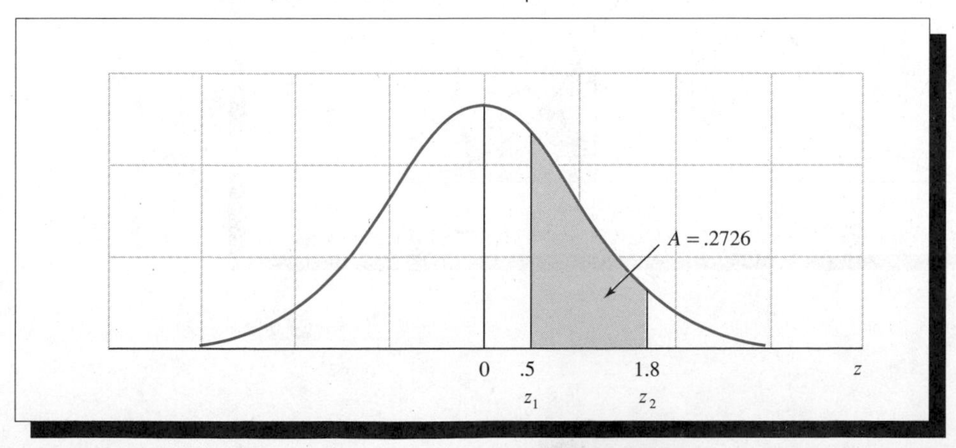

EXAMPLE 5.5 Studies show that gasoline use for compact cars sold in the United States is normally distributed, with a mean use of 25.5 miles per gallon (mpg) and a standard deviation of 4.5 mpg. What percentage of compacts obtain 30 or more mpg?

Solution The proportion of compacts obtaining 30 or more mpg is given by the shaded area in Figure 5.9.

We must find the z value corresponding to $x = 30$. Substituting into the formula for z, we obtain

$$z = \frac{x - \mu}{\sigma} = \frac{30 - 25.5}{4.5} = 1.0$$

The area A to the right of the mean, corresponding to $z = 1.0$, is .3413 (from Table 3). Then the proportion of compacts having an mpg ratio equal to or greater than 30 is equal to the entire area to the right of the mean, .5, minus the area A:

$$P(x \geq 30) = .5 - P(0 \leq z \leq 1) = .5 - .3413 = .1587$$

The percentage exceeding 30 mpg is

$$100(.1587) = 15.87\%$$ ∎

EXAMPLE 5.6 Refer to Example 5.5. In times of scarce energy resources, a competitive advantage is given to an automobile manufacturer who can produce a car obtaining substantially better fuel economy than the competitors' cars. If a manufacturer wishes to develop a compact car that outperforms 95% of the current compacts in fuel economy, what must the gasoline use rate for the new car be?

FIGURE 5.9 Area under the normal curve in Example 5.5

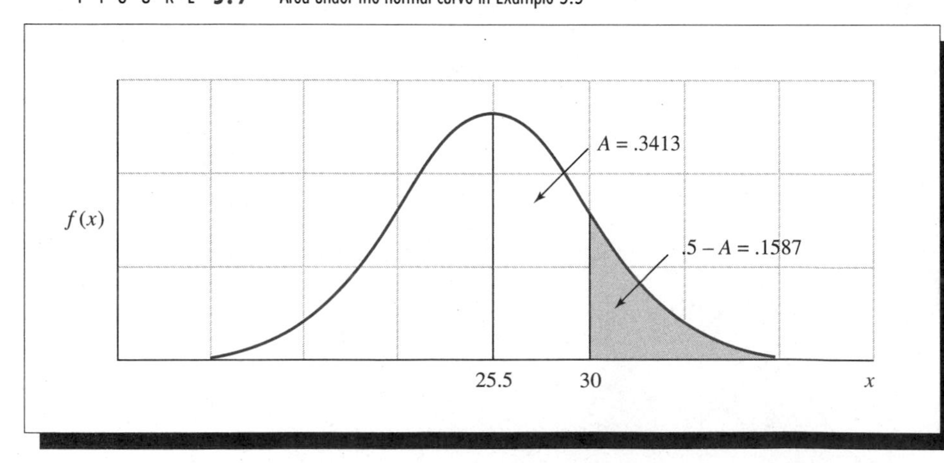

Solution Let x be a normally distributed random variable, with a mean of 25.5 and a standard deviation of 4.5. We want to find the value x_0 such that

$$P(x \leq x_0) = .95$$

As a first step, we find

$$z_0 = \frac{x_0 - \mu}{\sigma} = \frac{x_0 - 25.5}{4.5}$$

and note that our required probability is the same as the area to the left of z_0 for the standardized normal distribution. Therefore,

$$P(z \leq z_0) = .95$$

The area to the left of the mean is .5. The area to the right of the mean between z_0 and the mean is $.95 - .5 = .45$. Thus, from Table 3 we find that z_0 is between 1.64 and 1.65. Notice that the area .45 is exactly halfway between the areas for $z = 1.64$ and $z = 1.65$. Thus, z_0 is exactly halfway between 1.64 and 1.65; that is, $z_0 = 1.645$.

Substituting $z_0 = 1.645$ into the equation for z_0, we have

$$1.645 = \frac{x_0 - 25.5}{4.5}$$

Solving for x_0, we obtain

$$x_0 = (1.645)(4.5) + 25.5 = 32.9$$

The manufacturer's new compact car must therefore obtain a fuel economy of 32.9 mpg to outperform 95% of the compact cars currently available on the U.S. market. ■

Exercises

Basic Techniques

5.1 Using Table 3 in Appendix I, calculate the area under the normal curve between

a $z = 0$ and $z = 1.6$. **b** $z = 0$ and $z = 1.83$.

c $z = 0$ and $z = .90$. **d** $z = 0$ and $z = -.90$.

5.2 Repeat Exercise 5.1 for

a $z = -1.4$ and $z = 1.4$. **b** $z = -2.0$ and $z = 2.0$. **c** $z = -3.0$ and $z = 3.0$.

5.3 Repeat Exercise 5.1 for

a $z = -1.43$ and $z = .68$. **b** $z = .58$ and $z = 1.74$.

c $z = -1.55$ and $z = -.44$.

5.4 **a** Find a z_0 such that $P(z > z_0) = .025$. **b** Find a z_0 such that $P(z < z_0) = .9251$.

5.5 **a** Find a z_0 such that $P(z > z_0) = .9750$. **b** Find a z_0 such that $P(z > z_0) = .3594$.

5.6 Find a z_0 such that $P(-z_0 < z < z_0) = .8262$.

5.7 **a** Find a z_0 such that $P(z < z_0) = .9505$.
b Find a z_0 such that $P(z < z_0) = .05$.

5.8 **a** Find a z_0 such that $P(-z_0 < z < z_0) = .90$.
b Find a z_0 such that $P(-z_0 < z < z_0) = .99$.

5.9 A variable x is normally distributed with mean $\mu = 10$ and standard deviation $\sigma = 2$. Find the probability that
a $x > 13.5$. **b** $x < 8.2$. **c** $9.4 < x < 10.6$.

5.10 A variable x is normally distributed with mean $\mu = 1.20$ and standard deviation $\sigma = .15$. Find the probability that
a $1.00 < x < 1.10$. **b** $x > 1.38$. **c** $1.35 < x < 1.50$.

5.11 A variable is normally distributed with unknown mean μ and standard deviation $\sigma = 2$. If the probability that x exceeds 7.5 is .8023, find μ.

5.12 A variable is normally distributed with unknown mean μ and standard deviation $\sigma = 1.8$. If the probability that x exceeds 14.4 is .3015, find μ.

5.13 A variable is normally distributed with unknown mean and standard deviation. It is known that the probability that x exceeds 4 is .9772 and the probability that x exceeds 5 is .9332. Find μ and σ.

Applications

5.14 Studies indicate that drinking water supplied by some old lead-lined city piping systems may contain harmful levels of lead. Based on data presented by Karalekas, Ryan, and Taylor (1983), it appears that the distribution of lead content readings for individual water specimens has a mean and standard deviation equal (approximately) to .033 mg/l and .10 mg/l, respectively. Explain why you believe that this distribution is or is not normally distributed.

5.15 For a car traveling 30 miles per hour (mph), the distance required to brake to a stop is normally distributed with mean of 50 feet and a standard deviation of 8 feet. Suppose you are traveling 30 mph in a residential area and a car moves abruptly into your path at a distance of 60 feet.

a If you apply your brakes, what is the probability that you will brake to a stop within 40 feet or less? Within 50 feet or less?

b If the only way to avoid a collision is to brake to a stop, what is the probability that you will avoid the collision?

5.16 Suppose you must establish regulations concerning the maximum number of people who can occupy an elevator. A study of elevator occupancies indicates that, if eight people occupy the elevator, the probability distribution of the total weight of the eight people has a mean equal to 1200 pounds and a variance equal to 9800 $(\text{pounds})^2$. What is the probability that the total weight of eight people exceeds 1300 pounds? 1500 pounds? (Assume that the probability distribution is approximately normal.)

5.17 The discharge of suspended solids from a phosphate mine is normally distributed, with a mean daily discharge of 27 mg/l and a standard deviation of 14 mg/l. What proportion of days will the daily discharge exceed 50 mg/l?

5.18 Philatelists (stamp collectors) often buy stamps at or near retail prices, but, when they sell, the price is considerably lower. For example, it may be reasonable to assume that (depending on the mix of a collection, condition, demand, economic conditions, etc.) a collection will sell at x% of retail price, where x is normally distributed with a mean equal to 45% and a standard deviation of 4.5%. If a philatelist has a collection to sell that has a retail value of $30,000, what is the probability that the philatelist receives

a More than $15,000 for the collection?

b Less than $15,000 for the collection?

c Less than $12,000 for the collection?

5.19 In three western states, contaminated, undercooked hamburgers have been blamed in an outbreak of illness caused by *E. coli* bacteria (Gellene, 1993). A quick survey of fast-food restaurants reveals no universal agreement on which internal cooking temperature for cooked burgers is best. In fact, the minimum internal cooking temperatures ranged from a low of 141°F at McDonald's to a high of 165°F at Wendy's. Suppose this range represents a normal distribution, with a mean of 153°F and a standard deviation of 5°F.

a It is known that an internal temperature of 155°F destroys virtually all *E. coli* bacteria. What proportion of restaurants use cooking temperatures of 155°F or more?

b If a cooking temperature of 140°F or less destroys only 90% of the *E. coli* present, what proportion of restaurants use a cooking temperature of 140°F or less?

c Ninety-five percent of all restaurants use a minimum internal cooking temperature greater than what temperature?

5.20 The number of times x an adult human breathes per minute when at rest depends on the age of the human and varies greatly from person to person. Suppose the probability distribution for x is approximately normal, with the mean equal to 16 and the standard deviation equal to 4. If a person is selected at random and the number x of breaths per minute while at rest is recorded, what is the probability that x will exceed 22?

5.21 One method of arriving at economic forecasts is to use a consensus approach. A forecast is obtained from each of a large number of analysts, and the average of these individual forecasts is the consensus forecast. Suppose the individual 1995 January prime interest rate forecasts of all economic analysts are approximately normally distributed, with the mean equal to 10% and with the standard deviation equal to 1.3%. If a single analyst is randomly selected from among this group, what is the probability that the analyst's forecast of the prime interest rate will

a Exceed 13%?

b Be less than 9%?

5.22 The scores on a national achievement test were approximately normally distributed, with a mean of 540 and a standard deviation of 110.

a If you achieved a score of 680, how far, in standard deviations, did your score depart from the mean?

b What percentage of those who took the examination scored higher than you?

5.23 How does the IRS decide on the percentage of income tax returns to audit for each state? Suppose they do it by randomly selecting 50 values from a normal distribution with a mean equal to 1.55% and a standard deviation equal to .45%. (Computer programs are available for this type of sampling.)

a What is the probability that a particular state will have more than 2.5% of its income tax returns audited?

b What is the probability that a state will have less than 1% of its income tax returns audited?

5.24 Suppose the counts on the number of a particular type of bacteria in 1 ml of drinking water tend to be approximately normally distributed, with a mean of 85 and a standard deviation of 9. What is the probability that a given 1-ml sample will contain more than 100 bacteria?

5.25 A grain loader can be set to discharge grain in amounts that are normally distributed, with mean μ bushels and a standard deviation equal to 25.7 bushels. If a company wishes to use the loader to fill containers that hold 2000 bushels of grain and wants to overfill only one container in 100, at what value of μ should the company set the loader?

5.26 A publisher has discovered that the number of words contained in a new manuscript is normally distributed, with a mean equal to 20,000 words in excess of that specified in the author's contract and a standard deviation of 10,000 words. If the publisher wants to be almost certain (say, with a probability of .95) that the manuscript will be less than 100,000 words, what number of words should the publisher specify in the contract?

5.27 According to an Associated Press article that appeared in the *Press-Enterprise* ("Teachers Earn," 1992), average teachers' salaries reached a record high of \$34,213 in 1991–1992. Suppose the teacher salaries are normally distributed, with standard deviation of \$4000.

a What proportion of the teacher salaries are less than \$30,000?

b What proportion of the teacher salaries are between \$25,000 and \$30,000?

c What is the 90th percentile of this distribution?

5.28 A stringer of tennis rackets has found that the actual string tension achieved for any individual racket stringing will vary as much as 6 pounds per square inch from the desired tension set on the stringing machine. If the stringer wishes to string at a tension lower than that specified by a customer only 5% of the time, how much above or below the customer's specified tension should the stringer set the stringing machine? (NOTE: Assume that the distribution of string tensions produced by the stringing machine is normally distributed, with a mean equal to the tension set on the machine and a standard deviation equal to 2 pounds per square inch.)

5.29 Compared with their stay-at-home peers, employed women have higher levels of high-density lipoproteins (HDL), the "good" cholesterol associated with lower heart attack risk. A study of cholesterol levels in 2000 women, aged 25–64, living in Augsburg, Germany, was conducted by Ursula Haertel, Ulrich Keil, and colleagues (1989) at the GSF-Medis Institut in Munich. Of these 2000 women, the 48% who worked outside the home had HDL levels that were 2.5–3.6 milligrams per deciliter (mg/dl) higher than the HDL levels of their stay-at-home counterparts. If the difference in HDL levels is normally distributed, with a mean of 0 (indicating no difference between the two groups of women) and a standard deviation of 1.2 mg/dl, what is the probability of observing a difference in the HDL levels in a single pair of women between 2.5 and 3.6 mg/dl?

5.4 The Normal Approximation to the Binomial Probability Distribution

In Chapter 4, we considered several applications of the binomial probability distribution, all of which required us to calculate the probability that x, the number of successes in n trials, falls in a given region. Most examples involved small values of n because of the lengthy calculations involved in evaluating $p(x)$. However, when n was large and p was small, with $np < 7$, the Poisson probability distribution with $\mu = np$ produced satisfactory approximations to binomial probabilities. When n is large and the conditions for using the Poisson approximation are not met, another approximation based on normal curve areas is available.

The binomial probability histogram is symmetrical and bell-shaped for $p = .5$ and is relatively so for values of p not close to 0 or 1 when n is large. In such cases the binomial probability histogram is well approximated by a normal curve with mean $\mu = np$ and variance $\sigma^2 = npq$. Figures 5.10 and 5.11 show the binomial probability histograms for $n = 25$, $p = .5$, and $p = .1$, respectively, with the approximating normal curves superimposed. The correspondence between the areas for the binomial and normal distributions when $p = .5$ appears to be quite good. It is not so good for $p = .1$, for which the approximating symmetrical normal distribution poorly fits the nonsymmetrical binomial distribution. (However, when $n = 25$ and $p = .1$, $np = 2.5 < 7$, so that the Poisson approximation would be deemed appropriate in this particular case.)

Consider the binomial distribution for x when $n = 25$ and $p = .5$. For this distribution, $\mu = np = 25(.5) = 12.5$, and $\sigma = \sqrt{npq} = \sqrt{6.25} = 2.5$. Therefore, we choose an approximating normal distribution with a mean $\mu = 12.5$ and a standard

FIGURE 5.10 The binomial probability distribution for $n = 25$ and $p = .5$ and the approximating normal distribution with $\mu = 12.5$ and $\sigma = 2.5$

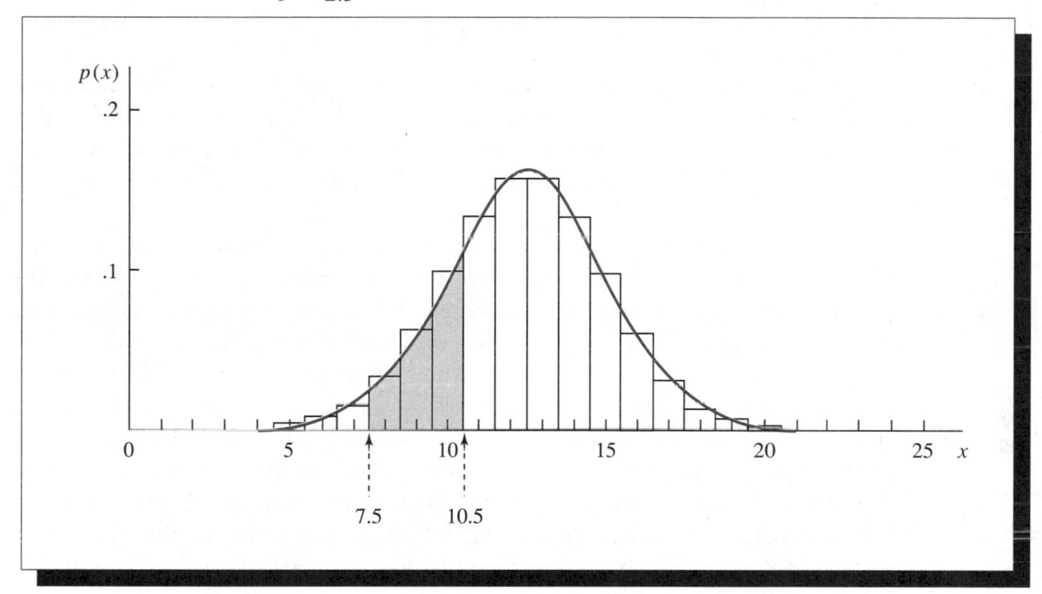

FIGURE 5.11 The binomial probability distribution and the approximating normal distribution for $n = 25$ and $p = .1$

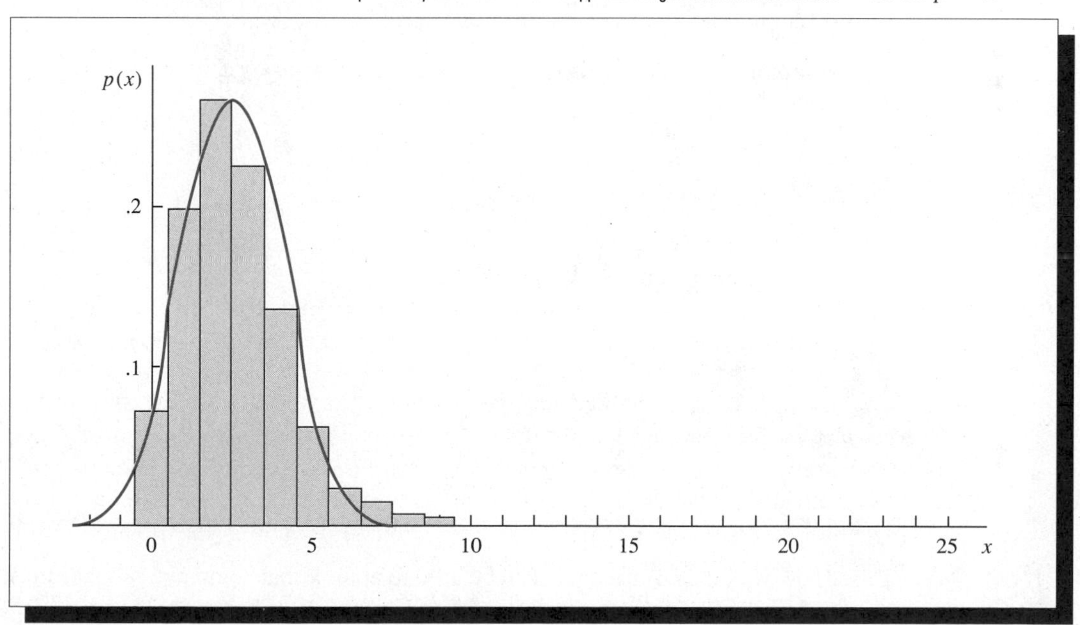

deviation $\sigma = 2.5$. The probability that $x = 8$, 9, or 10 is equal to the area of the three rectangles lying over $x = 8$, 9, and 10. This probability can be approximated by the area under the normal curve from $x = 7.5$ to $x = 10.5$, which is the shaded area in Figure 5.10.

Note that the area under the normal curve from 8 to 10 would not provide a good approximation to the probability that $x = 8$, 9, or 10 because it excludes half of the probability rectangles corresponding to $x = 8$ and $x = 10$. Thus, we must use the endpoints of the binomial probability rectangles, found by adding or subtracting .5 when calculating the approximating normal probabilities.

How can we tell whether the values of n and p are such that the normal approximation to binomial probabilities will be appropriate? From Section 5.2 (and from the Empirical Rule), we know that approximately 95% of the measurements associated with a normal curve lie within 2 standard deviations of the mean, and almost all lie within three. The binomial distribution would be nearly symmetrical if the distribution were able to spread out two standard deviations on both sides of the mean. Hence, to determine when the normal approximation is adequate, calculate $\mu = np$ and $\sigma = \sqrt{npq}$. **If the interval ($\mu \pm 2\sigma$) lies within the binomial bounds of 0 to n, the approximation is reasonably good. The approximation will be very good if the interval $\mu \pm 3\sigma$ lies in the interval 0 to n.** Note that this criterion is satisfied for the binomial probability distribution in Figure 5.10, but it is not satisfied for the distribution shown in Figure 5.11.

The Normal Approximation to the Binomial Probability Distribution

Appproximate the binomial probability distribution using a normal curve with

$$\mu = np$$
$$\sigma = \sqrt{npq}$$

where

$n =$ number of trials

$p =$ probability of success on a single trial

$q = 1 - p$

The approximation will be adequate when n is large and when the interval $\mu \pm 2\sigma$ falls between 0 and n.

EXAMPLE 5.7 To see how well the normal curve can be used to approximate binomial probabilities, refer to the binomial experiment illustrated in Figure 5.10, where $n = 25$ and $p = .5$. Calculate the probability that $x = 8$, 9, or 10, correct to three places, using Table 1 of binomial probabilities in Appendix I. Then calculate the corresponding normal approximation to this probability and compare the results.

Solution The exact probability can be calculated by using Table 1 for $n = 25$. Thus, we have

$$\sum_{x=8}^{10} p(x) = \sum_{x=0}^{10} p(x) - \sum_{x=0}^{7} p(x) = .212 - .022 = .190$$

As noted earlier in this section, the normal approximation requires that we calculate the area lying between $x_1 = 7.5$ and $x_2 = 10.5$, where $\mu = 12.5$ and $\sigma = 2.5$. Thus, we have

$$P(z_1 \leq z \leq z_2)$$

where

$$z_1 = \frac{x_1 - \mu}{\sigma} = \frac{7.5 - 12.5}{2.5} = -2.0 \qquad z_2 = \frac{x_2 - \mu}{\sigma} = \frac{10.5 - 12.5}{2.5} = -.8$$

This probability is shown in Figure 5.12. The area between $z = 0$ and $z = 2.0$ is .4772, and the area between $z = 0$ and $z = .8$ is .2881. From Figure 5.12,

$$\begin{aligned} P(z_1 \leq z \leq z_2) &= P(-2.0 \leq z \leq -.8) \\ &= .4772 - .2881 = .1891 \end{aligned}$$

The normal approximation is quite close to the binomial probability, .190, obtained from Table 1.

You must be careful not to exclude half of the two extreme probability rectangles when using the normal approximation to the binomial probability distribution. This means that the x values used to calculate z values will always have a 5 in the tenths decimal place. To be certain that you include all the probability rectangles in your approximation, always draw a sketch of the problem. ▪

FIGURE 5.12 Area under the normal curve for Example 5.7

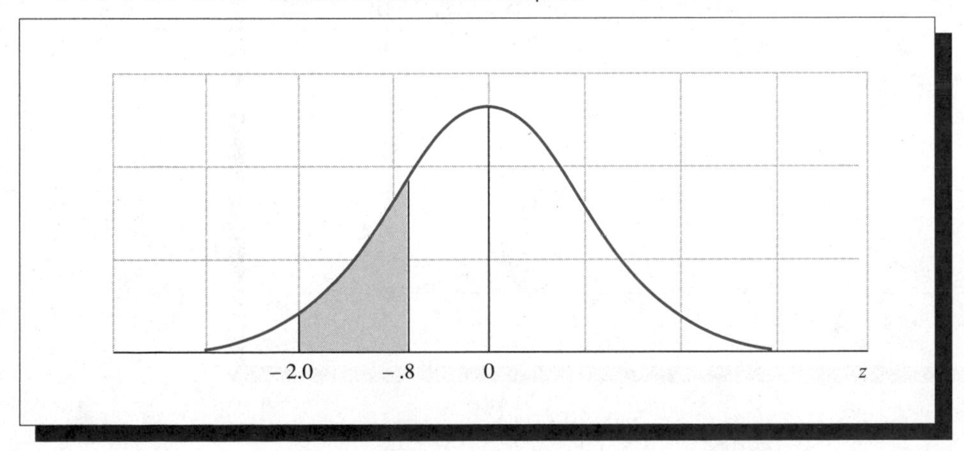

EXAMPLE 5.8 The reliability of an electrical fuse is the probability that a fuse, chosen at random from production, will function under the conditions for which it has been designed.

A random sample of 1000 fuses was tested and $x = 27$ defectives were observed. Calculate the approximate probability of observing 27 or more defectives, assuming that the fuse reliability is .98.

Solution The probability of observing a defective when a single fuse is tested is $p = .02$, given that the fuse reliability is .98. Then

$$\mu = np = 1000(.02) = 20$$
$$\sigma = \sqrt{npq} = \sqrt{1000(.02)(.98)} = 4.43$$

The probability of 27 or more defective fuses, given $n = 1000$, is

$$P = P(x \geq 27) = p(27) + p(28) + p(29) + \cdots + p(999) + p(1000)$$

The normal approximation to P would be the area under the normal curve to the right of $x = 26.5$ (NOTE: We must use $x = 26.5$ rather than $x = 27$ so as to include the entire probability rectangle associated with $x = 27$.) The z value corresponding to $x = 26.5$ is

$$z = \frac{x - \mu}{\sigma} = \frac{26.5 - 20}{4.43} = \frac{6.5}{4.43} = 1.47$$

and the area between $z = 0$ and $z = 1.47$ is equal to .4292, as shown in Figure 5.13. Since the total area to the right of the mean is equal to .5,

$$P(x \geq 27) \approx P(z \geq 1.47) = .5 - .4292 = .0708 \quad ■$$

FIGURE **5.13** Normal approximation to the binomial in Example 5.8

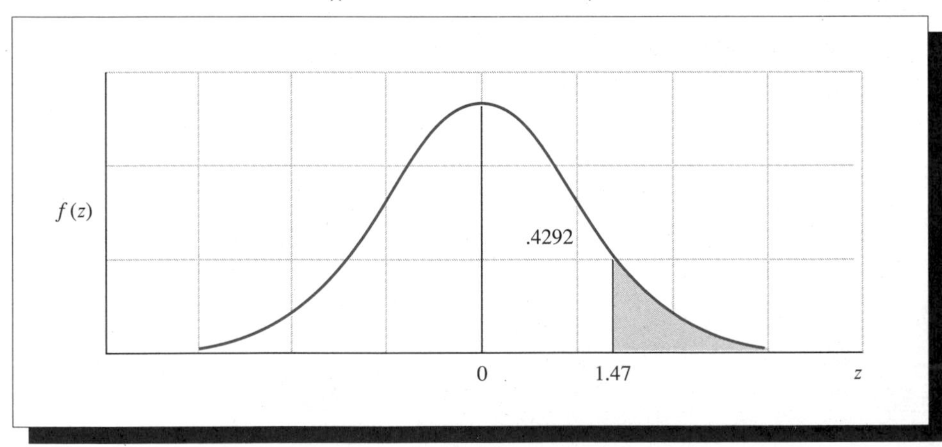

EXAMPLE **5.9** A producer of soft drinks was fairly certain that her brand had a 10% share of the soft drink market. In a market survey involving 2500 consumers of soft drinks, $x = 211$ expressed a preference for her brand. If the 10% figure is correct, find the probability of observing 211 or fewer consumers preferring her brand of soft drink.

Solution If the producer is correct, then the probability that a consumer prefers her brand of soft drink is $p = .10$. Then

$$\mu = np = 2500(.10) = 250$$
$$\sigma = \sqrt{npq} = \sqrt{2500(.10)(.90)} = \sqrt{225} = 15$$

The probability of observing 211 or fewer preferring her brand is

$$P(x \le 211) = p(0) + p(1) + \cdots + p(210) + p(211)$$

The normal approximation to this probability would be the area to the left of 211.5 under a normal curve with a mean of 250 and a standard deviation of 15.

$$z = \frac{x - \mu}{\sigma} = \frac{211.5 - 250}{15} = -2.57$$

and the area between $z = 0$ and $z = -2.57$ is equal to .4949. Since the area to the left of the mean is equal to .5,

$$P(x \le 211) \approx .5 - .4949 = .0051$$

The probability of observing a sample value of 211 or less when $p = .10$ is so small that we can conclude that one of two things has occurred; either we have observed an unusual sample even though $p = .10$, *or* the sample reflects that the actual value of p is less than .10 and perhaps closer to the value $211/2500 = .08$. ■

Exercises

Basic Techniques

5.30 Let x be a binomial random variable with $n = 25$, $p = .3$.

a Use Table 1 in Appendix I to find $P[8 \le x \le 10]$.

b Find μ and σ for the binomial distribution, and use the normal approximation to find $P[8 \le x \le 10]$. Compare the approximation with the exact value calculated in part (a).

5.31 Find the normal approximation to $P[x \ge 6]$ for a binomial probability distribution with $n = 10$, $p = .5$.

5.32 Find the normal approximation to $P[x > 6]$ for a binomial probability distribution with $n = 10$, $p = .5$.

5.33 Find the normal approximation to $P[x > 22]$ for a binomial probability distribution with $n = 100$, $p = .2$.

5.34 Find the normal approximation to $P[x \ge 22]$ for a binomial probability distribution with $n = 100$, $p = .2$.

5.35 Let x be a binomial random variable for $n = 25$, $p = .2$.

a Use Table 1 in Appendix I to calculate $P(4 \le x \le 6)$.

b Find μ and σ for the binomial probability distribution, and use the normal distribution to approximate the probability $P(4 \le x \le 6)$. Note that this value is a good approximation to the exact value of $P(4 \le x \le 6)$.

5.36 Consider a binomial experiment with $n = 20$, $p = .4$. Calculate $P(x \geq 10)$ using

a Table 1 in Appendix I.

b The normal approximation to the binomial probability distribution.

5.37 Find the normal approximation to $P[355 \leq x \leq 360]$ for a binomial probability distribution with $n = 400$, $p = .9$.

Applications

5.38 A Gallup survey, reported in *American Demographics* (Sept. 1989), indicated that the percentage of Americans who do not know the U.S. population is 56%. Does this hold true for you and your statistics classmates? Assume that it does and that your class contains 50 students. What is the probability that

a More than 25 do not know the U.S. population?

b Fewer than 15 do not know the U.S. population?

c Fewer than 30 *do know* the U.S. population?

5.39 Briggs and King developed the technique of nuclear transplantation, in which the nucleus of a cell from one of the later stages of the development of an embryo is transplanted into a zygote (a single-cell fertilized egg) to see if the nucleus can support normal development. If the probability that a single transplant from the early gastrula stage will be successful is .65, what is the probability that more than 70 transplants out of 100 will be successful?

5.40 To serve the changing lifestyles of children aged 2–5 years, the Children's Television Workshop has launched Sesame Street's Preschool Education Program. A Nielsen's survey ("Filling a Hole," 1992) revealed that 55% of children aged 2–5 years are cared for outside their homes; more than four out of every five are cared for in child-care centers. However, approximately 50% of the child-care centers have television that children can watch.

a If these survey results are correct and a random sample of $n = 100$ child-care centers are contacted, what are the mean and standard deviation of the number of child-care centers that have television sets that children can watch?

b What is the probability that 60 or more child-care centers in the sample have television sets that children can watch?

c If, in fact, you observe that 60 of the sampled child-care centers have television sets available for the children, what would you conclude about the 50% figure?

5.41 Data collected over a long period of time show that a particular genetic defect occurs in 1 out of every 1000 children. The records of a medical clinic show $x = 60$ children with the defect in a total of 50,000 examined. If the 50,000 children were a random sample from the population of children represented by past records, what is the probability of observing a value of x equal to 60 or more? Would you say that the observation of $x = 60$ children with genetic defects represents a rare event?

5.42 Does the appearance of an athlete or an athletic team on the cover of *Sports Illustrated* increase the athlete's (or the team's) performance? In a study of 271 randomly selected cover subjects from 1954 to 1983, researchers at the University of Southern California found that 58% of the athletes or teams improved their performance after appearing on the cover of *Sports Illustrated* (*USA Today*, July 2, 1984). Suppose that the probability that an athlete (or team) will improve is independent of exposure on the cover of *Sports Illustrated* and is, in fact, equal to .5.

a What is the probability that as many as 157 (58%) of a sample of 271 would show improvement?

b Is the value $x = 157$ (58%) a likely occurrence, assuming that there is no effect due to the athlete's (or team's) appearance on the cover of *Sports Illustrated*? What would you conclude regarding the effect of media coverage for an athlete?

5.43 Airlines and hotels often grant reservations in excess of capacity to minimize losses due to no-shows. Suppose the records of a motel show that, on the average, 10% of their prospective guests will not claim their reservation. If the motel accepts 215 reservations and there are only 200 rooms in the motel, what is the probability that all guests who arrive to claim a room will receive one?

5.44 Compilation of large masses of data on lung cancer shows that approximately 1 of every 40 adults acquires the disease. Workers in a certain occupation are known to work in an air-polluted environment that may cause an increased rate of lung cancer. A random sample of $n = 400$ workers shows 19 with identifiable cases of lung cancer. Do the data provide sufficient evidence to indicate a higher rate of lung cancer for these workers than for the national average?

5.45 Women are ten times more likely than men to die in the hospital following angioplasty, a procedure that uses a tiny balloon to reopen clogged heart arteries, according to an Associated Press article ("Angioplasty Much Riskier," 1993). Based on a sample of 1590 men who underwent the angioplasty procedure, 4 (or .25%) died before leaving the hospital.

a Assume that the .25% fatality rate also holds for women. Use the normal approximation to the binomial distribution to find the probability that 14 or more women among a group of 546 women who underwent angioplasty died before leaving the hospital.

b Based on the results of part (a), does it appear that the .25% figure is too low? What is the observed death rate for this group of women?

5.46 In Exercise 5.45, we used a normal distribution to approximate a binomial probability for the number of women who died before leaving the hospital following angioplasty.

a Was the normal approximation adequate? Explain.

b The Poisson distribution can also be used to approximate binomial probabilities when p (or q) is small and $np < 7$ (see Section 4.3). Use the Poisson approximation to find the probability that 14 or more women die before leaving the hospital following angioplasty.

c If, in fact, the actual death rate for women is 2.5%, 10 times that for men, reevaluate the probability in part (b). What is your conclusion concerning the in-hospital death rate for women following an angioplasty procedure?

5.5 Summary

Data on many continuous random variables observed in nature have relative frequency distributions that are bell-shaped and can be approximated by a normal distribution. In addition, many other random variables, such as the binomial and the Poisson, have distributions that can be approximated by a normal distribution when certain conditions are satisified. The binomial probability distribution can be approximated by a normal distribution when the n of trials is large and the probability p of success on a single trial is not too close to 0 or 1. The Poisson probability distribution can be approximated by a normal distribution when its mean becomes large. For example, if the mean of the Poisson distribution is as large as 9, the normal approximation to the Poisson distribution is reasonably good.

You might think that this is adequate reason to nominate the normal probability distribution as one of the most important distributions in statistics, but the most compelling reason is explained in Chapter 6. There you will learn that the probability distributions of many sample statistics, sample means, proportions, and so on are approximately normal when certain conditions are satisfied. This result plays an important role in statistical inference and explains why the normal distribution is the most important probability distribution in statistics.

Tips on Problem Solving

- Always sketch a normal curve and locate the probability areas pertinent to the exercise. If you are approximating a binomial probability distribution, sketch in the probability rectangles as well as the normal curve.
- Read each exercise carefully to see whether the data come from a binomial experiment or whether they possess a distribution that, by its very nature, is approximately normal. If you are approximating a binomial probability distribution using a normal curve, do not forget to make a half-unit correction, so that you will include the half-rectangles at the ends of the intervals. If the distribution is not binomial, *do not* make the half-unit corrections. If you make a sketch (as suggested in step 1), you will see why the half-unit correction is or is not needed.

Supplementary Exercises

5.47 Using Table 3 in Appendix I, calculate the area under the normal curve between

a $z = 0$ and $z = 1.2$. **b** $z = 0$ and $z = -.9$.

c $z = 0$ and $z = 1.46$. **d** $z = 0$ and $z = -.42$.

5.48 Repeat Exercise 5.47 using

a $z = .3$ and $z = 1.56$. **b** $z = .2$ and $z = -.2$.

5.49 **a** Find the probability that z is greater than $-.75$.

b Find the probability that z is less than 1.35.

5.50 Find z_0 such that $P(z > z_0) = .5$.

5.51 Find the probability that z lies between $z = -1.48$ and $z = 1.48$.

5.52 Find z_0 such that $P(-z_0 < z < z_0) = .5$.

5.53 The life span of oil-drilling bits depends on the types of rock and soil that the drill encounters, but it is estimated that the mean length of life is 75 hours. Suppose an oil exploration company purchases drill bits that have a life span that is approximately normally distributed, with a mean equal to 75 hours and a standard deviation equal to 12 hours.

a What proportion of the company's drill bits will fail before 60 hours of use?

b What proportion will last at least 60 hours?

c What proportion will have to be replaced after more than 90 hours of use?

5.54 The influx of new ideas into a college or university, introduced primarily by hiring new young faculty, is becoming a matter of concern because of the increasing ages of faculty members; that is, the distribution of faculty ages is shifting upward, due most likely to a shortage of vacant positions and an oversupply of Ph.Ds. Thus, faculty members are reluctant to move and give up a secure position. If the retirement age at most universities is 65, would you expect the distribution of faculty ages to be normal?

5.55 A machine operation produces bearings whose diameters are normally distributed, with mean and standard deviation equal to .498 and .002, respectively. If specifications require that the bearing diameter equal .500 inch ± .004 inch, what fraction of the production will be unacceptable?

5.56 A used-car dealership has found that the length of time before a major repair is required on the cars it sells is normally distributed, with a mean equal to ten months and a standard deviation of three months. If the dealer wants only 5% of the cars to fail before the end of the guarantee period, for how many months should the cars be guaranteed?

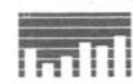

5.57 The daily sales total (excepting Saturday) at a small restaurant has a probability distribution that is approximately normal, with a mean μ equal to \$530 per day and a standard deviation σ equal to \$120.

a What is the probability that the sales will exceed \$700 on a given day?

b The restaurant must have at least \$300 in sales per day to break even. What is the probability that on a given day the restaurant will not break even?

5.58 The life span of a type of automatic washer is approximately normally distributed, with mean and standard deviation equal to 3.1 and 1.2 years, respectively. If this type of washer is guaranteed for 1 year, what fraction of original sales will require replacement?

5.59 Most users of automatic garage door openers activate their openers at distances that are normally distributed, with a mean of 30 feet and a standard deviation of 11 feet. To minimize interference with other radio-controlled devices, the manufacturer is required to limit the operating distance to 50 feet. What percentage of the time will users attempt to operate the opener outside its operating limit?

5.60 Consider a binomial experiment with $n = 25$, $p = .4$. Calculate $P(8 \leq x \leq 11)$ using

a The binomial probabilities given in Table 1 in Appendix I.

b The normal approximation to the binomial.

5.61 The average length of time required to complete a college achievement test was found to equal 70 minutes, with a standard deviation of 12 minutes. When should the test be terminated if we wish to allow sufficient time for 90% of the students to complete the test? (Assume that the time required to complete the test is normally distributed.)

5.62 The length of time required for the periodic maintenance of an automobile will usually have a probability distribution that is mound-shaped and, because some long service times will occur occasionally, is skewed to the right. Suppose the length of time required to run a 5000-mile check and to service an automobile has a mean equal to 1.4 hours and a standard deviation of .7 hour. Suppose that the service department plans to service 50 automobiles per 8-hour day and that, in order to do so, it must spend no more than an average of 1.6 hours per automobile. What proportion of all days will the service department have to work overtime?

5.63 An advertising agency has stated that 20% of all television viewers watch a particular program. In a random sample of 1000 viewers, $x = 184$ viewers were watching the program. Do these data present sufficient evidence to contradict the advertiser's claim?

5.64 One of the most valued rights in our democratic society is the right to vote. However, many citizens do not exercise this right. In a *Los Angeles Times* (July 8, 1990) poll of 2260 U.S. citizens living in California, 35% said that they would be encouraged to vote "if the issues were less confusing." Is this opinion true for citizens in the state of Iowa? Suppose that a random sample of 100 Iowans who had not registered to vote is taken and that the 35% figure is correct.

a What is the probability that 40 or more would be encouraged to vote if the issues were less confusing?

b What is the probability that 25 or fewer would be encouraged to vote if the issues were less confusing?

c If only 20 Iowans said that they would be more likely to vote if the issues were less confusing, what would you conclude about the 35% figure? Explain.

5.65 A researcher notes that senior corporation executives are not very accurate forecasters of their own annual earnings. He states that his studies of a large number of company executive forecasts "showed that the average estimate missed the mark by 15%."

a Suppose the distribution of these forecast errors has a mean of 15% and a standard deviation of 10%. Is it likely that the distribution of forecast errors is approximately normal?

b Suppose the probability is .5 that a corporate executive's forecast error exceeds 15%. If you were to sample the forecasts of 100 corporate executives, what is the probability that more than 60 would be in error by more than 15%?

5.66 A soft drink machine can be regulated to discharge an average of μ ounces per cup. If the ounces of fill are normally distributed, with standard deviation equal to .3 ounce, give the setting for μ so that 8-ounce cups will overflow only 1% of the time.

5.67 If you call the Internal Revenue Service with an income tax question, will you get the correct answer? According to an investigation conducted by the General Accounting Office, the IRS is giving incorrect answers to 39% of the questions posed by people who call for help with their taxes (*New York Times*, Feb. 23, 1988). If 65 taxpayers call the IRS with questions,

a What is the probability that more than 30 will receive incorrect advice?

b What is the probability that fewer than 20 will receive incorrect advice?

5.68 A manufacturing plant uses 3000 electric light bulbs whose life spans are normally distributed, with mean and standard deviation equal to 500 and 50 hours, respectively. In order to minimize the number of bulbs that burn out during operating hours, all the bulbs are replaced after a given period of operation. How often should the bulbs be replaced if we wish no more than 1% of the bulbs to burn out between replacement periods?

5.69 The admissions office of a small college is asked to accept deposits from a number of qualified prospective freshmen so that, with probability about .95, the size of the freshman class will be less than or equal to 120. Suppose the applicants constitute a random sample from a population of applicants, 80% of whom would actually enter the freshman class if accepted.

a How many deposits should the admissions counselor accept?

b If applicants in the number determined in part (a) are accepted, what is the probability that the freshman class size will be less than 105?

5.70 An airline finds that 5% of the persons making reservations on a certain flight will not show up for the flight. If the airline sells 160 tickets for a flight with only 155 seats, what is the probabililty that a seat will be available for every person holding a reservation and planning to fly?

5.71 It is known that 30% of all calls coming into a telephone exchange are long-distance calls. If 200 calls come into the exchange, what is the probability that at least 50 will be long-distance calls?

5.72 Suppose the random variable x has a binomial distribution corresponding to $n = 20$ and $p = .30$. Use Table 1 of Appendix I to calculate

a $P(x = 5)$. **b** $P(x \geq 7)$.

5.73 Refer to Exercise 5.72. Use the normal approximation to calculate $P(x = 5)$ and $P(x \geq 7)$. Compare with the exact values obtained from Table 1 in Appendix I.

5.74 A purchaser of electric relays buys from two suppliers, A and B. Supplier A supplies 2 of every 3 relays used by the company. If 75 relays are selected at random from those in use by the company, find the probability that at most 48 of these relays come from supplier A. Assume that the company uses a large number of relays.

5.75 Is television dangerous to your diet? Psychologists believe that excessive eating may be associated with emotional states (being upset, bored) and environmental cues (TV, reading, and so on). To test this theory, suppose we were to randomly select 60 overweight persons and match them by weight and sex in pairs. For a period of 2 weeks, one of each pair is required to spend evenings reading novels of interest to him or her. The other member of each pair spends each evening watching television. The calorie count for all snack and drink intake for the evenings is recorded for each person, and we record $x = 19$, the number of pairs for which the TV watchers' calorie intake exceeded the intake of the readers. If there is no difference in the effects of TV and reading on calorie intake, the probability p that the calorie intake of one member of a pair exceeds that of the other member is .5. Do these data provide sufficient evidence to indicate a difference between the effects of TV watching and reading on calorie intake? (HINT: Calculate the z-score for the observed value, $x = 19$.)

5.76 In Exercise 5.23 we suggested that the IRS assign auditing rates per state by randomly selecting 50 auditing percentages from a normal distribution with a mean equal to 1.55% and a standard deviation of .45%.

a What is the probability that a particular state would have more than 2% of its tax returns audited?

b What is the expected value of x, the number of states that will have more than 2% of their income tax returns audited?

c Is it likely that as many as 15 of the 50 states will have more than 2% of their income tax returns audited?

5.77 There is a difference in sports preferences between men and women, according to a Gallup poll (*American Demographics*, Sept. 1989). Among the ten most popular sports, men include competition-type sports—pool and billiards, basketball, and softball—whereas women include aerobics, running, hiking, and calisthenics. However, the top recreational activity for men was still the relaxing sport of fishing, with 41% of those surveyed indicating that they had fished during the year. Suppose 180 randomly selected men are asked whether they had fished in the last year.

a What is the probability that fewer than 50 had fished?

b What is the probability that between 50 and 75 had fished?

c If the 180 men selected for the interview were selected by the marketing department of a sporting-goods company based on information obtained from their mailing lists, what would you conclude about the reliability of their survey results?

CASE STUDY

The Long and the Short of It

If you were the boss, would height play a role in your selection of a successor for your job? In his *Fortune* column "Keeping Up" (1981), Daniel Seligman discussed his ideas concerning height as a factor in Deng Xiaoping's choice of Hu Yaobang as his replacement as Chairman of the Chinese Communist Party. As Seligman notes, the facts surrounding the case are enough to arouse suspicions when examined in the light of statistics.

Deng, it seemed, was only 5 foot tall, a height that is short even in China. Therefore, the choice of Hu Yaobang, who was also 5 feet tall, raised (or lowered) some eyebrows because, as Seligman notes, "the odds against a 'height-blind' decision producing a chairman as short as Deng are about 40 to 1." In other words, if we had the relative frequency distribution of the heights of all Chinese males, only 1 in 41 (i.e., 2.4%) would be 5 feet tall or less. To calculate these odds, Seligman notes that the Chinese equivalent of the U.S. Health Service does not exist and hence that health statistics on the current population of China are difficult to acquire. He says, however, that "it is generally held that a boy's length at birth represents 28.6% of his final height" and that, in prerevolutionary China, the average length of a Chinese boy at birth was 18.9 inches. From this, Seligman deduces that the mean height of mature Chinese males is

$$\frac{18.9}{.286} = 66.08 \text{ inches, or 5 feet 6.08 inches}$$

He then assumes that the distribution of the heights of males in China follows a normal distribution ("as it does in the U.S."), with a mean of 66 inches and a standard deviation equal to 2.7 inches, "a figure that looks about right for that mean."

1 Using Seligman's assumptions, calculate the probability that a single adult Chinese male, chosen at random, will be less than or equal to 5 feet tall, or equivalently, 60 inches tall.

2 Do the results in part 1 agree with Seligman's odds?

3 Comment on the validity of Seligman's assumptions. Are there any basic flaws in his reasoning?

4 Based on the results of parts 1 and 3, do you think that Deng Xiaoping took height into account in selecting his successor?

6

Sampling Distributions

Case Study

How would you like to try your hand at gambling without the risk of losing? You could do it by simulating the gambling process, making imaginary bets, and observing the results. This technique, called a **Monte Carlo procedure,** is the topic of the case study at the end of this chapter.

General Objectives

In the past several chapters, we studied *populations* and the *parameters* that describe them. These populations were either discrete or continuous, and we used *probability* as a tool for determining how likely certain sample outcomes might be. In this chapter, the focus of our discussion changes as we begin to study *samples* and the *statistics* that describe them. We will use these sample statistics to make inferences about the corresponding population parameters. Our objective in this chapter is to study sampling and **sampling distributions**, which describe the behavior of sample statistics in repeated sampling.

Specific Topics

1. Random samples (6.1)
2. Statistics and sampling distributions (6.2)
3. The Central Limit Theorem (6.3)
4. The sampling distribution of the sample mean, $\bar{x}$ (6.4)
5. The sampling distribution of the sample proportion, $\hat{p}$ (6.5)
6. Statistical process control: $\bar{x}$ and p charts (6.6)

6.1 Random Sampling

In earlier chapters, we studied random variables and their probability distributions. We presented several discrete and continuous probability distributions that are possible models for practical situations. These probability distributions depended upon descriptive measures called *parameters,* such as a population mean or standard deviation.

How do we apply these probability models in the practice of statistics? Usually, we are able to decide which type of probability distribution might serve as a model in a given situation; however, the values of the parameters that specify the distribution exactly are not available. For example, a pollster could be quite certain that the responses to an "agree/disagree" question on a survey could be modeled using a binomial distribution, without knowing the value of the parameter p. Similarly, an agronomist may be willing to assume that the yield per acre of a variety of wheat is normally distributed, without knowing the values of the population mean and standard deviation.

In situations such as these, we rely on the sample to provide information about these unknown population parameters. For example, the sample proportion of those who agree should reflect the actual value of p in the population. Similarly, the sample mean and standard deviation of yield per acre should provide information concerning the population mean and standard deviation. In both of these instances, sample information would be used to make inferences about population parameters.

The way a sample is selected is called the *sampling plan* or *experimental design* and determines the quantity of information in the sample. In addition, by knowing the sampling plan used in a particular situation, we can determine the probability of observing specific samples. These probabilities allow us to assess the reliability or goodness of the inferences that are based on these samples.

Simple random sampling is a commonly used sampling plan in which every sample of size n has the same chance of being selected. For example, suppose we want to select a sample of size $n = 2$ from a population containing $N = 4$ objects. If the four objects are identified by the symbols x_1, x_2, x_3, and x_4, there are six distinct pairs that could be selected:

Sample	Observations in Sample
1	x_1, x_2
2	x_1, x_3
3	x_1, x_4
4	x_2, x_3
5	x_2, x_4
6	x_3, x_4

If the sample of $n = 2$ observations is selected so that each of these six samples has the same chance of selection, given by 1/6, then the resulting sample would be called a **simple random sample**, or just a **random sample.**

It can be shown† that the number of ways of selecting a sample of size n elements from a population containing N elements is given by

$$C_n^N = \frac{N!}{n!(N-n)!}$$

†An explanation of this result along with examples and applications is given in the supplement to Chapter 3.

where $n! = n(n-1)\dots(3)(2)(1)$ and $0! = 1$. The symbol C_n^N stands for the number of distinct, unordered samples of size n selected *without replacement*. When $N = 4$ and $n = 2$, we have shown that there are

$$C_2^4 = \frac{4!}{2!2!} = \frac{4 \cdot 3 \cdot 2 \cdot 1}{(2 \cdot 1)(2 \cdot 1)} = 6$$

distinct samples. If we conduct an opinion poll of 5000 people based on a sample of size $n = 50$, there are C_{50}^{5000} different combinations of 50 people who could be selected in the sample. If each of these combinations has an equal chance of selection in the sampling plan, then the sample would be a *simple random sample*.

DEFINITION ▪ If a sample of n elements is selected from a population of N elements using a sampling plan in which each of the C_n^N samples has the same chance of selection, the sampling is said to be **random** and the resulting sample is a **simple random sample.** ▪

Perfect random sampling is difficult to achieve in practice. If the population is not too large, we might write each of the N numbers on a poker chip, mix the total, and select a sample of n chips. The numbers on the poker chips would specify the measurements to appear in the sample. In many situations, the population is conceptual, as in an observation made during a laboratory experiment. Here the population is envisioned to be the infinitely large number of measurements obtained when the experiment is repeated over and over again. If we want a sample of $n = 10$ measurements from this population, we repeat the experiment ten times and hope that the results represent, to a reasonable degree of approximation, a random sample.

The simplest and most reliable way to select a random sample of n elements from a large population is to use a table of random numbers (see Table 14, Appendix I). Random number tables are constructed so that integers occur randomly and with equal frequency. For example, suppose the population contains $N = 1000$ elements. Number the elements in sequence, 1 to 1000. Then turn to a table of random numbers such as the excerpt shown in Table 6.1.

TABLE **6.1** Portion of a table of random numbers

15574	35026	98924
45045	36933	28630
03225	78812	50856
88292	26053	21121

Select n of the random numbers in order. The population elements to be included in the random sample will be given by the first three digits of the random numbers (unless the first four digits are 1000). Thus if $n = 5$, we would include elements numbered 155, 450, 32, 882, and 350. To ensure against using the same sequence of random numbers over and over, select different starting points in Table 14 to begin selecting random numbers for different samples.

In addition to *simple random sampling,* there are other sampling plans that also involve random selection of the elements in the sample and thus also provide a probabilistic basis for inference making. Three such plans are based on *stratified, cluster,* and *systematic sampling*.

When the population of interest consists of two or more subpopulations, called **strata**, a sampling plan that ensures that each subpopulation is represented in the sample is called a **stratified random sample.**

DEFINITION ▪ **Stratified random sampling** involves selecting a simple random sample from each of a given number of subpopulations, or **strata**. ▪

A poll to determine citizens' opinions on the construction of a performing arts center could be implemented as a stratified random sample using city voting wards as strata. National polls usually involve some form of stratified random sampling with states as strata.

Another form of random sampling is used when the available sampling units are groups of elements, called **clusters.** For example, a household is a *cluster* of individuals living together. A city block or a neighborhood might constitute a convenient sampling unit and might be considered a *cluster* for a given sampling plan.

DEFINITION ▪ A **cluster sample** is a simple random sample of clusters from the available clusters in the population. ▪

When a cluster is selected for inclusion in the sample, a census of every element in the cluster is taken.

There are situations in which the population to be sampled is ordered, such as an alphabetized list of people with driver's licenses, a list of utility users arranged by service addresses, or a list of customers by account numbers. In these and other situations, one element is chosen at random from the first k elements, and then every kth element thereafter is included in the sample.

DEFINITION ▪ A **1-in-*k* systematic random sample** involves the random selection of one of the first k elements in an ordered population, and then the systematic selection of every kth element thereafter. ▪

Not all sampling plans, however, involve random selection. We have all heard of the nonrandom telephone polls in which those people who wish to express support for a question call one "900 number" and those opposed call a second "900 number." Each person must pay for his or her call. It is obvious that those people who call would not in general represent the population at large. This type of sampling plan is one form of a **convenience sample**, a sample that can be easily and simply obtained without random selection. Advertising for subjects who will be paid a fee for participating in an experiment produces a *convenience sample*. **Judgment sampling** allows the sampler to decide who will or will not be included in the sample. **Quota sampling**, in which the makeup of the sample must reflect the makeup of the population on some preselected characteristic, often has a nonrandom component in the selection process. **Remember that nonrandom samples can be described but cannot be used for inference purposes!**

6.2 Statistics and Sampling Distributions

Numerical descriptive measures calculated from a sample are called **statistics**. Since the values of these sample statistics are unpredictable and vary from sample to sample, they are *random variables* and have a *probability distribution* that describes their behavior in repeated sampling. This probability distribution, called the **sampling distribution of the statistic**, allows us to determine the goodness of any inferences based on this statistic.

DEFINITION ▪ The **sampling distribution** of a statistic is the probability distribution for all possible values of the statistic that results when random samples of size n are repeatedly drawn from the population. ▪

The sample mean, standard deviation, median, and other descriptive measures computed from sample values can be used not only to describe the sample but also to make inferences in the form of estimates or tests about corresponding population parameters. However, we must know the sampling distribution of the statistic in order to answer questions such as, Does the statistic consistently underestimate or overestimate the value of the parameter? Is this statistic less variable than other competitors, and hence more useful as an estimator?

The sampling distribution of a statistic may be derived mathematically or approximated empirically. Empirical approximations are found by drawing a large number of samples of size n from the specified population, calculating the value of the statistic for each sample, and tabulating the results in a relative frequency histogram. When the number of samples is large, the relative frequency histogram should closely approximate the theoretical sampling distribution.

Consider a random sample of size $n = 3$ drawn without replacement from a population of $N = 5$ elements. A simple random sample of size n is selected in such a way that every sample of size n has the same probability of being selected—equal to $1/C_n^N$. Let us begin with a population of $N = 5$ elements whose values are 3, 6, 9, 12, and 15. Since the five elements are distinct, the population probability distribution assigns a probability of 1/5 to each value of x, so that

$$p(x) = 1/5 \qquad \text{for } x = 3, 6, 9, 12, \text{ and } 15$$

FIGURE 6.1
Probability histogram for the $N = 5$ population values

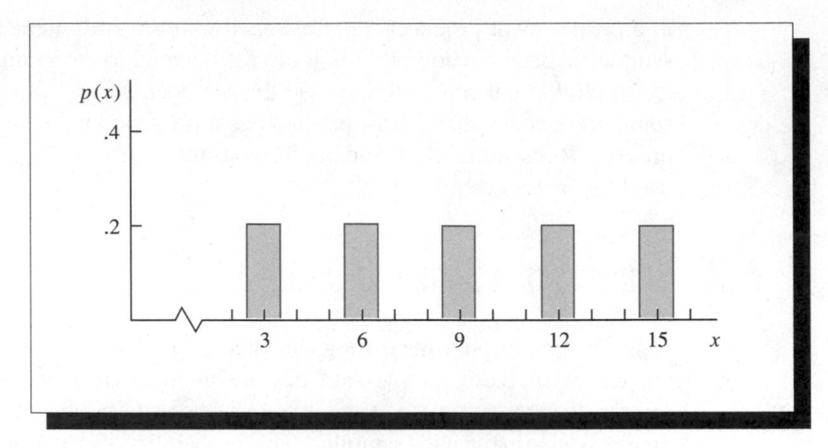

The probability histogram is given in Figure 6.1.

The population mean is found as

$$\mu = \frac{\sum_{i=1}^{N} x_i}{N} = \frac{45}{5} = 9$$

By inspection we see that the population median is $M = 9$ and that $\mu = M$. The number of possible samples is

$$C_3^5 = \frac{5!}{3!2!} = 10$$

and the ten possible samples are given in Table 6.2. For each sample we have calculated the sample mean $\bar{x}$ and the sample median m. Each of the ten samples in Table 6.2 is equally likely. Hence, the values of $\bar{x}$ and m associated with each sample are each assigned probability equal to 1/10. For example, we will observe a value of $\bar{x} = 6$ only if sample 1 is selected, and this occurs with probability .1. A value of $\bar{x} = 8$ will occur if sample 3 or sample 4 is drawn; therefore, the probability of observing $\bar{x} = 8$ is .2. Continuing in this manner, we can find the sampling distributions for $\bar{x}$ and m given in Figure 6.2.

TABLE 6.2
Values of $\bar{x}$ and m for simple random sampling when $n = 3$ and $N = 5$

Sample	Sample Values	$\bar{x}$	m
1	3, 6, 9	6	6
2	3, 6, 12	7	6
3	3, 6, 15	8	6
4	3, 9, 12	8	9
5	3, 9, 15	9	9
6	3, 12, 15	10	12
7	6, 9, 12	9	9
8	6, 9, 15	10	9
9	6, 12, 15	11	12
10	9, 12, 15	12	12

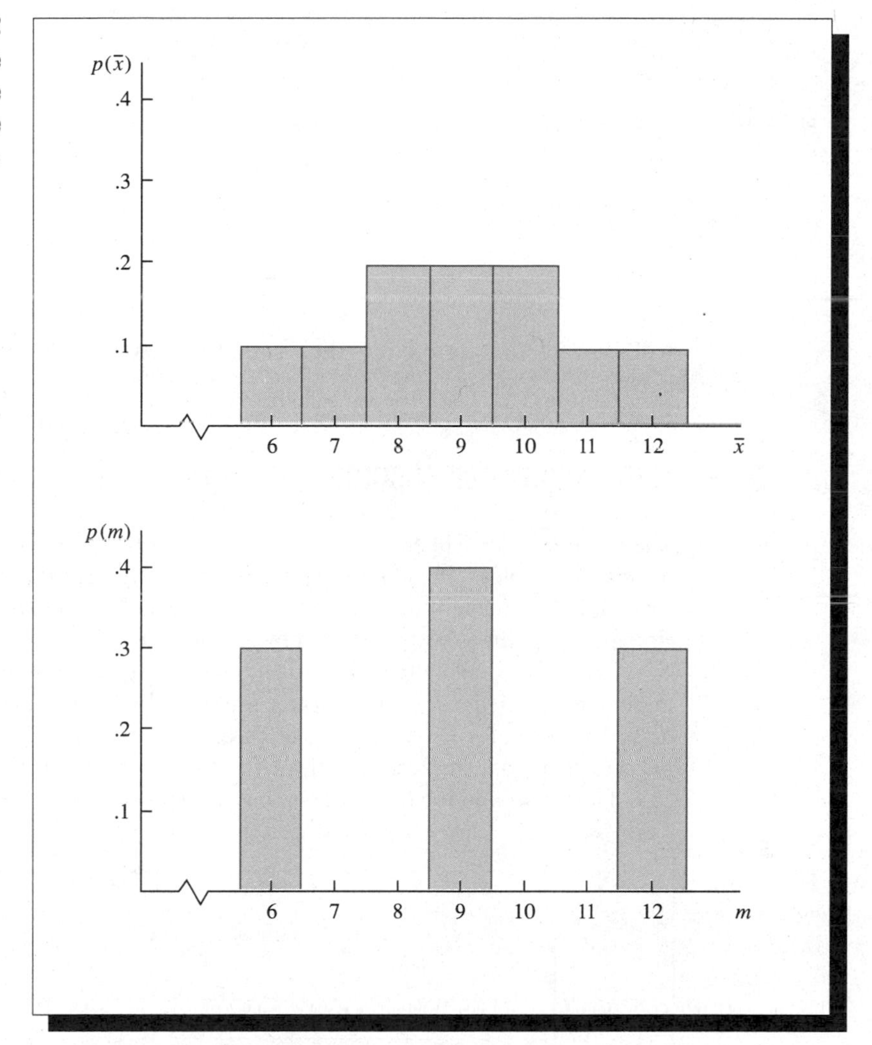

FIGURE 6.2 Probability histograms for the sampling distributions of the sample mean, $\bar{x}$, and the sample median, m

Since the population of $N = 5$ values is symmetric about the value $x = 9$, both the **population mean** and the **median** equal 9. It would seem reasonable, therefore, to consider using either $\bar{x}$ or m as possible estimators of $M = \mu = 9$. Which estimator would we choose? From Table 6.3, we see that, in using m as an estimator, we would be in error by $9 - 6 = 3$ with probability .3 or by $9 - 12 = -3$ with probability .3. That is, the error in estimation using m would be at least 3 with probability .6. In using $\bar{x}$, however, an error of at least 3 would occur only with probability .2. On these grounds alone, we may wish to use $\bar{x}$ as an estimator in preference to m.

When N is small, sampling distributions can be derived directly. However, when this is not the case, the sampling distribution can be approximated by repeatedly selecting samples of size n and calculating the value of the statistic for each sample. The resulting relative frequency histogram is an approximation to the unknown sampling distribution. Alternatively, **for certain statistics that are sums or means of the sample values, an important theorem that we introduce in the next section**

TABLE 6.3
Sampling distributions for (a) the sample mean and (b) the sample median

(a)

$\bar{x}$	$p(\bar{x})$
6	.1
7	.1
8	.2
9	.2
10	.2
11	.1
12	.1

(b)

m	$p(m)$
6	.3
9	.4
12	.3

will allow us to approximate their sampling distributions when the sample size is large.

6.3 The Central Limit Theorem

The Central Limit Theorem states that, under rather general conditions, sums and means of samples of random measurements drawn from a population tend to possess an approximately bell-shaped sampling distribution. The significance of this statement is perhaps best illustrated by an example.

Consider a population of die throws generated by tossing a balanced die infinitely many times, with the resulting **discrete uniform distribution** given in Figure 6.3. We draw a sample of $n = 5$ measurements from the population by tossing the balanced die five times and recording each of the five observations 3, 5, 1, 3, 2. We then calculate the sum of the five observations as well as the sample mean $\bar{x}$.

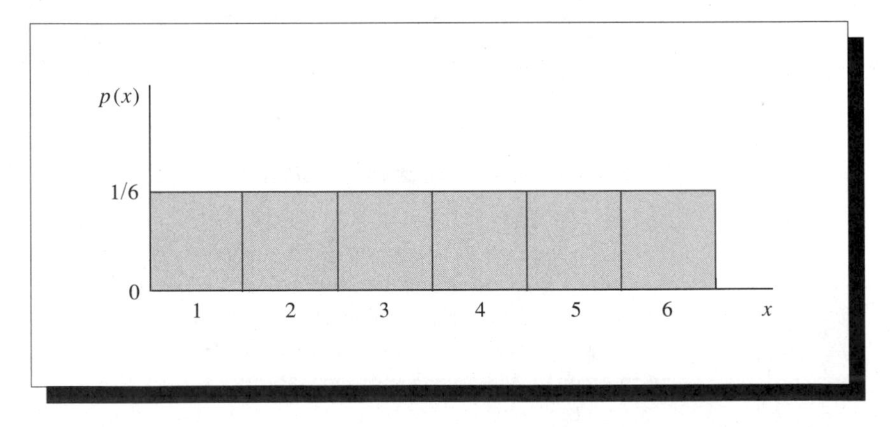

FIGURE 6.3
Probability distribution for x, the number appearing on a single toss of a die

For experimental purposes, repeat the sampling procedure 100 times (or preferably, an even larger number of times). The results for 100 samples are given in Table 6.4, along with the corresponding values of the sum $\sum_{i=1}^{5} x_i$ and the sample mean $\bar{x}$. A frequency histogram for the 100 values of $\bar{x}$ $\left(\text{or for } \sum_{i=1}^{5} x_i\right)$ is shown in Figure

TABLE **6.4**
Sampling from the population of die throws

Sample Number	x, Sample Measurements	$\sum x_i$	$\bar{x}$	Sample Number	x, Sample Measurements	$\sum x_i$	$\bar{x}$
1	3, 5, 1, 3, 2	14	2.8	51	2, 3, 5, 3, 2	15	3.0
2	3, 1, 1, 4, 6	15	3.0	52	1, 1, 1, 2, 4	9	1.8
3	1, 3, 1, 6, 1	12	2.4	53	2, 6, 3, 4, 5	20	4.0
4	4, 5, 3, 3, 2	17	3.4	54	1, 2, 2, 1, 1	7	1.4
5	3, 1, 3, 5, 2	14	2.8	55	2, 4, 4, 6, 2	18	3.6
6	2, 4, 4, 2, 4	16	3.2	56	3, 2, 5, 4, 5	19	3.8
7	4, 2, 5, 5, 3	19	3.8	57	2, 4, 2, 4, 5	17	3.4
8	3, 5, 5, 5, 5	23	4.6	58	5, 5, 4, 3, 2	19	3.8
9	6, 5, 5, 1, 6	23	4.6	59	5, 4, 4, 6, 3	22	4.4
10	5, 1, 6, 1, 6	19	3.8	60	3, 2, 5, 3, 1	14	2.8
11	1, 1, 1, 5, 3	11	2.2	61	2, 1, 4, 1, 3	11	2.2
12	3, 4, 2, 4, 4	17	3.4	62	4, 1, 1, 5, 2	13	2.6
13	2, 6, 1, 5, 4	18	3.6	63	2, 3, 1, 2, 3	11	2.2
14	6, 3, 4, 2, 5	20	4.0	64	2, 3, 3, 2, 6	16	3.2
15	2, 6, 2, 1, 5	16	3.2	65	4, 3, 5, 2, 6	20	4.0
16	1, 5, 1, 2, 5	14	2.8	66	3, 1, 3, 3, 4	14	2.8
17	3, 5, 1, 1, 2	12	2.4	67	4, 6, 1, 3, 6	20	4.0
18	3, 2, 4, 3, 5	17	3.4	48	2, 4, 6, 6, 3	21	4.2
19	5, 1, 6, 3, 1	16	3.2	69	4, 1, 6, 5, 5	21	4.2
20	1, 6, 4, 4, 1	16	3.2	70	6, 6, 6, 4, 5	27	5.4
21	6, 4, 2, 3, 5	20	4.0	71	2, 2, 5, 6, 3	18	3.6
22	1, 3, 5, 4, 1	14	2.8	72	6, 6, 6, 1, 6	25	5.0
23	2, 6, 5, 2, 6	21	4.2	73	4, 4, 4, 3, 1	16	3.2
24	3, 5, 1, 3, 5	17	3.4	74	4, 4, 5, 4, 2	19	3.8
25	5, 2, 4, 4, 3	18	3.6	75	4, 5, 4, 1, 4	18	3.6
26	6, 1, 1, 1, 6	15	3.0	76	5, 3, 2, 3, 4	17	3.4
27	1, 4, 1, 2, 6	14	2.8	77	1, 3, 3, 1, 5	13	2.6
28	3, 1, 2, 1, 5	12	2.4	78	4, 1, 5, 5, 3	18	3.6
29	1, 5, 5, 4, 5	20	4.0	79	4, 5, 6, 5, 4	24	4.8
30	4, 5, 3, 5, 2	19	3.8	80	1, 5, 3, 4, 2	15	3.0
31	4, 1, 6, 1, 1	13	2.6	81	4, 3, 4, 6, 3	20	4.0
32	3, 6, 4, 1, 2	16	3.2	82	5, 4, 2, 1, 6	18	3.6
33	3, 5, 5, 2, 2	17	3.4	83	1, 3, 2, 2, 5	13	2.6
34	1, 1, 5, 6, 3	16	3.2	84	5, 4, 1, 4, 6	20	4.0
35	2, 6, 1, 6, 2	17	3.4	85	2, 4, 2, 5, 5	18	3.6
36	2, 4, 3, 1, 3	13	2.6	86	1, 6, 3, 1, 6	17	3.4
37	1, 5, 1, 5, 2	14	2.8	87	2, 2,4, 3, 2	13	2.6
38	6, 6, 5, 3, 3	23	4.6	88	4, 4, 5, 4, 4	21	4.2
39	3, 3, 5, 2, 1	14	2.8	89	2, 5, 4, 3, 4	18	3.6
40	2, 6, 6, 6, 5	25	5.0	90	5, 1, 6, 4, 3	19	3.8
41	5, 5, 2, 3, 4	19	3.8	91	5, 2, 5, 6, 3	21	4.2
42	6, 4, 1, 6, 2	19	3.8	92	6, 4, 1, 2, 1	14	2.8
43	2, 5, 3, 1, 4	15	3.0	93	6, 3, 1, 5, 2	17	3.4
44	4, 2, 3, 2, 1	12	2.4	94	1, 3, 6, 4, 2	16	3.2
45	4, 4, 5, 4, 4	21	4.2	95	6, 1, 4, 2, 2	15	3.0
46	5, 4, 5, 5, 4	23	4.6	96	1, 1, 2, 3, 1	8	1.6
47	6, 6, 6, 2, 1	21	4.2	97	6, 2, 5, 1, 6	20	4.0
48	2, 1, 5, 5, 4	17	3.4	98	3, 1, 1, 4, 1	10	2.0
49	6, 4, 3, 1, 5	19	3.8	99	5, 2, 1, 6, 1	15	3.0
50	4, 4, 4, 4, 4	20	4.0	100	2, 4, 3, 4, 6	19	3.8

6.4. You will observe an interesting result: Although the values of x in the population ($x = 1, 2, 3, 4, 5, 6$) are equiprobable and hence possess a discrete uniform probability distribution, the distribution of the sample means (or sums) chosen from the population forms a **bell-shaped distribution.** We will make one additional comment without proof. If we should repeat the study outlined here by using larger samples of size $n = 10$, we would find that the distribution of the sample means tends to become more nearly bell-shaped.

FIGURE 6.4 Histogram of sample means for the die-tossing experiment

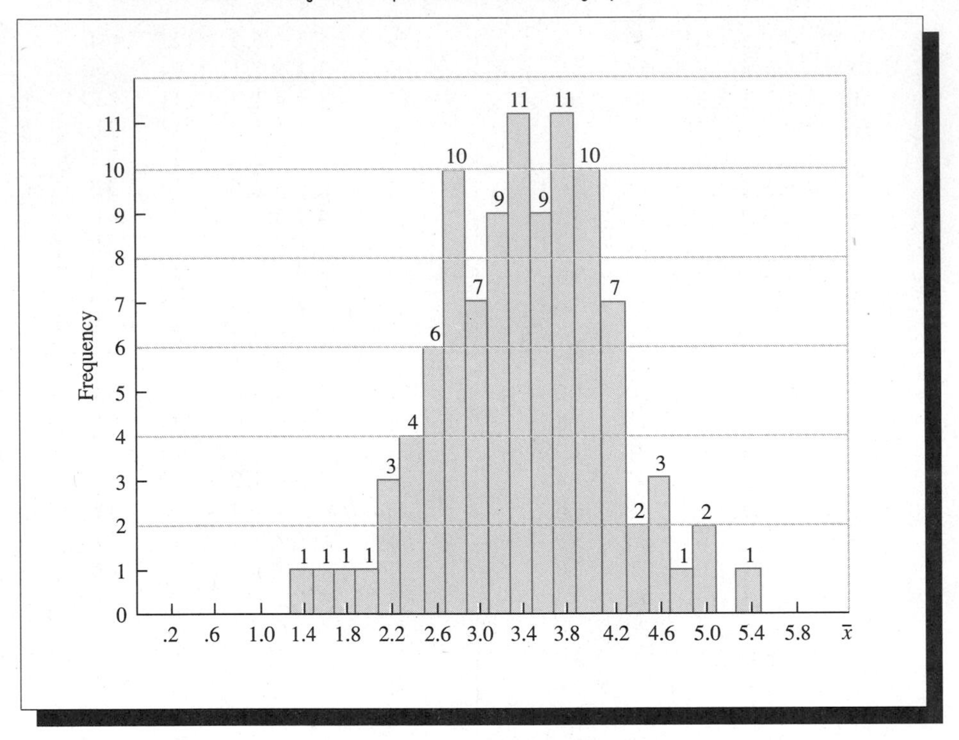

The relative frequency distribution (Figure 6.4) provides only a rough approximation to the sampling distribution of the sample mean. An accurate evaluation of the form of the sampling distribution would require an infinitely large number of samples or, at the very least, far more than the 100 samples contained in our experiment. Nevertheless, the relative frequency distribution constructed from the 100 samples provides a good indication of the form of the sampling distribution of the mean of samples of $n = 5$ observations for the die-tossing experiment and illustrates the basic idea involved in the Central Limit Theorem.

Central Limit Theorem If random samples of n observations are drawn from a nonnormal population with finite mean μ and standard deviation σ, then, when n is large, the sampling distribution of the sample mean $\bar{x}$ is approximately normally distributed, with mean and standard deviation

$$\mu_{\bar{x}} = \mu \qquad \text{and} \qquad \sigma_{\bar{x}} = \frac{\sigma}{\sqrt{n}}$$

The approximation will become more accurate as n becomes large. ▪

The Central Limit Theorem can be restated to apply to the **sum of the sample measurements**

$$\sum_{i=1}^{n} x_i$$

which, as n becomes large, would also tend to possess a normal distribution, in repeated sampling, with mean $n\mu$ and standard deviation $\sigma\sqrt{n}$.

It can be shown (proof omitted) that the mean and standard deviation of the sampling distribution of $\bar{x}$ are always related to the mean and standard deviation of the sampled population as well as to the sample size n. The two distributions have the same mean μ, and the standard deviation of the sampling distribution of $\bar{x}$ is equal to the population standard deviation σ divided by $\sqrt{n}$. (It can be shown that this relationship is true regardless of the sample size n.) Consequently, *the spread of the distribution of sample means is considerably less than the spread of the population distribution.*

The significance of the Central Limit Theorem is twofold. First, it explains the rather common occurrence of normally distributed random variables in nature. We might imagine the height of a person as stemming from a number of effects, each random, and being associated with such things as the height of the mother, the height of the father, the activity of a particular gland, the environment, and diet. If each effects tends to add to the others to yield the measurement of height, then height is the sum of a number of random variables, and according to the Central Limit Theorem, the distribution of heights is approximately normal.

The second and most important contribution of the Central Limit Theorem is in statistical inference. Many estimators that are used to make inferences about population parameters are sums or averages of the sample measurements. When sums or averages are used and the sample size n is sufficiently large, according to the Central Limit Theorem, we expect the estimator to possess (approximately) a normal probability distribution in repeated sampling. We can then use the normal distribution discussed in Chapter 5 to describe the behavior of the inference maker. This aspect of the Central Limit Theorem will be utilized in later chapters dealing with statistical inference.

One disturbing feature of the Central Limit Theorem, and of most approximation procedures, is that we need to have some idea of how large the sample size n must be for the approximation to be valid. Unfortunately, there is no clear-cut solution to this problem, since the appropriate value for n will depend on the population probability

FIGURE 6.5 Probability distributions and approximations of the sampling distributions for three populations (*Note:* vertical scale does not remain the same as n increases)

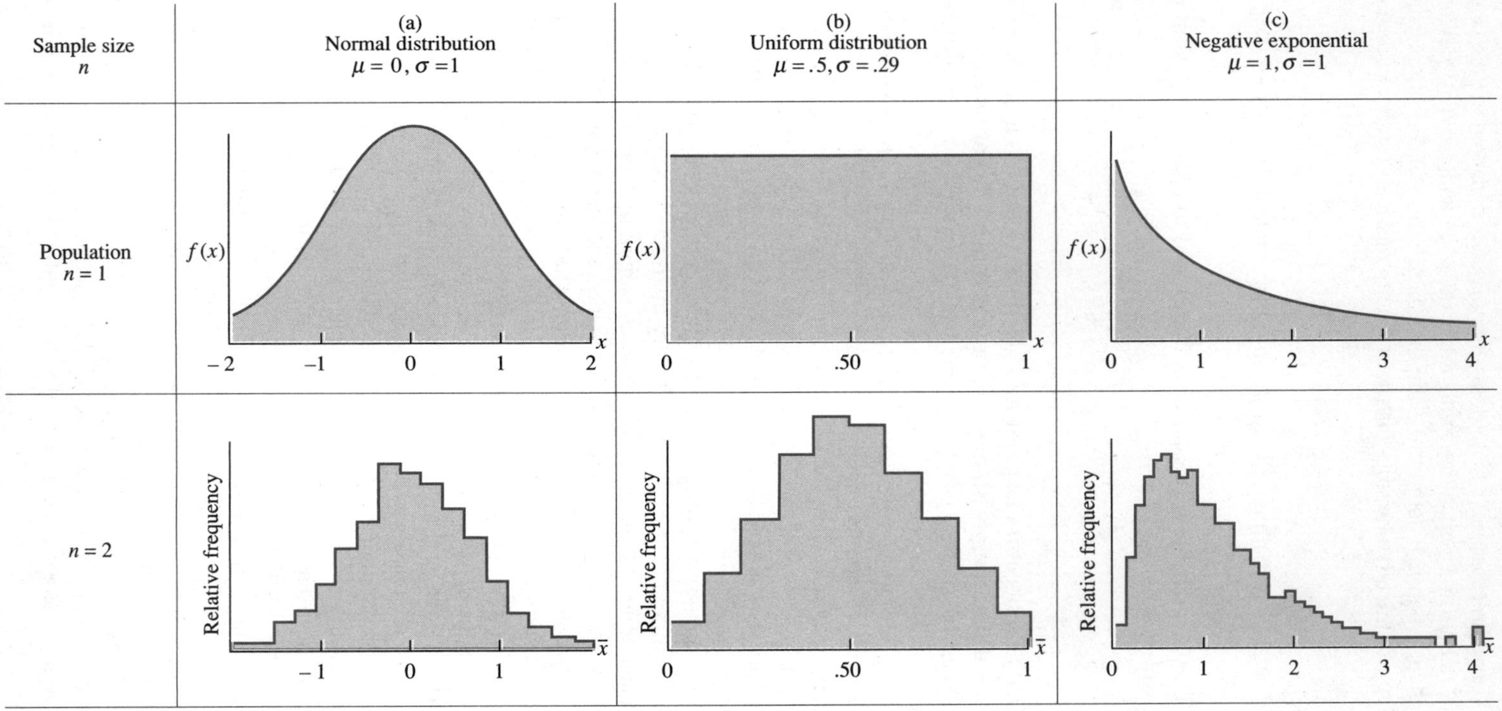

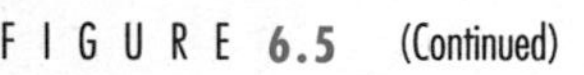

FIGURE 6.5 (Continued)

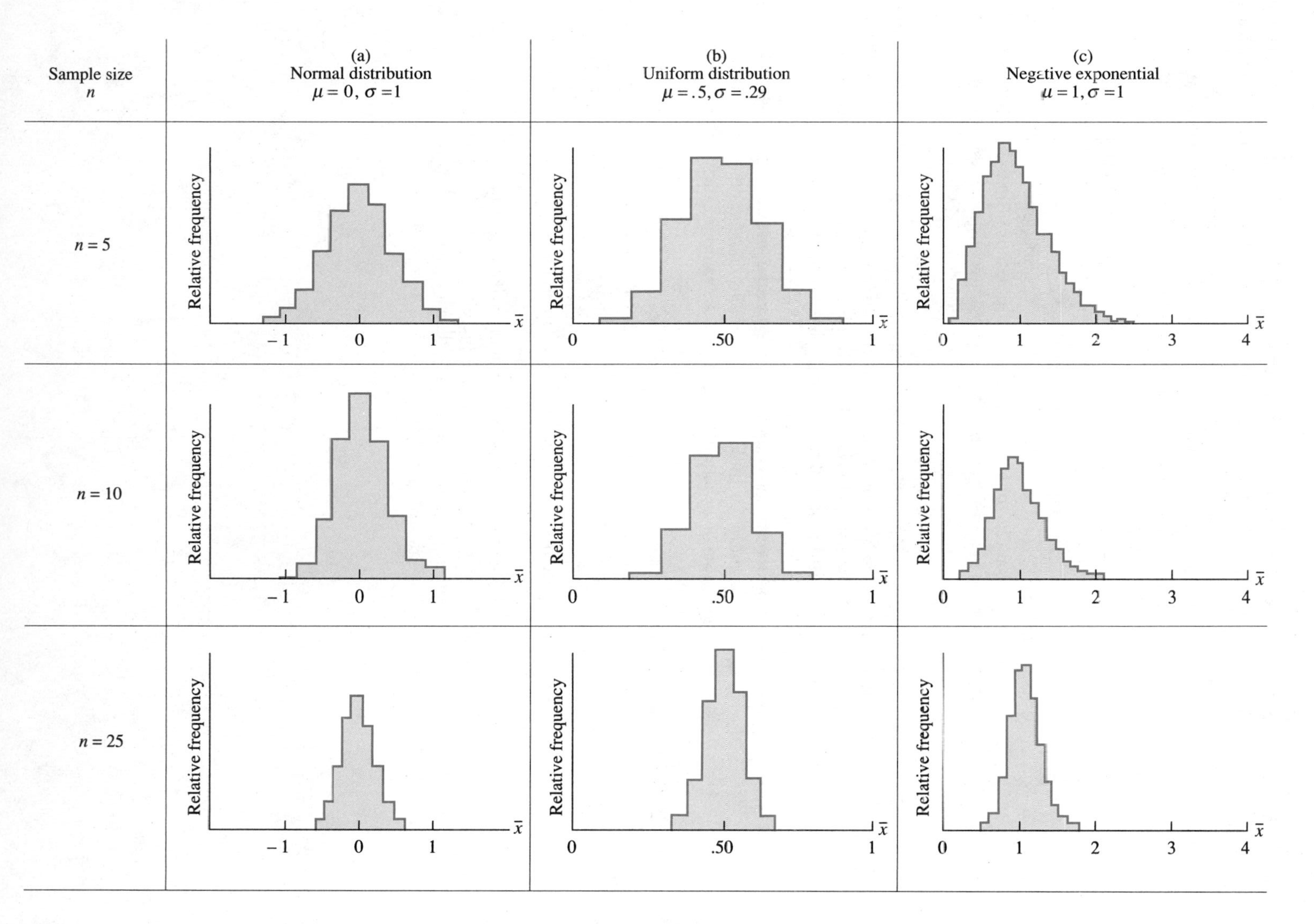

distribution as well as on how the approximation will be used. Although the preceding comment sidesteps the difficulty and suggests that we must rely solely on experience, we may take comfort in the results of the die-tossing experiment discussed earlier in this section. In repeated sampling, the distribution of $\bar{x}$, based on a sample of only $n = 5$ measurements, was approximately bell-shaped. This approximation would be even better for larger values of n.

Figure 6.5 gives the results of some other Monte Carlo sampling experiments. We programmed a computer to select random samples of size n, $n = 2, 5, 10$, and 25, from each of three populations, the first having a normal population probability distribution, the second a uniform probability distribution, and the third a negative exponential probability distribution. These population probability distributions are shown in the top row of Figure 6.5. The computer printouts of the approximations to sampling distributions of the sample mean $\bar{x}$ for sample sizes $n = 2, n = 5, n = 10$, and $n = 25$ are shown in rows 2, 3, 4, and 5 of Figure 6.5.

Figure 6.5 illustrates an important theorem of theoretical statistics. **The sampling distribution of the sample mean is exactly normally distributed (proof omitted), regardless of the sample size, when sampling is from a population that has a normal distribution.** In contrast, the sampling distributions of $\bar{x}$ for samples selected from populations with uniform and negative exponential probability distributions tend to become more nearly normal as the sample size n increases from $n = 2$ to $n = 25$, more rapidly for the uniform distribution, and more slowly for the highly skewed exponential distribution. But note that the sampling distribution of $\bar{x}$ is normal or approximately normal for sampling from either the uniform or the exponential probability distributions when the sample size is as large as $n = 25$. This suggests that, for many populations, the sampling distribution of $\bar{x}$ will be approximately normal for moderate sample sizes; an exception to this rule is the binomial experiment when either p or $(1 - p)$ is small. The appropriate sample size n will be given for specific applications of the Central Limit Theorem as they are encountered in this chapter and later in the text.

In using the sample mean $\bar{x}$ as a random variable, we must differentiate between the probability density function for a single observation x, which has mean μ and standard deviation σ, and the probability density function for $\bar{x}$, which has mean μ but standard deviation $\sigma/\sqrt{n}$.

6.4 The Sampling Distribution of the Sample Mean

Many estimators are available for estimating the population mean, including the median, the trimmed mean, and the midrange (the average of the largest and smallest observations in the set), as well as the sample mean $\bar{x}$. Each generates a sampling distribution in repeated sampling, and, depending on the population and the problem involved, each possesses certain advantages and disadvantages. Some statistics are easier to calculate than others; some statistics produce estimates that are consistently too large or too small; others produce estimates that are highly variable in repeated sampling. The sampling distributions for $\bar{x}$ and m with $n = 3$ for the population of $N = 5$ elements given in Section 6.2 showed that, when we use criteria such as these, the sample mean seemed to perform better than the median as an estimator of the population mean μ. In many situations, the sample mean $\bar{x}$ has desirable properties as an estimator that are not shared by other competing estimators; therefore, it is more widely used.

The Sampling Distribution of the Sample Mean $\bar{x}$

1 If a random sample of n measurements is selected from a population with mean μ and standard deviation σ, the sampling distribution of the sample mean $\bar{x}$ will have a mean

$$\mu_{\bar{x}} = \mu$$

and a standard deviation[†]

$$\sigma_{\bar{x}} = \frac{\sigma}{\sqrt{n}}$$

2 If the population has a *normal* distribution, the sampling distribution of $\bar{x}$ will be *exactly* normally distributed, *regardless of the sample size, n.*

3 If the population distribution is nonnormal, the sampling distribution of $\bar{x}$ will be, for large samples, approximately normally distributed (by the Central Limit Theorem). Figure 6.5 suggests that the sampling distribution of $\bar{x}$ will be approximately normal for sample sizes as small as $n = 25$ for most populations of measurements.

[†]When repeated samples of size n are randomly selected from a *finite* population with N elements whose mean is μ and whose variance is σ^2, the standard deviation of $\bar{x}$ is

$$\sigma_{\bar{x}} = \frac{\sigma}{\sqrt{n}}\sqrt{\frac{N-n}{N-1}}$$

where σ^2 is the population variance. When N is large relative to the sample size n, $\sqrt{(N-n)/(N-1)}$ is approximately equal to 1. Then

$$\sigma_{\bar{x}} = \frac{\sigma}{\sqrt{n}}$$

The standard deviation of a statistic used as an estimator of a population parameter is often called the **standard error of the estimator**, since it refers to the precision of the estimator. Therefore, the standard deviation of $\bar{x}$, given by $\sigma/\sqrt{n}$, is referred to as the **standard error of the mean.**

EXAMPLE 6.1 Suppose you select a random sample of $n = 25$ observations from a population with mean $\mu = 8$ and $\sigma = .6$. Find the approximate probability that the sample mean $\bar{x}$ will

a Be less than 7.9.

b Exceed 7.9.

c Lie within 0.1 of the population mean $\mu = 8$.

Solution **a** Regardless of the shape of the population relative frequency distribution, the sampling distribution of $\bar{x}$ will have a mean $\mu_{\bar{x}} = \mu = 8$ and a standard deviation

$$\sigma_{\bar{x}} = \frac{\sigma}{\sqrt{n}} = \frac{.6}{\sqrt{25}} = .12$$

And, for a sample as large as $n = 25$, it is likely (because of the Central Limit Theorem) that the sampling distribution of $\bar{x}$ is approximately normally distributed (we will assume that it is). Therefore, the probability that $\bar{x}$ will be less than 7.9 is approximated by the shaded area under the normal sampling distribution in Figure 6.6. To find this area, we need to calculate the value of z *corresponding to* $\bar{x} = 7.9$. The value of z = (variable − expected value)/(standard deviation of variable) measures the distance from the mean in units of standard deviations, or the distance between $\bar{x} = 7.9$ and $\mu_{\bar{x}} = \mu = 8.0$ expressed in units of

$$\sigma_{\bar{x}} = \frac{\sigma}{\sqrt{n}} = .12$$

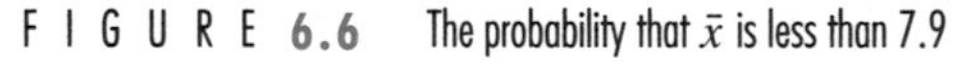

FIGURE 6.6 The probability that $\bar{x}$ is less than 7.9

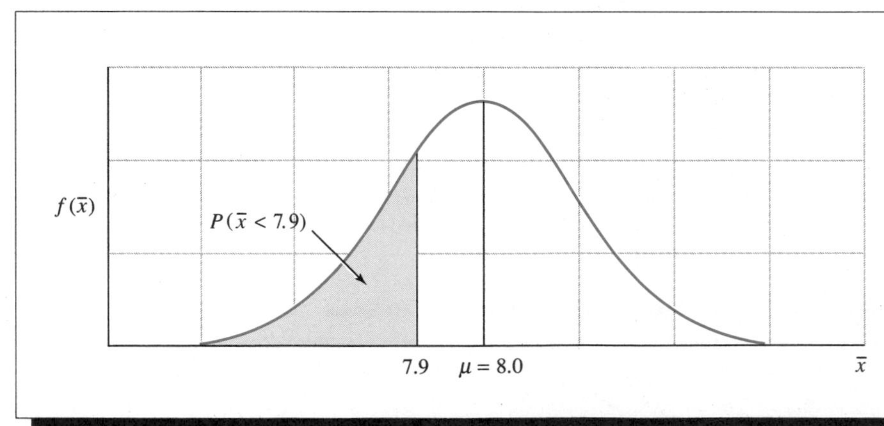

Thus,

$$z = \frac{\bar{x} - \mu}{\sigma_{\bar{x}}} = \frac{7.9 - 8.0}{.12} = -.83$$

From Table 3 in Appendix I, we find that the area corresponding to $z = .83$ is .2967. Therefore,

$$P(\bar{x} < 7.9) = .5 - .2967 = .2033$$

[NOTE: We must use $\sigma_{\bar{x}}$ (not σ) in the formula for z because we are finding an area under the sampling distribution for $\bar{x}$, not under the sampling distribution for x.]

b The event that $\bar{x}$ exceeds 7.9 is the complement of the event that $\bar{x}$ is less than 7.9. Thus, the probability that $\bar{x}$ exceeds 7.9 is

$$\begin{aligned} P(\bar{x} > 7.9) &= 1 - P(\bar{x} < 7.9) \\ &= 1 - .2033 \\ &= .7967 \end{aligned}$$

c The probability that $\bar{x}$ lies within 0.1 of $\mu = 8$ is the shaded area in Figure 6.7. We found in part (a) that the area between $\bar{x} = 7.9$ and $\mu = 8.0$ is .2967. Since the area under the normal curve between $\bar{x} = 8.1$ and $\mu = 8.0$ is identical to the area between $\bar{x} = 7.9$ and $\mu = 8.0$, it follows that

$$P(7.9 < \bar{x} < 8.1) = 2(.2967) = .5934 \quad \blacksquare$$

FIGURE 6.7 The probability that $\bar{x}$ lies within 0.1 of $\mu = 8$

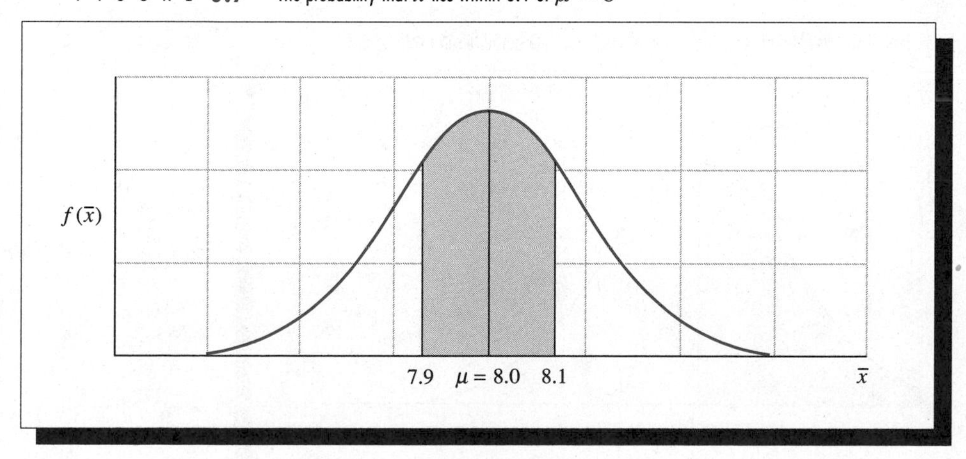

EXAMPLE 6.2 To avoid difficulties with the Federal Trade Commission or state and local consumer protection agencies, a beverage bottler must make reasonably certain that 12-ounce bottles actually contain 12 ounces of beverage. To determine whether a bottling machine is working satisfactorily, one bottler randomly samples ten bottles per hour and measures the amount of beverage in each bottle. The mean $\bar{x}$ of the ten fill measurements is used to decide whether to readjust the amount of beverage delivered per bottle by the filling machine. If records show that the amount of fill per bottle is normally distributed, with a standard deviation of .2 ounce, and if the bottling machine is set to produce a mean fill per bottle of 12.1 ounces, what is the approximate probability that the sample mean $\bar{x}$ of the ten test bottles is less than 12 ounces?

Solution The mean of the sampling distribution of the sample mean $\bar{x}$ is identical to the mean of the population of bottle fills—namely, $\mu = 12.1$ ounces—and the standard deviation (or standard error) of $\bar{x}$ is

$$\sigma_{\bar{x}} = \frac{\sigma}{\sqrt{n}} = \frac{.2}{\sqrt{10}} = .063$$

(NOTE: σ is the standard deviation of the population of bottle fills, and n is the number of bottles in the sample.) Since the amount of fill is normally distributed, $\bar{x}$ is also normally distributed. Then the probability distribution of $\bar{x}$ will appear as shown in Figure 6.8.

The probability that $\bar{x}$ will be less than 12 ounces will equal $(.5 - A)$, where A is the area between 12 and the mean $\mu = 12.1$. Expressing this distance in standard deviations, we have

$$z = \frac{\bar{x} - \mu}{\sigma_{\bar{x}}} = \frac{12 - 12.1}{.063} = -1.59$$

FIGURE 6.8 Probability distribution of $\bar{x}$, the mean of the $n = 10$ bottle fills in Example 6.2

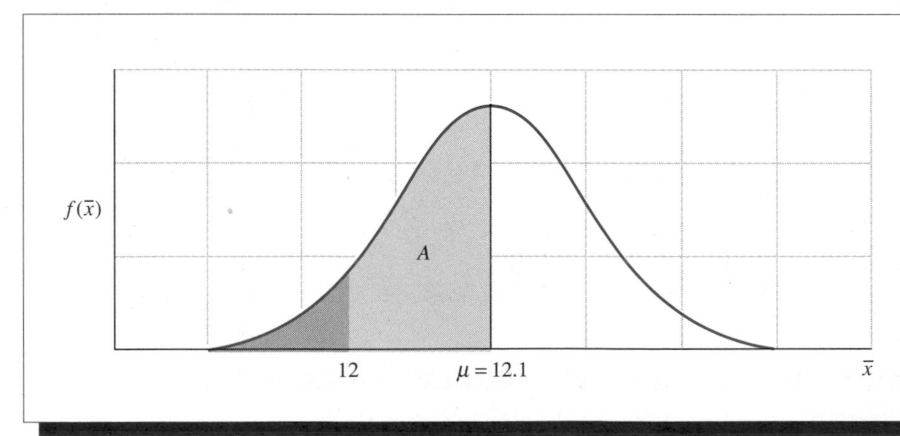

Then the area A over the interval $12 < \bar{x} < 12.1$, found in Table 3 of Appendix I, is .4441, and the probability that $\bar{x}$ will be less than 12 ounces is

$$P(\bar{x} < 12) = .5 - A = .5 - .4441 = .0559 \approx .056$$

Thus, if the machine is set to deliver an average fill of 12.1 ounces, the mean fill $\bar{x}$ of a sample of ten bottles will be less than 12 ounces with probability equal to .056. When this danger signal occurs ($\bar{x}$ is less than 12), the bottler takes a larger sample to recheck the setting of the filling machine. ▪

Tips on Problem Solving

Before attempting to calculate the probability that the statistic $\bar{x}$ falls in some interval, complete the following steps:

1. Calculate the mean and standard deviation of the sampling distribution of $\bar{x}$.
2. Sketch the sampling distribution. Show the location of the mean μ, and use the value of $\sigma_{\bar{x}}$ to locate the approximate location of the tails of the distribution.
3. Locate the interval on the sketch from part 2 and shade the area corresponding to the probability that you wish to calculate.
4. Find the z-score(s) associated with the value(s) of interest. Use Table 3 in Appendix I to find the probability.
5. When you have obtained your answer, look at your sketch of the sampling distribution to see whether your calculated answer agrees with the shaded area. This provides a very rough check on your calculations.

Exercises

Basic Techniques

6.1 Random samples of size n were selected from populations with the means and variances shown below. Find the mean and standard deviation (or standard error) of the sampling distribution of the sample mean for

a $n = 25, \mu = 10, \sigma^2 = 9$.

b $n = 100, \mu = 5, \sigma^2 = 4$.

c $n = 8, \mu = 120, \sigma^2 = 1$.

6.2 Refer to Exercise 6.1.

a If the sampled populations are normal, what is the sampling distribution of $\bar{x}$ for parts (a), (b), and (c)?

b According to the Central Limit Theorem, if the sampled populations are *not* normal, what can be said about the sampling distribution of $\bar{x}$ for parts (a), (b), and (c)?

6.3 Refer to Exercise 6.1, part (b).

a Sketch the sampling distribution for the sample mean and locate the mean and the interval $\mu \pm 2\sigma_{\bar{x}}$ along the $\bar{x}$-axis.

b Shade the area under the curve that corresponds to the probability that $\bar{x}$ lies within .15 unit of the population mean μ.

c Find the probability described in part (b).

6.4 Refer to the die-throwing experiment in Section 6.3 in which x is the number of dots observed when a single die is tossed. The probability distribution for x is given in Figure 6.3, and the relative frequency histogram for $\bar{x}$ is given in Figure 6.4 for 100 random samples of size $n = 5$.

a Verify that the mean and standard deviation of x are $\mu = 3.5$ and $\sigma = 1.71$, respectively.

b Look at the histogram in Figure 6.4. Guess the value of its mean and standard deviation. (HINT: The Empirical Rule states that approximately 95% of the measurements associated with a mound-shaped distribution will lie within two standard deviations of the mean.)

c What are the theoretical mean and standard deviation of the sampling distribution of $\bar{x}$? How do these values compare to the guessed values of part (b)?

6.5 Refer to Exercise 6.4. Suppose the sampling experiment in Section 6.3 was repeated over and over again an infinitely large number of times. Find the mean and standard deviation (or standard error) for the sampling distribution of $\bar{x}$ if each sample consists of

a $n = 10$ measurements.

b $n = 15$ measurements.

c $n = 25$ measurements.

6.6 Refer to Exercises 6.4 and 6.5. What effect does increasing the sample size have on the sampling distribution of $\bar{x}$?

6.7 Exercises 6.4 and 6.5 demonstrate that the standard deviation of the sampling distribution decreases as the sample size increases. To see this relationship more clearly, suppose that a random sample of n observations is selected from a population with standard deviation $\sigma = 1$. Calculate $\sigma_{\bar{x}}$ for $n = 1, 2, 4, 9, 16, 25$, and 100. Then plot $\sigma_{\bar{x}}$ versus the sample size n and connect the points with a smooth curve. Note the manner in which $\sigma_{\bar{x}}$ decreases as n increases.

6.8 Suppose a random sample of $n = 5$ observations is selected from a population that is normally distributed, with mean equal to 1 and standard deviation equal to .36.

a Give the mean and standard deviation of the sampling distribution of $\bar{x}$.

b Find the probability that $\bar{x}$ exceeds 1.3.

c Find the probability that the sample mean $\bar{x}$ will be less than .5.

d Find the probability that the sample mean will deviate from the population mean $\mu = 1$ by more than .4.

6.9 Suppose a random sample of $n = 25$ observations is selected from a population that is normally distributed, with mean equal to 106 and standard deviation equal to 12.

a Give the mean and the standard deviation of the sampling distribution of the sample mean $\bar{x}$.

b Find the probability that $\bar{x}$ exceeds 110.

c Find the probability that the sample mean will deviate from the population mean $\mu = 106$ by no more than 4.

Applications

6.10 The average household size in the United States is dropping, according to new figures released by the census bureau. This shrinkage has been occurring steadily over the last two decades, with the average household size in 1991 given as 2.63 people, down from 2.7 in 1980 (Dubin, 1992). Why would this very small decrease in the average household size be viewed as significant by the census bureau?

6.11 Explain why the weight of a package of one dozen tomatoes should be approximately normally distributed if the dozen tomatoes represent a random sample.

6.12 Use the Central Limit Theorem to explain why a Poisson random variable—say, the number of a particular type of bacteria in a cubic foot of water—has a distribution that can be approximated by a normal distribution when the mean μ is large. (HINT: One cubic foot of water contains 1728 cubic inches of water.)

6.13 Americans are taking better care of themselves, according to a Louis Harris survey conducted for *Prevention* magazine (*Press-Enterprise*, Riverside, Calif., May 24, 1989). The results of the survey were used to construct a Prevention Index—a composite score of personal health behaviors, including efforts to limit cholesterol, alcohol consumption, and cigarette smoking and measures relating to weight and stress levels. A random sample of 1250 American adults scored 65.4 out of a possible 100 points, nearly 4 points higher than the 1984 average.

a Explain why the Prevention Index scores should be approximately normally distributed.

b If we assume that the average score for all American adults has not changed since 1984 and is equal to 61.5, with a standard deviation of 12, what is the probability of observing a sample average of 65.4 or larger?

c Based on the results of part (b), is it reasonable to assume that the average score for all American adults has not changed since 1984? Explain.

6.14 An important expectation of the recent federal income tax reduction is that consumers will reap a substantial portion of the tax savings. Suppose estimates of the portion of total tax saved, based on a random sampling of 35 economists, had a mean of 26% and a standard deviation of 12%.

a What is the approximate probability that a sample mean, based on a random sample of $n = 35$ economists, will lie within 1% of the mean of the population of the estimates of all economists?

b Is it necessarily true that the mean of the population of estimates of all economists is equal to the percentage of tax savings that will actually be achieved?

6.15 A manufacturer of paper used for packaging requires a minimum strength of 20 pounds per square inch. To check on the quality of the paper, a random sample of ten pieces of paper is selected each hour from the previous hour's production and a strength measurement is recorded for each. The standard deviation σ of the strength measurements, computed by pooling the sum of squares of deviations of many samples, is known to equal 2 pounds per square inch.

a What is the approximate probability distribution of the sample mean of $n = 10$ test pieces of paper?

b If the mean of the population of strength samples is 21 pounds per square inch, what is the approximate probability that, for a random sample of $n = 10$ test pieces of paper, $\bar{x} < 20$?

c What value would you desire for the mean paper strength μ in order that $P(\bar{x} < 20)$ be equal to .001?

6.16 The normal daily human potassium requirement is in the range of 2000 to 6000 mg, with the larger amounts required during hot summer weather. The amount of potassium in food varies, depending on the food. For example, there are approximately 7 mg in a cola drink, 46 mg in a beer, 630 mg in a banana, 300 mg in a carrot, and 440 mg in a glass of orange juice. Suppose the distribution of potassium in a banana is normally distributed, with mean equal to 630 mg and standard deviation equal to 40 mg per banana. Suppose you eat $n = 3$ bananas per day, and T is the total number of milligrams of potassium you receive from them.

a Find the mean and standard deviation of T.

b Find the probability that your total daily intake of potassium from the three bananas will exceed 2000 mg.
(HINT: Note that T is the sum of three random variables x_1, x_2, and x_3, where x_1 is the amount of potassium in banana number 1, etc.)

6.17 The total daily sales, x, in the deli section of a local market is the sum of the sales generated by a fixed number of customers who make purchases on a given day.

a What kind of probability distribution would you expect the total daily sales to have? Explain.

b For this particular market, the average sale per customer in the deli section is \$8.50 with $\sigma = \$2.50$. If 30 customers make deli purchases on a given day, give the mean and standard deviation of the probability distribution of the total daily sales, x.

6.5 The Sampling Distribution of the Sample Proportion

Many sampling problems involve consumer preference or opinion polls, which are concerned with estimating the proportion p of people in the population who possess some specified characteristic. These and similar situations provide practical examples of binomial experiments, if the sampling procedure has been conducted in the appropriate manner. If a random sample of n persons is selected from the population and if x of these possess the specified characteristic, then the sample proportion

$$\hat{p} = x/n$$

is used to estimate the population proportion p.[†]

Since each distinct value of x results in a distinct value of $\hat{p} = x/n$, the probabilities associated with $\hat{p}$ are equal to the probabilities associated with the corresponding values of x. Hence, the sampling distribution of $\hat{p}$ will be the same shape as the binomial probability distribution for x. Like the binomial probability distribution, it can be approximated by a normal distribution when the sample size n is large. The mean of the sampling distribution of $\hat{p}$ is

$$\mu_{\hat{p}} = p$$

and its standard deviation is

$$\sigma_{\hat{p}} = \sqrt{\frac{pq}{n}}$$

where

$$q = 1 - p$$

[†]A "hat" placed over the symbol of a population parameter denotes a statistic used to estimate the population parameter. For example, the symbol $\hat{p}$ denotes the sample proportion.

Properties of the Sampling Distribution of the Sample Proportion $\hat{p}$

1. If a random sample of n observations is selected from a binomial population with parameter p, the sampling distribution of the sample proportion

$$\hat{p} = x/n$$

will have a mean

$$\mu_{\hat{p}} = p$$

and a standard deviation

$$\sigma_{\hat{p}} = \sqrt{\frac{pq}{n}}$$

where

$$q = 1 - p$$

2. When the sample size n is large, the sampling distribution of $\hat{p}$ can be approximated by a normal distribution. The approximation will be adequate if $\mu_{\hat{p}} - 2\sigma_{\hat{p}}$ and $\mu_{\hat{p}} + 2\sigma_{\hat{p}}$ fall in the interval 0 to 1.

EXAMPLE 6.3 In a survey, 500 mothers and fathers were asked about the importance of sports for boys and girls. Of the parents interviewed, 60% agreed that both sexes are equal and should have equal opportunities to participate in sports. Describe the sampling distribution of the sample proportion $\hat{p}$ of parents who agree that both sexes are equal and should have equal opportunities.

Solution We will assume that the 500 parents represent a random sample of the parents of all boys and girls in the United States and that the true proportion in the population is equal to some unknown value that we will call p. Then the sampling distribution of $\hat{p}$ can be approximated by a normal distribution,[†] with mean equal to p (see Figure 6.9) and standard deviation

$$\sigma_{\hat{p}} = \sqrt{\frac{pq}{n}}$$

Examining Figure 6.9, we see that the sampling distribution of $\hat{p}$ centers over its mean p. Even though we do not know the exact value of p (the sample proportion $\hat{p} = .60$ may be larger or smaller than p), we can calculate an approximate value

[†]Checking the conditions that allow the normal approximation to the distribution of $\hat{p}$, we find that $n = 500$ is adequate for values of p near .60, since $\hat{p} \pm 2\sigma_{\hat{p}} = .6 \pm 2\sqrt{.6(.4)/500} = .6 \pm .044$, or .556 to .644, falls in the interval 0 to 1.

FIGURE 6.9 The sampling distribution for $\hat{p}$ based on a sample of $n = 500$ parents

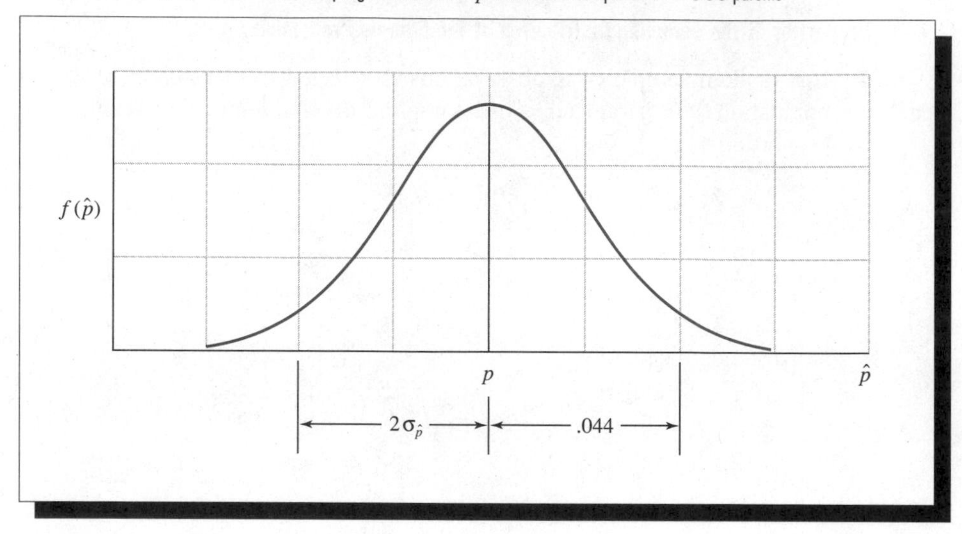

for the standard deviation of the sampling distribution using the sample proportion $\hat{p} = .60$ to approximate the unknown value of p. Thus

$$\sigma_{\hat{p}} = \sqrt{\frac{pq}{n}} \approx \sqrt{\frac{\hat{p}\hat{q}}{n}}$$
$$= \sqrt{\frac{(.60)(.40)}{500}}$$
$$= .022$$

Therefore, we know that, approximately 95% of the time, $\hat{p}$ will fall within $2\sigma_{\hat{p}} \approx .044$ of the (unknown) value of p. ▪

EXAMPLE 6.4 Refer to Example 6.3. Suppose the population proportion p of parents in the population is actually equal to .55. What is the probability of observing a sample proportion as large as or larger than the observed value $\hat{p} = .60$?

Solution Figure 6.10 shows the sampling distribution of $\hat{p}$ when $p = .55$ with the observed value $\hat{p} = .60$ located on the horizontal axis. From the figure, we can see that the probability of observing a sample proportion p equal to or larger than .60 can be approximated by the shaded area in the upper tail of a normal distribution with

$$\mu_{\hat{p}} = .55$$

and

$$\sigma_{\hat{p}} = \sqrt{\frac{pq}{n}} = \sqrt{\frac{(.55)(.45)}{500}} = .0222$$

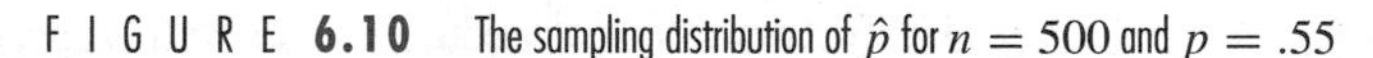

FIGURE **6.10** The sampling distribution of $\hat{p}$ for $n = 500$ and $p = .55$

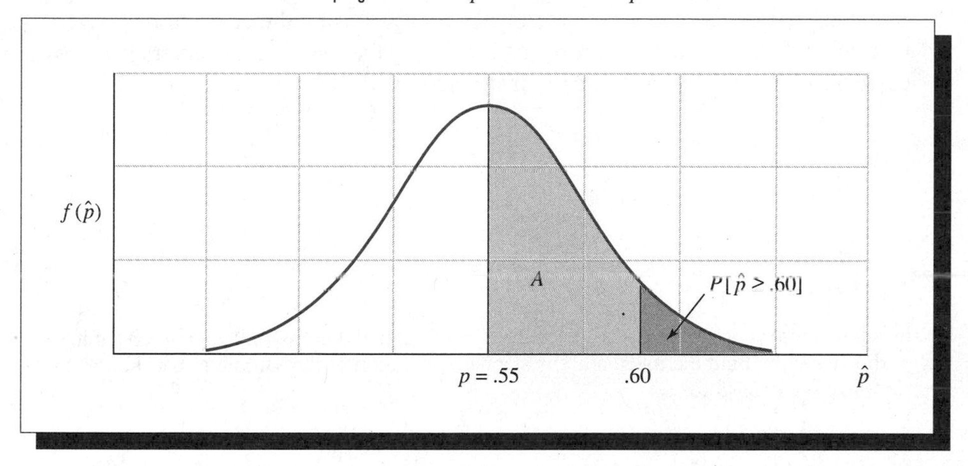

To find this shaded area, we need to know how many standard deviations the observed value $\hat{p} = .60$ lies away from the mean of the sampling distribution $p = .55$. This distance is given by the z value

$$z = \frac{\hat{p} - p}{\sigma_{\hat{p}}} = \frac{.60 - .55}{.0222} = 2.25$$

Table 3 in Appendix I gives the area A corresponding to $z = 2.25$ as

$$A = .4878$$

Therefore, the shaded area in the upper tail of the sampling distribution in Figure 6.10 is

$$P(\hat{p} > .60) = .5 - A = .5 - .4878 = .0122$$
$$\approx .01$$

This tells us that, if we were to select a random sample of $n = 500$ observations from a population with proportion p equal to .55, the probability that the sample proportion $\hat{p}$ would be as large as or larger than .60 is only .01. ▪

In employing the normal distribution to approximate the binomial probabilities associated with x, we used a correction of $\pm.5$ to improve the approximation. The equivalent correction here would be $\pm(1/2n)$. For example, for $\hat{p}$ the value of z with the correction would be

$$z_1 = \frac{(.60 - .001) - .55}{\sqrt{\frac{(.55)(.45)}{500}}} = 2.20$$

with $A = .4861$ and $(.5 - A) = .0139$. To two-decimal accuracy, this value agrees with our earlier result. When n is large, the effect of using the correction is generally negligible. You should solve problems in this and the remaining chapters *without* the correction factor unless specifically instructed to use it.

Exercises

Basic Techniques

6.18 Random samples of size n were selected from binomial populations with population parameters p shown below. Find the mean and the standard deviation of the sampling distribution of the sample proportion $\hat{p}$ for

a $n = 100, p = .3$.

b $n = 400, p = .1$.

c $n = 250, p = .6$.

6.19 Sketch each of the sampling distributions in Exercise 6.18. For each, locate the mean p and the interval $p \pm 2\sigma_{\hat{p}}$ along the $\hat{p}$-axis of the graph.

6.20 Refer to the sampling distribution in Exercise 6.18 (a).

a Sketch the sampling distribution for the sample proportion and shade the area under the curve that corresponds to the probability that $\hat{p}$ lies within .08 of the population proportion p.

b Find the probability described in part (a).

6.21 Random samples of size $n = 500$ were selected from a binomial population with $p = .1$.

a Is it appropriate to use the normal distribution to approximate the sampling distribution of $\hat{p}$? Check to make sure the necessary conditions are met.

Using the results of part (a), find the probability that

b $\hat{p} > .12$.

c $\hat{p} < .10$.

d $\hat{p}$ lies within .02 of p.

6.22 Calculate $\sigma_{\hat{p}}$ for $n = 100$ and

a $p = .01$. **b** $p = .10$. **c** $p = .30$. **d** $p = .50$.

e $p = .70$. **f** $p = .90$. **g** $p = .99$.

Plot $\sigma_{\hat{p}}$ versus p on graph paper and sketch a smooth curve through the points. For what value of p is the standard deviation of the sampling distribution of $\hat{p}$ a maximum? What happens to $\sigma_{\hat{p}}$ when p is near 0 or near 1.0?

6.23 **a** Is the normal approximation to the sampling distribution of $\hat{p}$ appropriate when $n = 400$ and $p = .8$?

b Use the results of part (a) to find the probability that $\hat{p}$ is greater than .83.

c Use the results of part (a) to find the probability that $\hat{p}$ lies between .76 and .84.

Applications

6.24 In a study to evaluate the attitudes of young adults aged 18–24 years toward marriage and the family, 85% of the 505 respondents felt that couples in their generation are more likely to get divorced than those in their parents' generation (*Time*, Special Fall Edition, 1990). Assume that the 505 respondents in the survey represent a random sample of 18–24-year-olds from the U.S. population.

a Describe the sampling distribution of $\hat{p}$, the proportion in the sample who think that divorce is more likely for their generation. (HINT: Use $\hat{p}$ to approximate p when calculating $\sigma_{\hat{p}}$.)

b Find the probability that $\hat{p}$ will lie within .02 of the proportion p of young adults in the population who think that divorce is more likely for their generation.

6.25 What are Americans' pet peeves: David Wessel (1989) reported that "Americans don't like anything that wastes their time." According to Wessel's report, four out of every ten people surveyed indicated that their pet peeve was staying home for deliverypersons or salespeople who don't show. As part of the survey, Peter Hart asked 1034 consumers (about half of the total sample) specifically about service complaints.

a What is the sampling distribution for $\hat{p}$, the sample proportion surveyed who indicated that their pet peeve was staying home for deliverypersons or salespeople who didn't show?

b What is the probability that the sample value of $\hat{p}$ lies within .02 of the true population value of p?

c How would the answer to part (b) change if the sample size were doubled?

6.26 Americans continue to buy and use more and more electronic conveniences and devices. In 1992, for example, 73% of American households had videocassette recorders (VCRs) and 60% of American homes had cable television (Wright, 1992, p. 222).

a In a random sample of $n = 1000$ American households, describe the sampling distribution of $\hat{p}$, the sample proportion of households having cable television, if in fact the 60% figure is correct.

b What is the probability that the sample proportion $\hat{p}$ of households having cable television differs from the true population proportion by at most 3%?

c If 73% of American households have a VCR, what is the probability that the sample proportion based on a random sample of $n = 1000$ lies within .03 of the true population proportion?

d Why are the answers to parts (b) and (c) different even though both samples consisted of $n = 1000$ households?

6.27 Teachers entering the classroom for the first time are optimistic about their careers but know that many problems await them. A random sample of 1002 beginning teachers felt that their training was adequate and that most teachers were dedicated to their work. However, when asked "Should all teachers take a national, standardized test to demonstrate their qualifications?" only 31% strongly agreed and 35% somewhat agreed (*Education Week*, Sept. 26, 1990).

a Describe the sampling distribution of $\hat{p}$, the proportion of beginning teachers in the sample who strongly agree that a national standardized test should be taken by all teachers. (Hint: Use $\hat{p}$ to approximate p when calculating $\sigma_{\hat{p}}$.)

b What is the probability that $\hat{p}$ will differ from p by more than .04?

6.6 A Sampling Application: Statistical Process Control

Statistical process control (SPC) methodology was developed to monitor, control, and improve products and services. Steel bearings must conform to size and hardness specifications, industrial chemicals must have a low prespecified level of impurities, and accounting firms must minimize and ultimately eliminate incorrect bookkeeping entries. It is often said that statistical process control consists of 10% statistics, and 90% engineering, and common sense. We can statistically monitor a process mean and tell when the mean falls outside preassigned limits, but we can't tell *why* it is out of control. Answering this last question requires knowledge of the process and problem-solving ability—the other 90%!

Product quality is usually monitored using statistical control charts. Measurements on a process variable to be monitored change over time. The cause of a change in the variable is said to be *assignable* if it can be found and corrected. Other variation—small haphazard changes due to alteration in the production environment—that is not controllable is regarded as *random variation*. If the variation in a process variable is solely random, the process is said to be *in control*. The first objective in statistical process control is to eliminate assignable causes of variation in the process variable and then get the process in control. The next step is to reduce variation and get the measurements on the process variable within *specification limits,* the limits within which the measurements on usable items or services must fall.

Once a process is in control and is producing a satisfactory product, the process variables are monitored by use of **control charts**. Samples of n items are drawn from the process at specified intervals of time, and a sample statistic is computed. These statistics are plotted on the control chart, so that the process can be checked for shifts in the process variable that might indicate control problems.

A Control Chart for the Process Mean: The $\bar{x}$ Chart

Assume that n items are selected from the production process at equal intervals and that measurements are recorded on the process variable. If the process is in control, the sample means should vary about the population mean μ in a random manner. Moreover, according to the Central Limit Theorem, the sampling distribution of $\bar{x}$ should be approximately normal, so that almost all (99.7%) of the values of $\bar{x}$ should fall in the interval $(\mu \pm 3\sigma_{\bar{x}}) = \mu \pm 3(\sigma/\sqrt{n})$. Although the exact values of μ and σ are unknown, we can obtain accurate estimates by using the sample measurements.

Every control chart has a *centerline* and *control limits*. The centerline is the estimate of μ, the grand average of all the sample statistics calculated from the measurements on the process variable. The upper and lower *control limits* are placed 3 standard deviations above and below the centerline. If we monitor the process mean based on k samples of size n taken at regular intervals, the centerline is $\bar{\bar{x}}$, the average of the sample means, and the control limits are at $\bar{\bar{x}} \pm 3(\sigma/\sqrt{n})$, with σ estimated by s, the standard deviation of the nk measurements.

EXAMPLE 6.5 A statistical process control monitoring system samples the inside diameters of $n = 4$ bearings each hour. Table 6.5 provides the data for $k = 25$ hourly samples. Construct an $\bar{x}$ chart for monitoring the process mean.

Solution The sample mean was calculated for each of the $k = 25$ samples. For example, the mean for sample 1 is

$$\bar{x} = \frac{.992 + 1.007 + 1.016 + .991}{4} = 1.0015$$

TABLE 6.5
25 hourly samples of bearing diameters, $n = 4$ bearings per sample, for Example 6.5

Sample	Sample Measurements				Sample Mean, $\bar{x}$
1	0.992	1.007	1.016	0.991	1.00150
2	1.015	0.984	0.976	1.000	0.99375
3	0.988	0.993	1.011	0.981	0.99325
4	0.996	1.020	1.004	0.999	1.00475
5	1.015	1.006	1.002	1.001	1.00600
6	1.000	0.982	1.005	0.989	0.99400
7	0.989	1.009	1.019	0.994	1.00275
8	0.994	1.010	1.009	0.990	1.00075
9	1.018	1.016	0.990	1.011	1.00875
10	0.997	1.005	0.989	0.001	0.99800
11	1.020	0.986	1.002	0.989	0.99925
12	1.007	0.986	0.981	0.995	0.99225
13	1.016	1.002	1.010	0.999	1.00675
14	0.982	0.995	1.011	0.987	0.99375
15	1.001	1.000	0.983	1.002	0.99650
16	0.992	1.008	1.001	0.996	0.99925
17	1.020	0.988	1.015	0.986	1.00225
18	0.993	0.987	1.006	1.001	0.99675
19	0.978	1.006	1.002	0.982	0.99200
20	0.984	1.009	0.983	0.986	0.99050
21	0.990	1.012	1.010	1.007	1.00475
22	1.015	0.983	1.003	0.989	0.99750
23	0.983	0.990	0.997	1.002	0.99300
24	1.011	1.012	0.991	1.008	1.00550
25	0.987	0.987	1.007	0.995	0.99400

The sample means are shown in column 6 of Table 6.5. The centerline is located at

$$\bar{\bar{x}} = \frac{99.87}{100} = .9987$$

The calculated value of s, the sample standard deviation of all $nk = 4(25) = 100$ observations, is $s = .011458$. The estimated standard error of the mean of $n = 4$ observations is then

$$\frac{s}{\sqrt{n}} = \frac{.011458}{\sqrt{4}} = .005729$$

The upper and lower control limits are found as

$$\text{UCL} = \bar{\bar{x}} + 3\frac{s}{\sqrt{n}} = .9987 + 3(.005729) = 1.015887$$

and

$$\text{LCL} = \bar{\bar{x}} - 3\frac{s}{\sqrt{n}} = .9987 - 3(.005729) = .981513$$

Figure 6.11 shows an EXECUSTAT printout of the $\bar{x}$ chart constructed from the data. Assuming that the samples used to construct the $\bar{x}$ chart were collected when the process was in control, the chart can now be used to detect changes in the process mean. Sample means are plotted periodically, and, if a sample mean falls outside the control limits, a warning should be conveyed. The process should be checked to locate the cause of the unusually large or small mean. ▪

FIGURE 6.11 EXECUSTAT $\bar{x}$ chart for Example 6.5

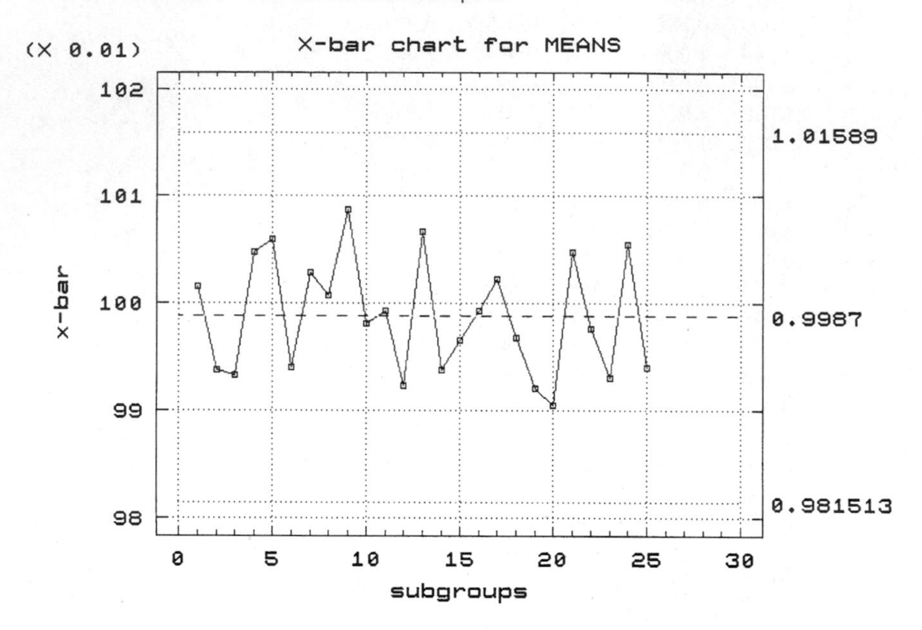

A Control Chart for the Proportion Defective: The p Chart

Sometimes the observation made on an item is simply whether or not it meets specifications; thus, it judged to be defective or nondefective. If the fraction defective produced by the process is p, then x, the number of defectives in a sample of n items, has a binomial distribution.

To monitor a process for defective items, samples of size n are selected at periodic intervals and the sample proportion $\hat{p}$ is calculated. When the process is in control, $\hat{p}$ should fall in the interval $p \pm 3\sigma_{\hat{p}}$, where p is the proportion of defectives in the population (or the process fraction defective) and

$$\sigma_{\hat{p}} = \sqrt{\frac{p(1-p)}{n}}$$

The process fraction defective is unknown but can be estimated by the average of the k sample proportions

$$\bar{p} = \frac{\sum_{i=1}^{k} \hat{p}_i}{k}$$

and $\sigma_{\hat{p}}$ is estimated by

$$\hat{\sigma}_{\hat{p}} = \sqrt{\frac{\bar{p}(1-\bar{p})}{n}}$$

The centerline for the ***p* chart** is located at $\bar{p}$, and the upper and lower control limits are

$$\text{UCL} = \bar{p} + 3\sqrt{\frac{\bar{p}(1-\bar{p})}{n}}$$

and

$$\text{LCL} = \bar{p} - 3\sqrt{\frac{\bar{p}(1-\bar{p})}{n}}$$

EXAMPLE 6.6 A manufacturer of ballpoint pens randomly samples 400 pens per day and tests each to see whether the ink flow is acceptable. The proportions of pens judged defective each day over a 40-day period are shown in Table 6.6. Construct a control chart for the proportion $\hat{p}$ defective in samples of $n = 400$ pens selected from the process.

TABLE 6.6 Proportions of defectives in samples of $n = 400$ pens for Example 6.6

Day	Proportion	Day	Proportion	Day	Proportion	Day	Proportion
1	.0200	11	.0100	21	.0300	31	.0225
2	.0125	12	.0175	22	.0200	32	.0175
3	.0225	13	.0250	23	.0125	33	.0225
4	.0100	14	.0175	24	.0175	34	.0100
5	.0150	15	.0275	25	.0225	35	.0125
6	.0200	16	.0200	26	.0150	36	.0300
7	.0275	17	.0225	27	.0200	37	.0200
8	.0175	18	.0100	28	.0250	38	.0150
9	.0200	19	.0175	29	.0150	39	.0150
10	.0250	20	.0200	30	.0175	40	.0225

Solution The estimate of the process proportion defective is the average of the $k = 40$ sample proportions of Table 6.6. Therefore, the centerline of the control chart is located at

$$\bar{p} = \frac{\sum_{i=1}^{k} p_i}{k} = \frac{.0200 + .0125 + \cdots + .0225}{40} = \frac{.7600}{40} = .019$$

An estimate of $\sigma_{\hat{p}}$, the standard deviation of the sample proportions, is

$$\hat{\sigma}_{\hat{p}} = \sqrt{\frac{\bar{p}(1-\bar{p})}{n}} = \sqrt{\frac{(.019)(.981)}{400}} = .00683$$

and $3\hat{\sigma}_{\hat{p}} = (3)(.00683) = .0205$. Therefore, the upper and lower control limits for the p chart are located at

$$\text{UCL} = \bar{p} + 3\hat{\sigma}_{\hat{p}} = .0190 + .0205 = .0395$$

and

$$\text{LCL} = \bar{p} - 3\hat{\sigma}_{\hat{p}} = .0190 - .0205 = -.0015$$

Or, since p cannot be negative, LCL = 0.

The $\hat{p}$ control chart is shown in Figure 6.12. Note that all 40 sample proportions fall within the control limits. If a sample proportion collected at some time in the future falls outside the control limits, the manufacturer will be warned of a possible increase in the value of the process proportion defective. Efforts will be initiated to seek possible causes for an increase in the value of the process proportion defective. ▪

FIGURE 6.12 EXECUSTAT p chart for Example 6.6

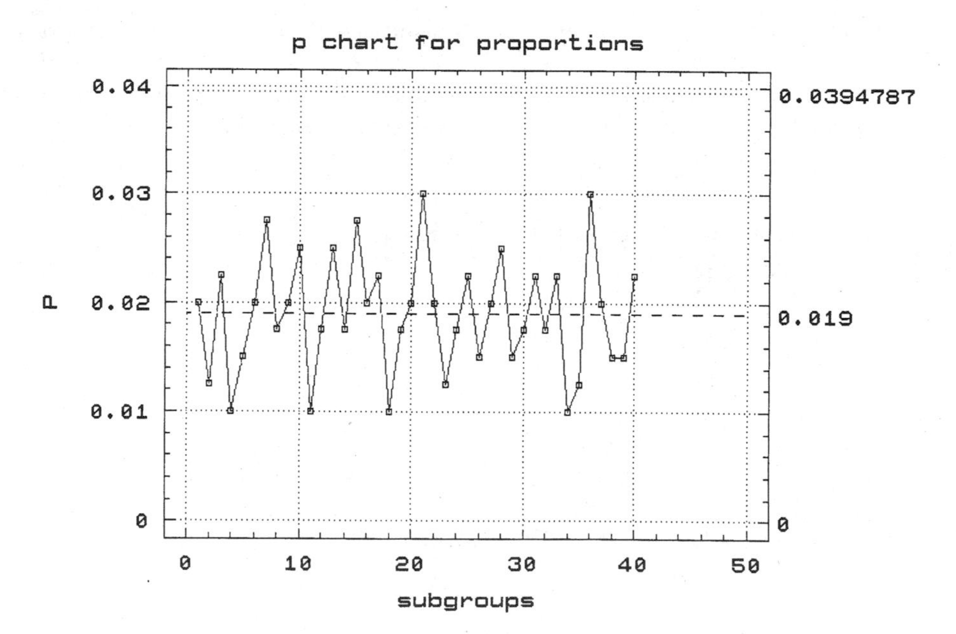

Other commonly used control charts are the *R chart*, which is used to monitor variation in the process variable by using the sample range, and the *c chart*, which is used to monitor the number of defects per item.

Exercises

Basic Techniques

6.28 The sample means were calculated for 30 samples of size $n = 10$ for a process that was judged to be in control. The means of the 30 $\bar{x}$ values and the standard deviation of the combined 300 measurements were $\bar{\bar{x}} = 20.74$ and $s = .87$.

a Use the data to determine the upper and lower control limits for an $\bar{x}$ chart.

b What is the purpose of an $\bar{x}$ chart?

c Construct an $\bar{x}$ chart for the process and explain how it can be used.

6.29 The sample means were calculated for 40 samples of size $n = 5$ for a process that was judged to be in control. The means of the 40 values and the standard deviation of the combined 200 measurements were $\bar{\bar{x}} = 155.9$ and $s = 4.3$.

a Use the data to determine the upper and lower control limits for an x chart.

b Construct an $\bar{x}$ chart for the process and explain how it can be used.

6.30 Explain the difference between an $\bar{x}$ chart and a p chart.

6.31 Samples of $n = 100$ items were selected hourly over a 100-hour period, and the sample proportion of defectives was calculated each hour. The mean of the 100 sample proportions was .035.

a Use the data to find the upper and lower control limits for a p chart.

b Construct a p chart for the process and explain how it can be used.

6.32 Samples of $n = 200$ items were selected hourly over a 100-hour period, and the sample proportion of defectives was calculated each hour. The mean of the 100 sample proportions was .041.

a Use the data to find the upper and lower control limits for a p chart.

b Construct a p chart for the process and explain how it can be used.

Applications

6.33 A gambling casino records and plots the mean of the daily gain or loss from five blackjack tables on an $\bar{x}$ chart. The overall mean of the sample means and the standard deviation of the combined data over 40 weeks were $\bar{\bar{x}} = \$10,752$ and $s = \$1605$.

a Construct an $\bar{x}$ chart for the mean daily gain per blackjack table.

b How might this $\bar{x}$ chart be of value to the manager of the casino?

6.34 A producer of brass rivets randomly samples 400 rivets each hour and calculates the proportion of defectives in the sample. The mean sample proportion, calculated from 200 samples, was equal to .021. Construct a control chart for the proportion of defectives in samples of 400 rivets. Explain how the control chart can be of value to a manager.

6.35 The manager of a building-supplies company randomly samples incoming lumber to see whether it meets quality specifications. From each shipment, 100 pieces of 2×4 lumber are inspected and judged according to whether they are first (acceptable) or second (defective) grade. The proportions of second-grade 2×4s recorded for 30 shipments were as follows:

.14	.21	.19	.18	.23	.20	.25	.19	.22	.17	.21	.15	.23	.12	.19
.22	.15	.26	.22	.21	.14	.20	.18	.22	.21	.13	.20	.23	.19	.26

Construct a control chart for the proportion of second-grade 2×4s in samples of 100 pieces of lumber. Explain how the control chart can be of use to the manager of the building-supplies company.

6.36 A coal-burning power plant tests and measures three specimens of coal each day to monitor the percentage of ash in the coal. The overall mean of 30 daily sample means and the combined standard deviation of all the data were $\bar{\bar{x}} = 7.24$ and $s = .07$. Construct an $\bar{x}$ chart for the process and explain how it can be of value to the manager of the power plant.

6.37 The data given in the following table are measures of the radiation in air particulates at a nuclear power plant. Four measurements were recorded at weekly intervals over a 26-week period. Use the data to construct an $\bar{x}$ chart and plot the 26 values of $\bar{x}$. Explain how the chart will be used.

Week	Radiation				Week	Radiation			
1	.031	.032	.030	.031	14	.029	.028	.029	.029
2	.025	.026	.025	.025	15	.031	.029	.030	.031
3	.029	.029	.031	.030	16	.014	.016	.016	.017
4	.035	.037	.034	.035	17	.019	.019	.021	.020
5	.022	.024	.022	.023	18	.024	.024	.024	.025
6	.030	.029	.030	.030	19	.029	.027	.028	.028
7	.019	.019	.018	.019	20	.032	.030	.031	.030
8	.027	.028	.028	.028	21	.041	.042	.038	.039
9	.034	.032	.033	.033	22	.034	.036	.036	.035
10	.017	.016	.018	.018	23	.021	.022	.024	.022
11	.022	.020	.020	.021	24	.029	.029	.030	.029
12	.016	.018	.017	.017	25	.016	.017	.017	.016
13	.015	.017	.018	.017	26	.020	.021	.020	.022

6.7 Summary

In a practical sampling situation, we draw a *single* random sample of n observations from a population, calculate a single value of a sample statistic, and use it to make an inference about a population parameter. But, to interpret the statistic—to know how close to the population parameter the computed statistic might be expected to fall—we need to observe the behavior of the statistic in repeated sampling. Thus, if we were to repeat the sampling process over and over again an infinitely large number of times, the distribution of values of the statistic produced by this enormous Monte Carlo experiment would be the sampling (or probability) distribution of the statistic.

This chapter described the properties of the sampling distributions for two useful statistics that we will employ in the following chapters to make inferences about population parameters: First, sample means and sample proportions have sampling distributions that can be approximated by a normal distribution when the sample sizes are large. Second, these distributions are centered over their respective population parameters. Thus, the mean of the sampling distribution of the sample mean $\bar{x}$ is the population mean μ, and the mean of the sampling distribution of the sample proportion $\hat{p}$ is the population proportion p. Third, the spread of the distributions, measured by their standard deviations, decreases as the sample size increases. As we will see in Chapter 7, this third characteristic is important when we wish to use a sample statistic to estimate its corresponding population parameter. By choosing a larger sample size, we can increase the probability that a sample statistic will fall close to the population parameter.

Supplementary Exercises

6.38 Review the die-tossing experiment in Section 6.3, where we simulated the selection of samples of $n = 5$ observations and obtained an approximation to the sampling distribution for the sample mean. Repeat this experiment, selecting 200 samples of size $n = 3$.

a Construct the sampling distribution for $\bar{x}$. Note that the sampling distribution of $\bar{x}$ for $n = 3$ does not achieve the bell shape that you observed for $n = 5$ (see Figure 6.4).

b The mean and standard deviation of the probability distribution for x, the number of dots that appear when a single die is tossed, are $\mu = 3.5$ and $\sigma = 1.71$. What are the theoretical values of the mean and standard deviation of the sampling distribution of $\bar{x}$ based on samples of $n = 3$?

c Calculate the mean and standard deviation of the simulated sampling distribution in part (a). Are these values close to the corresponding values obtained for part (b)?

6.39 Refer to the sampling experiment in Exercise 6.38. Calculate the median for each of the 200 samples of size $n = 3$.

a Use the 200 medians to construct a relative frequency histogram that approximated the sampling distribution of the sample median.

b Calculate the mean and standard deviation of the sampling distribution in part (a).

c Compare the mean and standard deviation of this sampling distribution with the mean and standard deviation calculated for the sampling distribution of $\bar{x}$, in Exercise 6.38(b). Which statistic, the sample mean or the sample median, appears to fall closer to μ?

6.40 Exercise 5.14 described part of a study of water samples taken from the Boston water supply system (Karalekas, Ryan, and Taylor, 1983). The researchers determined lead-level readings in drinking water for each of 23 days in 1977. Each daily reading was the average of the lead-level readings for a water specimen collected at each of 40 locations. The mean and standard deviation of the 23 daily lead-level readings were 0.033 and 0.016, respectively. The information given in Exercise 5.14 suggests that the distribution of the individual lead-level readings taken at each of the 40 locations in the piping system is highly skewed toward large values of lead concentration. What can you say about the distribution of the daily lead levels from which the sample of 23 days was selected?

6.41 A survey on the nutritional habits of Americans found that, in a sample of 1678 adults who had eaten dinner the night before, 71% had a dinner that was prepared at home and 4% had a commercially prepared frozen dinner as the main course (*New York Times,* Feb. 24, 1988). Suppose you drew another random sample of 1678 adults who had eaten dinner last night.

a If the proportion of adults who eat dinners prepared at home on any given night is .7, what is the probability that your sample percentage of adults who had a dinner that was prepared at home differs by as much as 4% from that obtained in the *New York Times* survey?

b If the sample percentages differ by as much as 10%, what might you suspect?

6.42 A finite population consists of the following four elements: 6, 1, 3, 2.

a How many different samples of size $n = 2$ can be selected from this population if we sample without replacement? (Sampling is said to be *without replacement* if an element cannot be selected twice for the same sample.)

b List the possible samples of size $n = 2$.

c Compute the sample mean for each of the samples given in part (b).

d Find the sampling distribution of $\bar{x}$. Use a probability histogram to graph the sampling distribution of $\bar{x}$.

e If all four population values are equally likely, calculate the value of the population mean μ. Do any of the samples listed in part (b) produce a value of $\bar{x}$ exactly equal to μ?

6.43 Refer to Exercise 6.42. Find the sampling distribution for $\bar{x}$ if random samples of size $n = 3$ are selected *without replacement.* Graph the sampling distribution of $\bar{x}$.

6.44 Find C_n^N for

a $N = 5, n = 3$. **b** $N = 10, n = 4$. **c** $N = 15, n = 5$.

6.45 Find C_n^N for

a $N = 7, n = 2$. **b** $N = 12, n = 5$. **c** $N = 100, n = 3$.

6.46 How many different samples of $n = 5$ observations can be selected from a population containing 500 observations?

6.47 If a random sample of four observations is selected from a population containing 100 observations, what is the probability that one particular sample of four observations will be selected?

6.48 A political analyst wishes to select a sample of $n = 20$ people from a population of 2000. Use the random number table to identify the people to be included in the sample.

6.49 A population contains 50,000 voters. Use the random number table to identify the voters to be included in a random sample of $n = 15$.

6.50 The total amount of vegetation held by the earth's forests is important to both ecologists and politicians. Since green plants absorb carbon dioxide, an underestimate of the earth's vegetative mass, or biomass, means that much of the carbon dioxide emitted by human activities (primarily fossil-burning fuels) will not be absorbed, and a climate-altering buildup of carbon dioxide will occur. New studies indicate that the biomass for tropical woodlands, thought to be about 35 kilograms per square meter (kg/m^2), may in fact be too high and that tropical biomass values vary regionally—from about 5 to 55 kg/m^2 (*Science News*, Aug. 19, 1989, p. 124). Suppose you measure the tropical biomass in 400 randomly selected square-meter plots.

a Approximate σ, the standard deviation of the biomass measurements.

b What is the probability that your sample average is within two units of the true average tropical biomass?

c If your sample average is $\bar{x} = 31.75$, what would you conclude about the overestimation about which the scientists are concerned?

6.51 The safety requirements for hard hats worn by construction workers and others, established by the American National Standards Institute (ANSI), specifies that each of three hats pass the following test. A hat is mounted on an aluminum head form. An 8-pound steel ball is dropped on the hat from a height of 5 feet, and the resulting force is measured at the bottom of the head form. The force exerted on the head form by each of the three hats must be less than 1000 pounds, and the average of the three must be less than 850 pounds. (The relationship between this test and actual human head damage is unknown.) Suppose the exerted force is normally distributed, and hence a sample mean of three force measurements is normally distributed. If a random sample of three hats is selected from a shipment with a mean equal to 900 and $\sigma = 100$, what is the probability that the sample mean will satisfy the ANSI standard?

6.52 The number of colleges seeing an increase in freshman applications is on the rise, despite the increasing costs of college education and the recession of the early 1990s. In a survey conducted by the National Association of College Admission Counselors (Kelly, 1992), 71% of the colleges surveyed reported an increase in applications, up from 50% in 1991. Suppose you were to conduct a similar survey using a random sample of 100 colleges.

a Describe the sampling distribution for $\hat{p}$, the sample proportion of colleges that reported an increase in applications in 1992.

b What would you use as an estimate of p, the population proportion?

c Suppose only 55 of the colleges in your sample report an increase in applications in 1992. Is this an unusual value, given that the 71% figure reported was accurate?

6.53 Refer to Exercise 6.52. In selecting your sample, suppose you had used a catalog, given to you by a counselor, that contained the names and addresses of all the *private* colleges and universities in the United States.

a Would your sample be representative of all colleges and universities in the United States?

b Will the results of your survey be similar to those of the survey noted in Exercise 6.52? Explain.

c What conclusions might you draw if only 55 colleges in your sample report an increase in applications in 1992?

6.54 Market research surveys depend to a large extent on responses to telephone interviews, door-to-door interviews, computer-generated telephone surveys, and so on. The techniques used by market researchers require representative random samples; however, "cooperation rates" are dropping to the point that the people who are responding to the surveys may no longer represent the rest of the population. Miami, Fort Lauderdale, Florida, and Los Angeles have the least cooperative residents, and about one-third of all Americans routinely refuse to be interviewed (Rothenberg, 1991). Suppose you were to take a random sample of 50 people and ask them if they would be willing to be interviewed for a market research survey.

a Describe the sampling distribution of $\hat{p}$, the proportion of people in the sample who refuse to be interviewed.

b If $p = 1/3$, what is the probability that your sample proportion will lie within .07 of the true value of p?

6.55 Suppose a telephone company executive wishes to select a random sample of $n = 20$ (a small number is used to simplify the exercise) out of 7000 customers for a survey of customer attitudes concerning service. If the customers are numbered for identification purposes, indicate the customers whom you will include in your sample. Use the random number table and explain how you selected your sample.

6.56 A small city contains 20,000 voters. Use the random number table to identify the voters to be included in a random sample of $n = 15$.

6.57 A random sample of public opinion in a small town was obtained by selecting every tenth person to pass by the busiest corner in the downtown area. Will this sample have the characteristics of a random sample selected from the town's citizenry? Explain.

6.58 The following notice, appearing in *USA Today* (Apr. 19, 1985), refers to one of the many telephone call-in opinion polls conducted by the major television networks: "NBC will conduct a phone vote via 900 numbers (50 cents a call) during its 1:00 P.M. EST baseball telecasts Saturday on whether a designated hitter should be used in both leagues. Bill Macatee will update the results until 4:00 P.M. from the studio."

Many television viewers will interpret the proportion of telephone calls responding "yes" to NBC's question as an estimate of the proportion of all baseball fans in favor of using a designated hitter in both leagues. Explain why the respondents will not represent a random sample of the opinions of all baseball fans. Explain the types of distortions that could creep into a call-in opinion poll.

6.59 The maximum load (with a generous safety factor) for the elevator in an office building is 2000 pounds. The relative frequency distribution of the weights of all of men and women using the elevator is mound-shaped (slightly skewed to the heavy weights), with mean μ equal to 150 pounds and standard deviation σ equal to 35 pounds. What is the largest number of people you can allow on the elevator if you want their total weight to exceed the maximum weight with a small probability (say near .01)? (HINT: If $x_1, x_2, \ldots, x_n$ are independent observations made on a random variable x, and if x has a mean μ and variance σ^2, then the mean and variance of $\sum_{i=1}^{n} x_i$ are $n\mu$ and $n\sigma^2$, respectively. This result was given in Section 6.3.)

6.60 The number of wiring packages that can be assembled by a company's employees has a normal distribution, with a mean equal to 16.4 per hour and a standard deviation of 1.3 per hour.

a What are the mean and standard deviation of the number x of packages produced per worker in an 8-hour day?

b Would you expect the probability distribution for x to be mound-shaped and approximately normal? Explain.

c What is the probability that a worker will produce at least 135 packages per 8-hour day?

6.61 Refer to Exercise 6.60. Suppose the company employs ten assemblers of wiring packages.

a Find the mean and standard deviations of the company's daily (8-hour day) production of wiring packages.

b What is the probability that the company's daily production would be less than 1280 wiring packages per day?

6.62 The following table lists the number of defective 60-watt light bulbs found in samples of 100 bulbs selected over 25 days from a manufacturing process. Assume that during these 25 days the manufacturing process was not producing an excessively large fraction of defectives.

Day	1	2	3	4	5	6	7	8	9	10	11	12	13
Defectives	4	2	5	8	3	4	4	5	6	1	2	4	3

Day	14	15	16	17	18	19	20	21	22	23	24	25
Defectives	4	0	2	3	1	4	0	2	2	3	5	3

a Construct a p chart to monitor the manufacturing process, and plot the data.

b How large must the fraction of defective items be in a sample selected from the manufacturing process before the process is assumed to be out of control?

c During a given day, suppose a sample of 100 items is selected from the manufacturing process and 15 defective bulbs are found. If a decision is made to shut down the manufacturing process in an attempt to locate the source of the implied controllable variation, explain how this decision might lead to erroneous conclusions.

6.63 A hardware store chain purchases large shipments of light bulbs from the manufacturer described in Exercise 6.62 and specifies that each shipment must contain no more than 4% defectives. When the manufacturing process is in control, what is the probability that the hardware store's specifications are met?

6.64 Refer to Exercise 6.62. During a given week the number of defective bulbs in each of 5 samples of 100 were found to be 2, 4, 9, 7, and 11. Is there reason to believe that the production process has been producing an excess proportion of defectives at any time during the week?

6.65 During long production runs of canned tomatoes, the average weights in ounces of samples of five cans of standard-grade tomatoes in puree form were taken at 30 control points during an 11-day period. These results are shown in the table. When the machine is performing normally, the average weight per can is 21 ounces with a standard deviation of 1.20 ounces.

a Compute the upper and lower control limits and the centerline for the $\bar{x}$ chart.

b Plot the sample data on the $\bar{x}$ chart and determine whether the performance of the machine is in control.

Sample Number	Average Weight	Sample Number	Average Weight
1	23.1	16	21.4
2	21.3	17	20.4
3	22.0	18	22.8
4	21.4	19	21.1
5	21.8	20	20.7
6	20.6	21	21.6
7	20.1	22	22.4
8	21.4	23	21.3
9	21.5	24	21.1
10	20.2	25	20.1
11	20.3	26	21.2
12	20.1	27	19.9
13	21.7	28	21.1
14	21.0	29	21.6
15	21.6	30	21.3

Source: Adapted from Hackl, (1991).

Exercises Using the Data Disks

Starred (*) exercises are optional.

6.66 The data disk contains a list of the *Fortune 500* largest U.S. industrial corporations in 1991. Use the random number table to select 25 samples of size $n = 10$ *Fortune 500* companies. Record the corporate sales for the companies in each sample, and calculate $\bar{x}$ for each sample. The true mean and standard deviation of the sales for the *Fortune 500* companies for 1991 are $\mu = 4525.17$ and $\sigma = 10574.9$.

a Find the interval $\mu \pm 2(\sigma/\sqrt{10})$. According to the Empirical Rule and the Central Limit Theorem, this interval should contain what proportion of the sample means?

b Find the mean and standard deviation of the 25 sample means. How do their values compare with the theoretical mean and standard deviation of $\bar{x}$ given by μ and $\sigma/\sqrt{10}$? What proportion of the sample means lie within the interval $\mu \pm 2(\sigma/\sqrt{10})$? Within $\mu \pm 3(\sigma/\sqrt{10})$?

6.67 A class Monte Carlo experiment: The data disk contains a set of 945 female systolic blood pressures. As an experiment, regard this data set as a population. Have each member of the class select a random sample of $n = 4$ observations from this population (using the random number table) and calculate the sample mean. Construct a relative frequency histogram for the sample means calculated by the class members. This provides an approximation to the sampling distribution of $\bar{x}$.

a The population relative frequency histogram for this data set is roughly mound-shaped, with range 75 to 145. How does your relative frequency histogram compare to this distribution?

b Calculate the theoretical mean and standard deviation of the sampling distribution of $\bar{x}$. (NOTE: The mean and standard deviation of the data set are $\mu = 110.390$ and $\sigma = 12.5753$.) Locate the mean μ and the interval $\mu \pm 2\sigma_{\bar{x}}$ along the horizontal axis of the relative frequency histogram you constructed in part (a). Does μ fall approximately in the center of the histogram? Does the interval $\mu \pm 2\sigma_{\bar{x}}$ include most of the sample means?

c Calculate the mean and standard deviation of the sample means used to construct the relative frequency histogram. Are these values close to the values found for μ and $\sigma_{\bar{x}}$ in part (a)?

6.68* If you have access to a computer and a computer program that generates random numbers, you can simulate sampling from a population that has a uniform probability distribution. The numbers produced by a random number generator are independent of one another, and the probability of observing any one number is the same as the probability of observing any other. Program the computer to generate a large number—say, 1000—of samples of $n = 2$ observations and calculate the sample mean for each. Use a computer program to arrange these 1000 sample means in a relative frequency histogram. The resulting histogram will provide a good approximation to the sampling distribution of $\bar{x}$ for samples selected from a population having a uniform relative frequency distribution. It should be similar to the sampling distribution shown in the third column of Figure 6.5 for $n = 2$.

6.69* Repeat Exercise 6.68 for $n = 5$, 10, and 25. Compare with the corresponding sampling distributions shown in Figure 6.5.

6.70* Repeat Exercise 6.68 for $n = 100$. Compare with the sampling distributions for $n = 2, 5, 10$, and 25.

CASE STUDY

Sampling the Roulette at Monte Carlo

The technique of simulating a process that contains random elements and repeating the process over and over to see how it behaves is called a **Monte Carlo procedure.** It is widely used in business and other fields to investigate the properties of an operation that is subject to a number of random effects, such as weather, human behavior, and so on. For example, you could model the behavior of a manufacturing company's inventory by creating, on paper, daily arrivals and departures of manufactured products from the company's warehouse. Each day a random number of items produced by the company would be received into inventory. Similarly, each day a random number of orders of varying random sizes would be shipped. Based on the input and output of items, you could calculate the inventory, that is, the number of items on hand at the end of each day. The values of the random variables, the number of items produced, the number of orders, and the number of items per order needed for each day's simulation would be obtained from theoretical distributions of observations that closely model the corresponding distributions of the variables that have been observed over time in the manufacturing operation. By repeating the simulation of the supply, the shipping, and the calculation of daily inventory for a large number of days (a sampling of what might really happen), you can observe the behavior of the plant's daily inventory. The Monte Carlo procedure is particularly valuable because it enables the manufacturer to see how the daily inventory would behave when certain changes are made in the supply pattern or in some other aspect of the operation that could be controlled.

In an article entitled "The Road to Monte Carlo," Daniel Seligman (1985) comments on the Monte Carlo method, noting that, although the technique is widely used in business schools to study capital budgeting, inventory planning, and cash flow management, no one seems to have used the procedure to study how well we might do if we were to gamble at Monte Carlo.

To follow up on this thought, Seligman programmed his personal computer to simulate the game of roulette. Roulette consists of a wheel whose rim is divided into 38 pockets. Thirty-six of the pockets are numbered 1 to 36 and are alternately colored red and black. The two remaining pockets are colored green and are marked 0 and 00. To play the game, you bet a certain amount of money on one or more pockets. The wheel is spun and turns until it stops. A ball falls into a slot on the wheel to indicate the winning number. If you have money on that number, you win a specified amount. For example, if you were to play the number 20, the payoff is 35 to 1. If the wheel does not stop at that number, you lose your bet. Seligman decided to see how his nightly gains (or losses) would fare if he were to bet $5 on each turn of the wheel and to repeat the process 200 times each night. He did this 365 times, thereby simulating the outcomes of 365 nights at the casino. Not surprisingly, the mean "gain" per $1000 evening for the 365 nights was a *loss* of $55, the average of the winnings retained by the gambling house. The surprise, according to Seligman, was the extreme variability of the nightly "winnings." Seven times out of the 365 evenings, the fictitious gambler lost the $1000 stake, and only once did he win a maximum of $1160. On 141 nights the loss exceeded $250.

1 To evaluate the results of Seligman's Monte Carlo experiment, first find the probability distribution of the gain x on a single $5 bet.

2 Find the expected value and variance of the gain x from part 1.

3 Find the expected value and variance for the evening's gain, the sum of the gains or losses for the 200 bets of $5 each.

4 Use the results of part 2 to evaluate the probability of 7 out of 365 evenings resulting in a loss of the total $1000 stake.

5 Use the results of part 3 to evaluate the probability that the largest evening's winnings were as large as $1160.

7

Large-Sample Estimation

Case Study

Do the national polls conducted by the Gallup and Harris organizations, the news media, and others really provide accurate estimates of the percentages of people in the United States who favor various propositions? The case study at the end of this chapter examines the reliability of a poll conducted by the *New York Times* using the theory of large-sample estimation.

General Objective

In previous chapters, we focused on the probability distributions of random variables and on the sampling distributions of several statistics that, for large sample sizes, can be approximated by a normal distribution according to the Central Limit Theorem. In this chapter, we present a method for estimating population parameters and illustrate the concept with practical examples. The Central Limit Theorem and the sampling distributions presented in Chapter 6 will play a key role in evaluating the reliability of the estimates.

Specific Topics

1 Types of estimators (7.3)
2 Evaluating the goodness of an estimator (7.4)
3 General formulas for large-sample estimation (7.4)
4 Estimation of a population mean (7.5)
5 Estimating the difference between two means (7.6)
6 Estimating a binomial proportion (7.7)
7 Estimating the difference between two binomial proportions (7.8)
8 Choosing the sample size (7.9)

7.1 A Brief Summary

The preceding six chapters set the stage for the objective of this text: an understanding of statistical inference and how it can be applied to the solution of practical problems. In Chapter 1, we stated that statisticians are concerned with making inferences about populations of measurements based on information contained in samples. We showed you how to phrase an inference—that is, how to describe a set of measurements—in Chapter 2. In Chapter 3, we discussed probability, the mechanism for making inferences, and we followed that with a general discussion of probability distributions.

Three useful discrete probability distributions were presented in Chapter 4—the binomial, the Poisson, and the hypergeometric. A continuous probability distribution—the normal—was presented in Chapter 5.

In Chapter 6, we noted that statistics, computed from the sample measurements, are used to make inferences about population parameters, and we found an important use for the normal probability distribution of Chapter 5. In particular, you learned that some of the most important statistics—sample means and proportions—have sampling distributions that can be approximated by a normal distribution when the sample sizes are large, owing to the Central Limit Theorem. These statistics will now be used to make inferences about population parameters, and their sampling distributions will provide a means of assessing the reliability of these inferences.

7.2 Inference: The Objective of Statistics

Inference—specifically decision making and prediction—is centuries old and plays a very important role in our individual lives. Each of us is faced daily with personal decisions and situations that require predictions concerning the future. The government is concerned with predicting short-term and long-term interest rates. The broker would like to forecast the behavior of the stock market. The metallurgist wishes to use the results of an experiment to infer whether a new type of steel is more resistant to temperature changes than another. The consumer wants to know whether detergent A is more effective than detergent B. Hopefully, these inferences will be based on relevant measurements, which are called **observations** or **data**.

In many practical situations, the relevant information is abundant, seemingly inconsistent, and, in many respects, overwhelming. As a result, a carefully considered decision or prediction is often little better than an outright guess. You need only refer to the "Market Views" section of the *Wall Street Journal* to observe the diversity of expert opinions concerning future stock market behavior. Similarly, a visual analysis of data by scientists and engineers often yields conflicting opinions regarding conclusions to be drawn from an experiment. Although many individuals tend to feel that their own built-in inference-making equipment is quite good, experience suggests that this may not be the case. It is the job of the mathematical statistician to provide inference-making techniques that are better than subjective guesses.

The objective of statistics is to make inferences about a population based on information contained in a sample. Since populations are characterized by numerical descriptive measures called **parameters**, statistical inference is concerned with making inferences about population parameters. Typical population parameters are the mean, the standard deviation, the area under the probability distribution above or below some value of the random variable, and the area between two values of the variable. Indeed, all the practical problems mentioned in the first paragraph of this section can be restated in the framework of a population with a specified parameter of interest.

Methods for making inferences about parameters fall into one of two categories. We may **make decisions** concerning the value of the parameter, or we may **estimate** the value of the parameter. For example, the circuits in computers and other electronic instruments consist of one or more printed circuit boards (PCBs); computer repairs often consist of simply replacing one or more defective PCBs. In an attempt to determine the proper setting of a plating process applied to one side of a PCB, a production supervisor might *estimate* the thickness of copper plating on PCBs using samples from several days of operation. However, if the supervisor were told by the plant owner that the thickness of the copper plating must not be less than .001 inch in order for the process to be in control, the supervisor might want to *decide* whether or not the average thickness of the copper plating is less than .001 inch. Although some statisticians view estimation as a decision-making problem, we will retain the two categories and concentrate separately on estimation and decision making using tests of hypotheses.

Which method of inference should be used? That is, should the parameter be estimated, or should we test a hypothesis concerning its value? The answer is dictated by the practical question posed and is often determined by personal preference. Some people like to test theories concerning parameters; others prefer to express their inference as an estimate. Inasmuch as both estimation and tests of hypotheses are used frequently in scientific literature, we will include both methods in our discussion.

A statistical problem, which involves planning, analysis, and inference making, would be incomplete without reference to a **measure of the goodness** of the inferential procedures. We may define numerous objective methods for making inferences in addition to our own individual procedures based on intuition. Therefore, a measure of goodness must be defined in such a way that one procedure may be compared with another. More than that, we wish to state the goodness of a particular inference in a given practical situation. Thus, to predict that the price of a stock will be $80 next Monday would be insufficient and would stimulate few of us to take action to buy or sell. We would also want to know whether the estimate was correct to within plus or minus $1, $2, or $10. **In summary, statistical inference contains two elements: (1) the inference and (2) a measure of its goodness.**

7.3 Types of Estimators

Estimation procedures can be divided into two types: point estimation and interval estimation. Suppose we wish to estimate the mean weight gain per month of four-month-old golden retriever pups that have been placed on a particular diet. The estimate might be given as a single number—for instance, 3.8 pounds—or we might estimate that the weight gain would fall in an interval such as 2.7 to 4.9 pounds. The first type of estimate is called a **point estimate** because the single number representing the estimate may be associated with a point on a line. The second type, involving two points and defining an interval on a line, is called an **interval estimate**. We will consider each of these methods of estimation.

In order to construct either a point or an interval estimate, we use information from the sample in the form of an estimator. Estimators are functions of sample observations and hence, by definition, are also **statistics**.

DEFINITION ▪ An **estimator** is a rule that tells us how to calculate an estimate based on information in the sample and that is generally expressed as a formula. ▪

For example, the sample mean

$$\bar{x} = \frac{\sum_{i=1}^{n} x_i}{n}$$

is an estimator of the population mean μ and explains exactly how the actual numerical value of the estimate can be obtained once the sample values $x_1, x_2, \ldots, x_n$ are known. The sample mean can be used to arrive at a single number to estimate μ or to construct an interval, two points that are intended to enclose the true value of μ.

DEFINITION ▪ A **point estimator** of a population parameter is a rule that tells us how to calculate a single number based on sample data. The resulting number is called a **point estimate.** ▪

DEFINITION ▪ An **interval estimator** of a population parameter is a rule that tells us how to calculate two numbers based on sample data, forming an interval within which the parameter is expected to lie. This pair of numbers is called an **interval estimate** or **confidence interval.** ▪

Both point and interval estimation procedures are developed using the sampling distribution of the best estimator of a specified population parameter. How do we decide which among several estimators of a specified population parameter is best? We will address this question in Section 7.4.

7.4 Evaluating the Goodness of an Estimator

The goodness of an estimator is evaluated by observing its behavior in repeated sampling. Let us consider the following analogy. In many respects, point estimation is similar to the firing of a revolver at a target. The estimator, which generates estimates, is analogous to the revolver; a particular estimate is analogous to the bullet; and the parameter of interest is analogous to the bull's-eye. Drawing a sample from the population and estimating the value of the parameter is equivalent to firing a single shot at the target.

Suppose a man fires a single shot at a target and the shot pierces the bull's-eye. Do we conclude that he is an excellent shot? The answer is no, because not one of us would consent to hold the target while a second shot is fired. On the other hand, if 1 million shots in succession hit the bull's-eye, we might acquire sufficient confidence in the marksman to hold the target for the next shot, if the compensation were adequate. The point we wish to make is that we cannot evaluate the goodness of an estimation procedure on the basis of a single estimate. Rather, we must observe the results when the estimation procedure is used over and over again, many, many times; then we observe how closely the shots are distributed about the bull's-eye. In fact, since the estimates are numbers, we would evaluate the goodness of the estimator by constructing a frequency distribution of the estimates obtained in repeated sampling and noting how closely the distribution centers about the parameter of interest. This relative frequency distribution would be the sampling distribution of the estimator.

As an illustration, consider the die-tossing experiment used in Chapter 6, where we generated 100 samples of $n = 5$ measurements each and calculated the mean for each sample. Since we know the mean value μ of the number showing on a die toss ($\mu = 3.5$), we can use the results of the die-tossing experiment to see how well the mean of a sample of $n = 5$ measurements estimates μ.

The frequency histogram of the 100 sample means (shown in Figure 7.1) is an approximation to the sampling distribution of the mean $\bar{x}$ of a sample of five observations, one for each die toss. Notice that the estimates group about the population mean $\mu = 3.5$ and that they range from 1.4 to 5.4. Surely this distribution of estimates tells us something about how good a new estimate of μ would be if we were to draw one more sample of $n = 5$ measurements and compute the sample mean $\bar{x}$.

Suppose we consider an estimator of a population parameter such as μ, σ, or p. What are some desirable properties of an estimator? Essentially, there are two, and they can be seen by observing the sampling distributions given in Figures 7.2 and 7.3.

First, we would like the sampling distribution to be centered over the true value of the parameter. **Thus, we would like the mean of the sampling distribution to equal the true value of the parameter.** Such an estimator is said to be **unbiased**.

DEFINITION ▪

An estimator of a parameter is said to be **unbiased** if the mean of its distribution is equal to the true value of the parameter. Otherwise, the estimator is said to be **biased**. ▪

The sampling distributions for an unbiased estimator and a biased estimator are shown in Figure 7.2. The sampling distribution for the biased estimator in Figure 7.2 is shifted to the right of the true value of the parameter. This biased estimator is more likely than an unbiased one to overestimate the value of the parameter.

The second desirable property of an estimator is that **the spread (as measured by the variance) of the sampling distribution should be as small as possible.** This ensures that, with a high probability, an individual estimate will fall close to the true value of the parameter. The sampling distributions for two unbiased estimators,

FIGURE 7.1 Histogram of sample means for the die-tossing experiment in Section 6.3

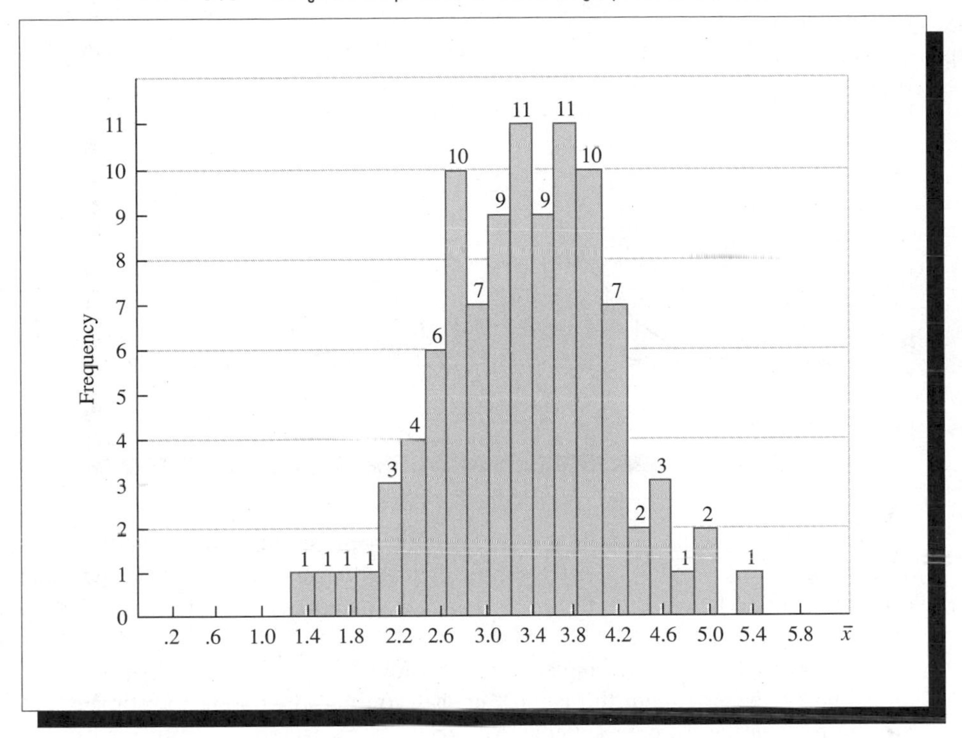

FIGURE 7.2 Distributions for biased and unbiased estimators

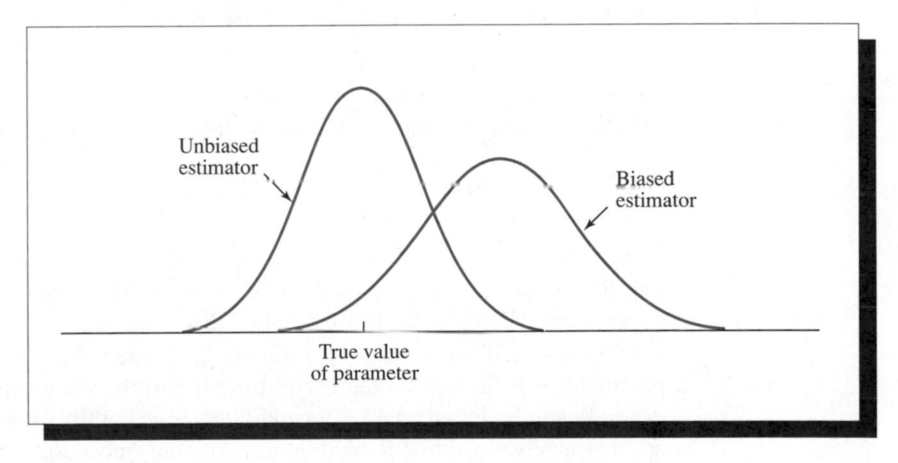

FIGURE 7.3
Comparison of estimator variability

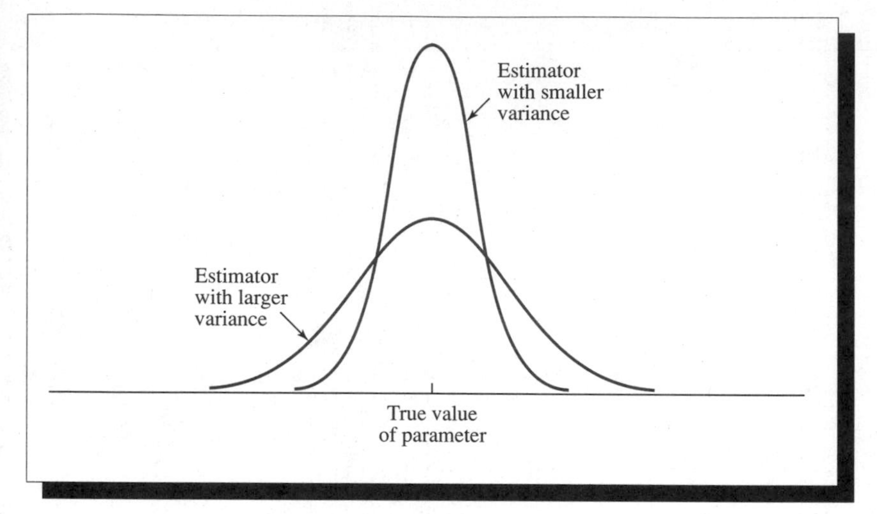

one with a small variance† and the other with a larger variance, are shown in Figure 7.3. Naturally, we would prefer the estimator with the smaller variance because the estimates tend to lie closer to the true value of the parameter than in the distribution with the larger variance.

In real-life sampling situations, you may know that the sampling distribution of an estimator centers about the parameter that you are attempting to estimate, but all you have is the estimate computed from the n measurements contained in the sample. How far from the true value of the parameter will your estimate lie? The distance between the estimate and the true value of the parameter is called the **error of estimation**.

DEFINITION ▪

The distance between an estimate and the estimated parameter is called the **error of estimation**. ▪

In this chapter we will assume that the sample sizes are always large and, therefore, that the *unbiased* estimators we will study have sampling distributions that can be approximated by a normal distribution (because of the Central Limit Theorem). Consequently, if we define the difference between a particular estimate and the parameter it estimates as the **error of estimation**, we would expect the error of estimation to be less than 1.96 standard deviations of the estimator, with probability approximately equal to .95 (see Figure 7.4). Thus, most estimates will be within 1.96 standard deviations of the true value of the parameter, and this quantity provides a practical upper limit for the error of estimation. This upper limit is called the **margin of error**. It is possible (with probability .05) that the error of estimation will exceed this margin of error, but it is very unlikely.

† Statisticians usually use the term *variance of an estimator* when in fact they mean the variance of the sampling distribution of the estimator. This contractive expression is used almost universally.

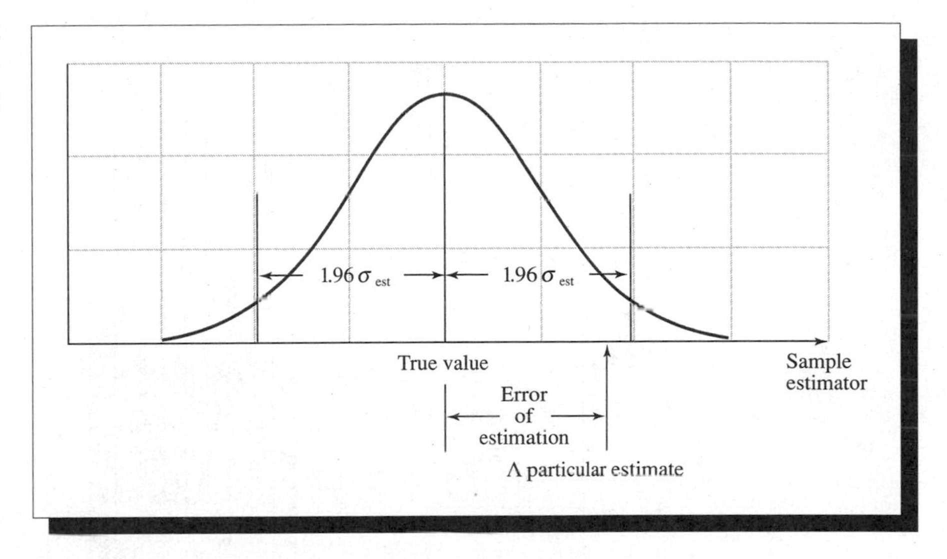

FIGURE 7.4
Sampling distribution of an unbiased estimator

Point Estimator for a Population Parameter

Point estimator: a statistic calculated using sample measurements

Margin of error: 1.96 × standard error of the estimator

The relative frequency distribution of the 100 values of $\bar{x}$ (Table 6.4), each calculated from a sample $n = 5$ die tosses, illustrates this concept. As noted in Chapter 6, the mean and standard deviation of the population of die tosses are $\mu = 3.5$ and $\sigma = 1.71$. Therefore,

$$1.96\sigma_{\bar{x}} = 1.96\frac{\sigma}{\sqrt{n}} = 1.96\frac{1.71}{\sqrt{5}} = 1.50$$

Even though this sample size is small and the sampling distribution of $\bar{x}$ is not exactly normally distributed, you can see in Figure 7.5 that all but 5 of the sample means fall within 1.50 of the mean $\mu = 3.5$—that is, in the interval 2.0 to 5.0.

The **goodness of an interval estimator** is analyzed in much the same way as that of a point estimator. Samples of the same size are repeatedly drawn from the population, and the interval estimate is calculated on each occasion. This process will generate a large number of intervals rather than points. **A good interval estimator would successfully enclose the true value of the parameter a large fraction of the time.** Constructing an interval estimate is like attempting to throw a lariat around a fence post. In this case the parameter that you wish to estimate corresponds to the post and the interval corresponds to the loop formed by the lariat. Each time you draw a sample, you construct a confidence interval for a parameter and you hope to

FIGURE 7.5
Histogram of sample means for the die-tossing experiment in Section 6.3

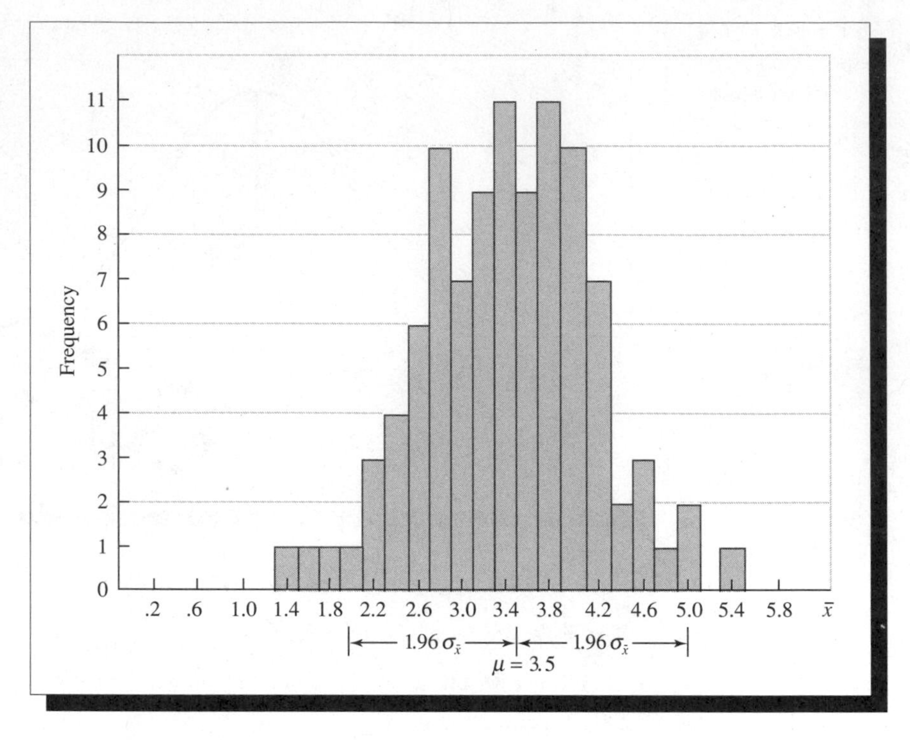

"rope it"—that is, include it in the interval. But you will not be successful for every sample. The "success rate" is referred to as the **confidence coefficient**.

DEFINITION ■

The probability that a confidence interval will enclose the estimated parameter is called the **confidence coefficient**. ■

To consider a practical example, suppose we want to estimate the mean number of bacteria per cubic centimeter in a polluted stream. If we were to draw ten samples, each containing $n = 30$ observations, and construct a confidence interval for the population mean μ for each sample, the intervals might appear as shown in Figure 7.6. The horizontal line segments represent the ten intervals, and the vertical line represents the location of the true mean number of bacteria per cubic centimeter. Note that the parameter is fixed and that the interval location and width may vary from sample to sample. **Thus we speak of "the probability that the interval encloses μ" not "the probability that μ falls in the interval," because μ is fixed. The *interval* is random.**

A good confidence interval is one that is as narrow as possible and has a large confidence coefficient, near 1. The narrower the interval, the more exactly we have located the estimated parameter. The larger the confidence coefficient, the more confidence we have that a particular interval encloses the estimated parameter. Remember that the confidence coefficient gives the probability that the interval estimator will

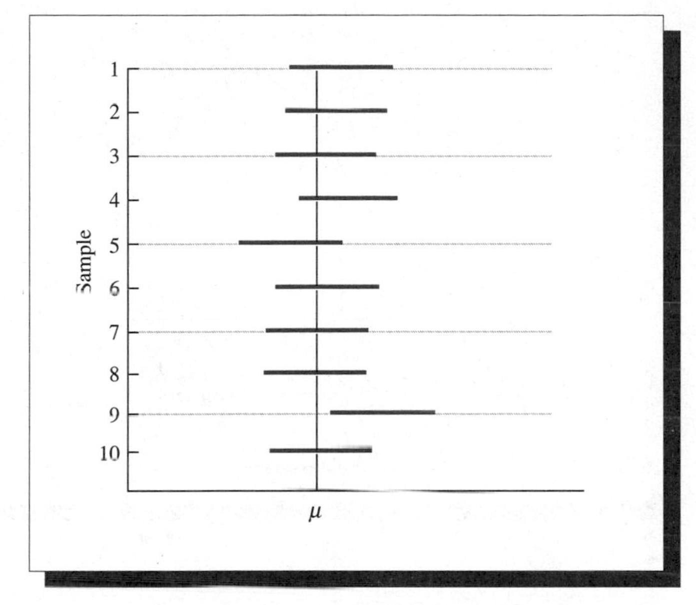

FIGURE 7.6
Ten confidence intervals for the mean number of bacteria per cubic centimeter (each based on a sample of $n = 30$ observations)

produce confidence limits that enclose the estimated parameter. It gives you a measure of the confidence you can place in the confidence limits constructed from the data contained in a sample. In that sense, the width of an interval and its associated confidence coefficient measure the goodness of the confidence interval.

We must remember that, once a sample is drawn and a confidence interval constructed, the resulting interval either encloses the true value of the parameter or doesn't. However, because the confidence coefficient is large and close to 1, we are confident that the interval just constructed does enclose the true value of the parameter.

A large-sample interval estimator, or **large-sample confidence interval**, for a population can be easily obtained when a sample estimator is normally distributed or approximately so due to the Central Limit Theorem. Since 95% of the point estimates will lie within 1.96 standard errors of the true value of the parameter, we can construct an interval estimate by measuring 1.96 × (standard error of the estimator) on either side of the point estimate. We know that 95% of the intervals constructed in this manner will enclose the true value of the unknown parameter (see Figure 7.7). Similarly, 90% of the intervals constructed by measuring 1.645 × (standard error of the estimator) on either side of the point estimate will enclose the unknown parameter. When the conditions specified above are met and the desired confidence coefficient is $(1 - \alpha)$, a $(1 - \alpha)$ 100% confidence interval estimate for the unknown population parameter is given in the following display.

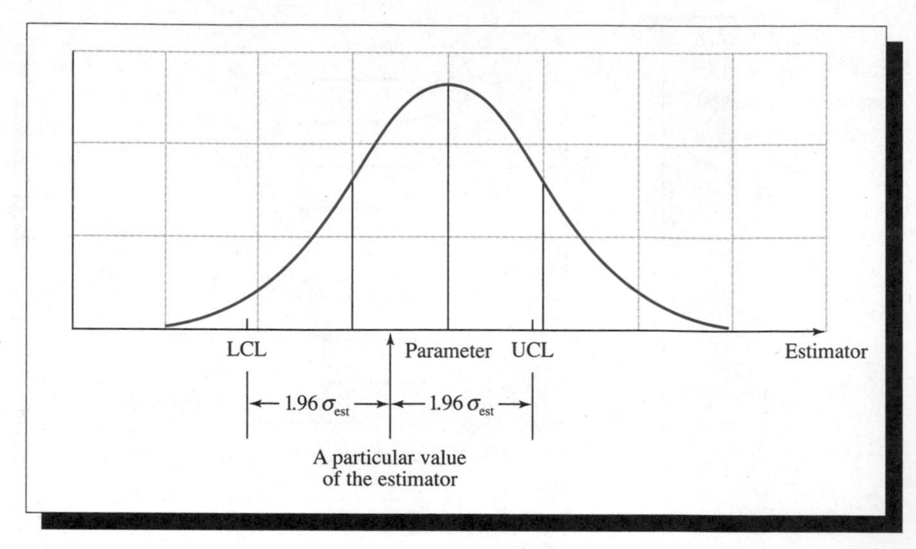

FIGURE 7.7
95% confidence limits for a population parameter

A $(1-\alpha)$ 100% Large-Sample Confidence Interval

$$(\text{Point estimator}) \pm z_{\alpha/2} \times (\text{standard error of the estimator})$$

where $z_{\alpha/2}$ is the z value corresponding to an area $\alpha/2$ in the upper tail of a standard normal distribution. This formula generates two values, the **lower confidence limit (LCL)** and the **upper confidence limit (UCL)**.

Throughout the remainder of this and succeeding chapters, populations and parameters of interest will be defined and the appropriate estimator indicated along with its mean and standard error. You will see how the general formulas for point and interval estimation apply for the specific situations described in the following sections.

7.5 Estimation of a Population Mean

Practical problems very often lead to the estimation of a population mean μ. We may be concerned with the average achievement of college students in a particular university, with the average strength of a new type of steel, with the average number of deaths per capita in a given social class, or with the average demand for a new product. Many estimators are available for estimating the population mean μ, including the sample median, the average of the largest and smallest measurements in the sample, and the sample mean $\bar{x}$. Each would have a sampling distribution and, depending on the population and practical problem involved, certain advantages and disadvantages. Although the sample median and the average of the sample extremes are easier to calculate, the sample mean $\bar{x}$ is usually superior in that, for some populations, its variance is a minimum and that, regardless of the population, it is always unbiased.

The sampling distribution of the sample mean, $\bar{x}$, discussed in Section 6.4, has four important properties:

- The sampling distribution of $\bar{x}$ will be *approximately normal* regardless of the probability distribution of the sampled population.
- If the sampled population is normal, the sampling distribution of $\bar{x}$ will be exactly *normal*.
- The *mean* of the sampling distribution of $\bar{x}$ will always equal μ. Thus, $\bar{x}$ is an unbiased estimator of μ.
- The standard deviation of the sampling distribution of $\bar{x}$, also called the *standard error of the mean*, is $\sigma_{\bar{x}} = \sigma/\sqrt{n}$.

The estimator $\bar{x}$ satisfies all the conditions given in Section 7.4, so that the general formulas can be applied for point and interval estimation.

Point Estimator of μ

Point estimator: $\bar{x}$

Margin of error: $1.96\sigma_{\bar{x}} = 1.96\sigma/\sqrt{n}$

If σ is unknown and n is 30 or larger, the sample standard deviation can be used to approximate σ.[†]

Assumption: $n \geq 30$

EXAMPLE **7.1** Suppose we wish to estimate the average daily yield of a chemical manufactured in a chemical plant. The daily yield, recorded for $n = 50$ days, produced a mean and standard deviation equal to

$$\bar{x} = 871 \text{ tons}$$
$$s = 21 \text{ tons}$$

Estimate the average daily yield μ.

Solution The estimate of the daily yield is $\bar{x} = 871$ tons. The margin of error is

$$1.96\sigma_{\bar{x}} = \frac{1.96\sigma}{\sqrt{n}} = \frac{1.96\sigma}{\sqrt{50}}$$

[†]When we sample a normal distribution, the statistic $(\bar{x} - \mu)/(s/\sqrt{n})$ has a t-distribution, which is discussed in Chapter 9. When the sample size is *large*, then this statistic is approximately normally distributed whether the sampled population is normal or nonnormal.

Although σ is unknown, the sample size is large, and we may approximate the value of σ by using s. Thus, the margin of error is approximately

$$1.96\frac{s}{\sqrt{n}} = 1.96\frac{(21)}{\sqrt{50}} = 5.82$$

We can feel fairly confident that our estimate of 871 tons is within 5.82 tons of the true average yield. ■

The standard deviation of the sampling distribution of $\bar{x}$ is $\sigma_{\bar{x}} = \sigma/\sqrt{n}$. The fact that this standard deviation is proportional to the population standard deviation σ and inversely proportional to the square root of the sample size n is intuitively reasonable. The more variable the population data, measured by σ, the more variable will be $\bar{x}$. On the other hand, more information will be available for estimating μ as n becomes larger. Therefore, the estimates should fall closer to μ, and $\sigma_{\bar{x}}$ should become smaller if we increase the sample size. The sampling distributions for $\bar{x}$ based on random samples of $n = 5$, $n = 20$, and $n = 80$ from a normal distribution are shown in Figure 7.8. Notice how these distributions center about μ and how the spread of the distributions decreases as n increases.

FIGURE 7.8
Sampling distributions for $\bar{x}$ based on random samples from a normal distribution, $n = 5$, 20, and 80

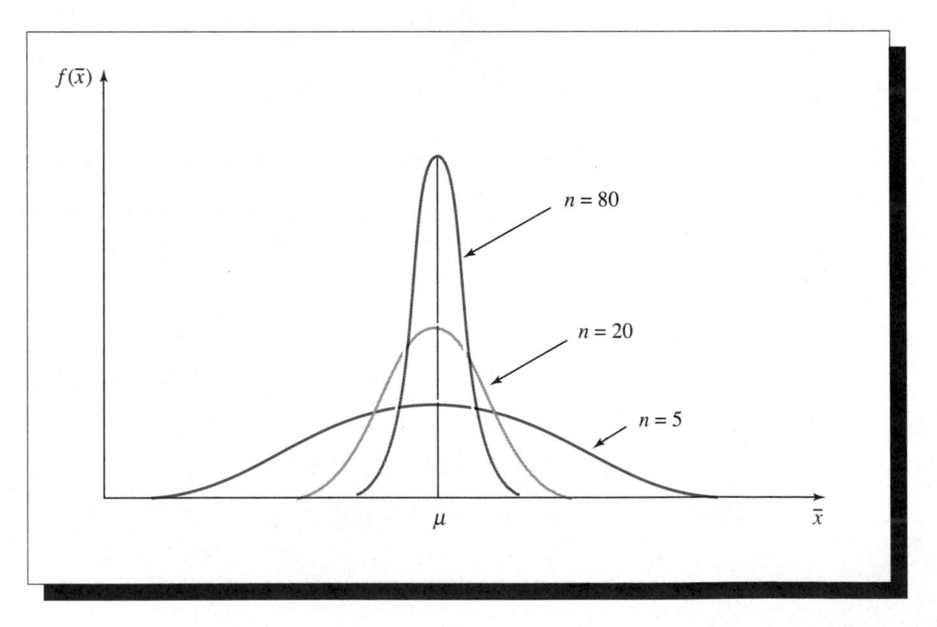

Similarly, we can construct confidence intervals for μ corresponding to any desired confidence coefficient—say, $(1 - \alpha)$—by using the following.

A $(1 - \alpha)$ 100% Large-Sample Confidence Interval for a Population Mean μ

$$\bar{x} \pm z_{\alpha/2} \frac{\sigma}{\sqrt{n}}$$

where $z_{\alpha/2}$ is the z value corresponding to an area $\alpha/2$ in the upper tail of a standard normal z-distribution.

n = sample size

σ = standard deviation of the sampled population

If σ is unknown, it can be approximated by the sample standard deviation s when the sample size is large.

Assumption: $n \geq 30$

The normal curve value $z_{\alpha/2}$, which appears in the formula for the confidence interval, is located as shown in Figure 7.9. For example, if you want a confidence coefficient $(1 - \alpha)$ equal to .95, then the tail-end area is $\alpha = .05$, and half of α (.025) is placed in each tail of the distribution. Then $z_{.025}$ is the table z value corresponding to an area of .475 to the right of the mean, or

$$z_{.025} = 1.96$$

Confidence limits corresponding to some of the commonly used confidence coefficients are shown in Table 7.1.

FIGURE 7.9 Location of $z_{\alpha/2}$

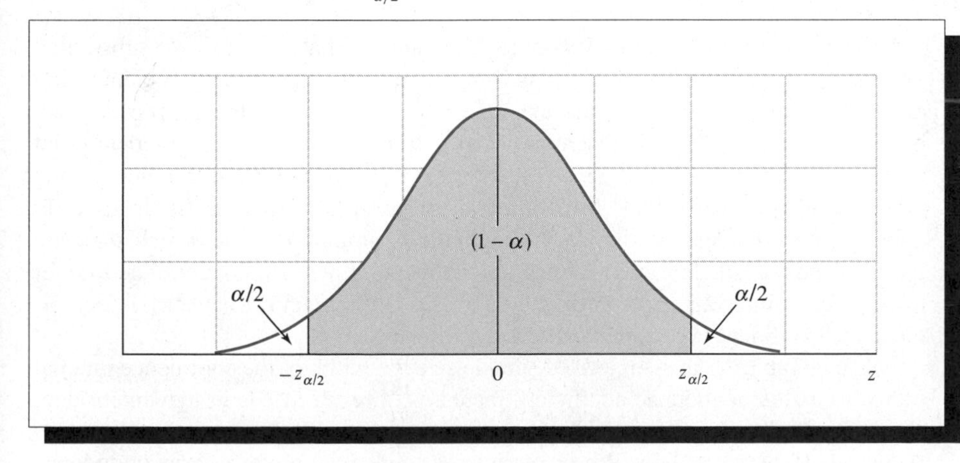

TABLE 7.1
Confidence limits for μ

Confidence coefficient, $(1-\alpha)$	α	$z_{\alpha/2}$	LCL	UCL
.90	.10	1.645	$\bar{x} - 1.645\frac{\sigma}{\sqrt{n}}$	$\bar{x} + 1.645\frac{\sigma}{\sqrt{n}}$
.95	.05	1.96	$\bar{x} - 1.96\frac{\sigma}{\sqrt{n}}$	$\bar{x} + 1.96\frac{\sigma}{\sqrt{n}}$
.99	.01	2.58	$\bar{x} - 2.58\frac{\sigma}{\sqrt{n}}$	$\bar{x} + 2.58\frac{\sigma}{\sqrt{n}}$

EXAMPLE 7.2 Find a 90% confidence interval for the population mean of Example 7.1. Recall that $\bar{x} = 871$ tons and $s = 21$ tons.

Solution The 90% confidence limits are

$$\bar{x} \pm 1.645\frac{\sigma}{\sqrt{n}}$$

Using s to estimate σ, we obtain

$$871 \pm (1.645)\frac{21}{\sqrt{50}} \quad \text{or} \quad 871 \pm 4.89$$

Therefore, we estimate that the average daily field μ lies in the interval from 866.11 to 875.89 tons. The confidence coefficient .90 implies that, in repeated sampling, 90% of the confidence intervals similarly formed would enclose μ. ▪

The confidence interval of Example 7.2 is approximate because we substituted s as an approximation for σ. That is, instead of the confidence coefficient being at .95, the value specified in the example, the true value of the coefficient may be .92, .94, or .97. But this discrepancy is of little concern from a practical point of view; as far as our "confidence" is concerned, there is little difference among these confidence coefficients. Most interval estimators employed in statistics yield approximate confidence intervals, because the assumptions upon which they are based are not satisfied exactly. Having made this point, we will not continue to refer to confidence intervals as "approximate." It is of little practical concern as long as the actual confidence coefficient is near the value specified.

Note in Table 7.1 that for a fixed sample size the width of the confidence interval increases as the confidence coefficient increases, a result that is in agreement with our intuition. Certainly, if we wish to be more confident that the interval will enclose μ, we would increase the width of the interval. Since we prefer narrow confidence intervals and large confidence coefficients, we must reach a compromise in choosing the confidence coefficient.

The choice of the confidence coefficient to be used in a given situation is made by the experimenter and depends on the degree of confidence the experimenter wishes to place in the estimate. Most confidence intervals are constructed by using one of the three confidence coefficients shown in Table 7.1. The most popular seem to be 95% confidence intervals. Use of 99% confidence intervals is less common because of the wider interval width that results. Of course, you can always decrease the width by increasing the sample size n.

Note the fine distinction between point estimators and interval estimators. Note also that, in placing bounds on the error of a point estimate, for all practical purposes we are constructing an interval estimate when a population mean is being estimated. Although this close relationship exists for most of the parameters estimated in this text, the two methods of estimation are not equivalent. For instance, it is not obvious that the best point estimator falls in the middle of the best interval estimator—in many cases it does not. Furthermore, it is not necessarily true that the best interval estimator is even a function of the best point estimator. Although these problems are of a theoretical nature, they are important and worth mentioning. From a practical point of view, the two methods are closely related, and **the choice between the point and the interval estimator in an actual problem depends on the preference of the experimenter.**

Exercises

Basic Techniques

7.1 Explain what is meant by "margin of error" in estimation.

7.2 Give the margin of error in estimating a population mean μ if

a $n = 20, \sigma^2 = 4$.
b $n = 100, \sigma^2 = .9$.
c $n = 50, \sigma^2 = 12$.

7.3 Find and interpret a 95% confidence interval for a population mean μ if

a $n = 36, \bar{x} = 13.1, s^2 = 3.42$.
b $n = 64, \bar{x} = 2.73, s^2 = .1047$.

7.4 Find a 90% confidence interval for a population mean μ if

a $n = 125, \bar{x} = .84, s^2 = .086$.
b $n = 50, \bar{x} = 21.9, s^2 = 3.44$.
c Interpret the intervals found in parts (a) and (b).

7.5 Find a $(1 - \alpha)$ 100% confidence interval for a population mean μ if

a $\alpha = .01, n = 38, \bar{x} = 34, s^2 = 12$.
b $\alpha = .10, n = 65, \bar{x} = 1049, s^2 = 51$.
c $\alpha = .05, n = 89, \bar{x} = 66.3, s^2 = 2.48$.

7.6 In Exercise 6.4 the mean and standard deviation for the die-toss population were $\mu = 3.5$ and $\sigma = 1.71$, respectively.

a If you were to toss a die $n = 5$ times, what is the approximate probability that the mean $\bar{x}$ of the sample would fall in the interval $2.5 \le \bar{x} \le 4.5$?

b If you were to toss a die $n = 10$ times, what is the approximate probability that $\bar{x}$ would fall in the interval $2.5 \le \bar{x} \le 4.5$?

c Note in part (b) that $P(2.5 \le \bar{x} \le 4.5)$ is large. If you have access to a die, toss it $n = 10$ times and calculate $\bar{x}$. Does $\bar{x}$ fall in the interval $2.5 \le \bar{x} \le 4.5$? Suppose the mean of the population $\mu = 3.5$ were unknown and that you were using your mean to estimate μ. What is your margin of error?

7.7 A random sample of n measurements is selected from a population with unknown mean μ and known standard deviation $\sigma = 10$. Calculate the width of a 95% confidence interval for μ when

a $n = 100$. **b** $n = 200$. **c** $n = 400$.

7.8 Compare the confidence intervals in Exercise 7.7. What is the effect on the width of a confidence interval when you

a Double the sample size?

b Quadruple the sample size?

7.9 Refer to Exercise 7.7.

a Calculate the width of a 90% confidence interval for μ when $n = 100$.

b Calculate the width of a 99% confidence interval for μ when $n = 100$.

c Compare the widths of 90, 95, and 99% confidence intervals for μ. What effect does increasing the confidence coefficient have on the width of the confidence interval?

Applications

7.10 Geologists are interested in shifts and movements of the earth's surface indicated by fractures (cracks) in the earth's crust. One of the most famous large fractures is the San Andreas fault (moving fracture) in California. A geologist attempting to study the movement of the relative shifts in the earth's crust at a particular location found many fractures in the local rock structure. In an attempt to determine the mean angle of the breaks, she sampled $n = 50$ fractures and found the sample mean and standard deviation to be 39.8° and 17.2°, respectively. Estimate the mean angular direction of the fractures and find the margin of error for your estimate.

7.11 Estimates of the earth's biomass, the total amount of vegetation held by the earth's forests, are important in determining the amount of unabsorbed carbon dioxide that we can expect to remain in the earth's atmosphere (*Science News*, August 19, 1989, p. 124). Suppose a sample of 75 1-square-meter plots, randomly chosen in North America's boreal (northern) forests, produced a mean biomass of 4.2 kilograms per square meter (kg/m^2), with a standard deviation of 1.5 kg/m^2. Estimate the average biomass for the boreal forests of North America and find the margin of error for your estimate.

7.12 An increase in the rate of consumer savings is frequently tied to a lack of confidence in the economy and is said to be an indicator of a recessional tendency in the economy. A random sampling of $n = 200$ saving accounts in a local community showed a mean increase in savings-account values of 7.2% over the past 12 months, with a standard deviation of 5.6%. Estimate the mean percent increase in savings-account values over the past 12 months for depositors in the community. Find the margin of error for your estimate.

7.13 A study of 392 healthy children living in the area of Tours, France, was designed to measure the serum levels of certain fat-soluble vitamins—namely, vitamin A, vitamin E, β-carotene, and cholesterol. Knowledge of the reference levels for children living in an industrial country, with normal food availability and feeding habits, would be important in establishing borderline levels for children living in other, less favorable conditions. Results of the study are shown in the following table.

	Mean $\pm$ SD ($n = 392$)	Boys ($n = 207$)	Girls ($n = 185$)
Retonol (μg/dl)	42.5 ± 12.0	43.0 ± 13.0	41.8 ± 10.7
β-carotene (μg/l)	572 ± 381	588 ± 406	553 ± 350
Vitamin E (mg/l)	9.5 ± 2.5	9.6 ± 2.7	9.5 ± 2.2
Cholesterol (g/l)	1.84 ± 0.42	1.84 ± 0.47	1.83 ± 0.38
Vitamin E/cholesterol (mg/g)	5.26 ± 1.04	5.26 ± 1.11	5.26 ± 0.25

Source: Malvy et al. (1989), p. 29.

a Describe the sampled population.

b Find a 99% confidence interval for the mean cholesterol level for the population of boys. Interpret this interval.

c Find a 99% confidence interval for the mean β-carotene level for the population of girls. Interpret this interval.

d Can the data collected by the researchers be used to estimate serum levels for all boys and girls in this age category? Why or why not?

7.14 Due to a variation in laboratory techniques, impurities in materials, and other unknown factors, the results of an experiment in a chemistry laboratory will not always yield the same numerical answer. In an electrolysis experiment, a class measured the amount of copper precipitated from a saturated solution of copper sulfate over a 30-minute period. The $n = 30$ students calculated a sample mean and standard deviation equal to .145 and .0051 mole, respectively. Find a 90% confidence interval for the mean amount of copper precipitated from the solution over a 30-minute period of time.

7.15 Intelligence tests are routinely administered by school guidance counselors and psychologists as screening devices for their students. However, are all of these tests really accurate indicators of a student's IQ? In a study to compare two such tests, the Slosson Intelligence Test (SIT) and the Wechsler Intelligence Scale for Children–Revised (WISC–R), the tests were administered to a sample of 72 children in a large urban school district in central Ohio. The mean age of the children was 8.5 years with a standard deviation of 16.6 months. Scores on the two tests for the 72 children were as follows:

Test	Mean	Standard Deviation
WISC–R Full Scale	86.11	15.65
SIT IQ	90.47	14.77

Source: Prewett and Fowler, (1992), p. 17.

a Assume that the scores of the 72 students represent a random sample from the population of scores for all students who might take the test. Find a point estimate for the average grade on the WISC–R for the population. What is the margin of error for this estimate?

b Find a 98% confidence interval for the mean grade on the SIT test.

c In fact, the sample taken by the experimenters was limited to students who were not making adequate academic progress in the regular classroom. What impact does this have on the inferences you can make in parts (b) and (c)?

7.16 The "Jobs and Careers" issue of the *National College Magazine* (Good, 1993) listed the average starting salaries of college graduates by majors. Chemical–engineering majors had the highest starting salary of $40,173, and journalism majors had the lowest starting salary of $19,114. In a random sample of 100 recent college graduates majoring in social science, the average starting salary was $20,450, with a standard deviation of $2390.

a What is the point estimate of average starting salaries for college graduates in the social sciences? What is the margin of error for this estimate?

b Construct a 99% confidence interval for the starting salary of college graduates majoring in a social science.

7.17 Acid rain, caused by the reaction of certain air pollutants with rainwater, appears to be a growing problem in the northeastern section of the United States. (Acid rain affects the soil and causes corrosion on exposed metal surfaces.) Pure rain falling through clean air registers a pH value of 5.7 (pH is a measure of acidity: 0 is acid; 14 is alkaline). Suppose water samples from 40 rainfalls are analyzed for pH and $\bar{x}$ and s are equal to 3.7 and 0.5, respectively. Find a 99% confidence interval for the mean pH in rainfall and interpret the interval. What assumption must be made for the confidence interval to be valid?

7.6 Estimating the Difference Between Two Means

A problem equally as important as the estimation of a single population mean is the comparison of two population means. For instance, we may wish to compare the difference in average scores on the Medical College Admission Test (MCAT) for students whose major was biochemistry and students whose major was biology. Or, we might wish to compare the average yield in a chemical plant using raw materials furnished by two suppliers, A and B. Samples of daily yield, one for each of the two suppliers, could be recorded and used to make inferences concerning the difference in mean yield.

For each of these examples there are two populations, the first with mean and variance μ_1 and σ_1^2 and the second with mean and variance μ_2 and σ_2^2. A random sample of n_1 measurements is drawn from population 1 and n_2 from population 2, where the samples are assumed to have been drawn independently of each other. Finally, the estimates of the population parameters are calculated from the sample data using the estimators $\bar{x}$, s_1^2, $\bar{x}_2$, and s_2^2.

Intuitively, the difference between two sample means would provide the maximum information about the actual difference between two population means, and this is in fact the case. The best point estimator of the difference $(\mu_1 - \mu_2)$ between the population means is $(\bar{x}_1 - \bar{x}_2)$. The sampling distribution of this estimator is not difficult to derive, but we state it here, without proof.

Properties of the Sampling Distribution of $(\bar{x}_1 - \bar{x}_2)$, the Difference Between Two Sample Means

When independent random samples of n_1 and n_2 observations have been selected from populations with means μ_1 and μ_2 and variances σ_1^2 and σ_2^2, respectively, the sampling distribution of the difference $(\bar{x}_1 - \bar{x}_2)$ will have the following properties:

1 The mean and the standard deviation of $(\bar{x}_1 - \bar{x}_2)$ will be

$$\mu_{(\bar{x}_1 - \bar{x}_2)} = \mu_1 - \mu_2$$

and

$$\sigma_{(\bar{x}_1 - \bar{x}_2)} = \sqrt{\frac{\sigma_1^2}{n_1} + \frac{\sigma_2^2}{n_2}}$$

2 **If the sampled populations are normally distributed**, then the sampling distribution of $(\bar{x}_1 - \bar{x}_2)$ is **exactly** normally distributed, regardless of the sample size.

3 **If the sampled populations are not normally distributed**, then the sampling distribution of $(\bar{x}_1 - \bar{x}_2)$ is **approximately normally distributed when n_1 and n_2 are large, due to the Central Limit Theorem.**

The sampling distribution of $(\bar{x}_1 - \bar{x}_2)$ is shown in Figure 7.10.

FIGURE 7.10 The distribution of $(\bar{x}_1 - \bar{x}_2)$ for large samples

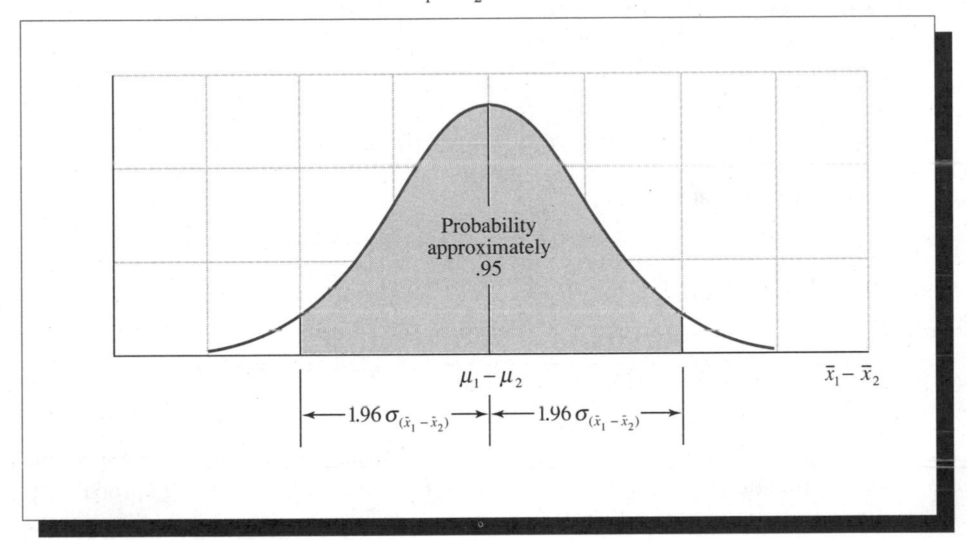

Since $\mu_1 - \mu_2$ is the mean of the sampling distribution, it follows that $(\bar{x}_1 - \bar{x}_2)$ is an unbiased estimator of $(\mu_1 - \mu_2)$. Hence, the general formulas of Section 7.4 can be used to construct point and interval estimates.

Point Estimation of $(\mu_1 - \mu_2)$

Estimator: $(\bar{x}_1 - \bar{x}_2)$

Margin of error: $1.96\sigma_{(\bar{x}_1 - \bar{x}_2)} = 1.96\sqrt{\dfrac{\sigma_1^2}{n_1} + \dfrac{\sigma_2^2}{n_2}}$

(NOTE: If σ_1^2 and σ_2^2 are unknown, but both n_1 and n_2 are 30 or more, you can use the sample variances s_1^2 and s_2^2 to estimate σ_1^2 and σ_2^2.)

EXAMPLE 7.3 A comparison of the wearing quality of two types of automobile tires was obtained by road-testing samples of $n_1 = n_2 = 100$ tires for each type. The number of miles until wearout was defined as a specific amount of tire wear. The test results are given below. Estimate $(\mu_1 - \mu_2)$, the difference in mean miles to wearout, and find the margin of error for your estimate.

Tire 1	Tire 2
$\bar{x}_1 = 26{,}400$ miles	$\bar{x}_2 = 25{,}100$ miles
$s_1^2 = 1{,}440{,}000$	$s_2^2 = 1{,}960{,}000$

Solution The point estimate of $(\mu_1 - \mu_2)$ is

$$(\bar{x}_1 - \bar{x}_2) = 26{,}400 - 25{,}100 = 1300 \text{ miles}$$

The standard error of $(\bar{x}_1 - \bar{x}_2)$ is

$$\sigma_{(\bar{x}-\bar{x}_2)} = \sqrt{\frac{\sigma_1^2}{n_1} + \frac{\sigma_2^2}{n_2}} \approx \sqrt{\frac{s_1^2}{n_1} + \frac{s_2^2}{n_2}}$$

$$= \sqrt{\frac{1{,}440{,}000}{100} + \frac{1{,}960{,}000}{100}} = \sqrt{34{,}000} = 184.4 \text{ miles}$$

where s_1^2 and s_2^2 are used to estimate σ_1^2 and σ_2^2, respectively. We would expect the error of estimation to be less than $(1.96)(184.4) = 361.4$ miles with high probability. If the point estimate of the difference in mean miles to wearout is 1300 miles and the margin of error is 361.4 miles, it seems fairly conclusive that there is a substantial difference in the mean miles to wearout for the two types of tires. If fact, tire 1 is apparently superior to tire 2 in wearing quality when subjected to the road test. ▪

Following the procedure of Section 7.4, we can construct a confidence interval for $(\mu_1 - \mu_2)$ with confidence coefficient $(1 - \alpha)$.

A $(1 - \alpha)$ 100% Confidence Interval for $(\mu_1 - \mu_2)$

$$(\bar{x}_1 - \bar{x}_2) \pm z_{\alpha/2}\sqrt{\frac{\sigma_1^2}{n_1} + \frac{\sigma_2^2}{n_2}}$$

If σ_1^2 and σ_2^2 are unknown, they can be approximated by the sample variances s_1^2 and s_2^2.

Assumption: n_1 and n_2 are both greater than or equal to 30.

EXAMPLE **7.4** Place a confidence interval on the difference in the mean miles to wearout for the problem described in Example 7.3. Use a confidence coefficient of .99.

Solution The confidence interval is

$$(\bar{x}_1 - \bar{x}_2) \pm 2.58\sqrt{\frac{\sigma_1^2}{n_1} + \frac{\sigma_2^2}{n_2}}$$

Using the results of Example 7.3, we find that the confidence interval is

$$1300 \pm (2.58)(184.4)$$

Therefore, LCL = 824.2, UCL = 1775.8, and the difference in the mean miles to wearout is estimated to lie between these two points. Note that the confidence interval is wider than the $\pm 1.96\sigma_{(\bar{x}_1 - \bar{x}_2)}$ used in Example 7.3 because we have chosen a larger confidence coefficient. ▪

Examples 7.3 and 7.4 deserve further comment with regard to using sample estimates in place of unknown parameters. The sampling distribution of

$$\frac{(\bar{x}_1 - \bar{x}_2) - (\mu_1 - \mu_2)}{\sqrt{\frac{\sigma_1^2}{n_1} + \frac{\sigma_2^2}{n_2}}}$$

has a standard normal distribution when both sampled populations are normal for all sample sizes, and an approximate standard normal distribution when the sampled populations are not normal but the sample sizes are large (greater than or equal to 30). When σ_1^2 and σ_2^2 are not known and are estimated by the sample estimates s_1^2 and s_2^2, the resulting statistic will also have an approximate standard normal distribution when the sample sizes are large. The behavior of this statistic when the population variances are unknown and the sample sizes are small will be discussed in Section 9.4.

Exercises

Basic Techniques

7.18 Independent random samples were selected from two populations 1 and 2. The sample sizes, means, and variances are as follows:

	Population	
	1	**2**
Sample size	35	49
Sample mean	12.7	7.4
Sample variance	1.38	4.14

Find the margin of error for estimating the difference in population means $(\mu_1 - \mu_2)$.

7.19 Independent random samples were selected from two populations 1 and 2. The sample sizes, means, and variances are as follows:

	Population	
	1	2
Sample size	64	64
Sample mean	2.9	5.1
Sample variance	0.83	1.67

a Find a 90% confidence interval for the difference in the population means and interpret your result.

b Find a 99% confidence interval for the difference in the population means and interpret your result.

Applications

7.20 A small amount of the trace element selenium, from 50–200 micrograms (μg) per day, is considered essential to good health. Suppose that random samples of $n_1 = n_2 = 30$ adults were selected from two regions of the United States and that a day's intake of selenium, from both liquids and solids, was recorded for each person. The mean and standard deviation of the selenium daily intakes for the 30 adults from region 1 were $\bar{x} = 167.1$ and $s_1 = 24.3$ μg, respectively. The corresponding statistics for the 30 adults from region 2 were $\bar{x}_2 = 140.9$ and $s_2 = 17.6$. Find a 95% confidence interval for the difference in the mean selenium intake for the two regions. Interpret this interval.

7.21 A study was conducted to compare the mean number of police emergency calls per 8-hour shift in two districts of a large city. Samples of 100 8-hour shifts were randomly selected from the police records for each of the two regions, and the number of emergency calls was recorded for each shift. The sample statistics are as follows:

	Region	
	1	2
Sample size	100	100
Sample mean	2.4	3.1
Sample variance	1.44	2.64

Find a 90% confidence interval for the difference in the mean number of police emergency calls per shift between the two districts of the city. Interpret the interval.

7.22 One method for solving the electric power shortage employs the construction of floating nuclear power plants located a few miles offshore. Because there is concern about the possibility of ships colliding with a floating (but anchored) plant, an estimate of the density of ship traffic in the area is needed. The number of ships passing within 10 miles of the proposed power-plant location per day, recorded for $n = 60$ days during July and August, has sample mean and variance equal to $\bar{x} = 7.2$ and $s^2 = 8.8$.

a Find a 95% confidence interval for the mean number of ships passing within 10 miles of the proposed power-plant location during a 1-day time period.

b The density of ship traffic was expected to decrease during the winter months. A sample of $n = 90$ daily recordings of ship sightings for December, January, and February gave a mean and variance of $\bar{x} = 4.7$ and $s^2 = 4.9$. Find a 90% confidence interval for the difference in mean density of ship traffic between the summer and winter months.

c What is the population associated with your estimate in part (b)? What could be wrong with the sampling procedure in parts (a) and (b)?

7.23 An experiment was conducted to compare two diets A and B designed for weight reduction. Two groups of 30 overweight dieters each were randomly selected. One group was placed on diet A and the other on diet B, and their weight losses were recorded over a 30-day period. The means and standard deviations of the weight-loss measurements for the two groups are shown in the table. Find a 95% confidence interval for the difference in mean weight loss for the two diets. Interpret your confidence interval.

Diet A	Diet B
$\bar{x}_A = 21.3$	$\bar{x}_B = 13.4$
$s_A = 2.6$	$s_B = 1.9$

7.24 Refer to Exercise 7.13.

a Find a 90% confidence interval for the difference in the mean serum cholesterol levels of boys versus girls.

b Find a 95% confidence interval for the difference in the mean β-carotene levels of boys versus girls.

c Interpret the intervals found in parts (a) and (b).

d Do the intervals in parts (a) and (b) contain the value $\mu_1 - \mu_2 = 0$? Why would this be of interest to the researcher?

7.25 Refer to Exercise 7.16. In an attempt to compare the starting salaries for college graduates majoring in education and social sciences, random samples of 50 recent college graduates in each major were selected and the following information was obtained:

Major	Mean	SD
Education	22,554	2225
Social Science	20,348	2375

a Find a point estimate for the difference in average starting salaries for college students majoring in education and the social sciences. What is the margin of error for your estimate?

b Based on the results of part (a), do you think that there is a significant difference in the means for the two groups in the general population? Explain.

7.26 In assessing students' leisure interests and stress ratings, Alice Chang and colleagues (1993) tested 559 high school students using the leisure interest checklist (LIC) inventory. The means and standard deviations for each of the seven LIC factor scales are given in the following table for both boys ($n_1 = 252$) and girls ($n_2 = 307$):

	Men		Women	
LIC Factor Scale Activity	**Mean**	**SD**	**Mean**	**SD**
Hobbies	10.06	6.47	13.64	7.46
Social fun	22.05	5.12	25.96	5.07
Sports	13.65	4.82	9.88	4.41
Cultural	11.48	5.69	13.21	5.31
Trips	6.90	3.41	6.49	2.97
Games	4.95	3.29	3.85	2.49
Church	2.15	2.10	3.00	2.26

a Provide a point estimate for the true mean LIC score for hobbies together with the margin of error for high school boys.

b What is the point estimate with a margin of error for the true mean hobby score for high school girls?

c Find a point estimate for the difference in mean scores between boys and girls for the hobby factor scale activity. What is the margin of error?

7.27 Refer to Exercise 7.26.

a Find a 95% confidence interval for the mean difference between boys and girls for the cultural factor scale activity. Interpret this interval.

b Find a 90% confidence interval for the mean difference between boys and girls for the social fun factor scale activity. Interpret this interval.

c Do there appear to be significant differences between boys and girls in the two LIC activity scores given in parts (a) and (b)? Explain.

7.7 Estimating a Binomial Proportion

Many surveys have as their objective the estimation of the proportion of people or objects in a large group that possess a particular attribute. Such a survey is a practical example of the binomial experiment discussed in Chapter 4. Estimating the proportion of sales that can be expected in a large number of customer contacts is a practical problem requiring the estimation of a binomial parameter p.

The best point estimator of the binomial parameter p is also the estimator that would be chosen intuitively. That is, the estimator of p, denoted by the symbol $\hat{p}$, is the total number x of successes divided by the total number n of trials:

$$\hat{p} = \frac{x}{n}$$

where x is the number of successes in n trials. [A "hat" ($\hat{\ }$) over a parameter is the symbol used to denote the estimator of the parameter.] By "best" we mean that $\hat{p}$ is unbiased and possesses a smaller variance than other possible estimators.

As noted in Chapter 6, the estimator $\hat{p}$ possesses a sampling distribution that can be approximated by a normal distribution because of the Central Limit Theorem. It is an unbiased estimator of the population proportion p, with mean and standard deviation given in the following display.

Mean and Standard Deviation of $\hat{p}$

$$E(\hat{p}) = p$$

$$\sigma_{\hat{p}} = \sqrt{\frac{pq}{n}}$$

Then, from Section 7.4, the point estimation procedure for p is as given in the following display.

Point Estimator for p

Estimator: $\hat{p} = \frac{x}{n}$

Margin of error: $1.96\sigma_{\hat{p}} = 1.96\sqrt{\frac{pq}{n}}$

Estimated margin of error: $1.96\sqrt{\frac{\hat{p}\hat{q}}{n}}$

The corresponding large-sample confidence interval with confidence coefficient $(1 - \alpha)$ is shown below.

A $(1 - \alpha)$ 100% Confidence Interval for p

$$\hat{p} \pm z_{\alpha/2}\sqrt{\frac{\hat{p}\hat{q}}{n}}$$

The sample size will be considered large when we can assume that the distribution of $\hat{p}$ can be approximated by a normal distribution—that is, when $p \pm 2\sigma_{\hat{p}}$ is contained in the interval 0 to 1.

The only difficulty encountered in our procedure will be in calculating $\sigma_{\hat{p}}$, which involves the unknown value of p (and $q = 1 - p$). You will note that we have substituted $\hat{p}$ for the parameter p in the standard deviation $\sqrt{pq/n}$. When n is large, little error will be introduced by this substitution. As a matter of fact, the standard deviation changes only slightly as p changes. This feature can be observed in Table 7.2, where $\sqrt{pq}$ is recorded for several values of p. Note that $\sqrt{pq}$ changes very little as p changes, especially when p is near .5.

TABLE **7.2**
Some calculated values of $\sqrt{pq}$

p	$\sqrt{pq}$
.5	.50
.4	.49
.3	.46
.2	.40
.1	.30

EXAMPLE **7.5** A random sample of $n = 100$ voters in a community produced $x = 59$ voters in favor of candidate A. Estimate the fraction of the voting population favoring A and find the estimated margin of error.

Solution The point estimate is

$$\hat{p} = \frac{x}{n} = \frac{59}{100} = .59$$

and the margin of error is

$$1.96\sigma_{\hat{p}} = 1.96\sqrt{\frac{pq}{n}} \approx 1.96\sqrt{\frac{(.59)(.41)}{100}} = .096^{\dagger}$$

A 95% confidence interval for p would be

$$\hat{p} \pm 1.96\sqrt{\frac{\hat{p}\hat{q}}{n}} \quad \text{or} \quad .59 \pm 1.96(.049)$$

Thus, we would estimate that p lies in the interval .494 to .686 with confidence coefficient .95. ▪

Exercises

Basic Techniques

7.28 A random sample of $n = 900$ observations from a binomial population produced $x = 655$ successes. Find a 99% confidence interval for p and interpret the interval.

7.29 A random sample of $n = 300$ observations from a binomial population produced $x = 263$ successes. Find a 90% confidence interval for p and interpret the interval.

7.30 Suppose the number of successes observed in $n = 500$ trials of a binomial experiment is 27. Find a 95% confidence interval for p. Why is the confidence interval narrower than the confidence interval in Exercise 7.28?

Applications

7.31 A survey concerning the health habits of Americans has been taken for over 10 years by Louis Harris & Associates ("Americans Round Retreating," March 14, 1993). Their latest findings have been labeled "terribly disturbing" by Dr. Todd Davis, the executive vice-president of the American Medical Association. The survey's sample of 1251 Americans showed that only 53% ate the recommended amounts of fibrous foods and vegetables, 51% avoided fat, and 46% avoided excess salt.

a Based on the survey results, construct a 90% confidence interval for the proportion of all Americans who eat the recommended amounts of fibrous foods and vegetables.

b Construct a 99% confidence interval for the proportion of all Americans who avoid fat in their diet.

c Find a point estimate for the proportion of all Americans who avoid salt in their diet. What is your margin of error?

† Checking the conditions that allow the normal approximation to the sampling distribution of $\hat{p}$, we find that $p \pm 2\sigma_{\hat{p}} \approx .59 \pm 2(.049) = .59 \pm .098$ falls in the interval 0 to 1.

d The survey reports a margin of sampling error of "plus or minus three percentage points." Does this agree with your results in part (c)?

7.32 Refer to Exercise 7.31. The survey conducted by Harris in 1992 for Baxter International, a manufacturer of health-care products, did provide some good news. The use of auto seatbelts has increased to 70%, from 19% in 1983, and the use of home smoke detectors has increased to 90%, from 67% in the same period.

a Find a 99% confidence interval for the proportion of Americans who used seatbelts in 1992. Interpret this interval.

b Find a 90% confidence interval for the proportion of Americans who used home smoke detectors in 1992. Interpret this interval.

7.33 Older Americans are more alienated politically than are young Americans, according to a poll by the Times Mirror Center for the People and the Press (Rosenbaum, 1992). Two-thirds of those polled who were over age 50 said that new leaders were needed in Washington, "even if they were not as effective as experienced politicians." If $n = 1752$ of those polled were over age 50, what is the point estimate of the true proportion of Americans over 50 who would agree that new leaders are needed in Washington? What is the margin of error?

7.34 "Athletes at major colleges graduated, on the whole, at virtually the same rate as other students," according to the National Collegiate Athletic Association (NCAA) ("Athletic Grade Rate," 1992). The survey involved students who had entered college before the NCAA Proposition 48 took effect in 1985. (It requires athletes to maintain at least a 2.0 grade point average on a scale of 4.0.) Suppose that, in a new poll of 500 athletes at major colleges, the number graduating was 268.

a Find a point estimate of the true proportion of college athletes at major colleges who graduate. Find the margin of error for your estimate.

b Find a 98% confidence interval for p. Does the confidence interval include the value $p = .51$, the graduation rate before Proposition 48? What might you conclude?

7.35 Radio and television stations often air controversial issues during broadcast time and ask viewers to indicate their agreement or disagreement with a given stand on the issue. A poll is conducted by asking those viewers who *agree* to call a certain 900 telephone number and those who *disagree* to call a second 900 telephone number (all respondents pay a fee for their calls).

a Does this polling technique result in a random sample?

b What can be said about the validity of the results of such a survey? Need we worry about a margin of error in this case?

7.36 Should professional athletes be tested for drugs? Reporting on this question, the *Orlando Sentinel* (Orlando, Fla., Oct. 31, 1985) noted that Donald Fehr, executive director of the Major League Baseball Player's Association, "criticized mandatory testing as being unnecessary."

Former baseball commissioner Peter Ueberroth and the public did not seem to agree. A *Washington Post*/ABC News survey found that 73% of 1506 people interviewed favored drug tests for professional athletes. Sixty-eight percent said that professional athletes using drugs for the first time should be banned or suspended from professional sports.

a Find a 95% confidence interval for the proportion of the public who favor drug tests for professional athletes.

b Upon what assumptions is your confidence interval in part (a) based?

7.37 Refer to Exercise 7.36, and find a 90% confidence interval for the proportion of the public who favor banning or suspending professional athletes who are found to be using drugs for the first time. What assumptions are required for your confidence interval to be valid?

7.38 A *New York Times*/CBS survey of 224 children aged 9 to 17, conducted shortly after the Jan. 28, 1986, space shuttle disaster, found that two-thirds of the children said they would like to travel in space (Orlando, Fla., *Orlando Sentinel*, Feb. 3, 1986). Find the margin of error for this estimate.

7.39 A survey of the college class of 1992 showed that more incoming college freshmen are smoking, their interest in business careers has declined, and they favor mandatory testing for AIDS and drugs. These findings were based on a survey by the American Council on Education (*New York Times*, Jan. 9, 1989) involving $n = 308{,}007$ freshmen entering 585 2- and 4-year colleges and universities. Specifically, 67% agreed that the best way to control AIDS was through widespread mandatory testing, and 71% agreed that employers should be allowed to test employees or job applicants for drugs.

a Find a 98% confidence interval for the proportion of college freshmen favoring mandatory AIDS testing.

b Find a 99% confidence interval for the proportion of college freshmen favoring drug testing for employees or job applicants. Interpret this interval.

7.40 Refer to Exercise 7.34. The results of this survey were reported to have a sampling error of no more than plus or minus 2 percentage points. If we interpret this to mean that the margin of error is at most two percentage points, determine whether this claim is true for the two estimates reported in the exercise.

7.8 Estimating the Difference Between Two Binomial Proportions

The fourth and final estimation problem considered in this chapter is the estimation of the difference between the parameters of two binomial populations—that is, the difference between two binomial proportions. We may be interested in the difference between the production rates for each of two production lines, or in the difference in the proportion of voters favoring two leading mayoral candidates. Intuitively, the difference between two sample proportions would provide the maximum information about the corresponding population proportions.

Assume that the two binomial populations 1 and 2 have parameters p_1 and p_2, respectively. Independent random samples consisting of n_1 and n_2 trials are drawn from the respective populations, and the estimates $\hat{p}_1$ and $\hat{p}_2$ are calculated. The sampling distribution of the difference $\hat{p}_1 - \hat{p}_2$ is stated without proof in the following display.

Properties of the Sampling Distribution of the Difference $(\hat{p}_1 - \hat{p}_2)$ Between Two Sample Proportions

Assume that independent random samples of n_1 and n_2 observations have been selected from binomial populations with parameters p_1 and p_2, respectively. The sampling distribution of the difference between sample proportions

$$(\hat{p}_1 - \hat{p}_2) = \left(\frac{x_1}{n_1} - \frac{x_2}{n_2}\right)$$

will have the following properties:

1 The mean and the standard deviation of $(\hat{p}_1 - \hat{p}_2)$ will be

$$\mu_{(\hat{p}_1 - \hat{p}_2)} = p_1 - p_2$$

and

$$\sigma_{(\hat{p}_1-\hat{p}_2)} = \sqrt{\frac{p_1q_1}{n_1} + \frac{p_2q_2}{n_2}}$$

2 The sampling distribution of $(\hat{p}_1 - \hat{p}_2)$ can be approximated by a normal distribution when n_1 and n_2 are large, due to the Central Limit Theorem.

When we use a normal distribution to approximate binomial probabilities, the interval $(\hat{p}_1 - \hat{p}_2) \pm 2\sigma_{(\hat{p}_1-\hat{p}_2)}$ should be contained within the range of $(\hat{p}_1 - \hat{p}_2)$, which varies from -1 to 1 and not from 0 to 1, as in the case of a single proportion.

Since $(\hat{p}_1 - \hat{p}_2)$ is an unbiased estimator of $(p_1 - p_2)$, the point estimator of $p_1 - p_2$ is given in the following display.

Point Estimator of $(p_1 - p_2)$

Estimator: $(\hat{p}_1 - \hat{p}_2)$

Margin of error: $1.96\sigma_{(\hat{p}_1-\hat{p}_2)} = 1.96\sqrt{\frac{p_1q_1}{n_1} + \frac{p_2q_2}{n_2}}$

(NOTE: The estimates $\hat{p}_1$ and $\hat{p}_2$ must be substituted for p_1 and p_2 to estimate the margin of error.)

The $(1 - \alpha)$ 100% confidence interval, appropriate when n_1 and n_2 are large, is shown here.

A $(1 - \alpha)$ 100% Large-Sample Confidence Interval for $(p_1 - p_2)$

$$(\hat{p}_1 - \hat{p}_2) \pm z_{\alpha/2}\sqrt{\frac{\hat{p}_1\hat{q}_1}{n_1} + \frac{\hat{p}_2\hat{q}_2}{n_2}}$$

Assumption: n_1 and n_2 must be sufficiently large so that the sampling distribution of $(\hat{p}_1 - \hat{p}_2)$ can be approximated by a normal distribution. The interval $(p_1 - p_2) \pm 2\sigma_{(\hat{p}_1-\hat{p}_2)}$ must be contained in the interval -1 to 1.

EXAMPLE **7.6** A bond proposal for school construction will be submitted to the voters during the next municipal election. A major portion of the money derived from this bond issue will be used to build schools in a rapidly developing section of the city, and the remainder will be used to renovate and update school buildings in the rest of the city.

To assess the viability of the bond proposal, a random sample of $n_1 = 50$ residents in the developing section and $n_2 = 100$ residents from the other parts of the city were asked whether they plan to vote for the proposal. The results are tabulated in the following table:

	Developing Section	**Rest of the City**
Sample size	50	100
Number favoring proposal	38	65
Proportion favoring proposal	.76	.65

a Estimate the difference in the true proportions favoring the bond proposal with a 99% confidence interval.

b If both samples were pooled into one sample of size $n = 150$, with 103 in favor of the proposal, provide a point estimate of the proportion of city residents who will vote for the bond proposal. What is the margin of error?

Solution **a** The best point estimate of the difference $p_1 - p_2$ is given by

$$\hat{p}_1 - \hat{p}_2 = .76 - .65 = .11$$

and the estimated standard deviation of $\hat{p}_1 - \hat{p}_2$ is

$$\sqrt{\frac{\hat{p}_1\hat{q}_1}{n_1} + \frac{\hat{p}_2\hat{q}_2}{n_2}} = \sqrt{\frac{(.76)(.24)}{50} + \frac{(.65)(.35)}{100}} = .0770$$

For a 99% confidence interval, $z_{.005} = 2.58$, and the approximate 99% confidence interval is found as

$$(\hat{p}_1 - \hat{p}_2) \pm z_{.005}\sqrt{\frac{\hat{p}_1\hat{q}_1}{n_1} + \frac{\hat{p}_2\hat{q}_2}{n_2}}$$

which is

$$.11 \pm (2.58)(.0770)$$
$$.11 \pm .199$$

or $(-.089, .309)$. Since this interval contains the value $p_1 - p_2 = 0$, it is highly possible that $p_1 = p_2$, which implies that there may be no difference in the proportions favoring the bond issue in the two sections of the city.

b If there is no difference in the two proportions, as suggested by the results of part (a), the two samples are not really different and might well be combined to obtain an overall estimate of the proportion of the city residents who will vote for the bond issue. If both samples are pooled, then $n = 150$ and

$$\hat{p} = \frac{103}{150} = .69$$

Therefore, the point estimate of the overall value of p is .69, with a margin of error given by

$$1.96\sqrt{\frac{(.69)(.31)}{150}} = 1.96(.0378) = .074$$

Notice that $.69 \pm .074$ produces the interval from .62 to .76, which includes only proportions larger than .5. Therefore, if voter attitudes do not change adversely prior to the election, the bond proposal should pass by a reasonable majority. ■

Exercises

Basic Techniques

7.41 Samples of $n_1 = 500$ and $n_2 = 500$ observations were selected from binomial populations 1 and 2, and $x_1 = 120$ and $x_2 = 147$ were observed. Find a bound on the error of estimating the difference in population proportions $(p_1 - p_2)$.

7.42 Samples of $n_1 = 800$ and $n_2 = 640$ observations were selected from binomial populations 1 and 2, and $x_1 = 337$ and $x_2 = 374$ were observed.

- **a** Find a 90% confidence interval for the difference $(p_1 - p_2)$ in the two population proportions. Interpret the interval.
- **b** What assumptions must you make for the confidence interval to be valid? Are these assumptions met?

7.43 Samples of $n_1 = 1265$ and $n_2 = 1688$ observations were selected from binomial populations 1 and 2, and $x_1 = 849$ and $x_2 = 910$ were observed.

- **a** Find a 99% confidence interval for the difference $(p_1 - p_2)$ in the two population proportions. Interpret the interval.
- **b** What assumptions must you make for the confidence interval to be valid? Are these assumptions met?

Applications

7.44 The survey on Americans' health habits that was conducted by Louis Harris & Associates (Exercise 7.31) not only reports the sample percentages in various categories for the 1992 survey but also give the results of a similar survey conducted in 1983 ("Americans Round Retreating," 1993). Although the number of participants in the 1983 survey is not reported in the article, we will assume that the 1983 survey is based on a sample of 1250 Americans. The sample results for several health categories are shown in the following table:

Health Issue	1983 Survey	1992 Survey
Ate recommended amount of fibrous foods	0.59	0.53
Avoided fat	0.55	0.51
Avoided excess salt	0.53	0.46
Used seatbelts	0.19	0.70
Used smoke detectors	0.67	0.90

a Construct a 95% confidence interval for the difference in the proportion of Americans who ate the recommended amount of fibrous foods in 1983 and 1992. Interpret this interval.

b Based on the results of part (a), do you think that there has been a significant decrease in the proportion of Americans eating the recommended amount of fibrous foods over the period reported in the article? Explain.

7.45 Refer to Exercise 7.44. Repeat the instructions in parts (a) and (b) for the other two categories relating to dietary health habits. Do you agree with the doctor quoted in the article as saying that the results are "terribly disturbing"? Explain.

7.46 Refer to Exercise 7.44. Based on the results of the surveys, there was some good news regarding the health behavior of Americans in areas enforced by laws, including drinking and driving, using automobile seatbelts, and installing smoke detectors at home.

a Construct a 95% confidence interval for the difference in the proportion of Americans who used seatbelts in 1983 and 1992.

b Construct a 95% confidence interval for the difference in the proportion of Americans who used smoke detectors in 1983 and 1992.

c Based on the results of parts (a) and (b), are these results indeed "good news"? Explain.

7.47 For several years now, aspirin-based products have been losing market share to substitute analgesics such as ibuprofen and acetaminophen. Finally, large companies such as Bayer, which had produced only aspirin-based products, are beginning to market acetaminophen-based pain relievers (Kraul, 1993). Although the proportion of acetaminophen users has remained constant at 41% between 1986 and 1991, the proportion of aspirin users has dropped from 45% to 34%, and the proportion of ibuprofen users has risen from 14% to 26%. Suppose these results were obtained using two independent random samples of size $n_1 = n_2 = 1000$ analgesic users.

a Find a point estimate of the difference in the proportion of aspirin users from 1986 to 1991. What is the margin of error of your estimate?

b Find a point estimate of the difference in the proportion of ibuprofen users from 1986 to 1991. What is the margin of error of your estimate?

c Based on the result of parts (a) and (b), what conclusions can you draw? Explain.

7.48 A sampling of political candidates—200 randomly chosen from the West and 200 from the East—was classified according to whether the candidate received backing by a national labor union and whether the candidate won. A summary of the data is as follows:

	West	East
Winners backed by union	120	142

Find a 95% confidence interval for the difference between the proportions of union-backed winners in the West versus the East. Interpret this interval.

7.49 In a study of the relationship between birth order and college success, an investigator found that 126 in a sample of 180 college graduates were firstborn or only children. In sample of

100 nongraduates of comparable age and socioeconomic background, the number of firstborn or only children was 54. Estimate the difference between the proportions of firstborn or only children in the two populations from which these samples were drawn. Use a 90% confidence interval and interpret your results.

7.9 Choosing the Sample Size

The design of an experiment is essentially a plan for purchasing a quantity of information. This information, like any other commodity, may be acquired at varying prices depending on the manner in which the data are obtained. Some measurements contain a large amount of information concerning the parameter of interest; others may contain little or none. Since the sole product of research is information, we should try to purchase it at minimum cost.

The **sampling procedure**, or **experimental design** (as it is usually called), affects the quantity of information per measurement. This procedure, along with the sample size n, controls the total amount of relevant information in a sample. With a few exceptions, we will be concerned with the simplest sampling situation—random sampling from a relatively large population—and will focus our attention on the selection of the sample size n.

The researcher makes little progress in planning an experiment before encountering the problem of selecting the sample size. Indeed, perhaps one of the questions most frequently asked of the statistician is, **How many measurements should be included in the sample?** Unfortunately, the statistician cannot answer this question without knowing how much information the experimenter wishes to buy. Certainly, the total amount of information in the sample will affect the measure of goodness of the method of inference and must be specified by the experimenter. Referring specifically to estimation, we would like to know how accurate the experimenter wishes the estimate to be. This accuracy may be stated by specifying a bound on the margin of error for the estimator.

For instance, suppose we wish to estimate the average daily yield μ of a chemical (Example 7.1) and wish the margin of error to be less than 4 tons, with a probability of .95. Since approximately 95% of the sample means will lie within $1.96\sigma_{\bar{x}}$ of μ in repeated sampling, we are asking that $1.96\sigma_{\bar{x}}$ equal 4 tons (see Figure 7.11). Then,

$$1.96\sigma_{\bar{x}} = 4 \qquad \text{or} \qquad 1.96\frac{\sigma}{\sqrt{n}} = 4$$

Solving for n, we obtain

$$n = \left(\frac{1.96}{4}\right)^2 \sigma^2 \qquad \text{or} \qquad n = .24\sigma^2$$

You can see that we cannot obtain a numerical value for n unless the population standard deviation σ is known. And, certainly, this is exactly what we would expect because the variability of $\bar{x}$ depends on the variability of the population from which the sample was drawn. Lacking an exact value for σ, we would use the best approximation available, such as an estimate s obtained from a previous sample or knowledge of the range in which the measurements will fall. Since the range is approximately equal to

FIGURE 7.11 Approximate sampling distribution of $\bar{x}$ for large samples

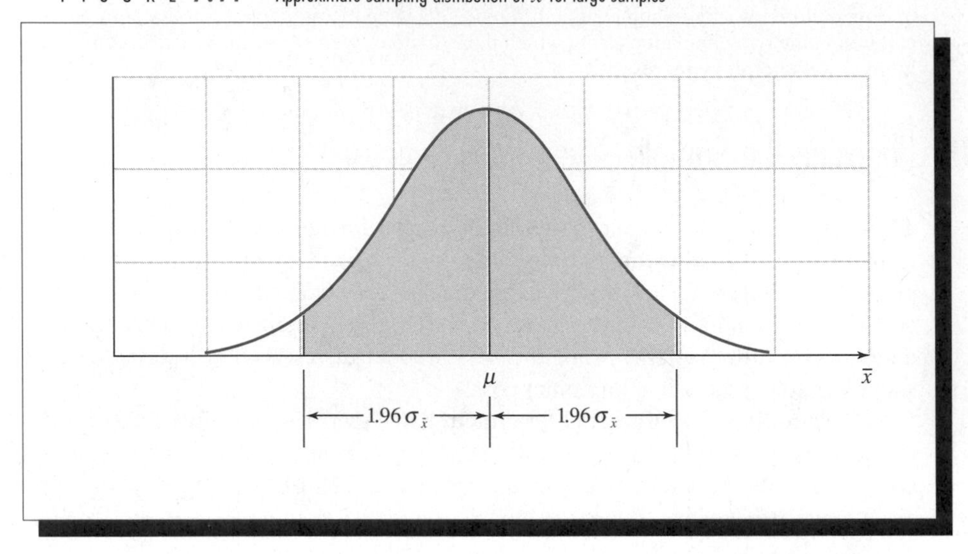

4σ (the Empirical Rule), one-fourth of the range will provide an approximate value for σ. For our example, we would use the results of Example 7.1, which provided a reasonably accurate estimate of σ equal to $s = 21$. Then,

$$n = .24\sigma^2 = .24(21)^2 = 105.8$$

Using a sample of size $n = 106$ or larger, we could be reasonably certain (with probability approximately equal to .95) that our estimate will be within $1.96\sigma_{\bar{x}} = 4$ tons of the true average daily yield.

Actually, we should expect the margin of error to be much less than 4 tons. According to the Empirical Rule, the probability is approximately equal to .68 that the margin of error is less that $\sigma_{\bar{x}} = 2$ tons. You will note that the probabilities .95 and .68 used in these statements are inexact, since s was substituted for σ. Although this method of choosing the sample size is only approximate for a specified desired accuracy of estimation, it is the best available and is certainly better than selecting the sample size on the basis of our intuition.

The method of choosing the sample size for all the large-sample estimation procedures discussed in preceding sections is identical to that described above.

Procedure for Choosing the Sample Size

Determine the parameter to be estimated and the standard error of its point estimator. Then proceed as follows:

1 Choose B, the bound on the margin of error, and a confidence coefficient $(1 - \alpha)$.

2 Solve the following equation for the sample size n:

$$z_{\alpha/2} \times (\text{standard error of the estimator}) = B$$

where $z_{\alpha/2}$ is the value of z having area $\alpha/2$ to its right.

[NOTE: For most estimators (all presented in this text), the standard error is a function of the sample size n.]

We will illustrate with examples.

EXAMPLE 7.7 The reaction of an individual to a stimulus in a psychological experiment can take one of two forms: either an event A will occur, or it will not occur. If an experimenter wishes to estimate the probability p that a person will react in favor of A, how many people must be included in the experiment? Assume that he or she will be satisfied if the margin of error is less than .04, with probability equal to .90. Assume also that he or she expects p to lie somewhere in the neighborhood of .6.

Solution For this particular example, the bound B on the margin of error is .04. Since the confidence coefficient is $1 - \alpha = .90$, α must equal .10 and $\alpha/2 = .05$. The z value corresponding to an area equal to .05 in the upper tail of the z distribution is $z_{\alpha/2} = 1.645$.† We then require

$$1.645\sigma_{\hat{p}} = .04 \qquad \text{or} \qquad 1.645\sqrt{\frac{pq}{n}} = .04$$

Since the variability of $\hat{p}$ is dependent on p, which is unknown, we must use the guessed value of $p = .6$ provided by the experimenter as an approximation. Then,

$$1.645\sqrt{\frac{(.6)(.4)}{n}} = .04 \qquad \text{or} \qquad \sqrt{n} = \frac{1.645}{.04}\sqrt{(.6)(.4)}$$

and

$$n = 406$$

Remember that we used an approximate value of p to obtain this value of n. Consequently, the value $n = 406$ is not exact; it is only an approximation to the required sample size. Therefore, it is reasonable to say that a sample of approximately $n = 406$ people will allow you to estimate p with a margin of error less than .04 with probability equal to .90. ∎

---

†NOTE: It is necessary to carry three decimal places on $z_{\alpha/2}$ for these calculations. The calculated value of n will be very approximate because we are using an approximate value for p in the calculations.

In the preceding example, the value of p was known to be approximately .6. When the value of p is unknown, using $p = .5$ produces the maximum value of $pq = .25$ (see Table 7.2). Hence, the standard deviation of $\hat{p}$ can never be larger than $\sqrt{.25/n}$. **Therefore, using $p = .5$ always produces a value of n satisfying the desired bound.**

EXAMPLE **7.8** An experimenter wishes to compare the effectiveness of two methods of training industrial employees to perform a certain assembly operation. A number of employees are to be divided into two equal groups, the first receiving training method 1 and the second training method 2. Each employee will perform the assembly operation, and the length of assembly time will be recorded. The measurements for both groups are expected to have a range of approximately 8 minutes. If the estimate of the difference in mean times to assemble is desired to be correct to within 1 minute, with probability equal to .95, how many workers must be included in each training group?

Solution Equating $1.96\sigma_{(\bar{x}_1 - \bar{x}_2)}$ to $B = 1$ minute, we obtain

$$1.96\sqrt{\frac{\sigma_1^2}{n_1} + \frac{\sigma_2^2}{n_2}} = 1$$

Or, since we desire n_1 to equal n_2, we may let $n_1 = n_2 = n$ and obtain the equation

$$1.96\sqrt{\frac{\sigma_1^2}{n} + \frac{\sigma_2^2}{n}} = 1$$

As noted above, the variability of each method of assembly is approximately the same, and hence $\sigma_1^2 = \sigma_2^2 = \sigma^2$. Since the range, equal to 8 minutes, is approximately equal to 4σ, then $4\sigma \approx 8$ and $\sigma \approx 2$. Substituting this value for σ_1 and σ_2 in the above equation, we obtain

$$1.96\sqrt{\frac{(2)^2}{n} + \frac{(2)^2}{n}} = 1$$

$$1.96\sqrt{8/n} = 1$$

$$\sqrt{n} = 1.96\sqrt{8}$$

Solving, we have $n = 31$. Thus, each group should contain approximately $n = 31$ members. ▪

Exercises

Basic Techniques

7.50 Suppose you wish to estimate a population mean based on a random sample of n observations, and prior experience suggests that $\sigma = 12.7$. If you wish to estimate μ correct to within 1.6, with probability equal to .95, how many observations should be included in your sample?

7.51 Suppose you wish to estimate a binomial parameter p correct to within .04, with probability equal to .95. If you suspect that p is equal to some value between .1 and .3 and you want to be certain that your sample is large enough, how large should n be? (HINT: When calculating $\sigma_{\hat{p}}$, use the value of p in the interval $.1 < p < .3$ that will give the largest sample size.)

7.52 Independent random samples of $n_1 = n_2 = n$ observations are to be selected from each of two populations 1 and 2. If you wish to estimate the difference between the two population means correct to within .17, with probability equal to .90, how large should n_1 and n_2 be? Assume that you know $\sigma_1^2 \approx \sigma_2^2 \approx 27.8$.

7.53 Independent random samples of $n_1 = n_2 = n$ observations are to be selected from each of two binomial populations 1 and 2. If you wish to estimate the difference in the two population binomial proportions correct to within .05, with probability equal to .98, how large should n be? Assume that you have no prior information on the values of p_1 and p_2, but you want to make certain that you have an adequate number of observations in the samples.

Applications

7.54 According to the Sheriff's Office in Marion County, Florida, an automobile salesman who rolled back the mileage on a used car by 60,000 miles increased the car's value by $1375 (*New York Times*, Apr. 14, 1985). Fraud involving the rollback of odometer readings in used cars is prevalent and costly to society. Suppose that it is possible to positively identify an automobile with an altered odometer reading. If you were to accept an appraiser's statement of the amount of the increase in the car's value produced by the odometer alteration, how would you proceed to estimate the mean loss to odometer fraud per used automobile sold?

a Explain how you would select your sample.

b Explain how you would decide on the number of used automobile purchases to be included in your sample.

7.55 Exercise 7.36 discussed a *Washington Post*/ABC News poll conducted to determine the public's attitudes concerning drug use by professional athletes. Suppose you were designing a poll of this type.

a How would you collect your sample?

b If you wanted to estimate the percentage of the population favoring a particular proposition correctly to within 1%, with probability equal to .95, approximately how many people would have to be polled? (NOTE: Use a conservative approximation for p, $p = .5$.)

7.56 If a wildlife service wishes to estimate the mean number of days of hunting per hunter for all hunters licensed in the state during a given season, with a bound on the error of estimation equal to 2 hunting days, how many hunters must be included in the survey? Assume that data collected in earlier surveys have shown σ to be approximately equal to 10.

7.57 Suppose you wish to estimate the mean pH of rainfalls in an area that suffers heavy pollution due to the discharge of smoke from a power plant. You know that σ is in the neighborhood of .5 pH, and you wish your estimate to lie within .1 of μ, with a probability near .95. Approximately how many rainfalls must be included in your sample (one pH reading per rainfall)? Would it be valid to select all of your water specimens from a single rainfall? Explain.

7.58 Refer to Exercise 7.57. Suppose you wish to estimate the difference between the mean acidity for rainfalls at two different locations, one in a relatively unpolluted area along the ocean and the other in an area subject to heavy air pollution. If you wish your estimate to be correct to the nearest .1 pH, with probability near .90, approximately how many rainfalls (pH values) would have to be included in each sample? (Assume that the variance of the pH measurements is approximately .25 at both locations and that the samples will be of equal size.)

7.59 You want to estimate the difference in grade point average between two groups of college students accurate to within .2 grade point, with probability approximately equal to .95. If the standard deviation of the grade point measurement is approximately equal to .6, how many students must be included in each group? (Assume that the groups will be of equal size.)

7.60 Refer to the comparison of the daily adult intake of selenium in two different regions of the United States in Exercise 7.20. Suppose you wish to estimate the difference in the mean daily intake between the two regions correct to within 5 micrograms, with probability equal to .90. If you plan to select an equal number of adults from the two regions (i.e, $n_1 = n_2$), how large should n_1 and n_2 be?

7.10 Summary

This chapter presented the basic concepts of statistical estimation and demonstrated how these concepts can be applied to the solution of some practical problems.

Estimators are rules (usually formulas) that tell how to calculate a parameter estimate based on sample data. Point estimators produce a single number (point) that estimates the value of a population parameter. The properties of a point estimator are contained in its sampling distribution. Thus, we prefer a point estimator that is unbiased (i.e., the mean of its sampling distribution is equal to the estimated parameter) and that has a small, preferably a minimum, variance.

The reliability of a point estimator is usually measured by a 1.96 standard deviation (i.e., the standard deviation of the sampling distribution of the estimator) margin on the error of estimation. When we use sample data to calculate a particular estimate, the probability that the margin of error will be less than the bound is approximately .95.

An interval estimator uses the sample data to calculate two points—a confidence interval—that we hope will enclose the estimated parameter. Since we will want to know the probability that the interval will enclose the parameter, we need to know the sampling distribution of the statistic used to calculate the interval.

Four estimators—a sample mean, a sample proportion, the difference between two sample means, and the difference between two sample proportions—were used to estimate their population equivalents and to demonstrate the concepts of estimation developed in this chapter. These estimators were chosen for a particular reason. Generally, they are "good" estimators for the respective population parameters in a wide variety of applications. Fortunately, they all have, for large samples, sampling distributions that can be approximated by a normal distribution. This enabled us to use the exact same procedure to construct confidence intervals for the four population parameters μ, p, $(\mu_1 - \mu_2)$, and $(p_1 - p_2)$. Thus, we were able to demonstrate an important role that the Central Limit Theorem (Chapter 6) plays in statistical inference.

Tips on Problem Solving

In solving the exercises in this chapter, you will be required to answer practical questions of interest to a businessperson, a professional person, a scientist, or a layperson. To find the answers to the questions, you will need to make an inference about one or more population parameters. Consequently, the first step in solving a problem is to decide on the objective of the exercise. What parameters do you wish to make an inference about? Answering the following two questions will help you reach a decision.

- What *type of data* is involved? This will help you decide which type of parameters you want to make inferences about: binomial proportions (p's) or population means (μ's). Check to see if the data are of the yes/no (two-possibility) variety. If they are, the data are probably binomial, and you will be interested in proportions. If not, the data probably represent measurements on one or more quantitative random variables, and you will be interested in means. Look for key words such as "proportions," "fractions," and so on, which indicate binomial data. Binomial data often (but not exclusively) evolve from a "sample survey" or an opinion poll.
- Do you wish to make an inference about a single parameter p or μ or about the *difference between two parameters* $(p_1 - p_2)$ or $(\mu_1 - \mu_2)$? This is an easy question to answer. Check on the number of samples involved. One sample implies an inference about a single parameter; two samples imply a comparison of two parameters. The answers to the first two questions identify the parameter.
- After identifying the parameter(s) involved in the exercise, you must identify the exercise objective. It will be either (a) choosing the sample size required to estimate a parameter with a specified bound on the margin of error or (b) estimating a parameter (or difference between two parameters). The objective will be very clear if it is (a) because the question will ask for or direct you to find the "sample size." Objective (b) will be clear because the exercise will specifically direct you to estimate a parameter (or the difference between two parameters).
- Check the conditions required for the sampling distribution of the parameter to be approximated by a normal distribution. For quantitative data, the sample size or sizes must be 30 or more. For binomial data, a large sample size will ensure that $p \pm 2\sigma_{\hat{p}}$ is contained in the interval 0 to 1 [$(p_1 - p_2) \pm 2\sigma_{(\hat{p}_1 - \hat{p}_2)}$ contained in the interval -1 to 1 for the two-sample case].

To summarize these tips, your thought process should follow the decision tree shown in Figure 7.12.

FIGURE 7.12
Decision tree

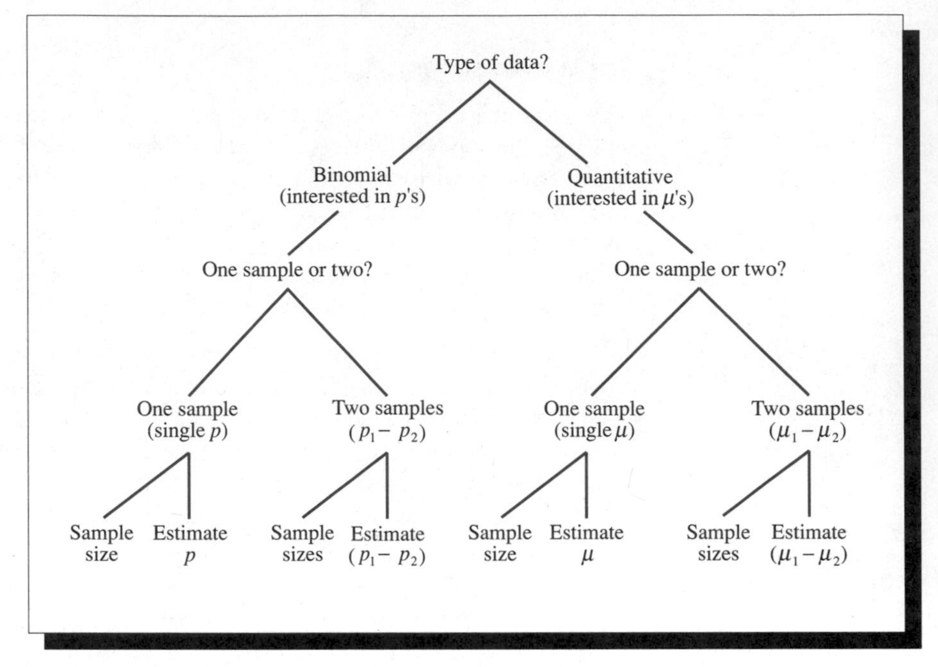

Supplementary Exercises

7.61 State the Central Limit Theorem. Of what value is the Central Limit Theorem in statistical inference?

7.62 A random sample of $n = 64$ observations has a mean $\bar{x} = 29.1$ and a standard deviation $s = 3.9$. Give the point estimate for the population mean μ and find the margin of error for your estimator.

7.63 Refer to Exercise 7.62 and find a 90% confidence interval for μ. Interpret this interval.

7.64 Refer to Exercise 7.62. How many observations would be required if you wish to estimate μ with a bound on the margin of error equal to .5, with probability equal to .95?

7.65 Independent random samples of $n_1 = 50$ and $n_2 = 60$ observations were selected from populations 1 and 2, respectively. The sample sizes and computed sample statistics are as follows:

	Population	
	1	**2**
Sample size	50	60
Sample mean	100.4	96.2
Sample standard deviation	0.8	1.3

Find a 90% confidence interval for the difference in population means and interpret the interval.

7.66 Refer to Exercise 7.65. Suppose you wish to estimate $(\mu_1 - \mu_2)$ correct to within .2, with probability equal to .95. If you plan to use equal sample sizes, how large should n_1 and n_2 be?

7.67 A random sample of $n = 500$ observations from a binomial population produced $x = 240$ successes.

a Find a point estimate for p, and find the margin of error for your estimator.

b Find a 90% confidence interval for p. Interpret this interval.

7.68 Refer to Exercise 7.67. How large a sample would be required if you wish to estimate p correct to within .025, with probability equal to .90?

7.69 Independent random samples of $n_1 = 40$ and $n_2 = 80$ observations were selected from binomial populations 1 and 2, respectively. The number of successes in the two samples were $x_1 = 17$ and $x_2 = 23$. Find a 90% confidence interval for the difference between the two binomial population proportions. Interpret this interval.

7.70 Refer to Exercise 7.69. Suppose you wish to estimate $(p_1 - p_2)$ correct to within .06, with probability equal to .90, and that you plan to use equal sample sizes—that is, $n_1 = n_2$. How large should n_1 and n_2 be?

7.71 An experiment was conducted to estimate the effect of smoking on the blood pressure of a group of 25 college-age cigarette smokers. The difference for each participant was obtained by taking the difference in the blood pressure readings at the time of graduation and again 5 years later. The sample mean increase, measured in millimeters of mercury, was $\bar{x} = 9.7$. The sample standard deviation was $s = 5.8$. Estimate the mean increase in blood pressure that one would expect for cigarette smokers over the time span indicated by the experiment. Find the margin of error. Describe the population associated with the mean that you have estimated.

7.72 Using a confidence coefficient equal to .90, place a confidence interval on the mean increase in blood pressure for Exercise 7.71.

7.73 An unusual survey conducted in Washington, D.C., and Moscow, reported in the *Tampa Tribute-Times* (Tampa, Fla., Jan. 17, 1988), found Soviets more optimistic but less informed than their American counterparts concerning bilateral issues. The poll, which was conducted by telephone, involved 352 people in Washington and 305 nationally. In Moscow, 1000 randomly selected Moscovites were interviewed by telephone. The news article cited a maximum error for a sample of 1000 as ±3 percentage points; for 500, ±5 percentage points; and for 350, ±6 percentage points. Do you agree with these error margins? If not, what are the maximal error rates for samples of size $n = 1000$, $n = 500$, and $n = 350$?

7.74 Based on repeated measurements of the iodine concentration in a solution, a chemist reports the concentration as 4.614, with an "error margin of .006."

a How would you interpret the chemist's "error margin"?

b If the reported concentration is based on a random sample $n = 30$ measurements, with a sample standard deviation $s = .017$, would you agree that the chemist's "error margin" is .006?

7.75 If it is assumed that the heights of men are normally distributed, with a standard deviation of 2.5 inches, how large a sample should be taken to be fairly sure (probability .95) that the sample mean does not differ from the true mean (population mean) by more than .50 in absolute value?

7.76 In measuring reaction time, a psychologist estimates that the standard deviation is 0.05 second. How large a sample of measurements must he take in order to be 90% confident that the margin of error will not exceed 0.01 second?

7.77 In an experiment, 320 out of 400 seeds germinate. Find a confidence interval for the true fraction of germinating seeds; use a confidence coefficient of .98.

7.78 To estimate the proportion of unemployed workers in Panama, an economist selected 400 people at random from the working class. Of these, 25 were unemployed.

a Estimate the true proportion of unemployed workers. What is the margin of error for your estimate?

b How many people must be sampled to reduce the margin of error to .02?

7.79 A random sample of 36 cigarettes of a certain brand were tested for nicotine content. The sample gave a mean of 22 and a standard deviation of 4 milligrams. Find a 98% confidence interval for μ, the true mean nicotine content of the brand. Interpret this interval.

7.80 America's public schools are under fire from several directions. Teachers are looking for competitive salaries; parents and children are concerned about quality education; and school administrators are trying to balance quality education and rising educational costs. A Gallup poll conducted for Phi Delta Kappa, a national educational fraternity (*Press-Enterprise*, Riverside, Calif., Aug. 25, 1989), found that 60% of the people surveyed supported the controversial idea of allowing parents and students to choose which public school to attend, and nearly two-thirds favored giving school principals broader authority in running their schools in return for making them responsible for school performance.

a If the survey sample size was 3000, find a 95% confidence interval for the true proportion of Americans who support the idea of allowing parents and children to choose which public school to attend.

b Estimate the true proportion of Americans who favor broader authority for school principals and provide the margin of error for your estimate.

7.81 Refer to Exercise 7.80. If the pollster would like the estimates in the education survey to be accurate to within 2 percentage points of the actual percentages, how large a sample should she take? Is the sample size given in Exercise 7.80 large enough to achieve the desired bound?

7.82 In a controlled pollination study involving *Phlox drummondii*, a spring-flowering annual plant common along roadsides and sandy fields in central Texas, Karen Pittman and Donald Levin (1989) found that seed survival rates were not affected by water or nutrient deprivation. In the experiment, flowers on plants were counted as males when donating pollen and as females when pollinated by donor pollen in crosses involving plants in the three treatment groups: control, low water, and low nutrient. The following table, which reflects one aspect of the results of the experiment, records the number of seeds surviving to maturity for each of the three groups for both male parents and female parents.

	Male		Female	
Treatment	n	**Number Surviving**	n	**Number Surviving**
Control	585	543	632	560
Low water	578	522	510	466
Low nutrient	568	510	589	546

a Find the sample estimate of the proportion of surviving seeds and the standard error of the estimate for each of the six treatment–sex combinations.

b Find a 99% confidence interval for the difference between survival proportions for low water versus low nutrients for the male category.

c Find a 99% confidence interval for the difference between survival proportions for low water versus low nutrients for the female category.

d Find a 99% confidence interval for the difference between survival proportions for the male control group compared to the female control group.

e Interpret the intervals calculated in parts (a), (b), (c), and (d).

7.83 How large a sample is necessary in order to estimate a binomial parameter p to within .01 of its true value with probability .95? Assume the value of p is approximately .5.

7.84 A dean of men wishes to estimate the average cost of the freshman year at a particular college correct to within \$500, with a probability of .95. If a random sample of freshmen is to be selected, and each asked to keep financial data, how many must be included in the sample? Assume that the dean knows only that the range of expenditure will vary from approximately \$4800 to \$13,000.

7.85 An experimenter wishes to estimate the difference $(p_1 - p_2)$ for two binomial populations to within .1 of the true difference. Both p_1 and p_2 are expected to assume values between .3 and .7. If samples of equal size are to be selected from the populations to estimate $(p_1 - p_2)$, how large should they be?

7.86 An experimenter fed different rations to two groups of 100 chicks each. Assume that all factors other than rations are the same for both groups. A record of mortality for each group is as follows:

Chicks	Ration A	Ration B
n	100	100
Number died	13	6

Construct a 98% confidence interval for the true difference in mortality rates for the two rations. Interpret this interval.

7.87 You want to estimate the mean hourly yield for a process manufacturing an antibiotic. You observe the process for 100 hourly periods chosen at random, with the results $\bar{x} = 34$ oz/hr and $s = 3$. Estimate the mean hourly yield for the process using a 95% confidence interval.

7.88 A quality-control engineer wants to estimate the fraction of defectives in a large lot of light bulbs. From previous experience, he feels that the actual fraction of defectives should be somewhere around .2. How large a sample should he take if he wants to estimate the true fraction to within .01 using a 95% confidence interval?

7.89 Samples of 400 printed circuit boards were selected from each of two production lines A and B. The number of defectives in each sample was as follows:

Line	Number of Defectives
A	40
B	80

Estimate the difference between the actual fractions of defectives for the two lines with a confidence coefficient of .90.

7.90 Refer to Exercise 7.89. Suppose ten samples of $n = 400$ printed circuit boards were tested and a confidence interval was constructed for p for each of the ten samples. What is the probability that exactly one of the intervals will not enclose the true value of p? That at least one interval will not enclose the true value of p?

7.91 The ability to accelerate rapidly is an important attribute for an ice hockey player. G. Wayne Marino (1983) investigated some of the variables related to the acceleration and speed of a hockey player from a stopped position. Sixty-nine hockey players, varsity and intramural, from the University of Illinois were included in the experiment. Each player was required to move as rapidly as possible from a stopped position to cover a distance of 6 meters. The means and standard deviations of some of the variables recorded for each of the 69 skaters are shown in the following table:

	Mean	Standard Deviation
Weight (kilograms)	75.270	9.470
Stride length (meters)	1.110	0.205
Stride rate (strides/second)	3.310	0.390
Average acceleration (meters/sec^2)	2.962	0.529
Instantaneous velocity (meters/sec)	5.753	0.892
Time to skate (seconds)	1.953	0.131

a Give the formula that you would use to construct a 95% confidence interval for one of the population means (e.g., mean time to skate the 6-meter distance).

b Construct a 95% confidence interval for the mean time to skate. Interpret this interval.

7.92 Exercise 7.91 presented statistics from a study of fast starts by ice hockey skaters. The mean and standard deviation of the 69 individual average acceleration measurements over the 6-meter distance were 2.962 and 0.529 meters per second, respectively (Marino, 1983).

a Find a 95% confidence interval for this population mean. Interpret the interval.

b Suppose you were dissatisfied with the width of this confidence interval and wanted to cut the interval in half by increasing the sample size. How many skaters (total) would have to be included in the study?

7.93 The mean and standard deviation of the speeds of the sample of 69 skaters at the end of the 6-meter distance in Exercise 7.91 were 5.753 and 0.892 meters per second, respectively (Marino, 1983).

a Find a 95% confidence interval for the mean velocity at the 6-meter mark. Interpret the interval.

b Suppose that you wanted to repeat the experiment and that you wanted to estimate this mean velocity correct to within .1 second, with probability .99. How many skaters would have to be included in your sample?

7.94 The following statistics are the result of an experiment conducted by P. I. Ward (1984) to investigate a theory concerning the molting behavior of the male *Gammarus pulex*, a small crustacean. If a male has to molt while paired with a female, he must release her and so loses her. The theory is that the male *G. pulex* is able to postpone the time to molt and thereby reduce the possibility of losing his mate. Ward randomly assigned 100 pairs of males and females to two groups of 50 each. Pairs in the first group were maintained together (Normal); those in the second group were separated (Split). The length of time to molt was recorded for both males and females, and the means, standard deviations, and sample sizes are shown in the table. (The number of crustaceans in each of the four samples is less than 50 because some in each group did not survive until molting time.) Find a 99% confidence interval for the difference in mean molt time for "Normal" males versus those "Split" from their mates. Interpret the interval.

	Time to Molt (days)		
	Mean	s	n
Males			
Normal	24.8	7.1	34
Split	21.3	8.1	41
Females			
Normal	8.6	4.8	45
Split	11.6	5.6	48

7.95 Refer to the data in Exercise 7.94, and find a 99% confidence interval for the difference in mean molt time for "Normal" females versus those "Split" from their mates. Interpret the interval.

Exercises Using the Data Disks

7.96 Class experiment: In our analysis of the blood pressure data on the data disk, we found that the mean and standard deviation of the 965 systolic male blood pressures given are 118.71 and 14.23, respectively. Regarding this data set as a population, we will demonstrate the concept of a confidence interval. Each student should select a random sample of $n = 40$ observations from the data set, calculate the sample mean $\bar{x}$ and variance s^2, and then calculate a 95% confidence interval for the population mean $\mu = 118.71$. Note whether the confidence interval encloses μ. Record the confidence intervals for all members of the class. Notice how they shift in a random manner. Theoretically, if there were millions of class members, approximately 95% of the confidence intervals should enclose μ. Calculate the percentage of the confidence intervals constructed by the class that enclose μ. It is unlikely that this percentage will equal 95%

(because the number of members in the class is not large enough), but most of the confidence intervals should enclose μ.

7.97 In our descriptive analysis of the NIH blood pressure data given on the data disk, we found the means and sample standard deviations for the samples of systolic blood pressures to be as follows:

	Sample Size	Mean	Standard Deviation
Male	964	118.71	14.23
Female	945	110.39	12.58

Find a 99% confidence interval for the difference in population means $(\mu_1 - \mu_2)$. Interpret the interval.

7.98 Refer to the list of the *Fortune 500* companies given on the data disk. The true mean and standard deviation of the sales for the *Fortune 500* companies for 1991 are $\mu = 4525.17$ and $\sigma = 10{,}574.9$.

a Consider the sales for the 500 companies to be the population of interest and draw a random sample of size 30 from this population (use the random number table in Appendix I).

b Use the sample from part (a) to construct a 99% confidence interval for μ. Does this confidence interval enclose the true value, $\mu = 4525.17$?

c Draw another random sample of size 30 and construct another 99% confidence interval for μ. Does this interval enclose μ?

d In general, what proportion of intervals constructed in this way would you expect to enclose the value $\mu = 4525.17$?

CASE STUDY

How Reliable Is That Poll?

It is almost impossible to read a daily newspaper or listen to the radio or watch TV without hearing about some opinion poll or economic survey. Many of us ask, How reliable are the percentages derived from these samples of public opinion? Do the national polls conducted by the Gallup and Harris organizations, the news media, and so on really provide accurate estimates of the percentages of people in the United States who favor various propositions?

A report of the results of a *New York Times*/CBS poll provides a clue to these answers ("Polls Say Clinton," 1992). The object of the poll was to determine voter preferences in the 1992 presidential election, focusing especially on the preference of *likely voters*. Nested in the middle of the report was a box entitled "How the Poll Was Conducted," part of which is reproduced below.

> The latest *New York Times*/CBS News poll is based on telephone interviews conducted Tuesday through Friday with 2374 adults around the United States, excluding Alaska and Hawaii, of whom 1912 said they were registered voters.
>
> The sample of telephone exchanges called was selected by a computer from a complete list of exchanges in the country. The exchanges were chosen to ensure that each region of the country was represented in proportion to its population. For each exchange, the telephone numbers were formed by random digits, thus permitting access to both listed and unlisted numbers.

Describing the reliability of the poll results, the box states, "In theory, in 19 cases out of 20, the results based on such samples will differ by no more than a few percentage points in either direction from what would have been obtained by seeking out all adults in the country. For results based on all four days, the potential sampling error is plus or minus two percentage points; for results based on either two-day period separately, it is three percentage points."

The potential sampling error for smaller subgroups is larger. For example, for results based on registered voters who called themselves independents, during all 4 days of interviewing, it is plus or minus 4 percentage points.

1 Verify the margin of error given by the pollster.

2 Do the calculations in part 1 confirm the *New York Times* statement that the sample percentages will vary less than 2% from the actual population percentages? What assumptions may have been violated in the *New York Times* sampling procedure?

3 From the potential bounds given for a 2-day period, and for the independents in the sample, can you determine approximately how many people were interviewed in each of these subgroups?

8

Large-Sample Tests of Hypotheses

Case Study

Will an aspirin a day reduce the risk of heart attack? A very large study of U.S. physicians showed that a single aspirin taken every other day reduced the risk of heart attack in men by one-half. However, 3 days later, a British study reported a completely opposite conclusion. How could this be? The case study at the end of this chapter looks at how the studies were conducted, and you will analyze the data using large-sample techniques.

General Objective

In this chapter, the concept of a statistical test of a hypothesis will be formally introduced. The sampling distributions of statistics presented in Chapters 6 and 7 will be used to construct large-sample tests concerning the values of population parameters of interest to the experimenter.

Specific Topics

8.1 Testing Hypotheses about Population Parameters

Some practical problems require that we estimate the value of a population parameter; others require that we make decisions concerning the value of the parameter. For example, if a pharmaceutical company were fermenting a vat of antibiotic, it would want to test the potency of samples of the antibiotic and use those samples to estimate the mean potency μ of the antibiotic in the vat. In contrast, suppose that there were no concern that the potency of the antibiotic would be too high; the company's only concern was that the mean potency exceed some government-specified minimum in order that the vat be declared acceptable for sale. In this case, the company would not wish to estimate the mean potency. Rather, it would want to show that the mean potency of the antibiotic in the vat exceeded the minimum specified by the government. Thus, the company would want to decide whether or not the mean potency exceeded the minimum allowable potency. The pharmaceutical company's problem illustrates a **statistical test of hypothesis**.

The reasoning employed in testing an hypothesis bears a striking resemblance to the procedure used in a court trial. In trying a person for theft, the court assumes the accused innocent until proved guilty. The prosecution collects and presents all available evidence in an attempt to contradict the "not guilty" hypothesis and hence to obtain a conviction. However, if the prosecution fails to disprove the "not guilty" hypothesis, this does not prove that the accused is "innocent" but merely that there is not sufficient evidence to conclude that the accused is "guilty."

The statistical problem portrays the potency of the antibiotic as the accused. The hypothesis to be tested, called the **null hypothesis**, is that the potency does not exceed the minimum government standard. The evidence in this case is contained in the sample of specimens drawn from the vat. The pharmaceutical company, playing the role of the prosecutor, believes that an **alternative hypothesis** is true—namely, that the potency of the antibiotic does exceed the minimum standard. Hence, the company attempts to use the evidence in the sample to reject the null hypothesis (potency does not exceed minimum standard), thereby supporting the alternative hypothesis (potency exceeds minimum standard). You will recognize this procedure as an essential feature of the scientific method, in which all proposed theories must be compared with reality.

In this chapter, we will explain the basic concepts of a test of an hypothesis and demonstrate the concepts with some very useful large-sample statistical tests of the values of a population mean, a population proportion, the difference between a pair of population means, and the difference between two proportions. We will employ the four point estimators discussed in Chapter 7—$\bar{x}$, $(\bar{x}_1 - \bar{x}_2)$, $\hat{p}$, and $(\hat{p}_1 - \hat{p}_2)$—as test statistics and, in doing so, will obtain a unity in these four statistical tests. All four test statistics will, for large samples, have sampling distributions that are normal or can be approximated by a normal distribution.

8.2 A Statistical Test of Hypothesis

A statistical test of hypothesis consists of four parts:

- A null hypothesis, denoted by the symbol H_0
- An alternative hypothesis, denoted by the symbol H_a
- A test statistic
- A rejection region

The specification of these four elements defines a particular test; changing one or more of the parts creates a new test.

The **alternative hypothesis** is the hypothesis that the researcher wishes to support. The **null hypothesis** is a contradiction of the alternative hypothesis; that is, if the null hypothesis is false, the alternative hypothesis must be true. For reasons you will subsequently see, it is easier to show support for the alternative hypothesis by presenting evidence (sample data) that indicates that the null hypothesis is false. Thus, we are building a case in support of the alternative hypothesis by using a method that is analogous to proof by contradiction.

Even though we wish to gain evidence in support of the alternative hypothesis, the null hypothesis is the hypothesis to be tested. Thus, H_0 will specify hypothesized values for one or more population parameters.

EXAMPLE **8.1** We wish to show that the average hourly wage of construction workers in the state of California is different from \$14, which is the national average. This is the alternative hypothesis, written as

$$H_a : \mu \neq 14$$

The null hypothesis is written as

$$H_0 : \mu = 14$$

We would like to reject the null hypothesis, thus concluding that the California mean is not equal to \$14. ▪

EXAMPLE **8.2** A milling process currently produces an average of 3% defectives. We are interested in showing that a simple adjustment on a machine will decrease p, the proportion of defectives produced in the milling process. Thus, we write the alternative hypothesis as

$$H_a : p < .03$$

and the null hypothesis as

$$H_0 : p = .03$$

If we can reject H_0, we can conclude that the adjusted process produces fewer defectives. ▪

There is a difference in the forms of the alternative hypotheses given in Examples 8.1 and 8.2. In Example 8.1, there is no directional difference suggested for the value of μ; that is, μ might be either larger or smaller than \$14 if H_a is true. This type of test is called a **two-tailed test of hypothesis.** In Example 8.2, however, we are specifically interested in detecting a directional difference in the value of p; that is, if H_a is true, the value of p will be smaller than .03. This type of test is called a **one-tailed test of hypothesis.**

The decision to reject or accept the null hypothesis is based on information contained in a sample drawn from the population of interest. The sample values are used to compute a single number, corresponding to a point on a line, which operates as a decision maker. This decision maker is called the **test statistic**. The entire set of values that the test statistic may assume is divided into two sets, or regions. One set, consisting of values that support the alternative hypothesis, is called the **rejection region.** The other, consisting of values that support the null hypothesis, is called the **acceptance region.**

The acceptance and rejection regions are separated by a **critical value** of the test statistic. If the test statistic computed from a particular sample assumes a value in the rejection region, the null hypothesis is rejected, and the alternative hypothesis H_a is accepted. If the test statistic falls in the acceptance region, either the null hypothesis is accepted or the test is judged to be inconclusive. The circumstances leading to this latter decision are explained subsequently.

EXAMPLE 8.3

For the test of hypothesis given in Example 8.1, the average wage $\bar{x}$ for a random sample of 100 California construction workers might provide a good test statistic for testing

$$H_0 : \mu = 14 \qquad \text{vs.} \qquad H_a : \mu \neq 14$$

Since the sample mean is the best estimator of the corresponding population mean, we would be inclined to reject H_0 in favor of H_a if the sample mean $\bar{x}$ is either much smaller than \$14 or much larger than \$14. Hence, the rejection region would consist of both large and small values of $\bar{x}$ as shown in Figure 8.1. ▪

The decision procedure described above is subject to two types of errors, which are prevalent in a two-choice decision problem.

DEFINITION ▪

A **type I error** for a statistical test is the error made by rejecting the null hypothesis when it is true. The probability of making a type I error is denoted by the symbol α.

A **type II error** for a statistical test is the error made by accepting (not rejecting) the null hypothesis when it is false and some alternative hypothesis is true. The probability of making a type II error is denoted by the symbol β. ▪

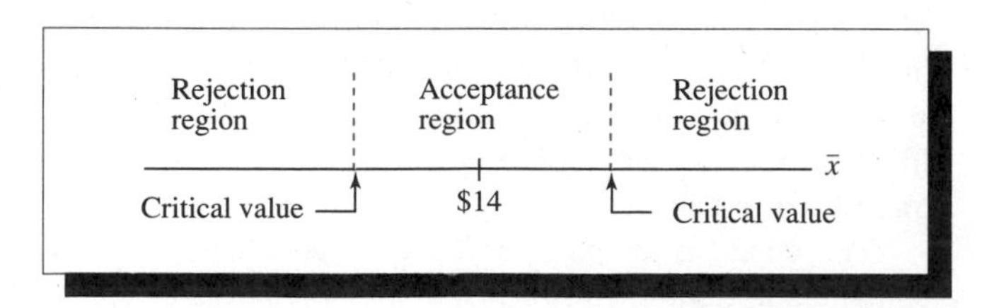

FIGURE 8.1
Rejection and acceptance regions for Example 8.1

The two possibilities for the null hypothesis—that is, true or false—along with the two decisions the experimenter can make, are indicated in the two-way table, Table 8.1. The occurrences of the type I and type II errors are indicated in the appropriate cells.

TABLE **8.1**
Decision table

	Null Hypothesis	
Decision	**True**	**False**
Reject H_0	Type I error	Correct decision
Accept H_0	Correct decision	Type II error

The goodness of a statistical test of an hypothesis is measured by the probabilities of making a type I or a type II error, denoted by the symbols α and β, respectively.

Since α is the probability of rejecting H_0 when it is true, this is a measure of the chance of *incorrectly rejecting* H_0. Since β is the probability of accepting H_0 when it is false, its complement, $1 - \beta$, is the probability of rejecting H_0 when it is false, and measures the chance of *correctly rejecting* H_0. This probability, $1 - \beta$, is called the **power of the test**, the chance that the test performed as it should.

DEFINITION ■

The **power of a statistical test**, given as

$$1 - \beta = P[\text{ reject } H_0 \text{ when } H_0 \text{ is false}]$$

measures the ability of the test to perform as required. ■

A graph of $1 - \beta$, the probability of rejecting H_0 when, in fact, H_0 is false, as a function of the true value of the parameter of interest is called the **power curve** for the statistical test. Ideally, we would like α to be small and the *power* $(1 - \beta)$ to be large. The experimenter should be able to specify values of α and β, measuring the risks of the respective errors he or she is willing to tolerate, as well as some deviation from the hypothesized value of the parameter he or she considers of practical importance and wishes to detect. The rejection region for the test will be located in accordance with the specified value of α; the sample size will be chosen large enough to achieve an acceptable value of β for the specified deviation the experimenter wishes to detect. This choice could be made by consulting the power curves, corresponding to various sample sizes, for the chosen test.

The language involved in a statistical test of hypothesis is summarized in the following display.

Parts of a Statistical Test

- **Null hypothesis:** The hypothesis that is assumed true until proven false, the negation of the alternative hypothesis
- **Alternative hypothesis:** The hypothesis that the researcher wishes to support or prove true
- **One-tailed test of hypothesis:** A test that assumes a one-directional difference for the parameter of interest if the alternative hypothesis is true
- **Two-tailed test of hypothesis:** A test that assumes a two-directional difference (either larger or smaller) for the alternative hypothesis
- **Test statistic:** A statistic calculated from sample measurements that will be used as a decision maker
- **Rejection region:** Values of the test statistic for which H_0 will be rejected
- **Acceptance region:** Values of the test statistic for which H_0 will be accepted
- **Critical values of the test statistic:** The value(s) of the test statistic that separate the rejection and acceptance regions
- **Type I error (with probability α):** Rejecting H_0 when H_0 is true
- **Type II error (with probabilty β):** Accepting H_0 when H_0 is false
- **Power of the test ($1 - \beta$):** The probability of correctly rejecting the null hypothesis

Large-sample tests of hypotheses concerning population means and proportions are similar. The similarity lies in the fact that all of the point estimators discussed in Chapter 7 are unbiased and have sampling distributions that, for large samples, can be approximated by a normal distribution. Therefore, we can use the point estimators as test statistics to test hypotheses about the respective parameters. We will consider tests of hypotheses concerning the four population parameters μ, p, $\mu_1 - \mu_2$, and $p_1 - p_2$ separately in the following sections.

8.3 A Large-Sample Test of Hypothesis about a Population Mean

Consider a random sample of n measurements drawn from a population that has mean μ and standard deviation σ. We would like to test a hypothesis of the form[†]

$$H_0 : \mu = \mu_0$$

[†]Note that if the test rejects the null hypothesis $\mu = \mu_0$ in favor of the alternative hypothesis $\mu > \mu_0$, then it will certainly reject a null hypothesis of the form $\mu < \mu_0$, since this is even more contradictory to the alternative hypothesis. For this reason, in this text we state the null hypothesis for a one-tailed test as $\mu = \mu_0$ rather than $\mu \leq \mu_0$.

where μ_0 is some hypothesized value for μ versus a one-tailed alternative hypothesis

$$H_a : \mu > \mu_0$$

The sample mean $\bar{x}$ is the best estimate of the actual value of μ, which is presently in question. What values of $\bar{x}$ would lead us to believe that H_0 is false and μ is, in fact, greater than the hypothesized value? Those values of $\bar{x}$ that are extremely *large* would imply that μ is larger than hypothesized. Hence, we will reject H_0 if $\bar{x}$ is "too large."

The next problem is to define what we mean by "too large." Values of $\bar{x}$ that lie too many standard deviations to the right of the mean are not very likely to occur. Hence, we can define "too large" as being too many standard deviations away from μ_0. Remember that the standard deviation or standard error of $\bar{x}$ is calculated as

$$\sigma_{\bar{x}} = \frac{\sigma}{\sqrt{n}}$$

Since the sampling distribution of the sample mean $\bar{x}$ is approximately normal when n is large, we can measure the number of standard deviations that $\bar{x}$ lies from μ_0 using the **standardized test statistic**

$$z = \frac{\bar{x} - \mu_0}{\sigma/\sqrt{n}}$$

which has a standard normal distribution when H_0 is true and $\mu = \mu_0$.

The probability of rejecting the null hypothesis, assuming it to be true, is equal to the area under the normal curve lying above the rejection region. Thus, if we want $\alpha = .05$, we would reject H_0 when $\bar{x}$ is more than 1.645 standard deviations to the right of μ_0. Equivalently, we would reject H_0 if the standardized test statistic z, defined above, is greater than 1.645 (Figure 8.2).

If we wish to detect departures either greater or less than μ_0, the alternative hypothesis would be *two-tailed*, written as

$$H_a : \mu \neq \mu_0$$

which implies either $\mu > \mu_0$ or $\mu < \mu_0$. Values of $\bar{x}$ that are either "too large" or "too small" in terms of their distance from μ_0 will be placed in the rejection region. Since we still want $\alpha = .05$, the area in the rejection region is equally divided between the two tails of the normal distribution, as shown in Figure 8.3. Using the standardized test statistic z, we will reject H_0 if $z > 1.96$ or $z < -1.96$. For different values of α, the critical values of z that separate the rejection and acceptance regions will change.

FIGURE 8.2 Distribution of $z = \dfrac{\bar{x} - \mu_0}{\sigma/\sqrt{n}}$ when H_0 is true

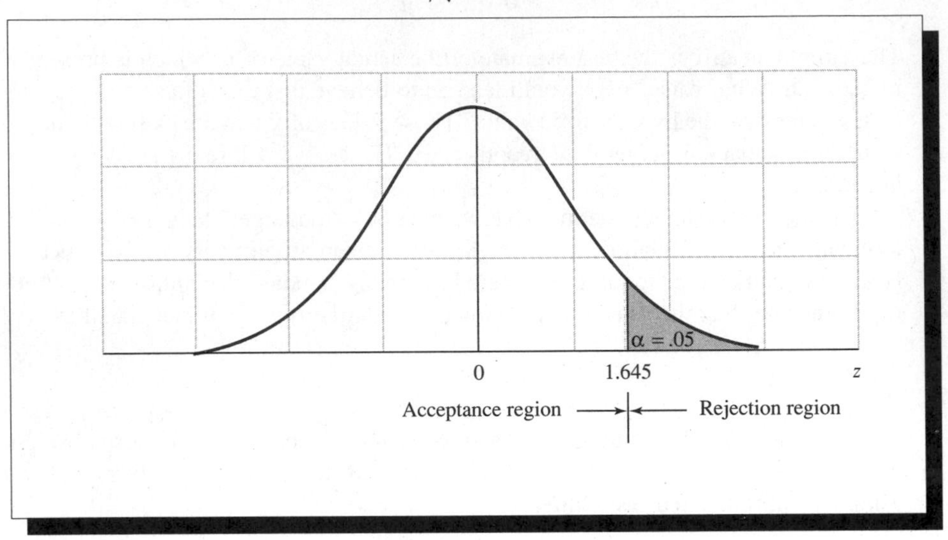

FIGURE 8.3 The rejection region for a two-tailed test with $\alpha = .05$

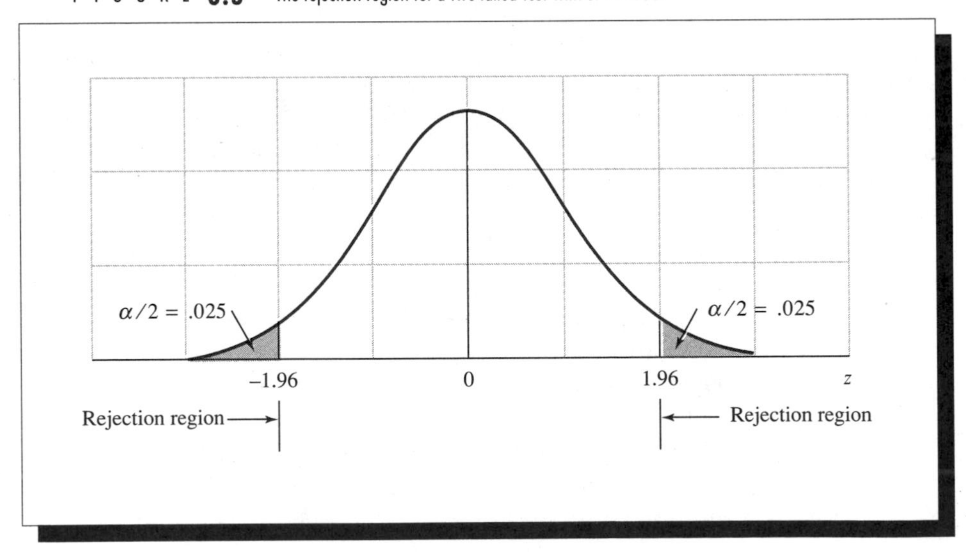

Large-Sample Statistical Test for μ

1 Null Hypothesis: $H_0 : \mu = \mu_0$

2 Alternative Hypothesis:

One-Tailed Test	Two-Tailed Test
$H_a : \mu > \mu_0$ (or, $H_a : \mu < \mu_0$)	$H_a : \mu \neq \mu_0$

3 Test Statistic: $z = \dfrac{\bar{x} - \mu_0}{\sigma_{\bar{x}}} = \dfrac{\bar{x} - \mu_0}{\sigma/\sqrt{n}}$

If σ is unknown (which is usually the case), substitute the sample standard deviation s for σ.

4 Rejection Region:

One-Tailed Test	Two-Tailed Test
$z > z_\alpha$ (or, $z < -z_\alpha$ when the alternative hypothesis is $H_a : \mu < \mu_0$)	$z > z_{\alpha/2}$ or $z < -z_{\alpha/2}$

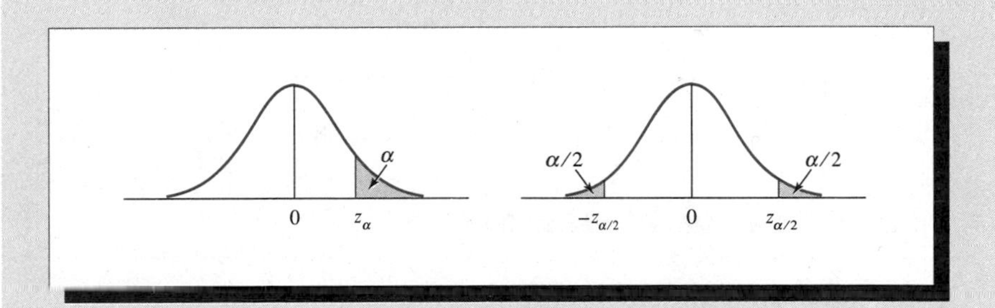

Assumptions: The n observations in the sample were randomly selected from the population and n is large, say, $n \geq 30$.

EXAMPLE **8.4** The average weekly earnings for men in managerial and professional positions is \$725. Do women in the same positions have average weekly earnings that are less than those for men? A random sample of $n = 40$ women in managerial and professional positions showed $\bar{x} = \$670$ and $s = \$102$. Test the appropriate hypothesis using $\alpha = .01$.

Solution We would like to show that the average weekly earnings are less than \$725, the men's average. Hence, if μ is the average weekly earnings in managerial and professional positions for women, the hypothesis to be tested is

$$H_0 : \mu = 725 \qquad \text{vs.} \qquad H_a : \mu < 725$$

The rejection region for this one-tailed test consists of small values of $\bar{x}$, or equivalently, values of the *standardized test statistic* z in the left tail of the standard normal distribution, with $\alpha = .01$. This value is found in Table 3 of Appendix I to be $z = -2.33$, as shown in Figure 8.4. The observed value of the test statistic, using s as an estimate of the population standard deviation, is

$$z = \frac{\bar{x} - 725}{s/\sqrt{n}} = \frac{670 - 725}{102/\sqrt{40}} = -3.41$$

Since the observed value of the test statistic falls in the rejection region, H_0 is rejected, and we conclude that the average weekly earnings for women in managerial and professional positions are significantly less than those for men. The probability that we have made an incorrect decision is $\alpha = .01$. ▪

EXAMPLE 8.5 Refer to Example 7.1. Test the hypothesis that the average daily yield of the chemical is $\mu = 880$ tons/day against the alternative that μ is either greater or less than 880 tons/day. The sample (Example 7.1), based on $n = 50$ measurements, yielded $\bar{x} = 871$ and $s = 21$ tons.

Solution The null and alternative hypotheses are

$$H_0 : \mu = 880 \qquad \text{and} \qquad H_a : \mu \neq 880$$

FIGURE **8.4** Rejection region for Example 8.4

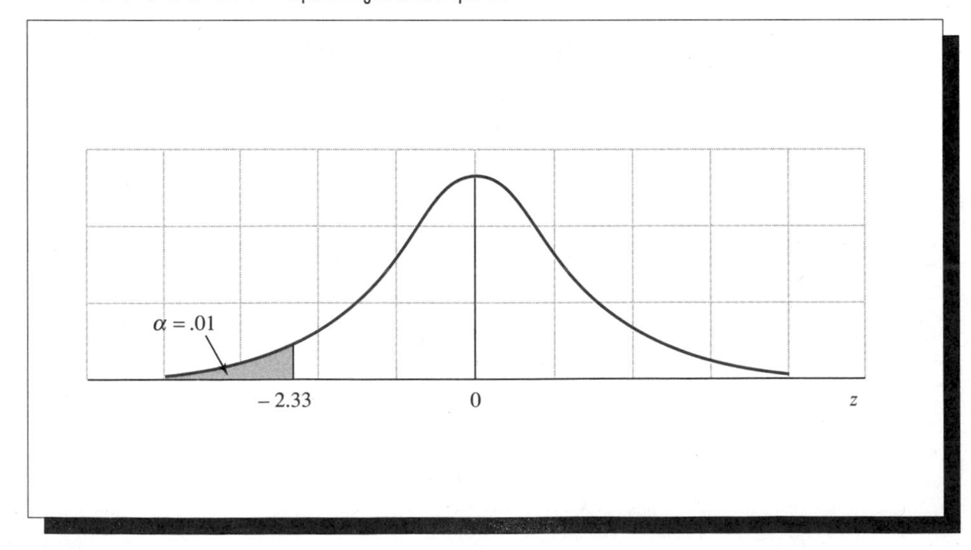

The point estimate for μ is $\bar{x}$. Therefore, the test statistic is

$$z = \frac{\bar{x} - \mu_0}{\sigma_{\bar{x}}} = \frac{\bar{x} - \mu_0}{\sigma/\sqrt{n}}$$

Using s to approximate σ, we obtain

$$z = \frac{871 - 880}{21/\sqrt{50}} = -3.03$$

For $\alpha = .05$ the rejection region consists of values of $z > 1.96$ and values of $z < -1.96$. Since -3.03, the calculated value of z, falls in the rejection region, we reject the hypothesis that $\mu = 880$ tons and conclude that it is less. The probability of rejecting H_0 when H_0 is true is $\alpha = .05$. Hence, we are reasonably confident that our decision is correct. ▪

The statistical test based on a normally distributed test statistic, with a given α, and the $(1 - \alpha)$ 100% confidence interval of Section 7.5 are related. The interval $[\bar{x} \pm (1.96\sigma/\sqrt{n})]$, or approximately (871 ± 5.82) for Example 8.4 is constructed so that in repeated sampling $(1 - \alpha)$ 100% of the intervals will enclose μ. Noting that $\mu = 880$ does not fall in the interval, we would be inclined to reject $\mu = 880$ as a likely value and conclude that the mean daily yield was, indeed, less.

There is another similarity between this test and the confidence interval of Section 7.5. The test is "approximate" because we substituted s, an approximate value, for σ. That is, the probability α of a type I error selected for the test is not .05. It is close to .05, close enough for all practical purposes, but it is not exact. This will be true for most statistical tests because one or more assumptions will not be satisfied exactly.

EXAMPLE 8.6 Refer to Example 8.5. Calculate β and the power of the test, $1 - \beta$, when μ is actually equal to 870 tons.

Solution The acceptance region for the test of Example 8.5 is located in the interval $(\mu_0 \pm 1.96\sigma_{\bar{x}})$. Substituting numerical values, we obtain

$$880 \pm 1.96\frac{21}{\sqrt{50}} \quad \text{or} \quad 874.18 \text{ to } 885.82$$

The probability of accepting H_0, given $\mu = 870$, is equal to the area under the frequency distribution for the test statistic $\bar{x}$ in the interval from 874.18 to 885.82. Since $\bar{x}$ is normally distributed with a mean of 870 and $\sigma_{\bar{x}} \approx 21/\sqrt{50} = 2.97$, β is equal to the area under the normal curve located between 874.18 and 885.82 (see Figure 8.5).

Calculating the z values corresponding to 874.18 and 885.82, we obtain

$$z_1 = \frac{\bar{x} - \mu}{\sigma/\sqrt{n}} \approx \frac{874.18 - 870}{21/\sqrt{50}} = 1.41$$

FIGURE 8.5 Calculating β in Example 8.6

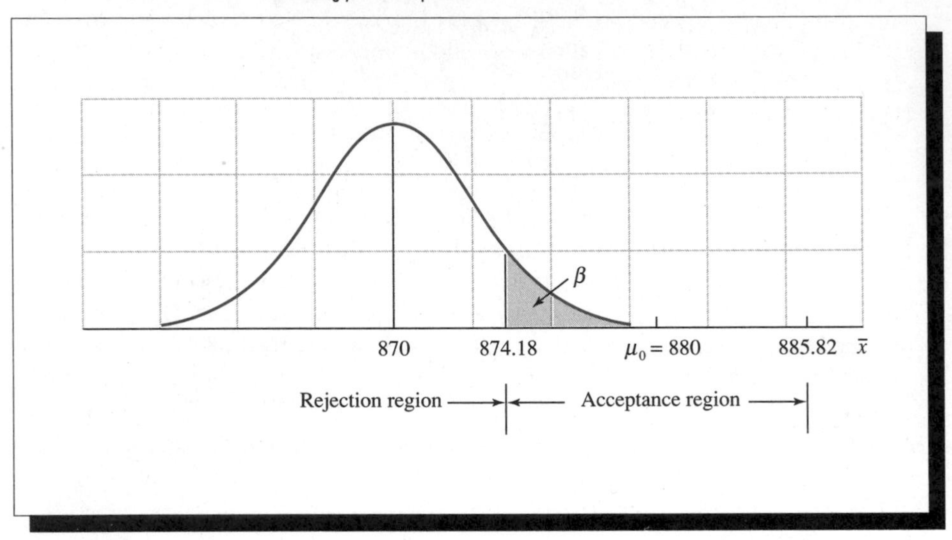

$$z_2 = \frac{\bar{x} - \mu}{\sigma/\sqrt{n}} \approx \frac{885.82 - 870}{21/\sqrt{50}} = 5.33$$

Then

$$\beta = P(\text{accept } H_0 \text{ when } \mu = 870) = P(874.18 < \bar{x} < 885.82 \text{ when } \mu = 870)$$
$$= P(1.41 < z < 5.33)$$

You can see from Figure 8.5 that the area under the normal curve above $\bar{x} = 885.82$ (or $z = 5.33$) is negligible. Therefore,

$$\beta = P(z > 1.41)$$

From Table 3 in Appendix I we find that the area between $z = 0$ and $z = 1.41$ is .4207 and

$$\beta = .5 - .4207 = .0793$$

Hence, the power of the test is

$$1 - \beta = 1 - .0793 = .9207$$

The probability of correctly rejecting H_0, given that μ is really equal to 870, is .9207, or approximately 92 chances in 100.

Values of $1 - \beta$ can be calculated for various values of μ_a different from $\mu_0 = 880$ to measure the power of the test. For example, if $\mu_a = 885$,

$$\beta = P(874.18 < \bar{x} < 885.82 \text{ when } \mu = 885)$$
$$= P(-3.64 < z < .28) = .5 + .1103 = .6103$$

and the power is $1 - \beta = .3897$. Table 8.2 shows the power of the test for various vales of μ_a, and a power curve is graphed is Figure 8.6. Note that the power of the test increases as the distance between μ_a and μ_0 increases. The result is a U-shaped curve for this two-tailed test. ■

TABLE 8.2 Values of $1 - \beta$ for various values of μ_a, Example 8.6

μ_a	$1-\beta$	μ_a	$1-\beta$
865	.9990	883	.1726
870	.9207	885	.3897
872	.7673	888	.7673
875	.3897	890	.9207
877	.1726	895	.9990
880	.0500		

Tips on Problem Solving: Calculating β

1. Find the critical value or values of $\bar{x}$ used to separate the acceptance and rejection regions.
2. Using one or more values for μ consistent with the alternative hypothesis, H_a, calculate the probability that the sample mean $\bar{x}$ falls in the *acceptance region*. This will produce the value $\beta = P[\text{accept } H_0 \text{ when } \mu = \mu_a]$.

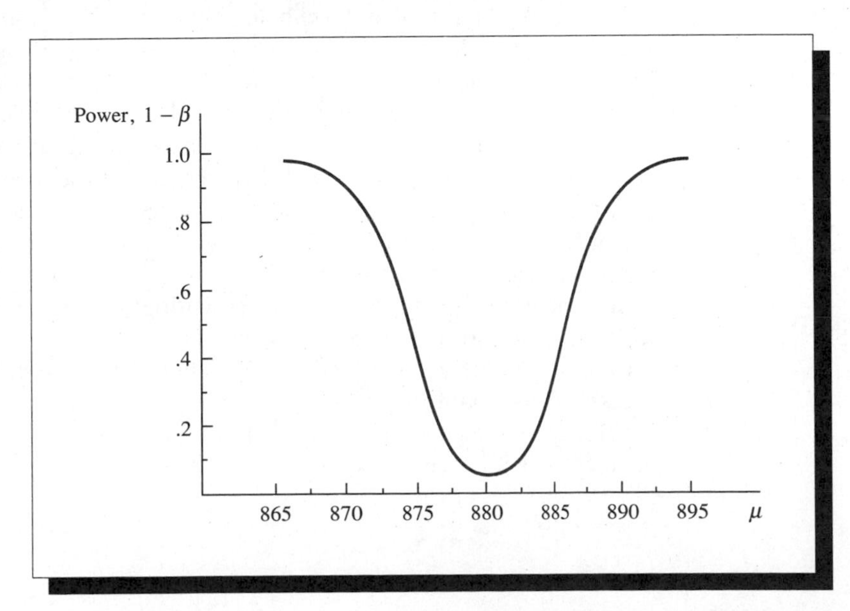

FIGURE 8.6 Power curve for Example 8.6

Comments Concerning Error Rates

Since α is the probability that the test statistic falls in the rejection region, assuming H_0 to be true, **an increase in the size of the rejection region increases α** and, at the same time, decreases β for a fixed sample size. Reducing the size of the rejection region decreases α and increases β. If the sample size n is increased, more information is available on which to base the decision, and for fixed α, β will decrease.

The probability β of making a type II error varies depending on the true value of the population parameter. For instance, suppose that we wish to test the null hypothesis that the binomial parameter p is equal to $p_0 = .4$. (We will use a subscript 0 to indicate the parameter value specified in the null hypothesis H_0.) Furthermore, suppose that H_0 is false and that p is really equal to an alternative value, say, p_a. Which will be more easily detected, a $p_a = .4001$ or a $p_a = 1.0$? Certainly, if p is really equal to 1.0, every single trial will result in success and the sample results will produce strong evidence to support a rejection of $H_0 : p_0 = .4$. On the other hand, $p_a = .4001$ lies so close to $p_0 = .4$ that it would be extremely difficult to detect p_a without a very large sample. In other words, the probability β of accepting H_0 will vary depending on the difference between the true value of p and the hypothesized value p_0. Ideally, the further p_a lies from p_0, the higher the probability is of rejecting H_0. This probability is measured by $1 - \beta$, which is called the **power** of the test.

For constant values of n and α, **the power of a test should increase as the distance between the true and hypothesized values of the parameter increases.** An increase in the sample size n will increase the power, $1 - \beta$, for all alternative values of the parameter being tested. Thus, we can create a power curve corresponding to each sample size.

In practice, β is often unknown, either because it was never computed before the test was conducted or because it may be extremely difficult to compute for the test. Then rather than accept the null hypothesis when the test statistic falls in the acceptance region, you should withhold judgment. That is, you should not accept the null hypothesis unless you know the risk (measured by β) of making an incorrect decision. Notice that you will never be faced with this "no conclusion" situation when the test statistic falls in the rejection region. Then you can reject the null hypothesis (and accept the alternative hypothesis) because you always know the value of α, the probability of rejecting the null hypothesis when it is true. The fact that β is often unknown explains why we attempt to support the alternative hypothesis by rejecting the null hypothesis. When we reach this decision, the probability α that such a decision is incorrect is known.

In conclusion, keep in mind that **"accepting" a particular hypothesis means deciding in its favor.** Regardless of the outcome of a test, you are never *certain* that the hypothesis you "accept" is true. **There is always a risk of being wrong (measured by α and β).** Consequently, you never "accept" H_0 if β is unknown or its value is unacceptable to you. When this situation occurs, you should withhold judgment and collect more data.

Exercises

Basic Techniques

8.1 A random sample of $n = 35$ observations from a population produced a mean $\bar{x} = 2.4$ and standard deviation equal to 0.29. Suppose your research objective is to show that the population mean μ exceeds 2.3.

a Give the null and alternative hypotheses for the test.

b Describe the probability of a type I error, α, in terms of the null and alternative hypotheses.

c Locate the rejection region for the test if $\alpha = .05$.

d Before you conduct the test, glance at the data and use your intuition to decide whether the sample mean $\bar{x} = 2.4$ implies that $\mu > 2.3$. Now conduct the test. Do the data provide sufficient evidence to indicate that $\mu > 2.3$?

8.2 Refer to Exercise 8.1. Suppose that your research objective is to show that the population mean is less than 2.9. Give the null and alternative hypotheses for the test. Is the test one- or two-tailed? Explain.

8.3 Refer to Exercise 8.1. If the research objective is to show that μ differs from 2.9 (i.e., is either greater than or less than 2.9), state the null and alternative hypotheses. Is the test one- or two-tailed?

8.4 Refer to Exercise 8.1. In testing $H_0 : \mu = 2.3$ against $H_a : \mu > 2.3$,

a Find the critical value of $\bar{x}$ necessary for rejection of H_0.

b Calculate $\beta = P$[accept H_0 when $\mu = 2.4$]. Repeat the calculation for $\mu = 2.3$, 2.5, and 2.6.

c Use the value of β calculated in part (b) to graph the power curve for the test.

Applications

8.5 High airline occupancy rates on scheduled flights are essential to profitability. Suppose a scheduled flight must average at least 60% occupancy in order to be profitable and an examination of the occupancy rate for 120 10:00 A.M. flights from Atlanta to Dallas showed a mean occupancy rate per flight of 58% and a standard deviation of 11%.

a If μ is the mean occupancy per flight and if the company wishes to determine whether or not this scheduled flight is unprofitable, give the alternative and the null hypotheses for the test.

b Does the alternative hypothesis in part (a) imply a one- or two-tailed test? Explain.

c Do the occupancy data for the 120 flights suggest that this scheduled flight is unprofitable? Test using $\alpha = .10$.

8.6 Many computer buyers have discovered that they can save a considerable amount by purchasing a personal computer (PC) from a mail-order company—an average of $900 by their estimates ("Whos's Tops," 1992). In an attempt to test this claim, a random sample of 35 customers who recently bought a PC through a mail-order company were contacted and asked to estimate the amount that they saved by purchasing by mail. The average and standard deviation of these 35 estimates were $885 and $50, respectively.

a State the null and alternative hypotheses to be tested.

b Find the appropriate rejection region using $\alpha = .01$.

c Calculate the observed value of the test statistic.

d Based on the results of part (c), is there sufficient evidence to indicate that the average savings is different from the $900 claimed by mail-order PC buyers?

8.7 A drug manufacturer claimed that the mean potency of one of its antibiotics was 80%. A random sample of $n = 100$ capsules were tested and produced a sample mean of $\bar{x} = 79.7\%$, with a standard deviation $s = .8\%$. Do the data present sufficient evidence to refute the manufacturer's claim? Let $\alpha = .05$.

a State the null hypothesis to be tested.

b State the alternative hypothesis.

c Conduct a statistical test of the null hypothesis and state your conclusion.

8.8 In Exercise 6.13, we described the Prevention Index—a composite index designed to measure fitness based on 21 key health- and safety-promoting activities. A random sample of 1250 American adults (*Press-Enterprise*, Riverside, Calif., May 24, 1989) scored 65.4 of a possible 100 points, nearly 4 points higher than the 1984 average. If the mean of the Prevention Index scores in 1984 was 61.5 and the standard deviation of the scores was approximately equal to 12, do the data provide sufficient evidence to indicate that the mean score for all Americans is higher now than it was in 1984? Test using $\alpha = .05$.

8.9 As more and more Americans enter the work force, it is becoming increasingly difficult for the present transportation and communications systems to handle the increasing demand for services during the traditional work hours. Hence, many companies are becoming involved in *flextime*, in which the worker schedules his or her own work hours, or compresses work weeks. A company that was contemplating the installation of a flextime schedule estimated that it needed a minimum mean of 7 hours/day per assembly worker in order to operate effectively. Each of a random sample of 80 of the company's assemblers was asked to submit a tentative flextime schedule. If the mean of the number of hours per day for Monday was 6.7 hours and the standard deviation was 2.7 hours, do the data provide sufficient evidence to indicate that the mean number of hours worked per day on Mondays, for all of the company's assemblers, will be less than 7 hours? Test using $\alpha = .05$.

8.10 In Exercise 1.41, we examined an advertising flyer for the *Princeton Review* (Irvine, Calif., 1993), a review course designed for high school students taking the SAT tests. The flyer claimed that the average score improvements for students who have taken the *Princeton Review* course is between 110 and 160 points. Are the claims made by the *Princeton Review* advertisers exaggerated? That is, is the average score improvement less than 110, the minimum claimed in the advertising flyer? A random sample of 100 students who took the *Princeton Review* course achieved an average score improvement of 107 points with a standard deviation of 13 points. Test the *Princeton Review* claim using $\alpha = .01$.

8.4 A Large-Sample Test of Hypothesis for the Difference in Two Population Means

In many situations, the statistical question to be answered involves a comparison of two population means. For example, the U.S. Postal Service is interested in reducing its massive 350 million gallons/year gasoline bill by replacing gasoline-powered trucks with electric-powered trucks. To determine if there are significant savings in operating costs achieved by changing to electric-powered trucks, a pilot study should be undertaken using, say, 100 conventional gasoline-powered mail trucks and 100 electric-powered mail trucks operated under similar conditions. The statistic that summarizes the sample information regarding the difference in population means $\mu_1 - \mu_2$ is the difference in sample means $\bar{x}_1 - \bar{x}_2$. Therefore, in testing whether the difference in sample means indicates that the true difference in population means differs from a specified value, $\mu_1 - \mu_2 = D_0$, we would use the number of standard deviations that $\bar{x}_1 - \bar{x}_2$ lie from the hypothesized difference D_0. The formal testing procedure is given in the following display.

Large-Sample Statistical Test for $(\mu_1 - \mu_2)$

1 Null Hypothesis: $H_0 : (\mu_1 - \mu_2) = D_0$ where D_0 is some specified difference that you wish to test. For many tests, you will wish to hypothesize that there is no difference between μ_1 and μ_2; that is, $D_0 = 0$.

2 Alternative Hypothesis:

One-Tailed Test	*Two-Tailed Test*
$H_a : (\mu_1 - \mu_2) > D_0$ [or, $H_a : (\mu_1 - \mu_2) < D_0$]	$H_a : (\mu_1 - \mu_2) \neq D_0$

3 Test Statistic: $z = \dfrac{(\bar{x}_1 - \bar{x}_2) - D_0}{\sigma_{(\bar{x}_1 - \bar{x}_2)}} = \dfrac{(\bar{x}_1 - \bar{x}_2) - D_0}{\sqrt{\dfrac{\sigma_1^2}{n_1} + \dfrac{\sigma_2^2}{n_2}}}$

If σ_1^2 and σ_2^2 are unknown (which is usually the case), substitute the sample variances s_1^2 and s_2^2 for σ_1^2 and σ_2^2, respectively.

4 Rejection Region:

One-Tailed Test	*Two-Tailed Test*
$z > z_\alpha$ [or $z < -z_\alpha$ when the alternative hypothesis is $H_a : (\mu_1 - \mu_2) < D_0$]	$z > z_{\alpha/2}$ or $z < -z_{\alpha/2}$

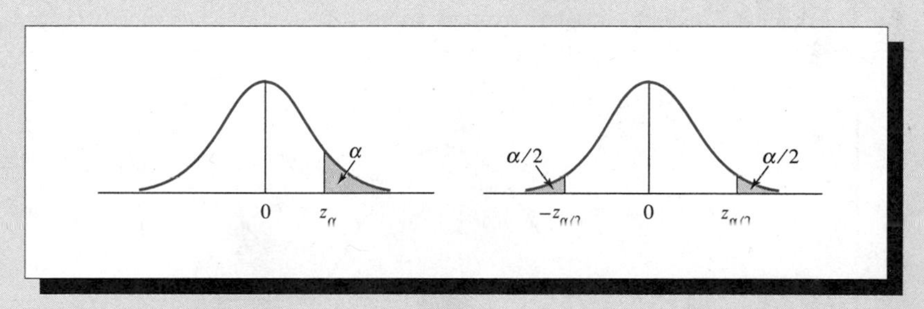

Assumptions: The samples were randomly and independently selected from the two populations and $n_1 \geq 30$ and $n_2 \geq 30$.

EXAMPLE **8.7** A university investigation, conducted to determine whether car ownership affects academic achievement, was based on two random samples of 100 male students, each drawn from the student body. The grade point average for the $n_1 = 100$ nonowners of cars had an average and variance equal to $\bar{x}_1 = 2.70$ and $s_1^2 = .36$, as opposed to $\bar{x}_2 = 2.54$ and $s_2^2 = .40$ for the $n_2 = 100$ car owners. Do the data present sufficient evidence to indicate a difference in the mean achievement between car owners and nonowners of cars? Test using $\alpha = .10$.

Solution Since we wish to detect a difference, if it exists, between the mean academic achievement for nonowners of cars μ_1 and car owners μ_2, we will wish to test the null hypothesis that there is no difference between the means, against the alternative hypothesis that $(\mu_1 - \mu_2) \neq 0$; that is,

$$H_0 : (\mu_1 - \mu_2) = D_0 = 0 \qquad \text{and} \qquad H_a : (\mu_1 - \mu_2) \neq 0$$

Substituting into the formula for the test statistic, we obtain

$$z = \frac{(\bar{x}_1 - \bar{x}_2 - D_0)}{\sqrt{\dfrac{\sigma_1^2}{n_1} + \dfrac{\sigma_2^2}{n_2}}} \approx \frac{2.70 - 2.54}{\sqrt{\dfrac{.36}{100} + \dfrac{.40}{100}}} = 1.84$$

Using a two-tailed test with $\alpha = .10$, we will place $\alpha/2 = .05$ in each tail of the z distribution and reject H_0 if $z > 1.645$ or $z < -1.645$. (See Figure 8.7.) Since $z = 1.84$ exceeds $z_{\alpha/2} = 1.645$, it falls in the rejection region. Therefore, we would reject the null hypothesis that there is no difference in the average academic achievement of car owners versus nonowners of cars. The chance of rejecting H_0, assuming H_0 is true, is only $\alpha = .10$, and hence we would be inclined to think that we have made a reasonably good decision. ▪

FIGURE 8.7 Location of the rejection region in Example 8.7

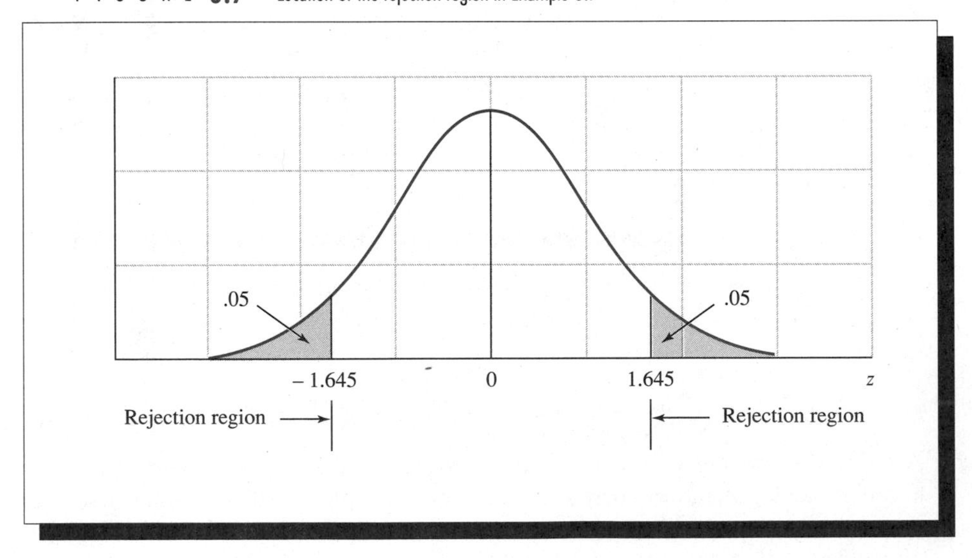

Exercises

Basic Techniques

8.11 Independent random samples of $n_1 = 80$ and $n_2 = 80$ were selected from populations 1 and 2, respectively. The population parameters and the sample means and variance are shown below:

	Population	
	1	**2**
Population mean	μ_1	μ_2
Population variance	σ_1^2	σ_2^2
Sample size	80	80
Sample mean	11.6	9.7
Sample variance	27.9	38.4

a If your research objective is to show that μ_1 is larger than μ_2, state the alternative and the null hypotheses that you would choose for a statistical test.

b Is the test in part (a) a one- or a two-tailed test?

c Give the test statistic that you would use for the test in parts (a) and (b) and give the rejection region for $\alpha = .10$.

d Look at the data. Based on your intuition, do you think that the data provide sufficient evidence to indicate that μ_1 is larger than μ_2? [We will employ a statistical test to reach this decision in part (e).]

e Conduct the test and draw your conclusions. Do the data present sufficient evidence to indicate that $\mu_1 > \mu_2$?

8.12 Independent random samples of $n_1 = 36$ and $n_2 = 45$ were selected from populations 1 and 2, respectively. The population parameters and the sample means and variances are shown in the accompanying table:

	Population	
	1	**2**
Population mean	μ_1	μ_2
Population variance	σ_1^2	σ_2^2
Sample size	36	45
Sample mean	1.24	1.31
Sample variance	0.0560	0.0540

a If your research objective is to show that μ_1 and μ_2 are different, state the alternative and the null hypotheses that you would choose for a statistical test.

b Is the test in part (a) a one- or a two-tailed test?

c Give the test statistic that you would use for the test in parts (a) and (b) and give the rejection region for $\alpha = .05$.

d Look at the data. Based on your intuition, do you think that the data provide sufficient evidence to indicate that μ_1 differs from μ_2? [We will employ a statistical test to reach this decision in part (e).]

e Conduct the test and draw your conclusions. Do the data present sufficient evidence to indicate that $\mu_1 \neq \mu_2$?

8.13 Suppose that you wish to detect a difference between μ_1 and μ_2 (either $\mu_1 > \mu_2$ or $\mu_1 < \mu_2$) and that, instead of running a two-tailed test using $\alpha = .10$, you employ the following test procedure. You wait until you have collected the sample data and have calculated $\bar{x}_1$ and $\bar{x}_2$. If $\bar{x}_1$ is larger than $\bar{x}_2$, you choose the alternative hypothesis $H_a : \mu_1 > \mu_2$ and run a one-tailed test placing $\alpha_1 = .10$ in the upper tail of the z distribution. If, on the other hand, $\bar{x}_2$ is larger than $\bar{x}_1$, you reverse the procedure and run a one-tailed test, placing $\alpha_2 = .10$ in the lower tail of the z distribution. If you use this procedure and if μ_1 actually equals μ_2, what is the probability α that you will conclude that μ_1 is not equal to μ_2 (i.e., what is the probability α that you will incorrectly reject H_0 when H_0 is true)? This exercise demonstrates why statistical tests should be formulated *prior* to observing the data.

Applications

8.14 An experiment was planned to compare the mean time (in days) required to recover from a common cold for persons given a daily dose of 4 milligrams of vitamin C versus those who were not given a vitamin supplement. Suppose that 35 adults were randomly selected for each treatment category and that the mean recovery times and standard deviations for the two groups were as follows:

	Treatment	
	No Vitamin Supplement	**4 mg Vitamin C**
Sample size	35	35
Sample mean	6.9	5.8
Sample standard deviation	2.9	1.2
Population mean	μ_1	μ_2

a Suppose your research objective is to show that the use of vitamin C reduces the mean time required to recover from a common cold and its complications. Give the null and alternative hypotheses for the test. Is this a one- or a two-tailed test?

b Conduct the statistical test of the null hypothesis in part (a) and state your conclusion. Test using $\alpha = .05$.

8.15 Analyses of drinking water samples for 100 homes in each of two different sections of a city gave the following means and standard deviations of lead levels (in parts per million):

	Section of City	
	1	**2**
Sample size	100	100
Mean	34.1	36.0
Standard deviation	5.9	6.0

Do the data provide sufficient evidence to indicate a difference in the mean lead levels in the drinking water for the two sections of the city? Use $\alpha = .05$.

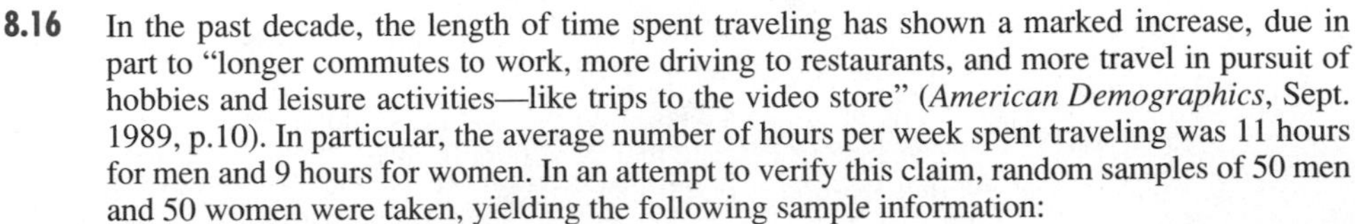

8.16 In the past decade, the length of time spent traveling has shown a marked increase, due in part to "longer commutes to work, more driving to restaurants, and more travel in pursuit of hobbies and leisure activities—like trips to the video store" (*American Demographics*, Sept. 1989, p.10). In particular, the average number of hours per week spent traveling was 11 hours for men and 9 hours for women. In an attempt to verify this claim, random samples of 50 men and 50 women were taken, yielding the following sample information:

	Men	Women
Mean	11.5	8.5
Standard deviation	1.25	1.20

a If the population means given in the *American Demographics* article are correct, then the difference between the two population means is $\mu_1 - \mu_2 = 11 - 9 = 2$. You suspect that the difference in commute times between men and women is greater than that claimed in the article. State the null and alternative hypotheses for the test.

b Give the rejection region for the test in part (a) using $\alpha = .01$.

c Calculate the test statistic and state your conclusions.

8.17 In an attempt to compare the starting salaries for college graduates majoring in education and social sciences (Exercise 7.25), random samples of 50 recent college graduates in each major were selected and the following information was obtained:

Major	Mean	Standard Deviation
Education	22,554	2225
Social science	20,348	2375

a Do the data provide sufficient evidence to indicate a difference in average starting salaries for college graduates in education and the social sciences? Test using $\alpha = .05$.

b Compare your conclusions in part (a) with the results of part (b), Exercise 7.25. Are your conclusions the same? Explain.

8.18 In comparing high school and college students' leisure interests and stress ratings (Exercise 7.26), Alice Chang and colleagues (1993) tested 559 high school students using the leisure interest checklist (LIC) inventory. The means and standard deviations for each of the seven LIC factor scales are given in the accompanying table for 252 boys and 307 girls.

LIC Factor	Boys		Girls	
Scale Activity	Mean	SD	Mean	SD
Hobbies	10.06	6.47	13.64	7.46
Social fun	22.05	5.12	25.96	5.07
Sports	13.65	4.82	9.88	4.41
Cultural	11.48	5.69	13.21	5.31
Trips	6.90	3.41	6.49	2.97
Games	4.95	3.29	3.85	2.49
Church	2.15	2.10	3.00	2.26

a Without looking at the sample data, would you expect that boys would have a greater or lesser interest in sports as a leisure time activity than girls? Do the data provide sufficient evidence to indicate that the mean LIC score for sports is greater for boys than the mean for girls? Test using $\alpha = .01$.

b Does the data provide sufficient evidence to indicate a difference in mean LIC scores for church between boys and girls? Test using $\alpha = .01$.

8.19 In an article entitled "A Strategy for Big Bucks," Charles Dickey (1980) discusses studies of the habits of white-tailed deer that indicate that they live and feed within very limited ranges—approximately 150 to 205 acres. To determine whether there was a difference between the ranges of deer located in two different geographical areas, 40 deer were caught, tagged, and fitted with small radio transmitters. Several months later, the deer were tracked and identified, and the distance x from the release point was recorded. The mean and standard deviation of the distances from the release point were as follows:

	Location	
	1	2
Sample size	40	40
Sample mean	2980 ft	3205 ft
Sample standard deviation	1140 ft	963 ft
Population mean	μ_1	μ_2

a If you have no preconceived reason for believing one population mean to be larger than another, what would you choose for your alternative hypothesis? Your null hypothesis?

b Would your alternative hypothesis in part (a) imply a one- or a two-tailed test? Explain.

c Do the data provide sufficient evidence to indicate that the mean distances differ for the two geographical locations? Test using $\alpha = .10$.

8.5 A Large-Sample Test of Hypothesis for a Binomial Proportion

When a random sample of n identical trials is drawn from a binomial population, the sample proportion $\hat{p}$ has an approximately normal distribution when n is large, with mean p and standard deviation

$$\sigma_{\hat{p}} = \sqrt{\frac{pq}{n}}$$

In testing an hypothesis about p, the proportion in the population possessing a certain attribute, the test follows the same general form as the large-sample tests in Sections 8.3 and 8.4. To test an hypothesis of the form

$$H_0 : p = p_0$$

versus a one- or a two-tailed alternative

$$H_a : p \neq p_0 \quad \text{or} \quad H_a : p > p_0 \quad \text{or} \quad H_a : p < p_0$$

the test statistic is constructed using $\hat{p}$, the best estimator of the true population proportion p. The sample proportion $\hat{p}$ is standardized, using the hypothesized mean and standard deviation, to form a test statistic z, which has a standard normal distribution if H_0 is true. This large-sample test is summarized in the following display.

Large-Sample Test for a Population Proportion p

1 Null Hypothesis: $H_0 : p = p_0$.

2 Alternative Hypothesis:

One-Tailed Test	*Two-Tailed Test*
$H_a: p > p_0$ (or, $H_a: p < p_0$)	$H_a: p \neq p_0$

3 Test Statistic: $z = \dfrac{\hat{p} - p_0}{\sigma_{\hat{p}}} = \dfrac{\hat{p} - p_0}{\sqrt{\dfrac{p_0 q_0}{n}}}$, with $\hat{p} = \dfrac{x}{n}$

where x is the number of successes in n binomial trials.†

4 Rejection Region:

One-Tailed Test	*Two-Tailed Test*
$z > z_\alpha$ (or $z < -z_\alpha$ when the alternative hypothesis is $H_a: p < p_0$)	$z > z_{\alpha/2}$ or $z < -z_{\alpha/2}$

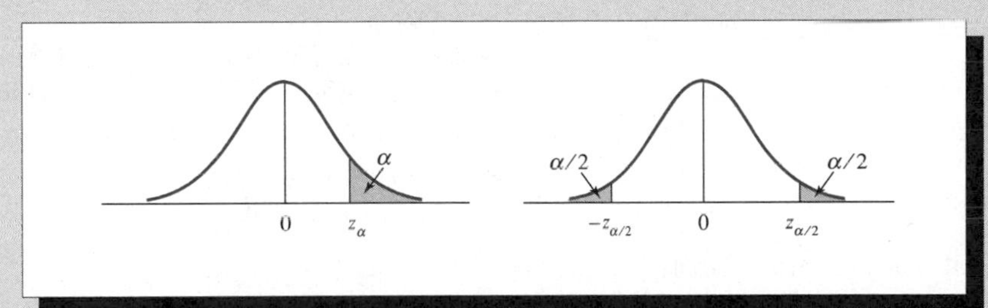

Assumption: The sampling satisfies the assumptions of a binomial experiment (Section 4.2), and n is large enough so that the sampling distribution of $\hat{p}$ can be approximated by a normal distribution. The interval $p \pm 2\sigma_{\hat{p}}$ must be contained in the interval 0 to 1.

EXAMPLE **8.8** Regardless of age, about 20% of American adults participate in fitness activities at least twice a week. However, the fitness activities in which people participate change as they get older, and occasional participants become nonparticipants as they age. In a local survey of $n = 100$ adults over 40 years of age, a total of 15 people indicated that they participated in a fitness activity at least twice a week. Do these data indicate

†An equivalent test statistic can be found by multiplying the numerator and denominator of z by n to obtain

$$z = \frac{x - np_0}{\sqrt{np_0 q_0}}$$

that the participation rate for adults over 40 years of age is significantly less than the 20% figure?

Solution It is assumed that the sampling procedure satisfies the requirements of a binomial experiment. An answer to the question posed can be determined by testing the hypothesis

$$H_0 : p = .2$$

against the alternative

$$H_a : p < .2$$

A one-tailed test is used because we wish to detect whether the value of p is less than .2.

The point estimator of p is $\hat{p} = x/n$, and the test statistic is

$$z = \frac{\hat{p} - p_0}{\sqrt{p_0 q_0 / n}}$$

When H_0 is true, the value of p is $p_0 = .2$, and the sampling distribution of $\hat{p}$ has a mean equal to p_0 and a standard deviation of $\sqrt{p_0 q_0/n}$. **Hence, $\sqrt{\hat{p}\hat{q}/n}$ is not used to estimate the standard error of $\hat{p}$ in this case because the test statistic is calculated under the assumption that H_0 is true.** (When estimating the value of p using the estimator $\hat{p}$, the standard error of $\hat{p}$ is not known and is *estimated* by $\sqrt{\hat{p}\hat{q}/n}$.)

FIGURE 8.8 Location of the rejection region in Example 8.8

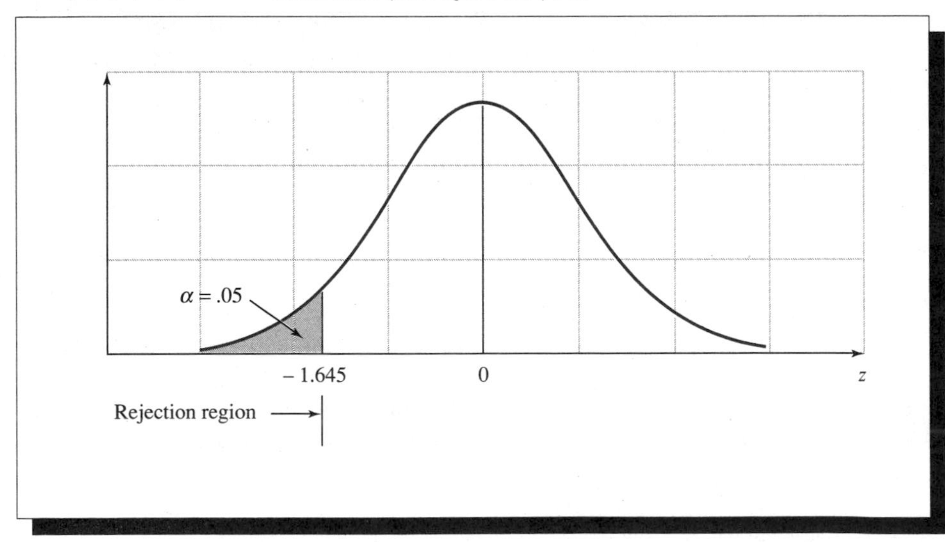

With $\alpha = .05$, we would reject H_0 when $z < -1.645$ (Figure 8.8). With $\hat{p} = 15/100 = .15$, the value of the test statistic is

$$z = \frac{\hat{p} - p_0}{\sqrt{\frac{p_0 q_0}{n}}} = \frac{.15 - .20}{\sqrt{\frac{(.20)(.80)}{100}}} = -1.25$$

The calculated value of the test statistic does not fall in the rejection region, and hence *we do not reject* H_0.

Do we accept H_0? No, not until we have stated alternative values of p different from $p_0 = .2$ that are of *practical significance*. The probability of a type II error should be calculated using these alternative values. If β is sufficiently small, we would accept H_0 with full awareness of the risk of an erroneous decision. ■

Tips on Problem Solving

When testing an hypothesis concerning p, use p_0 (not $\hat{p}$) to calculate $\sigma_{\hat{p}}$ in the denominator of the z statistic. The reason for this is that the rejection region is determined by the distribution of $\hat{p}$ when the null hypothesis is true, namely, when $p = p_0$.

Example 8.5 illustrates an important point. **If the data present sufficient evidence to reject H_0, the probability of an erroneous conclusion α is known in advance because α is used in locating the rejection region. Since α is usually small, we are fairly certain that we have made a correct decision.** On the other hand, if the data present insufficient evidence to reject H_0, the conclusions are not so obvious. Ideally, following the statistical test procedure outlined in Section 8.2, we would have specified a practically significant alternative p_a in advance and chosen n such that β would be small. Unfortunately, many experiments are not conducted in this ideal manner. Someone chooses a sample size, and the experimenter or statistician is left to evaluate the evidence.

The calculation of β is not too difficult for the statistical test procedure outlined in this section but may be extremely difficult, if not beyond the capability of the beginner, in other test situations. **A much simpler procedure is to *not reject* H_0 rather than to accept it and then to estimate using a confidence interval.** The interval will give you a range of plausible values for p.

Exercises

Basic Techniques

8.20 A random sample of $n = 1000$ observations from a binomial population produced $x = 279$.

a If your research hypothesis is that p is less than .3, what should you choose for your alternative hypothesis? Your null hypothesis?

b Does your alternative hypothesis in part (a) imply a one- or a two-tailed statistical test? Explain.

c Do the data provide sufficient evidence to indicate that p is less than .3? Test using $\alpha = .05$.

8.21 A random sample of $n = 1400$ observations from a binomial population produced $x = 529$.

a If your research hypothesis is that p differs from .4, what should you choose for your alternative hypothesis? Your null hypothesis?

b Does your alternative hypothesis in part (a) imply a one- or a two-tailed statistical test? Explain.

c Do the data provide sufficient evidence to indicate that p differs from .4? Test using $\alpha = .10$.

8.22 A random sample of 120 observations was selected from a binomial population, and 72 successes were observed. Do the data provide sufficient evidence to indicate that p is larger than .5? Test using $\alpha = .05$.

Applications

8.23 In a survey designed to monitor the spending patterns of employed high school seniors (Exercise 3.24), 70% of the boys surveyed spent none or only a little of their earnings on long-term savings (Schmittroth, 1991, p. 335). In an attempt to determine whether the 70% figure is accurate, a random sample of 200 employed high school boys in the senior class is taken, and 132 of the 200 indicate that they spent none or only a little of their earnings on long-term savings. Do the data suggest that the 70% figure is not correct? Test using $\alpha = .01$.

8.24 Voters' reactions to proposed tax increases are often tempered by the ways in which the tax monies are to be used. A report ("What Americans Are Saying," 1989) claims that 51% of the voting public view the problem of "providing adequate medical care for all who need it but can't afford it" as one that requires government action even if new taxes are needed. In a random sample of $n = 500$ U.S. voters, 271 agreed that adequate medical care for all requires government action even if new taxes are needed. Use a test of hypothesis to determine if there is sufficient sample evidence to indicate that the true proportion in favor of government action even if taxes are needed differs from $p = .51$.

8.25 An Associated Press article (Bovee, 1991) claimed that one in four college students is aged 30 years or more, indicating that older Americans are returning to school at a higher rate than in previous years. Do you think that the 25% figure is accurate? A random sample of 300 college students at a small community college included 98 who were 30 or older.

a Assume that the random sample chosen at this small community college is representative of all college students in the United States. Test the accuracy of the 25% figure using $\alpha = .05$.

b Is it likely that the assumption made in part (a) is correct? Does this help to explain the results of your test in part (a)? Explain.

8.26 More than ever before, Americans are working at two jobs, according to a Labor Department survey reported in the *Wall Street Journal* (Nov. 7, 1989). According to the survey, the proportion of employed Americans holding two or more jobs is 6.2% compared with 5.4% in 1985. Assume that the current survey was based on a random sample of 850 employed Americans. If you wish to show that the proportion of Americans holding two or more jobs is greater than the 1985 figure, a State the null and alternative hypotheses to be tested.

b Locate the rejection region for $\alpha = .01$.

c Conduct the test and state your conclusions.

8.27 Although more than one in ten public and commerical buildings in the United States contain damaged asbestos that could cause cancer, the Environmental Protection Agency (EPA) has declined to order a nationwide cleanup program (*Press-Enterprise*, Riverside, Calif., Mar. 1, 1988). An EPA administrator conceded that removal is attractive in concept but that improperly performed removal could result in high levels of exposure. The survey concluded that 20% of the nation's 3.6 million public or commercial buildings contain asbestos in a form that could be crushed or damaged with simple hand pressure.

a In a small-scale survey involving 50 randomly selected public and commercial buildings within the county, 15 buildings were found to contain asbestos that could be crushed or damaged by simple hand pressure. Perform a test to determine whether this county rate differs significantly from the postulated national figure of 20%. Use $\alpha = .05$.

b Find a 95% confidence interval for the proportion of public or commercial buildings in the county that have asbestos that can be damaged by simple hand pressure. Does this interval confirm or contradict the results of part (a)?

8.28 An article in the *Washington Post* ("Seeing the World," 1993, p. 5) stated that nearly 45% of the U.S. population is born with brown eyes, although they don't necessarily stay that way. Many Americans change the hue of their eyes with tinted contact lenses, some for corrective reasons and others just for fun. To test the newspaper's claim, a random sample of 80 people was selected, and 32 had brown eyes. Is there sufficient evidence to dispute the newspaper's claim regarding the proportion of brown-eyed people in the United States? Use $\alpha = .01$.

8.29 Refer to Exercise 8.28. Contact lenses, worn by about 26 million Americans, come in many styles and colors. Most Americans wear the soft-lens version, with the most popular colors being the blue varieties (25%), followed by greens (24%) and then hazel or brown. A random sample of 80 tinted–contact lens wearers was checked for the color of their lenses. Of these people, 22 wore blue lenses, and only 15 wore green lenses.

a Do the sample data provide sufficient evidence to indicate that the proportion of tinted–contact lens wearers who wear blue lenses is different from 25%? Use $\alpha = .05$.

b Do the sample data provide sufficient evidence to indicate that the proportion of tinted–contact lens wearers who wear green lenses is different from 24%? Use $\alpha = .05$.

c Would there be any reason to conduct a one-tailed test for either parts (a) or (b)? Explain.

8.6 A Large-Sample Test of Hypothesis for the Difference Between Two Binomial Proportions

When the focus of an experiment or study is the difference in the proportion of individuals or items possessing a specified characteristic, the pivotal statistic for testing hypotheses about $p_1 - p_2$ is the difference in the sample proportions $\hat{p}_1 - \hat{p}_2$. The formal testing procedure concerning the difference in population proportions is given in the following display.

A Large-Sample Statistical Test for $(p_1 - p_2)$

1 Null Hypothesis: $H_0 : (p_1 - p_2) = D_0$ where D_0 is some specified difference that you wish to test. For many tests, you will wish to hypothesize that there is no difference between p_1 and p_2; that is, $D_0 = 0$.

2 Alternative Hypothesis:

One-Tailed Test	*Two-Tailed Test*
$H_a : (p_1 - p_2) > D_0$	$H_a : (p_1 - p_2) \neq D_0$
(or $H_a : (p_1 - p_2) < D_0$)	

3 Test Statistic: $z = \dfrac{(\hat{p}_1 - \hat{p}_2) - D_0}{\sigma_{(\hat{p}_1 - \hat{p}_2)}} = \dfrac{(\hat{p}_1 - \hat{p}_2) - D_0}{\sqrt{\dfrac{p_1 q_1}{n_1} + \dfrac{p_2 q_2}{n_2}}}$

where $\hat{p}_1 = x_1/n_1$ and $\hat{p}_2 = x_2/n_2$. Since p_1 and p_2 are unknown, we will need to approximate their values in order to calculate the standard deviation of $\hat{p}_1 - \hat{p}_2$ that appears in the denominator of the z statistic. Approximations are available for two cases.

Case I: If we hypothesize that p_1 equals p_2, that is,

$$H_0 : p_1 = p_2$$

or equivalently, that

$$p_1 - p_2 = 0$$

then $p_1 = p_2 = p$ and the best estimate of p is obtained by pooling the data from both samples. Thus, if x_1 and x_2 are the numbers of successes obtained from the two samples,

$$\hat{p} = \frac{x_1 + x_2}{n_1 + n_2}$$

The test statistic would be

$$z = \frac{(\hat{p}_1 - \hat{p}_2) - 0}{\sqrt{\dfrac{\hat{p}\hat{q}}{n_1} + \dfrac{\hat{p}\hat{q}}{n_2}}} \quad \text{or} \quad z = \frac{\hat{p}_1 - \hat{p}_2}{\sqrt{\hat{p}\hat{q}\left(\dfrac{1}{n_1} + \dfrac{1}{n_2}\right)}}$$

Case II: On the other hand, if we hypothesize that D_0 is *not* equal to zero; that is,

$$H_0 : (p_1 - p_2) = D_0$$

where $D_0 \neq 0$, then the best estimates of p_1 and p_2 are $\hat{p}_1$ and $\hat{p}_2$, respectively. The test statistic would be

$$z = \frac{(\hat{p}_1 - \hat{p}_2) - D_0}{\sqrt{\dfrac{\hat{p}_1\hat{q}_1}{n_1} + \dfrac{\hat{p}_2\hat{q}_2}{n_2}}}$$

4 Rejection Region:

One-Tailed Test	*Two-Tailed Test*
$z > z_\alpha$ [or $z < -z_\alpha$ when the alternative hypothesis is $H_a : (p_1 - p_2) < D_0$]	$z > z_{\alpha/2}$ or $z < -z_{\alpha/2}$

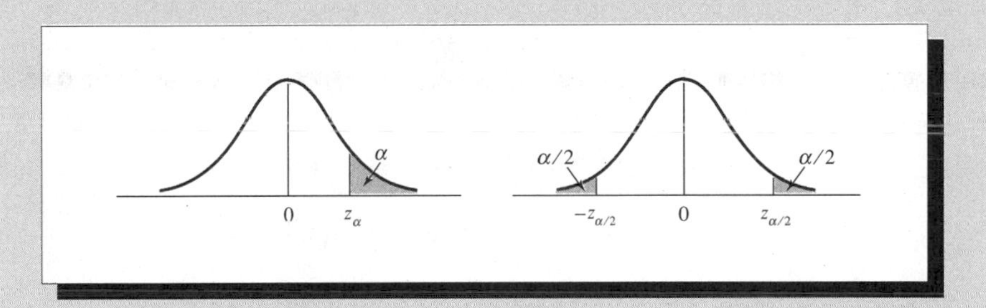

Assumptions: Samples were selected in a random and independent manner from two binomial populations, and n_1 and n_2 are large enough so that the sampling distribution of $\hat{p}_1 - \hat{p}_2$ can be approximated by a normal distribution. The interval $(p_1 - p_2) \pm 2\sigma_{(\hat{p}_1 - \hat{p}_2)}$ must be contained in the interval -1 to 1.

EXAMPLE 8.9 The records of a hospital show that 52 men in a sample of 1000 men versus 23 women in a sample of 1000 women were admitted because of heart disease. Do these data present sufficient evidence to indicate a higher rate of heart disease among men admitted to the hospital?

Solution Assume that the number of patients admitted for heart disease will follow approximately a binomial probability distribution for both men and women with parameters p_1 and p_2, respectively. Then, since we wish to determine whether $p_1 > p_2$, we will test the null hypothesis that $p_1 = p_2$—that is, $H_0 : (p_1 - p_2) = 0$—against the alternative hypothesis $H_a : p_1 > p_2$ or, equivalently, $H_a : (p_1 - p_2) > 0$. To conduct this test, we use the z test statistic and approximate the value of $\sigma_{(\hat{p}_1 - \hat{p}_2)}$ using the pooled estimate of p described in Case I. Since H_a implies a one-tailed test, we reject H_0

FIGURE 8.9 Location of the rejection region in Example 8.9

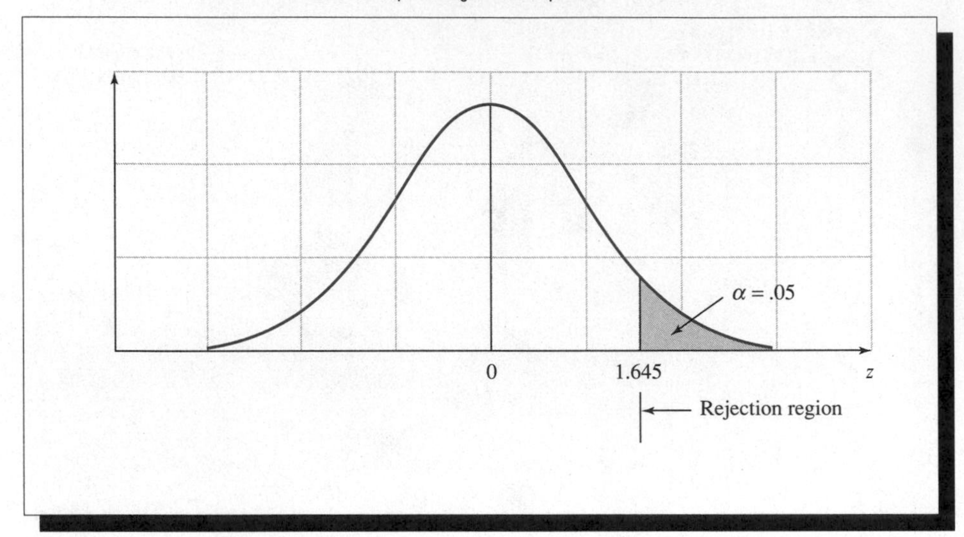

only for large values of z. Thus, for $\alpha = .05$, we reject H_0 if $z > 1.645$ (see Figure 8.9).

The pooled estimate of p required for $\sigma_{(\hat{p}_1 - \hat{p}_2)}$ is

$$\hat{p} = \frac{x_1 + x_2}{n_1 + n_2} = \frac{52 + 23}{1000 + 1000} = .0375$$

The test statistic is

$$z = \frac{\hat{p}_1 - \hat{p}_2}{\sqrt{\hat{p}\hat{q}\left(\frac{1}{n_1} + \frac{1}{n_2}\right)}} = \frac{.052 - .023}{\sqrt{(.0375)(.9625)\left(\frac{1}{1000} + \frac{1}{1000}\right)}} = 3.41$$

Since the computed value of z falls in the rejection region, we reject the hypothesis that $p_1 = p_2$ and conclude that the data present sufficient evidence to indicate that the percentage of men entering the hospital because of heart disease is higher than that of women. (NOTE: This does not imply that the *incidence* of heart disease is higher in men. Perhaps fewer women enter the hospital when afflicted with the disease!) ▪

Exercises

Basic Techniques

8.30 Independent random samples of $n_1 = 140$ and $n_2 = 140$ observations were randomly selected from binomial populations 1 and 2, respectively. The number of successes in the samples and the population parameters are shown below:

	Population	
	1	2
Sample size	140	140
Number of successes	74	81
Binomial parameter	p_1	p_2

a Suppose you have no preconceived theory concerning which parameter, p_1 or p_2, is the larger and you only wish to detect a difference between the two parameters if one exists. What should you choose as the alternative hypothesis for a statistical test? As the null hypothesis?

b Does your alternative hypothesis in part (a) imply a one- or two-tailed test? Explain.

c Conduct the test and state your conclusions. Test using $\alpha = .05$.

8.31 Refer to Exercise 8.30. Suppose, for practical reasons, you know that p_1 cannot be larger than p_2.

a Given this knowledge, what should you choose as the alternative hypothesis for your statistical test? Your null hypothesis?

b Will your alternative hypothesis in part (a) imply a one- or two-tailed test? Explain.

c Conduct the test and state your conclusions. Test using $\alpha = .10$.

8.32 Independent random samples of $n_1 = 280$ and $n_2 = 350$ observations were randomly selected from binomial populations 1 and 2, respectively. The number of successes in the samples and the population parameters are shown below:

	Population	
	1	2
Sample size	280	350
Number of successes	132	178
Binomial parameter	p_1	p_2

a Suppose you know that p_1 can never be larger than p_2 and you want to know if p_1 is less than p_2. What should you choose for your null and alternative hypotheses?

b Does your alternative hypothesis in part (a) imply a one- or two-tailed test? Explain.

c Conduct the test and state your conclusions. Test using $\alpha = .05$.

Applications

8.33 An experiment was conducted to test the effect of a new drug on a viral infection. The infection was induced in 100 mice and the mice were randomly split into two groups of 50. The first group, the *control group*, received no treatment for the infection. The second group received the drug. After a 30-day period, the proportion of survivors, $\hat{p}_1$ and $\hat{p}_2$, in the two groups were found to be $\hat{p}_1 = .36$ and $\hat{p}_2 = .60$. Is there sufficient evidence to indicate that the drug is effective in treating the viral infection? Use $\alpha = .10$.

8.34 Childless women over 50 years of age are more likely to have heart attacks, according to Evelyn Talbot, a University of Pittsburgh researcher, who reported her results at a meeting of the American Heart Association (*Florida Times-Union,* Jacksonville, Fla., Nov. 20, 1986). A study of the medical records of women who died of heart attacks in Allegheny County, Pennsylvania, found that 12 out of 51 heart attack victims were childless. In a similar group of 47 women who survived a heart attack, only two were childless. Do these data present sufficient evidence to conclude at the $\alpha = .05$ level of significance that a difference exists between the fatality rates for heart attacks among childless women versus women of the same age who were not childless? (HINT: Form a 2×2 table to enumerate the number of childless and nonchildless women and their reaction to the heart attack.)

8.35 For several years now, aspirin-based products have been losing market share to substitute analgesics such as ibuprofen and acetaminophen (Exercise 7.47). Finally, large companies such as Bayer, which had produced only aspirin-based products, are beginning to market acetaminophen-based pain relievers (Kraul, 1993). The following table gives the proportion of people who preferred each of the three major pain relievers for the years 1986 and 1991:

Pain Reliever	1986	1991
Aspirin	45%	34%
Acetaminophen	41%	41%
Ibuprofen	14%	26%

a If these results were based on two independent random samples of size $n_1 = 1000$ and $n_2 = 1000$, test the hypothesis of no difference in the proportion of aspirin users in 1986 and 1991. Use $\alpha = .05$.

b Use a test of hypothesis to determine whether ibuprofen has significantly increased its market share between 1986 and 1991. Use $\alpha = .01$.

c Are the tests in parts (a) and (b) related? Explain.

8.36 A survey concerning the health habits of people in the United States, which was conducted by the Louis Harris & Associates (Exercise 7.31), gives not only the sample percentages in various categories for the 1992 survey but also the results of a similar survey conducted in 1983 ("Americans Found Retreating," 1993). Although the number of participants in the 1983 survey is not given in the article, we will assume that the 1983 survey was based on a sample of 1250 people. The sample results for several health categories are shown below:

Health Issue	1983 Survey	1992 Survey
Ate recommended amount of fibrous foods	0.59	0.53
Avoided fat	0.55	0.51
Avoided excess salt	0.53	0.46

Physicians who examined these survey results are concerned that Americans are being less careful about their dietary health habits than they were in 1983. Do the sample data indicate that there has been a significant decrease in the proportions who eat the recommended amount of fibrous foods between 1983 and 1992? Use $\alpha = .05$.

8.37 Refer to Exercise 8.36. Do the sample data indicate that there has been a significant decrease in the proportions who avoid fat and those who avoid salt between 1983 and 1992? Conduct two separate tests using $\alpha = .05$. Are the three tests conducted in this exercise and Exercises 8.36 independent? Explain.

8.38 A survey conducted by the National Association of College Admission Counselors examined enrollment in colleges and universities for the years 1991 and 1992 (Kelly, 1992). The 1992 survey was based on 1232 questionnaires, and we will assume that the 1991 figures were based on an identical number of responses. Based on these samples, 71% of the colleges reported increases in applications, compared with 50% in 1991. Do the data indicate a significant difference in the proportion of colleges who report an increase in applications between 1991 and 1992? Use $\alpha = .01$.

8.39 Refer to Exercise 8.38. The survey also reported that 58% of the colleges said minority applications had increased, compared with 52% in 1991. Do the data indicate a significant difference in the proportion of colleges who report an increase in minority applications between 1991 and 1992? Use $\alpha = .01$.

8.40 In a comparison study of homeless and vulnerable meal-program users, Michael Sosin has investigated determinants that account for a transition from having a home (domiciled) but utilizing meal programs to becoming homeless. The following information provides the study data:

	Homeless Men	Domiciled Men
Sample size	112	260
Number currently working	34	98
Sample proportion	0.30	0.38

Source: Sosin (1992), Table 1.

Test for a significant difference in the population proportions who are currently working for homeless versus domiciled men. Use $\alpha = .01$.

8.7 Another Way to Report the Results of Statistical Tests: p-Values

The probability α of making a type I error is often called the significance level of the statistical test, a term that originated in the following way. The probability of the observed value of the test statistic, or some value even more contradictory to the null hypothesis, measures, in a sense, the weight of evidence favoring rejection of H_0. Some experimenters report test results as being significant (we would reject H_0) at the 5% significance level but not at the 1% level. This means that we would reject H_0 if α were .05 but not if α were .01.

The smallest value of α for which test results are statistically significant is often called the ***p*-value**, or the **observed significance level**, for the test. Some statistical computer programs compute p-values for statistical tests correct to four or five decimal places. But if you are using statistical tables to determine a p-value, you will only be able to approximate its value. This is because most statistical tables give the critical values of test statistics only for large differential values of α (e.g., .01, .025, 0.5, .10). Consequently, the p-value reported by most experimenters is the largest tabulated value of α for which the test remains statistically significant. For example, if a test result is statistically significant for $\alpha = .10$, but not for $\alpha = .05$, then the p-value for the test would be given as p-value = .10 or, more precisely, as p-value $< .10$.

Many scientific journals require researchers to report the p-values associated with statistical tests because these values provide a reader with *more information* than simply stating that a null hypothesis is or is not to be rejected for some value of α chosen by the experimenter. In a sense, it allows the reader of published research to evaluate the extent to which the data disagree with the null hypothesis. In particular, it enables each reader to choose his or her own personal value for α and then decide whether or not the data lead to rejection of the null hypothesis.

The procedure for finding the p-value for a test will be illustrated by the following examples.

EXAMPLE 8.10 Find the p-value for the statistical test in Example 8.5. Interpret your results.

Solution Example 8.5 presents a test of the null hypothesis $H_0 : \mu = 880$ against the alternative hypothesis $H_a : \mu \neq 880$. The value of the test statistic, computed from the sample data, was $z = -3.03$. Therefore, the p-value for this two-tailed test is the probability that $z \leq -3.03$ or $z \geq 3.03$ (the shaded areas in Figure 8.10).

From Table 3 in Appendix I, you can see that the tabulated area under the normal curve between $z = 0$ and $z = 3.03$ is .4988 and the area to the right of $z = 3.03$ is $.5 - .4988 = .0012$. Then, since this was a two-tailed test, the area corresponding to the z values, $z > 3.03$ or $z < -3.03$, is $2(.0012) = .0024$. Consequently, we would report the p-value for the test as p-value $= .0024$. ■

EXAMPLE 8.11 Find the p-value for the statistical test in Example 8.8. Interpret your results.

Solution Example 8.8 presented a one-tailed test of the null hypothesis $H_0 : p = .20$ against the alternative hypothesis $H_a : p < .20$, and the observed value of the test statistic was $z = -1.25$. Therefore, the p-value for the test is the probability of observing a

FIGURE 8.10 Locating the p-value for the test in Example 8.5

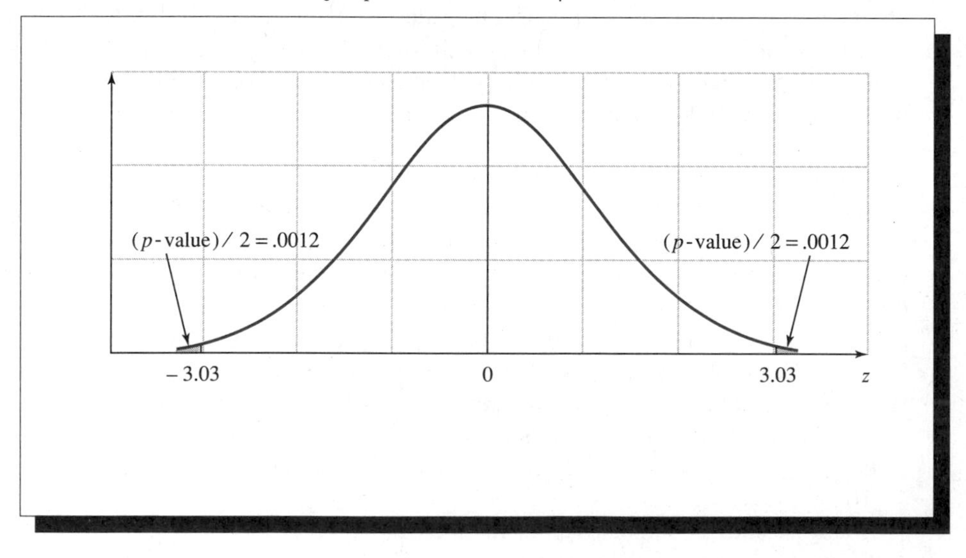

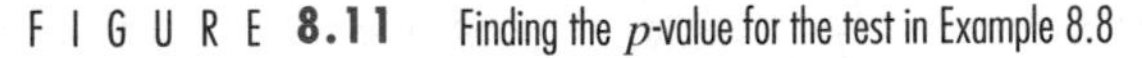

FIGURE 8.11 Finding the p-value for the test in Example 8.8

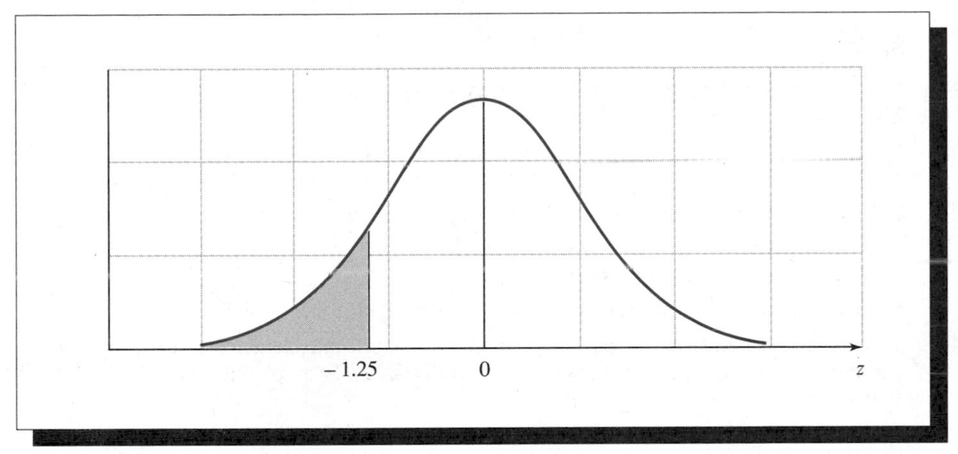

value of the z statistic less than -1.25. This value is the area under the normal curve to the left of $z = -1.25$ (the shaded area in Figure 8.11).

From Table 3 in Appendix I, the area under the normal curve between $z = 0$ and $z = -1.25$ is .3944. Therefore, the area under the normal curve to the left of $z = -1.25$, the p-value for the test, is $p\text{-value} = .5 - .3944 = .1056$.

Advocating that a researcher report the p-value for a test and leave its interpretation to the reader does not violate the traditional statistical test procedure described in the preceding sections. It simply leaves the decision of whether or not to reject the null hypothesis (with the potential for a type I or type II error) to the reader. Thus, it shifts the responsibility for choosing the value of α and, possibly, the problem of evaluating the probability β of making a type II error, to the reader. ▪

Exercises

Basic Techniques

8.41 Suppose you tested the null hypothesis $H_0 : \mu = 94$ against the alternative hypothesis $H_a : \mu < 94$. For a random sample of $n = 52$ observations, $\bar{x} = 92.9$, and $s = 4.1$.

a Give the observed significance level for the test. Interpret this value.

b If you wish to conduct your test using $\alpha = .05$, what would be your test conclusions?

8.42 Suppose you tested the null hypothesis $H_0 : \mu = 94$ against the alternative hypothesis $H_a : \mu \neq 94$. For a random sample of $n = 52$ observations, $\bar{x} = 92.1$, and $s = 4.1$.

a Give the observed significance level for the test.

b If you wish to conduct your test using $\alpha = .05$, what would be your test conclusions?

8.43 Suppose you tested the null hypothesis $H_0 : \mu = 15$ against the alternative hypothesis $H_a : \mu \neq 15$. For a random sample of $n = 38$ observations, $\bar{x} = 15.7$, and $s = 2.4$.

a Give the observed significance level for the test.

b If you wish to conduct your test using $\alpha = .05$, what would be your test conclusions?

Applications

8.44 Refer to Example 8.4. What is the p-value associated with this one-tailed test in the left tail? The observed value of z is -3.41.

8.45 In Example 8.7, we conducted a two-tailed test of the equality of the grade point averages for nonowners and owners of cars. What is the p-value of the two-tailed test for which $z = 1.84$?

8.46 In Example 8.8 where we tested whether participation in fitness activities decreased for people over 40 years of age, the value of z was -1.25.

a What is the p-value for this test?

b What is your conclusion based on this observed significance level?

8.47 When testing whether the proportion of men was significantly larger than the proportion of women admitted to a hospital because of heart disease in Example 8.9, the value of the test statistic was $z = 3.41$.

a What is the p-value associated with this test?

b If your personal value of a type I error was $\alpha = .01$, what would be your decision in this case?

8.48 Refer to Exercise 8.6.

a What is the p-value for this two-tailed test?

b Based on this p-value, what would your decision be?

8.49 Refer to Exercise 8.10.

a What is the p-value for this one-tailed test?

b If you considered a value of $\alpha = .05$ significantly small, what would be your decision?

8.50 The percentage of Americans who smoke did not decline for the first time in a quarter of a century in 1991; 25.7% of the population smoked in 1991, unchanged from 25.5% in 1990 (Neergaard, 1993). According to the Centers for Disease Control (CDC), this means that the United States will not meet a national health objective of having only 15% of Americans smoking by the year 2000. Suppose that in a random sample of 1000 Americans, 230 were found to be smokers. Test the hypothesis that the percentage of smokers has decreased from the 1990 figure of 25.5%.

a What is the p-value for this test?

b Would you agree with the conclusion of the CDC?

8.51 Of those women who are diagnosed to have early-stage breast cancer, one-third eventually die of the disease. Suppose a community public health department instituted a screening program to provide for the early detection of breast cancer and, thus, a consequent increase in the survival rate p of those diagnosed to have the disease. A random sample of 200 women was selected from among those who were periodically screened by the program and who eventually were diagnosed to have the disease. Let x represent the number of those in the sample who survive the disease.

a If we wish to detect whether the community screening program has been effective, state the null hypothesis that should be tested.

b State the alternative hypothesis.

c If the number of women in the sample of 200 who survive the disease is 164, would you conclude that the community screening program was effective? Test using $\alpha = .05$ and explain the practical conclusions to be derived from your test.

d Find the p-value for the test and interpret it.

8.8 Some Comments on the Theory of Tests of Hypotheses

As outlined in Section 8.2, the theory of a statistical test of an hypothesis is indeed a very clear-cut procedure, enabling the experimenter to either reject or accept the null hypothesis with measured risks α and β. Unfortunately, as we noted, the theoretical framework does not suffice for all practical situations.

The crux of the theory requires that we be able to specify a meaningful alternative hypothesis that permits the calculation of the probability β of a type II error for all alternative values of the parameter(s). This indeed can be done for many statistical tests, including the large-sample test discussed in Section 8.3, although the calculation of β for various alternatives and sample sizes may, in some cases, be difficult. On the other hand, in some test situations it is extremely difficult to clearly specify alternatives to H_0 that have practical significance. This may occur when we wish to test an hypothesis concerning the values of a set of parameters, a situation that we will encounter in Chapter 12 in analyzing enumerative data.

The obstacle that we mention does not invalidate the use of statistical tests. Rather, it urges caution in drawing conclusions when insufficient evidence is available to reject the null hypothesis. The difficulty of specifying meaningful alternatives to the null hypothesis, together with the difficulty in calculating and tabulating β for other than the simplest statistical tests, justifies skirting this issue in an introductory text. Hence, we can adopt one of two procedures. We can present the p-value associated with a statistical test and leave the interpretation to you. Or, we can agree to adopt the procedure described in Example 8.8 when tabulated values of β are unavailable for the test. When the test statistic falls in the acceptance region, we will *not reject*, rather than *accept*, the null hypothesis. Further conclusions may be made by calculating an interval estimate for the parameter or by consulting one of several published statistical handbooks for tabulated values of β. We will not be too surprised to learn that these tabulations are inaccessible, if not completely unavailable, for some of the more complicated statistical tests.

Finally, we might comment on the choice between a one- or a two-tailed test for a given situation. We emphasize that this choice is dictated by the practical aspects of the problem and will depend on the alternative value of the parameter that the experimenter is trying to detect. If, for example, we were to sustain a large financial loss if μ were greater than μ_0, but not if it were less, we would concentrate our attention on the detection of values of μ greater than μ_0. Hence, we would reject in the upper tail of the distribution for the test statistics previously discussed. On the other hand, if we are equally interested in detecting values of μ that are either less than or greater than μ_0, we would employ a two-tailed test.

Tips on Problem Solving

When conducting a statistical test of an hypothesis, it is important to follow the same basic procedure for each problem.

1. **Determine the type of data** involved (quantitative or binomial) and the number of samples involved (one or two). This will allow you to identify the parameter of interest in the experiment.
2. **Check the conditions** required for the sampling distribution of the parameter to be approximated by a normal distribution. For quantitative data, the sample size (or sizes) must be 30 or more. For binomial data, a large sample size will ensure that $p \pm 2\sigma_{\hat{p}}$ is contained in the interval 0 to 1 [$(p_1 - p_2) \pm 2\sigma_{(\hat{p}_1 - \hat{p}_2)}$ contained in the interval -1 to 1 for the two-sample case].
3. **State the null and alternative hypotheses (H_0 and H_a).** The alternative hypothesis is the hypothesis that the researcher wishes to support; the null hypothesis is a contradiction of the alternative hypothesis.
4. **State the test statistic** to be used in the test of the hypothesis.
5. **Locate the rejection region** for the test. In this chapter, the rejection region will be found in the tail areas of the standard normal (z) distribution. The exact rejection region will be determined by the desired value of α and the form of the alternative hypothesis (one- or two-tailed).
6. **Conduct the test**, calculating the observed value of the test statistic based on the sample data.
7. **Draw conclusions** based on the observed value of the test statistic. If the test statistic falls in the rejection region, the null hypothesis is rejected in favor of the alternative hypothesis. The probability of an incorrect decision is α. However, if the test statistic does not fall in the rejection region, we cannot reject the null hypothesis. There is insufficient evidence to show that the alternative hypothesis is true. Judgment is withheld until more data can be collected.

8.9 Summary

In this chapter, we have presented the basic concepts of a statistical test of an hypothesis and have demonstrated the procedure with four separate tests. All of the statistical tests described in this chapter are based on the Central Limit Theorem and hence apply to large samples. When n is large, each of the respective test statistics possesses a sampling distribution that can be approximated by the normal distribution. This result, along with the properties of the normal distribution studied in Chapter 5, permits the calculation of α, β, and p-values for the statistical tests.

In addition to the test statistics presented in this chapter, many other test statistics have sampling distributions that are approximately normal when the sample sizes are large (you will encounter several in Chapter 14). But when the sample sizes are small, the sampling distributions of most test statistics are not normal. One of these nonnormal sampling distributions—the t distribution—will be employed in Chapter 9 to obtain confidence intervals and tests of hypotheses for a single population mean and the difference between two population means.

Supplementary Exercises

Starred (*) exercises are optional.

8.52 Define α and β for a statistical test of an hypothesis.

8.53 What is the observed significance level of a test?

8.54 The daily wages in a particular industry are normally distributed with a mean of \$43.20 and a standard deviation of \$9.50. Suppose a company in this industry employing 40 workers pays these workers \$41.20 on the average. Based on this sample mean, could these workers be viewed as a random sample from among all workers in the industry?

a Find the observed significance level of the test.

b If you planned to conduct your test using $\alpha = .01$, what would be your test conclusions?

8.55 Refer to Exercise 7.17 and the collection of water samples to estimate the mean acidity (in pH) of rainfalls in the northeastern United States. As noted, the pH for pure rain falling through clean air is approximately 5.7. The sample of $n = 40$ rainfalls produced pH readings with $\bar{x} = 3.7$ and $s = .5$. Do the data provide sufficient evidence to indicate that the mean pH for rainfalls is more acidic ($H_a : \mu < 5.7$ pH) than pure rainwater? Test using $\alpha = .05$. Note that this inference is appropriate only for the area in which the rainwater specimens were collected.

8.56 A manufacturer of automatic washers provides a particular model in one of three colors: A, B, or C. Of the first 1000 washers sold, it is noted that 400 were of color A. Would you conclude that more than one-third of all customers have a preference for color A?

a Find the observed significance level of the test.

b If you planned to conduct your test using $\alpha = .05$, what would be your test conclusions?

8.57 The pH factor is a measure of the acidity or alkalinity of water. A reading of 7.0 is neutral; values in excess of 7.0 indicate alkalinity; those below 7.0 imply acidity. Loren Hill (1980) states that the best chance of catching bass occurs when the pH of the water is in the range of 7.5 to 7.9. Suppose you suspect that acid rain is lowering the pH of your favorite fishing spot and you wish to determine whether the pH is less than 7.5.

a State the alternative and null hypotheses that you would choose for a statistical test.

b Would the alternative hypothesis in part (a) imply a one- or a two-tailed test? Explain.

c Suppose that a random sample of 30 water specimens gave pH readings with $\bar{x} = 7.3$ and $s = .2$. Just glancing at the data, do you think that the difference $\bar{x} - 7.5 = -.2$ is large enough to indicate that the mean pH of the water samples is less than 7.5? (Do *not* conduct the test.)

d Now conduct a statistical test of the null hypothesis in part (a) and state your conclusions. Test using $\alpha = .05$. Compare your statistically based decision with your intuitive decision in part (c).

8.58 What conditions must be met so that the z test can be used to test an hypothesis concerning a population mean μ?

8.59 In Exercise 7.82, we presented data from a controlled pollination study involving seed survival rates for plants subjected to water or nutrient deprivation. The data, representing the number of seeds surviving to maturity, are reproduced below:

	Male		Female	
Treatment	n	**Number Surviving**	n	**Number Surviving**
Control	585	543	632	560
Low water	578	522	510	466
Low nutrient	568	510	589	546

Source: Pittman and Levin (1989).

a Is there a significant difference between the proportions of surviving seeds for low water versus low nutrients in the male category? Use $\alpha = .01$.

b Do the data provide sufficient evidence to indicate a difference between survival proportions for low water versus low nutrients for the female category? Use $\alpha = .01$.

c Look at the data for the other four treatment–sex combinations. Does it appear that there are any significant differences due to low levels of water or nutrients? Explain.

8.60 In Exercise 4.32, we noted that 16 out of every 100 doctors in any given year are subject to malpractice claims (*New York Times*, Feb. 15, 1985). A hospital that has a staff consisting of 300 physicians seems to have an unusually large number of doctors involved in malpractice claims—58 cases in 1 year. Does this number provide sufficient evidence to indicate that the doctor–malpractice claim rate at the hospital differs from the national rate? Test using $\alpha = .05$.

8.61 In a study to assess various effects of using a female model in automobile advertising, each of 100 male subjects was shown photographs of two automobiles matched for price, color, and size, but of different makes. One of the automobiles was shown with a female model to 50 of the subjects (group A), and both automobiles were shown without the model to the other 50 subjects (group B). In group A, the automobile shown with the model was judged as more expensive by 37 subjects, in group B the same automobile was judged as the more expensive by 23 subjects. Do these results indicate that using a female model influences the perceived expensiveness of an automobile? Use a one-tailed test with $\alpha = .05$.

8.62 Random samples of 200 bolts manufactured by machine A and 200 bolts manufactured by machine B showed 16 and 8 defective bolts, respectively. Do these data present sufficient evidence to suggest a difference in the performance of the machines? Use $\alpha = .05$.

8.63 In Exercise 6.50, we reported that the biomass for tropical woodlands, thought to be about 35 kilograms per square meter (kg/m^2), may in fact be too high and that tropical biomass values vary regionally—from about 5 to 55 kg/m^2 (*Science News*, Aug. 19, 1989, p. 124). Suppose you measure the tropical biomass in 400 randomly selected square-meter plots and obtain $\bar{x} = 31.75$ and $s = 10.5$. Do the data present sufficient evidence to indicate that scientists are overestimating the mean biomass for tropical woodlands and that the mean is in fact lower than estimated?

a State the null and alternative hypotheses to be tested.

b Locate the rejection region for the test with $\alpha = .01$.

c Conduct the test and state your conclusions.

8.64 A social scientist believes that the fraction p_1 of Republicans in favor of the death penalty is greater than the fraction p_2 of Democrats in favor of the death penalty. She acquired independent random samples of 200 Republicans and 200 Democrats and found 46 Republicans and 34 Democrats favoring the death penalty. Do these data support the social scientist's belief?

a Find the observed significance level of the test.

b If you planned to conduct your test using $\alpha = .05$, what would be your test conclusions?

8.65* Refer to Exercise 8.64. Some thought should have been given to designing a test for which β is tolerably low when p_1 exceeds p_2 by an important amount. For example, find a common sample size n for a test with $\alpha = .05$ and $\beta \leq .20$, when in fact p_1 exceeds p_2 by .1. [HINT: The maximum value of $p(1 - p)$ is .25.]

8.66 In comparing the mean weight loss for two diets, the following sample data were obtained:

	Diet I	Diet II
Sample size n	40	40
Sample mean $\bar{x}$	10 lb	8 lb
Sample variance s^2	4.3	5.7

Do the data provide sufficient evidence to indicate that diet I produces a greater mean weight loss than diet II? Use $\alpha = .05$.

8.67 A test of the breaking strengths of two different types of cables was conducted using samples of $n_1 = n_2 = 100$ pieces of each type of cable.

Cable I	Cable II
$\bar{x}_1 = 1925$	$\bar{x}_2 = 1905$
$s_1 = 40$	$s_2 = 30$

Do the data provide sufficient evidence to indicate a difference between the mean breaking strengths of the two cables? Use $\alpha = .10$.

8.68 The braking ability was compared for two types of 1993 automobiles. Random samples of 64 automobiles were tested for each type. The recorded measurement was the distance (in feet) required to stop when the brakes were applied at 40 miles per hour. The computed sample means and variances were the following:

$$\bar{x}_1 = 118 \qquad \bar{x}_2 = 109$$
$$s_1^2 = 102 \qquad s_2^2 = 87$$

Do the data provide sufficient evidence to indicate a difference between the mean stopping distances for the two types of automobiles?

8.69 A fruit grower wishes to test a new spray that a manufacturer claims will *reduce* the loss due to damage by a certain insect. To test the claim, the grower sprays 200 trees with the new spray and 200 other trees with the standard spray. The following data were recorded:

	New Spray	Standard Spray
Mean yield per tree $\bar{x}$ (lb)	240	227
Variance s^2	980	820

a Do the data provide sufficient evidence to conclude that the new spray is better than the old? Use $\alpha = .05$.

b Construct a 95% confidence interval for the difference between the mean yields for the two sprays.

8.70* Refer to Example 8.6. Use the procedure described in Example 8.6 to calculate β for several alternative values of μ. (For example, $\mu = 873, 875$, and 877.) Use the three computed values of β along with the value computed in Example 8.6 to construct a power curve for the statistical test.

8.71* Repeat the procedure described in Exercise 8.70 for a sample size $n = 25$ (as opposed to $n = 50$ used there) and compare the two power curves.

8.72 In Exercise 7.13, we described a study of 392 healthy children living in the area of Tours, France. The table below shows some results of the study, measuring the serum levels of certain fat-soluble vitamins—namely, vitamin A (retinol) and β-carotene—in both boys and girls:

	Boys ($n = 207$)	Girls ($n = 185$)
Retinol (μg/dl)	$\bar{x}_1 = 43.0$	$\bar{x}_2 = 41.8$
	$s_1 = 13.0$	$s_2 = 10.7$
β-carotene (μg/l)	$\bar{x}_1 = 588$	$\bar{x}_2 = 553$
	$s_1 = 406$	$s_2 = 350$

Source: Malvy, J., et al. (1989), p. 29.

a Do the data provide sufficient evidence to indicate a difference between the mean retinol levels for boys versus girls? Use $\alpha = .01$.

b Do the data indicate a significant difference between the mean β-carotene levels for boys versus girls? Use $\alpha = .01$.

c Would the results of parts (a) and (b) change if you had used $\alpha = .05$? if $\alpha = .10$?

8.73 A biologist hypothesizes that high concentrations of actinomysin D inhibit RNA synthesis in cells and hence the production of proteins as well. An experiment conducted to test this theory compared the RNA synthesis in cells treated with two concentrations of actinomysin D, .6 and .7 microgram per milliliter, respectively. Cells treated with the lower concentration (.6) of actinomysin D showed that 55 out of 70 developed normally, whereas only 23 out of 70 appeared to develop normally for the higher concentration (.7). Do these data provide sufficient evidence to indicate a difference between the rates of normal RNA synthesis for cells exposed to the two different concentrations of actinomysin D?

a Find the observed significance level of the test.

b If you planned to conduct your test using $\alpha = .10$, what would be your test conclusions?

8.74 In Exercise 7.94, we described an experiment conducted by Paul Ward (1984) to investigate a theory concerning the molting of the male *Gammarus pulex*, a small crustacean. If a male has to molt while paired with a female, he must release her and so lose her. The theory is that the male *Gammarus pulex* is able to postpone the time to molt and thereby reduce the possibility of losing his mate. Ward randomly assigned 100 pairs of males and females to two groups of 50 each. Pairs in the first group were maintained together (Normal); those in the second group were separated (Split). The length of time to molt was recorded for both males and females, and the means, standard deviations, and sample sizes are shown in the table. (The numbers of crustaceans in each of the four samples are less than 50 because some did not survive until molting time.)

	Time to Molt (days)		
	Mean	s	n
Males			
Normal	24.8	7.1	34
Split	21.3	8.1	41
Females			
Normal	8.6	4.8	45
Split	11.6	5.6	48

a Do the data present sufficient evidence to indicate that the mean molt time for "Normal" males exceeds the mean time for those "Split" from their mates? State the null and alternative hypotheses that you would use for the test.

b Give the rejection region for the test using $\alpha = .05$.

c Conduct the test and state your results.

8.75 Refer to the time to molt data for the female *Gammarus pulex* crustaceans in Exercise 8.74. Do the data present sufficient evidence to indicate that the mean molt time for "Normal" females exceeds the mean time for those "Split" from their mates?

8.76 *Psychology Today* reports on a study by environmental psychologist Karen Frank and two colleagues, which was conducted to determine whether it is more difficult to make friends in a large city than in a small town. Two groups of graduate students were used for the study: 45 new arrivals at a private university in New York City and 47 at a prestigious university located in a town of 31,000 in an upstate rural area. The number of friends (more than just acquaintances) made by each student was recorded for each student at the end of 2 months and then again at the end of 7 months. The 7-month sample means are shown below:

	Location	
	New York City	Upstate Rural Small Town
Sample size	45	47
Sample mean (7 months)	5.1	5.3
Population mean (7 months)	μ_1	μ_2

Source: Reprinted from *Psychology Today*. Copyright ©1981 Ziff-Davis Publishing Company.

a If your theory is that it is more difficult to make friends in a city than in a small town, state the alternative hypothesis that you would use for a statistical test. State your null hypothesis.

b Suppose that $\sigma_1 = 2.2$ and $\sigma_2 = 2.3$ for the populations of "numbers of friends" made by new graduate students after 7 months at a new location. Do the data provide sufficient evidence to indicate that the mean number of friends acquired after 7 months differs for the two locations? Test using $\alpha = .05$.

Exercises Using the Data Disk

8.77 Refer to the *Fortune 500* sales data on the data disk. Draw a random sample of size $n = 30$ (you may use one of the samples you used in Exercise 7.98).

a Since the mean of the population from which you are sampling is, in fact, $\mu = 4592$, a test of hypothesis $H_0 : \mu = 4592$ should not be rejected. Conduct this test using a two-tailed alternative, with $\alpha = .01$. Do you arrive at the correct decision?

b A test of the hypothesis $H_0 : \mu = 3000$ should be rejected (since we know that $\mu = 4592$ for this population). Conduct this test using your sample information and a two-tailed alternative, with $\alpha = .01$. Do you arrive at the correct decision?

c Is it likely that you will be able to reject a test of the hypothesis $H_0 : \mu = 4000$ even though we know that this is not the true population mean? Explain.

8.78 In our descriptive analysis of the NIH blood pressure data, we noted that it appears that the distribution of systolic blood pressures for males appears to be shifted more than 8 pressure units above the corresponding distribution for females. Is this difference real, or are we just observing a difference that can be explained by the variation within the data? In other words, do the data present sufficient evidence to indicate a difference between mean systolic blood pressures for males versus females in the 15-to-20 age group? The sample sizes, means, and standard deviations for the two data sets are shown below. Test the null hypothesis that no difference exists between the means against the two-sided alternative hypothesis that the means differ. Use $\alpha = .01$. Interpret your results.

	Sample Size	Mean	Standard Deviation
Male	965	118.728	14.2343
Female	945	110.390	12.5753

CASE STUDY

An Aspirin a Day . . . ?

On Wednesday, January 27, 1988, the front page of the *New York Times* read "heart attack risk found to be cut by taking aspirin: Lifesaving effects seen." A very large study of U.S. physicians showed that a single aspirin tablet taken every other day reduced by one-half the risk of heart attack in men (Greenhouse and Greenhouse, 1988). Three days later, a headline in the *Times* read "Value of daily aspirin disputed in British study of heart attacks." How could two seemingly similar studies, both involving doctors as participants, result in such opposite conclusions?

The U.S. physicians' health study consisted of two randomized clinical trials in one. The first tested the hypothesis that 325 milligrams (mg) of aspirin taken every other day reduces mortality from cardiovascular disease. The second tested whether 50 mg of beta-carotene taken on alternate days decreases the incidence of cancer. Using an American Medical Association computer tape, 261,248 male physicians between the ages of 40 and 84 were invited to participate in the trial. Of those responding, 59,285 were willing to participate. After excluding those physicians who had a history of medical disorders, or who were currently taking aspirin or had negative reactions to aspirin, 22,071 physicians were randomized into one of four treatment groups: (1) buffered aspirin and beta-carotene; (2) buffered aspirin and a beta-carotene placebo; (3) aspirin placebo and beta-carotene; and (4) aspirin placebo and beta-carotene placebo. Thus, half were assigned to receive aspirin and half to receive beta-carotene.

The study was conducted as a double-blind study, in which neither the participants nor the investigators responsible for following the participants knew to which group a participant belonged. The results of the American study concerning myocardial infarctions (the technical name for heart attacks) are given in the following table:

American Study

	Aspirin ($n = 11{,}037$)	Placebo ($n = 11{,}034$)
Myocardial infarction		
Fatal	5	18
Nonfatal	99	171
Total	104	189

The objective of the British study was to determine whether 500 mg of aspirin taken daily would reduce the incidence of and mortality from cardiovascular disease. In 1978 all male physicians in the United Kingdom were invited to participate. After the usual exclusions, 5139 doctors were randomly allocated to take aspirin, unless some problem developed, and one-third were randomly allocated to *avoid* aspirin. Placebo tablets were not used, and so the study was not blind! The results of the British study are given in the following table:

British Study

	Aspirin ($n = 3429$)	**Control** ($n = 1710$)
Myocardial infarction		
Fatal	89 (47.3)	47 (49.6)
Nonfatal	80 (42.5)	41 (43.3)
Total	169 (89.8)	88 (92.9)

To account for unequal sample sizes, the British study reported rates per 10,000-subject-years-alive (given in parentheses).

1 Test to see whether the American study does in fact indicate that the rate of heart attacks for physicians taking 325 mg of aspirin every other day was significantly different from the rate for those on the placebo. Is the American claim justified?

2 Repeat the analysis using the data from the British study in which one group took 500 mg of aspirin every day and the control group took none. Based on their data, is the British claim justified?

3 Can you think of some possible explanations as to why the results of these two studies, which were alike in some respects, produced such different conclusions?

9

Inference from Small Samples

Case Study

Will the price of books continue to rise? If more books are on the market, will competition cause the price of books to drop? According to the information given in an article by Chandler Grannis (1992), the number of new books in 1991 increased over the number of new books in 1990. Does this imply that the price per volume of hardback books decreased? The case study at the end of this chapter examines this question using small-sample inferential techniques.

General Objective

The basic concepts of statistical estimation and tests of hypotheses were presented in Chapters 7 and 8, along with summaries of the methodologies for some large-sample estimation and test procedures. Large-sample estimation and test procedures for population means and proportions were used to illustrate concepts as well as to give you some useful tools for solving some practical problems. Because all these techniques rely on the Central Limit Theorem to justify the normality of the estimators and test statistics, they apply only when the sample sizes are large. Consequently, the objective of this chapter is to supplement the results of Chapters 7 and 8 by presenting small-sample statistical test and estimation procedures for population means and variances. These techniques differ from those of Chapters 7 and 8 because they require the relative frequency distributions of the sampled populations to be normal or approximately so.

Specific Topics

1. Student's t distribution (9.2)
2. Small-sample inferences concerning a population mean (9.3)
3. Small-sample inferences concerning the difference in two means (9.4)
4. Paired difference test (9.5)
5. Inference concerning a population variance (9.6)
6. Comparing two population variances (9.7)
7. Assumptions (9.8)

9.1 Introduction

Large-sample methods for making inferences concerning population means and the difference between two means were discussed with examples in Chapters 7 and 8. Cost, available time, and other factors frequently limit the size of the sample that can be acquired. In this case, large-sample procedures are inappropriate, and other test and estimation procedures must be used. In this chapter, we will study several small-sample inferential procedures that are closely related to the large-sample methods presented in Chapters 7 and 8. Specifically, we will consider methods for estimating and testing hypotheses about population means, the difference between two means, a population variance, and a comparison of two population variances. Small-sample tests and confidence intervals for binomial parameters will be omitted from our discussion.[†]

9.2 Student's t Distribution

We introduce our topic by considering the following problem: A very costly experiment has been conducted to evaluate a new process for producing synthetic diamonds. Six diamonds have been generated by the new process with recorded weights .46, .61, .52, .48, .57, and .54 karat.

A study of the process costs indicates that the average weight of the diamonds must be greater than .5 karat in order for the process to be operated at a profitable level. Do the six diamond-weight measurements present sufficient evidence to indicate that the average weight of the diamonds produced by the process is in excess of .5 karat? That is, we wish to test the null hypothesis that $\mu = .5$ against the alternative hypothesis that $\mu > .5$.

According to the Central Limit Theorem,

$$z = \frac{\bar{x} - \mu}{\sigma/\sqrt{n}}$$

has approximately a normal distribution in repeated sampling when n is large. For $\alpha = .05$, we could use a one-tailed statistical test and reject H_0 when $z > 1.645$. This procedure assumes that σ is known or that a good estimate s is available and is based on a reasonably large sample (we have suggested $n \geq 30$). Unfortunately, this latter requirement will not be satisfied for the $n = 6$ diamond-weight measurements. How, then, may we test the hypothesis that $\mu = .5$ against the alternative that $\mu > .5$ when we have a small sample?

The problem that we pose is not new; it is one that received serious attention from statisticians and experimenters at the turn of the century. If a sample standard deviation s were substituted for σ in z, would the resulting quantity have approximately a standardized normal distribution in repeated sampling? More specifically, is the rejection region $z > 1.645$ appropriate? That is, do approximately 5% of the values of the test statistic, computed in repeated sampling, exceed 1.645 when H_0 is true?

The answers to these questions, not unlike many of the problems encountered in the sciences, may be resolved by experimentation. That is, we could draw a small sample—say, $n = 6$ measurements—and compute the value of the test statistic. Then we would repeat this process over and over again a very large number of times and construct a frequency distribution for the computed values of the test statistic. The

[†]A small-sample test for the binomial parameter p will be presented in Chapter 14.

general shape of the distribution and the location of the rejection region would then be evident.

The sampling distribution of the test statistic

$$t = \frac{\bar{x} - \mu}{s/\sqrt{n}}$$

for samples drawn from a normally distributed population was discovered by W. S. Gosset and published in 1908 under the pen name of "Student." He referred to the quantity under study as t, and ever since it has been known as **Student's *t***. We omit the complicated mathematical expression for the density function for t, but we will describe some of its characteristics.

In repeated sampling, the distribution of the test statistic

$$t = \frac{\bar{x} - \mu}{s/\sqrt{n}}$$

is, like z, mound-shaped and perfectly symmetrical about $t = 0$. Unlike z, it is much more variable, tailing rapidly out to the right and left, a phenomenon that can be readily explained. The variability of z in repeated sampling is due solely to $\bar{x}$; the other quantities appearing in z (n and σ) are nonrandom. On the other hand, the variability of t is contributed by *two* random quantities, $\bar{x}$ and s, that can be shown to be independent of each other. Thus, when $\bar{x}$ is very large, s may be very small, and vice versa. As a result, t will be more variable than z in repeated sampling (Figure 9.1). Finally, as we might surmise, the variability of t decreases as n increases because the estimate s of σ is based on more and more information. When n is infinitely large, the t and z distributions are identical. Thus, Gosset discovered that the distribution of t depended on the sample size n.

FIGURE 9.1 Standard normal z and the t distribution based on $n = 6$ measurements (5 d.f.)

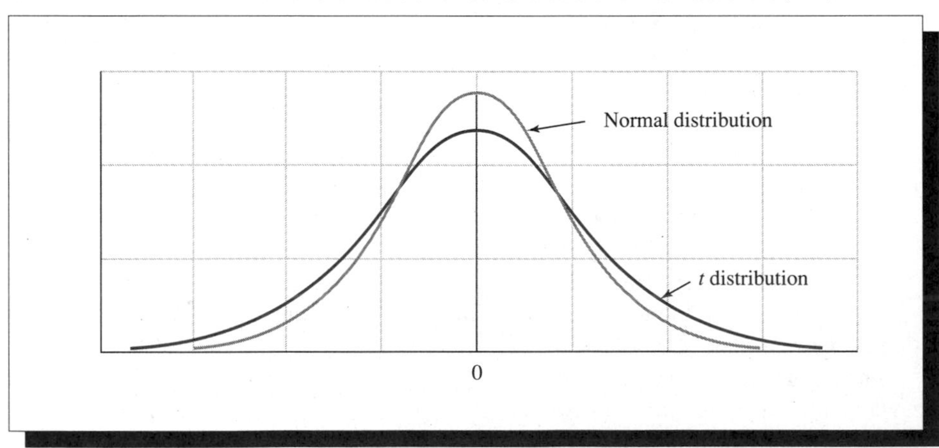

The divisor $(n - 1)$ of the sum of squares of deviations that appears in the formula for s^2 is called the **number of degrees of freedom associated with s^2**. The origin of the term *degrees of freedom* is linked to the statistical theory underlying the probability distribution of s^2 and refers to the number of independent squared deviations available for estimating σ^2. We will not pursue this point further except to say that the test statistic t, based on a sample of n measurements, possesses $(n - 1)$ degrees of freedom (d.f.).

The critical values of t that separate the rejection and acceptance regions for the statistical test are presented in Table 4 in Appendix I. Table 4 is partially reproduced in Table 9.1. The tabulated value t_a records the value of t such that an area a lies to its right, as shown in Figure 9.2. The degrees of freedom associated with t are shown in the first and last columns of the table, and the values of t_a corresponding to various values of a appear in the top row. Thus, if we wish to find the value of t, such that 5% of the area lies to its right, we would use the column marked $t_{.05}$. The critical value of t for our example, found in the $t_{.05}$ column opposite the row corresponding to d.f. $= (n - 1) = (6 - 1) = 5$, is $t = 2.015$ (shaded in Table 9.1). Thus, we would reject $H_0 : \mu = .5$ in favor of $H_a : \mu > .5$ when $t > 2.015$. Since the distribution of t is symmetric about $t = 0$, a left-tailed critical value of t is simply the negative of the corresponding right-tailed value. For example, with 5 d.f., the area to the left of $t = -2.015$ is equal to .05, the area to the right of $t = 2.015$.

TABLE **9.1**
Format of the Student's t table from Table 4 in Appendix I

d.f.	$t_{.100}$	$t_{.050}$	$t_{.025}$	$t_{.010}$	$t_{.005}$	d.f.
1	3.078	6.314	12.706	31.821	63.657	1
2	1.886	2.920	4.303	6.965	9.925	2
3	1.638	2.353	3.182	4.541	5.841	3
4	1.533	2.132	2.776	3.747	4.604	4
5	1.476	2.015	2.571	3.365	4.032	5
6	1.440	1.943	2.447	3.143	3.707	6
7	1.415	1.895	2.365	2.998	3.499	7
8	1.397	1.860	2.306	2.896	3.355	8
9	1.383	1.833	2.262	2.821	3.250	9
⋮	⋮	⋮	⋮	⋮	⋮	⋮
26	1.315	1.706	2.056	2.479	2.779	26
27	1.314	1.703	2.052	2.473	2.771	27
28	1.313	1.701	2.048	2.467	2.763	28
29	1.311	1.699	2.045	2.462	2.756	29
∞	1.282	1.645	1.960	2.326	2.576	∞

Note that the critical value of t is always larger than the corresponding critical value of z for a specified α. For example, when $a = .05$, the critical value of t for $n = 2$ (d.f. $= 1$) is $t = 6.314$, which is very large when compared with the corresponding $z_{.05} = 1.645$. Proceeding down the $t_{.05}$ column, we note that the critical value of t decreases, reflecting the effect of a larger sample size (more degrees of freedom) on the estimation of σ. Finally, when n is infinitely large, the critical value of t equals $z_{.05} = 1.645$.

The reason for choosing $n = 30$ (an arbitrary choice) as the dividing line between large and small samples is apparent. For $n = 30$ (d.f. $= 29$), the critical value of $t_{.05} = 1.699$ is numerically quite close to $z_{.05} = 1.645$. For a two-tailed test based

FIGURE 9.2 Tabulated values of Student's t

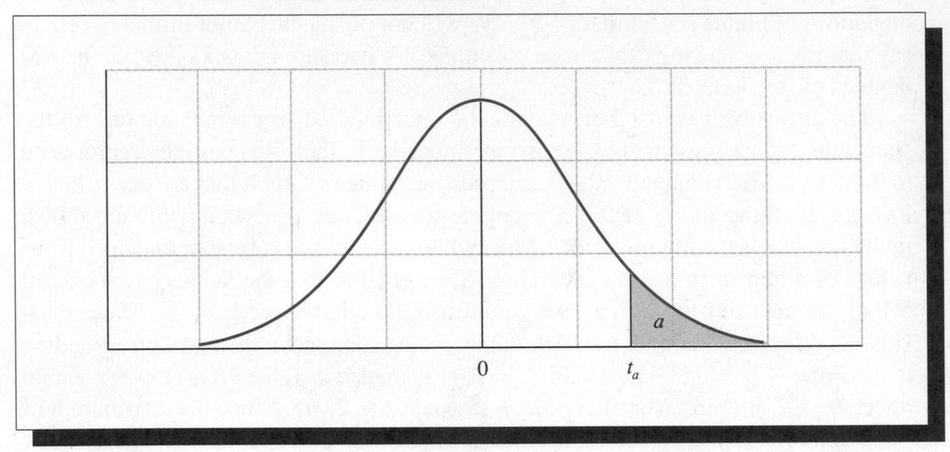

on $n = 30$ measurements and $\alpha = .05$, we would place .025 in each tail of the t distribution and reject $H_0 : \mu = \mu_0$ when $t > 2.045$ or $t < -2.045$. Note that this is very close to $z_{.025} = 1.96$ used in the z test.

Note that the Student's t and corresponding tabulated critical values are based on the assumption that the sampled population has a normal probability distribution. This indeed is a very restrictive assumption because, in many sampling situations, the properties of the population are completely unknown and may be nonnormal. If this restriction were to seriously affect the distribution of the t statistic, the application of the t test would be very limited. Fortunately, this point is of little consequence because **it can be shown that the distribution of the t statistic has nearly the same shape as the theoretical t distribution for populations that are nonnormal but have a mound-shaped probability distribution.** This property of the t statistic and the common occurrence of mound-shaped distributions of data in nature enhance the value of Student's t for use in statistical inference.

Having discussed the origin of Student's t and the tabulated critical values from Table 4 in Appendix I, we now return to the problem of making an inference about the mean diamond weight based on our sample of $n = 6$ measurements. Prior to considering the solution, you may wish to test your built-in inference-making equipment by glancing at the six measurements and arriving at a conclusion concerning the significance of the data.

9.3 Small-Sample Inferences Concerning a Population Mean

The statistical test of hypothesis concerning a population mean may be stated as follows:

1. Null Hypothesis: $H_0 : \mu = \mu_0$
2. Alternative Hypothesis:

One-Tailed Test	*Two-Tailed Test*
$H_a : \mu > \mu_0$ (or, $H_a : \mu < \mu_0$)	$H_a : \mu \neq \mu_0$

3. Test Statistic: $t = \dfrac{\bar{x} - \mu_0}{s/\sqrt{n}}$
4. Rejection Region:

One-Tailed Test	*Two-Tailed Test*
$t > t_\alpha$ (or, $t < -t_\alpha$ when the alternative hypothesis is $H_a : \mu < \mu_0$)	$t > t_{\alpha/2}$ or $t < -t_{\alpha/2}$

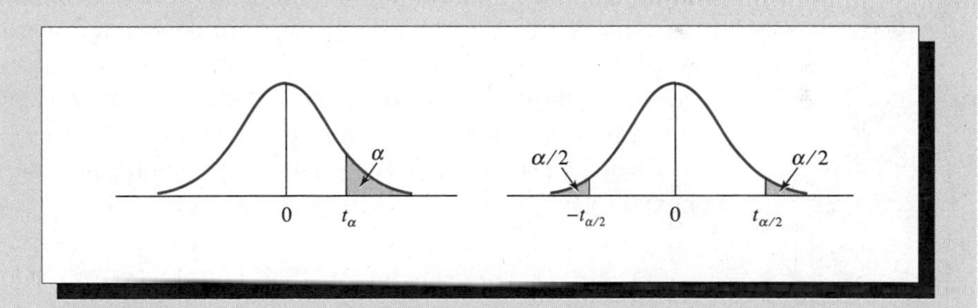

The critical values of t, t_α, and $t_{\alpha/2}$ are based on $(n - 1)$ d.f. These tabulated values can be found in Table 4 of Appendix I.

Assumption: The sample has been randomly selected from a normally distributed population.

To apply this test to the diamond-weight data, we must first calculate the sample mean $\bar{x}$ and standard deviation s. You can verify that the mean and standard deviation for the six diamond weights are .53 and .0559, respectively.

We wish to test the null hypothesis that the mean diamond weight is .5 against the alternative hypothesis that it is greater than .5. The elements of the test as defined above are

$$H_0 : \mu = .5$$
$$H_a : \mu > .5$$
$$\text{Test statistic: } t = \frac{\bar{x} - \mu_0}{s/\sqrt{n}} = \frac{.53 - .5}{.0559/\sqrt{6}} = 1.32$$

Rejection region: The rejection region for $\alpha = .05$ and $(n - 1) = (6 - 1) = 5$ d.f. is $t > 2.015$.

Since the calculated value of the test statistic does not fall in the rejection region, we cannot reject H_0. Thus, the data do not present sufficient evidence to indicate that the mean diamond weight exceeds .5 karat.

The calculation of the probability β of a type II error for the t test is very difficult and is beyond the scope of this text. Therefore, we will avoid this problem and obtain an interval estimate for μ. The large-sample confidence interval for μ, given in Chapter 7, is

$$\bar{x} \pm z_{\alpha/2} \frac{\sigma}{\sqrt{n}}$$

where $z_{\alpha/2} = 1.96$ for a confidence coefficient equal to .95. This result assumes that σ is known and simply involves a measurement of $1.96\sigma_{\bar{x}}$ (or approximately $2\sigma_{\bar{x}}$) on either side of $\bar{x}$ in conformity with the Empirical Rule. When σ is unknown and must be estimated by a small-sample standard deviation s, the large-sample confidence interval will not enclose μ 95% of the time in repeated sampling. To account for the added variability introduced through s, the estimate of σ, the appropriate confidence interval for μ is obtained by replacing the critical value $z_{\alpha/2}$ with the critical value $t_{\alpha/2}$ based on $(n - 1)$ d.f. The small-sample confidence interval for μ is given in the next display.

Small-Sample $(1 - \alpha)$ 100% Confidence Interval for μ

$$\bar{x} \pm t_{\alpha/2} \frac{s}{\sqrt{n}}$$

where $s/\sqrt{n}$ is the estimated standard error of $\bar{x}$ but is often referred to as the **standard error of the mean.**

Assumption: The sampled population is approximately normally distributed.

For our example, a 95% confidence interval for μ is

$$\bar{x} \pm t_{\alpha/2} \frac{s}{\sqrt{n}} \quad \text{or} \quad .53 \pm 2.571 \frac{.0559}{\sqrt{6}} \quad \text{or} \quad .53 \pm .059$$

Therefore, the interval estimate for μ is .471 to .589, with confidence coefficient of .95. If the experimenter wishes to detect a small increase in mean diamond weight in excess of .5 carat, the width of the interval must be reduced by obtaining more diamond-weight measurements. Additional measurements will decrease both $1/\sqrt{n}$ and $t_{\alpha/2}$ and thereby decrease the width of the interval. From the standpoint of a statistical test of hypothesis, an increase in n increases the available information on which to base a decision and decreases the probability of making a type II error.

EXAMPLE 9.1 Labels on gallon cans of paint usually indicate the drying time and the size of the area that can be covered in one coat. Most brands of paint indicate that, in one coat, 1 gallon will cover between 250 and 500 square feet depending upon the texture of the surface to be painted. One particular brand of paint claimed that 1 gallon of its paint would cover 400 square feet. A random sample of ten 1-gallon cans of white paint of this brand were used to paint ten identical areas using the same kind of equipment. The actual areas covered by these 10 gallons of paint are given in Table 9.2. Do the data present sufficient evidence to indicate that the average coverage differs from 400 square feet?

TABLE 9.2 Data for Example 9.1

Area Covered (Square Feet)	
310	311
412	368
447	376
303	410
365	350

Solution Testing the null hypothesis that $\mu = 400$ square feet against the alternative that μ is either greater than or less than 400 square feet results in a two-tailed statistical test. Thus,

$$H_0 : \mu = 400 \quad \text{and} \quad H_a : \mu \neq 400$$

Using $\alpha = .05$ and placing .025 in each tail of the t distribution, we find that the critical value of t for $n = 10$ measurements, or $(n - 1) = 9$ d.f. is $t_{.025} = 2.262$. Hence, we will reject H_0 if $t > 2.262$ or $t < -2.262$. (Remember, the distribution of t is symmetric about $t = 0$.)

The sample mean and standard deviation for the recorded data are

$$\bar{x} = 365.2 \quad \text{and} \quad s = 48.42.$$

Then,

$$t = \frac{\bar{x} - \mu}{s/\sqrt{n}} = \frac{365.2 - 400}{48.417/\sqrt{10}} = -2.27$$

Since the observed value of t falls in the rejection region, we reject H_0 and conclude that the average area covered by 1 gallon of paint is less than 400 square feet. We can be reasonably confident that we have made the correct decision. Using our procedure, we should falsely reject H_0 only $\alpha = .05$ of the time in repeated applications of the statistical test. ▪

EXAMPLE 9.2 Find a 95% confidence interval for the mean in Example 9.1.

Solution Substituting $\bar{x} = 365.2$, $t_{.025} = 2.262$, $s = 48.417$, and $n = 10$ in the formula

$$\bar{x} \pm t_{\alpha/2} \frac{s}{\sqrt{n}}$$

we obtain

$$365.2 \pm (2.262) \frac{48.417}{\sqrt{10}} \quad \text{or} \quad 365.2 \pm 34.63$$

Thus, we estimate that the average area covered by 1 gallon of this brand of paint lies in the interval 330.6 to 399.8. A more accurate interval estimate (a shorter interval) can generally be obtained by increasing the sample size. Notice that the upper limit of this interval is very close to the value of 400 square feet, the coverage claimed on the label. This coincides with the fact that the observed value of $t = -2.27$ was just slightly smaller than the left-tail critical value of $t_{.025} = -2.262$, and hence we were able to reject the null hypothesis that the mean was 400. ▪

EXAMPLE 9.3 If you planned to report the results of the statistical test in Example 9.1, what p-value would you report?

Solution The p-value for this test is the probability of observing a value of the t statistic as contradictory to the null hypothesis as the one observed for this set of data, namely, $t = -2.27$. Since this is a two-tailed test, the p-value is the probability that either $t \leq -2.27$ or $t \geq 2.27$.

Unlike the table of areas under the normal curve (Table 3 of Appendix I), the table for t (Table 4 of Appendix I) does not give the areas corresponding to various values of t. Rather, it gives the values of t corresponding to upper-tail areas equal to .10, .05, .025, .010, and .005. Consequently, we can only approximate the upper-tail area that corresponds to the probability that $t > 2.27$. Since the t statistic for this test is based on 9 d.f., we refer to the row corresponding to d.f. $= 9$ in Table 4 and find that 2.27 falls between the $t_{.025} = 2.262$ and $t_{.010} = 2.821$. Therefore, the right-tail area corresponding to the probability that $t > 2.27$ lies between .025 and .010. Since this value represents only half of the required area, the actual p-value lies between $2(.025) = .05$ and $2(.010) = .020$. Using the tabulated values of t, we could reject H_0 with $\alpha = .05$, but not with $\alpha = .02$. Therefore, the p-value for this test would be reported as p-value $= .05$. Since most researchers would be using a t table similar or identical to Table 4, the reader of a published research report would realize that for a reported p-value of .05, the exact p-value for the test was probably less than .05 and was between .05 and .02. In fact, some researchers report these results as $.02 < p$-value $< .05$, indicating significance at the .05 level but not at the .02 level. ▪

The MINITAB package contains commands that can be used to implement a small-sample test of a population mean or to produce a small-sample confidence interval for a population mean. The command TTEST requires the user to provide the value of the population mean to be tested, together with the column number in which the data are stored. The subcommand ALTERNATIVE, followed by a 1, -1, or 0, will implement a right-tailed, left-tailed, or two-tailed test procedure. If this subcommand is not used, a two-tailed test is implemented by default.

The results of using TTEST to implement the test in Example 9.1 are given in Table 9.3. Besides the observed value of $t = -2.27$, the output gives the sample mean $\bar{x} = 365.2$, the sample standard deviation $s = 48.417$, and the standard error of the mean (SE MEAN), $s/\sqrt{n} = 15.311$. When compared with our results in Example 9.1, the only noticeable difference is in the reported decimal accuracy of the results. In addition, the exact p-value for the test appears in the printout and is given as P VALUE $= .049$. In using Table 4 to approximate this value, we found that the p-value was between .05 and .02, consistent with the MINITAB result.

TABLE **9.3**
MINITAB printout for the data in Table 9.2

```
MTB > TTEST 400 C1

TEST OF MU = 400.000 VS MU N.E. 400.000

            N      MEAN     STDEV   SE MEAN         T     P VALUE
C1         10   365.200    48.417    15.311     -2.27       0.049

MTB > TINTERVAL C1

            N      MEAN     STDEV   SE MEAN    95.0 PERCENT C.I.
C1         10     365.2      48.4      15.3     (330.6,   399.8)

MTB >
```

The MINITAB command TINTERVAL is available for constructing a confidence interval for a population mean. The command requires that the user supply the confidence coefficient and the column location of the data. A 95% confidence interval for μ using the data in Example 9.1 also appears in Table 9.3. Apart from the reported decimal accuracy, the results agree with those found in Example 9.2.

Exercises

Basic Techniques

9.1 Find the following:

a $t_{.05}$ for 5 d.f. **b** $t_{.025}$ for 8 d.f.

c $t_{.10}$ for 18 d.f. **d** $t_{.025}$ for 30 d.f.

9.2 Find t_α given that $P(t > t_\alpha) = \alpha$ for

a $\alpha = .10$, 12 d.f. **b** $\alpha = .01$, 25 d.f.

c $\alpha = .05$, 16 d.f. **d** $\alpha = .025$, 7 d.f.

9.3 A random sample of $n = 12$ observations from a normal population produced $\bar{x} = 47.1$ and $s^2 = 4.7$.

a Test the hypothesis $H_0: \mu = 48$ against $H_a: \mu \neq 48$. Use $\alpha = .10$.

b What is the observed level of significance for the test in part (a)?

c Find a 90% confidence interval for the population mean. Interpret this interval.

9.4 The following $n = 10$ observations resulted when sampling from a normal population:

7.4 7.1 6.5 7.5 7.6 6.3 6.9 7.7 6.5 7.0

a Find the mean and the standard deviation of these data.

b Find a 99% confidence interval for the population mean μ.

c Test $H_0: \mu = 7.5$ versus $H_a: \mu < 7.5$. Use $\alpha = .01$.

d Do the results of part (b) support your conclusion in part (c)?

Applications

9.5 One family of fish native to tropical Africa can breathe both in air and in water. Thus, the fish can live in water with a relatively low-oxygen content by taking a portion of their oxygen from the air. Michael Pettit and Thomas Beitinger (1985) conducted a study to investigate the effects of some variables on oxygen partitioning by reedfish— that is, the proportions of oxygen taken by reedfish from the surrounding air and from the water in which they lived. The data shown below represent the oxygen uptake readings from air and water for 11 reedfish in water at 25°C. The first column of the table gives the weights of the fish. The second and third columns give the averages of repeated measurements of oxygen uptake from air and water, respectively, for each fish. The fourth column gives the percentage of total oxygen uptake attributable to air.

Weight (grams)	Oxygen Uptake ($\text{ml } O_2 g^{-1} h^{-1}$)		Percentage†	
	Air	Water	Air	Water
22.1	0.043	0.045	49	51
21.1	0.034	0.066	34	66
13.4	0.026	0.084	24	76
19.5	0.048	0.045	52	48
21.9	0.009	0.056	14	86
19.5	0.028	0.072	28	72
16.5	0.032	0.081	28	72
18.5	0.045	0.051	47	53
18.0	0.061	0.041	60	40
15.4	0.028	0.060	32	68
13.4	0.024	0.113	18	82

†Percentages shown were calculated by the author based on Pettit and Beitinger's data.

Assume that the reedfish included in this experiment represent a random sample of mature reedfish.

a Find a 90% confidence interval for the mean weight of mature reedfish. Interpret the interval.

b Find a 90% confidence interval for the mean oxygen uptake from the air for reedfish living in an environment similar to that used in this experiment. Interpret your interval.

c Find a 90% confidence interval for the mean percentage of oxygen uptake from air in reedfish living in an environment similar to that employed in this experiment. Interpret your interval.

9.6 Scholastic Aptitude Test (SAT) scores, falling slowly since the inception of the tests, have now begun to rise. Originally, a score of 500 was intended to be "average." The mean scores for 1991 are approximately 422 for the verbal test and 474 for the mathematics test. A sampling of the test scores of 20 seniors at an urban high school produced the following mean scores and standard deviations:

	Verbal	Mathematics
Sample mean	419	455
Sample standard deviation	57	69

a Do the data provide sufficient evidence to indicate that the mean verbal SAT score for the urban high school seniors is less than the 1991 national mean of 422? Test using $\alpha = .05$.

b Find the approximate observed significance level for the test in part (a) and interpret its value.

c Do the data provide sufficient evidence to indicate that the mean mathematics SAT score for the urban high school seniors is less than the 1991 national mean of 474? Test using $\alpha = .05$.

d Find the observed significance level for the test in part (c) and interpret its value.

9.7 Industrial wastes and sewage dumped into our rivers and streams absorb oxygen and thereby reduce the amount of dissolved oxygen available for fish and other forms of aquatic life. One state agency requires a minimum of 5 parts per million (ppm) of dissolved oxygen in order for the oxygen content to be sufficient to support aquatic life. Six water specimens taken from a river at a specific location during the low-water season (July) gave readings of 4.9, 5.1, 4.9, 5.0, 5.0, and 4.7 ppm of dissolved oxygen. Do the data provide sufficient evidence to indicate that the dissolved oxygen content is less than 5 ppm? Test using $\alpha = .05$.

9.8 In Exercise 0.7, we presented some data from a study of the infestation of the *T. orientalis* lobster by two types of barnacles, *O. tridens* and *O. lowei*. The carapace lengths of the ten lobsters are reproduced below. Find a 95% confidence interval for the mean carapace length of *T. orientalis* lobsters caught in the seas in the vicinity of Singapore.

Lobster Field Number	A061	A062	A066	A070	A067	A069	A064	A068	A065	A063
Carapace Length (mm)	78	66	65	63	60	60	58	56	52	50

Source: Jeffries, Voris, and Yang (1984).

9.9 It is recognized that cigarette smoking has a deleterious effect on lung function. In their study of the effect of cigarette smoking on the carbon monoxide diffusing capacity (DL) of the lung, Ronald Knudson, Walter Kaltenborn, and Benjamin Burrows found that current smokers had DL readings significantly lower than either exsmokers or nonsmokers. The carbon monoxide diffusing capacities for a random sample of $n = 20$ current smokers follows:

103.768	88.602	73.003	123.086
91.052	92.295	61.675	90.677
84.023	76.014	100.615	88.017
71.210	82.115	89.222	102.754
108.579	73.154	106.755	90.479

Do these data indicate that the mean DL reading for current smokers is significantly lower than 100 DL, the average for nonsmokers? Use $\alpha = .01$.

9.10 Refer to Exercise 9.9 and find a 95% confidence interval estimate for the mean DL reading for current smokers.

9.11 Organic chemists often purify organic compounds by a method known as fractional crystallization. An experimenter wanted to prepare and purify 4.85 grams (g) of aniline. Ten 4.85-g quantities of aniline were individually prepared and purified to acetanilide. The following dry yields were recorded:

3.85	3.80
3.88	3.85
3.90	3.36
3.62	4.01
3.72	3.82

Estimate the mean number of grams of acetanilide that could be recovered from an initial amount of 4.85 g of aniline. Use a 95% confidence interval.

9.12 Refer to Exercise 9.11. Approximately how many 4.85-g specimens of aniline would be required if you wished to estimate the mean number of grams of acetanilide correct to within .06 g with probability equal to .95?

9.13 Researchers are calling for closer scrutiny of milk after reports show that vitamin D levels did not meet or far exceeded federal standards (Haney, 1992). Vitamin D deficiency causes rickets in the young; among the elderly, it leads to weak bones and fractures. In a test of 42 containers of 13 different brands of milk, 62% fell at least 20% below the 400 units required by the U.S. Food and Drug Administration, and in 10% it was at least 20% above 400 units. Suppose, in an independent study in which 20 containers were tested, the average and standard deviation of the number of units of vitamin D were $\bar{x} = 365$ and $s = 46$.

a Use the data from the independent study to determine if there is sufficient evidence to conclude that the mean of the population sampled is not 400 units of vitamin D.

b Find a 95% confidence interval estimate of μ, the true value of the vitamin D content for the population sampled.

9.14 Operators of gasoline-fueled vehicles complain about the price of gasoline at the pump. Contrary to what you might think, the federal tax per gallon is constant (14.45 cents as of January 1, 1993 ["Pump Prices," 1993]), whereas state and local taxes vary from .50 cent to 25.45 cents for $n = 18$ metropolitan areas. The following table lists the amount of taxes (in cents) at the pump at $n = 18$ "key" locations around the country:

42.89	40.45	35.09
53.91	39.65	35.04
48.55	38.65	34.95
47.90	37.95	33.45
47.73	36.80	28.99
46.61	35.95	27.45

a Suppose that we consider these $n = 18$ key locations around the country as a random sample. Use a test of hypothesis to determine whether the average tax at the pump is significantly less than .45 cents.

b Construct a 95% confidence interval for the average tax at the gas pump in the United States.

9.15 Although there are many treatments for *bulimia nervosa*, some subjects fail to benefit from treatment. In a study to determine which factors predict who will benefit from treatment, Wendy Baell and E. H. Wertheim (1992) found that self-esteem was one of these important predictors. The following table gives the mean and standard deviation of self-esteem scores prior to treatment, posttreatment, and during a follow-up:

	Pretreatment	Posttreatment	Follow-Up
Sample mean $\bar{x}$	20.3	26.6	27.7
Standard deviation s	5.0	7.4	8.2
Sample size n	21	21	20

a Use a test of hypothesis to determine if there is sufficient evidence to conclude that the true pretreatment mean is less than 25.

b Find a 95% confidence interval estimate for the true posttreatment mean.

c In Section 9.4, we will introduce small-sample techniques for making inferences about the difference between two population means. Without the formality of a statistical test, what would you be willing to conclude about the differences among the three sampled population means represented by the results in the accompanying table?

9.4 Small-Sample Inferences Concerning the Difference Between Two Means

The physical setting for the problem we consider here is identical to that discussed in Section 7.6. Independent random samples of n_1 and n_2 measurements, respectively, are drawn from two populations, which possess means and variances μ_1, σ_1^2 and μ_2, σ_2^2. Our objective is to make inferences concerning the difference $(\mu_1 - \mu_2)$ between the two population means.

The following small-sample methods for testing hypotheses and placing a confidence interval on the difference between two means are, like the case for a single mean, founded on assumptions regarding the probability distributions of the sampled populations.

Assumptions for Small-Sample Inferences Concerning $\mu_1 - \mu_2$

1. Random and independent samples are drawn from each of two normal populations.
2. The variability of the measurements in the two populations is the same and can be measured by a common variance σ^2. That is, $\sigma_1^2 = \sigma_2^2 = \sigma^2$.

Although these assumptions may seem to be restrictive, they are reasonable for many sampling situations.

The point estimator of $(\mu_1 - \mu_2)$ is $(\bar{x}_1 - \bar{x}_2)$, the difference between the sample means. In Section 7.6, we found that in repeated sampling $(\bar{x}_1 - \bar{x}_2)$ was an unbiased estimator of $(\mu_1 - \mu_2)$, with a standard deviation

$$\sigma_{(\bar{x}_1 - \bar{x}_2)} = \sqrt{\frac{\sigma_1^2}{n_1} + \frac{\sigma_2^2}{n_2}}$$

This result was used in calculating the margin of error for estimation, in constructing a large-sample confidence interval, and in testing an hypothesis concerning the difference between two population means. The large-sample test of the hypothesis $H_0 : \mu_1 - \mu_2 = D_0$, where D_0 is the hypothesized difference between the means, was based on the z statistic given by

$$z = \frac{(\bar{x}_1 - \bar{x}_2) - D_0}{\sqrt{\frac{\sigma_1^2}{n_1} + \frac{\sigma_2^2}{n_2}}}$$

When we sample two normal populations with equal variances, and $\sigma_1^2 = \sigma_2^2 = \sigma^2$, then the z statistic can be simplified as follows:

$$z = \frac{(\bar{x}_1 - \bar{x}_2) - D_0}{\sqrt{\frac{\sigma_1^2}{n_1} + \frac{\sigma_2^2}{n_2}}} = \frac{(\bar{x}_1 - \bar{x}_2) - D_0}{\sigma\sqrt{\frac{1}{n_1} + \frac{1}{n_2}}}$$

For small-sample tests of the hypothesis $H_0 : \mu_1 - \mu_2 = D_0$, it seems reasonable to use the test statistic given next.

Small-Sample Test Statistic t for the Difference Between Two Means

$$t = \frac{(\bar{x}_1 - \bar{x}_2) - D_0}{s\sqrt{\frac{1}{n_1} + \frac{1}{n_2}}}$$

That is, we use a sample standard deviation s as an estimator of σ. This test statistic possesses a Student's t distribution in repeated sampling when the stated assumptions are satisfied, a fact that can be proved mathematically or verified by experimental sampling from two normal populations.

The estimate s used in the t statistic could be either s_1 or s_2, the standard deviation for each of the two samples. However, the use of either one alone would be wasteful, since both sample standard deviations are independent estimators of σ. The information from both sample variances can be combined by using a weighted average in which the weights reflect the relative amount of information in s_1^2 and s_2^2. Since s_1^2

is based on $(n_1 - 1)$ d.f. and s_2^2 is based on $(n_2 - 1)$ d.f., a **pooled estimator of σ^2** using the degrees of freedom as weights is

$$s^2 = \frac{(n_1 - 1)s_1^2 + (n_2 - 1)s_2^2}{(n_1 - 1) + (n_2 - 1)}$$

with $(n_1 - 1) + (n_2 - 1) = n_1 + n_2 - 2$ d.f.

The pooled estimator s^2 can also be expressed as a sum of the squared deviations in each sample. If x_{1i} and x_{2i} represent the ith observations in samples 1 and 2, then

$$s_1^2 = \frac{\sum_{i=1}^{n_1}(x_{1i} - \bar{x}_1)^2}{(n_1 - 1)} \quad \text{and} \quad s_2^2 = \frac{\sum_{i=1}^{n_2}(x_{2i} - \bar{x}_2)^2}{(n_2 - 1)}$$

Using these expressions, we can write the pooled estimator of σ^2 in either of the following two ways.

Pooled Estimator of σ^2

$$s^2 = \frac{\sum_{i=1}^{n_1}(x_{1i} - \bar{x}_1)^2 + \sum_{i=1}^{n_2}(x_{2i} - \bar{x}_2)^2}{(n_1 - 1) + (n_2 - 1)}$$

$$s^2 = \frac{(n_1 - 1)s_1^2 + (n_2 - 1)s_2^2}{(n_1 - 1) + (n_2 - 1)}$$

The first form for calculating s^2 shows that the numerator is the pooled sum of squared deviations from each sample, and the denominator is the pooled sum of the degrees of freedom from each sample. Regardless of the form used for calculation, it can be proved that s^2 is an unbiased estimator of σ^2 and hence is an unbiased estimator of the common population variance. If the pooled estimator s^2 is used to estimate σ^2 and the samples are randomly and independently drawn from normal populations with a common variance, then the statistic

$$t = \frac{(\bar{x}_1 - \bar{x}_2) - (\mu_1 - \mu_2)}{s\sqrt{\frac{1}{n_1} + \frac{1}{n_2}}}$$

will have a Student's t distribution with $n_1 + n_2 - 2$ d.f.

The small-sample test for the difference between two means is given in the next display.

Test of an Hypothesis Concerning the Difference Between Two Means

1. Null Hypothesis: $H_0 : (\mu_1 - \mu_2) = D_0$, where D_0 is some specified difference that you wish to test. For many tests, you will wish to hypothesize that there is no difference between μ_1 and μ_2; that is, $D_0 = 0$.
2. Alternative Hypothesis:

   One-Tailed Test
   $H_a : (\mu_1 - \mu_2) > D_0$
   (or, $H_a : (\mu_1 - \mu_2) < D_0$)

   Two-Tailed Test
   $H_a : (\mu_1 - \mu_2) \neq D_0$
3. Test Statistic: $t = \dfrac{(\bar{x}_1 - \bar{x}_2) - D_0}{s\sqrt{\dfrac{1}{n_1} + \dfrac{1}{n_2}}}$ where

$$s^2 = \frac{\sum_{i=1}^{n_1}(x_{1i} - \bar{x}_1)^2 + \sum_{i=1}^{n_2}(x_{2i} - \bar{x}_2)^2}{n_1 + n_2 - 2}$$

4. Rejection Region:

   One-Tailed Test
   $t > t_\alpha$
   (or, $t < -t_\alpha$ when the alternative hypothesis is $H_a : (\mu_1 - \mu_2) < D_0$)

   Two-Tailed Test
   $t > t_{\alpha/2}$ or $t < -t_{\alpha/2}$

The critical values of t, t_α, and $t_{\alpha/2}$ will be based on $(n_1 + n_2 - 2)$ d.f. The tabulated values can be found in Table 4 in Appendix I.

Assumptions: The samples were randomly and independently selected from normally distributed populations. The variances of the populations σ_1^2 and σ_2^2 are equal.

EXAMPLE **9.4** An assembly operation in a manufacturing plant requires approximately a 1-month training period for a new employee to reach maximum efficiency. A new method of training was suggested and a test was conducted to compare the new method with the standard procedure. Two groups of nine new employees were trained for a period of 3 weeks, one group using the new method and the other following standard training procedure. The length of time in minutes required for each employee to assemble the device was recorded at the end of the 3-week period. These measurements appear in Table 9.4. Do the data present sufficient evidence to indicate that the mean time to assemble at the end of a 3-week training period is less for the new training procedure?

TABLE 9.4

Standard Procedure	New Procedure
32	35
37	31
35	29
28	25
41	34
44	40
35	27
31	32
34	31

Solution Let μ_1 and μ_2 be the mean time to assemble for the standard and the new assembly procedures, respectively. Then, since we seek evidence to support the theory that $\mu_1 > \mu_2$, we will test the null hypothesis $H_0 : \mu_1 = \mu_2$ (i.e., $\mu_1 - \mu_2 = 0$) against the alternative hypothesis $H_a : \mu > \mu_2$ (i.e., $\mu_1 - \mu_2 > 0$). To conduct this test, assume that the population distributions of measurements are normal, and that the variability for the two populations of measurements is the same.

The sample means and sums of squared deviations are

$$\bar{x}_1 = 35.22 \quad \text{and} \quad \sum_{i=1}^{9}(x_{1i} - \bar{x}_1)^2 = 195.5556$$

$$\bar{x}_2 = 31.56 \quad \text{and} \quad \sum_{i=1}^{9}(x_{2i} - \bar{x}_2)^2 = 160.2222$$

Then the pooled estimate of the common variance is

$$s^2 = \frac{\sum_{i=1}^{9}(x_{1i} - \bar{x}_1)^2 + \sum_{i=1}^{9}(x_{2i} - \bar{x}_2)^2}{n_1 + n_2 - 2} = \frac{195.5556 + 160.2222}{9 + 9 - 2} = 22.2361$$

and the standard deviation is $s = 4.716$.

The alternative hypothesis $H_a : \mu_1 > \mu_2$, or, equivalently, $\mu_1 - \mu_2 > 0$, implies that we should use a one-tailed statistical test and that the rejection region for the test is located in the upper tail of the t distribution. Referring to Table 4 in Appendix I, we find that the critical value of t for $\alpha = .05$ and $(n_1 + n_2 - 2) = 16$ d.f. is 1.746. Therefore, we will reject the null hypothesis when the calculated value of t is greater than 1.746.

The calculated value of the test statistic is

$$t = \frac{(\bar{x}_1 - \bar{x}_2)}{s\sqrt{\frac{1}{n_1} + \frac{1}{n_2}}} = \frac{35.22 - 31.56}{4.716\sqrt{\frac{1}{9} + \frac{1}{9}}} = 1.65$$

Comparing this value with the critical value $t_{.05} = 1.746$, we note that the calculated value does not fall in the rejection region. Therefore, we must conclude that there is insufficient evidence to indicate that the new method of training is superior at the .05 level of significance. ■

EXAMPLE 9.5 Find the p-value that would be reported for the statistical test of Example 9.4.

Solution The observed value of t for this one-tailed test was $t = 1.65$. Therefore, the p-value for the test would be the probability that $t > 1.65$. Since we cannot obtain this probability from Table 4 of Appendix I, we would report the p-value for the test as the smallest tabulated value of α that leads to the rejection of H_0. Consulting the row in Table 4 corresponding to 16 d.f., we see that the observed value, $t = 1.65$, lies between $t_{.10} = 1.337$ and $t_{.05} = 1.746$. Therefore, the probability that $t > 1.65$ is between .05 and .10, and the p-value for this test would be reported as $.05 < p\text{-value} < .10$. ■

The small-sample confidence interval for $(\mu_1 - \mu_2)$ is based on the same assumptions as the statistical test procedure. This confidence interval, with confidence coefficient $(1 - \alpha)$, is given by the formula in the following display.

Small-Sample $(1 - \alpha)$ 100% Confidence Interval for $(\mu_1 - \mu_2)$ Based on Independent Random Samples

$$(\bar{x}_1 - \bar{x}_2) \pm t_{\alpha/2} s \sqrt{\frac{1}{n_1} + \frac{1}{n_2}}$$

where s is obtained from the pooled estimate of σ^2.

Assumptions: The samples were randomly and independently selected from normally distributed populations. The variances of the populations σ_1^2 and σ_2^2 are equal.

Note the similarity in the procedures for constructing the confidence intervals for a single mean in Section 9.3 and the difference between two means. In both cases, the interval is constructed by using the appropriate point estimator and then adding and subtracting an amount equal to $t_{\alpha/2}$ times the standard error of the point estimator.

EXAMPLE 9.6 Find an interval estimate for $(\mu_1 - \mu_2)$ in Example 9.4 using a confidence coefficient equal to .95.

Solution Substituting into the formula

$$(\bar{x}_1 - \bar{x}_2) \pm t_{\alpha/2} s \sqrt{\frac{1}{n_1} + \frac{1}{n_2}}$$

we find that the interval estimate (or 95% confidence interval) is

$$(35.22 - 31.56) \pm (2.120)(4.716)\sqrt{\frac{1}{9} + \frac{1}{9}} \quad \text{or} \quad 3.66 \pm 4.71$$

Thus, we estimate that the difference $(\mu_1 - \mu_2)$ in mean time to assemble falls in the interval from -1.05 to 8.37. Note that the interval width is considerable; it seems advisable to increase the size of the samples and reestimate $(\mu_1 - \mu_2)$ by using this additional information. ■

To implement a two-sample t procedure with a pooled estimate of variance using the MINITAB package, enter the main command TWOSAMPLE, the columns identifying the two sets of sample values, and a semicolon. The subcommand POOLED, followed by a period, completes the instructions. The calculated value of t, its p-value, and its degrees of freedom are given as output together with a 95% confidence interval estimate of $(\mu_1 - \mu_2)$. Using the subcommand ALTERNATIVE, you may also specify whether the test is to be left-tailed (−1), right-tailed (1), or two-tailed (0). The test is performed as two-tailed unless another alternative is specified.

The MINITAB output for the TWOSAMPLE command using the data in Table 9.4 appears in Table 9.5. Notice that the entry SE MEAN (standard error of the mean), given for each column, is calculated as $s/\sqrt{n}$. For example, the standard error of the mean for C1 is $4.94/\sqrt{9} = 1.646$, or 1.6. The remaining entries are self-explanatory and can be compared with the results of Examples 9.4–9.6.

Before concluding our discussion, we should comment on the two assumptions on which our inferential procedures are based. **Moderate departures from the assumption that the populations possess normal probability distributions do not seriously affect the distribution of the test statistic and the confidence coefficient for the corresponding confidence interval. On the other hand, the population variances should be nearly equal to ensure that the procedures given above are valid.**

If there is reason to believe that the population variances are far from being equal, two changes must be made in the testing and estimation procedures. Since the pooled estimator s^2 is no longer appropriate when the population variances are not equal,

TABLE **9.5**
MINITAB output for two independent samples, using the data in Table 9.4

```
MTB > PRINT  C1  C2
 ROW     C1      C2

   1     32      35
   2     37      31
   3     35      29
   4     28      25
   5     41      34
   6     44      40
   7     35      27
   8     31      32
   9     34      31

MTB  >  TWOSAMPLE  C1  C2;
SUBC >  POOLED;
SUBC >  ALTERNATIVE 1.

TWOSAMPLE T FOR CI VS C2
     N       MEAN         STDEV       SE MEAN
C1   9      35.22          4.94          1.6
C2   9      31.56          4.48          1.5

95 PCT CI FOR MU C1 - MU C2: (-1.0, 8.4)
TTEST MU C1 = MU C2 (VS GT): T = 1.65  P = 0.059  DF = 16.0
```

the sample variances s_1^2 and s_2^2 are used as estimators of σ_1^2 and σ_2^2. The resulting test statistic is

$$\frac{(\bar{x}_1 - \bar{x}_2) - D_0}{\sqrt{\dfrac{s_1^2}{n_1} + \dfrac{s_2^2}{n_2}}}$$

When the sample sizes are *small*, critical values for this statistic are found in Table 4 of Appendix I, using degrees of freedom approximated by the formula

$$\text{d.f.} \approx \frac{\left(\dfrac{s_1^2}{n_1} + \dfrac{s_2^2}{n_2}\right)^2}{\dfrac{\left(\frac{s_1^2}{n_1}\right)^2}{(n_1 - 1)} + \dfrac{\left(\frac{s_2^2}{n_2}\right)^2}{(n_2 - 1)}}$$

Obviously, this result must be rounded to the nearest integer. In the MINITAB package, the TWOSAMPLE t command without the subcommand POOLED implements this procedure.

In Section 9.7, we will present a procedure for testing an hypothesis concerning the equality of two population variances that can be used to determine whether or not the underlying population variances are equal.

If there is reason to believe that the normality assumptions have been violated, you can test for a shift in location of two population distributions using the nonparametric Mann-Whitney U test of Chapter 14. This test procedure, which requires fewer assumptions concerning the nature of the population probability distributions, is almost as sensitive in detecting a difference in population means when the conditions necessary for the t test are satisfied. It may be more sensitive when the assumptions are not satisfied.

Exercises

Basic Techniques

9.16 Give the number of degrees of freedom for s^2, the pooled estimator of σ^2, if

a $n_1 = 16, n_2 = 8$.

b $n_1 = 10, n_2 = 12$.

c $n_1 = 15, n_2 = 3$.

9.17 Calculate s^2, the pooled estimator for σ^2, if

a $n_1 = 10, n_2 = 4, s_1^2 = 3.4, s_2^2 = 4.9$.

b $n_1 = 12, n_2 = 21, s_1^2 = 18, s_2^2 = 23$.

9.18 Two independent random samples of size $n_1 = 4$ and $n_2 = 5$ were selected from each of two normal populations. The data are as follows:

Population	
1	**2**
12	14
3	7
8	7
5	9
	6

a Calculate s^2, the pooled estimator of σ^2.

b Find a 90% confidence interval for $(\mu_1 - \mu_2)$, the difference between the two population means.

c Test $H_0 : (\mu_1 - \mu_2) = 0$ against $H_a : (\mu_1 - \mu_2) > 0$ for $\alpha = .05$. State your conclusions.

9.19 Independent random samples of $n_1 = 16$ and $n_2 = 13$ observations were selected from two normal populations with equal variances. The sample means and variances are shown below:

	Population	
	1	**2**
Sample size	16	13
Sample mean	34.6	32.2
Sample variance	4.8	5.9

a Suppose you wish to detect a difference between the population means. State the null and alternative hypotheses that you would use for the test.

b Find the rejection region for the test in part (a) for $\alpha = .10$.

c Find the value of the test statistic.

d Find the approximate observed significance level for the test.

e Conduct the test and state your conclusions.

9.20 Refer to Exercise 9.19. Find a 90% confidence interval for $(\mu_1 - \mu_2)$.

Applications

9.21 Jan Lindhe (1984) conducted a study on the effect of an oral antiplaque rinse on plaque buildup on teeth. Fourteen subjects, whose teeth were thoroughly cleaned and polished, were randomly assigned to two groups of seven subjects each. Both groups were assigned to use oral rinses (no brushing) for a 2-week period. Group 1 used a rinse that contained an antiplaque agent. Group 2, the control group, received a similar rinse except that, unknown to the subjects, the rinse contained no antiplaque agent. A plaque index x, a measure of plaque buildup, was recorded at 4, 7, and 14 days. The mean and standard deviation for the 14-day plaque measurements are shown below for the two groups:

	Control Group	Antiplaque Group
Sample size	7	7
Mean	1.26	.78
Standard deviation	.32	.32

a State the null and alternative hypotheses that should be used to test the effectiveness of the antiplaque oral rinse.

b Do the data provide sufficient evidence to indicate that the oral antiplaque rinse is effective? Test using $\alpha = .05$.

c Find the p-value for the test.

9.22 Refer to Exercise 9.21. Find a 95% confidence interval for the mean difference between the plaque indexes for the control group and the oral antiplaque rinse group. Interpret the interval.

9.23 Refer to the description of the oxygen partition study in reedfish conducted by Pettit and Beitinger (1985), Exercise 9.5. The table below gives oxygen uptake readings for reedfish exposed to two temperature environments, one group of $n_1 = 11$ reedfish in water at 25°C and the other group of $n_2 = 12$ reedfish in water at 33°C.

	Weight (grams)	Oxygen Uptake (ml O_2 $g^{-1}h^{-1}$)		Percentage	
		Air	Water	Air	Water
25°C	22.1	0.043	0.045	49	51
	21.1	0.034	0.066	34	66
	13.4	0.026	0.084	24	76
	19.5	0.048	0.045	52	48
	21.9	0.009	0.056	14	86
	19.5	0.028	0.072	28	72
	16.5	0.032	0.081	28	72
	18.5	0.045	0.051	47	53
	18.0	0.061	0.041	60	40
	15.4	0.028	0.060	32	68
	13.4	0.024	0.113	18	82

	Weight (grams)	Oxygen Uptake (ml O_2 $g^{-1}h^{-1}$)		Percentage	
		Air	Water	Air	Water
33°C	22.1	0.020	0.051	28	72
	21.1	0.044	0.036	55	45
	19.5	0.035	0.043	45	55
	21.9	0.051	0.049	51	49
	19.5	0.035	0.051	41	59
	16.5	0.027	0.073	27	73
	18.5	0.043	0.054	44	56
	18.9	0.053	0.057	48	52
	15.4	0.055	0.046	54	46
	21.8	0.075	0.037	67	33
	22.1	0.042	0.049	46	54
	13.2	0.055	0.039	59	41

Use the MINITAB printout given below to answer the following questions.

```
TWOSAMPLE T FOR 25-DEG VS 33-DEG
         N       MEAN      STDEV    SE MEAN
25-DEG  11       35.1       14.9        4.5
33-DEG  12       47.1       11.6        3.4

95 PCT CI FOR MU 25-DEG - MU 33-DEG: (-23.5, -0.5)

TTEST MU 25-DEG = MU 33-DEG (VS NE) : T= -2.17  P=0.042  DF= 21

POOLED STDEV =          13.3
```

a Do the data in the table present sufficient evidence to indicate a difference between the percentages of oxygen uptake by air for reedfish exposed to water at 25°C and those at 33°C? Test using $\alpha = .10$.

b Find the approximate p-value for the test.

9.24 Chronic anterior compartment syndrome is a condition characterized by exercise-induced pain in the lower leg. Swelling and impaired nerve and muscle function also accompany this exercise-induced pain, which is relieved by rest. Susan Beckham and colleagues (1993) conducted an experiment involving ten healthy runners and ten healthy cyclists to determine if there are significant differences in pressure measurements within the anterior muscle compartment for runners and cyclists. The data summary—compartment pressure in millimeters of mercury (Hg)—is as follows:

	Runners		**Cyclists**	
Condition	**Mean**	**Standard Deviation**	**Mean**	**Standard Deviation**
Rest	14.5	3.92	11.1	3.98
80% maximal O_2 consumption	12.2	3.49	11.5	4.95
Maximal O_2 consumption	19.1	16.9	12.2	4.47

Source: Beckham et al. (1993), Table 1.

a Test for a significant difference in compartment pressure between runners and cyclists under the resting condition. Use $\alpha = .05$.

b Construct a 95% confidence interval estimate of the difference in means for runners and cyclists under the condition of exercising at 80% of maximal oxygen consumption.

9.25 Is there a difference in SAT scores for high school students, depending on their intended field of study? Fifteen students who intend to major in engineering were compared with 15 students intending to major in language and literature by recording their verbal and math scores on the SAT tests. The results are as follows:

	Verbal		**Math**	
Engineering	$\bar{x} = 446$	$s = 42$	$\bar{x} = 548$	$s = 57$
Language/literature	$\bar{x} = 534$	$s = 45$	$\bar{x} = 517$	$s = 52$

Source: "SAT Scores" (1993).

a Is there a significant difference in the average verbal scores for students majoring in engineering and language/literature? Test using $\alpha = .01$.

b Is there a significant difference in the average math scores for students majoring in engineering and language/literature? Test using $\alpha = .01$.

9.26 Refer to Exercise 9.25. Suppose that the verbal and math scores for the two groups of students were not reported separately, but instead the total scores were recorded, along with the standard deviation of the scores. The results are as follows:

Engineering	Language/Literature
$\bar{x} = 994$	$\bar{x} = 1051$
$s = 71$	$s = 69$
$n = 15$	$n = 15$

a Is there a significant difference in the average total scores for students majoring in engineering and language/literature? Test using $\alpha = .01$.

b What is the p-value for the test in part (a)?

c Construct a 99% confidence interval for the difference in the average total scores for students intending to major in engineering and language/literature. Interpret this interval.

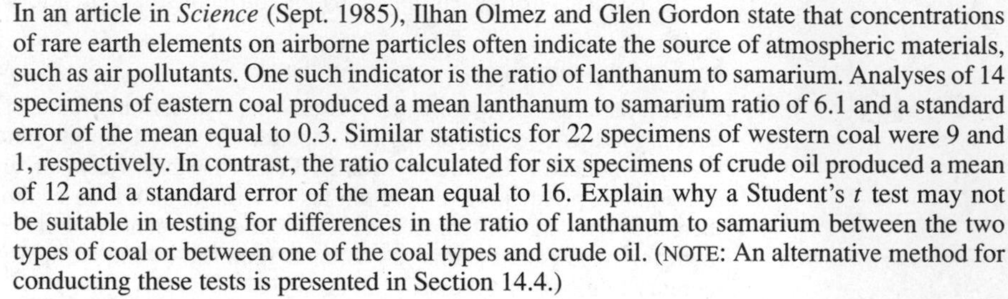

9.27 In an article in *Science* (Sept. 1985), Ilhan Olmez and Glen Gordon state that concentrations of rare earth elements on airborne particles often indicate the source of atmospheric materials, such as air pollutants. One such indicator is the ratio of lanthanum to samarium. Analyses of 14 specimens of eastern coal produced a mean lanthanum to samarium ratio of 6.1 and a standard error of the mean equal to 0.3. Similar statistics for 22 specimens of western coal were 9 and 1, respectively. In contrast, the ratio calculated for six specimens of crude oil produced a mean of 12 and a standard error of the mean equal to 16. Explain why a Student's t test may not be suitable in testing for differences in the ratio of lanthanum to samarium between the two types of coal or between one of the coal types and crude oil. (NOTE: An alternative method for conducting these tests is presented in Section 14.4.)

9.28 Refer to Exercise 9.7, where we measured the dissolved oxygen content in river water to determine whether a stream had sufficient oxygen to support aquatic life. A pollution control inspector suspected that a river community was releasing amounts of semitreated sewage into a river. To check his theory, he drew five randomly selected specimens of river water at a location above the town, and another five below. The dissolved oxygen readings, in parts per million, are as follows:

Above town	4.8	5.2	5.0	4.9	5.1
Below town	5.0	4.7	4.9	4.8	4.9

a Do the data provide sufficient evidence to indicate that the mean oxygen content below the town is less than the mean oxygen content above? Test using $\alpha = .05$.

b Suppose you prefer estimation as a method of inference. Estimate the difference in mean dissolved oxygen content between locations above and below the town. Use a 95% confidence interval.

9.5 A Paired-Difference Test

A manufacturer wishes to compare the wearing qualities of two different types of automobile tires, A and B. For the comparison, a tire of type A and one of type B are randomly assigned and mounted on the rear wheels of each of five automobiles. The automobiles are then operated for a specified number of miles, and the amount of wear is recorded for each tire. These measurements appear in Table 9.6. Do the data present sufficient evidence to indicate a difference in the average wear for the two tire types?

Analyzing the data, we note that the differences between the two sample means is $(\bar{x}_1 - \bar{x}_2) = .48$, a rather small quantity, considering the variability of the data and the small number of measurements involved. At first glance, it would seem that there

TABLE 9.6

Automobile	Tire A	Tire B
1	10.6	10.2
2	9.8	9.4
3	12.3	11.8
4	9.7	9.1
5	8.8	8.3
	$\bar{x}_1 = 10.24$	$\bar{x}_2 = 9.76$

is little evidence to indicate a difference between the population means, a conjecture that we can check by the method outlined in Section 9.4.

The pooled estimate of the common variance σ^2 is

$$s^2 = \frac{\sum_{i=1}^{n_1}(x_{1i} - \bar{x}_1)^2 + \sum_{i=1}^{n_2}(x_{2i} - \bar{x}_2)^2}{n_1 + n_2 - 2} = \frac{6.932 + 7.052}{5 + 5 - 2} = 1.748$$

and $s = 1.32$. The calculated value of t used to test the hypothesis that $\mu_1 = \mu_2$ is

$$t = \frac{\bar{x}_1 - \bar{x}_2}{s\sqrt{\frac{1}{n_1} + \frac{1}{n_2}}} = \frac{10.24 - 9.76}{1.32\sqrt{\frac{1}{5} + \frac{1}{5}}} = .57$$

a value that is not nearly large enough to reject the hypothesis that $\mu_1 = \mu_2$.

The corresponding 95% confidence interval is

$$(\bar{x}_1 - \bar{x}_2) \pm t_{\alpha/2} s \sqrt{\frac{1}{n_1} + \frac{1}{n_2}}$$

$$(10.24 - 9.76) \pm (2.306)(1.32)\sqrt{\frac{1}{5} + \frac{1}{5}}$$

or -1.45 to 2.41. This interval is quite wide, considering the small difference between the sample means.

A second glance at the data reveals a marked inconsistency with this conclusion. We note that the wear measurement for type A is larger than the corresponding value for type B for *each* of the five automobiles. These differences, recorded as $d = A - B$, are shown in Table 9.7.

TABLE 9.7
Differences in tire wear, using the data of Table 9.6

Automobile	A	B	$d = A - B$
1	10.6	10.2	.4
2	9.8	9.4	.4
3	12.3	11.8	.5
4	9.7	9.1	.6
5	8.8	8.3	.5
			$\bar{d} = .48$

If there is no difference in mean tire wear for the two tire types, then the probability that tire A shows more wear than tire B is equal to $p = .5$, and the five automobiles correspond to $n = 5$ independent binomial trials. A two-tailed test of the null hypothesis $p = .5$ would include a rejection region consisting of $x = 0$ and $x = 5$ and $\alpha = P(x = 0) + P(x = 5) = 2(1/2)^5 = 1/16 = .0625$. Since five of the differences are positive ($x = 5$), we have evidence to indicate that a difference exists in the mean wear of the two tire types.

You will note that we have used two different statistical tests to test the same hypothesis. Isn't it peculiar that the t test, which uses more information (the actual sample measurements) than the binomial test, fails to supply sufficient evidence for rejection of the hypothesis $\mu_1 = \mu_2$?

There is an explanation for this inconsistency. The t test described in Section 9.4 is not the proper statistical test to be used for our example. The statistical test procedure of Section 9.4 requires that the two samples be *independent and random*. Certainly, the independence requirement was violated by the manner in which the experiment was conducted. The (pair of) measurements, an A and a B tire, for a particular automobile are definitely related. A glance at the data shows that the readings have approximately the same magnitude for a particular automobile but vary markedly from one automobile to another. This, of course, is exactly what we might expect. Tire wear is largely determined by driver habits, the balance of the wheels, and the road surface. Since each automobile has a different driver, we would expect a large amount of variability in the data from one automobile to another.

The familiarity we have gained with interval estimation has shown us that the width of the large-sample confidence intervals depends on the magnitude of the standard deviation of the point estimator of the parameter. The smaller its value, the better is the estimate and the more likely it is that the test statistic will provide evidence to reject the null hypothesis if it is, in fact, false. Knowledge of this phenomenon was utilized in *designing* the tire-wear experiment. The experimenter realized that the wear measurements would vary greatly from auto to auto and that this variability could not be separated from the data if the tires were assigned to the ten wheels in a random manner. (A random assignment of the tires would have implied that the data be analyzed according to the procedure of Section 9.4.) Instead, a comparison of the wear between tire types A and B made on each automobile resulted in the five difference measurements. This design eliminates the effect of the car-to-car variability and yields more information on the mean difference in the wearing quality for the two tire types.

A proper analysis of the data would use the five difference measurements to test the hypothesis that the average difference μ_d is equal to 0 or, equivalently, to test the null hypothesis $H_0 : \mu_d = \mu_1 - \mu_2 = 0$ against the alternative hypothesis $H_a : \mu_d = (\mu_1 - \mu_2) \neq 0$.

Paired-Difference Test for $(\mu_1 - \mu_2) = \mu_d$

1 Null Hypothesis: $H_0 : \mu_d = 0$

2 Alternative Hypothesis:

One-Tailed Test	*Two-Tailed Test*
$H_a : \mu_d > 0$ (or, $H_a : \mu_d < 0$)	$H_a : \mu_d \neq 0$

3 Test Statistic: $t = \dfrac{\bar{d} - 0}{s_d/\sqrt{n}} = \dfrac{\bar{d}}{s_d/\sqrt{n}}$

where n = number of paired differences

$$s_d = \sqrt{\frac{\sum_{i=1}^{n}(d_i - \bar{d})^2}{n-1}}$$

4 Rejection Region:

One-Tailed Test	*Two-Tailed Test*
$t > t_\alpha$ (or, $t < -t_\alpha$ when the alternative hypothesis is $H_a : \mu_d < 0$)	$t > t_{\alpha/2}$ or $t < -t_{\alpha/2}$

The critical values of t, t_α, and $t_{\alpha/2}$ are based on $(n - 1)$ d.f. These tabulated values are given in Table 4 of Appendix I.

Assumptions: The n paired observations are randomly selected from normally distributed populations.

EXAMPLE **9.7** Do the data in Table 9.6 provide sufficient evidence to indicate a difference in mean wear for tire types A and B? Test using $\alpha = .05$.

Solution You can verify that the average and standard deviation of the five difference measurements are

$$\bar{d} = .48 \quad \text{and} \quad s_d = .0837$$

Then,

$$H_0 : \mu_d = 0 \quad \text{and} \quad H_a : \mu_d \neq 0$$

and

$$t = \frac{\bar{d} - 0}{s_d/\sqrt{n}} = \frac{.48}{.0837/\sqrt{5}} = 12.8$$

The critical value of t for a two-tailed statistical test, $\alpha = .05$ and 4 d.f., is 2.776. Certainly, the observed value of $t = 12.8$ is extremely large and highly significant.

Hence, we would conclude that there is a difference in the mean amount of wear for tire types A and B. ▪

$(1 - \alpha)$ 100% Small-Sample Confidence Interval for $(\mu_1 - \mu_2) = \mu_d$, Based on a Paired-Difference Experiment

$$\bar{d} \pm t_{\alpha/2} \frac{s_d}{\sqrt{n}}$$

where n = number of paired differences and

$$s_d = \sqrt{\frac{\sum_{i=1}^{n} (d_i - \bar{d})^2}{n - 1}}$$

Assumptions: The n paired observations are randomly selected from normally distributed populations.

EXAMPLE 9.8 Find a 95% confidence interval for $(\mu_1 - \mu_2) = \mu_d$ using the data in Table 9.6.

Solution A 95% confidence interval for the difference between the mean wear would be

$$\bar{d} \pm t_{\alpha/2} \frac{s_d}{\sqrt{n}} = .48 \pm (2.776) \frac{.0837}{\sqrt{5}}$$

or $.48 \pm .10$. ▪

When the units used to compare two or more procedures exhibit marked variability before any experimental procedures are implemented, the effect of this variability can be minimized by comparing the procedures *within* groups of relatively homogeneous units called **blocks**. In this way, the effects of the procedures are not masked by the initial variability among the units in the experiment. An experiment conducted in this manner is called a **randomized block design**. In an experiment involving daily sales, blocks may represent days of the week; in an experiment involving product marketing, blocks may represent geographic areas. (Randomized block designs are discussed in more detail in Section 13.6.)

The statistical design of the tire experiment is a simple example of a randomized block design, and the resulting statistical test is often called a *paired-difference test*. You will note that the pairing occurred when the experiment was planned and not after the data were collected. Comparisons of tire wear were made within relatively homogeneous blocks (automobiles), with the tire types randomly assigned to the two automobile wheels.

The amount of information gained by blocking the tire experiment may be measured by comparing the calculated confidence interval for the unpaired (and incorrect) analysis with the interval obtained for the paired-difference analysis. The confidence interval for $(\mu_1 - \mu_2)$ that might have been calculated, had the tires been randomly assigned to the ten wheels (unpaired), is unknown but probably would have been of the same magnitude as the interval -1.45 to 2.41, which was calculated by analyzing the observed data in an unpaired manner. Pairing the tire types on the automobiles (blocking) and the resulting analysis of the differences produced the interval estimate .38 to .58. Note the difference in the widths of the intervals, which indicates the very sizable increase in information obtained by blocking in this experiment.

Although blocking proved to be very beneficial in the tire experiment, it may not always be. We observe that the degrees of freedom available for estimating σ^2 are less for the paired than for the corresponding unpaired experiment. If there were actually no differences among the blocks, the reduction in the degrees of freedom would produce a moderate increase in the $t_{\alpha/2}$ employed in the confidence interval and hence would increase the width of the interval. This, of course, did not occur in the tire experiment because the large reduction in the standard error of $\bar{d}$ more than compensated for the loss in degrees of freedom.

Except for notation, the analysis of a paired-difference experiment is the same as that for a single sample presented in Section 9.3. This similarity enables us to use the MINITAB commands TTEST and TINTERVAL to analyze the differences in a paired-difference experiment.

Before concluding, we want to reemphasize a point. Once you have used a paired design for an experiment, you no longer have the option of using the unpaired analysis of Section 9.4. The assumptions on which that test is based have been violated. Your only alternative is to use the correct method of analysis, the paired-difference test (and associated confidence interval) of this section.

Exercises

Basic Techniques

9.29 A paired-difference experiment was conducted using $n = 10$ pairs of observations. Test the null hypothesis $H_0: (\mu_1 - \mu_2) = 0$ against $H_a: (\mu_1 - \mu_2) \neq 0$ for $\alpha = .05$, $\bar{d} = .3$, and $s_d^2 = .16$. Give the approximate p-value for the test.

9.30 Find a 95% confidence interval for $(\mu_1 - \mu_2)$ in Exercise 9.29.

9.31 How many pairs of observations would you need if you wished to estimate $(\mu_1 - \mu_2)$ in Exercise 9.29, correct to within .1 with probability equal to .95?

9.32 A paired-difference experiment consists of $n = 18$ pairs, $\bar{d} = 5.7$, and $s_d^2 = 256$. Suppose we wish to detect $\mu_d > 0$.

a Give the null and alternative hypotheses for the test.

b Conduct the test and state your conclusions.

9.33 A paired-difference experiment was conducted to compare the means of two populations. The data are as follows:

	Pairs				
Population	**1**	**2**	**3**	**4**	**5**
1	1.3	1.6	1.1	1.4	1.7
2	1.2	1.5	1.1	1.2	1.8

a Do the data provide sufficient evidence to indicate that μ_1 differs from μ_2? Test using $\alpha = .05$.

b Find the approximate observed significance level for the test and interpret its value.

c Find a 95% confidence interval for $(\mu_1 - \mu_2)$. Compare your interpretation of the confidence interval with your test results in part (a).

d What assumptions must you make for your inferences to be valid?

Applications

9.34 Persons submitting computing jobs to a computer center are usually required to estimate the amount of computer time required to complete the job. This time is measured in CPUs, the amount of time that a job will occupy a portion of the computer's central processing unit's memory. A computer center decided to perform a comparison of the estimated versus actual CPU times for a particular customer. The corresponding times were available for 11 jobs. The sample data are shown below:

	Job Number										
CPU Time (minutes)	**1**	**2**	**3**	**4**	**5**	**6**	**7**	**8**	**9**	**10**	**11**
Estimated	.50	1.40	.95	.45	.75	1.20	1.60	2.6	1.30	.85	.60
Actual	.46	1.52	.99	.53	.71	1.31	1.49	2.9	1.41	.83	.74

a Why would you expect these pairs of data to be correlated?

b Do the data provide sufficient evidence to indicate that, *on the average*, the customer tends to underestimate CPU time required for computing jobs? Test using $\alpha = .10$.

c Find the observed significance level for the test and interpret its value.

d Find a 90% confidence interval for the difference in mean estimated CPU time versus mean actual CPU time.

9.35 Refer to Exercise 9.24. In addition to the compartment pressures, the level of creatine phosphokinase (CPK) in blood samples, a measure of muscle damage, was also determined for each of ten runners and ten cyclists before and after exercise. The data summary—CPK values in units/liter—is as follows:

	Runners		Cyclists	
Condition	**Mean**	**Standard Deviation**	**Mean**	**Standard Deviation**
Before exercise	255.63	115.48	173.8	60.69
After exercise	284.75	132.64	177.1	64.53
Difference	29.13	21.01	3.3	6.85

Source: Beckham et al. (1993), Table 2.

a Test for a significant difference in mean CPK values for runners and cyclists before exercise under the assumption that $\sigma_1^2 \neq \sigma_2^2$; use $\alpha = .05$. Find a 95% confidence interval estimate for the corresponding difference in means.

b Test for a significant difference in mean CPK values for runners and cyclists after exercise under the assumption that $\sigma_1^2 \neq \sigma_2^2$; use $\alpha = .05$. Find a 95% confidence interval estimate for the corresponding difference in means.

c Test for a significant difference in mean CPK values for runners before and after exercise. Use $\alpha = .05$.

d Find a 95% confidence interval estimate for the difference in mean CPK values for cyclists before and after exercise. Does your estimate indicate that there is no significant difference in mean CPK levels for cyclists before and after exercise?

9.36 In an advertisement for Albertsons, a supermarket chain in the western United States, the advertiser claims that Albertsons has been consistently lower than four other full-service supermarkets. As part of a survey conducted by an "independent market basket price-checking company," the average weekly total, based on the prices of approximately 95 items, is given for two different supermarket chains recorded during four consecutive weeks in December 1992 and January 1993.

Week	Albertsons	Ralphs
1	254.26	256.03
2	240.62	255.65
3	231.90	255.12
4	234.13	261.18

a Is there a significant difference in the average prices for these two different supermarket chains? Test using $\alpha = .05$.

b What is the approximate p-value for the test conducted in part (a)?

c Construct a 99% confidence interval for the difference in the average prices for the two supermarket chains. Interpret this interval.

9.37 An experiment was conducted to compare mean reaction time to two types of traffic signs, prohibitive (No Left Turn) and permissive (Left Turn Only). Ten subjects were included in the experiment. Each subject was presented with 40 traffic signs, 20 prohibitive and 20 permissive, in random order. The mean time to reaction and the number of correct actions were recorded for each subject. The mean reaction times to the 20 prohibitive and 20 permissive traffic signs are shown below for each of the ten subjects:

	Mean Reaction Times (ms) for 20 Traffic Signs	
Subject	**Prohibitive**	**Permissive**
1	824	702
2	866	725
3	841	744
4	770	663
5	829	792
6	764	708
7	857	747
8	831	685
9	846	742
10	759	610

a Explain why this is a paired-difference experiment and give reasons why the pairing should be useful in increasing information on the difference between the mean reaction times to prohibitive and permissive traffic signs.

b Do the data present sufficient evidence to indicate a difference in mean reaction times to prohibitive and permissive traffic signs? Test using $\alpha = .05$.

c Find and interpret the approximate p-value for the test in part (b).

d Find a 95% confidence interval for the difference in mean reaction times to prohibitive and permissive traffic signs.

9.38 Two computers are often compared by running a collection of different "benchmark" programs and recording the difference in CPU time required to complete the same program. Six benchmark programs, run on two computers, produced the following CPU times (in minutes):

	Benchmark Program					
Computer	**1**	**2**	**3**	**4**	**5**	**6**
1	1.12	1.73	1.04	1.86	1.47	2.10
2	1.15	1.72	1.10	1.87	1.46	2.15

a Do the data provide sufficient evidence to indicate a difference in mean CPU times required for the two computers to complete a job? Test using $\alpha = .05$.

b Find the approximate observed significance level for the test and interpret its value.

c Find a 95% confidence interval for the difference in mean CPU times required for the two computers to complete a job.

9.39 Exercise 9.21 describes a dental experiment conducted to investigate the effectiveness of an oral rinse used to deter the growth of plaque on teeth. Subjects were divided into two groups: one group used a rinse with an antiplaque ingredient, and the control group used a rinse containing inactive ingredients. Suppose that the plaque growth on each person's teeth was measured after using the rinse after 4 hours and then again after 8 hours. If you wish to estimate the difference in plaque growth from 4 to 8 hours, should you use a confidence interval based on a paired or an unpaired analysis? Explain.

9.40 A study conducted by nutritionists at the University of Toronto has determined that people who nibble all day instead of eating their ordinary-size meals have significantly lower blood cholesterol levels (*Press-Enterprise*, Riverside, Calif., Oct. 5, 1989). In the study, seven men ate 2500 calories/day. For 2 weeks, they ate ordinary-size meals, but for another 2 weeks, they got the same amount of calories in 17 snacks eaten once an hour. The total cholesterol level, as well as the hazardous low-density lipoprotein cholesterol (LDL), was measured for each subject after each 2-week period. Is this a paired or an unpaired experiment? Explain.

9.41 The earth's temperature (which affects seed germination, crop survival in bad weather, and many other aspects of agricultural production) can be measured using either ground-based sensors or infrared-sensing devices mounted in aircraft or space satellites. Ground-based sensoring is tedious, requiring many replications to obtain an accurate estimate of ground temperature. On the other hand, airplane or satellite sensoring of infrared waves appears to introduce a bias in the temperature readings. To determine the bias, readings were obtained at six different locations using both ground- and air-based temperature sensors. The readings, in degrees Celsius, were as follows:

Location	Ground	Air
1	46.9	47.3
2	45.4	48.1
3	36.3	37.9
4	31.0	32.7
5	24.7	26.2

a Do the data present sufficient evidence to indicate a bias in the air-based temperature readings? Explain. (Use $\alpha = .05$.)

b Estimate the difference in mean temperature between ground- and air-based sensors using a 95% confidence interval.

9.42 Refer to Exercise 9.41. How many paired observations would be required to estimate the difference between mean temperatures for ground- versus air-based sensors correct to within .2°C, with probability approximately equal to .95?

9.43 In response to a complaint that a particular tax assessor (A) was biased, an experiment was conducted to compare the assessor named in the complaint with another tax assessor (B) from the same office. Eight properties were selected, and each was assessed by both assessors. The assessments (in thousands of dollars) are shown in the table:

Property	Assessor A	Assessor B
1	76.3	75.1
2	88.4	86.8
3	80.2	77.3
4	94.7	90.6
5	68.7	69.1
6	82.8	81.0
7	76.1	75.3
8	79.0	79.1

Use the MINITAB printout given below to answer the following questions.

```
MTB > TTEST 0 C3

TEST OF MU = 0.000 VS MU N.E. 0.000

         N       MEAN      STDEV   SE MEAN         T      P VALUE
C3       8      1.487      1.491     0.527      2.82        0.026
```

a Do the data provide sufficient evidence to indicate that assessor A tends to give higher assessments than assessor B? Test using $\alpha = .05$.

b Estimate the difference in mean assessments for the two assessors.

c What assumptions must you make in order for inferences (a) and (b) to be valid?

d Suppose that assessor A had been compared with a more stable standard, say the average $\bar{x}$, of the assessments given by four assessors selected from the tax office. Thus, each property would be assessed by A and also by each of the four other assessors, and $x_A - \bar{x}$ would be calculated. If the test in part (a) is valid, could you use the paired-difference t test to test the hypothesis that the bias, the mean difference between A's assessments and the mean of the assessments of the four assessors, is equal to 0? Explain.

9.44 In a study of the well-being of cancer patients, investigators matched each of a group of 104 cancer survivors (those who have lived 1 or more years past their last cancer treatment) with a control, that is, a person who was considered physically healthy and who had not had cancer (Schmale et al., 1983). Patients and controls were matched according to sex, age, and education. Each person in the study was then administered a questionnaire, "The General Well-Being Measure," which purports to measure a person's well-being. The following table gives the means for 104 cancer survivors (CS) and 104 controls for each of a number of factors associated with well-being and for two composite indices of well-being. Also shown are the t values and associated p-values for testing an hypothesis of "no difference" between the means of the cancer survivors and the controls.

Factors	CS	Control	t	2-Tailed p
	$n_1 = n_2 = 104$ Means			
Anxiety	12.68	12.39	0.46	0.64
Depression	5.27	5.49	0.74	0.46
Self-control	15.34	16.29	3.30	0.00*
Positive well-being	17.53	16.99	1.09	0.27
Vitality	17.70	17.38	0.68	0.49
General health	13.95	15.20	3.67	0.00*

*Asterisks refer to statistical significance.

Composites	CS	Control	t	2-Tailed p
	$n_1 = n_2 = 104$ Means			
Mental health index	70.79	71.39	0.41	0.67
General well-being	102.14	103.98	0.88	0.37

a Describe the type of experimental design used for this study.

b Describe the practical conclusions to be derived from the test results.

c Calculate one of the p-values, say, the approximate p-value for positive well-being. Does it appear to agree with the p-value given in the author's table?

9.6 Inferences Concerning a Population Variance

We have seen in the preceding sections that an estimate of the population variance σ^2 is fundamental to procedures for making inferences about population means. Moreover, there are many practical situations where σ^2 is the primary objective of an experimental investigation; thus, it may assume a position of far greater importance than that of the population mean.

Scientific measuring instruments must provide unbiased readings with a very small error of measurement. An aircraft altimeter that measured the correct altitude on the *average* would be of little value if the standard deviation of the error of measurement were 5000 feet. Indeed, bias in a measuring instrument can often be corrected, but the precision of the instrument, measured by the standard deviation of the error of measurement, is usually a function of the design of the instrument itself and cannot be controlled.

Machined parts in a manufacturing process must be produced with minimum variability in order to reduce out-of-size and hence defective products. And, in general, it is desirable to maintain a minimum variance in the measurements of the quality characteristics of an industrial product in order to achieve process control and therefore minimize the percentage of poor-quality product.

The sample variance

$$s^2 = \frac{\sum_{i=1}^{n}(x_i - \bar{x})^2}{n - 1}$$

is an unbiased estimator of the population variance σ^2. Thus, the distribution of sample variances generated by repeated sampling will have a probability distribution that commences at $s^2 = 0$ (since s^2 cannot be negative), with a mean equal to σ^2. Unlike the distribution of $\bar{x}$, the sampling distribution of s^2 is nonsymmetric, the exact form being dependent on the probability distribution of the population.

Assumptions: For the methodology that follows we will assume that the sample is drawn from a normal population and that s^2 is based on a random sample of n measurements.

Or, using the terminology of Section 9.2, we would say that s^2 has $(n - 1)$ d.f.

The next and obvious step would be to consider the distribution of s^2 in repeated sampling from a specified normal distribution—one with a specific mean and variance—and to tabulate the critical values of s^2 for some of the commonly used tail areas. If this is done, we will find that the distribution of s^2 is independent of the population mean μ, but has a different distribution for each sample size and each value of σ^2. This task would be quite laborious, but fortunately it may be simplified by *standardizing*, as was done by using z in the normal tables. The quantity

$$\chi^2 = \frac{(n - 1)s^2}{\sigma^2}$$

called a **chi-square variable** by statisticians, admirably suits our purposes. Its distribution in repeated sampling is called, as we might suspect, a **chi-square probability distribution**. The equation of the density function for the chi-square distribution is well known to statisticians who have tabulated critical values corresponding to various tail areas of the distribution. These values are presented in Table 5 in Appendix I.

The shape of the chi-square distribution, like that of the t distribution, will vary with the sample size or, equivalently, with the degrees of freedom associated with s^2. Thus, Table 5 in Appendix I is constructed in exactly the same manner as the t table, with the degrees of freedom shown in the first and last columns. A partial reproduction of Table 5 in Appendix I is shown in Table 9.8. The symbol χ_a^2 indicates that the tabulated χ^2 value is such that an area a lies to its right. (See Figure 9.3.) Stated in probabilistic terms,

$$P(\chi^2 > \chi_a^2) = a$$

Thus, 99% of the area under the χ^2 distribution would lie to the right of $\chi_{.99}^2$. We note that the extreme values of χ^2 must be tabulated for both the lower and upper tail of the distribution because it is nonsymmetric.

You can check your ability to use the table by verifying the following statements. The probability that χ^2, based on $n = 16$ measurements (d.f. $= 15$), will exceed 24.9958 is .05. For a sample of $n = 6$ measurements (d.f. $= 5$), 95% of the area under the χ^2 distribution will lie to the right of $\chi^2 = 1.145476$. These values of

TABLE **9.8** Format of the chi-square table from Table 5 in Appendix I

d.f.	$\chi^2_{0.995}$	...	$\chi^2_{0.950}$	$\chi^2_{0.900}$	$\chi^2_{0.100}$	$\chi^2_{0.050}$	...	$\chi^2_{0.005}$	d.f.
1	0.0000393		0.0039321	0.0157908	2.70554	3.84146		7.87944	1
2	0.0100251		0.102587	0.210720	4.60517	5.99147		10.5966	2
3	0.0717212		0.351846	0.584375	6.25139	7.81473		12.8381	3
4	0.206990		0.710721	1.063623	7.77944	9.48773		14.8602	4
5	0.411740		1.145476	1.61031	9.23635	11.0705		16.7496	5
6	0.0675727		1.63539	2.20413	10.6446	12.5916		18.5476	6
⋮	⋮		⋮	⋮	⋮	⋮		⋮	⋮
15	4.60094		7.26094	8.54675	22.3072	24.9958		32.8013	15
16	5.14224		7.96164	9.31223	23.5418	26.2962		34.2672	16
17	5.69724		8.67176	10.0852	24.7690	27.5871		35.7185	17
18	6.26481		9.39046	10.8649	25.9894	28.8693		37.1564	18
19	6.84398		10.1170	11.6509	27.2036	30.1435		38.5822	19
⋮	⋮		⋮	⋮	⋮	⋮		⋮	⋮

FIGURE **9.3**
A chi-square distribution

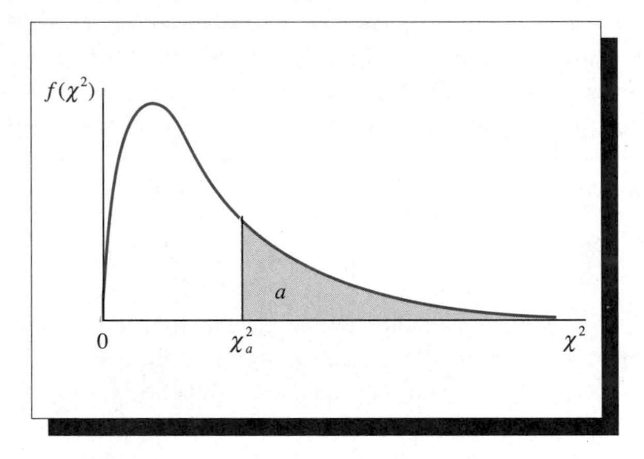

χ^2 are shaded in Table 9.8. The statistical test of a null hypothesis concerning a population variance

$$H_0 : \sigma^2 = \sigma_0^2$$

will employ the test statistic

$$\chi^2 = \frac{(n-1)s^2}{\sigma_0^2}$$

Notice that when H_0 is true, s^2/σ_0^2 should be near 1, so that χ^2 should be close to $(n-1)$, the degrees of freedom. If σ^2 is really greater than the hypothesized value σ_0^2, the test statistic will tend to be larger than $(n-1)$ and will probably fall toward the upper tail of the distribution. If $\sigma^2 < \sigma_0^2$, the test statistic will tend to be smaller than $(n-1)$ and will probably fall toward the lower tail of the chi-square distribution. As in other testing situations, we may use either a one- or a two-tailed statistical test, depending on the alternative hypothesis under test.

Test of Hypothesis Concerning a Population Variance

1. Null Hypothesis: $H_0 : \sigma^2 = \sigma_0^2$
2. Alternative Hypothesis:

One-Tailed Test	*Two-Tailed Test*
$H_a : \sigma^2 > \sigma_0^2$	$H_a : \sigma^2 \neq \sigma_0^2$
(or, $H_a : \sigma^2 < \sigma_0^2$)	

3. Test Statistic: $\chi^2 = \dfrac{(n-1)s^2}{\sigma_0^2}$
4. Rejection Region:

One-Tailed Test	*Two-Tailed Test*
$\chi^2 > \chi^2_\alpha$ (or, $\chi^2 < \chi^2_{(1-\alpha)}$ when the alternative hypothesis is $H_a : \sigma^2 < \sigma_0^2$), where χ^2_α and $\chi^2_{(1-\alpha)}$ are, respecttively, the upper- and lower-tail values of χ^2 that place α in the tail areas.	$\chi^2 > \chi^2_{\alpha/2}$ or $\chi^2 < \chi^2_{(1-\alpha/2)}$, where $\chi^2_{\alpha/2}$ and $\chi^2_{(1-\alpha/2)}$ are, respectively, the upper- and lower-tail values of χ^2 that place $\alpha/2$ in the tail areas.

The critical values of χ^2 are based on $(n-1)$ d.f. These tabulated values are given in Table 5 in Appendix I.

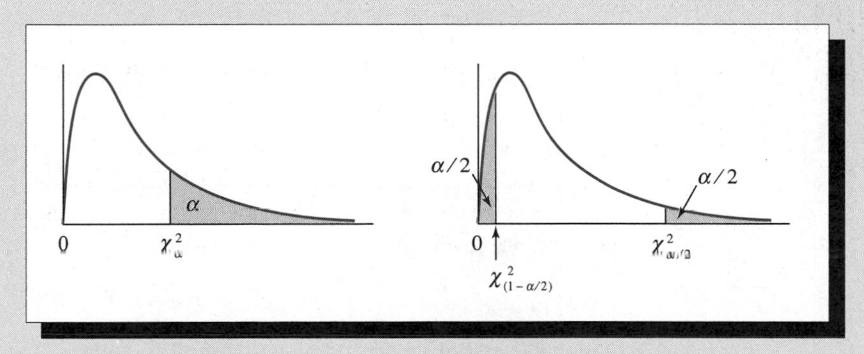

Assumption: The sample has been randomly selected from a normal population.

EXAMPLE **9.9** A cement manufacturer claimed that concrete prepared from his product would have a relatively stable compressive strength and that the strength measured in kilograms per square centimeter (kg/cm^2) would lie within a range of 40 kg/cm^2. A sample of $n = 10$ measurements produced a mean and variance equal to, respectively,

$$\bar{x} = 312$$
$$s^2 = 195$$

Do these data present sufficient evidence to reject the manufacturer's claim?

Solution As stated, the manufacturer claimed that the range of the strength measurements would equal 40 kg/cm^2. We will suppose that he meant that the measurements would lie within this range 95% of the time and, therefore, that the range would equal approximately 4σ and that $\sigma = 10$. We would then wish to test the null hypothesis

$$H_0 : \sigma^2 = (10)^2 = 100$$

against the alternative

$$H_a : \sigma^2 > 100$$

The alternative hypothesis would require a one-tailed statistical test with the entire rejection region located in the upper tail of the χ^2 distribution. The critical value of χ^2 for $\alpha = .05$ and $(n - 1) = 9$ d.f. is $\chi^2 = 16.9190$, which implies that we will reject H_0 if the test statistic exceeds this value.

Calculating, we obtain

$$\chi^2 = \frac{(n-1)s^2}{\sigma_0^2} = \frac{1755}{100} = 17.55$$

Since the value of the test statistic falls in the rejection region, we conclude that the null hypothesis is false and that the range of concrete-strength measurements will exceed the manufacturer's claim. ▪

EXAMPLE 9.10 Find the approximate observed significance level for the test in Example 9.9.

Solution Examining the row corresponding to 9 d.f. in Table 5 of Appendix I, you will see that the observed value of chi-square, $\chi^2 = 17.55$, is larger than the tabulated value, $\chi^2_{.05} = 16.9190$, and less than $\chi^2_{.025} = 19.0228$. Therefore, the observed significance level (p-value) for the test lies between .05 and .025, and we would report the observed significance level for the test as

$$.025 < p\text{-value} < .05$$

This tells us that we would reject the null hypothesis for any value of α equal to or greater than .05. ▪

A confidence interval for σ^2 with a $(1 - \alpha)$ confidence coefficient can be shown to be

$(1 - \alpha)$ 100% Confidence Interval for σ^2

$$\frac{(n-1)s^2}{\chi^2_{\alpha/2}} < \sigma^2 < \frac{(n-1)s^2}{\chi^2_{(1-\alpha/2)}}$$

where $\chi^2_{\alpha/2}$ and $\chi^2_{(1-\alpha/2)}$ are the upper and lower χ^2 values, which would locate one-half of α in each tail of the chi-square distribution.

Assumption: The sample has been randomly selected from a normal population.

EXAMPLE 9.11 Find a 90% confidence interval for σ^2 in Example 9.9.

Solution The first step would be to obtain the tabulated values of $\chi^2_{.95}$ and $\chi^2_{.05}$ corresponding to $(n - 1) = 9$ d.f.

$$\chi^2_{(1-\alpha/2)} = \chi^2_{.95} = 3.32511$$
$$\chi^2_{\alpha/2} = \chi^2_{.05} = 16.9190$$

Substituting these values and $s^2 = 195$ into the formula for the confidence interval,

$$\frac{(n-1)s^2}{\chi^2_{\alpha/2}} < \sigma^2 < \frac{(n-1)s^2}{\chi^2_{(1-\alpha/2)}}$$

The interval estimate for σ^2 would be

$$\frac{9(195)}{16.9190} < \sigma^2 < \frac{9(195)}{3.32511}$$

or $103.73 < \sigma^2 < 527.80$. ▪

EXAMPLE 9.12 An experimenter was convinced that her measuring equipment had a variability measured by a standard deviation $\sigma = 2$. During an experiment, she recorded the measurements 4.1, 5.2, and 10.2. Do these data disagree with her assumption? Test the hypothesis $H_0 : \sigma = 2$ or $\sigma^2 = 4$. Then place a 90% confidence interval on σ^2.

Solution The calculated sample variance is $s^2 = 10.57$. Since we wish to detect $\sigma^2 > 4$ as well as $\sigma^2 < 4$, we should employ a two-tailed test. When we use $\alpha = .10$ and place .05 in each tail, we will reject H_0 when $\chi^2 > 5.99147$ or $\chi^2 < .102587$. The calculated value of the test statistic is

$$\chi^2 = \frac{(n-1)s^2}{\sigma_0^2} = \frac{2(10.57)}{4} = 5.29$$

Since the test statistic does not fall in the rejection region, the data do not provide sufficient evidence to reject the null hypothesis $H_0 : \sigma^2 = 4$.

The corresponding 90% confidence interval is

$$\frac{(n-1)s^2}{\chi^2_{\alpha/2}} < \sigma^2 < \frac{(n-1)s^2}{\chi^2_{(1-\alpha/2)}}$$

The values of $\chi^2_{(1-\alpha/2)}$ and $\chi^2_{\alpha/2}$ are

$$\chi^2_{(1-\alpha/2)} = \chi^2_{.95} = .102587$$
$$\chi^2_{\alpha/2} = \chi^2_{.05} = 5.99147$$

Substituting these values into the formula for the interval estimate, we obtain

$$\frac{2(10.57)}{5.99147} < \sigma^2 < \frac{2(10.57)}{.102587} \quad \text{or} \quad 3.53 < \sigma^2 < 206.07$$

Thus, we estimate the population variance to fall in the interval 3.53 to 206.07. This very wide confidence interval indicates how little information on the population variance is obtained in a sample of only three measurements. Consequently, it is not surprising that there was insufficient evidence to reject the null hypothesis $\sigma^2 = 4$. To obtain more information on σ^2, the experimenter needs to increase the sample size. ■

Exercises

Basic Techniques

9.45 A random sample of $n = 25$ observations from a normal population produced a sample variance equal to 21.4. Do these data provide sufficient evidence to indicate that $\sigma^2 > 15$? Test using $\alpha = .05$.

9.46 A random sample of $n = 15$ observations was selected from a normal population. The sample mean and variance were $\bar{x} = 3.91$ and $s^2 = .3214$. Find a 90% confidence interval for the population variance σ^2.

9.47 A random sample of size $n = 7$ from a normal population produced the following measurements:

1.4 3.6 1.7 2.0 3.3 2.8 2.9

a Calculate the sample variance, s^2.

b Construct a 95% confidence interval for the population variance, σ^2.

c Test $H_0 : \sigma^2 = .8$ against $H_a : \sigma^2 \neq .8$, using $\alpha = .05$. State your conclusions.

d What is the approximate p-value for the test in part (c)?

Applications

9.48 Refer to the comparison of the estimated versus actual CPU times for computer jobs submitted by the customer in Exercise 9.35.

CPU Time (minutes)	Job Number 1	2	3	4	5	6	7	8	9	10	11
Estimated	.50	1.40	.95	.45	.75	1.20	1.60	2.6	1.30	.85	.60
Actual	.46	1.52	.99	.53	.71	1.31	1.49	2.9	1.41	.83	.74

Find a 90% confidence interval for the variance of the differences between estimated and actual CPU times.

9.49 Refer to Exercise 9.48. Suppose that the computer center expects customers to estimate CPU time correct to within .2 minute most of the time (say, 95% of the time).

a Do the data provide sufficient evidence to indicate that the customer's error in estimating job CPU time exceeds the computer center's expectation? Test using $\alpha = .10$.

b Find the approximate observed significance level for the test and interpret its value.

9.50 A precision instrument is guaranteed to read accurately to within 2 units. A sample of four instrument readings on the same object yielded the measurements 353, 351, 351, and 355. Test the null hypothesis that $\sigma = .7$ against the alternative $\sigma > .7$. Use $\alpha = .05$.

9.51 Find a 90% confidence interval for the population variance in Exercise 9.50.

9.52 To properly treat patients, drugs prescribed by physicians must have a potency that is accurately defined. Consequently, the distribution of potency values for shipments of a drug must not only have a mean value as specified on the drug's container, but the variation in potency must be small. Otherwise, pharmacists would be distributing drug prescriptions that could be harmfully potent or have a low potency that would be ineffective. A drug manufacturer claims that his drug is marketed with a potency of $5 \pm .1$ milligram per cubic centimeter (mg/cc). A random sample of four containers gave potency readings equal to 4.94, 5.09, 5.03, and 4.90 mg/cc.

a Do the data present sufficient evidence to indicate that the mean potency differs from 5 mg/cc?

b Do the data present sufficient evidence to indicate that the variation in potency differs from the error limits specified by the manufacturer? [HINT: It is sometimes difficult to determine exactly what is meant by limits on potency as specified by a manufacturer. Since he implies that the potency values will fall in the interval $5.0 \pm .1$ mg/cc with very high probability—the implication is *always*—let us assume that the range .2, i.e., (4.9 to 5.1) represents 6σ, as suggested by the Empirical Rule. Note that letting the range equal 6σ rather than 4σ places a stringent interpretation on the manufacturer's claim. We want the potency to fall in the interval $5.0 \pm .1$ with very high probability.]

9.53 Refer to Exercise 9.52. Testing of 60 additional randomly selected containers of the drug gave a sample mean and variance equal to 5.04 and .0063 (for the total of $n = 64$ containers). Using a 95% confidence interval, estimate the variance of the manufacturer's potency measurements.

9.54 A manufacturer of hard safety hats for construction workers is concerned about the mean and the variation of the forces helmets transmit to wearers when subjected to a standard external force. The manufacturer desires the mean force transmitted by helmets to be 800 pounds (or less), well under the legal 1000-pound limit, and σ to be less than 40. A random sample of $n = 40$ helmets was tested, and the sample mean and variance were found to be equal to 825 pounds and 2350 pounds^2, respectively.

a If $\mu = 800$ and $\sigma = 40$, is it likely that any helmet, subjected to the standard external force, will transmit a force to a wearer in excess of 1000 pounds? Explain.

b Do the data provide sufficient evidence to indicate that when subjected to the standard external force, the mean force transmitted by the helmets exceeds 800 pounds?

9.55 Refer to Exercise 9.54. Do the data provide sufficient evidence to indicate that σ exceeds 40?

9.56 A manufacturer of industrial light bulbs likes its bulbs to have a mean life that is acceptable to its customers and a variation in life that is relatively small. If some bulbs fail too early in their life, customers become annoyed and shift to competitive products. Large variations above the mean reduce replacement sales, and variation in general disrupts customers' replacement schedules. A sample of 20 bulbs tested produced the following lengths of life (in hours):

2100, 2302, 1951, 2067, 2415, 1883, 2101, 2146, 2278, 2019,
1924, 2183, 2077, 2392, 2286, 2501, 1946, 2161, 2253, 1827

The manufacturer wishes to control the variability in length of life so that σ is less than 150 hours. Do the data provide sufficient evidence to indicate that the manufacturer is achieving this goal? Test using $\alpha = .01$.

9.7 Comparing Two Population Variances

The need for statistical methods to compare two population variances is readily apparent from the discussion in Section 9.6. We may frequently wish to compare the precision of one measuring device with that of another, the stability of one manufacturing process with that of another, or even the variability in the grading procedure of one college professor with that of another.

Intuitively, we might compare two population variances σ_1^2 and σ_2^2 using the ratio of the sample variances s_1^2/s_2^2. If s_1^2/s_2^2 is nearly equal to 1, we would find little evidence to indicate that σ_1^2 and σ_2^2 are unequal. On the other hand, a very large or very small value for s_1^2/s_2^2 would provide evidence of a difference in the population variances.

How large or small must s_1^2/s_2^2 be for sufficient evidence to exist to reject the following null hypothesis?

$$H_0 : \sigma_1^2 = \sigma_2^2$$

The answer to this question may be found by studying the distribution of s_1^2/s_2^2 in repeated sampling.

When independent random samples are drawn from two normal populations with equal variances, that is, $\sigma_1^2 = \sigma_2^2$, then s_1^2/s_2^2 has a probability distribution in repeated sampling that is known to statisticians as an ***F* distribution**.

Assumptions for s_1^2/s_2^2 to Have an F Distribution

1. Random and independent samples are drawn from each of two normal populations.
2. The variability of the measurements in the two populations is the same and can be measured by a common variance, σ^2. That is, $\sigma_1^2 = \sigma_2^2 = \sigma^2$.

We need not concern ourselves with the equation of the density function for F except to state that, as we might surmise, it is reasonably complex. For our purposes, it will suffice to accept the fact that the distribution is well known and that critical values have been tabulated. These appear in Table 6 in Appendix I.

The shape of an F distribution as shown in Figure 9.4 is nonsymmetric and will depend on the number of degrees of freedom associated with s_1^2 and s_2^2. We will represent these quantities as ν_1 and ν_2, respectively. This fact complicates the tabulation of critical values of the F distribution and necessitates the construction of a table to accommodate differing values of ν_1, ν_2, and a.

FIGURE **9.4**
An F distribution with $\nu_1 = 10$ and $\nu_2 = 10$

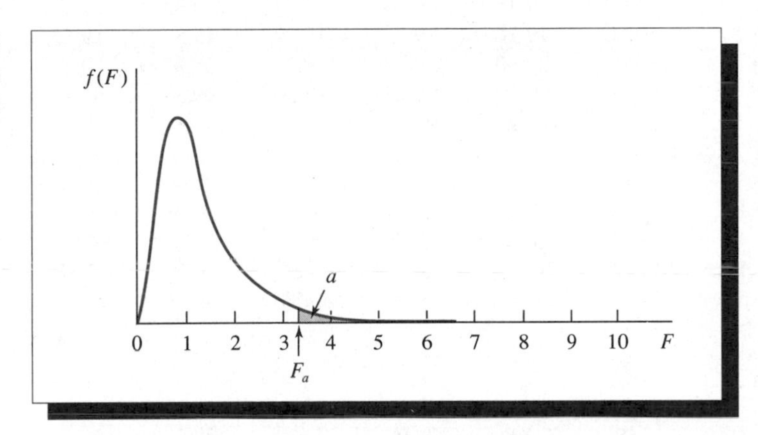

In Table 6, Appendix I, critical values of F for right-tailed areas corresponding to $a = .10, .05, .025, .010$, and $.005$ are tabulated for various combinations of ν_1 numerator degrees of freedom and ν_2 denominator degrees of freedom. A portion of Table 6 is reproduced in Table 9.9. The numerator degrees of freedom ν_1 are listed across the top margin, and the denominator degrees of freedom ν_2 are listed along both side margins. The values of a are listed in the second column from the left as well as from the right. For a fixed combination of ν_1 and ν_2 the appropriate critical values of F are found in the line indexed by the value of a required.

Referring to Table 9.9, we note that $F_{.05}$ for sample sizes $n_1 = 7$ and $n_2 = 10$ (i.e., $\nu_1 = 6$, $\nu_2 = 9$) is 3.37. Similarly, the critical value $F_{.05}$ for samples sizes $n_1 = 9$ and $n_2 = 12$ ($\nu_1 = 8$, $\nu_2 = 11$) is 2.95. These values of F are shaded in Table 9.9.

In a similar manner, the critical values for a tail area, $a = .01$, are presented in Table 6 in Appendix I. Thus, if $\nu_1 = 6$ and $\nu_2 = 9$,

$$P(F > F_{.01}) = P(F > 5.80) = .01$$

The statistical test of the null hypothesis

$$H_0 : \sigma_1^2 = \sigma_2^2$$

uses the test statistic

$$F = \frac{s_1^2}{s_2^2}$$

TABLE 9.9 Format of the F table from Table 6 in Appendix I

		ν_1								
ν_2	a	1	2	3	4	5	6	7	8	9
1	.100	39.86	49.50	53.59	55.83	57.24	58.20	58.91	59.44	59.86
	.050	161.4	199.5	215.7	224.6	230.2	234.0	236.8	238.9	240.5
	.025	647.8	799.5	864.2	899.6	921.8	937.1	948.2	956.7	963.3
	.010	4052	4999.5	5403	5625	5764	5859	5928	5982	6022
	.005	16211	20000	21615	22500	23056	23437	23715	23925	24091
2	.100	8.53	9.00	9.16	9.24	9.29	9.33	9.35	9.37	9.38
	.050	18.51	19.00	19.16	19.25	19.30	19.33	19.35	19.37	19.38
	.025	38.51	39.00	39.17	39.25	39.30	39.33	39.36	39.37	39.39
	.010	98.50	99.00	99.17	99.25	99.30	99.33	99.36	99.37	99.39
	.005	198.5	199.0	199.2	199.2	199.3	199.3	199.4	199.4	199.4
3	.100	5.54	5.46	5.39	5.34	5.31	5.28	5.27	5.25	5.24
	.050	10.13	9.55	9.28	9.12	9.01	8.94	8.89	8.85	8.81
	.025	17.44	16.04	15.44	15.10	14.88	14.73	14.62	14.54	14.47
	.010	34.12	30.82	29.46	28.71	28.24	27.91	27.67	27.49	27.35
	.005	55.55	49.80	47.47	46.19	45.39	44.84	44.43	44.13	43.88
⋮	⋮			⋮			⋮			⋮
9	.100	3.36	3.01	2.81	2.69	2.61	2.55	2.51	2.47	2.44
	.050	5.12	4.26	3.86	3.63	3.48	3.37	3.29	3.23	3.18
	.025	7.21	5.71	5.08	4.72	4.48	4.32	4.20	4.10	4.03
	.010	10.56	8.02	6.99	6.42	6.06	5.80	5.61	5.47	5.35
	.005	13.61	10.11	8.72	7.96	7.47	7.13	6.88	6.69	6.54
10	.100	3.29	2.92	2.73	2.61	2.52	2.46	2.41	2.38	2.35
	.050	4.96	4.10	3.71	3.48	3.33	3.22	3.14	3.07	3.02
	.025	6.94	5.46	4.83	4.47	4.24	4.07	3.95	3.85	3.78
	.010	10.04	7.56	6.55	5.99	5.64	5.39	5.20	5.06	4.94
	.005	12.83	9.43	8.08	7.34	6.87	6.54	6.30	6.12	5.97
11	.100	3.23	2.86	2.66	2.54	2.45	2.39	2.34	2.30	2.27
	.050	4.84	3.98	3.59	3.36	3.20	3.09	3.01	2.95	2.90
	.025	6.72	5.26	4.63	4.28	4.04	3.88	3.76	3.66	3.59
	.010	9.65	7.21	6.22	5.67	5.32	5.07	4.89	4.74	4.63
	.005	12.23	8.91	7.60	6.88	6.42	6.10	5.86	5.68	5.54
12	.100	3.18	2.81	2.61	2.48	2.39	2.33	2.28	2.24	2.21
	.050	4.75	3.89	3.49	3.26	3.11	3.00	2.91	2.85	2.80
	.025	6.55	5.10	4.47	4.12	3.89	3.73	3.61	3.51	3.44
	.010	9.33	6.93	5.95	5.41	5.06	4.82	4.64	4.50	4.39
	.005	11.75	8.51	7.23	6.52	6.07	5.76	5.52	5.35	5.20

When the alternative hypothesis implies a one-tailed test, that is,

$$H_a : \sigma_1^2 > \sigma_2^2$$

we can use the tables directly. However, when the alternative hypothesis requires a two-tailed test,

$$H_a : \sigma_1^2 \neq \sigma_2^2$$

we note that the rejection region will be divided between the lower and upper tails of the F distribution and that tables of critical values for the lower tail are conspicuously missing. The reason for their absence is explained as follows: We are at liberty to identify either of the two populations as population I. If the population with the larger sample variance is designated as population II, then $s_2^2 > s_1^2$ and we will be concerned with rejection in the lower tail of the F distribution. Since the identification of the populations was arbitrary, we can avoid this difficulty by designating the population with the larger sample variance as population I. In other words, always place the larger sample variance in the numerator of

$$F = \frac{s_1^2}{s_2^2}$$

and designate that population as I. Then, since the area in the right-hand tail will represent only $\alpha/2$, we double this value to obtain the correct value for the probability of a type I error α.

Test of an Hypothesis Concerning the Equality of Two Population Variances

1 Null Hypothesis: $H_0 : \sigma_1^2 = \sigma_2^2$

2 Alternative Hypothesis:

One-Tailed Test	*Two-Tailed Test*
$H_a : \sigma_1^2 > \sigma_2^2$ (or, $H_a : \sigma_2^2 > \sigma_1^2$)	$H_a : \sigma_1^2 \neq \sigma_2^2$

3 Test Statistic:

One-Tailed Test	*Two-Tailed Test*
$F = \frac{s_1^2}{s_2^2}$ $\left(\text{or, } F = \frac{s_2^2}{s_1^2} \text{ for } H_a : \sigma_2^2 > \sigma_1^2\right)$	$F = \frac{s_1^2}{s_2^2}$ where s_1^2 is the larger sample variance.

4 Rejection Region:

One-Tailed Test	*Two-Tailed Test*
$F > F_\alpha$	$F > F_{\alpha/2}$

When $F = s_1^2/s_2^2$, the critical values of F, F_α, and $F_{\alpha/2}$ are based on $\nu_1 = n_1 - 1$ and $\nu_2 = n_2 - 1$ d.f. These tabulated values, for $\alpha = .10, .05, .025, .01$, and $.005$, can be found in Table 6 in Appendix I.

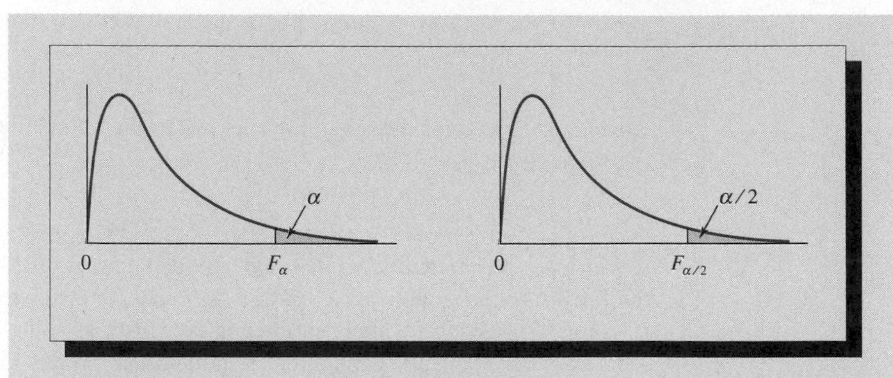

Assumptions: The samples were randomly and independently selected from normally distributed populations.

We will illustrate with examples.

EXAMPLE 9.13 Two samples consisting of 10 and 8 measurements each were observed to have sample variances equal to $s_1^2 = 7.14$ and $s_2^2 = 3.21$, respectively. Do the sample variances present sufficient evidence to indicate that the population variances are unequal?

Solution Assume that the populations have probability distributions that are reasonably mound-shaped and hence will satisfy, for all practical purposes, the assumption that the populations are normal.

We wish to test the null hypothesis

$$H_0 : \sigma_1^2 = \sigma_2^2$$

against the alternative

$$H_a : \sigma_1^2 \neq \sigma_2^2$$

Using Table 6 in Appendix I for $\alpha/2 = .025$, we will reject H_0 when $F > 4.82$ with $\alpha = .05$.

The calculated value of the test statistic is

$$F = \frac{s_1^2}{s_2^2} = \frac{7.14}{3.21} = 2.22$$

Because the test statistic does not fall in the rejection region, we do not reject $H_0 : \sigma_1^2 = \sigma_2^2$. Thus, there is insufficient evidence to indicate a difference in the population variances. ■

The confidence interval, with confidence coefficient $(1-\alpha)$, for the ratio of two population variances can be shown to equal

A Confidence Interval for σ_1^2/σ_2^2

$$\frac{s_1^2}{s_2^2}\frac{1}{F_{\nu_1,\nu_2}} < \frac{\sigma_1^2}{\sigma_2^2} < \frac{s_1^2}{s_2^2}F_{\nu_2,\nu_1}$$

where $\nu_1 = n_1 - 1$ and $\nu_2 = n_2 - 1$.

Assumptions: The samples were randomly and independently selected from normally distributed populations.

F_{ν_1,ν_2} is the tabulated critical value of F corresponding to ν_1 and ν_2 degrees of freedom in the numerator and denominator of F, respectively. Similar to the two-tailed test, α will be twice the tabulated value.

EXAMPLE 9.14 Refer to Example 9.13 and find a 90% confidence interval σ_1^2/σ_2^2.

Solution The 90% confidence interval for σ_1^2/σ_2^2 in Example 9.13 is

$$\frac{s_1^2}{s_2^2}\frac{1}{F_{\nu_1,\nu_2}} < \frac{\sigma_1^2}{\sigma_2^2} < \frac{s_1^2}{s_2^2}F_{\nu_2,\nu_1}$$

where

$$s_1^2 = 7.14 \qquad s_2^2 = 3.21$$
$$\nu_1 = (n_1 - 1) = 9 \qquad \nu_2 = (n_2 - 1) = 7$$
$$F_{\nu_1,\nu_2} = F_{9,7} = 3.68 \qquad F_{\nu_2,\nu_1} = F_{7,9} = 3.29$$

Substituting these values into the formula for the confidence interval, we obtain

$$\left(\frac{7.14}{3.21}\right)\frac{1}{3.68} < \frac{\sigma_1^2}{\sigma_2^2} < \frac{(7.14)(3.29)}{3.21} \quad \text{or} \quad .60 < \frac{\sigma_1^2}{\sigma_2^2} < 7.32$$

The calculated interval estimate .60 to 7.32 is observed to include 1.0, the value hypothesized in H_0. This indicates that it is quite possible that $\sigma_1^2 = \sigma_2^2$ and therefore agrees with our test conclusions. Do not reject $H_0 : \sigma_1^2 = \sigma_2^2$. ■

EXAMPLE 9.15 The variability in the amount of impurities present in a batch of chemical used for a particular process depends on the length of time the process is in operation. A manufacturer using two production lines 1 and 2 has made a slight adjustment to process 2, hoping to reduce the variability as well as the average amount of impurities in the chemical. Samples of $n_1 = 25$ and $n_2 = 25$ measurements from the two batches yield means and variances as follows:

$$\bar{x}_1 = 3.2 \quad s_1^2 = 1.04$$
$$\bar{x}_2 = 3.0 \quad s_2^2 = 0.51$$

Do the data present sufficient evidence to indicate that the process variability is less for process 2? Test the null hypothesis $H_0 : \sigma_1^2 = \sigma_2^2$.

Solution The practical implications of this example are illustrated in Figure 9.5. We believe that the mean levels of impurities in the two production lines are nearly equal (in fact, they may be equal) but that there is a possibility that the variation in the level of impurities is substantially less for line 2. Then distributions of impurity measurements for the two production lines would have nearly the same mean level, but they would differ in their variation. A large variance for the level of impurities increases the probability of producing shipments of chemical with an unacceptably high level of impurities. Consequently, we hope to show that the process change in line 2 has made σ_2^2 less than σ_1^2.

Testing the null hypothesis

$$H_0 : \sigma_1^2 = \sigma_2^2$$

against the alternative

$$H_a : \sigma_1^2 > \sigma_2^2$$

at an $\alpha = .05$ significance level, we will reject H_0 when F is greater than $F_{.05} = 1.98$; that is, we will employ a one-tailed statistical test.

We readily observe that the calculated value of the test statistic

$$F = \frac{s_1^2}{s_2^2} = \frac{1.04}{.51} = 2.04$$

falls in the rejection region, and hence we conclude that the variability of process 2 is less than that of process 1.

FIGURE 9.5
Distributions of impurity measurements for two production lines

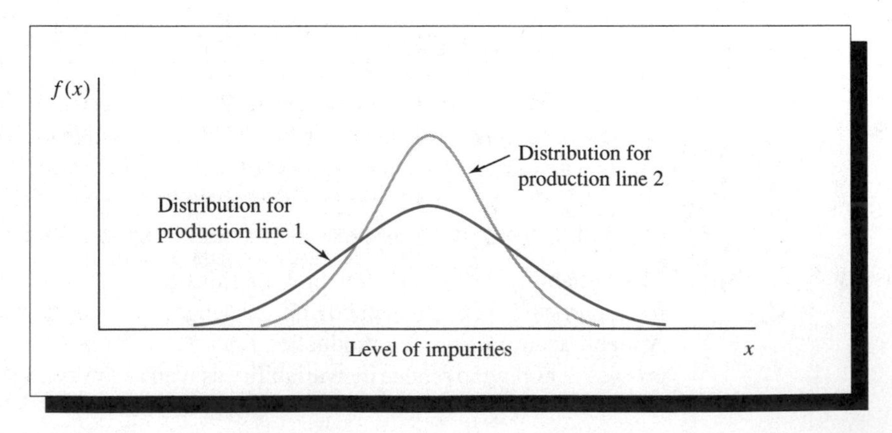

The 90% confidence interval for the ratio σ_1^2/σ_2^2 is

$$\frac{s_1^2}{s_2^2}\frac{1}{F_{\nu_1,\nu_2}} < \frac{\sigma_1^2}{\sigma_2^2} < \frac{s_1^2}{s_2^2}F_{\nu_2,\nu_1}$$

$$\frac{1.04}{(.51)(1.98)} < \frac{\sigma_1^2}{\sigma_2^2} < \frac{(1.04)(1.98)}{.51}$$

$$\text{or } 1.03 < \frac{\sigma_1^2}{\sigma_2^2} < 4.04$$

Thus, we estimate the reduction in the variance of the amount of impurities to be as large as 4.04 to 1 or as small as 1.03 to 1. The actual reduction would likely be somewhere between these two extremes. This would suggest that the adjustment is effective in reducing the variation in the amount of impurities in the chemical. ▪

Exercises

Basic Techniques

9.57 Independent random samples from two normal populations produced the following variances:

Sample Size	Sample Variance
16	55.7
20	31.4

a Do the data provide sufficient evidence to indicate that σ_1^2 differs from σ_2^2? Test using $\alpha = .05$.

b Find the approximate observed significance level for the test and interpret its value.

9.58 Refer to Exercise 9.57 and find a 95% confidence interval for σ_1^2/σ_2^2.

9.59 Independent random samples from two normal populations produced the following variances:

Sample Size	Sample Variance
13	18.3
13	7.9

a Do the data provide sufficient evidence to indicate that $\sigma_1^2 > \sigma_2^2$? Test using $\alpha = .05$.

b Find the approximate observed significance level for the test and interpret its value.

Applications

9.60 Refer to Exercise 9.26, in which the average total SAT scores (verbal plus math) were recorded for two groups of students, one group planning to major in engineering and one group planning to major in language/literature. The data are reproduced below.

Engineering	Language/Literature
$\bar{x} = 994$	$\bar{x} = 1051$
$s = 71$	$s = 69$
$n = 15$	$n = 15$

To use the two-sample t test with a pooled estimate of σ^2, we assumed that the two population variances were equal. Test this assumption using the F test for equality of variances. What is the approximate p-value for the test? Were we correct in using the two-sample analysis in Exercise 9.26?

9.61 The stability of measurements of the characteristics of a manufactured product is important in maintaining product quality. In fact, it is sometimes better to have small variation in the measured value of some important characteristic of a product and have the process mean be slightly off target than to suffer wide variation with a mean value that perfectly fits requirements. The latter situation may produce a higher percentage of defective products than the former. A manufacturer of light bulbs suspected that one of her production lines was producing bulbs with a high variation in length of life. To test this theory, she compared the lengths of life for $n = 50$ bulbs randomly sampled from the suspect line and $n = 50$ from a line that seemed to be "in control." The sample means and variances for the two samples were as follows:

"Suspect Line"	Line "in Control"
$\bar{x}_1 = 1{,}520$	$\bar{x}_2 = 1{,}476$
$s_1^2 = 92{,}000$	$s_2^2 = 37{,}000$

a Do the data provide sufficient evidence to indicate that bulbs produced by the "suspect line" have a larger variance in length of life than those produced by the line that is assumed to be in control? Test using $\alpha = .05$.

b Find the approximate observed significance level for the test and interpret its value.

9.62 Use the method of Section 9.7 to obtain a 90% confidence interval for the variance ratio in Exercise 9.61.

9.63 In Exercise 9.23, we conducted a test to detect a difference between the percentages of oxygen uptake by air in reedfish exposed to water at 25°C versus those at 33°C.

a What assumption had to be made concerning the population variances in order that the test be valid?

b Do the data provide sufficient evidence to indicate that they violate the assumption in part (a)? Test using $\alpha = .05$.

9.64 Refer to Exercise 9.24. Susan Beckham and colleagues conducted an experiment involving ten healthy runners and ten healthy cyclists to determine if there are significant differences in pressure measurements within the anterior muscle compartment for runners and cyclists. The data—compartment pressure in millimeters of mercury (Hg)—are reproduced below:

	Runners		Cyclists	
Condition	Mean	Standard Deviation	Mean	Standard Deviation
Rest	14.5	3.92	11.1	3.98
80% maximal O_2 consumption	12.2	3.49	11.5	4.95
Maximal O_2 consumption	19.1	16.9	12.2	4.47

Source: Beckham et al. (1993), Table 1.

For each of the three variables measured in this experiment, test to see if there is a significant difference in the variances for runners versus cyclists. Find the approximate p-values for each of these tests. Will a two-sample t test with a pooled estimate of σ^2 be appropriate for all three of these variables? Explain.

9.65 A pharmaceutical manufacturer purchases a particular material from two different suppliers. The mean level of impurities in the raw material is approximately the same for both suppliers, but the manufacturer is concerned about the variability of the impurities from shipment to shipment. If the level of impurities tends to vary excessively for one source of supply, it could affect the quality of the pharmaceutical product. To compare the variation in percentage impurities for the two suppliers, the manufacturer selects ten shipments from each of the two suppliers and measures the percentage of impurities in the raw material for each shipment. The sample means and variances are shown in the table:

Supplier A	Supplier B
$\bar{x}_1 - 1.89$	$\bar{x}_2 - 1.85$
$s_1^2 = .273$	$s_2^2 = .094$
$n_1 = 10$	$n_2 = 10$

a Do the data provide sufficient evidence to indicate a difference in the variability of the shipment impurity levels for the two suppliers? Test using $\alpha = .10$. Based on the results of your test, what recommendation would you make to the pharmaceutical manufacturer?

b Find a 90% confidence interval for σ_2^2 and interpret your results.

9.66 In Exercise 9.27 we raised questions about the validity of the Student's t test for making a comparison of population means. Use the F test to show that the data present sufficient evidence to indicate differences in population variances for the comparison questioned in Exercise 9.27.

9.8 Assumptions

As noted earlier, the tests and confidence intervals based on the Student's t, the chi-square, and the F statistic require that the data satisfy specific assumptions in order that the error probabilities (for the tests) and the confidence coefficients (for the confidence intervals) be equal to the values we have specified. For example, if the assumptions are violated by selecting a sample from a nonnormal population, and the data are used to construct a 95% confidence interval for μ, the actual confidence coefficient might, unbeknownst to us, be equal to .85 instead of .95. The assumptions are summarized next for your convenience.

Assumptions

1. For all tests and confidence intervals described in this chapter, it is assumed that **samples are randomly selected from normally distributed populations.**
2. When two samples are selected, we assume that they are **selected in an independent manner** except in the case of the paired-difference experiment.
3. For tests or confidence intervals concerning **the difference between two population means** μ_1 and μ_2 based on independent random samples, we **assume that** $\sigma_1^2 = \sigma_2^2$.

In a practical sampling situation, you never know everything about the probability distribution of the sampled population. If you did, there would be no need for sampling or statistics. Second, it is highly unlikely that a population would have exactly the characteristics described above. Consequently, to be useful, the inferential methods described in this chapter must give good inferences when moderate departures from the assumptions are present. For example, if the population has a mound-shaped distribution that is nearly normal, we would like a 95% confidence interval constructed for μ to be one with a confidence coefficient close to .95. Similarly, if we conduct a t test of the null hypothesis $\mu_1 = \mu_2$ based on independent random samples from normal populations where σ_1^2 and σ_2^2 are not exactly equal, we want the probability of incorrectly rejecting the null hypothesis α to be approximately equal to the value we used in locating the rejection region.

A statistical method that is insensitive to departures from the assumptions on which the method is based is said to be robust. The t tests are quite robust to moderate departures from normality. In contrast, the chi-square and F tests are sensitive to departures from normality. The t test for comparing two means is moderately robust to departures from the assumption $\sigma_1^2 = \sigma_2^2$, when $n_1 = n_2$. However, the test becomes sensitive to departures from this assumption as n_1 becomes large relative to n_2 (or vice versa).

If you are concerned that your data do not satisfy the assumptions prescribed for one of the statistical methods described in this chapter, you may be able to use a nonparametric statistical method to make your inference. These methods, which require few or no assumptions about the nature of the population probability distributions, are particularly useful for testing hypotheses, and some nonparametric methods have been developed for estimating population parameters. Tests of hypotheses concerning the location of a population distribution or a test for the equivalence of two population distributions are presented in Chapter 14. If you can select relatively large samples, you can usually use one of the large-sample estimation or test procedures of Chapters 7 and 8.

9.9 Summary

It is important to note that the t, χ^2, and F statistics employed in the small-sample statistical methods discussed in the preceding sections are based on the assumption that the sampled populations have a normal probability distribution. This requirement will be adequately satisfied for many types of experimental measurements.

You will observe the very close relationship connecting Student's t and the z statistic and, therefore, the similarity of the methods for testing hypotheses and the construction of confidence intervals. The χ^2 and F statistics employed in making inferences concerning population variances do not, of course, follow this pattern, but the reasoning employed in the construction of the statistical tests and confidence intervals is identical for all the methods we have presented.

Tips on Problem Solving

To help you decide whether the techniques of this chapter are appropriate for the solution of a problem, ask yourself the following questions:

1. Does the problem imply that an inference should be made about a population mean or the difference between two means? Are the samples small—say, $n < 30$? If the answers to both questions are yes, you may be able to use one of the methods of Sections 9.3, 9.4, or 9.5. If the sample sizes are large, $n \geq 30$, you can use the methods of Chapters 7 and 8. In practice, you would also need to verify that the assumptions underlying each procedure are satisfied. Are the population distributions nearly normal, and have the sampling procedures conformed to those prescribed for the statistical method?
2. When comparing population means, were the observations from the two populations selected in a paired manner? If they were, you must use the paired-difference analysis of Section 9.5. If the samples were selected independently and in a random manner, use the methods of Section 9.4.
3. Is data variation the primary objective of the problem? If it is, you may be required to make an inference about a population variance σ^2 (Section 9.6) or to compare two population variances σ_1^2 and σ_2^2 (Section 9.7).

(NOTE: The tips on problem solving following Section 7.11 will also be helpful in solving problems in this chapter.)

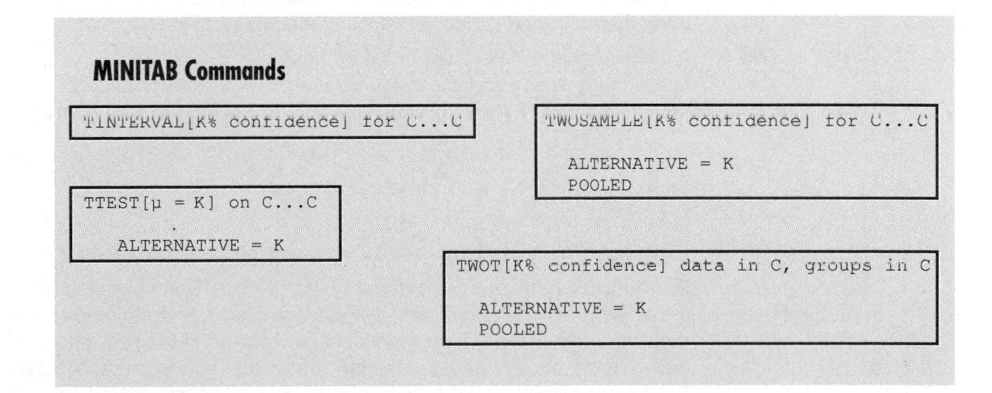

MINITAB Commands

```
TINTERVAL[K% confidence] for C...C
```

```
TWOSAMPLE[K% confidence] for C...C

  ALTERNATIVE = K
  POOLED
```

```
TTEST[μ = K] on C...C
      .
  ALTERNATIVE = K
```

```
TWOT[K% confidence] data in C, groups in C

  ALTERNATIVE = K
  POOLED
```

Supplementary Exercises

9.67 What assumptions are made when Student's t test is employed to test an hypothesis concerning a population mean?

9.68 What assumptions are made about the populations from which independent random samples are obtained when using the t distribution in making small-sample inferences concerning the difference in population means?

9.69 Why use paired observations to estimate the difference between two population means rather than estimation based on independent random samples selected from the two populations? Is a paired experiment always preferable? Explain.

9.70 A manufacturer can tolerate a small amount [.05 milligrams per liter (mg/l)] of impurities in a raw material needed for manufacturing its product. Because the laboratory test for the impurities is subject to experimental error, the manufacturer tests each batch ten times. Assume that the mean value of the experimental error is 0 and hence that the mean value of the ten test readings is an unbiased estimate of the true amount of the impurities in the batch. For a particular batch of the raw material, the mean of the ten test readings is .058 mg/l, with a standard deviation of .012 mg/l. Do the data provide sufficient evidence to indicate that the amount of impurities in the batch exceeds .05 mg/l? Find the p-value for the test and interpret its value.

9.71 The main stem growth measured for a sample of 17 four-year-old red pine trees produced a mean and standard deviation equal to 11.3 and 3.4 inches, respectively. Find a 90% confidence interval for the mean growth of a population of four-year-old red pine trees subjected to similar environmental conditions.

9.72 The object of a general chemistry experiment is to determine the amount (in milliliters) of sodium hydroxide (NaOH) solution needed to neutralize 1 gram of a specified acid. This will be an exact amount, but when run in the laboratory, variation will occur as the result of experimental error. Three titrations are made using phenolphthalein as an indicator of the neutrality of the solution (pH equals 7 for a neutral solution). The three volumes of NaOH required to attain a pH of 7 in each of the three titrations are as follows: 82.10, 75.75, and 75.44 milliliters. Use a 90% confidence interval to estimate the mean number of milliliters required to neutralize 1 gram of the acid.

9.73 Measurements of water intake, obtained from a sample of 17 rats that had been injected with a sodium chloride solution, produced a mean and standard deviation of 31.0 and 6.2 cubic centimeters (cm^3), respectively. Given that the average water intake for noninjected rats observed over a comparable period of time is 22.0 cm^3, do the data indicate that injected rats drink more water than noninjected rats? Test at the 5% level of significance. Find a 90% confidence interval for the mean water intake for injected rats.

9.74 An experimenter, interested in determining the mean thickness of the cortex of the sea urchin egg, employed an experimental procedure developed by Sakai. The thickness of the cortex was measured for $n = 10$ sea urchin eggs. The following measurements were obtained.

4.5	6.1	3.2	3.9	4.7
5.2	2.6	3.7	4.6	4.1

Estimate the mean thickness of the cortex using a 95% confidence interval.

9.75 A production plant has two extremely complex fabricating systems; one system is twice as old as the other. Both systems are checked, lubricated, and maintained once every 2 weeks. The number of finished products fabricated daily by each of the systems is recorded for 30 working days. The results are given in the table. Do these data present sufficient evidence to conclude that the variability in daily production warrants increased maintenance of the older fabricating system? Use a 5% level of significance.

New System	Old System
$\bar{x}_1 = 246$	$\bar{x}_2 = 240$
$s_1 = 15.6$	$s_2 = 28.2$

9.76 The data in the table give the diameters and heights of ten fossil specimens of a species of small shellfish, *Rotularia (Annelida) fallax*, that were unearthed in a mapping expedition near the Antarctic Peninsula (Macellari, 1984). The first column gives Macellari's identification symbol for the fossil specimen. The second and third columns give the fossil's diameter and height, respectively, in millimeters (mm). The fourth column gives the ratio of diameter to height.

Specimen	Diameter	Height	D/H
OSU 36651	185	78	2.37
OSU 36652	194	65	2.98
OSU 36653	173	77	2.25
OSU 36654	200	76	2.63
OSU 36655	179	72	2.49
OSU 36656	213	76	2.80
OSU 36657	134	75	1.79
OSU 36658	191	77	2.48
OSU 36659	177	69	2.57
OSU 36660	199	65	3.06
$\bar{x}$:	184.5	73	2.54
s :	21.5	5	3.7

Find a 95% confidence interval for

a The mean diameter of the specimens.

b The mean ratio of diameter to height.

c The mean height of the specimens.

9.77 Refer to Exercise 9.76. Suppose you want to estimate the mean diameter of the fossil specimens correct to within 5 mm with probability equal to .95. How many fossils would you have to include in your sample?

9.78 The length of time to recovery was recorded for patients randomly assigned and subjected to two different surgical procedures. The data, recorded in days, are as follows:

Procedure I	Procedure II
$n_1 = 21$	$n_2 = 23$
$\bar{x}_1 = 7.3$	$\bar{x}_2 = 8.9$
$s_1^2 = 1.23$	$s_2^2 = 1.49$

a Do the data present sufficient evidence to indicate a difference between the mean recovery times for the two surgical procedures? Test using $\alpha = .05$.

b Find the approximate observed significance level for the test and interpret its value.

9.79 In behavior analysis, self-control is most often defined as the choice of a large, delayed reinforcer over the choice of a small, immediate reinforcer. In an experiment by Stephen Flora and fellow researchers (1992) to determine the effects of noise on human self-control, subjects pressed two buttons for points that were exchangeable for money. Pressing one of the buttons produced 2 points over 4 seconds; pressing the other button, the self-control choice, produced 10 points over 4 seconds after a delay of 16 seconds. Seven subjects were asked to make 30 choices in the presence of aversive noise and in the absence of noise. Another six subjects were asked to make 30 choices first in the absence of noise and then in the presence of aversive noise. The number of correct choices for each subject for both conditions is given in the following table:

Subject	Noise	Quiet	Subject	Quiet	Noise
1	17	11	8	26	30
2	24	27	9	26	30
3	11	13	10	10	25
4	16	30	11	19	24
5	25	27	12	20	26
6	12	17	13	26	30
7	17	12			
Means	17.43	19.57		21.17	27.5

a Test for a significant difference in the mean number of self-control choices made by the first seven subjects and the second six subjects under the *quiet* condition at the $\alpha = .05$ level of significance. Construct a 95% confidence interval estimate of the difference in means under the *quiet* condition.

b Test for a significant difference in the mean number of self-control choices made by the first seven subjects and the second six subjects under the *noise* condition at the $\alpha = .05$ level of significance.

c Since each subject was tested under both conditions (either *noise* followed by *quiet* or *quiet* followed by *noise*), these within-subject observations are, in fact, paired. Using the data for the first seven subjects, test for a significant difference in the average number of self-control choices between the noise and quiet conditions, with $\alpha = .05$. Construct a 95% confidence interval estimate of the mean difference for the noise-versus-quiet condition using these seven subjects.

d Using the data for the six subjects numbered 8–13, use a paired t test to determine if there is a significant mean difference between the quiet and noise conditions. Use $\alpha = .05$.

9.80 A stringer of tennis racquets claims to be able to string racquets to within 3 pounds per square inch (lb/in.^{2}) of the tension requested by her customers. Laboratory tests of a sample of $n = 10$ of the stringer's racquets produced the following data:

	Racquet									
	1	2	3	4	5	6	7	8	9	10
Requested tension	70	65	65	55	60	68	65	60	65	60
Test tension	68	62	66	54	62	72	64	62	69	60

a Assume that the stringer's claim to be within 3 lb/in.^{2} of the requested tension means that three standard deviations of the difference between requested and test tensions is 3 lb/in.^{2}. Do the data present sufficient evidence to indicate that the stringer's deviations from requested tension exceed the claim? Test using $\alpha = .05$.

b Do the data present sufficient evidence to indicate bias in the stringing machine, that is, a tendency to string above (or below) the requested tension? Test using $\alpha = .05$.

9.81 Do we lose our memory capability as we get older? In a study of the effect of glucose on memory in elderly men and women, C. A. Manning, J. L. Hall, and P. E. Gold (1990) tested 16 volunteers (5 men and 11 women) for long-term word memory, recording the number of words recalled from a list read to the volunteer. The volunteer was reminded of those words missed and was asked to recall as many words as possible from the total list. The mean and standard deviation of the long-term word-memory scores were $\bar{x} = 79.47$ and $s = 25.25$. Find a 99% confidence interval for μ, the average long-term word-memory scores for elderly men and women. Interpret this interval.

9.82 An experiment was conducted to compare the mean lengths of time required for the bodily absorption of two drugs A and B. Ten people were randomly selected and assigned to each drug treatment. Each person received an oral dosage of the assigned drug, and the length of time (in minutes) for the drug to reach a specified level in the blood was recorded. The means and variances for the two samples are as follows:

Drug A	Drug B
$\bar{x}_1 = 27.2$	$\bar{x}_2 = 33.5$
$s_1^2 = 16.36$	$s_2^2 = 18.92$

a Do the data provide sufficient evidence to indicate a difference in mean times to absorption for the two drugs? Test using $\alpha = .10$.

b Find the approximate observed significance level for the test and interpret its value.

9.83 Refer to Exercise 9.82. Find a 95% confidence interval for the difference in mean times to absorption.

9.84 Refer to Exercise 9.83. Suppose you wish to estimate the difference in mean times to absorption correct to within 1 minute with probability approximately equal to .95.

a Approximately how large a sample would be required for each drug (assume that the sample sizes are equal)?

b To conduct the experiment using the sample sizes of part (a) will require a large amount of time and money. Can anything be done to reduce the sample sizes and still achieve the 1-minute margin of error for estimation?

9.85 Insects hovering in flight expend enormous amounts of energy for their size and weight. The data shown below were taken from a much larger body of data collected by T. M. Casey, M. L. May, and K. R. Morgan (1985). They show the wing stroke frequencies, in hertz, for two different species of bees, $n_1 = 4$ *Euglossa mandibularis* Friese and $n_2 = 6$ *Eublossa imperialis* Cockerell.

Wing Stroke Frequencies

Euglossa mandibularis Friese	*Euglossa imperialis* Cockerell
235	180
225	169
190	180
188	185
	178
	182

Do these data provide sufficient evidence to indicate a difference in the variances of the populations of wing stroke frequencies corresponding to the two species of bees?

a Just look at the data. Based on your intuition, do you think that a difference exists between the two population variances?

b Test to determine whether a difference exists. Use $\alpha = .05$.

c Give the approximate p-value for the test in part (b).

d Explain why a Student's t test with a pooled estimator s^2 would be unsuitable for comparing the mean wing stroke frequencies for the two species of bees.

9.86 Refer to Exercise 9.81. In the same study, the 16 elderly volunteers were asked to perform a variety of tasks on each of two mornings, after ingesting 8-ounces of lemon-flavored beverage. One morning, the drink was sweetened with glucose (50 grams); on the other morning, it was sweetened with saccharin (23.7 milligrams). The results of selected tests are shown below:

Test	Glucose Mean (SD)	Saccharin Mean (SD)	Difference (SEM)
Quick test	108	106	2.00
	(14.60)	(13.17)	(2.39)
Attention test	64.65	68.94	−4.29
	(11.87)	(13.00)	(2.96)
Finger oscillation	32.18	33.90	−1.72
	(5.15)	(6.10)	(1.02)

Source: Manning et al. (1990), Table 1.

a What type of analysis (paired or unpaired) should be used to test for a significant difference between responses using glucose-versus-saccharin treatments? Explain.

b The statistics reported in the last column of the table represent the sample mean of the difference in responses for glucose-versus-saccharin treatments and (in parentheses) the standard error of the mean for these differences. Is there a significant difference in the average response for the quick test for glucose-versus-saccharin treatments? Test using $\alpha = .05$.

c Construct a 95% confidence interval for the difference in average responses for the attention test using the two treatments. Is there a significant difference in the treatment means? Explain.

d Is there a significant difference in the average response for the finger oscillation test for glucose versus saccharin treatments? Test using $\alpha = .05$.

9.87 Karl Niklas and T. G. Owens (1989) examined the differences in a particular plant, *Plantago Major L.*, when grown in full sunlight versus shade conditions. In this study, shaded plants were those receiving direct sunlight for less than 2 hours each day, whereas full-sun plants were never shaded. A partial summary of the data based on $n_1 = 16$ full-sun plants and $n_2 = 15$ shade plants is shown below:

	Full-Sun		Shade	
	$\bar{x}$	s	$\bar{x}$	s
Leaf area (cm^2)	128	43	78.7	41.7
Overlap area (cm^2)	46.8	2.21	8.1	1.26
Leaf number	9.75	2.27	6.93	1.49
Thickness (mm)	0.9	0.03	0.5	0.02
Length (cm)	8.70	1.64	8.91	1.23
Width (cm)	5.24	0.98	3.41	0.61

a What assumptions are required in order to use the small-sample procedures given in this chapter in comparing full-sun versus shade plants? From the summary presented, do you think that any of these assumptions have been violated?

b Do the data present sufficient evidence to indicate a difference in mean leaf area for full-sun versus shade plants? Test using $\alpha = .05$.

c Do the data present sufficient evidence to indicate a difference in mean overlap area for full-sun versus shade plants? Find the observed level of significance for the test.

9.88 Refer to Exercise 9.87.

a Find a 95% confidence interval for the difference in mean leaf thickness for full-sun versus shade plants.

b Find a 95% confidence interval for the difference in mean leaf width for full-sun versus shade plants.

c Use the confidence intervals from parts (a) and (b) to determine whether there is a significant difference in mean leaf thickness and width for the two populations.

9.89 Most working parents in the United States pay large amounts of money for child care; however, they are not always satisfied with the results. In fact, a Louis Harris survey indicated that only 8% of parents of preschoolers believe the child care system is working very well (*American Demographics*, Sept. 1989, p. 14). Suppose that 12 single mothers and 12 married couples are randomly selected. Their monthly child care costs are recorded below.

Married Couple	Single Mother
$\bar{x}_1 = \$185$	$\bar{x}_2 = \$211$
$s_1 = 13$	$s_2 = 15$

a Do the data provide sufficient evidence to indicate that the population variances are different? Test using $\alpha = .05$.

b If the population variances are not significantly different, use the Student's t test to determine whether there is a significant difference between the average monthly child-care payments for married couples and single mothers. Determine the approximate p-value for the test.

9.90 American adolescents who espouse traditional values, such as respect for authority, allegiance to a church, or strong belief in a god, may be a minority in the population at large, and as a result, may feel alienated from their peers. In their study of junior and senior high school students in the Midwest, Raymond Calabrese and Edgar Raymond (1989) used the Dean Alienation Scale to assess alienation, isolation, normlessness, and powerlessness. The following table displays means and standard deviations for the measure of isolation for various groups as reported by Calabrese and Raymond. (The sample sizes are not those actually used in their study.)

	Alienation		
Group	**Mean**	**Standard Deviation**	**Sample Size**
Religious affiliation			
Church	68.3	10.1	18
No church	69.8	9.9	12
Religious commitment			
Strong	71.0	13.3	10
Moderate	69.7	12.1	8
None	67.7	9.2	12
Family organization			
Traditional	70.1	9.8	21
Nontraditional	68.9	10.5	9

a Test for a significant difference in mean isolation scores for those students with a church affiliation and those without a church affiliation. Use $\alpha = .05$.

b Is there a significant difference in variances for those students with a moderate religious commitment and those with no commitment? Use $\alpha = .05$.

c Find a 95% confidence interval estimate of the actual difference in mean alienation scores for students from homes having a traditional family organization and those from nontraditional homes.

9.91 Scott Powers and colleagues (1983) conducted a study to investigate the relationship between several variables and a competitive runner's finish time in a 10-kilometer (km) race. The finish time (in minutes) for the 10-km race for each of the nine runners is shown below:

Runner	1	2	3	4	5	6	7	8	9
Finish time	33.15	33.33	33.50	33.55	33.73	33.86	33.90	34.15	34.90

Assume that these experienced competitive runners' times to complete a 10-km race represent a random sample of the running times for all competitive runners of 10-km races. Find a 90% confidence interval for the mean time to complete a 10-km race. Interpret your interval.

9.92 Four sets of identical twins (pairs A, B, C, and D) were selected at random from a population of identical twins. One child was selected at random from each pair to form an "experimental group." These four children were sent to school. The other four children were kept at home as a control group. At the end of the school year the following IQ scores were obtained:

Pair	Experimental Group	Control Group
A	110	111
B	125	120
C	139	128
D	142	135

Does this evidence justify the conclusion that lack of school experience has a depressing effect on IQ scores? Use $\alpha = .10$.

9.93 In the study of the physical characteristics of outstanding female junior tennis players, Powers and Walker (1983) give statistics on the circumferences of the forearms of ten junior tennis players. The circumference was measured on the forearms of both the preferred hand (the one that the player uses most often to grip the racquet) and the nonpreferred hand. The means and standard errors of the means are shown below. If there is any difference in the mean circumference of the forearms in tennis players, the larger circumference should be in the forearm most likely to have the greater buildup in muscle, namely the one for the preferred hand.

	Preferred Hand	Nonpreferred Hand
Mean	23.88	22.33
Standard error of the mean	0.82	0.41

a Suppose that you wish to detect this difference, assuming that it exists for female junior tennis players. State the null and alternative hypotheses that you would use for the test.

b Give the rejection region for $\alpha = .01$.

c Conduct the test and state your conclusions.

d What type of experimental design was used for this experiment?

e Based on your answer to part (d), was your test in parts (a), (b), and (c) the correct procedure for analyzing the data? Explain.

9.94 Seven obese persons were placed on a diet for 1 month, and the weights, at the beginning and at the end of the month, were recorded. They are shown in the table:

	Weights	
Subjects	Initial	Final
1	310	263
2	295	251
3	287	249
4	305	259
5	270	233
6	323	267
7	277	242
8	299	265

Estimate the mean weight loss for obese persons when placed on the diet for a 1-month period. Use a 95% confidence interval and interpret your results. What assumptions must you make in order that your inference be valid?

9.95 A comparison of reaction times for two different stimuli in a psychological word-association experiment produced the following results when applied to a random sample of 16 people.

Stimulus	Reaction Time (sec)							
1	1	3	2	1	2	1	3	2
2	4	2	3	3	1	2	3	3

Do the data present sufficient evidence to indicate a difference in mean reaction times for the two stimuli? Test using $\alpha = .05$.

9.96 Refer to Exercise 9.95. Suppose that the word-association experiment had been conducted using eight people as blocks and making a comparison of reaction time within each person; that is, each person would be subjected to both stimuli in a random order. The data for the experiment are as follows:

	Reaction Time (sec)	
Person	Stimulus 1	Stimulus 2
1	3	4
2	1	2
3	1	3
4	2	1
5	1	2
6	2	3
7	3	3
8	1	3

Do the data present sufficient evidence to indicate a difference in mean reaction time for the two stimuli? Test using $\alpha = .05$.

9.97 Obtain a 90% confidence interval for $(\mu_1 - \mu_2)$ in Exercise 9.95.

9.98 Obtain a 95% confidence interval for $(\mu_1 - \mu_2)$ in Exercise 9.96.

9.99 Analyze the data in Exercise 9.96 as though the experiment had been conducted in an unpaired manner. Calculate a 95% confidence interval for $(\mu_1 - \mu_2)$ and compare with the answer to Exercise 9.98. Does it appear that blocking increased the amount of information available in the experiment?

9.100 The following data give readings in foot-pounds of the impact strength of two kinds of packaging material. Use the MINITAB printout shown below to determine whether there is evidence of a difference in mean strength between the two kinds of material. Test using $\alpha = .10$.

A	B
1.25	.89
1.16	1.01
1.33	.97
1.15	.95
1.23	.94
1.20	1.02
1.32	.98
1.28	1.06
1.21	.98

```
TWOSAMPLE T FOR A VS B
   N        MEAN       STDEV      SE MEAN
A  9      1.2367      0.0644        0.021
B  9      0.9778      0.0494        0.016

95 PCT CI FOR MU A - MU B: (0.201, 0.316)

TTEST MU A = MU B (VS NE): T= 9.56  P=0.0000  DF=  16

POOLED STDEV =        0.0574
```

9.101 Would the amount of information extracted from the data in Exercise 9.100 be increased by pairing successive observations and analyzing the differences? Calculate 90% confidence intervals for $(\mu_1 - \mu_2)$ for the two methods of analysis (unpaired and paired) and compare the widths of the intervals.

9.102 When should one employ a paired-difference analysis in making inferences concerning the difference between two means?

9.103 An experiment was conducted to compare the density of cakes prepared from two different cake mixes A and B. Six cake pans received batter A, and six received batter B. Expecting a variation in oven temperature, the experimenter placed an A and a B side by side at six different locations in the oven. The six paired observations are as follows:

Mix	Density (oz/in.^{3})					
A	.135	.102	.098	.141	.131	.144
B	.129	.120	.112	.152	.135	.163

a Do the data present sufficient evidence to indicate a difference between the average densities of cakes prepared using the two types of batter? Test using $\alpha = .05$.

b Place a 95% confidence interval on the difference between the average densities for the two mixes.

9.104 Under what assumptions may the F distribution be used in making inferences about the ratio of population variances?

9.105 A dairy is in the market for a new bottle-filling machine and is considering models A and B manufactured by company X and company Y, respectively. If ruggedness, cost, and convenience are comparable in the two models, the deciding factor is the variability of fills (the model producing fills with the smaller variance being preferred). Let σ_1^2 and σ_2^2 be the fill variances for models A and B, respectively, and consider various tests of the null hypothesis $H_0 : \sigma_1^2 = \sigma_2^2$. Obtaining samples of fills from the two machines and using the test statistic s_1^2/s_2^2, one could set up as the rejection region an upper-tail area, a lower-tail area, or a two-tailed area of the F distribution, depending on the point of view. Which type of rejection region would be most favored by

a The manager of the dairy? Why?

b A salesperson for company X? Why?

c A salesperson for company Y? Why?

9.106 The closing prices of two common stocks were recorded for a period of 15 days. The means and variances are

$$\bar{x}_1 = 40.33 \quad \bar{x}_2 = 42.54$$
$$s_1^2 = 1.54 \quad s_2^2 = 2.96$$

a Do these data present sufficient evidence to indicate a difference between the variabilities of the closing prices of the two stocks for the populations associated with the two samples? Give the p-value for the test and interpret its value.

b Place a 90% confidence interval on the ratio of the two population variances.

9.107 Refer to Exercise 9.105. Wishing to demonstrate that the variability of fills is less for model A than for model B, a salesperson for company X acquired a sample of 30 fills from a machine of model A and a sample of 10 fills from a machine of model B. The sample variances were

$s_1^2 = .027$ and $s_2^2 = .065$, respectively. Does this result provide statistical support at the .05 level of significance for the salesperson's claim?

9.108 A chemical manufacturer claims that the purity of his product never varies more than 2%. Five batches were tested and given purity readings of 98.2, 97.1, 98.9, 97.7, and 97.9%. Do the data provide sufficient evidence to contradict the manufacturer's claim? (HINT: To be generous, let a range of 2% equal 4σ.)

9.109 Refer to Exercise 9.108. Find a 90% confidence interval for σ^2.

9.110 A cannery prints "weight 16 ounces" on its label. The quality-control supervisor selects nine cans at random and weighs them. She finds $\bar{x} = 15.7$ and $s = .5$. Do the data present sufficient evidence to indicate that the mean weight is less than that claimed on the label? Test using $\alpha = .05$.

9.111 A psychologist wishes to verify that a certain drug increases the reaction time to a given stimulus. The following reaction times in tenths of a second were recorded before and after injection of the drug for each of four subjects:

	Reaction Time	
Subject	**Before**	**After**
1	7	13
2	2	3
3	12	18
4	12	13

Test at the 5% level of significance to determine whether the drug significantly increases reaction time.

9.112 At a time when energy conservation is so important, some scientists think we should give closer scrutiny to the cost (in energy) of producing various forms of food. One study compares the mean amount of oil required to produce 1 acre of different types of crops. For example, suppose we wish to compare the mean amount of oil required to produce 1 acre of corn versus 1 acre of cauliflower. The readings in barrels of oil per acre, based on 20-acre plots, seven for each crop, are shown in the table. Use these data to find a 90% confidence interval for the difference between the mean amounts of oil required to produce these two crops.

Corn	Cauliflower
5.6	15.9
7.1	13.4
4.5	17.6
6.0	16.8
7.9	15.8
4.8	16.3
5.7	17.1

9.113 The effect of alcohol consumption on the body appears to be much greater at high altitudes than at sea level. To test this theory, a scientist randomly selects 12 subjects and randomly divides them into two groups of six each. One group is transported to an altitude of 12,000 feet, where each subject ingests a drink containing 100 cc of alcohol. The second group receives the same drink at sea level. After two hours, the amount of alcohol in the blood (grams per 100 cc) for each subject is measured. The data are shown in the table. Do the data provide sufficient evidence to support the theory that retention of alcohol in the blood is greater at high altitudes? Test using $\alpha = .10$.

Sea Level	12,000 Feet
.07	.13
.10	.17
.09	.15
.12	.14
.09	.10
.13	.14

9.114 An experiment is conducted to compare two new automobile designs. Twenty people are randomly selected, and each person is asked to rate each design on a scale of 1 (poor) to 10 (excellent). The resulting ratings will be used to test the null hypothesis that the mean level of approval is the same for both designs against the alternative hypothesis that one of the automobile designs is preferred. Would these data satisfy the assumptions required for the Student's t test of Section 9.4? Explain.

9.115 The following data were collected on lost-time accidents (the figures given are mean work-hours lost per month over a period of 1 year) before and after an industrial safety program was put into effect. Data were recorded for six industrial plants. Do the data provide sufficient evidence to indicate whether the safety program was effective in reducing lost-time accidents? Test using $\alpha = .10$.

	Plant Number					
	1	2	3	4	5	6
Before program	38	64	42	70	58	30
After program	31	58	43	65	52	29

9.116 To compare the demand for two different entrées, the manager of a cafeteria recorded the number of purchases for each entrée on seven consecutive days. The data are shown in the table. Do the data provide sufficient evidence to indicate a greater mean demand for one of the entrées? Use the MINITAB printout shown below.

Day	A	B
Monday	420	391
Tuesday	374	343
Wednesday	434	469
Thursday	395	412
Friday	637	538
Saturday	594	521
Sunday	679	625

```
TEST OF MU = 0.000 VS MU N.E.  0.000

        N      MEAN     STDEV   SE MEAN          T    P VALUE
C3      7    33.429    47.504    17.955       1.86       0.11
```

9.117 The EPA limit on the allowable discharge of suspended solids into rivers and streams is 60 milligrams per liter (mg/l) per day. A study of water samples selected from the discharge at a phosphate mine shows that over a long period, the mean daily discharge of suspended solids is 48 mg/l, but day-to-day discharge readings are variable. State inspectors measured the discharge rates of suspended solids for $n = 20$ days and found $s^2 = 39$ (mg/l)^2. Find a 90% confidence interval for σ^2. Interpret your results.

9.118 Exercise 9.44 discussed a study conducted by Arthur Schmale and colleagues (1983) of the well-being of cancer patients who had survived at least 1 year beyond their last cancer treatment and who appeared to be in good health. Each of 104 of these cancer survivors was administered a questionnaire called "The General Well-Being Measure." The table below gives the mean scores when the 104 patients are split into groups according to sex and some other sociodemographic variables. The group sample sizes for each of these sociodemographic variables, along with t values and p-values for testing the null hypothesis of "no difference" between a pair of means, are shown in the table.

	Sample Size†	Mean	t	2-Tailed p
Sex				
Male	44	100.2		
Female	60	103.5	1.15	0.25
Marital status				
Married	72	104.1		
Other	32	97.7	2.19	0.03
Living arrangements				
Not alone	85	103.2		
Alone	19	95.6	1.94	0.06
Change in occupation since cancer diagnosed				
No	88	103.4		
Yes	16	95.0	2.23	0.03

†Sample sizes for the groups were calculated from percentages given in the reference.

a Explain the test results.

b Using the value of $t = 1.94$ given in the table for comparing the mean well-being scores for patients' living arrangements, calculate the approximate p-value of the test. Does your value appear to agree with the value $p = .06$ given in the table?

9.119 Are Americans putting in more time at work than they once did? John Robinson (1989) discusses a survey of men and women who work at least 10 hours per week. Respondents kept a record of the number of hours per week spent at paid work, including the commute. The average work week for men, broken down by educational background, follows. (Only the means are reported by Robinson.)

Education	Mean	Standard Deviation	Sample Size
Grade school	45	10.0	25
High school graduate	42	8.9	25
College graduate	40	8.1	25
Postgraduate	38	9.7	25

a Is there a significant difference between mean hours per work week for men with a high school education compared with men having only a grade school education? Test using $\alpha = .01$.

b Can we conclude that a man with postgraduate training works significantly fewer hours per week than a man who is a high school graduate? Test using $\alpha = .01$.

Exercises Using the Data Disk

9.120 Refer to the batting averages for National and American League batting champions, Data Set B on the data disk. Draw a random sample of ten champions from each of the two leagues and calculate the means and standard deviations of the samples.

a Is there sufficient evidence to indicate that there is a difference in the mean batting averages for the champions of the National and American Leagues?

b What is the p-value for the test in part (a)?

c What assumptions must be made in order that the test in part (a) be valid? these assumptions are met? Explain.

9.121 A random sample of 15 women was selected from among those listed in Data Set A on the data disk. The systolic blood pressures for these 15 women are shown below:

84	100	110
110	110	114
112	120	104
112	90	112
130	110	88

a Construct a 95% confidence interval for the population variance, σ^2. Interpret this interval.

b If we consider the 945 female systolic blood pressures to be the population of interest, the variance of this population is given on the data disk as $\sigma^2 = 158.2654$. Does this value fall in the confidence interval constructed in part (a)?

9.122 Refer to Exercise 9.121. A random sample of $n = 15$ men is now selected from the population in Data Set A on the data disk. The systolic blood pressures are shown below:

112	120	110
90	120	115
128	130	126
140	106	110
134	90	120

a Do the data provide sufficient evidence to indicate that there is a difference in the test variability of systolic blood pressures between men and women? Test using $\alpha = .01$.

b Construct a 95% confidence interval for the ratio of the two population variances.

c The ratio of the two population variances, given on the data disk, is $(14.23)^2/(12.58)^2$. Does this value fall within the confidence interval constructed in part (b)? Explain.

CASE STUDY

Will Your Bill for College Textbooks Continue to Rise?

The number of new U.S. book titles increased from almost 47,000 in 1990 to over 48,000 in 1991. However, this was still below the historic high of about 56,000 titles attained in 1987 (Grannis, 1992). Can we expect an increase or decrease in the price of books, especially hardbacks, if there are more competitors on the market? The following table gives the number of titles and the average price of hardback books classified according to 23 standard subject groups representing one or more specific Dewey Decimal Classification numbers.

	1990		1991	
Category	**Volumes**	**Average Price**	**Volumes**	**Average Price**
Agriculture	359	$54.24	371	$57.73
Art	759	42.18	717	44.99
Biography	1,337	28.58	1,416	27.52
Business	748	45.48	790	43.38
Education	562	38.72	556	41.26
Fiction	1,962	19.83	2,062	21.30
General works	1,035	54.77	1,071	51.74
History	1,450	36.43	1,442	39.87
Home economics	357	23.80	341	24.23
Juveniles	3,675	13.01	3,705	16.64
Language	312	42.98	240	51.71
Law	596	60.78	240	63.89
Literature	1,312	35.80	1,265	35.76
Medicine	2,215	72.24	2,078	71.44
Music	184	41.86	173	41.04
Philosophy/Psychology	963	40.58	945	42.74
Poetry/drama	486	32.19	511	33.29
Religion	977	31.31	958	32.33
Science	2,028	74.39	958	80.14
Sociology/Economics	4,504	42.10	4,306	48.83
Sports/recreation	403	30.52	440	30.68
Technology	1,521	76.48	1,620	76.40
Travel	181	30.41	156	33.50
Total	27,926	$42.12	26,361	$43.93

Consider the number of volumes and average price per volume in 1990 and 1991 as paired samples for two randomly chosen years for each of the 23 categories of books. Although there was an increase in the total number of books in 1991, the number of hardbacks seems relatively unchanged and the average price per volume seems to have increased over the average 1990 price.

1 Determine whether the difference in the average number of volumes per category for 1991 differs significantly from the 1990 average, using a significance level of 5%.

2 Determine whether the change in the average price of a hardback book per category in 1991 differs significantly from that in 1990 at the 5% level of significance.

3 Summarize your results concerning the difference in the number and price of books per category in 1991 compared with 1990.

10

Linear Regression and Correlation

Case Study

The phrase "made in the U.S.A." has become a battle cry in the last few years, as American workers try to protect their jobs from overseas competition. In the case study at the end of this chapter we explore the changing attitudes of American consumers toward automobiles made outside of the United States, using a simple linear regression analysis.

General Objective

We have made inferences about population means when random samples consisting of measurements on a single random variable were selected from the populations of interest. In this chapter, we consider the case in which the mean value of a random variable y is related to another variable, say, x. By measuring both y and x on each experimental unit, thus generating *bivariate* measurements, we can use information provided by x to estimate the mean value of y and to predict particular values of y for preassigned values of x.

Specific Topics

1. Linear probabilistic models (10.2)
2. Method of least squares (10.3)
3. Inferences concerning β_1 (10.4)
4. Estimating $E(y)$ for a given value of x (10.5)
5. Predicting y for a particular value of x (10.5)
6. Coefficient of correlation, r (10.7)
7. Assumptions (10.8)

10.1 Introduction

An estimation problem of importance to high school seniors, freshman entering college, their parents, and a university administration concerns the expected academic achievements of a particular student after he or she has enrolled in a university. For example, we might wish to estimate a student's grade point average (GPA) at the end of the freshman year *before* the student has been accepted or enrolled in the university. At first glance, this would seem to be a difficult task.

The statistical approach to this problem is, in many respects, a formalization of the procedure we might follow intuitively. If data were available giving the high school academic grades, psychological and sociological information, and the grades attained at the end of the college freshman year for a large number of students, we might categorize the students into groups having similar characteristics. Highly motivated students who have had a high rank in their high school class, have graduated from a high school with known superior academic standards, and so on, should achieve, on the average, a high GPA at the end of the college freshman year. On the other hand, students who lack proper motivation or who achieved only moderate success in high school would not be expected, on the average, to do as well. Carrying this line of thought to the ultimate and idealistic extreme, we would expect the GPA of a student to be a **function** of the many variables that define the psychological and physical characteristics of the individual, as well as those characteristics that define the academic and social environment to which the student will be exposed. Ideally, we would like to have a mathematical equation that would relate a student's GPA to all relevant variables so that the equation could be used for prediction.

The problem we have defined is of a very general nature. We are interested in a random variable y that is related to a number of independent variables $x_1, x_2, x_3, \ldots$. The variable y for our example would be the student's GPA, and the independent variables might be

x_1 = rank in high school class

x_2 = score on a mathematics achievement test

x_3 = score on a verbal achievement test

and so on. The ultimate objective is to measure $x_1, x_2, x_3, \ldots, x_k$ for a particular student, substitute these values into the prediction equation, and thereby predict the student's GPA. To accomplish this objective, we must first locate the related variables $x_1, x_2, x_3, \ldots, x_k$ and obtain a measure of the strength of their relationship to y. Then we must construct a good prediction equation that expresses y as a function of the selected independent predictor variables.

Practical examples of our prediction problem are numerous in business, industry, and the sciences. The stockbroker wishes to predict stock market behavior as a function of a number of "key indices" that are observable and serve as the independent predictor variables $x_1, x_2, x_3, \ldots$. The manager of a manufacturing plant would like to relate yield of a chemical to a number of process variables. The manager would then use the prediction equation to find settings for the controllable process variables that would provide the maximum yield of the chemical. The biologist would like to relate body characteristics to the amounts of various glandular secretions. The political scientist may wish to relate success in a political campaign to the characteristics of a candidate, the nature of the opposition, and various campaign issues, campaign expenditures, and promotional techniques. All these prediction problems involve the same general concept.

In this chapter, we will restrict our attention to the simple problem of predicting y as a **linear function** of a **single** variable. The solution for the multivariable problem—for example, predicting a student's GPA—will be the subject of Chapter 11.

10.2 A Simple Linear Probabilistic Model

We introduce our topic by considering the problem of predicting a student's final grade in a college freshman calculus course based on his score on a mathematics achievement test administered prior to college entrance. As noted in Section 10.1, we wish to determine whether the achievement test is really worthwhile—that is, whether the achievement test score is actually related to a student's grade in calculus—and to obtain an equation that will be useful for predicting y as a function of x. The evidence, presented in Table 10.1, represents a sample of the achievement test scores and calculus grades for ten college freshmen. Since two random variables (achievement test score and calculus grade) have been measured on each experimental unit (college freshman), the resulting data is called **bivariate data**, as defined in Section 1.7. We will assume that ten students constitute a random sample drawn from the population of freshmen who have already entered the university or will do so in the immediate future.

TABLE 10.1
Mathematics achievement test scores and final calculus grades for college freshmen

Student	Mathematics Achievement Test Score	Final Calculus Grade
1	39	65
2	43	78
3	21	52
4	64	82
5	57	92
6	47	89
7	28	73
8	75	98
9	34	56
10	52	75

Our initial approach to the analysis of the data in Table 10.1 would be to plot the data on a two-dimensional **scatterplot** as presented in Section 1.7, representing a student's calculus grade as y and the corresponding achievement test score as x. Figure 10.1 shows a scatterplot generated by the EXECUSTAT software program; you can see that y appears to increase as x increases. Could this arrangement of the points have occurred due to chance even if x and y were unrelated?

One method of obtaining a prediction equation relating y to x is to place a ruler on the graph and move it about until it seems to pass through the points, thus providing what we might regard as the "best fit" to the data. Indeed, if we were to draw a line through the points, it would appear that our prediction problem had been solved. Certainly, we can now use the graph to predict a student's calculus grade as a function of his score on the mathematics achievement test. Furthermore, we have chosen a **mathematical model** that expresses the supposed functional relation between y and x.

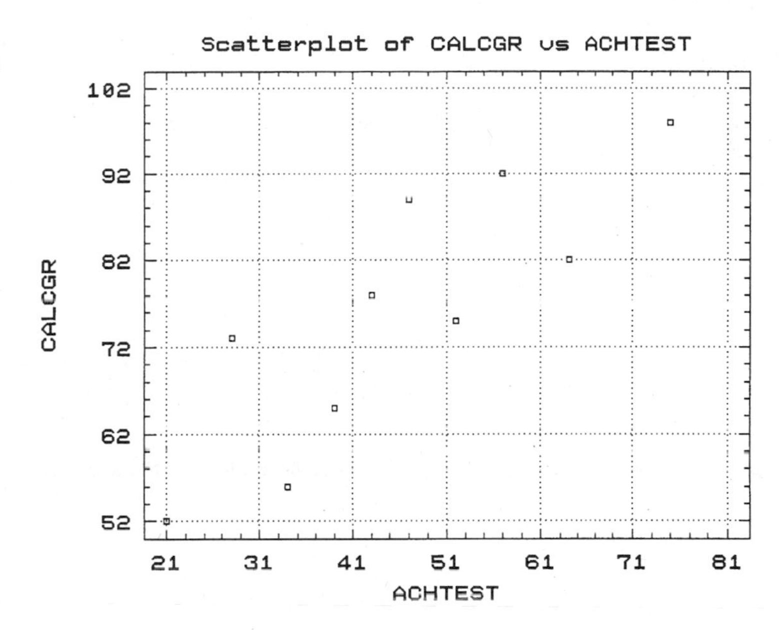

FIGURE 10.1
EXECUSTAT scatterplot of the data in Table 10.1

Let us review several facts concerning the graphing of mathematical functions. First, the mathematical **equation of a straight line is**

$$y = \beta_0 + \beta_1 x$$

where β_0 is the ***y* intercept**, the value of y when $x = 0$, and β_1 is the **slope** of the line, the change in y for a one unit change in x (Figure 10.2). Second, the line that we graph corresponding to any linear equation is unique. Each equation corresponds to only one line and vice versa. Thus, when we draw a line through the points, we automatically choose a mathematical equation

$$y = \beta_0 + \beta_1 x$$

where β_0 and β_1 have unique numerical values.

Equation of a Straight Line

$$y = \beta_0 + \beta_1 x$$

where $\beta_0 = y$ intercept
$= \text{value of } y \text{ when } x = 0$
and $\beta_1 = \text{slope of the line}$
$= \text{change in } y \text{ for a 1-unit increase in } x$

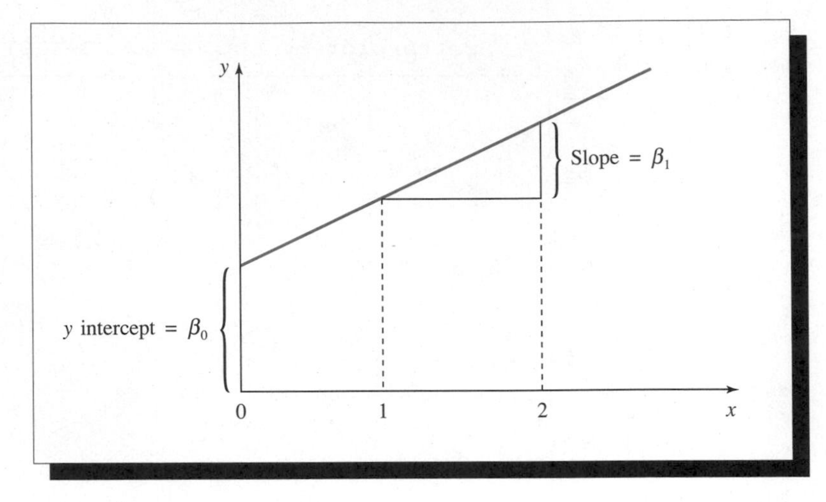

FIGURE 10.2
The y intercept and slope for a line

The linear model

$$y = \beta_0 + \beta_1 x$$

is said to be a **deterministic mathematical model** because, when a value of x is substituted into the equation, the value of y is determined and no allowance is made for error. Fitting a straight line through a set of points by eye produces a deterministic model. Many other examples of deterministic mathematical models can be found by leafing through the pages of elementary chemistry, physics, or engineering textbooks.

Deterministic models are suitable for explaining physical phenomena and predicting when the error of prediction is negligible for practical purposes. Thus, Newton's Law, which expresses the relation between the force F imparted by a moving body with mass m and acceleration a, given by the deterministic model

$$F = ma$$

predicts force with very little error for most practical applications. "Very little" is, of course, a relative concept. An error of .1 inch in forming an I-beam for a bridge is extremely small but would be impossibly large in the manufacture of parts for a wristwatch. Thus, in many physical situations, the error of prediction cannot be ignored. Indeed, consistent with our stated philosophy, we would be hesitant to place much confidence in a prediction unaccompanied by a measure of its goodness. For this reason, a visual choice of a line to relate the calculus grade and achievement test score would have limited value.

In contrast to the deterministic model, we might employ a **probabilistic mathematical model**, which is a simple modification of the deterministic model. Rather than saying that y and x are related by the deterministic model

$$y = \beta_0 + \beta_1 x$$

we say that the **mean (or expected) value of y for a given value of x**, denoted by the symbol $E(y \mid x)$, has a graph that is a straight line. That is, we let

$$E(y \mid x) = \beta_0 + \beta_1 x$$

For any given value of x, y values will vary in random manner about the mean $E(y \mid x)$.

So to summarize, we write the probabilistic model for any particular observed value of y as

$$\begin{aligned} y &= (\text{mean value of } y \text{ for given value of } x + (\text{random error}) \\ &= \overbrace{\beta_0 + \beta_1 x}^{E(y \mid x)} + \epsilon \end{aligned}$$

where ϵ is a **random error**, the difference between an observed value of y and the mean value of y for a given value of x. Thus, we assume that for any given value of x the observed value of y varies in a random manner and possesses a probability distribution with a mean value $E(y \mid x)$. For assistance in conveying this idea, the probability distributions of y about the line of means $E(y \mid x)$ are shown in Figure 10.3 for three hypothetical values of x—x_1, x_2, and x_3—although, as noted, we imagine a distribution of y values for every value of x.

FIGURE 10.3
Linear probabilistic model

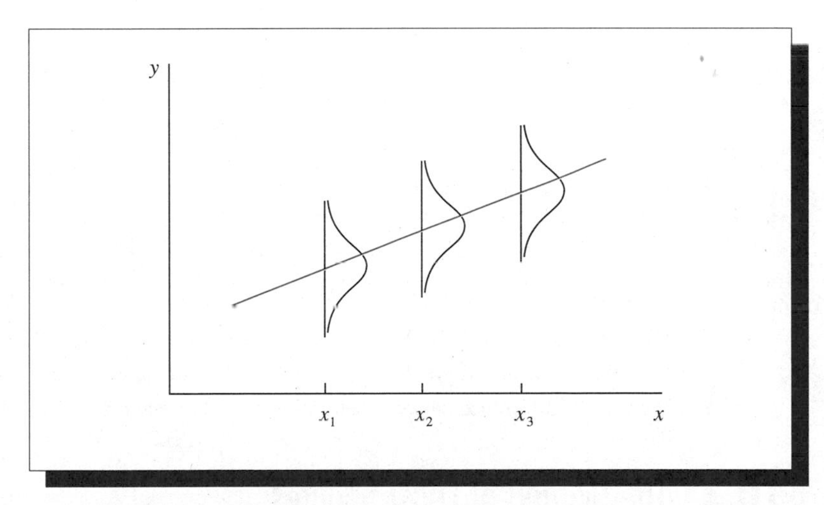

What properties will we assume for the probability distribution of y for a given value of x? Assumptions that seem to fit many practical situations and, upon which the methodology of Sections 10.4–10.7 is based, are given as follows:

Assumptions for the Probabilistic Model

For any given value of x, y possesses a normal distribution, with a mean value given by the equation

$$E(y \mid x) = \beta_0 + \beta_1 x$$

and with a variance of σ^2. Furthermore, any one value of y is independent of every other value.

These assumptions permit us to construct tests of hypotheses and confidence intervals for β_0, β_1, and $E(y \mid x)$. In addition, we are able to give a prediction interval for y when we use the fitted model (the line that we find to fit a particular set of data) to predict some new value of y for a particular value of x. The practical problems we can solve by using these procedures will become apparent when we have worked some examples.

In the next section, we consider the problem of finding the "best-fitting" line for a given set of data. This line will give us a prediction equation that is also called a **regression line**.

Exercises

Basic Techniques

10.1 Graph the line corresponding to the equation $y = 2x + 1$ by graphing the points corresponding to $x = 0$, 1, and 2. Give the y intercept and slope for the line.

10.2 Graph the line corresponding to the equation $y = -2x + 1$ by graphing the points corresponding to $x = 0$, 1, and 2. Give the y intercept and slope for the line. How is this line related to the line $y = 2x + 1$ of Exercise 10.1?

10.3 Give the equation and graph for a line with a y intercept equal to 3 and a slope equal to -1.

10.4 Give the equation and graph for a line with a y intercept equal to -3 and a slope equal to 1.

10.5 What is the difference between deterministic and probabilistic mathematical models?

10.3 The Method of Least Squares

The statistical procedure for finding the best-fitting straight line for a set of points is, in many respects, a formalization of the procedure employed when we fit a line by eye. For instance, when we visually fit a line to a set of data, we move the ruler until we think that we have minimized the **deviations** of the points from the prospective line. If we denote the predicted value of y obtained from the fitted line as $\hat{y}$, the prediction equation will be

$$\hat{y} = \hat{\beta}_0 + \hat{\beta}_1 x$$

where $\hat{\beta}_0$ and $\hat{\beta}_1$ represent estimates of the parameters β_0 and β_1. This line for the data of Table 10.1 is shown in Figure 10.4. The vertical lines drawn from the prediction line to each point represent the deviations of the points from the predicted value of y.

FIGURE **10.4** EXECUSTAT graph of the least-squares line, $\hat{y} = \hat{\beta}_0 + \hat{\beta}_1 x$, and data points from Table 10.1

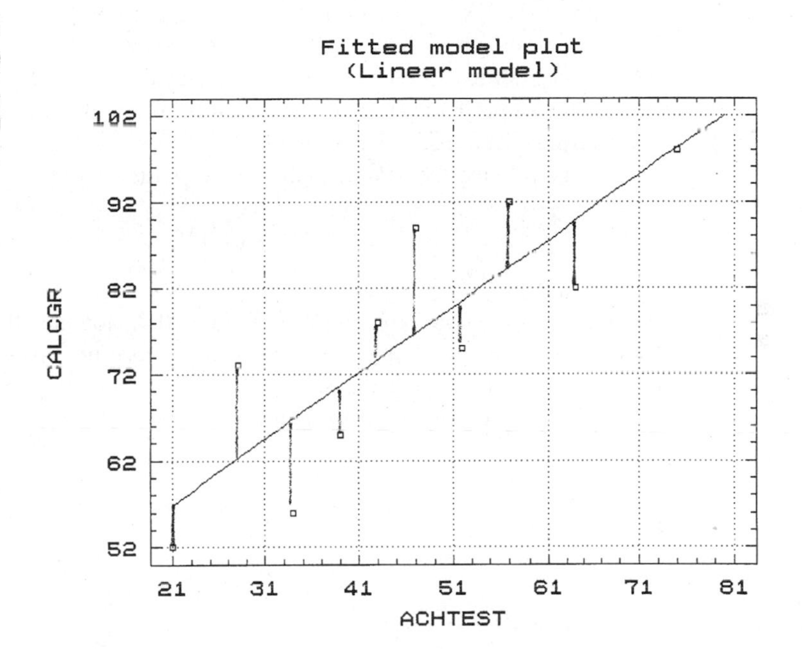

Having decided that in some manner or other we will attempt to minimize the deviations of the points in choosing the best-fitting line, we must now define what we mean by "best." That is, we wish to define a criterion for "best fit" that will seem intuitively reasonable, that is objective, and that, under certain conditions, will give the best prediction of y for a given value of x.

We will employ a criterion of goodness that is known as the **principle of least squares**, which may be stated as follows: **Choose as the "best-fitting" line the line that minimizes the sum of squares of the deviations of the observed values of y from those predicted.** This sum of squares of deviations, commonly called the **sum of squares for error (SSE)** and defined as

$$\text{SSE} = \sum_{i=1}^{n} (y_i - \hat{y}_i)^2$$

is the sum of the squared distances represented by the vertical lines in Figure 10.4.

The method for finding the numerical values of β_0 and β_1 that minimize SSE uses differential calculus and hence is beyond the scope of this text. However, if you have a calculator with a statistics function or have access to a statistical software package, you can easily obtain the least squares estimates, $\hat{\beta}_0$ and $\hat{\beta}_1$, once you have entered the data using the appropriate commands. Each calculator and software program provides slightly different output, but they are all similar in format.

EXAMPLE 10.1 Obtain the least squares prediction line for the data in Table 10.1 using an appropriate statistical software package. Predict a student's calculus grade if the student scored $x = 50$ on the achievement test.

Solution A portion of the EXECUSTAT output for a simple linear regression using the data in Table 10.1 is shown in Table 10.2. The least squares line is given on the first line of the printout as **linear model**, and the least squares estimates $\hat{\beta}_0$ and $\hat{\beta}_1$ are given in the **table of estimates** in the column labeled **estimate** and the rows labeled **intercept** and **slope**, respectively. Then, according to the principle of least squares, the best-fitting straight line relating the calculus grade to the achievement test score is

$$\begin{aligned}\hat{y} &= \hat{\beta}_0 + \hat{\beta}_1 x \\ &= 40.784 + .766x\end{aligned}$$

The graph of this equation is shown in Figure 10.4. Note that the y intercept, 40.784, is the value of y when $x = 0$. The slope of the line, .766, gives the estimated change in y for a 1-unit change in x. Finally, if a student scores $x = 50$ on the achievement test, his or her predicted calculus grade would be

$$\begin{aligned}\hat{y} &= \hat{\beta}_0 + \hat{\beta}_1 x \\ &= 40.7842 + (.7655618)(50) = 79.062\end{aligned}$$

TABLE 10.2 EXECUSTAT printout for Example 10.1

```
              Simple Regression Analysis for IPS10-1

=============================================================
Linear model: CALCGR = 40.7842 + 0.765562*ACHTEST

                      Table of Estimates
=============================================================
                             Standard         t          P-
               Estimate        Error        Value      Value
=============================================================
Intercept     40.7842        8.50686        4.79       0.001
Slope          0.765562      0.174985       4.38       0.002
=============================================================
```
■

In general, the least-squares estimates will either be labeled "estimate" or "coefficient," and the intercept and slope may be labeled "constant" and "x," respectively. We will present computer printouts from both MINITAB and EXECUSTAT software packages throughout this chapter so that you will become familiar with their formats.

Exercises

Basic Techniques

10.6 Given five points whose coordinates are

x	−2	−1	0	1	2
y	1	1	3	5	5

a computer software package produces the following information about the least-squares regression line:

```
SLOPE = 1.2000          INTERCEPT = 3.0000
```

a Find the least-squares line for the data.

b Plot the five points and graph the line in part (a). Does the line appear to provide a good fit to the data points?

10.7 Given the points whose coordinates are

x	1	2	3	4	5	6
y	5.6	4.6	4.5	3.7	3.2	2.7

a computer printout gives the least-squares estimates as

```
PREDICTOR     COEF
CONSTANT      6.0000
X             -.557143
```

a Find the least-squares line for the data.

b Plot the six points and graph the line. Does the line appear to provide a good fit to the data points?

c Use the least-squares line to predict the value of y when $x = 3.5$.

Applications

10.8 In Exercise 1.27, we presented data on the productivity of Professor Isaac Asimov, one of the most prolific writers of all time. Prior to his death in 1992, he wrote nearly 500 books during a 40-year career. In fact, as his career progressed, he became even more productive in terms of the number of books written within a given period of time (Ohlsson, 1992). The data below give the time in months required to write his books in increments of 100:

Number of books x	100	200	300	400	490
Time in months y	237	350	419	465	507

```
-----------------------------------------------------------
The regression equation is
y = 196 + 0.670 x

Predictor        Coef          Stdev       t-Ratio        p
Constant       195.90          26.40          7.42    0.005
x             0.67013        0.08034          8.34    0.004

s = 24.90        R-sq = 95.9%        R-sq(adj) = 94.5%
-----------------------------------------------------------
```

a Use the MINITAB printout given above to find the least-squares line relating the time in months y to the number of books written x.

b Plot the time as a function of the number of books written, using a scatterplot, and graph the least-squares line on the same paper. Does it seem to provide a good fit to the data points?

10.9 A new and booming area of specialty sales is the gourmet coffee market. Gourmet coffee stores—often small, informal, and quirky—are gaining a bigger share of the total coffee market, whereas the conventional coffee market, not including the market for decaffeinated coffee, continues to decline. The following table gives retail sales and forecasts for gourmet coffee (y), in millions of dollars, for the years 1983–1992:

Year	Gourmet Coffee
1983	210
1984	270
1985	340
1986	420
1987	500
1988	600
1989	675
1990	750
1991	825
1992	900

Source: Robichaux (1989), p. B1.

```
          Simple Regression Analysis for IPS10-9
============================================================
Linear model: COFFEE = -156162 + 78.8485*YEAR

                   Table of Estimates
============================================================
                            Standard          t          P
               Estimate       Error         Value      Value
============================================================
Intercept  -156162         2229.63         -70.04      0.000
Slope            78.8485      1.12183        70.29      0.000
============================================================
```

a Find the least-squares line using the EXECUSTAT printout given above.

b Plot the data points and graph the least-squares line.

10.4 Inferences Concerning the Slope of the Line, β_1

The initial inference of importance in studying the relationship between y and x concerns the existence of the relationship. Does x contribute information for the prediction of y; that is, **do the data present sufficient evidence to indicate that y increases (or decreases) linearly as x increases over the region of observation?** Or, is it probable that the points would fall on the graph in a manner similar to that observed in Figure 10.1 when y and x are completely unrelated?

Any inference concerning the least-squares line requires that we first estimate σ^2, the variability of the points about the line. It seems reasonable to use SSE, the sum of the squared deviations, about the predicted line for this purpose. Indeed, it can be shown that the formula given in the display provides an estimator for σ^2 that is unbiased, based on $(n - 2)$ degrees of freedom (d.f).

An Estimator for σ^2

$$\hat{\sigma}^2 = s^2 = \frac{\text{SSE}}{n-2}$$

The sum of squares of deviations, SSE, can be calculated directly by using the prediction equation to calculate $\hat{y}$ for each point, then calculating the deviations $(y_i - \hat{y}_i)$, and finally calculating

$$\text{SSE} = \sum_{i=1}^{n}(y_i - \hat{y}_i)^2$$

This procedure tends to be tedious and is rather poor from a computational point of view because the numerous subtractions tend to introduce computational rounding errors. An easier and computationally better procedure is to use the formula given in Section 10.6 or to obtain SSE and s^2 from a computer printout.

Table 10.3 shows the MINITAB output for a linear regression analysis using the data in Table 10.1. Notice the clear specification of the regression equation in lines 1 and 2, and that the least-squares estimates $\hat{\beta}_0$ and $\hat{\beta}_1$ are given in the column labeled **coef** in lines 4 and 5. The estimator of σ is clearly labeled as $\mathbf{s = 8.704}$ in the last line of the printout. In an EXECUSTAT printout, this estimator will be labeled **Standard error of estimation.**

The practical interpretation that can be given to s ultimately rests on the meaning of σ. Since σ measures the spread of the y values about the line of means $E(y \mid x) = \beta_0 + \beta_1 x$ (see Figure 10.3), we would expect (from the Empirical Rule) approximately 95% of the y values fall within 2σ of that line. Since we do not know σ, $2s$ provides an approximate value for the half width of this interval. Now return to Figure 10.4 and note the location of the data points about the least-squares line. Since we used the $n = 10$ data points to fit the least-squares line, you would not be too surprised to find that most of the points fall within $2s = 2(8.7) = 17.4$ of the line. If you check Figure 10.4, you will see that all ten points fall within $2s$ of the

TABLE 10.3
MINITAB regression printout for the data in Table 10.1

```
-----------------------------------------------------------------
The regression equation is
y = 40.8 + 0.766 x

Predictor      Coef       Stdev    t-Ratio        p
Constant     40.784       8.507       4.79    0.000
x            0.7656      0.1750       4.38    0.002

s = 8.704        R-sq = 70.5%       R-sq(adj) = 66.8%
-----------------------------------------------------------------
```

least-squares line. (You will find that, in general, most of the data points used to fit the least-squares line will fall within $2s$ of the line. This estimate provides you with a rough check for your calculated value of s.)

The initial question we pose concerns the value of β_1, which is the average change in y for a 1-unit increase in x. Stating that y does not increase (or decrease) linearly as x increases is equivalent to saying that $\beta_1 = 0$. Thus, we would wish to test an hypothesis that $\beta_1 = 0$ against the alternative that $\beta_1 \neq 0$. As we might suspect, the estimator $\hat{\beta}_1$ is extremely useful in constructing a test statistic to test this hypothesis. Therefore, we wish to examine the distribution of estimates $\hat{\beta}_1$ that would be obtained when samples, each containing n points, are repeatedly drawn from the population of interest. If we assume that the random error ϵ is normally distributed, in addition to the previously stated assumption, it can be shown that the test statistic

$$t = \frac{\hat{\beta}_1 - \beta_1}{s/\sqrt{S_{xx}}}, \qquad \text{where } S_{xx} = \sum_{i=1}^{n}(x_i - \bar{x})^2$$

follows a Student's t distribution in repeated sampling with $(n - 2)$ d.f. Note that the number of degrees of freedom associated with s^2 determines the number of degrees of freedom associated with t. Thus, we observe that the test of an hypothesis that β_1 equals some particular numerical value, say β_{10}, is the familiar t test encountered in Chapter 9.

Test of Hypothesis Concerning the Slope of a Line

1 Null Hypothesis: $H_0 : \beta_1 = \beta_{10}$

2 Alternative Hypothesis:

One-Tailed Test	*Two-Tailed Test*
$H_a : \beta_1 > \beta_{10}$ (or, $\beta_1 < \beta_{10}$)	$H_a : \beta_1 \neq \beta_{10}$

3 Test Statistic: $t = \dfrac{\hat{\beta}_1 - \beta_{10}}{s/\sqrt{S_{xx}}}$

When the assumptions given in Section 10.2 (and 10.8) are satisfied, the test statistic will have a Student's t distribution with $(n - 2)$ d.f.

4 Rejection Region:

One-Tailed Test	*Two-Tailed Test*
$t > t_\alpha$ (or, $t < -t_\alpha$ when the alternative hypothesis is $H_a : \beta_1 < \beta_{10}$)	$t > t_{\alpha/2}$ or $t < -t_{\alpha/2}$

The values of t_α and $t_{\alpha/2}$ are given in Table 4 in Appendix I. Use the values of t corresponding to $(n-2)$ d.f.

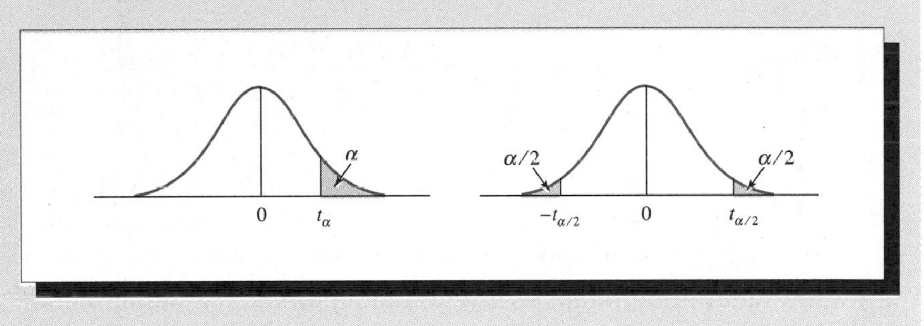

EXAMPLE **10.2** Use the data in Table 10.1 to determine whether the value of $\hat{\beta}_1$ provides sufficient evidence to indicate that β_1 differs from zero; that is, does a linear relationship exist between a freshman's mathematics achievement score x and his or her final calculus grade y?

Solution We wish to test the null hypothesis

$$H_0 : \beta_1 = 0 \qquad \text{against} \qquad H_a : \beta_1 \neq 0$$

for the calculus grade-achievement test score data in Table 10.1. ▪

Table 10.4 gives the EXECUSTAT regression analysis printout for the data in Table 10.1 and contains the estimated line, the estimates of the intercept and the slope, together with their standard errors, and the value of the t statistic in testing $H_0 : \beta_1 = 0$ versus the two-tailed alternative $H_a : \beta_1 \neq 0$. The value of the t statistic is found in the column labeled ***t* Value** in the row corresponding to the **Slope** and is equal to **Estimate/(Standard Error) = 0.765562/0.174985 = 4.38**. The p-value associated with this test is the second entry in the column labeled ***p*-Value** and is equal to **.002**. Therefore, if we were using a value of α greater than .002, we would reject $H_0 : \beta_1 = 0$. The value of $s = \sqrt{\text{MSE}}$ is given in the second part of the printout as the **Standard error of estimation = 8.70363**.

Once we have decided that x and y are linearly related, we are interested in examining this relationship in detail. If x increases by 1 unit, what is the estimated

TABLE 10.4
EXECUSTAT printout for Example 10.2

```
                  Simple Regression Analysis for IPS10-1
==================================================================
Linear Model: CALCGR = 40.7842 + 0.765562*ACHTEST

                         Table of Estimates
==================================================================
                               Standard           t            p-
                Estimate          Error        Value        Value
==================================================================
Intercept    40.7842          8.50686          4.79         0.001
Slope           0.765562      0.174985         4.38         0.002
==================================================================
R-squared = 70.52%
Correlation coeff. = 0.840
Standard error of estimation = 8.70363
Durbin-Watson statistic = 1.17368
Mean absolute error = 6.97801
Sample size (n) = 10
==================================================================
```

change in y, and how much confidence can be placed in the estimate? In other words, we require an estimate of the slope β_1. You will not be surprised to observe a continuity in the procedures of Chapters 9 and 10; that is, the confidence interval for β_1, with confidence coefficient $(1 - \alpha)$, can be shown to be

$$\hat{\beta}_1 \pm t_{\alpha/2}(\text{estimated } \sigma_{\hat{\beta}_1})$$

or

A $100(1 - \alpha)\%$ Confidence Interval for β_1

$$\hat{\beta}_1 \pm t_{\alpha/2}(\text{standard error of } \hat{\beta}_1)$$

where $t_{\alpha/2}$ is based on $(n - 2)$ d.f.

EXAMPLE 10.3 Find a 95% confidence interval for β_1 based on the data in Table 10.2.

Solution The regression analysis printout in Table 10.4 contains the estimate of the slope and its standard error. Therefore, the 95% confidence interval for β_1 is $\hat{\beta}_1 \pm t_{.025}$ (standard error of $\hat{\beta}_1$)

$$.766 \pm 2.306(.175) \qquad \text{or} \qquad .766 \pm .404$$

Several points concerning the interpretation of our results deserve particular attention. As we have noted, β_1 is the slope of the assumed line over the region of observation and indicates the linear change in $E(y \mid x)$ for a 1-unit change in x. **Even**

if we do not reject the null hypothesis that the slope of the line β_1 equals zero, it does not necessarily mean that x and y are unrelated. In the first place, we must be concerned with the probability of committing a type II error—that is, of accepting the null hypothesis that the slope equals zero when this hypothesis is false. Second, it is possible that x and y might be perfectly related in a curvilinear, but not linear, manner. For example, Figure 10.5 depicts a curvilinear relationship between y and x over the domain of $x : a \le x \le f$. We note that a straight line would provide a good predictor of y if fitted over a small interval in the x domain, say, $b \le x \le c$. The resulting line is line 1. On the other hand, if we attempt to fit a line over the region $c \le x \le d$, then β_1 equals zero and the best fit to the data is the horizontal line 2. This result would occur even though all the points fell perfectly on the curve and y and x possessed a functional relation as defined in Section 10.2. We must take care in drawing conclusions if we do not find evidence to indicate that β_1 differs from zero. Perhaps we have chosen the wrong type of probabilistic model for the physical situation.

FIGURE 10.5
Curvilinear relation

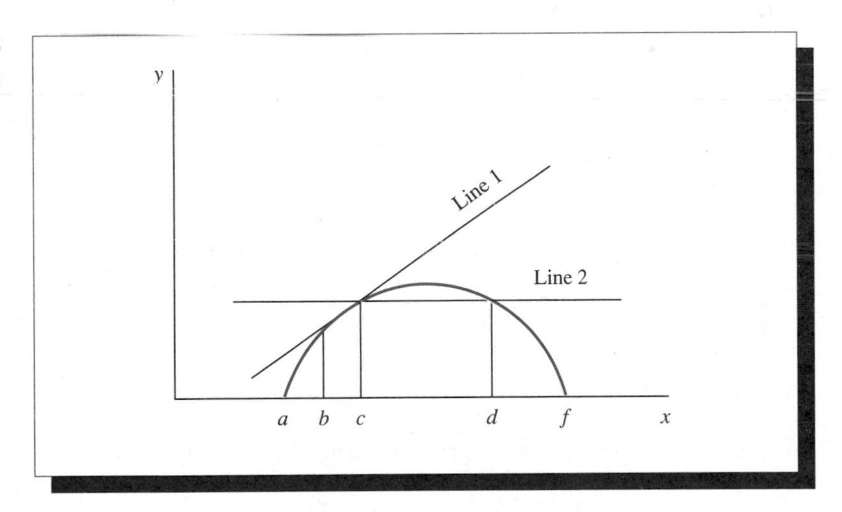

Note that the comments contain a second implication. **If the data provide values of x in an interval $b \le x \le c$, then the calculated prediction equation is appropriate only over this region. Extrapolation in predicting y for values of x outside the region $b \le x \le c$ for the situation indicated in Figure 10.5 would result in a serious prediction error.**

If the data present sufficient evidence to indicate that β_1 differs from zero, we do not conclude that the true relationship between y and x is linear. Undoubtedly, y is a function of a number of variables, which demonstrate their existence to a greater or lesser degree in terms of the random error ϵ that appears in the model. This random error, of course, is why we have been obliged to use a probabilistic model in the first place. Large errors of prediction imply curvatures in the true relation between y and x, the presence of other important variables that do not appear in the model, or both, as most often is the case. **All we can say is that we have evidence to indicate that y changes as x changes and that we may obtain a better prediction of y by**

using x and the linear predictor than by simply using $\bar{y}$ and ignoring x. Note that this statement does not imply a causal relationship between x and y. Some third variable may have caused the change in both x and y, producing the relationship that we have observed.

Exercises

Basic Techniques

10.10 The EXECUSTAT printout below gives the least-squares regression analysis for the data in Exercise 10.6. Do the data present sufficient evidence to indicate the y and x are linearly related? (Test the hypothesis that $\beta_1 = 0$; use $\alpha = .05$.)

```
              Simple Regression Analysis for IPS10-6
================================================================
Linear Model: y = 3 + 1.2*x

                        Table of Estimates
================================================================
                           Standard             t            p-
              Estimate       Error            Value        Value
================================================================
Intercept         3        0.326599           9.19        0.0027
Slope           1.2        0.23094            5.20        0.0138
================================================================
R-squared = 90.00%
Correlation coeff. = 0.949
Standard error of estimation = 0.730297
Durbin-Watson statistic = 2.6
Mean absolute error = 0.48
Sample size (n) = 5
================================================================
```

10.11 Refer to Exercise 10.10. Find a 90% confidence interval for the slope of the line. Interpret this interval estimate.

10.12 Do the data in Exercise 10.7 present sufficient evidence to indicate that y and x are linearly related? Use the MINITAB printout provided below:

```
----------------------------------------------------------------
The regression equation is
y = 6.00 - 0.557 x

Predictor         Coef        Stdev       t-ratio          p
Constant        6.0000       0.1759         34.10        0.000
x             -0.55714      0.04518        -12.33        0.000

s = 0.1890          R-sq = 97.4%           R-sq(adj) = 96.8%
----------------------------------------------------------------
```

10.13 Refer to Exercise 10.12. Find a 90% confidence interval for the slope of the line. Interpret this interval estimate.

Applications

10.14 Refer to the data in Exercise 10.8, relating x, the number of books written by Professor Issac Asimov, to y, the number of months required to write his books in increments of 100. Do the data support the hypothesis that $\beta_1 \neq 0$? Use the computer printout given Exercise 10.8 and $\alpha = .05$.

10.15 Refer to the computer printout given in Exercise 10.9. Suppose you wish to determine whether the gourmet coffee sales are increasing with time.

a State the null and alternative hypotheses to be tested.

b Conduct the test using the results of the printout in Exercise 10.9 and state your conclusions. Give the observed level of significance for the test.

10.16 A study was conducted to determine the effects of sleep deprivation on subjects' ability to solve problems. The amount of sleep deprivation varied over 8, 12, 16, 20, and 24 hours without sleep. A total of ten subjects participated in the study, two at each sleep-deprivation level. After his or her specified sleep-deprivation period, each subject was administered a set of simple addition problems, and the number of errors was recorded. The following results were obtained:

Number of errors y	8, 6	6, 10	8, 14	14, 12	16, 12
Number of hours without sleep x	8	12	16	20	24

```
------------------------------------------------------------
The regression equation is
Y = 3.00 + 0.475 X

Predictor       Coef       Stdev      t-ratio         p
Constant       3.000       2.127         1.41     0.196
X             0.4750      0.1253         3.79     0.005

s = 2.242        R-sq = 64.2%        R-sq(adj) = 59.8%

Analysis of Variance

SOURCE        DF         SS         MS         F         p
Regression     1     72.200     72.200     14.37     0.005
Error          8     40.200      5.025
Total          9    112.400
------------------------------------------------------------
```

Use the MINITAB printout to answer the following questions:

a Find the least-squares line appropriate to these data.

b Plot the points and graph the least-squares line as a check on your calculations.

c Calculate s^2.

10.17 Do the data in Exercise 10.16 present sufficient evidence to indicate that number of errors is linearly related to number of hours without sleep? (Test using $\alpha = .05$.) Would you expect the relation between y and x to be linear if x were varied over a wider range (say, $x = 4$ to $x = 48$)?

10.18 Find a 95% confidence interval for the slope of the line in Exercise 10.17. Give a practical interpretation to this interval estimate.

10.19 Daily readings of air temperatures and soil moisture percentages were collected by William Reiners and colleagues (1984) at four elevations, 670, 825, 1145, and 1379 meters, in the White Mountains of New Hampshire. Among the analyses of the data are simple linear regressions relating air temperature y (in °C) in a given month to altitude x in meters. For example, the regression analysis for July produced the following statistics: $\hat{\beta}_0 = 21.4$, $\hat{\beta}_1 = -0.0063$, $r^2 = .99$, and the estimate of the standard deviation of $\hat{\beta}_1$ is 0.0007.

a Give the least-squares prediction equation relating air temperature y (in °C) to altitude x in meters for the month of July.

b How many degrees of freedom would be associated with SSE and s^2 in this analysis? (Assume that readings were taken each day during the month of July at all four elevations.)

c Do the data present sufficient evidence to indicate that altitude x contributes information for predicting air temperature y? Test using $\alpha = .01$.

10.5 Estimation and Prediction Using the Fitted Line

In Chapters 7 and 9, we studied methods for estimating a population mean μ and encountered numerous practical applications of these methods in the examples and exercises. Now we will consider a generalization of this problem.

Estimating the mean value of y for a given value of x—that is, estimating $E(y|x)$—can be a very important practical problem. If a corporation's profit y is linearly related to advertising expenditures x, the marketing director may wish to estimate the mean profit for a given expenditure x. Similarly, a research physician might wish to estimate the mean response of a human to a specific drug dosage x, and an educator might wish to know the mean grade expected of calculus students who acquired a mathematics achievement test score of $x = 50$. The least-squares prediction equation can be used to obtain these estimates.

Assume that x and y are linearly related according to the probabilistic model defined in Section 10.2 and therefore that $E(y \mid x) = \beta_0 + \beta_1 x$ represents the expected value of y for a given value of x. **Since the fitted line**

$$\hat{y} = \hat{\beta}_0 + \hat{\beta}_1 x$$

attempts to estimate the line of means $E(y \mid x)$ (i.e., we estimate β_0 and β_1), then $\hat{y}$ can be used to estimate the expected value of y as well as to predict some value of y that might be observed in the future. It seems reasonable to assume that the errors of estimation and prediction differ for these two cases. Consequently, the two estimation procedures differ.

Observe the two lines in Figure 10.6. The first line represents the line of means

$$E(y \mid x) = \beta_0 + \beta_1 x$$

and the second is the fitted prediction equation

$$\hat{y} = \hat{\beta} + \hat{\beta}_1 x$$

We observe from the figure that the error in estimating the expected value of y when $x = x_p$ is the deviation between the two lines above the point x_p. Also, this error increases as we move to the endpoints of the interval over which x has been measured. Although the expected value of y for a particular value of x is of interest for our example in Table 10.1, we are primarily interested in *using* the prediction equation $\hat{y} = \hat{\beta}_0 + \hat{\beta}_1 x$ based on our observed data to predict the final calculus grade for some prospective student selected from the population of interest. That is, we want to use the prediction equation obtained for the ten measurements in Table 10.1 to predict the final calculus grade for a new student selected from the population. If the student's achievement test score was x_p, we intuitively see that the error of prediction (the deviation between $\hat{y}$ and the actual grade y that the student will obtain) is composed of two elements. Since the student's grade will equal

$$y = \beta_0 + \beta_1 x_p + \epsilon$$

$(y - \hat{y})$ equals the deviation between $\hat{y}$ and the expected value of y, shown in Figure 10.6, *plus* the random amount ϵ that represents the deviation of the student's grade from the expected value (Figure 10.7). **Thus, the variability in the error for predicting a single value of y exceeds the variability for estimating the expected value of y.**

FIGURE 10.6
Estimating $E(y \mid x)$ when $x = x_p$

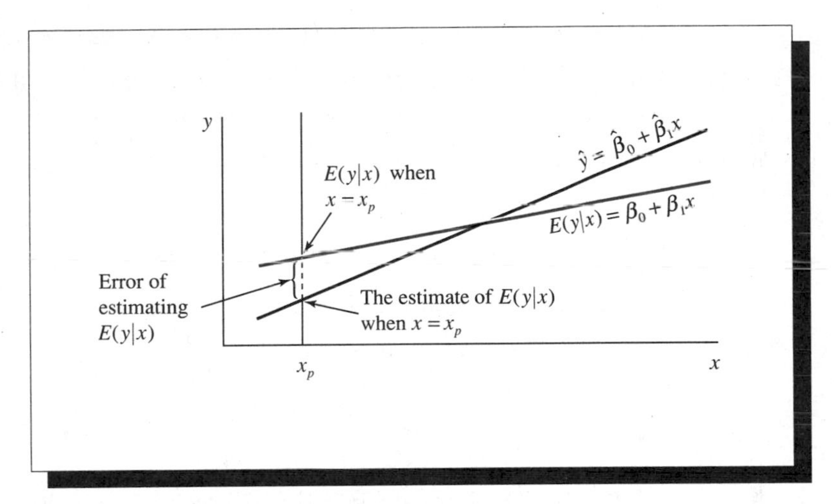

FIGURE 10.7
Error in predicting a particular value of y

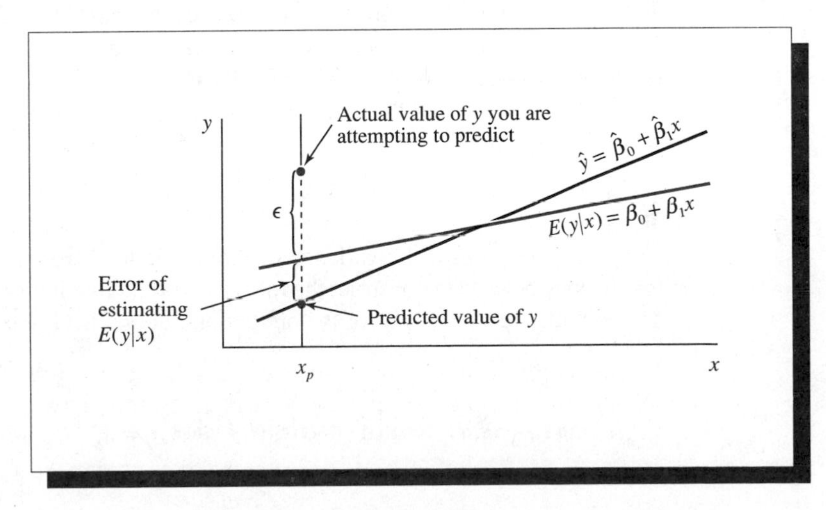

A $100(1-\alpha)\%$ confidence interval for the expected value of y, given $x = x_p$, is

A $100(1-\alpha)\%$ Confidence Interval for $E(y \mid x)$ when $x = x_p$

$$\hat{y} \pm t_{\alpha/2} s_{\hat{y}}$$

where $t_{\alpha/2}$ is based on $(n-2)$ d.f. and $s_{\hat{y}}$, given in Section 10.6, is the standard error in estimating $E(y \mid x)$.

EXAMPLE 10.4 Find a 95% confidence interval for the expected value of y, the final calculus grade, given that the mathematics achievement test score is $x = 50$.

Solution To estimate the mean calculus grade for students whose achievement test score was $x_p = 50$, we would use

$$\hat{y} = \hat{\beta}_0 + \hat{\beta}_1 x_p$$

to calculate $\hat{y}$, the estimate of $E(y \mid x = 50)$. Then, using values calculated previously

$$\hat{y} = 40.784 + (.765562)(50) = 79.062 \quad \blacksquare$$

The MINITAB printout in Table 10.5 shows that the **REGRESS** command followed by the subcommand **PREDICT** will produce an estimate of $E(y)$ for a given value of x, labeled **Fit**; the standard error of the prediction, $s_{\hat{y}}$ labeled **Stdev.Fit**; and the resulting 95% confidence interval estimate, **95% C.I.** Therefore, the estimate of $E(y \mid x = 50)$ is **79.06**, with a standard error of **2.84**, and the resulting 95% confidence interval estimate is **(72.51, 85.61).**

It can be shown that the variance of the error of predicting a particular value of y when $x = x_p$, that is, $(y - \hat{y})$, is

$$\sigma^2_{(y-\hat{y})} = \sigma^2_y + \sigma^2_{\hat{y}}$$

When n is very large, the variance of the prediction error will approach σ^2. These results may be used to construct the following prediction interval for y, given $x = x_p$. The confidence coefficient for the prediction interval is $(1-\alpha)$.

A $100(1-\alpha)\%$ Prediction Interval for y when $x = x_p$

$$\hat{y} \pm t_{\alpha/2} s_{(y-\hat{y})}$$

where $t_{\alpha/2}$ is based on $(n-2)$ d.f. and $s^2_{(y-\hat{y})} = s^2 + s^2_{\hat{y}}$

TABLE 10.5
MINITAB printout for Example 10.4

```
MTB > REGRESS C1 1 C2;
SUBC> PREDICT 50.

The regression equation is
Y = 40.8 + 0.766 X

Predictor        Coef        Stdev     t-ratio        p
Constant       40.784        8.507        4.79    0.000
X              0.7656       0.1750        4.38    0.002

s = 8.704            R-sq = 70.5%        R-sq(adj) = 66.8%
   .                    .                    .          .
   .                    .                    .          .
   .                    .                    .          .

  Fit    Stdev.Fit        95% C.I.            95% P.I.
79.06         2.84    (72.51,   85.61)     (57.94, 100.18)
```

EXAMPLE 10.5 Refer to Example 10.1 and predict the final calculus grade for some new student who scored $x = 50$ on the mathematics achievement test.

Solution The predicted value of y would be

$$\begin{aligned}\hat{y} &= \hat{\beta}_0 + \hat{\beta}_1 x_p \\ &= 40.784 + (.75562)(50)\end{aligned}$$

and the 95% prediction interval for the final calculus grade would be

$$79.06 \pm (2.306)\sqrt{75.754 + (2.84)^2}$$

or

$$79.06 \pm 21.11$$

You can generate the EXECUSTAT printout shown in Table 10.6 by using a subcommand in the RELATE/SIMPLE REGRESSION menu of the program. The least-squares estimate and the prediction limits for $x = 50$ are given in the third, fourth, and fifth columns of the table and agree with the hand calculations.

TABLE 10.6
EXECUSTAT printout for Example 10.5

Table of Predicted Values

Row	ACHTEST	Predicted CALCGR	95.00% Prediction Limits Lower	Upper	95.00% Confidence Limit Lower	Upper
1	50	79.0622	57.9502	100.174	72.5133	85.611

Note that in a practical situation we would probably have the grades and achievement test scores for many more than the $n = 10$ students indicated in Table 10.1 and that this would reduce somewhat the width of the prediction interval. In fact, when n

is large, the prediction interval approaches $\hat{y} \pm z_{\alpha/2}s$ or, for a 95% prediction interval, $\hat{y} \pm 1.96s$. ▪

Again, note the distinction between the confidence interval for $E(y \mid x)$ and the prediction interval for y presented in this section. $E(y \mid x)$ is a mean, a parameter of a population of y values, and y is a random variable that oscillates in a random manner about $E(y \mid x)$. The mean value of y when $x = 50$ is vastly different from some value of y chosen at random from the set of all y values for which $x = 50$. To make this distinction when making inferences, **we always *estimate* the value of a parameter and *predict* the value of a random variable.** As noted in our earlier discussion and as shown in Figures 10.6 and 10.7, the error of predicting y is different from the error of estimating $E(y \mid x)$. This is evident in the difference in widths of the two prediction and confidence intervals.

A graph of the confidence interval for $E(y \mid x)$ and the prediction interval for a particular value of y for the data in Table 10.1 is shown in Figure 10.8. The plot of the confidence interval is shown by dotted lines, and the prediction interval is identified by dashed lines. Note how the widths of the intervals increase as you move to the right or left of $\bar{x} = 46$. In particular, see the confidence interval and prediction interval for $x = 50$ calculated in Examples 10.4 and 10.5.

FIGURE **10.8** Confidence intervals for $E(y \mid x)$ and prediction intervals for y based on data in Table 10.2

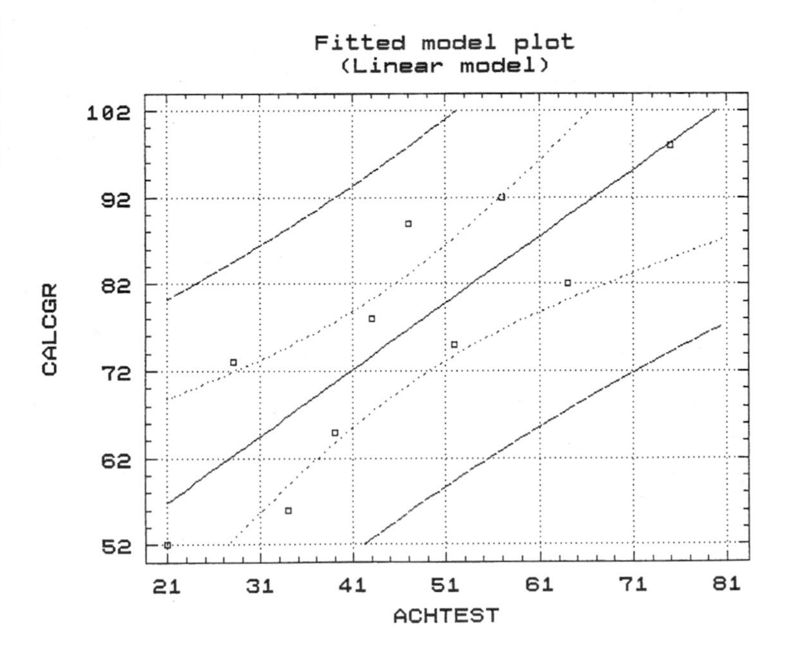

Exercises

Basic Techniques

10.20 Refer to Exercise 10.6. The EXECUSTAT printout shown below was generated using the **ESTIMATE** command in the **RELATE/SIMPLE REGRESSION** menu:

Table of Predicted Values

		Predicted	90.00% Prediction Limits		90.00% Confidence Limits	
Row	x	y	Lower	Upper	Lower	Upper
1	-2	0.6	-1.57396	2.77396	-0.731273	1.93127
2	-1	1.8	-0.15958	3.75958	0.858648	2.74135
3	0	3	1.1173	4.8827	2.23139	3.76861
4	1	4.2	2.24042	6.15958	3.25865	5.14135
5	2	5.4	3.22604	7.57396	4.06873	6.73127

a Estimate the expected value of y when $x = 1$, using a 90% confidence interval.

b Find a 90% prediction interval for some value of y to be observed in the future when $x = 1$.

10.21 Refer to Exercise 10.7. If $s_{\hat{y}} = .102685$, find a 90% confidence interval for the mean value of y when $x = 2$.

10.22 Refer to Exercise 10.7 and use the computer printout in Exercise 10.12.

a Find the value of s_y in the computer printout.

b If $s_{\hat{y}} = .102685$, find $s_{y-\hat{y}}$.

c Find a 95% prediction interval for some value of y to be observed in the future when $x = 2$.

Applications

10.23 Refer to the simple linear regression analysis in Exercise 10.8, relating the time in months y to write his books to x, the number of books written. Part of an EXECUSTAT printout for the data is given below:

Table of Predicted Values

		Predicted	95.00% Prediction Limits		95.00% Confidence Limits	
Row	x	y	Lower	Upper	Lower	Upper
1	250	363.434	275.759	451.109	325.928	400.939
2	550	564.473	456.365	672.58	490.94	638.005

a Find a 95% confidence interval for the average time taken to write $x = 250$ books.

b If Professor Asimov had written 550 books, predict the amount of time it would have taken him using a 95% prediction interval.

10.24 A marketing research experiment was conducted to study the relationship between the length of time necessary for a buyer to reach a decision and the number of alternative package designs of a product presented. Brand names were eliminated from the packages to reduce the effects of brand preferences. The buyers made their selections using the manufacturer's product descriptions on the packages as the only buying guide. The length of time necessary to reach a decision is recorded for 15 participants in the marketing research study.

Length of decision time y (sec)	5, 8, 8, 7, 9	7, 9, 8, 9, 10	10, 11, 10, 12, 9
Number of alternatives x	2	3	4

```
-----------------------------------------------------------
The regression equation is
y = 4.30 + 1.50 x

Predictor        Coef       Stdev     t-ratio         p
Constant        4.300       1.216        3.53     0.004
x              1.5000      0.3913        3.83     0.002

s = 1.237         R-sq = 53.1%     R-sq(adj) = 49.5%

   Fit  Stdev.Fit         95% C.I.           95% P.I.
 8.800      0.319    (8.110, 9.490)    (6.039, 11.561)
-----------------------------------------------------------
```

Use the MINITAB printout given above to answer the following questions.

a Find the least-squares line appropriate for these data.

b Plot the points and graph the line as a check on your calculations.

c Calculate s^2.

d Do the data present sufficient evidence to indicate that the length of decision time is linearly related to the number of alternative package designs? (Test at the $\alpha = .05$ level of significance.)

e Find the approximate observed significance level for the test and interpret its value.

f Estimate the average length of time necessary to reach a decision when three alternatives are presented, using a 95% confidence interval.

10.25 If you try to rent an apartment or buy a house, you find that real estate representatives establish apartment rents and house prices on the basis of square footage of heated floor space. The data in the table give the square footages and sales prices of $n = 12$ houses randomly selected from those sold in a small city.

Square Feet x	**Price** y
1460	$ 88,700
2108	109,300
1743	101,400
1499	91,100
1864	102,400
2391	114,900
1977	105,400
1610	97,000
1530	92,400
1759	98,200
1821	104,300
2216	111,700

```
-------------------------------------------------------------
The regression equation is
y = 51206 + 27.4 x

Predictor       Coef        Stdev      t-ratio        p
Constant       51206         3389        15.11    0.000
x             27.406        1.828        14.99    0.000

s = 1793          R-sq = 95.7%         R sq(adj) = 95.3%

     Fit    Stdev.Fit          95% C.I.             95% P.I.
  106018          602   (104676, 107360)    (101803, 110233)
   99989          526   ( 98816, 101161)    ( 95825, 104153)
-------------------------------------------------------------
```

Use the MINITAB printout given above to answer the following questions:

a Estimate the mean increase in the price for an increase of 1 square foot for houses sold in the city. Use a 90% confidence interval. Interpret your estimate.

b Suppose that you are a real estate salesperson and you desire an estimate of the mean sales price of houses with a total of 2000 square feet of heated space. Use a 95% confidence interval and interpret your estimate.

c Calculate the price per square foot for each house and then calculate the sample mean. Why is this estimate of the mean cost per square foot not equal to the answer in part (a)? Should it be? Explain.

d Suppose that a house containing 1780 square feet of heated floor space is offered for sale. Give a 90% prediction interval for the price at which the house will sell. Interpret this prediction. (HINT: Modify the 95% prediction interval given in the printout.)

10.26 An experiment was conducted to determine the effect of soil applications of various levels of phosphorus on the inorganic phosphorus levels in a particular plant. The following data represent the levels of inorganic phosphorus in micromoles (μM) per gram dry weight of sudan grass roots grown in the greenhouse for 28 days, in the absence of zinc.

Phosphorus Applied x	Phosphorus in Plant y
.5 μM	204
	195
	247
	245
.25 μM	159
	127
	95
	144
.10 μM	128
	192
	84
	71

```
-------------------------------------------------------------
The regression equation is
y = 80.9 + 271 x

Predictor        Coef       Stdev    t-ratio        p
Constant        80.85       22.40       3.61    0.005
x              270.82       68.31       3.96    0.003

s = 39.04          R-sq = 61.1%     R-sq(adj) = 57.2%

     Fit    Stdev.Fit         95% C.I.            95% P.I.
   135.0         12.6     (106.9, 163.2)      (43.6, 226.5)
-------------------------------------------------------------
```

Use the MINITAB printout given above to answer the following questions:

a Plot the data. Do the data appear to exhibit a linear relationship?

b Find the least-squares line relating the plant phosphorus levels y to the amount of phosphorus applied to the soil x. Graph the least-squares line as a check on your calculations.

c Do the data provide sufficient evidence to indicate that the amount of phosphorus present in the plant is linearly related to the amount of phosphorus applied to the soil? Test using $\alpha = .05$.

d Find the approximate p-value for the test in part (c).

e Estimate the mean amount of phosphorus in the plant if .20 μM of phosphorus is applied to the soil, in the absence of zinc. Use a 90% confidence interval.

10.27 In Exercise 10.19, another simple linear regression analysis presented in the paper by William Reiners and colleagues (1984) involved the relationship between the number y of degree-days (days when the temperature is above freezing) per year and the altitude x in meters. The analysis is based on the recorded number of degree-days for two winters, 1976–1977 and 1977–1978, at each of the four altitudes. Statistics presented for this analysis are $\hat{\beta}_0 = 3393.93$, $\hat{\beta}_1 = -1.319$, and $r^2 = .99$, and the estimated standard deviation of $\hat{\beta}_1$ is 0.091.

a How many degrees of freedom are associated with SSE and s^2 in this analysis?

b Do the data provide sufficient evidence to indicate that altitude x provides information for the prediction of degree-days y? Test using $\alpha = .01$.

c Give the least-squares prediction equation.

d Predict the number of degree-days at an altitude of 825 meters. Do the statistics above enable you to construct a prediction interval for your prediction? Explain.

10.6 Calculational Formulas

There are six basic quantities that are used in regression analysis calculations:

1 n **2** $\bar{y} = \dfrac{\sum_{i=1}^{n} y_i}{n}$ **3** $\bar{x} = \dfrac{\sum_{i=1}^{n} x_i}{n}$

4 $S_{yy} = \sum_{i=1}^{n} (y_i - \bar{y})^2 = \sum_{i=1}^{n} y_i^2 - \dfrac{\left(\sum_{i=1}^{n} y_i\right)^2}{n}$

5 $S_{xy} = \sum_{i=1}^{n}(x_i - \bar{x})(y_i - \bar{y}) = \sum_{i=1}^{n} x_i y_i - \frac{\left(\sum_{i=1}^{n} x_i\right)\left(\sum_{i=1}^{n} y_i\right)}{n}$

6 $S_{xx} = \sum_{i=1}^{n}(x_i - \bar{x})^2 = \sum_{i=1}^{n} x_i^2 - \frac{\left(\sum_{i=1}^{n} x_i\right)^2}{n}$

Relevant formulas for commonly used estimators and their standard errors follow:

1 Slope: $\hat{\beta} = \frac{S_{xy}}{S_{xx}}$; intercept: $\hat{\beta}_0 = \bar{y} - \hat{\beta}_1\bar{x}$; least-squares line: $\hat{y} = \hat{\beta}_0 + \hat{\beta}_1 x$

2 $\text{SSE} = S_{yy} - \frac{(S_{xy})^2}{S_{xx}}$; $s^2 = \frac{\text{SSE}}{n-2}$; $s = \sqrt{s^2}$

3 Standard error of $\hat{\beta}_1$: $s_{\hat{\beta}_1} = \sqrt{\frac{s^2}{S_{xx}}}$

4 Standard error of $\hat{\beta}_0$: $s_{\hat{\beta}_0} = \sqrt{s^2\left[\frac{1}{n} + \frac{\bar{x}^2}{S_{xx}}\right]}$

5 Standard error in estimating $E(y)$: $s_{\hat{y}} = \sqrt{s^2\left[\frac{1}{n} + \frac{(x_p - \bar{x})^2}{S_{xx}}\right]}$

6 Standard error of prediction: $s_{(y-\hat{y})} = \sqrt{s^2 + s_{\hat{y}}^2} = \sqrt{s^2\left[1 + \frac{1}{n} + \frac{(x_p - \bar{x})^2}{S_{xx}}\right]}$

Remember that, in general, a small-sample test statistic is constructed as

$$t = \frac{(\text{estimator}) - (\text{parameter} \mid H_0)}{\text{SE(estimator)}}$$

based upon $n - 2$ d.f. A $100(1 - \alpha)\%$ interval estimator is found using (estimator) $\pm$ $t_{\alpha/2}$SE(estimator).

Tips on Problem Solving

1 Be careful of rounding errors, which can greatly affect the answer you obtain in calculating S_{xx}, S_{xy}, and S_{yy}. If you must round a number, it is recommended that you carry six or more significant figures in the calculations. (Note also that in working exercises, rounding errors might cause some slight discrepancies between your answers and the answers given in the back of the text.)

2 Always plot the data points and graph your least-squares line. If the line does not provide a reasonable fit to the data points, you may have made an error in your calculations.

10.7 A Coefficient of Correlation

Sometimes we wish to obtain an indicator of the strength of the linear relationship between two variables y and x that is independent of their respective scales of measurement. We call this a measure of the **linear correlation between y and x.**

The measure of the linear correlation commonly used in statistics is called the **Pearson product moment coefficient of correlation** between y and x. This quantity, denoted by the symbol r, is computed as follows:

Pearson Product Moment Coefficient of Correlation

$$r = \frac{S_{xy}}{\sqrt{S_{xx}S_{yy}}}$$

where S_{xy}, S_{xx}, and S_{yy} were defined in Section 10.6.

We will show you how to compute the Pearson product moment coefficient of correlation for the GPA data in Table 10.1, and then we will explain how it measures the strength of the relationship between y and x.

EXAMPLE **10.6** Calculate the coefficient of correlation for the calculus grade–achievement test score data in Table 10.1.

Solution The coefficient of correlation for the calculus grade–achievement test score data in Table 10.1 can be obtained by using the formula for r and the quantities

$$S_{xy} = 1894$$
$$S_{xx} = 2474$$
$$S_{yy} = 2056$$

Then,

$$r = \frac{S_{xy}}{\sqrt{S_{xx}S_{yy}}} = \frac{1894}{\sqrt{2474(2056)}} = .84$$

Most regression programs will provide the value of r or r^2. The value of **R-sq** is found on the MINITAB printout, and the **correlation coeff.** is found on the EXECUSTAT simple regression analysis printout. ▪

A study of the coefficient of correlation r yields rather interesting results and explains the reason for its selection as a measure of linear correlation. We note that the denominators used in calculating r and $\hat{\beta}_1 = S_{xy}/S_{xx}$ will always be positive since they both involve sums of squares of numbers. Since the numerator used in calculating r is identical to the numerator of the formula for the slope $\hat{\beta}_1$, the coefficient of correlation r will assume exactly the same sign as $\hat{\beta}_1$ and will equal zero when $\hat{\beta}_1 = 0$. **Thus, $r = 0$ implies no linear correlation between y and x. A positive value for r implies that the line slopes upward to the right; a negative value indicates that it slopes downward to the right.**

Figure 10.9 shows four typical scatterplots and their associated correlation coefficients. Note that **$r = 0$ implies no linear correlation**, not simply "no correlation." A pronounced curvilinear pattern may exist, as in Figure 10.9(d), but its linear correlation coefficient may equal 0. In general, we can say that r measures the linear association of the two variables y and x. When $r = 1$ or -1, all the points fall on a straight line; when $r = 0$, they are scattered and give no evidence of a *linear* relationship. Any other value of r suggests the degree to which the points tend to be linearly related.

FIGURE 10.9
Some typical scatterplots with approximate values of r

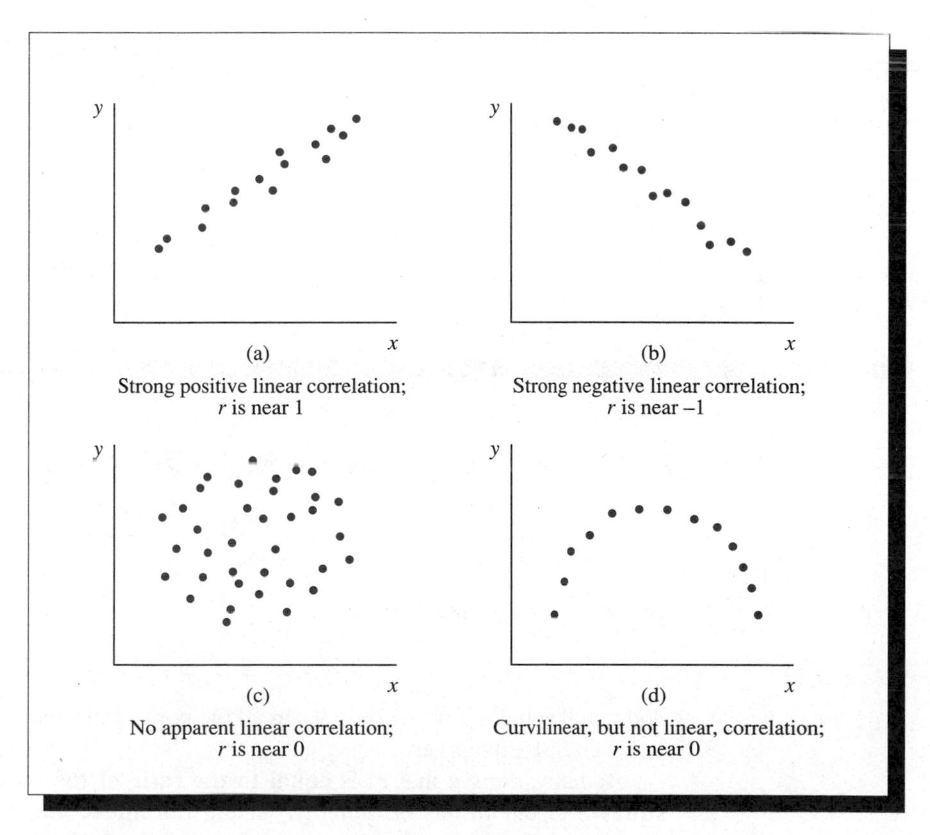

The interpretation of nonzero values of r can be obtained by comparing the errors of prediction for the prediction equation

$$\hat{y} = \hat{\beta}_0 + \hat{\beta}_1 x$$

with the predictor of y, $\bar{y}$, that would be employed if x were ignored. Figure 10.10(a) and (b) shows the lines $\hat{y} = \hat{\beta}_0 + \hat{\beta}_1 x$ and $\hat{y} = \bar{y}$ fit to the same set of data. Certainly, if x is of any value in predicting y, then SSE, the sum of squares of deviations of y about the linear model, should be less than the sum of squares of deviations about the predictor $\bar{y}$, which is

$$S_{yy} = \sum_{i=1}^{n} (y_i - \bar{y})^2$$

FIGURE 10.10 Two models fit to the same data

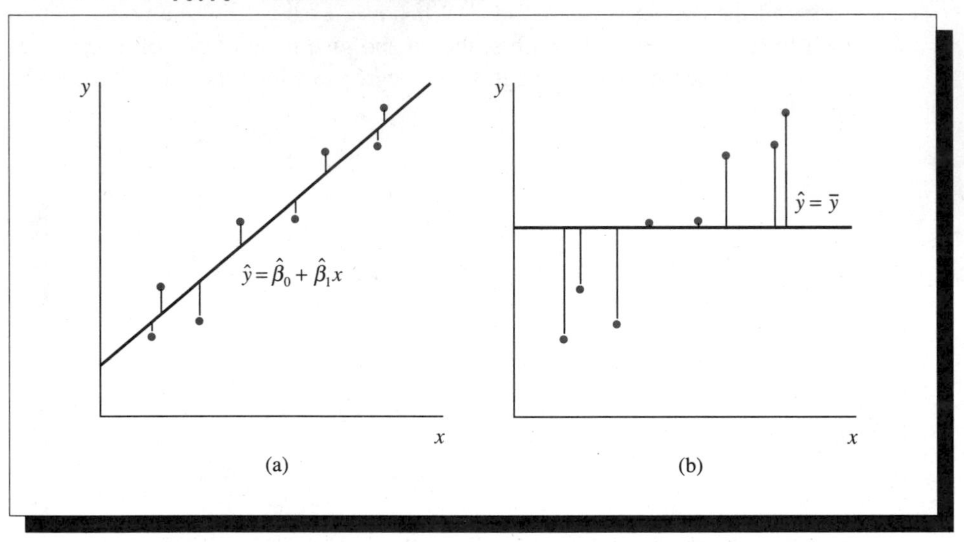

With the aid of a bit of algebraic manipulation, we can show that

$$r^2 = 1 - \frac{\text{SSE}}{S_{yy}} = \frac{S_{yy} - \text{SSE}}{S_{yy}}$$

In other words, r^2 lies in the interval

$$0 \leq r^2 \leq 1$$

and r will equal $+1$ or -1 only when all the points fall exactly on the fitted line, that is, when SSE equals zero.

Actually, we see that **r^2 is equal to the ratio of the reduction in the sum of squares of deviations obtained by using the linear model to the total sum of squares of deviations about the sample mean $\bar{y}$, which would be the predictor of y if x were ignored.** Thus r^2, called the *coefficient of determination*, would seem to give a more meaningful interpretation of the strength of the relation between y and x than would the correlation coefficient r.

Coefficient of Determination

$$r^2 = \frac{S_{yy} - \text{SSE}}{S_{yy}}$$

You will observe that the sample correlation coefficient r is an estimator of a population correlation coefficient ρ (Greek letter rho), which would be obtained if the coefficient of correlation were calculated by using all the points in the population.

A test of the null hypothesis that no correlation exists between y and x, $H_0: \rho = 0$, is exactly equivalent to a test of the hypothesis $H_0: \beta_1 = 0$. The test statistic is given as

$$t = \frac{r\sqrt{n-2}}{\sqrt{1-r^2}} = \frac{\hat{\beta}_1}{s/\sqrt{S_{xx}}}, \qquad \text{with } n-2 \text{ d.f.}$$

For the data in Table 10.1, with $r = .84$, the value of the test statistic is

$$t = \frac{r\sqrt{n-2}}{\sqrt{1-r^2}} = \frac{.84\sqrt{8}}{\sqrt{1-(.84)^2}} = 4.38$$

which is identical to the value calculated for testing $H_0: \beta_1 = 0$ in Example 10.3.

Although r gives a rather nice measure of the goodness of fit of the least-squares line to the fitted data, its use in making inferences concerning ρ may be of dubious practical value in many situations. Ordinarily, we would be interested in testing the null hypothesis that $\rho = 0$ and, since this is algebraically equivalent to testing the hypothesis that $\beta_1 = 0$, we have already considered this problem. If the evidence in the sample suggests that y and x are related, it would seem that we would redirect our attention to the ultimate objective of our data analysis, using the prediction equation to obtain interval estimates for $E(y \mid x)$ and prediction intervals for y.

If the linear coefficients of correlation between y and each of two variables x_1 and x_2 were calculated to be .4 and .5, respectively, it does not follow that a predictor using both variables would account for $[(.4)^2 + (.5)^2] = .41$, or a 41% reduction in the sum of squares of deviations. Actually, x_1 and x_2 might be highly correlated and therefore contribute virtually the same information for the prediction of y.

Finally, remember that r is a measure of **linear correlation** and that x and y could be perfectly related by some **curvilinear** function when the observed value of r is equal to 0.

Exercises

Basic Techniques

10.28 How does the coefficient of correlation measure the strength of the linear relationship between two variables y and x?

10.29 Describe the significance of the algebraic sign and the magnitude of r.

10.30 What value does r assume if all the data points fall on the same straight line and

a The line has positive slope?

b The line has negative slope?

10.31 Given the following data,

x	−2	−1	0	1	2
y	2	2	3	4	4

a Plot the data points. Based on your graph, what will be the sign of the sample correlation coefficient?

b If $S_{xx} = 10$, $S_{yy} = 4$, and $S_{xy} = 6$, calculate r and r^2 and interpret their values.

10.32 For the following data,

x	1	2	3	4	5	6
y	7	5	5	3	2	0

a Plot the six points on graph paper.

b If $S_{xx} = 17.5$, $S_{yy} = 31.3333$, and $S_{xy} = -23$, calculate the sample coefficient of correlation r and interpret.

c By what percentage was the sum of squares of deviations reduced by using the least-squares predictor $\hat{y} = \hat{\beta}_0 + \hat{\beta}_1 x$, rather than $\bar{y}$ as a predictor of y?

10.33 Reverse the slope of the line in Exercise 10.32 by reordering the y observations, as follows:

x	1	2	3	4	5	6
y	0	2	3	5	5	7

Repeat the steps of Exercise 10.32 with $S_{xy} = 23$. Notice the change in sign of r and the relation between the values of r^2 of Exercise 10.32 and this exercise.

Applications

10.34 In Exercise 0.7, we presented some data from a study of the infestation of the *T. orientalis* lobster by two types of barnacles, *O. tridens* and *O. lowei* (Jeffries, Vorhis, and Yang, 1984). The data from Exercise 1.9 are reproduced below. Examine the columns of the table that give the numbers of *O. tridens* and *O. lowei* barnacles on each lobster. Does it appear that the barnacles compete for space on the surface of a lobster?

		Number of Barnacles		
Field Number	**Carapace Length (mm)**	***O. tridens***	***O. lowei***	**Total**
AO61	78	645	6	651
AO62	66	320	23	343
AO66	65	401	40	441
AO70	63	364	9	373
AO67	60	327	24	351
AO69	60	73	5	78
AO64	58	20	86	106
AO68	56	221	0	221
AO65	52	3	109	112
AO63	50	5	350	355

a If they do compete, would you expect the number x of *O. tridens* and the number y of *O. lowei* barnacles to be positively or negatively correlated? Explain.

b If you want to test the theory that the two types of barnacles compete for space by conducting a test of the null hypothesis "the population correlation coefficient ρ equals 0," what would you use for your alternative hypothesis?

c A computer program is run to calculate $r = -.5519$. Conduct the test in part (b) and state your conclusions. Use $\alpha = .05$.

10.35 A social skills training program was implemented with seven mildly handicapped students in a study to determine whether the program caused improvement in pre/post measures and behavior ratings. For one such test (Exercise 1.28), the pre- and posttest scores for the seven students are given below:

	SSRS-T (Standard Score)	
Subject	**Pretest**	**Posttest**
Earl	101	113
Ned	89	89
Jasper	112	121
Charlie	105	99
Tom	90	104
Susie	91	94
Lori	89	99

Source: Torrey et al. (1992), p. 248.

a What type of correlation, if any, would you expect to see between the pre- and posttest scores? Plot the data. Does the correlation appear to be positive or negative?

b The correlation between the pre- and posttest scores is $r = .760$. Is there a significant positive correlation? Test using $\alpha = .05$.

10.36 In Exercise 7.91, we described a study by G. W. Marino (1983) to investigate the variables related to a hockey player's ability to make a fast start from a stopped position. In the experiment, each skater started from a stopped position and attempted to move as rapidly as possible over a 6-meter distance. The correlation coefficient r between a skater's stride rate (number of strides per second) and the length of time to cover the 6-meter distance for the sample of 69 skaters was -0.37.

a Do the data provide sufficient evidence to indicate a correlation between stride rate and time to cover the distance? Test using $\alpha = .05$.

b Find the approximate p-value for the test.

c What are the practical implications of the test in part (a)?

10.37 Refer to Exercise 7.91. Marino (1983) calculated the sample correlation coefficient r for the stride rate and the average acceleration rate for the 69 skaters to be .36.

a Do the data provide sufficient evidence to indicate a correlation between stride rate and average acceleration for the skaters? Test using $\alpha = .05$.

b Find the approximate p-value for the test in part (a).

10.38 Geothermal power is an important source of energy. Since the amount of energy contained in 1 pound of water is a function of its temperature, you might wonder whether water obtained from deeper wells contains more energy per pound. The data in the table are reproduced from an article on geothermal systems by A. J. Ellis (1975).

Location of Well	Average (max.) Drill Hole Depth (m)	Average (max.) Temperature (°C)
El Tateo, Chile	650	230
Ahuachapan, El Salvador	1000	230
Namafjall, Iceland	1000	250
Larderello (region), Italy	600	200
Matsukawa, Japan	1000	220
Cerro Prieto, Mexico	800	300
Wairakei, New Zealand	800	230
Kizildere, Turkey	700	190
The Geysers, United States	1500	250

Use the EXECUSTAT printout to answer the questions that follow.

```
             Simple Regression Analysis for IPS10-38
=============================================================
Linear model: TEMP = 198.925 + 0.0384686*DEPTH

                      Table of Estimates
=============================================================
                                 Standard         t           p-
                  Estimate          Error     Value        Value
=============================================================
Intercept     198.925           38.7696        5.13       0.0014
Slope            0.0384686       0.0416386     0.92       0.3863
=============================================================
R-squared = 10.87%
Correlation coeff. = 0.330
Standard error of estimation = 32.3128
Durbin-Watson statistic = 2.26465
Mean absolute error = 19.8391
Sample size (n) = 9
=============================================================
```

a Find the least-squares line.

b Sketch the least-squares line on graph paper and plot the nine points.

c Calculate the sample coefficient of correlation r and interpret it.

d By what percentage was the sum of squares of deviations reduced by using the least-squares predictor $\hat{y} = \hat{\beta}_0 + \hat{\beta}_1 x$, rather than $\bar{y}$?

10.8 Assumptions

The assumptions for a regression analysis are given in the accompanying display.

Assumptions for a Regression Analysis

1 The response y can be represented by the probabilistic model

$$y = \beta_0 + \beta_1 x + \epsilon$$

2 x is measured without error.

3 ϵ is a random variable such that, for a given value of x,

$$E(\epsilon) = 0 \qquad \text{and} \qquad \sigma_{\epsilon}^2 = \sigma^2$$

and all pairs, ϵ_i, ϵ_j, are independent in a probabilistic sense.

4 ϵ possesses a normal probability distribution.

At first glance, you might fail to understand the significance of the first assumption. Models, deterministic or otherwise, are, as the name implies, only models for real relationships that occur in nature. Consequently, model misspecification is always a possibility. Even if you have obtained a good fit to the data, a large error in prediction is possible if you use the model to predict y for some value of x outside the range of values used to fit the least-squares equation. Of course, this problem will always occur if x is time and you attempt to forecast y at some point in the future. This problem occurs with any model for the future time predictions; consequently, you make the forecast but keep the model limitations in mind.

The assumption that the variance of the ϵ's is constant and equal to σ^2 will not be true for all types of data. Furthermore, if x is time, it is possible that y values measured over adjacent time periods will tend to be dependent (an overly large value of y in 1993 might signal a large value of y in 1994). Substantial departure from either of these assumptions will affect the confidence coefficients of interval estimates and significance levels and tests described in this chapter.

Like unequal variances and correlation of the random errors, if the normality assumption (4) is not satisfied, the confidence coefficients and significance levels for interval estimates and tests will not be what we expect them to be. However, modest departures from normality will not seriously disturb these values.

10.9 Summary

Although it was not stressed, you will observe that predicting the value of a random variable y was considered for the most elementary situation in Chapters 7 and 8. Thus, if we possessed no information concerning variables related to y, the sole information available for predicting y would be provided by its probability distribution. If we were to select one value as representative of the population, we would most likely choose μ, or some other measure of central tendency. The estimation of the mean was considered in Chapters 7 and 8.

In this chapter, we were concerned with the problem of predicting y when auxiliary information is available on other variables—say, $x_1, x_2, x_3, \ldots, x_k$—which are related to y and hence assist in its prediction. We have concentrated primarily on the problem of predicting y as a linear function of a single variable x, which provides the simplest extension of the prediction problem beyond that considered in Chapters 7 and 8. The more interesting case, where y is a linear function of a set of independent variables, is the subject of Chapter 11.

MINITAB Commands

```
REGRESS C on K pred. C...C [store st. res. in C [fits in C]]

  MSE          put into K
  COEFFICIENTS put into C
```

```
CORRELATION for C...C [put in M]
```

Supplementary Exercises

10.39 An experiment was conducted to observe the effect of an increase in temperature on the potency of an antibiotic. Three 1-ounce portions of the antibiotic were stored for equal lengths of time at each of the following temperatures: 30°, 50°, 70°, and 90°. The potency readings observed at the temperature of the experimental period were

Potency readings y	38, 43, 29	32, 26, 33	19, 27, 23	14, 19, 21
Temperature x	30°	50°	70°	90°

Use an appropriate computer program to answer the following questions:

a Find the least-squares line appropriate for these data.

b Plot the points and graph the line as a check on your calculations.

c Calculate s^2.

d Estimate the change in potency for a 1-unit change in temperature. Use a 90% confidence interval.

e Estimate the mean potency corresponding to a temperature of 50°. Use a 90% confidence interval.

f Suppose that a batch of the antibiotic were stored at 50° for the same length of time as the experimental period. Predict potency of the batch at the end of the storage period. Use a 90% confidence interval.

10.40 The following data are estimates of the number of reported robberies and arrests in the New York City subway system each month during the first half of 1988.

Month	Robberies	Robbery Arrests
January	475	180
February	465	155
March	470	160
April	500	190
May	550	225
June	600	220
July	602	223

Source: Data estimated from graph in the *New York Times* (Oct. 2, 1988).

a Without looking at the data, do you think that the correlation between number of robberies and number of robbery arrests should be positive or negative?

b Calculate the coefficient of correlation and interpret its value.

c Calculate the coefficient of determination and interpret its value.

d Do the data provide sufficient evidence to indicate that a positive correlation exists between number of robberies and number of robbery arrests in the New York City subway system? Test using $\alpha = .01$.

10.41 Refer to Exercise 1.12. In the following table, the U.S. census bureau details the federal government's income through federal income taxes (on a per capita basis) as well as the amount spent per capita in federal aid for each of the 50 states in fiscal year 1991:

State	Taxes	U.S Aid	State	Taxes	U.S. Aid
AL	964	566	MT	1007	755
AK	3168	1239	NE	1103	467
AZ	1256	396	NV	1312	330
AR	997	502	NH	565	399
CA	1477	578	NJ	1500	543
CO	951	466	NM	1347	595
CT	1514	646	NY	1567	816
DE	1713	532	NC	1165	454
FL	1040	345	ND	1189	772
GA	1080	510	OH	1056	507
HI	2325	644	OK	1216	475
ID	1159	516	OR	1037	623
IL	1151	416	PA	1088	465
IN	1102	460	RI	1251	737
IA	1233	518	SC	1104	575
KS	1120	453	SD	796	713
KY	1358	597	TN	870	551
LA	1013	668	TX	923	438
ME	1261	610	UT	1051	551
MD	1317	457	VT	1207	795
MA	1615	673	VA	1090	374
MI	1185	555	WA	1592	551
MN	1588	560	WV	1292	596
MS	948	704	WI	1416	532
MO	968	432	WY	1385	1218

Source: U.S. Department of Commerce (1992).

a If the correlation between federal per capita taxes and per capita U.S. aid for the 50 states is $r = .5215$, calculate the coefficient of determination and interpret its value.

b Is it true that the more a state pays in federal taxes, the more financial aid they will receive? Test using $\alpha = .01$.

10.42 As the United States becomes more aware of environmental problems, many cities are instituting curbside recycling programs in an attempt to conserve space in local landfills. These investments in recycling programs are more prevalent in some regions of the country than in others. The table below (reproduced from Exercise 1.29) shows the number of curbside recycling programs and the number of landfills in each of seven regions of the United States.

Region	Curbside Recycling Programs	Landfills
West	569	1374
Rocky Mountain	44	661
Midwest	108	1402
Great Lakes	1148	531
South	402	1007
Mid-Atlantic	1379	334
New England	305	503

Source: EPA Journal (July/August 1992).

a Would you expect to find a correlation between the number of landfills and the number of curbside recycling programs? If so, do you think the correlation will be positive or negative?

b Plot the data using a scatterplot. Does the correlation appear to be positive or negative?

c If the correlation coefficient relating the number of landfills to the number of curbside recycling programs is $r = -.5203$, is the correlation significantly different from zero? Use $\alpha = .05$.

10.43 An experiment was conducted to investigate the effect of a training program on the length of time for a typical male college student to complete the 100-yard dash. Nine students were placed in the program. The reduction y in time to complete the 100-yard dash was measured for three students at the end of 2 weeks, for three at the end of 4 weeks, and for three at the end of 6 weeks of training. The data are shown below:

Reduction in time y (sec)	1.6, .8, 1.0	2.1, 1.6, 2.5	3.8, 2.7, 3.1
Length of training x (wk)	2	4	6

Use an appropriate computer software package to analyze these data. State any conclusions you can draw.

10.44 It is generally thought that the temperature on football fields with synthetic turf surfaces is higher than the temperature on a natural grass surface. Jerry Ramsey (1982) made 81 daily temperature readings of the surface temperature of the synthetic turf at the Texas Tech University football stadium and also at a nearby natural grass surface. Figure 10.11 shows Ramsey's computer printout for a simple linear regression analysis.

a Give the least-squares prediction equation relating the dry-bulb synthetic surface temperature (DBS), the dependent variable, to the dry-bulb natural surface temperature (DBN).

b Do Ramsey's data present sufficient evidence to indicate that the dry-bulb natural grass temperature (DBN) provides information for the prediction of the dry-bulb synthetic turf temperature (DBS)? Explain.

FIGURE 10.11
Prediction of dry-bulb temperature on synthetic turf—linear regression

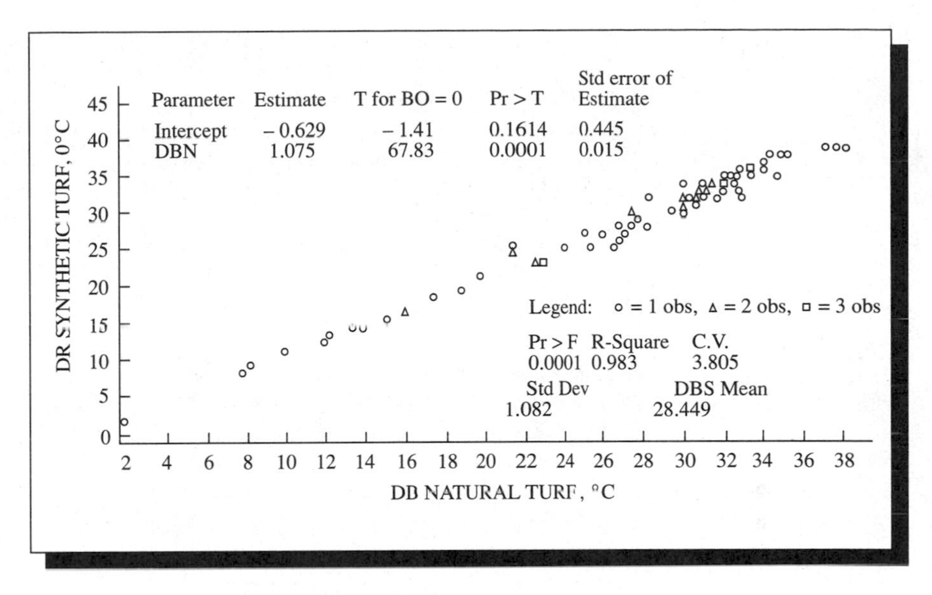

10.45 Some varieties of nematodes, round worms that live in the soil and frequently are so small as to be invisible to the naked eye, feed upon the roots of lawn grasses and other plants. This pest, which is particularly troublesome in warm climates, can be treated by the application of nematicides. Data collected on the percentage kill of nematodes for various rates of application (dosages given in pounds per acre of active ingredient) are as follows:

Rate of application x	2	3	4	5
Percent kill y	50, 56, 48	63, 69, 71	86, 82, 76	94, 99, 97

Use an appropriate computer printout to answer the following questions:

a Calculate the coefficient of correlation r between rates of application x and percent kill y.

b Calculate the coefficient of determination r^2 and interpret.

c Fit a least-squares line to the data.

d Suppose that you wish to estimate the mean percent kill for an application of 4 pounds of the nematicide per acre. Do the data satisfy the assumptions that are required for the confidence intervals of Section 10.5?

10.46 If you play tennis, you know that tennis racquets vary in their physical characteristics. The data shown in the accompanying table give measures of bending stiffness and twisting stiffness as measured by engineering tests for 12 tennis racquets:

Racquet	Bending Stiffness x	Twisting Stiffness y	Racquet	Bending Stiffness x	Twisting Stiffness y
1	419	227	7	424	384
2	407	231	8	359	194
3	363	200	9	346	158
4	360	211	10	556	225
5	257	182	11	474	305
6	622	304	12	441	235

a If a racquet has bending stiffness, is it also likely to have twisting stiffness? If $r = .5470$, do the data provide evidence that x and y correlated? (HINT: Test $H_0: \rho = 0$.) Find the p-value for the test and interpret its value.

b Calculate the coefficient of determination r^2 and interpret its value.

10.47 Movement of avocados into the United States from certain areas is prohibited because of the possibility of bringing fruit flies into the country with the avocado shipments. However, certain avocado varieties supposedly are resistant to fruit fly infestation before they soften as a result of ripening. The following data resulted from an experiment in which avocados ranging from 1 to 9 days after harvest were exposed to Mediterranean fruit flies. Penetrability of the avocados was measured on the day of exposure, and the percentage of the avocado fruit infested was assessed.

Days After Harvest	Penetrability	Percentage Infected
1	.91	30
2	.81	40
4	.95	45
5	1.04	57
6	1.22	60
7	1.38	75
9	1.77	100

Use the following MINITAB printout of the regression of penetrability (y) on days after harvest (x) to analyze the relationship between these two variables. Explain all pertinent parts of the printout and interpret the results of any tests.

```
-------------------------------------------------------------------
MTB > regress c2 1 c1

The regression equation is
y = 0.616 + 0.111 x


Predictor        Coef        Stdev      t-ratio        p
Constant       0.6159       0.1088         5.66    0.002
x             0.11085      0.01977         5.61    0.002


s = 0.1353        R-sq = 86.3%      R-sq(adj)  = 83.5%


Analysis of Variance


SOURCE       DF          SS           MS           F          p
Regression    1     0.57581      0.57581       31.44      0.002
Error         5     0.09157      0.01831
Total         6     0.66737
-------------------------------------------------------------------
```

10.48 Refer to Exercise 10.47. Suppose that the experimenter wants to examine the relationship between the percentage of infected fruit and the number of days after harvest. Does the method of linear regression discussed in this chapter provide an appropriate method of analysis? If not, what assumptions have been violated?

Exercises Using the Data Disk

10.49 Is there any relationship between blood pressure and age? To answer this question, select a random sample of 15 male subjects from Data Set A on the data disk. For each subject, record the subject's age, his systolic blood pressure, and his diastolic blood pressure.

a Plot systolic blood pressure y as a function of age x on a scatterplot. Does there appear to be a linear relationship between x and y?

b Use a computer program to find the least-squares line relating blood pressure and age. Is there a significant linear relationship between x and y? Use $\alpha = .05$.

c Use the program to predict the systolic blood pressure for a subject who is 21 years old. Use a 95% prediction interval.

10.50 Refer to Exercise 10.49. Using Data Set A on the data disk, select a random sample of 15 female subjects, recording their ages and systolic and diastolic blood pressures. Follow the instructions in parts (a), (b), and (c) of Exercise 10.49 for these 15 subjects. Are the results similar for men and women? Explain.

CASE STUDY

Is Your Car "Made in the U.S.A."?

The phrase "made in the U.S.A." has become a battle cry in the last few years, as U.S. workers try to protect their jobs from overseas competition (*Automotive News*, 1992). For the last decade, a major trade imbalance in the United States has been caused by a flood of imported goods that enter the country and are sold at lower cost than comparable American-made goods. One prime source of competition is in the automotive industry, in which the number of imported cars steadily increased during the 1970s and 1980s. The U.S. automobile industry has been besieged with complaints about product quality, worker layoffs, and high prices and has spent billions in advertising and research to produce an American-made car that will satisfy consumer demands. Have they been successful in stopping the flood of imported cars being purchased by American consumers? The data shown below represent the number of imported cars y sold in the United States (in millions) for the years 1969–1991. To simplify the analysis, we have coded the year using the coded variable $x =$ Year—1969.

Year	(Year—1969) x	Number of Imported Cars y	Year	(Year—1969) x	Number of Imported Cars y
1969	0	1.1	1981	12	2.3
1970	1	1.3	1982	13	2.2
1971	2	1.6	1983	14	2.4
1972	3	1.6	1984	15	2.4
1973	4	1.8	1985	16	2.8
1974	5	1.4	1986	17	3.2
1975	6	1.6	1987	18	3.1
1976	7	1.5	1988	19	3.1
1977	8	2.1	1989	20	2.8
1978	9	2.0	1990	21	2.5
1979	10	2.3	1991	22	2.1
1980	11	2.4			

1 Using a scatterplot, plot the data using the years 1969–1988. Does there appear to be a linear relationship between the number of imported cars and the year?

2 Use a computer software package to find the least-squares line for predicting the number of imported cars as a function of year.

3 Is there a significant linear relationship between the number of imported cars and the year? Test using $\alpha = .05$.

4 Use the computer program to predict the number of cars that will be imported using 95% prediction intervals for each of the years 1989, 1990, and 1991.

5 Now add the actual data points for the years 1989, 1990, and 1991 to your scatterplot in step 1. Do the predictions obtained in step 4 provide accurate estimates of the actual values observed in these years? Explain.

6 Given the form of the scatterplot for the years 1969–1991, does it appear that a straight line provides an accurate model for the data? What other types of model might be more appropriate?

11

Multiple Regression Analysis

Case Study

In Chapter 10, we used simple linear regression analysis to try to predict the number of imported cars over a period of years. Unfortunately, the number of imported cars does not really follow a linear trend pattern, and our predictions were far from accurate. We reexamine the same data at the end of this chapter, using the techniques of multiple regression analysis.

General Objective

In Chapter 10, we introduced the concept of simple linear regression and correlation, the ultimate goal being to estimate the mean value of y or to predict a value of y using information contained in a single independent (predictor) variable x. In this chapter, we will extend this concept and relate the mean value of y to one or more independent variables—$x_1, x_2, \ldots, x_k$—in models that are more flexible than the straight-line model of Chapter 10. The process of finding the least-squares prediction equation, testing the adequacy of the model, and conducting tests about and estimating the values of the model parameters is called a multiple regression analysis.

Specific Topics

1 The objectives of a multiple regression analysis (11.1)
2 The multiple regression model (11.2)
3 The least-squares prediction equation (11.3)
4 Estimation and testing of the parameters (11.3)
5 Multiple coefficient of determination (11.3)
6 Qualitative and quantitative variables in model building (11.4)
7 Testing sets of model parameters (11.5)
8 Residual analysis (11.6)
9 Misinterpretations in a regression analysis (11.7)

11.1 The Objectives of a Multiple Regression Analysis

The objective of a multiple regression analysis is to relate a response variable y to a set of predictor variables by using a multiple regression model. Ultimately, we want to be able to estimate the mean value of y and/or predict particular values of y to be observed in the future when the predictor variables assume specific values.

To illustrate, we might want to relate a company's regional sales y of a product to the amount x_1 of the company's television advertising expenditures, to the amount x_2 of newspaper advertising expenditures, and to the number x_3 of sales representatives assigned to the region. Thus, we would want to use data collected on y, x_1, x_2, and x_3 to obtain a mathematical prediction equation relating y to x_1, x_2, and x_3. We would use this prediction equation to predict the regional sales of a product for specific advertising expenditures (values of x_1 and x_2) and a specific number of sales representatives. If we are successful in developing a good prediction equation, one that makes accurate sales predictions, we can (by examining the equation) obtain a better understanding of the manner in which these controllable predictor variables x_1, x_2, and x_3 affect the product's sales.

This chapter is intended to be a brief introduction to multiple regression analysis, to help you understand what it is and aid you in interpreting the results of a multiple regression computer printout. We do not intend to cover all of the topics usually covered in a discussion of multiple regression analysis, nor do we intend to spend much time discussing the difficult task of formulating the regression model.

11.2 The Multiple Regression Model and Associated Assumptions

The **general linear model** for a multiple regression analysis will take the following form.

The General Linear Model and Assumptions

$$y = \beta_0 + \beta_1 x_1 + \beta_2 x_2 + \cdots + \beta_k x_k + \varepsilon$$

where the assumptions are as follows:

1. y is the response variable that you want to predict.
2. $\beta_0, \beta_1, \beta_2, \ldots, \beta_k$ are unknown constants.
3. $x_1, x_2, \ldots, x_k$ are independent **predictor variables** that are measured without error.
4. ε is a random error that for any given set of values for $x_1, x_2, \cdots, x_k$ is normally distributed with mean of zero and variance equal to σ^2.
5. The random errors—say, ε_i and ε_j—associated with any pair of y values are independent.

With these assumptions, it follows that the mean value of y for a given set of values for $x_1, x_2, \ldots, x_k$ is equal to

$$E(y) = \beta_0 + \beta_1 x_1 + \beta_2 x_2 + \cdots + \beta_k x_k$$

Although the variables $x_1, x_2, \ldots, x_k$ that appear in the general linear model need not represent *different* predictor variables, we will assume initially that they do. In any case, the development that follows requires that when we observe a value of y, the variables $x_1x_2, \ldots, x_k$ can be recorded without error. Furthermore, the random error ε associated with an observation y has mean of zero and a constant variance σ^2, independent of the values of the predictor variables $x_1, x_2, \ldots, x_k$.

Suppose we want to relate y, the listed selling price of a home, as a function of several independent variables such as

$$x_1 = \text{square footage of living space}$$
$$x_2 = \text{number of bedrooms}$$
$$x_3 = \text{number of bathrooms}$$

The multiple linear regression model relating y to x_1, x_2, and x_3 is

$$y = \beta_0 + \beta_1 x_1 + \beta_2 x_2 + \beta_3 x_3 + \varepsilon$$

with

$$E(y) = \beta_0 + \beta_1 x_1 + \beta_2 x_2 + \beta_3 x_3$$

which describes a plane in four-dimensional space. The parameter β_0 is called the **intercept** and represents the average value of y when x_1, x_2, and x_3 are each zero. The parameters β_1, β_2, and β_3 are called the **partial slopes** or **partial regression coefficients** to distinguish their values from the **total slopes** obtained in the three simple linear regression equations relating y to x_1, y to x_2, and y to x_3. In the multiple linear regression equation relating y to x_1, x_2, and x_3, the partial slope β_1 represents the average increase in y for a 1-unit increase in x_1 *when x_2 and x_3 are held constant.* The partial regression coefficients β_2 and β_3 have similar interpretations. For example, if x_2 and x_3 are held constant with $x_2 = 3$ and $x_3 = 2$, then

$$\begin{aligned} E(y) &= \beta_0 + \beta_1 x_1 + \beta_2(3) + \beta_3(2) \\ &= \beta_0^* + \beta_1 x_1 \end{aligned}$$

with $\beta_0^* = \beta_0 + 3\beta_2 + 2\beta_3$. In this form, we can see that β_1 is the increase in y for a 1-unit increase in x_1 in the presence of x_2 and x_3 in the model. In general, the value of β_1 with x_2 and x_3 in the model **will not be the same** as the slope found when fitting x_1 alone. Notice that the effect of changing the value of x_2 or x_3 is to produce a line parallel to the first with a new intercept β_0^*.

The regression analysis methodology that we present in Section 11.3 is appropriate for the multiple linear regression model in which terms linear in $x_1, x_2, \ldots, x_k$ appear; however, it is also appropriate for models in which polynomial or cross product terms such as x_1^3 or x_1x_2 appear, since x_1^3 and x_1x_2 are simply treated as new predictor variables. In this case, however, care must be taken when interpreting the fitted regression coefficients.

Exercises

Basic Techniques

11.1 Suppose that $E(y)$ is related to two predictor variables x_1 and x_2 by the equation

$$E(y) = 3 + x_1 - 2x_2$$

a Graph the relationship between $E(y)$ and x_1 when $x_2 = 2$. Repeat for $x_2 = 1$ and for $x_2 = 0$.

b What relationship do the lines in part (a) have to each other?

11.2 Refer to Exercise 11.1

a Graph the relationship $E(y)$ and x_2 when $x_1 = 0$. Repeat for $x_1 = 1$ and for $x_1 = 2$.

b What relationship do the lines in part (a) have to each other?

c Suppose, in a practical situation, you want to model the relationship between $E(y)$ and two predictor variables x_1 and x_2. What would be the implication of using the first-order model $E(y) = \beta_0 + \beta_1 x_1 + \beta_2 x_2$?

11.3 Suppose $E(y)$ is related to two predictor variables x_1 and x_2 by the equation

$$E(y) = 3 + x_1 - 2x_2 + x_1 x_2$$

a Graph the relationship between $E(y)$ and x_1 when $x = 0$. Repeat for $x_2 = 2$ and for $x_2 = -2$.

b Note that the equation for $E(y)$ is exactly the same as the equation in Exercise 11.1 except that we have added the term $x_1 x_2$. How does the addition of the $x_1 x_2$ term affect the graphs of the three lines?

c What flexibility is added to the first-order model $E(y) = \beta_0 + \beta_1 x_1 + \beta_2 x_2$ by the addition of the term $\beta_3 x_1 x_2$, using the model $E(y) = \beta_0 + \beta_1 x_1 + \beta_2 x_2 + \beta_3 x_1 x_2$?

11.3 A Multiple Regression Analysis

A multiple regression analysis is performed somewhat like a simple linear regression analysis. A multiple regression model, say,

$$E(y) = \beta_0 + \beta_1 x_1 + \beta_2 x_2 + \cdots + \beta_k x_k$$

is fitted to a set of data by using the method of least squares, a procedure that finds the prediction equation

$$\hat{y} = \hat{\beta}_0 + \hat{\beta}_1 x_1 + \hat{\beta}_2 x_2 + \cdots + \hat{\beta}_k x_k$$

that minimizes SSE, the sum of squares of deviations of the observed values of y from their predicted values. This procedure is usually implemented using one of several regression programs available in the MINITAB, SAS, SPSS^x, or other computer packages. In this section, we will present two sets of data and formulate models for each. We will then present and discuss the printout of one or more regression programs. Although the printouts differ in format (selection and placement of relevant information), in general they contain the same essential information.

EXAMPLE 11.1 A large number of residents living in a community located 30 miles south of a major metropolitan area are commuters who work in the city. Table 11.1 gives the listed selling price y in thousands of dollars, the square feet of living area x_1 in thousands of square feet, the number of floors x_2, bedrooms x_3, and bathrooms x_4, for $n = 29$ randomly selected residences on the market during the summer of 1993. Use a multiple linear regression model relating the listed selling price y and the independent variables x_1, x_2, x_3, and x_4.

TABLE 11.1 Listed selling prices for 29 residential properties

Observation	y LPRICE	x_1 SQFT	x_2 NUMFLRS	x_3 BDRMS	x_4 BATHS
1	69.0	6	1	2	1.0
2	11.5	8	1	2	1.0
3	118.5	10	1	2	2.0
4	104.0	11	1	3	2.0
5	116.5	10	1	3	2.0
6	121.5	10	1	3	2.0
7	125.0	11	1	3	2.0
8	128.0	15	2	3	2.5
9	129.9	13	1	3	1.7
10	133.0	13	2	3	2.5
11	135.0	13	2	3	2.5
12	137.5	15	2	3	2.5
13	139.9	13	1	3	2.0
14	143.9	14	2	3	2.5
15	147.9	17	2	3	2.5
16	154.9	15	2	3	2.5
17	160.0	19	2	3	2.0
18	169.0	15	1	3	2.0
19	169.9	18	1	3	2.0
20	125.0	13	1	4	2.0
21	134.9	13	1	4	2.0
22	139.9	17	1	4	2.0
23	147.0	18	1	4	2.0
24	159.0	14	1	4	2.0
25	169.9	17	2	4	3.0
26	178.9	19	1	4	2.0
27	194.5	20	2	4	3.0
28	219.9	21	1	4	2.5
29	269.0	25	2	4	3.0

Solution The multiple linear regression model to be fitted is

$$E(y) = \beta_0 + \beta_1 x_1 + \beta_2 x_2 + \beta_3 x_3 + \beta_4 x_4$$

The computer printout in Table 11.2 resulted when the REGRESS command in the MINITAB package was used to regress y, stored in column 8 on the four predictor variables x_1, x_2, x_3, and x_4 stored in C1, C2, C3, and C4. ▪

TABLE 11.2 MINITAB regression analysis for the data in Example 11.1

```
MTB > REGRESS C8 4 C1-C4;
SUBC > PREDICT 10 1 3 2;
SUBC > PREDICT 14 2 3 2.5.
                                        (1)
The regression equation is
LPRICE = -16.6 + 7.84 SQFT - 34.4 NUMFLRS - 7.99 BDRMS + 54.9 BATHS

                   (2)          (3)            (4)         (5)
Predictor    Coef       Stdev      t-ratio          p
Constant   -16.58       18.88        -0.88      0.389
SQFT        7.839       1.234         6.35      0.000
NUMFLRS    -34.39       11.15        -3.09      0.005
BDRMS      -7.990       8.249        -0.97      0.342
BATHS       54.93       13.52         4.06      0.000

s = 16.58 (6)   R-sq = 88.2% (7)   R-sq(adj) = 86.2% (8)

Analysis of Variance                                        (9)

SOURCE         DF       SS       MS        F        p
Regression      4    49359    12340    44.88    0.000
Error          24     6599      275
Total          28    55958

SOURCE      DF    SEQ SS   (10)
SQFT         1     44444
NUMFLRS      1        59
BDRMS        1       321
BATHS        1      4536

Unusual Observations
Obs.   SQFT   LPRICE     Fit   Stdev.Fit   Residual   St.Resid
  1     6.0    69.00   35.02        9.45      33.98      2.49R
  2     8.0    11.50   50.70        9.45     -39.20     -2.88R

R denotes an obs. with a large st. resid.

    Fit  Stdev.Fit        95% C.I.              95% P.I.        (11)
 113.32       5.80   (101.34, 125.30)    ( 77.05, 149.59)
 137.75       5.48   (126.44, 149.07)    (101.70, 173.81)
```

Estimation of Parameters

The fitted regression equation, found in area (1), is given by

$$\hat{y} = -16.6 + 7.84x_1 - 34.4x_2 - 7.99x_3 + 54.9x_4$$

Information about the individual parameters of the model is found in the upper portion of the printout. The column listed as PREDICTOR identifies each parameter, with either the actual column number containing that variable or the column's assigned name. The estimated coefficients are found in area (2) and their estimated standard deviations are found in area (3). For example, $\hat{\beta}_1 = 7.839$ with $s_{\hat{\beta}_1} = 1.234$, where $\hat{\beta}_1$ represents the estimated average increase in listed selling price for an increase of 1000 square feet of living area **when the other predictor variables are held constant.**

In area (6), the estimate of σ is given by $s = 16.58$, where

$$s^2 = \text{MSE} = \frac{\text{SSE}}{n - (k+1)}$$

The degrees of freedom (d.f.) associated with s^2 is equal to $n - (k+1)$, where $(k+1)$ is the number of parameters (including β_0) in the model. In this case, s^2 has $29 - (4+1) = 24$ d.f. The values of SSE and MSE are found in the ERROR line of the analysis of variance table in area (9).

Estimation and Testing of Individual Parameters

Confidence interval estimates and tests of hypotheses concerning individual partial slopes (β_i) are based on the estimated partial regression coefficients ($\hat{\beta}_i$) and their estimated standard errors. **A $(1-\alpha)$ 100% confidence interval for β_i is given by**

$$\hat{\beta}_i \pm t_{\alpha/2} s_{\hat{\beta}_i}$$

where $t_{\alpha/2}$ is the value of t based on $n - (k+1)$ d.f. having an area of $\alpha/2$ to its right. For example, a 95% confidence interval for β_1 is

$$\hat{\beta}_1 \pm t_{.025} s_{\hat{\beta}_1}$$
$$7.839 \pm 2.064(1.234)$$
$$7.839 \pm 2.547$$

or from 5.292 to 10.386.

Tests of significance for individual partial regression coefficients are based on the Student's t statistic given by

$$t = \frac{\hat{\beta}_i - \beta_i}{s_{\hat{\beta}_i}}$$

The procedure is identical to the one used to test an hypothesis about the slope β_1 in a simple linear regression model.† In our example, testing $H_0 : \beta_1 = 0$ versus $H_a : \beta_1 \neq 0$ uses the statistic

$$t = \frac{\hat{\beta}_i - 0}{s_{\hat{\beta}_i}} = \frac{7.839}{1.234} = 6.35$$

†Some packages use the t statistic just described, whereas others use the equivalent F statistic ($F = t^2$), since the square of a t statistic with v degrees of freedom is equal to an F statistic with 1 d.f. in the numerator and v degrees of freedom in the denominator.

which is the t ratio given in area (4). This value exceeds the critical value of t with $\alpha = .05$ and 24 d.f. given by $t_{.025} = 2.064$. The p-value of 0.000 for this test is found in the column labeled p in area (5).

Assessing the Utility of the Model

The portion of the printout labeled (9) presents an **analysis of variance** for the data. In regression analysis, an analysis of variance is the technique that partitions the total variation in the response y into one portion associated with random error and another portion associated with the variability accounted for by regression. The **Error** row in this portion of the printout provides the degrees of freedom for error in the **DF** column, the value of SSE in the **SS** column, and value of s^2 in the **MS** column. The value of the standard deviation s, found in area (6) on the printout, is the square root of s^2. The value found in the **Total** row and **SS** column is the total variation in the response y, given as

$$S_{yy} = \sum (y_i - \bar{y})^2 = 55{,}958$$

The amount of the total variation that can be accounted for by regression is SSR = 49,359, the difference between the total variation and the error variation, SSE. The *analysis of variance* will be discussed in detail in Chapter 13.

In area (9), the analysis of variance table has two sources of variation: one due to *regression* with $k = 4$ d.f. (corresponding to the four fitted parameters, $\hat{\beta}_1, \hat{\beta}_2, \hat{\beta}_3$, and $\hat{\beta}_4$) and the other due to *error* with $n - (k + 1) = 29 - (4 + 1) = 24$ d.f. The adequacy of the model using x_1, x_2, x_3, and x_4 is assessed by testing the hypothesis

$$H_0 : \beta_1 = \beta_2 = \beta_3 = \beta_4 = 0$$

against the alternative $H_a : \beta_i \neq 0$ for at least one value of $i = 1, 2, 3$, or 4, using the statistic

$$F = \frac{\text{MSR}}{\text{MSE}} = \frac{12{,}340}{275} = 44.88$$

which when compared to the critical value of F with 4 numerator and 24 denominator d.f. given by $F_{.05} = 2.78$ is significant with $\alpha = .05$ and, in fact, has a p-value reported as 0.000.

In assessing the strength of the relationship between y and the predictor variables x_1, x_2, x_3, and x_4, we use the fact that the total sum of squared deviations of y about $\bar{y}$, S_{yy}, is equal to

$$S_{yy} = \text{SSR} + \text{SSE}$$

where SSR, the sum of squares due to regression, is the portion of S_{yy} explained by regression, while SSE, the sum of squares due to error, is the unexplained portion of S_{yy}. Hence, the larger the value of the ratio SSR/S_{yy}, the stronger the relationship between y and the predictor variables. The **coefficient of determination** is defined as

$$R^2 = \left(\frac{\text{SSR}}{S_{yy}}\right) 100\% \qquad \text{or} \qquad R^2 = \left(\frac{S_{yy} - \text{SSE}}{S_{yy}}\right) 100\%$$

For this example,

$$R^2 = \left(\frac{49{,}359}{55{,}958}\right) 100\% = 88.2\%$$

and 88.2% of the variation in y is explained by linear regression of y on x_1, x_2, x_3, and x_4. This value is found in area (7) of the printout. The value of the F statistic used for testing $H_0 : \beta_1 = \beta_2 = \beta_3 = \beta_4 = 0$ is related to R^2 in the following way:

$$F = \frac{R^2/k}{(1 - R^2)/[n - (k + 1)]}$$

so that when R^2 is large, F is large and vice versa.

The quantity R (the positive square foot of R^2) is called the **multiple correlation coefficient** and measures the correlation between y and its predicted value $\hat{y}$. Notice that the value of R^2 can never decrease with the addition of one or more variables into the regression model. Hence R^2 *can be artificially inflated by the inclusion of more and more predictor variables.*

An alternative measure of the strength of the relationship between y and x_1, x_2, x_3, and x_4, adjusted for degrees of freedom by using mean squares rather than sums of squares, is defined as

$$R^2(\text{adj}) = \left(1 - \frac{\text{MSE}}{s_y^2}\right) 100\%$$

where $s_y^2 = S_{yy}/(n - 1)$. In this example,

$$R^2(\text{adj}) = \left(1 - \frac{275}{55{,}958/28}\right) 100\% = (1 - .138)100\% = 86.2\%$$

which agrees with the value in area (8). The value of R^2(adj) is mainly used to compare two or more regression models using different numbers of independent predictor variables.

The decomposition of the sum of squares due to regression appears in area (10), in which the conditional contribution of each predictor variable *given the variables already entered into the model* is given with the order specified in the REGRESS command. For example, the sum of squares due to regression on x_1 is 44,444; the conditional sum of squares due to regression on x_2, *given that x_1 is already in the model*, is 59. The conditional sum of squares due to regression on x_3, *given that x_1 and x_2 are already in the model*, is 321, and so on. In this example, it is interesting to note that the predictor variable x_1 alone accounts for $44{,}444/55{,}958 = .794$, or 79.4% of the total variation compared with 88.2% using the variables x_1, x_2, x_3, and x_4.

Estimating $E(y)$ and Predicting y

The subcommand PREDICT in the MINITAB package, followed by fixed values of x_1, x_2, x_3, and x_4, implements the calculation of the estimated values $\hat{y}$ (FIT), which is the point estimator for both $E(y)$ and y. This value is shown in area (11) together with $s_{\hat{y}}$, its estimated standard error (Stdev. Fit) and both a 95% confidence interval for $E(y)$ and a 95% prediction interval for y. (Recall from Section 10.5 that the prediction interval is always *wider* than the confidence interval). For our example, using the values from the first PREDICT command, we can verify that when $x_1 = 10$, $x_2 = 1$, $x_3 = 3$, and $x_4 = 2$,

$$\begin{aligned}\hat{y} &= -16.58 + 7.84(10) - 34.39(1) - 7.99(3) + 54.93(2)\\ &= 113.32\end{aligned}$$

and the 95% confidence interval for $E(y)$, based on 24 d.f., is

$$\hat{y} \pm t_{.025} s_{\hat{y}}$$

$$113.32 \pm 2.064(5.80)$$

$$113.32 \pm 11.97$$

or from 101.35 to 125.29, which to one decimal place agrees with that given in area (11).

EXAMPLE 11.2 In a study of variables affecting productivity in the retail grocery trade, W. S. Good (1984) uses value added per work-hour to measure the productivity of retail grocery outlets. He defines "value added" as "the surplus [money generated by the business] available to pay for labor, furniture and fixtures, and equipment." Data consistent with the relationship between value added per work-hour y and the size x of a grocery outlet described in Good's article are shown in Table 11.3 for ten fictitious grocery outlets. Choose a linear model to relate y to x.

TABLE 11.3 Value added per work-hour versus size of store

	Store									
Variable	**1**	**2**	**3**	**4**	**5**	**6**	**7**	**8**	**9**	**10**
Value added per work-hour (dollars) y	4.08	3.40	3.51	3.09	2.92	1.94	4.11	3.16	3.75	3.60
Size of store (thousands of square feet) x	21.0	12.0	25.2	10.4	30.9	6.8	19.6	14.5	25.0	19.1

Solution The first step in choosing a model to describe the data is to plot the data points (Figure 11.1). The relationship suggested by the data is one that depicts productivity rising as the size of a grocery outlet increases, until an optimal size is reached. Above that size, productivity tends to decrease. Since the relationship that we have described suggests curvature, we will fit a quadratic model

$$E(y) = \beta_0 + \beta_1 x + \beta_2 x^2$$

to the data.

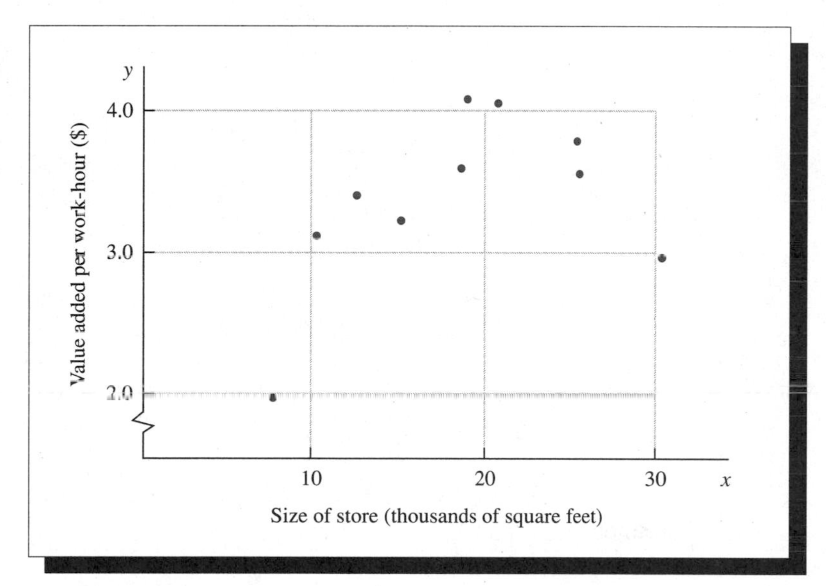

FIGURE 11.1 Plot of the data points for Example 11.2

Keep in mind that in choosing a quadratic model to fit to our data, we are not saying that the true relationship between the mean value of y and the value x is defined by an equation of the type

$$E(y) = \beta_0 + \beta_1 x + \beta_2 x^2$$

Rather, we have chosen this equation to *model* the relationship. Presumably, it will provide a better description of the relationship than a simple linear model that graphs as a straight line, and it will enable us to estimate the mean productivity of a grocery retail outlet as a function of the outlet size. ▪

EXAMPLE 11.3 Refer to the data on grocery retail outlet productivity and outlet size given in Example 11.2.

1. Fit a quadratic model to the data, using the MINITAB computer program package and discuss the adequacy of the fitted model.
2. Graph the quadratic prediction curve, along with the plotted data points.

Solution From the printout in Table 11.4, we see that the regression equation is

$$\hat{y} = -.159 + .392x - .00949x^2$$

TABLE 11.4
MINITAB regression analysis for the data in Example 11.2

```
The regression equation is
Y = -0.159 + 0.392 X -0.00949 XSQ

Predictor        Coef         Stdev          t-ratio          P
Constant      -0.1594        0.5006            -0.32      0.760
X             0.39193       0.05801             6.76      0.000
XSQ         -0.009495      0.001535            -6.19      0.000

s = 0.2503      R-sq = 87.9%      R-sq(adj) = 84.5%

Analysis of Variance

SOURCE        DF        SS        MS         F        P
Regression     2    3.1989    1.5994     25.53    0.001
Error          7    0.4385    0.0626
Total          9    3.6374

SOURCE        DF    SEQ SS
X              1    0.8003
XSQ            1    2.3986
```

The graph of this quadratic equation together with the data points is shown in Figure 11.2.

To assess the adequacy of the quadratic model, the test of

$$H_0 : \beta_1 = \beta_2 = 0$$

versus $H_a : \beta_i \neq 0$ for $i = 1$ or 2 is given in the analysis of variance section of the printout as

$$F = \frac{\text{MSR}}{\text{MSE}} = 25.53$$

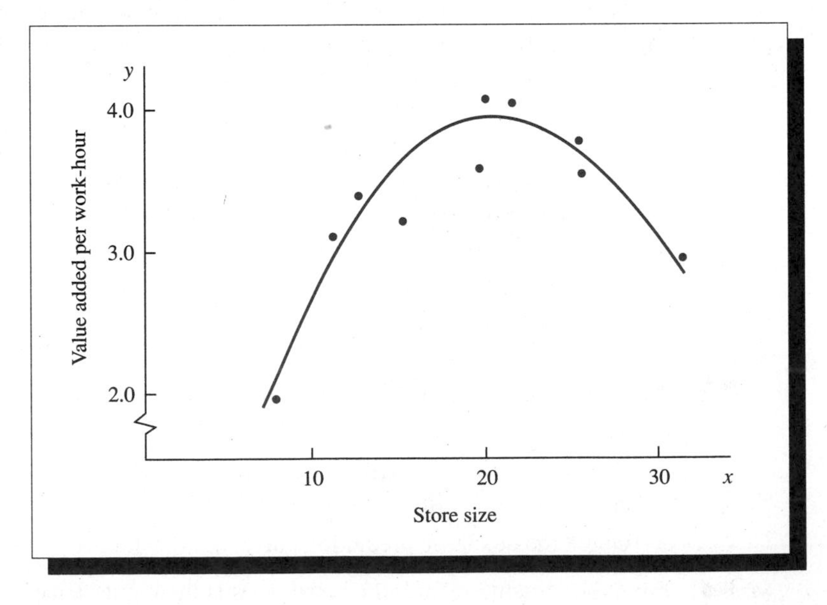

FIGURE 11.2
A graph of the least-squares prediction equation for the quadratic model of Example 11.3

with p-value 0.001. Hence, at least one of the two regression coefficients is highly significant. Quadratic regression accounts for $R^2 = 87.9\%$ of the variation in y ($R^2(\text{adj}) = 84.5\%$).

In examining the individual analysis of variables section of the printout, we see that both $\hat{\beta}_1$ and $\hat{\beta}_2$ are highly significant with p-values equal to 0.000. It is interesting to note that, from the sequential sum of squares section, the sum of squares for linear regression is .8003 with an additional sum of squares of 2.3986 when adding the quadratic term. It is apparent from this analysis that a simple linear regression model would be inadequate in describing the data. ■

Exercises

Basic Techniques

11.4 Suppose that you were to fit the model

$$E(y) = \beta_0 + \beta_1 x_1 + \beta_2 x_2 + \beta_3 x_3$$

to 15 data points and found R^2 equal to 94.

a Interpret the value of R^2.

b Do the data provide sufficient evidence to indicate that the model contributes information for the prediction of y? Test using $\alpha = .05$.

11.5 The computer output for the multiple regression analysis for Exercise 11.4 provides the following information:

$$\hat{\beta}_0 = 1.04 \qquad \begin{array}{l}\hat{\beta}_1 = 1.29 \\ s_{\hat{\beta}_1} = .42\end{array} \qquad \begin{array}{l}\hat{\beta}_2 = 2.72 \\ s_{\hat{\beta}_2} = .65\end{array} \qquad \begin{array}{l}\hat{\beta}_3 = .41 \\ s_{\hat{\beta}_3} = .17\end{array}$$

Which, if any, of the independent variables x_1, x_2, and x_3 contribute information for the prediction of y? Test using $\alpha = .05$.

11.6 Refer to Exercise 11.5.

a Give the least-squares prediction equation.

b On the same sheet of graph paper, graph y versus x_1 when $x_2 = 1$ and $x_3 = 0$ and when $x_2 = 1$ and $x_3 = .5$. What relationship do the two lines have to each other?

c What is the practical interpretation of the parameter β_1?

11.7 Refer to Exercise 11.5. Find a 90% confidence interval for β_1 and interpret it.

11.8 Suppose that you were to fit the model

$$E(y) = \beta_0 + \beta_1 x + \beta_2 x^2$$

to 20 data points and found $R^2 = .762$.

a What type of model have you chosen to fit the data?

b How well does the model fit the data? Explain.

c Do the data provide sufficient evidence to indicate that the model contributes information for the prediction of y? Test using $\alpha = .05$.

11.9 The computer output for the multiple regression analysis for Exercise 11.8 provides the following information:

$$\hat{\beta}_0 = 1.21 \qquad \hat{\beta}_1 = 7.60 \qquad \hat{\beta}_2 = -.94$$
$$s_{\hat{\beta}_0} = .62 \qquad s_{\hat{\beta}_1} = 1.97 \qquad s_{\hat{\beta}_1} = .33$$

a Give the prediction equation.

b Graph the prediction equation over the interval $0 \le x \le 6$.

11.10 Refer to Exercise 11.9.

a What is your estimate of the mean value of y when $x = 0$?

b Do the data provide sufficient evidence to indicate that the mean value of y differs from 0 when $x = 0$? Test using $\alpha = .10$.

c Find a 90% confidence interval for $E(y)$ when $x = 0$.

11.11 Refer to Exercise 11.9.

a Suppose that the relationship between $E(y)$ and x is a straight line. What would you know about the value of β_2?

b Do the data provide sufficient evidence to indicate curvature in the relationship between y and x?

11.12 Refer to Exercise 11.9. Suppose that y is the profit for some business and x is the amount of capital invested, and you know that the rate of increase in profit for a unit increase in capital invested can only decrease as x increases. Do the data provide sufficient evidence to indicate a decreasing rate of increase in profit as the amount of capital invested increases?

a The circumstances that we have described imply a one-tailed statistical test. Why?

b Conduct the test for $\alpha = .05$. State your conclusions.

Applications

11.13 A publisher of college textbooks conducted a study to relate profit per text to cost of sales over a 6-year period when its sales force (and sales costs) were growing rapidly. The following inflation-adjusted data (in thousands of dollars) were collected:

Profit per text, y	16.5	22.4	24.9	28.8	31.5	35.8
Sales cost per text, x	5.0	5.6	6.1	6.8	7.4	8.6

Expecting profit per book to rise and then plateau, the publisher fitted the model $E(y) = \beta_0 + \beta_1 x + \beta_2 x^2$ to the data.

```
DEPENDENT VARIABLE:  Y

SOURCE            DF   SUM OF SQUARES    MEAN SQUARE    F VALUE    PR > F    R-SQUARE       C.V.
MODEL              2     234.95514252   117.47757126     332.53    0.0003    0.995509     2.2303
ERROR              3       1.05985748     0.35328583             ROOT MSE                 Y MEAN
CORRECTED TOTAL    5     236.01500000                          0.59437852             26.6500000

SOURCE     DF      TYPE I SS     F VALUE     PR > F    DF     TYPE IV SS     F VALUE     PR > F
X           1   227.81864814      644.86     0.0001     1    15.20325554       43.03     0.0072
X*X         1     7.13649437       20.20     0.0206     1     7.13649437       20.20     0.0206

                              PARAMETER ESTIMATES

                                 T FOR HO:
VARIABLE          ESTIMATE     PARAMETER = 0    PR > |T|     STD ERROR
INTERCEPT     -44.19249551             -5.33      0.0129    8.28688218
X              16.33386317              6.56      0.0072    2.48991042
X*X            -0.81976920             -4.49      0.0206    0.18239471
```

OBSERVATION	OBSERVED VALUE	PREDICTED VALUE	RESIDUAL	LOWER 95% CL FOR MEAN	UPPER 95% CL FOR MEAN
1	16.50000000	16.98259044	-0.48259044	15.33066719	18.63451369
2	22.40000000	21.56917626	0.83082374	20.56714686	22.57120566
3	24.90000000	24.94045805	-0.04045805	23.93483233	25.94608377
4	28.80000000	28.97164643	-0.17164643	27.81528749	30.12800537
5	31.50000000	31.78753079	-0.28753079	30.65188830	32.92317327
6	35.80000000	35.64859804	0.15140196	33.81460946	37.48258661
7*		27.34236657		26.23203462	28.45269852

*OBSERVATION WAS NOT USED IN THIS ANALYSIS

SUM OF RESIDUALS	-0.00000000
SUM OF SQUARED RESIDUALS	1.05985748
SUM OF SQUARED RESIDUALS-ERROR SS	-0.00000000
PRESS STATISTIC	12.11621807
FIRST ORDER AUTOCORRELATION	-0.39797399
DURBIN-WATSON D	2.55457960

a What sign would you expect the actual value of β_2 to assume? The SAS computer printout is shown in the preceding table. Find the value of $\hat{\beta}_2$ on the printout and see whether the sign agrees with your answer.

b Find SSE and s^2 on the printout.

c How many degrees of freedom do SSE and s^2 possess? Show that

$$s^2 = \frac{\text{SSE}}{\text{degrees of freedom}}$$

d Do the data provide sufficient evidence to indicate that the model contributes information for the prediction of y? Test using $\alpha = .05$.

e Find the observed significance level for the test in part (d) and interpret its value.

f Do the data provide sufficient evidence to indicate curvature in the relationship between $E(y)$ and x (i.e., evidence to indicate that β_2 differs from 0)? Test using $\alpha = .05$.

g Find the observed significance level for the test in part (f) and interpret its value.

h Find the prediction equation and graph the relationship between $\hat{y}$ and x.

i Find R^2 on the printout and interpret its value.

j Use the prediction equation to estimate the mean profit per text when the sales cost per text is $6500. (Express the sales cost in thousands of dollars before substituting into the prediction equation.) We instructed the SAS program to print this confidence interval. The confidence interval when $x = 6.5$, 26.23203462 to 28.45269852 is shown at the bottom of the printout.

11.14 Refer to the printout given in Exercise 11.13.

a Verify that $S_{yy} = \text{SSR} + \text{SSE}$.

b Use the values of SSR and S_{yy} to calculate R^2. Compare this value with the value R-SQUARE given in the printout.

c Use the values given in the printout to calculate $R^2(\text{adj})$. When would it be appropriate to use this value to measure the strength of the relationship between y and x and x^2?

11.15 The federal government has spent billions of dollars in bailout costs for failed savings and loan institutions across the country. Following the passage of federal legislation enacted in 1989 requiring 1.5% of a savings and loan's assets to be tangible capital, the performance of savings and loan institutions with headquarters in Riverside and San Bernardino counties (California) was evaluated. At the conclusion of the 3-month period from March 31 to June 30, 1990, the following data were recorded:

Savings and Loan	**Total Assets x_1 (in millions)**	**Net income y (in thousands)**	**Tangible Capital x_2 (% of assets)**
Redlands Federal	$791.8	$2700	5.92
Hemet Federal	556.3	1261	5.69
Provident Federal	518.0	68	5.47
Palm Springs Federal	137.0	251	3.57
Inland Savings	111.1	748	5.14
Secure Savings	56.4	195	6.87
Mission Savings	35.6	68	5.53

Source: Press-Enterprise, Riverside, Calif., Oct 12, 1990.

The MINITAB command REGRESS was used to fit the model

$$y = \beta_0 + \beta_1 x_1 + \beta_2 x_2 + \varepsilon$$

to this data. The printout is shown in the following table:

```
MTB > REGRESS C1 2 C2-C3

The regression equation is
Y = -383 + 2.42 X1 + 69 X2

Predictor       Coef     Stdev      t-ratio         P
Constant        -383      1700        -0.23     0.833
X1             2.418     1.027         2.35     0.078
X2              69.1     311.4         0.22     0.835

s = 748.1       R-sq = 59.7%     R-sq(adj) = 39.5%

Analysis of Variance

SOURCE        DF          SS          MS          F          P
Regression     2     3312151     1656075       2.96      0.163
Error          4     2238508      559627
Total          6     5550659

SOURCE        DF      SEQ SS
X1             1     3284613
X2             1       27538
```

a Find the values of SSE and s^2.

b Find the prediction equation.

c Find R^2 and interpret its value.

d Find the value of R^2(adj). If the value of R^2(adj) for a linear regression was 51.0%, what conclusions might you draw about the value of x_2 in the model?

e Do the data provide sufficient evidence to indicate that the model contributes information for the prediction of y? Test using $\alpha = .10$.

f Note that the observed significant level for the test $H_0 : \beta_2 = 0$ is large. Does this mean that there is little evidence to indicate that x_2 contributes information for the prediction of y?

g What would you conclude regarding the adequacy of the model that we have fit in this exercise?

11.16 R. C. Curham, W. J. Salmon, and R. D. Buzzell (1983) describe a study of supermarket sales and profitability in the sales of health and beauty aids (HBA) and small-ticket general merchandise (GM) for two types of firms, those that use service merchandisers to supply HBA products and service to their stores and those (direct buyers) that buy HBA products, warehouse them, and service the stores themselves. A portion of this study involved multiple regression analyses on data from $n = 15$ supermarkets for each of four dependent variables. The results are shown in the accompanying table:

Dependent Variable	Intercept	Average per Store Weekly Customer Count	Category Annual Sales per Firm (in thousands)	Per Store Average Linear Feet of Shelf Space	Category Percentage of Sales Purchased Direct	R^2	F-Ratio
HBA sales per linear foot	555	+.31	−\$.003	−10.42	−6.90%	.65	4.590*
		(2.708)†	(.104)	(1.028)	(.842)		
HBA contribution per linear foot	220	+.06	\$.007	−3.98	−1.73%	.72	6.345‡
		(2.847)*	(1.401)	(2.143)†	(1.157)		
GM sales per linear foot	1124	+.06	+\$.02	−4.36	−8.07%	.69	4.904‡
		(1.605)§	(1.525)§	(2.174)†	(1.394)§		
GM contribution per linear foot	466	+.02	−\$.009	−1.90	−2.88%	.71	5.389‡
		(1.187)	(1.811)§	(2.589)†	(1.357)§		

Source: Curhan, Salmon, and Buzzell (1983).
NOTE: t statistics are shown in parentheses.
*$.01 < p < .05$.
†$.05 < p < .10$.
‡$p < .01$.
§$.10 < p < .20$.

The four dependent variables are listed in column 1. The four independent variables

x_1 = average per store weekly customer count
x_2 = category annual sales per firm (\$1000)
x_3 = per store average linear feet of shelf space
x_4 = category percentage of sales purchased directly

appear in columns 3, 4, 5, and 6. A first-order model was used to relate each dependent variable y to these four independent variables. The numbers appearing in the four independent variable columns are the estimates of the regression coefficients. The y intercept for each of the four multiple regressions is shown in column 2 of the table.

a Write the equation of the multiple regression model used in the regression analysis.

b Examine the regression analysis for y = HBA sales per linear foot. Find the F value in the table for testing the complete model. Do the data provide sufficient evidence to indicate that the model contributes information of the prediction of y? Test using $\alpha = .05$.

c Find the approximate p-value for the test in part (b).

d Which, if any, of the independent variables appear to contribute information for the prediction of y? Explain.

e Refer to part (b). Use the value of R^2 to calculate the value of the F statistic used to test the complete model. Does your calculated value agree with the value shown in the table and used in the test in part (b)?

f Interpret the value of R^2 given for the analysis.

g Does the first-order model used in this regression analysis provide a good fit to the data? Explain. Can you suggest possible improvements to the model?

11.17 Answer the questions of Exercise 11.16, except refer to the multiple regression analysis for y = HBA contribution (to profit) per linear foot.

11.18 Answer the questions of Exercise 11.16, except refer to the multiple regression analysis for y = GM sales per linear foot.

11.19 Answer the questions of Exercise 11.16, except refer to the multiple regression analysis for y = GM contribution (to profit) per linear foot.

11.4 The Use of Quantitative and Qualitative Variables in Linear Regression Models

The variables in a regression analysis can be one of two types, quantitative or qualitative. For example, the age of a person is a **quantitative variable** because its values express the quantity or amount of something—in this case, age. In contrast, the nationality of a person is a **qualitative variable** that varies from person to person, but the values of the variable cannot be quantified; they can only be classified. For the methods we present, the response variable y must always be (according to the assumptions of Section 11.2) a quantitative variable. In contrast, predictor variables may be either quantitative or qualitative. For example, suppose we want to predict the annual income of a person. Both the age and the nationality of the person could be important predictor variables, and there are probably many others.

DEFINITION ■ A **quantitative variable** is one whose values correspond to the quantity or the amount of something. ■

DEFINITION ■ A **qualitative variable** is one that assumes values that cannot be quantified. It can only be categorized. ■

The variables $x_1, x_2, \ldots, x_k$ that appear in the general linear model need not represent *different* predictor variables. Our assumption only requires that when we observe a value of y, the values of $x_1, x_2, \ldots, x_k$ can be recorded without error. For example, if we want to relate the mean measure $E(y)$ of worker absenteeism to two quantitative predictor variables, say,

$$x_1 = \text{worker age}$$
$$x_2 = \text{worker hourly wage rate}$$

we can construct any of a number of models. One possibility is

$$E(y) = \beta_0 + \beta_1 x_1 + \beta_2 x_2$$

This model graphs as a *plane*, a surface in the three-dimensional space defined by y, x_1, and x_2. Or if we suspect that the response surface relating $E(y)$ to x_1 and x_2 possesses curvature, we might use the model

$$E(y) = \beta_0 + \beta_1 x_1 + \beta_2 x_2 + \beta_3 x_1 x_2$$

or

$$E(y) = \beta_0 + \beta_1 x_1 + \beta_2 x_2 + \beta_3 x_1 x_2 + \beta_4 x_1^2 + \beta_5 x_2^2$$

Besides containing the *first-order terms*, those involving only x_1 or x_2, these models include *second-order terms*, such as x_1^2, x_2^2, and the two-variable cross product $x_1 x_2$.[†] Thus, all three of these models relate $E(y)$ to only two predictor variables, x_1 and x_2, but the number of terms and interpretations of the models differ.

Choosing a good model relating y to a set of predictor variables is a difficult and important task, much more difficult than fitting the model to a set of data (this is usually done automatically by a computer). Even if your model includes all of the important predictor variables, it still may provide a poor fit to your data if the form of the model is not properly specified.

For example, suppose that $E(y)$ is *perfectly* related to a single predictor variable x_1 by the relation

$$E(y) = \beta_0 + \beta_1 x_1 + \beta_2 x_1^2$$

If you fit the first-order (straight-line) model

$$E(y) = \beta_0 + \beta_1 x_1$$

to the data, you will obtain the *best-fitting, least-squares line*, but it may still provide a poor fit to your data and may be of little value for estimation or prediction [see Figure 11.3(a)]. In contrast, if you fit the model

$$E(y) = \beta_0 + \beta_1 x_1 + \beta_2 x_1^2$$

you obtain a perfect fit to the data [see Figure 11.3(b)].

The lesson is quite clear. Including all of the important predictor variables in the model as first-order terms—that is, $x_1, x_2, \ldots, x_k$—may not (and probably will not) produce a model that provides a good fit to your data. You may have to include second-order terms such as $x_1^2, x_2^2, x_3^2, x_1 x_2, x_1 x_3$. Plots of y versus x_1, y versus x_2, and y versus x_3 are helpful tools in determining whether to include quadratic or other higher-order terms in the model.

In contrast to quantitative predictor variables, qualitative predictor variables are entered into a model by using **dummy (indicator) variables**. For example, suppose you are attempting to relate the mean salary of a group of employees to a set of predictor variables and one of the variables that you want to include is the employee's ethnic background. If each employee included in your study belongs to one of three

[†]The *order* of a term is determined by the sum of the exponents of variables making up that term. Terms involving x_1 or x_2 are first-order. Terms involving x_1^2, x_2^2, or $x_1 x_2$ are second-order.

FIGURE 11.3 Two models, each using one predictor variable

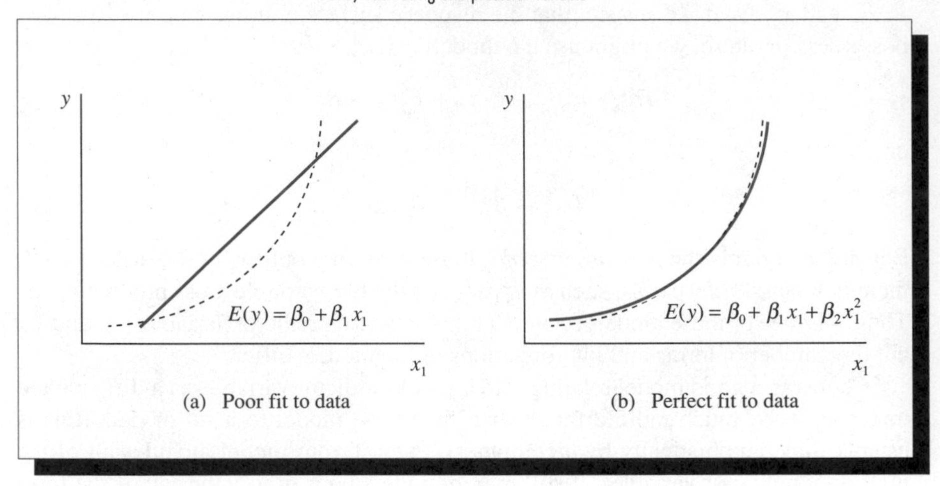

ethnic groups, say A, B, or C, you will want to enter the qualitative predictor variable "ethnicity" into your model as follows:

$$E(y) = \beta_0 + \beta_1 x_1 + \beta_2 x_2$$

where

$$x_1 = \begin{cases} 1, & \text{if group B} \\ 0, & \text{if not} \end{cases} \qquad x_2 = \begin{cases} 1, & \text{if group C} \\ 0, & \text{if not} \end{cases}$$

If you want to find $E(y)$ for group A, you examine the coding for the dummy variables x_1 and x_2 and note that $x_1 = 0$ and $x_2 = 0$. Therefore, for group A

$$E(y) = \beta_0 + \beta_1 x_1 + \beta_2 x_2 = \beta_0 + \beta_1(0) + \beta_2(0) = \beta_0$$

The value of $E(y)$ for group B is obtained by letting $x_1 = 1$ and $x_2 = 0$; that is, for group B

$$E(y) = \beta_0 + \beta_1 x_1 + \beta_2 x_2 = \beta_0 + \beta_1(1) + \beta_2(0) = \beta_0 + \beta_1$$

Similarly, the value of $E(y)$ for group C is obtained by letting $x_1 = 0$ and $x_2 = 1$. Therefore, the group C

$$E(y) = \beta_0 + \beta_1 x_1 + \beta_2 x_2 = \beta_0 + \beta_1(0) + \beta_2(1) = \beta_0 + \beta_2$$

The coefficient β_0 is the mean for group A, β_1 is the mean difference between groups B and A, and β_2 is the mean difference between groups C and A.

Models with qualitative predictor variables that may assume, say, k values can be constructed in a similar manner, by using $(k - 1)$ terms involving dummy variables. These terms can be added to models containing other predictor variables, quantitative or qualitative, and you can also include terms involving the cross products (interaction terms) of the dummy variables with other variables that appear in the model.

This section introduced two important aspects of model formulation: the concept of predictor variable interaction (and how to cope with it) and, second, the method for introducing qualitative predictor variables into a model. Clearly, this is not enough to make you proficient in model formulation, but it will help you to understand why and how some terms are included in regression models, and it may help you to avoid some pitfalls encountered in model construction.

EXAMPLE 11.4 A study was conducted to examine the relationship between university salary y, the number of years of experience of the faculty member, and the sex of the faculty member. If we expect a straight line relationship between mean salary and years of experience for both men and women, write the model relating mean salary to the two predictor variables:

1 Years of experience (quantitative)

2 Sex of the professor (qualitative)

Solution Since we may suspect the mean salary lines for females and males to be different, we want to construct a model for mean salary $E(y)$ that may appear as shown in Figure 11.4.

A straight-line relationship between $E(y)$ and years of experience x_1 implies the model

$$E(y) = \beta_0 + \beta_1 x_1 \quad \textbf{(graphs as a straight line)}$$

The qualitative variable sex can only assume two "values," male and female. Therefore, we can enter the predictor variable sex into the model by using one **dummy** (or **indicator variable**), x_2, as

$$E(y) = \beta_0 + \beta_1 x_1 + \beta_2 x_2 \quad \textbf{(graphs as two parallel lines)}$$

$$x_2 = \begin{cases} 1, & \text{if male} \\ 0, & \text{if female} \end{cases}$$

The fact that we want to allow the slopes of the two lines to differ means that we think that the two predictor variables **interact**—that is, the change in $E(y)$

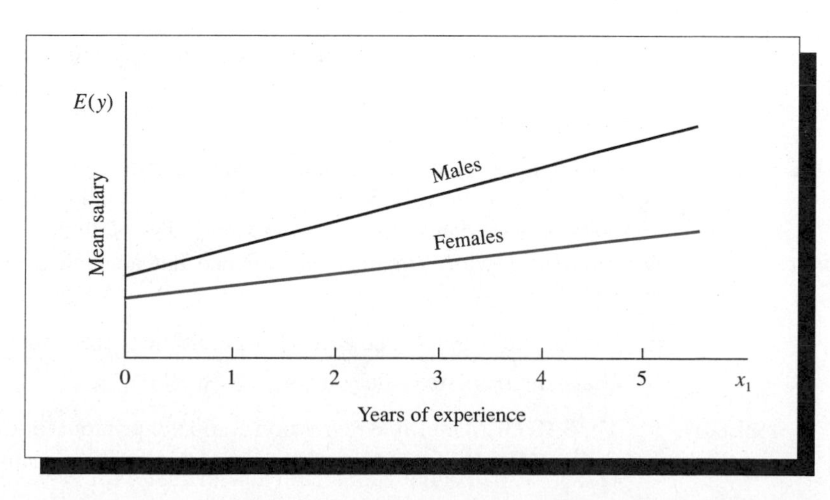

FIGURE 11.4 Hypothetical relationship between mean salary $E(y)$, years of experience (x_1), and sex (x_2) for Example 11.4

corresponding to a change in x_1 depends on whether the professor is a man or a woman. To allow for this interaction (difference in slopes), we introduce the interaction term x_1x_2 into the model. The complete model that characterizes the graph in Figure 11.4 is

$$E(y) = \beta_0 + \beta_1 \underset{\substack{\uparrow \\ \text{years of} \\ \text{experience}}}{x_1} + \beta_2 \overset{\substack{\text{dummy variable} \\ \text{for sex} \\ \downarrow}}{x_2} + \beta_3 \underset{\substack{\uparrow \\ \text{interaction}}}{x_1x_2}$$

where

$$x_1 = \text{years of experience}$$

$$x_2 = \begin{cases} 1, & \text{if a male} \\ 0, & \text{if a female} \end{cases}$$

The interpretation of the model parameters can be seen by assigning values to the dummy variable x_2. Thus, when you want to acquire the line for females, the dummy variable $x_2 = 0$ (according to our coding), and

$$E(y) = \beta_0 + \beta_1 x_1 + \beta_2(0) + \beta_3 x_1(0) = \beta_0 + \beta_1 x_1$$

Therefore, β_0 is the y intercept for the females' line, and β_1 is the slope of the line relating expected salary to years of experience for *females only.*

Similarly, the line for males is obtained by letting $x_2 = 1$. Then

$$\begin{aligned} E(y) &= \beta_0 + \beta_1 x_1 + \beta_2(1) + \beta_3 x_1(1) \\ &= \underbrace{(\beta_0 + \beta_2)}_{y \text{ intercept}} + \underbrace{(\beta_1 + \beta_3)}_{\text{slope}} x_1 \end{aligned}$$

The y intercept of the males' line is $(\beta_0 + \beta_2)$, and the slope, the coefficient of x_1, is equal to $(\beta_1 + \beta_3)$.

Because the slope of the males' line is $(\beta_1 + \beta_3)$ and the slope of the females' line is β_1, it follows that $(\beta_1 + \beta_3) - \beta_1 = \beta_3$ is the difference in the slopes of the two lines. Similarly, β_2 is equal to the difference in the y intercepts for the two lines. ■

EXAMPLE 11.5 Random samples of six female and six male assistant professors were selected from among the assistant professors in a college of arts and sciences. The data on salary and years of experience are shown in Table 11.5. Note that both samples contained two professors with 3 years of experience but no male professor had 2 years of experience.

a Explain the output of the SAS GLM multiple regression computer printout.

b Graph the predicted salary lines.

Solution **a** The SAS GLM multiple regression computer printout for the data in Table 11.5 is shown in Table 11.6. The explanation of this printout is similar to the explanation

TABLE 11.5 Salary versus sex and years of experience

Years of experience x_1	1	2	3	4	5
Salary y (males)	20,710		23,160 23,210	24,140	25,760 25,590
Salary y (females)	19,510	20,440	21,340 21,760	22,750	23,200

TABLE 11.6 The SAS GLM multiple regression analysis computer printout for Example 11.5

```
DEPENDENT VARIABLE: Y
                                                               (2)               (1)
     SOURCE DF       SUM OF SQUARES         MEAN SQUARE      F VALUE PR > F    R-SQUARE         C.V.

       MODEL 3    42108777.02898556   14036259.00966185      346.24 0.0001    0.992357        0.8897

       ERROR 8      324314.63768142      40539.32971018 (7)      ROOT MSE                     Y MEAN
                                                                              (8)
 CORRECTED TOTAL 11  42433091.66666698                         201.34380971              22630.83333333

                   (9)
 SOURCE DF         TYPE I SS   F VALUE   PR > F   D        TYPE IV SS   F VALUE   PR > F
    X1 1   33294036.23595509    821.28   0.0001   1  9389610.00000008    231.62   0.0001
    X2 1    8452796.51598297    208.51   0.0001   1   326808.74399183      8.06   0.0218
 X1*X2 1     361944.27704750      8.93   0.0174   1   361944.27704750      8.93   0.0174

                            PARAMETER ESTIMATES
                                   (4)
                 (3)           T FOR HO:        (5)          (6)
 VARIABLE         ESTIMATE    PARAMETER = 0    PR > |T|     STD ERROR
 INTERCEPT  18593.00000000            89.41     0.0001    207.94699250
        X1    969.00000000            15.22     0.0001     63.67050315
        X2    866.71014493             2.84     0.0218    305.25678646
     X1*X2    260.13043478             2.99     0.0174     87.05798112
```

Note: The star is used to indicate multiplication. Thus x_1x_2 is denoted by X1*X2.

of the MINITAB computer printout given in Example 11.2. The relevant portions of the printout are boxed and numbered.

The value of R^2, the **multiple coefficient of determination**, provides a measure of how well the model fits the data. As you can see, 99.2% of the sum of squares of deviations of the y values about $\bar{y}$ is explained by all of the terms in the model.

If the model contributes information for the prediction of y, at least one of the model parameters, β_1, β_2, or β_3, will differ from 0. Consequently, we want to test the null hypothesis $H_0 : \beta_1 = \beta_2 = \beta_3 = 0$ against the alternative hypothesis H_a : at least one of the parameters, β_1, β_2, or β_3, differs from 0. The test statistic for this test, an F statistic with v_1 = (the number of parameters in H_0) = 3 and $v_2 = n -$ (number of parameters in the model) $= 12 - 4 = 8$, is shown in area (2) to be 346.24. Since this value exceeds the tabulated value for F—with $v_1 = 3$, $v_2 = 8$, and $\alpha = .05$, $F = 4.07$ (from Table 6 of Appendix I)—we reject H_0 and conclude that at least one of the parameters β_1, β_2, β_3 differs from 0. The observed significance level (p-value) for the test, PR > F, shown to the right of the value of the F statistic, is equal to .0001.

The estimates of the four model parameters are shown under the heading ESTIMATE. Rounding these estimates, we obtain the prediction equation

$$\hat{y} = 18593.0 + 969.0x_1 + 866.7x_2 + 260.1x_1x_2$$

The computed value of the t statistic for each of the model parameters is shown in area (4) under the column T FOR H_0 : PARAMETER = 0, and the observed significance levels (p-values) are shown in area (5) under the column headed PR > |T|. These values are calculated for two-tailed tests. The p-value for a one-tailed test is half of the value shown in the printout. Looking at the observed significance levels, we see that all of the regression coefficients are significantly different from zero, with p-values less than or equal to .03.

The estimated standard deviations of the estimators, used in constructing confidence intervals for the regression coefficients, are shown in area (6) under the heading STD ERROR. The values of SSE = 324314.638 and s^2 = MSE = 40539.330 are shown in area (7), and the standard deviation $s = \sqrt{\text{MSE}}$ = 201.334 is shown in area (8), labeled ROOT MSE. Area (9), labeled TYPE I SS, are the sequential sum of squares given in the MINITAB printout and discussed in Example 11.2.

b A graph of the two salary lines is shown in Figure 11.5. Note that the salary line corresponding to the male faculty members appears to be rising at a more rapid rate than the salary line for females. Is this real, or could it be due to chance? The following example will answer this question. ■

EXAMPLE **11.6** Refer to Example 11.5. Do the data provide sufficient evidence to indicate that the annual rate of increase in male junior faculty salaries exceeds the annual rate of increase for women junior faculty salaries? Thus, we want to know whether the data provide sufficient evidence to indicate that the slope of the men's faculty salary line exceeds the slope of the women's faculty salary line.

Solution Since β_3 measures the difference in slopes, the slopes of the two lines will be identical if $\beta_3 = 0$. Therefore, we want to test

$$H_0 : \beta_3 = 0$$

—that is, the slopes of the two lines are identical—against the alternative hypothesis

$$H_a : \beta_3 > 0$$

—that is, the slope of the male faculty salary line is greater than the slope of the female faculty salary line.

The calculated value of t corresponding to β_3, shown in area (4) of the computer printout (Table 11.6), is 2.99. Since we want to detect values $\beta_3 > 0$, we will conduct a one-tailed test and reject H_0 if $t > t_\alpha$. The tabulated t value from Table 4 of Appendix I, for $\alpha = .05$ and $v = 8$ d.f., is $t_{.05} = 1.860$. The calculated value of t exceeds this value, and thus there is evidence to indicate that the annual rate of increase in male

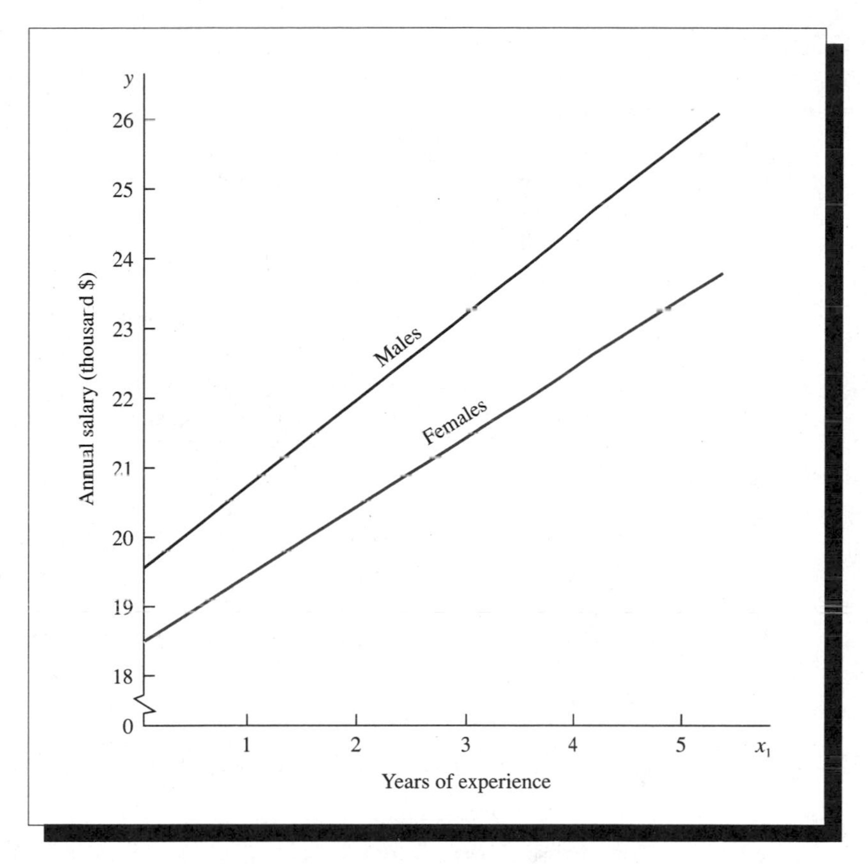

FIGURE 11.5
A graph of the faculty salary prediction lines for Example 11.5

faculty salaries exceeds the corresponding annual rate of increase in faculty salaries for women.[†] ■

Most popular statistical program packages contain multiple regression programs. Some programs can be used only on specific computers; others are more versatile. If you plan to conduct a multiple regression analysis, you will want to determine the statistical program packages available at your computer center so that you can become familiar with their output.

The computer printouts for the MINITAB, SAS, SPSS^x, and other multiple regression computer program packages are very similar. Once you can read one, it is likely that you will be able to read the output of the computer program available at your computer center. If you have difficulty understanding the output, consult the package instruction manual.

[†] If we want to determine whether the data provide sufficient evidence to indicate that the male faculty members start at higher salaries, we test $H_0 : \beta_2 = 0$ against the alternative hypothesis $H_a : \beta_2 > 0$.

Exercises

Basic Techniques

11.20 Suppose you wish to predict production yield y as a function of several independent predictor variables. Indicate whether each of the following independent variables is qualitative or quantitative.

a The prevailing interest rate in the area

b The price per pound of one item used in the production process

c The plant (A or B) at which the production yield is measured

d The length of time that the production machine has been in operation

e The shift (night or day) in which the yield is measured

11.21 A multiple linear regression model, involving one qualitative and one quantitative independent variable, produced the following prediction equation:

$$\hat{y} = 12.6 + .54x_1 - 1.2x_1x_2 + 3.9x_2^2$$

a Which of the two variables is the quantitative variable? Explain.

b If x_1 can take only the values 0 or 1, find the two possible prediction equations for this experiment.

c Graph the two equations found in part (b). Compare the shapes of the two curves.

Applications

11.22 Data concerning the median annual income of full-time workers aged 25 years or older by the number of years of schooling completed, is given in the accompanying table.

Years schooling	Salary (dollars) Men	Women
8	18,000	11,500
10	20,500	13,000
12	25,000	16,100
14	28,100	18,300
16	34,500	22,100
19	39,700	27,600

Source: Data adapted from "Upward Mobility" (1989), p. 13.

The MINITAB command REGRESS was used to fit the model

$$y = \beta_0 + \beta_1 x_1 + \beta_2 x_2 + \varepsilon$$

to this data, where

$$\begin{aligned} y &= \text{salary} \\ x_1 &= \text{years of schooling} \\ x_2 &= 0 \quad \text{if female} \\ & 1 \quad \text{if male} \end{aligned}$$

The printout is shown in the accompanying table:

```
MTB> REGRESS C1 2 C2 C3;
SUBC> PREDICT 18 0;
SUBC> PREDICT 18 1.

The regression equation is
Y = -5125 + 1764 X1 + 9533 X2

Predictor     Coef   Stdev  t-ratio       P
Constant     -5125    1678    -3.05   0.014
X1          1763.9   118.5    14.88   0.000
X2          9533.3   870.1    10.96   0.000

s = 1507   R-sq = 97.4%   R-sq(adj) = 96.9%

Analysis of Variance

SOURCE       DF         SS          MS        F       P
Regression    2  775663808   387831904   170.74   0.000
Error         9   20442852     2271428
Total        11  796106688

SOURCE       DF     SEQ SS
X1            1  503010496
X2            1  272653344

Unusual Observations

Obs.    X1       Y   Fit  Stdev.Fit   Residual   St.Resid
   7   8.0   11500  8986        868       2514      2.04R

R denotes an obs. with a large st. resid.

  Fit   Stdev.Fit        95% C.I.          95% P.I.
26626         841   (24723,28528)    (22721,30531)
36159         841   (34257,38061)    (32254,40064)
```

a Find the prediction equation for these data.

b Test the adequacy of the complete model using the F statistic. What is the observed level of significance of the test?

c Which of the independent variables are significant? Test using $\alpha = .05$.

d Use the prediction equation in part (a) to predict the salary of a woman who has 18 years of schooling. Compare with the prediction given in the printout.

e Use the prediction equation in part (a) to predict the salary of a man who has 18 years of schooling. Compare with the prediction given in the printout.

11.23 The German system of industrial relations is based upon condetermination, or, equivalently, worker participation in management. In a study of the German worker's perception of this system, J. B. Dworkin and associates interviewed a sample of 135 workers, 113 men and 22 women. Each worker responded to ten job satisfaction statements, rating each statement from 1 (least satisfaction) to 5 (most satisfaction). One aspect of the study involved an examination of some variables that might be predictors of job satisfaction. The dependent variable y used to measure a worker's job satisfaction was the average of a worker's responses to the ten job-satisfaction questions. These observations, along with the recorded values of nine independent variables (listed in the accompanying table), were recorded for 135 workers. The results obtained after fitting a first-order model to the data are also shown in the table.

Independent Variable	Description	Parameter Estimate $\hat{\beta}_i$	Standard Error $s_{\hat{\beta}_i}$
x_1	Sex of worker	−1.53	.201
x_2	Job rank of worker	.398	.142
x_3	Level of education	−.026	.080
x_4	Union member (yes or no)	.239	.166
x_5	Union official (yes or no)	−.002	.120
x_6	Worker active in codetermination system (yes or no)	−.264	.189
x_7	(Identification omitted)	−.308	.131
x_8	(Identification omitted)	.072	.062
x_9	Monthly salary of worker	.110	.052

Source: Dworkin, et al. (1983).

a Give the equation of the model used in the regression analysis.

b How many degrees of freedom are associated with SSE and s^2?

c Examine the table and identify those variables that were entered into the model by using dummy variables.

d The value of R^2 for the regression analysis was .281. Interpret the value of R^2 and comment on its relevance to the regression analysis.

e Do the data provide sufficient information to indicate that at least one variable in the model contributes information for predicting worker job satisfaction? (HINT: Test the usefulness of the complete model with an F test. Test using $\alpha = .05$.)

11.24 Refer to the worker job satisfaction study in Exercise 11.23. Use the information in the table above to determine which, if any, of the variables contribute information for the prediction of worker job satisfaction. Test each and indicate those with p-values less than or equal to .10.

11.5 Testing Sets of Model Parameters

In the preceding sections, we found it useful to test a complete linear model to determine whether the model contributes information for the prediction of y. We also were able to test an hypothesis about an individual β parameter, using a Student's t test. Besides these two important tests, we may want to test hypotheses about sets of parameters.

For example, suppose a company suspects that the demand y for some product could be related to as many as five independent variables x_1, x_2, x_3, x_4, and x_5. However, the cost of obtaining measurements on the variables x_3, x_4, and x_5 is very high. If, in a small pilot study, the company could show that these three variables contribute little or no information for the prediction of y, they can be eliminated from the study at great savings to the company.

If all five variables x_1, x_2, x_3, x_4, and x_5 are used to predict y, the multiple linear regression model would be written as

$$y = \beta_0 + \beta_1 x_1 + \beta_2 x_2 + \beta_3 x_3 + \beta_4 x_4 + \beta_5 x_5 + \varepsilon$$

However, if x_3, x_4, and x_5 contribute no information for the prediction of y, then they would not appear in the model—that is, $\beta_3 = \beta_4 = \beta_5 = 0$—and the reduced model would be

$$y = \beta_0 + \beta_1 x_1 + \beta_2 x_2 + \varepsilon$$

Hence, we want to test

$$H_0 : \beta_3 = \beta_4 = \beta_5 = 0$$

—that is, the independent variables x_3, x_4, and x_5 contribute no information for the prediction of y—against the alternative hypothesis

$$H_a : \text{ at least one of the parameters } \beta_3, \beta_4, \text{ or } \beta_5 \text{ differs from } 0$$

—that is, at least one of the variables x_3, x_4, or x_5 contributes information for the prediction of y. Thus, in deciding whether the complete model is preferable to the reduced model in predicting demand, we are led to a test of an hypothesis about a set of three parameters, β_3, β_4, and β_5.

To explain how to test an hypothesis concerning a set of model parameters, we will define two models:

Model 1 (reduced model)

$$E(y) = \beta_0 + \beta_1 x_1 + \beta_2 x_2 + \cdots + \beta_r x_r$$

Model 2 (complete model)

$$E(y) = \underbrace{\beta_0 + \beta_1 x_1 + \beta_2 x_2 + \cdots + \beta_r x_r}_{\text{terms in model 1}} + \underbrace{\beta_{r+1} x_{r+1} + \beta_{r+2} x_{r+2} + \cdots + \beta_k x_k}_{\text{additional terms in model 2}}$$

Suppose we were to fit both models to the data set and calculate the sum of squares for error for both regression analyses. If model 2 contributes more information for the prediction of y than model 1, then the errors of prediction for model 2 should be smaller than the corresponding errors for model 1, and SSE_2 should be smaller than SSE_1. In fact, the greater the difference between SSE_1 and SSE_2, the greater is the evidence to indicate that model 2 contributes more information for the prediction of y than model 1.

The test of the null hypothesis

$$H_0 : \beta_{r+1} = \beta_{r+2} = \cdots = \beta_k = 0$$

against the alternative hypothesis

$$H_a : \text{ at least one of the parameters } \beta_{r+1}, \beta_{r+2}, \ldots, \beta_k \text{ differs from zero}$$

uses the test statistic

$$F = \frac{(\text{SSE}_1 - \text{SSE}_2)/(k - r)}{\text{MSE}_2}$$

where F is based on $v_1 = k - r$ and $v_2 = n - (k + 1)$ d.f. Note that the $(k - r)$ parameters involved in H_0 are those added to model 1 to obtain model 2. The numerator degrees of freedom v_1 of will always equal $(k - r)$, the number of parameters involved in H_0. The denominator degrees of freedom v_2 is the number of degrees of freedom associated with the sum of squares for error, SSE_2, for the complete model.

The rejection region for the test is identical to the rejection region for all of the analysis of variance F tests, namely

$$F > F_\alpha$$

EXAMPLE 11.7 Refer to the real estate data of Example 11.1 in which the listed selling price y is related to the square feet of living area x_1, the number of floors x_2, the number of bedrooms x_3, and the number of bathrooms, x_4. The realtor suspects that the square footage of living area is the most important predictor variable and that the other variables might be eliminated from the model without loss of much prediction information. Test this claim with $\alpha = .05$.

Solution The hypothesis to be tested is

$$H_0 : \beta_2 = \beta_3 = \beta_4 = 0$$

versus the alternative that at least one of β_2, β_3, or β_4 is different from zero. The **complete model (2)**, given as

$$y = \beta_0 + \beta_1 x_1 + \beta_2 x_2 + \beta_3 x_3 + \beta_4 x_4 + \varepsilon$$

was fitted in Example 11.1 and the MINITAB printout from Table 11.2 is reproduced in Table 11.7 along with the MINITAB printout for the simple linear regression analysis of the **reduced model (1)**, given as

$$y = \beta_0 + \beta_1 x_1 + \varepsilon$$

Then $SSE_1 = 11514$ from Table 11.7(b) and $SSE_2 = 6599$ and $MSE_2 = 275$ from Table 11.7(a). The test statistic is

$$F = \frac{(SSE_1 - SSE_2)/(k - r)}{MSE_2}$$
$$= \frac{(11514 - 6599)/(4 - 1)}{275} = 5.96$$

The critical value of F with $\alpha = .05$, $v_1 = 3$, and $v_2 = n - (k + 1) = 29 - (4 + 1) = 24$ d.f. is $F_{.05} = 3.01$. Hence, H_0 is rejected. There is evidence to indicate that at least one of the three variables, number of floors, bedrooms, or bathrooms is contributing significant information for the prediction of listed selling price. ▪

TABLE 11.7
MINITAB printouts for complete (a) and reduced (b) models for Example 11.7

```
(a) Complete Model

The regression equation is
LPRICE = -16.6 + 7.84 SQFT - 34.4 NUMFLRS - 7.99 BDRMS + 54.9 BATHS

Predictor      Coef       Stdev    t-ratio       p
Constant     -16.58       18.88      -0.88   0.389
SQFT          7.839       1.234       6.35   0.000
NUMFLRS      -34.39       11.15      -3.09   0.005
BDRMS        -7.990       8.249      -0.97   0.342
BATHS         54.93       13.52       4.06   0.000

s = 16.58     R-sq = 88.2%     R-sq(adj) = 86.2%

Analysis of Variance

SOURCE        DF         SS         MS          F        P
Regression     4      49359      12340      44.88    0.000
Error         24       6599        275
Total         28      55958

SOURCE        DF     SEQ SS
SQFT           1      44444
NUMFLRS        1         59
BDRMS          1        321
BATHS          1       4536

(b) Reduced Model

The regression equation is
LPRICE = 3.0 + 9.61 SQFT

Predictor        Coef    Stdev    t-ratio       P
Constant         3.00    14.26       0.21   0.835
SQFT           9.6121   0.9416      10.21   0.000

s = 20.65     R-sq = 79.4%     R-sq(adj) = 78.7%

Analysis of Variance

SOURCE        DF         SS         MS          F        p
Regression     1      44444      44444     104.22    0.000
Error         27      11514        426
Total         28      55958
```

11.6 Residual Analysis

The deviations between the observed values of y and their predicted values are called **residuals**. For example, the first three columns of Table 11.8 reproduce the data of Table 10.2, and the ten predicted values of y obtained from the prediction equation (Example 11.1)

$$\hat{y} = 40.78 + .77x$$

are shown in column 4. The predicted value of y for $x_1 = 39$ is $\hat{y}_1 = 40.78 + .77x = 40.78 + (.77)(39) = 70.81$, and the residual is $e_1 = y_1 - \hat{y}_1 = 65 - 70.81 = -5.81$.

TABLE 11.8
Management data of Table 10.2 with predicted values and residuals

Student	Mathematics Achievement Test Scores x	Final Calculus Grade y	$\hat{y} = 40.78 + .77x$	$e = y - \hat{y}$
1	39	65	70.81	−5.81
2	43	78	73.89	4.11
3	21	52	56.95	−4.95
4	64	82	90.06	−8.06
5	57	92	84.67	7.33
6	47	89	76.97	12.03
7	28	73	62.34	10.66
8	75	98	98.53	−0.53
9	34	56	66.96	−10.96
10	52	75	80.82	−5.82

FIGURE 11.6
Residuals for the least-squares line $\hat{y} = 40.78 + .77x$

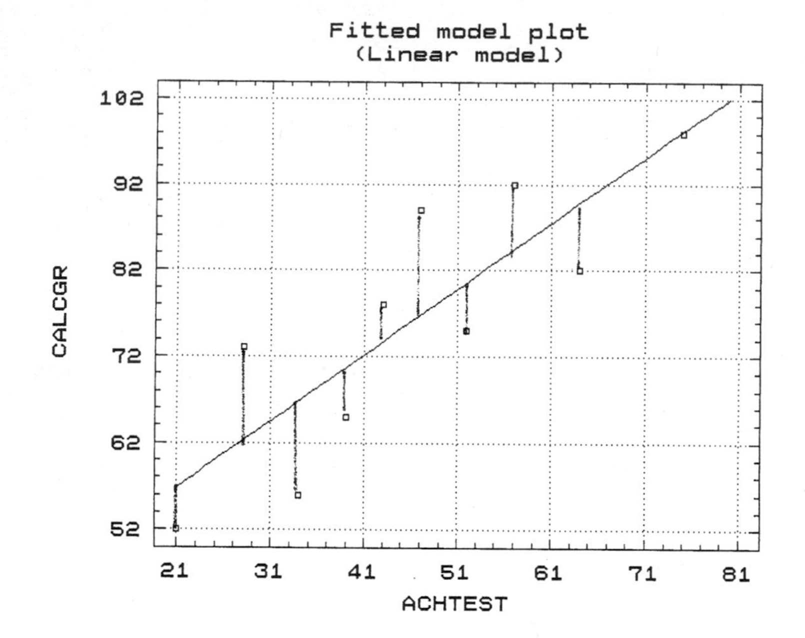

The ten residuals, one for each x value, appear in column 5 of Table 11.8 and are displayed graphically as the vertical line segments in Figure 11.6.

DEFINITION ▪

The **residual** corresponding to the data point (x_i, y_i) is

$$e_i = y_i - \hat{y}_i \quad ▪$$

In **residual analysis**, plots of the residuals against $\hat{y}$ or against the individual independent variables often indicate departures from the assumptions required for an analysis of variance, and they also may suggest changes in the underlying model.

Plots of the residuals against $\hat{y}$ are particularly useful for detecting nonuniformity in the variance of y for different values of $E(y)$.

For example, the variance of some types of data—Poisson data, in particular—increases with the mean. A plot of the residuals for this type of data might appear as shown in Figure 11.7(a). Note that the range of the residuals increases as $\hat{y}$ increases, thus indicating that the variance of y is increasing as the mean value of y, $E(y)$, increases.

The variances for percentages and proportions calculated from binomial data also increase for values of $p = 0$ to $p = .5$ and then decrease from $p = .5$ to $p = 1.0$. Plots of residuals versus $\hat{y}$ for this type of data would appear as shown in Figure 11.7(b). Therefore, if the data were small percentages, the plot of residuals would show the range of the residuals *increasing* as $\hat{y}$ increases. If the data were large percentages, the range of the residuals would appear to *decrease* as $\hat{y}$ increases.

FIGURE 11.7
Plots of residuals against $\hat{y}$

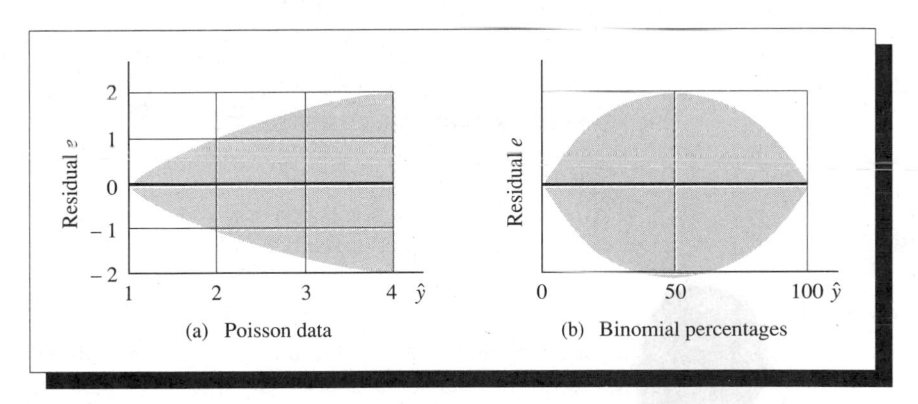

If the range of the residuals increases as $\hat{y}$ increases and you know that the data are measurements on Poisson variables, you can stabilize the variance of the response by running the regression analysis on $y^* = \sqrt{y}$. Or if percentages are calculated from binomial data, you can use the arcsin transformation, $y^* = \sin^{-1} \sqrt{y}$.[†]

If you have no prior reason to explain why the range of the residuals increases as $\hat{y}$ increases, you can still use a transformation of y that affects larger values of y more than smaller values, say, $y^* = \sqrt{y}$ or $y^* = \ln y$. These transformations have a tendency to both stabilize the variance of y^* and to make the distribution of y^* more nearly normal when the distribution of y is highly skewed.

Plots of the residuals against the individual independent variables often indicate the problems with models selection. In theory, the residuals should vary in size and sign (positive or negative) in a random manner about $\hat{y}$ if the equation that you have selected for $E(y)$ is a good approximation to the true relationship between $E(y)$ and

[†]In Chapter 9 and earlier chapters we represented the response variable by the symbol x. In the chapters on regression analysis, Chapters 10 and 11, the response variable is represented by the symbol y.

the independent variables. For example, if $E(y)$ and a single independent variable x are linearly related; that is,

$$E(y) = \beta_0 + \beta_1 x$$

and you fit a straight line to the data, then the observed y values should vary in a random manner about $\hat{y}$, and a plot of the residuals against x will appear as shown in Figure 11.8.

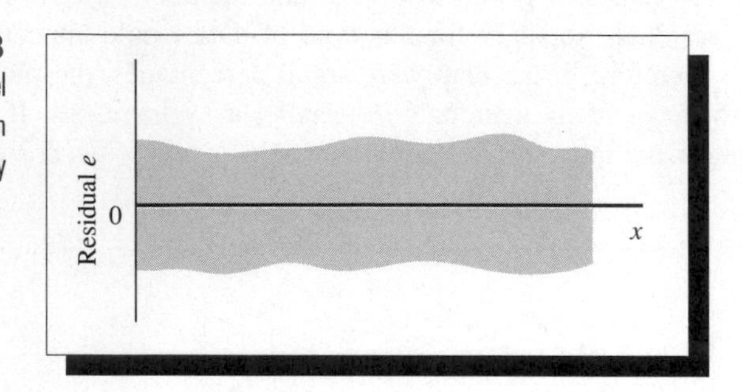

FIGURE **11.8**
Residual plot when the model provides a good approximation to reality

The residuals from the regression of assistant professors' salaries (y) on years of experience (x_1) and sex (x_2) described in Example 11.5 using the model

$$E(y) = \beta_0 + \beta_1 x_1 + \beta_2 x_2 + \beta_2 x_1 x_2$$

are plotted against the values of $\hat{y}$ in Figure 11.9. (Recall that this model plots as two straight lines, one for male salaries and one for female salaries.) Notice that there does not appear to be a pattern in the residuals from the model. On the other hand, when a common regression line is fitted for both male and female assistant professors using the model

$$E(y) = \beta_0 + \beta_1 x_1$$

the plot of the residuals versus $\hat{y}$ given in Figure 11.10 reveals one distinct set of positive residuals corresponding to the salaries of the men and a second distinct set of negative residuals corresponding to the salaries of the women. This nonrandom grouping indicates the need for the variable x_2, which was not included in the model.

In Example 11.3, we fitted a quadratic regression model relating the value added per work-hour (y) as a function of the size of the store (x). These data were fitted using the simple linear regression model

$$E(y) = \beta_0 + \beta_1 x$$

The residuals plotted against the values of $\hat{y}$ in Figure 11.11 display a distinct quadratic trend, indicating the omission of a quadratic term in the model. However, not all residual plots give so clear an indication of which terms or variables are missing, as we observed in these two examples.

Most computer multiple regression analysis printouts give the predicted values of y and the residuals for each value of x when requested. In addition, you can

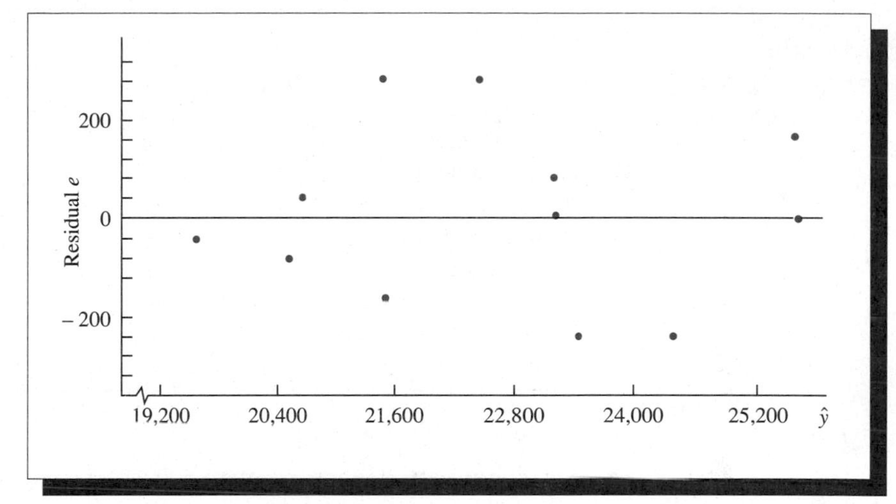

FIGURE 11.9 Residual plot based on two regression lines, data of Example 11.5

FIGURE 11.10 Residual plot based on a common regression line, data of Example 11.5

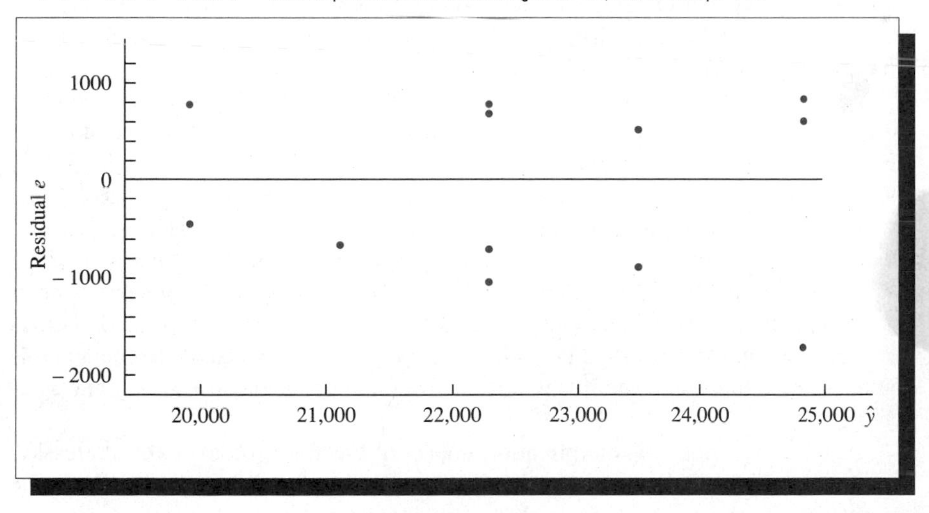

request plots of the residuals versus $\hat{y}$ or versus the individual independent variables. **Examine the plots and look for nonrandomness in the behavior of the residuals. If nonrandomness is observed, seek an explanation for the behavior and attempt to correct it.**

11.7 Misinterpretations in a Regression Analysis

Several misinterpretations of the output of a regression analysis are common. We have already mentioned the importance of model selection. If a model does not fit a set of data, it does not mean that the variables included in the model contribute little or no information for the prediction of y. The variables may be very important contributors of information, but you may not have entered the variables into the model in an

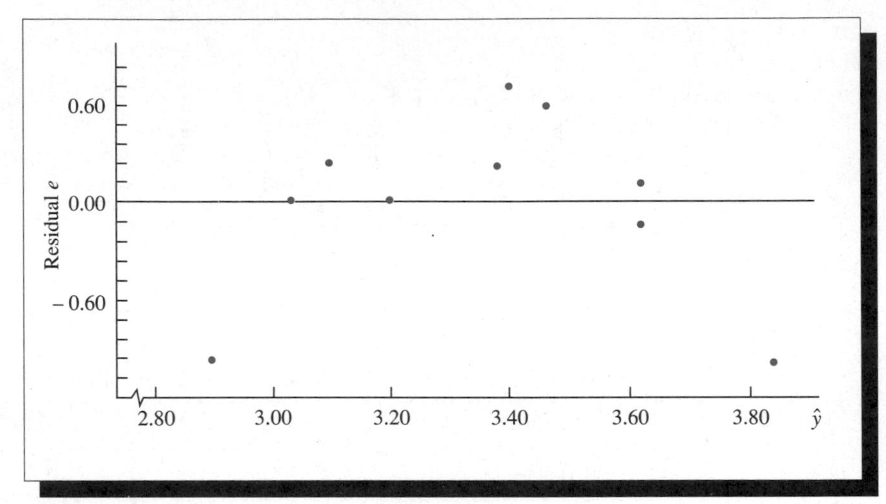

FIGURE 11.11
Residual plot from a simple linear regression model, data of Example 11.3

appropriate way. For example, a second-order model in the variables might provide a very good fit to the data when a first-order model appears to be completely useless in describing the response variable y.

Second, you must be careful not to deduce that a causal relationship exists between a response y and a variable x. Just because a variable x contributes information for the prediction of y, it does not imply that changes in x *cause* changes in y. It is possible to *design* an experiment to detect causal relationships. For example, if you randomly assign experimental units to each of two levels of a variable x—say, $x = 5$ and $x = 10$—and the data show that the mean value of y is larger when $x = 10$, then you could say that the change in the level of x caused a change in the mean value of y. But in most regression analyses, where the experiments are not designed, there is no guarantee that an important predictor variable—say, x_1—caused y to change. It is quite possible that some variable that is not even in the model cause *both* y and x_1 to change.

A third common misinterpretation concerns the magnitude of the regression coefficients. Neither the size of a regression coefficient nor its t value indicates the importance of the associated variable as an information contributor. For example, suppose you wish to predict a college student's calculus grade y as a function of the student's high school average mathematics grade x_1 and the student's score x_2 on a college mathematics placement test. A regression analysis, using the first-order model $E(y) = \beta_0 + \beta_1 x_1 + \beta_2 x_2$, would likely show that both x_1 and x_2 contribute information for the prediction of y. However, it is conceivable that the t value associated with one of the regression coefficients would not be statistically significant because much of the information contained in x_1 is the same information contained in x_2. When this occurs, the one-variable model

$$E(y) = \beta_0 + \beta_1 x_1 \qquad \text{or} \qquad E(y) = \beta_0 + \beta_2 x_2$$

may be almost as useful in predicting y as the model

$$E(y) = \beta_0 + \beta_1 x_1 + \beta_2 x_2$$

Multicollinearity

One or more of the predictor variables in a regression analysis may be highly correlated with one another. This situation and the problems that it causes in a regression analysis is referred to as **multicollinearity**. This problem arises when two predictor variables are linearly related and hence share the same predictive information, or when successive powers of $x (x, x^2, x^3$, etc.) are used as predictor variables in polynomial regression.

When multicollinearity is present in a regression analysis, the estimated regression coefficients have standard errors that are quite large and thereby produce interval estimates of $E(y)$ or prediction intervals for y that are wide and hence imprecise. Furthermore, when multicollinearity is present, the addition or deletion of a predictor variable causes significant changes in the value of the other regression coefficients.

What are some obvious indications that multicollinearity may be present, and therefore is distorting the results of a regression analysis? One easily recognized indication is a large value of R^2 and/or the presence of a highly significant F test when testing $H_0 : \beta_1 = \beta_2 = \cdots = \beta_k = 0$, together with nonsignificant t tests for individual regression coefficients. Another not so obvious indication is having signed values of the estimated regression coefficients that are contrary to the values dictated by theory or common sense. Alternatively, examining a matrix of correlations among the predictors $x_1, x_2, \ldots, x_k$ and the dependent variable y will allow you to determine which predictors are highly correlated with each other, as well as which predictors are most highly correlated with y.

The matrix of correlation coefficients for the independent variables square feet of living area x_1, the number of floors x_2, bedrooms x_3, bathrooms x_4, and listed selling price y for the data of Example 11.1 is given in Table 11.9. The elements in the diagonal positions are always equal to 1, and elements in symmetric locations with respect to the diagonal elements are always equal. The correlation between any two variables is found by cross-indexing the row corresponding to the first variable

TABLE 11.9
The matrix of correlation coefficients for the variables in Example 11.1

	x_1	x_2	x_3	x_4	y
x_1	1.00000	0.39351	0.69002	0.67142	0.89120
x_2	0.39351	1.00000	0.03924	0.71765	0.32092
x_3	0.69002	0.03924	1.00000	0.51601	0.67447
x_4	0.67142	0.71765	0.51601	1.00000	0.74095
y	0.89120	0.32092	0.67447	0.74095	1.00000

and the column corresponding to the second (or vice versa). For example, notice that the correlation between x_2 and x_4 is $r_{24} = .71765$; the next highest correlation among the predictor variables, between x_1 and x_3, is $r_{13} = .69002$, and the correlation between x_1 and x_4 is $r_{14} = .67142$, and so on. Therefore, x_1, x_2, x_3, and x_4 share a lot of information in common. Furthermore, we see that x_1 would be the best one-variable predictor of y since $r_{1y} = .89120$. Taken together, this information suggests the possible presence of multicollinearity in these data.

In examining the regression printout for Example 11.1 given in Table 11.2, we see that the signs of the coefficients of x_2 and x_3 are negative, although we would expect them to be positive. Also, t tests for individual βs are nonsignificant for both of these variables. In summary, we could conclude that some degree of multicolinearity may be distorting the regression results and that one or more of the predictor variables could be eliminated from the regression equation.

Since multicollinearity exists to some degree in most regression analyses, the individual terms in the model should be viewed as information contributors. The primary decision to be made is whether one or more of the terms contribute sufficient information for the prediction of y and whether they should be retained in the model.

11.8 Summary

This chapter provided an introduction to an extremely useful area of statistical methodology, one that enables us to relate the mean value of a response variable y to a set of predictor variables. We start by postulating a model that expresses $E(y)$ as the sum of a number of terms, each term involving the product of a single unknown parameter and a function of one or more of the predictor variables. Values of y are observed when the predictor variables assume specified values, and these data are used to estimate the unknown parameters in the model by using the method of least squares. The procedure for estimating the unknown parameters of the model, testing its utility, or testing hypotheses about sets of the model parameters is known as a multiple regression analysis.

Perhaps the most important result of a multiple regression analysis is the use of the fitted model—the prediction equation—to estimate the mean value of y or to predict a particular value of y for a given set of values of the predictor variables. Scientists, sociologist, engineers, and business managers all face the same problem: forecasting some quantitative response based on the values of a set of predictor variables that describe current or future conditions. When the appropriate assumptions are satisfied, a multiple regression analysis can be used to solve this important problem.

Supplementary Exercises

Starred (*) exercises are optional.

11.25 Utility companies, which must plan the operation and expansion of electricity generation, are vitally interested in predicting customer demand over both short and long periods of time. A short-term study was conducted to investigate the effect of mean monthly daily temperature x_1, and cost per kilowatt x_2 on the mean daily consumption (kilowatt-hours, kWh) per household. The company expected the demand for electricity to rise in cold weather (due to heating), fall when the weather was moderate, and rise again when the temperature rose and there was need for air-conditioning. They expected demand to decrease as the cost per kilowatt-hour increased, reflecting greater attention to conservation. Data were available for 2 years, a period in which the cost per kilowatt-hour x_2 was increased owing to the increasing cost of fuel. The company fitted the model

$$E(y) = \beta_0 + \beta_1 x_1 + \beta_2 x_1^2 + \beta_3 x_2 + \beta_4 x_1 x_2 + \beta_5 x_1^2 x_2$$

Price per kWh x_2	Daily Temperature and Consumption	Mean Daily Consumption (kWh) per Household											
8¢	Mean daily temperature (°F) x_1	31	34	39	42	47	56	62	66	68	71	75	78
	Mean daily consumption y	55	49	46	47	40	43	41	46	44	51	62	73
10¢	Mean daily temperature x_1	32	36	39	42	48	56	62	66	68	72	75	79
	Mean daily consumption y	50	44	42	42	38	40	39	44	40	44	50	55

to the data shown in the table above. The SAS multiple regression computer printout is shown in the table below.

```
DEPENDENT VARIABLE: Y
SOURCE            DF   SUM OF SQUARES    MEAN SQUARE   F VALUE   PR > F   R-SQUARE         C.V.
MODEL              5    1346.44751546   269.28950309     31.85   0.0001   0.898455       6.2029
ERROR             18     152.17748454     8.45430470   ROOT MSE                   Y MEAN
CORRECTED TOTAL   23    1498.62500000                2.90762871              46.87500000

SOURCE      DF      TYPE I SS   F VALUE   PR > F   DF    TYPE IV SS   F VALUE   PR > F
X1           1   140.71101071     16.64   0.0007    1  104.43456772     12.35   0.0025
X1*X1        1   892.77914933    105.60   0.0001    1  125.58263370     14.85   0.0012
X2           1   192.44266992     22.76   0.0002    1   46.78392523      5.53   0.0302
X1*X2        1    57.83521130      6.84   0.0175    1   50.03623552      5.92   0.0256
X1*X1*X2     1    62.67947420      7.41   0.0140    1   62.67947420      7.41   0.0140

                  PARAMETER ESTIMATES

                                  T FOR HO:
VARIABLE        ESTIMATE     PARAMETER = 0   PR > |T|    STD ERROR
INTERCEPT   325.60644505              3.92     0.0010  83.06412820
X1          -11.38255606             -3.51     0.0025   3.23859476
X1*X1         0.11319735              3.85     0.0012   0.02944828
X2          -21.69920900             -2.35     0.0302   9.22432344
X1*X2         0.87302921              2.43     0.0256   0.35886030
X1*X1*X2     -0.00886945             -2.72     0.0140   0.00325742
```

a Find SSE and s^2 on the printout.

b How many degrees of freedom do SSE and s^2 possess? Show that s^2 =SSE/(degrees of freedom).

c Do the data provide sufficient evidence to indicate that the model contributes information for the prediction of mean daily kilowatt-hour consumption per household? Test using $\alpha = .10$.

d Use Table 6 in Appendix I to find the approximate observed significance level for the test in part (c).

e Find the Total SS, S_{yy}, on the printout. Find the value of R^2 and interpret its value. Show that $R^2 = 1 - (\text{SSE}/S_{yy})$.

f Use the prediction equation to predict mean daily consumption when the mean daily temperature is 60° F and the price per kilowatt-hour is 8¢.

g If you have access to a computer, use it to perform a multiple regression analysis for the data in Exercise 11.25. Compare your computer output with the SAS output shown in the accompanying table.

```
DEPENDENT VARIABLE: Y
SOURCE            DF    SUM OF SQUARES   MEAN SQUARE   F VALUE        PR > F   R-SQUARE          C.V.
MODEL              2     1033.49016004  516.74908002     23.33        0.0001    0.68926       10.0401
ERROR             21      465.13483996   22.14927809                ROOT MSE                   Y MEAN
CORRECTED TOTAL   23     1498.62500000                            4.70630196              46.87500000

SOURCE            DF         TYPE I SS    F VALUE   PR > F  DF      TYPE IV SS  F VALUE   PR > F
X1                 1      140.71101071       6.35   0.0199   1    810.33211666    36.59   0.0001
X1*X1              1      892.77914933      40.31   0.0001   1    892.77914933    40.31   0.0001

                        PARAMETER ESTIMATE

                            T FOR HO:
VARIABLE        ESTIMATE  PARAMETER = 0    PR >|T|      STD ERROR
INTERCEPT   130.00929147           8.74     0.0001    14.87580615
X1           -3.50171859          -6.05     0.0001     0.57893460
X1*X1         0.03337133           6.35     0.0001     0.00525631
```

11.26* Refer to Exercise 11.25.

a Graph the curve depicting $\hat{y}$ as a function of temperature x_1 when the cost per kilowatt-hour is $x_2 = 8¢$. Construct a similar graph for the case when $x_2 = 10¢$ per kilowatt-hour. Does it appear that the consumption curves differ?

b If cost per kilowatt-hour is unimportant in predicting use, then we do not need the terms involving x_2 in the model. Therefore, the null hypothesis H_0 : "x_2 does not contribute information for the prediction of y" is equivalent to the hypothesis $H_0 : \beta_3 = \beta_4 = \beta_5 = 0$ (if $\beta_3 = \beta_4 = \beta_5 = 0$, the terms involving x_2 disappear from the model). This hypothesis is tested by using the procedure of Section 11.5. The SAS multiple regression computer printout, obtained by fitting the reduced model

$$E(y) = \beta_0 + \beta_1 x_1 + \beta_2 x_1^2$$

to the data, is shown in the table in Exercise 11.25. The following steps enable you to conduct a test to determine whether the data provide sufficient evidence to indicate that price per kilowatt-hour x_2 contributes information for the prediction of y. Test using $\alpha = .05$.

i Find SSE_1 and SSE_2.

ii Give the values for v_1 and v_2.

iii Calculate the value of F.

iv Find the tabulated value of $F_{.05}$.

v State your conclusions.

11.27 Refer to Example 11.5. The t value for testing the hypothesis $H_0 : \beta_1 = 0$ is

$$t = \frac{\hat{\beta} - 0}{s_{\hat{\beta}_1}} = \frac{\hat{\beta}}{s_{\hat{\beta}_1}}$$

where $\hat{\beta}_1$ is the estimate of β_1 and $s_{\hat{\beta}_1}$ is the estimated standard deviation (or standard error) of $\hat{\beta}_1$. Both of these quantities are shown on the SAS, MINITAB, and SPSS^x printouts in Tables 11.6 and 11.7. Examine the printouts and verify that the corresponding printed values for $\hat{\beta}_1$ and $s_{\hat{\beta}_1}$ are identical except for rounding errors. Calculate $t = \hat{\beta}_1/s_{\hat{\beta}_1}$. Verify that your calculated value of t is equal to the value printed on the printouts; that is, $t = 15.22$.

11.28 The t test of Exercise 11.27 can also be conducted by using an F test (the test is described in Section 11.5). We can use an F test because the square of a t statistic with v degrees of freedom is equal to an F statistic with $v_1 = 1$ and $v_2 = v$ degrees of freedom; that is,

$$F_{1,v} = t_v^2$$

a To convince yourself that this relationship is valid, find the value $t_{.025}$ (Table 4 in Appendix I) that is used to locate the rejection regions for the two-tailed ($\alpha = .05$)t tests of Exercise 11.27.

b Show that $t^2_{.025}$ [found in part (a)] is equal to the tabulated value $F_{.05}$ (Table 6 in Appendix I) for $v_1 = 1$ and $v_2 = 8$ d.f.

11.29 When will an F statistic equal the square of a t statistic?

11.30 A department store conducted an experiment to investigate the effects of advertising expenditures on the weekly sales for its men's wear, children's wear, and women's wear departments. Five weeks for observation were randomly selected from each department, and an advertising budget x_1 (hundreds of dollars) was assigned for each. The weekly sales (thousands of dollars) are shown in the accompanying table for each of the 15 1-week sales periods. If we expect weekly sales $E(y)$ to be linearly related to advertising expenditure x_1 and if we expect the slopes of the lines corresponding to the three departments to differ, then an appropriate model for $E(y)$ is

$$E(y) = \beta_0 + \underbrace{\beta_1 x_1}_{\substack{\text{quantitative}\\ \text{variable}\\ \text{“advertising}\\ \text{expenditure”}}} + \underbrace{\beta_2 x_2 + \beta_3 x_3}_{\substack{\text{dummy variables}\\ \text{used to introduce}\\ \text{the qualitative}\\ \text{variable “department”}\\ \text{into the model}}} + \underbrace{\beta_4 x_1 x_2 + \beta_5 x_1 x_3}_{\substack{\text{interaction terms that introduce}\\ \text{differences in slopes}}}$$

where

$$x_1 = \text{advertising expenditure}$$

$$x_2 = \begin{cases} 1, & \text{if children's wear department B} \\ 0, & \text{if not} \end{cases}$$

$$x_3 = \begin{cases} 1, & \text{if women's wear department C} \\ 0, & \text{if not} \end{cases}$$

Department	Advertising Expenditure (hundreds of dollars)				
	1	**2**	**3**	**4**	**5**
Men's wear A	5.2	5.9	7.7	7.9	9.4
Children's wear B	8.2	9.0	9.1	10.5	10.5
Women's wear C	10.0	10.3	12.1	12.7	13.6

a Find the equation of the line relating $E(y)$ to advertising expenditure x_1 for the men's wear department A. [HINT: According to the coding used for the dummy variables, the model represents mean sales $E(y)$ for the men's wear department A when $x_2 = x_3 = 0$. Substitute $x_2 = x_3 = 0$ into the equation for $E(y)$ to find the equation of this line.]

b Find the equation of the line relating $E(y)$ to x_1 for the children's wear department B. [HINT: According to the coding, the model represents $E(y)$ for the children's wear department when $x_2 = 1$ and $x_3 = 0$.]

c Find the equation of the line relating $E(y)$ to x_1 for the women's wear department C.

d Find the difference between the intercepts of the $E(y)$ lines corresponding to the children's wear B and men's wear A departments.

e Find the difference in slopes between $E(y)$ lines corresponding to the women's wear C and men's wear A departments.

f Refer to part (e). Suppose that you want to test the null hypothesis that the slopes of the lines corresponding to the three departments are equal. Express this as a test of an hypothesis about one or more of the model parameters.

11.31* If you have access to a computer and a multiple regression computer program package, perform a multiple regression analysis for the data of Exercise 11.30.

a Verify that SSE = 1.2190, that $s^2 = .1354$, and that $s = .368$.

b How many degrees of freedom will SSE and s^2 possess?

c Verify that Total SS = $S_{yy} = 76.0493$.

d Calculate R^2 and interpret its value.

e Verify that the parameter estimates (values rounded) are approximately equal to

$$\hat{\beta}_0 \approx 4.10 \qquad \hat{\beta}_1 \approx 1.04 \qquad \hat{\beta}_2 \approx 3.53$$
$$\hat{\beta}_3 \approx 4.76 \qquad \hat{\beta}_4 \approx -.43 \qquad \hat{\beta}_5 \approx -.08$$

f Find the prediction equation, and graph the three department sales lines.

g Examine the graphs in part (f). Do the slopes of the lines corresponding to the children's wear B and men's wear A departments appear to differ? Test the null hypothesis that the slopes do not differ (i.e., $H_0 : \beta_4 = 0$) against the alternative hypothesis that the slopes do differ (i.e., $H_a : \beta_4 \neq 0$). Test using $\alpha = .05$. (NOTE: Verify on your computer printout that the estimated standard deviation of $\hat{\beta}_4$ is $s_{\hat{\beta}_4} \approx .165$.)

h Find a 95% confidence interval for the difference between the slopes of lines for B and A.

i Do the data provide sufficient evidence to indicate a difference in the slopes between lines C and A? Test using $\alpha = .05$. (HINT: $s_{\hat{\beta}_5} \approx .165$.)

11.32 Groups of 10-day-old chicks were randomly assigned to seven treatment groups in which a basal diet was supplemented with 0, 50, 100, 200, 250, or 300 μg/kg d-biotin. The following data give the average biotin intake (x) in micrograms per day (μg/d) and the average weight gain (y) in grams per day (g/d):

Added Biotin	Biotin Intake x	Weight Gain y
0	0.14	8.0
50	2.01	17.1
100	6.06	22.3
150	6.34	24.4
200	7.15	26.5
250	9.65	23.4
300	12.50	23.3

Source: Blair and Miser (1989), pp. 55–58.

In the MINITAB printout that follows, the second-order polynomial model

$$E(y) = \beta_0 + \beta_1 x + \beta_2 x^2$$

was fitted to the data. Use the printout to answer the questions that follow.

```
MTB > REGRESS C3 2 C1 C2

THE REGRESSION EQUATION IS
Y = 8.59 + 3.82 X - 0.217 X*X

PREDICTOR      COEF      STDEV    T-RATIO       P
CONSTANT      8.585      1.641       5.23   0.006
X            3.8208     0.5683       6.72   0.003
X*X        -0.21663    0.04390      -4.93   0.008

S = 1.833     R-SQ = 94.4%     R-SQ(ADJ) = 91.5%
```

```
ANALYSIS OF VARIANCE

SOURCE         DF        SS         MS        F        P
REGRESSION      2    224.75     112.37    33.44    0.003
ERROR           4     13.44       3.36
TOTAL           6    238.19

SOURCE         DF    SEQ SS
X               1    142.92
X*X             1     81.83

MTB >
```

a What is the fitted least-squares line?

b Find R^2 and interpret its value.

c Do the data provide sufficient evidence to conclude that the model contributes significant information for predicting y?

d Find the results of the test of $H_0: \beta_2 = 0$. Is there sufficient evidence to indicate that the quadratic model provides a better fit to the data than a simple linear model does?

11.33 The waiting time y that elapses between the time a computing job is submitted to a large computer and the time at which the job is initiated (computing commences) is a function of many variables, including the priority assigned to the job, the number and sizes of the jobs already on the computer, the size of the job being submitted, and so on. A study was initiated to investigate the relationship between waiting time y (in hours) for a job and x_1, the estimated CPU time (in seconds) for the job, and x_2, the CPU utilization factor. The estimated CPU time x_1 is an estimate of the amount of time that a job will occupy a portion of the computer's central processing unit's memory. The CPU utilization factor x_2 is the percentage of the memory bank of the central processing unit that is occupied at the time the job is submitted. We would expect the waiting time y to increase as the size of the job x_1 increases and as the CPU utilization factor x_2 increases. To conduct the study, 15 jobs of varying sizes were submitted to the computer at randomly assigned times throughout the day. The job waiting time y, estimated CPU time x_1, and CPU utilization factor x_2 were recorded for each job. The data[†] are shown below.

	Job 1	2	3	4	5	6	7	8	9	10	11	12	13	14	15
x_1	2.0	9.3	5.6	3.7	12.4	18.1	13.5	26.6	34.2	38.8	56.1	60.3	4.4	2.6	20.9
x_2	45	80	23	25	67	30	55	21	79	40	22	37	50	66	42
y	0.001	1.140	0.030	0.001	0.780	0.300	0.600	0.200	2.240	0.440	0.001	0.320	0.160	0.290	0.490

A second-order model, $E(y) = \beta_0 + \beta_1 x_1 + \beta_2 x_2 + \beta_3 x_1 x_2 + \beta_4 x_1^2 + \beta_5 x_2^2$, was selected to model mean waiting time $E(y)$. The SAS multiple regression analysis for the data is shown in the following printout:

[†]Waiting time data frequently violate the assumptions required for significance tests and confidence intervals in a regression analysis. The probability distribution for waiting times is often skewed, and its variance increases as the mean waiting time increases. Methods are available for coping with this problem, but we will ignore it for the purposes of this introductory discussion.

```
DEPENDENT VARIABLE: Y

SOURCE              DF     SUM OF SQUARES    MEAN SQUARE      F VALUE      PR>F   R-SQUARE       C.V.
MODEL                5         4.74848837     0.94969767       159.23    0.0001   0.988822    16.5655
ERROR                9         0.05367803     0.00596423                ROOT MSE                Y MEAN
CORRECTED TOTAL 14             4.80216640                            0.077722840            0.46620000

SOURCE              DF        TYPE I SS        F VALUE     PR > F  DF   TYPE III SS    F VALUE    PR> F
X1                   1       0.08032055          13.47     0.0052   1    0.00448637       0.75   0.4083
X2                   1       3.21553051         539.14     0.0001   1    0.13905895      23.32   0.0009
X1*X2                1       0.96272106         161.42     0.0001   1    0.48988534      82.14   0.0001
X1*X1                1       0.22455767          37.65     0.0002   1    0.16181969      27.13   0.0006
X2*X2                1       0.26535859          44.49     0.0001   1    0.26535859      44.49   0.0001

                            T FOR HO:          PR > ITI     STD ERROR OF
PARAMETER     ESTIMATE     PARAMETER = 0                     ESTIMATE
INTERCEPT   0.43806816              2.71         0.0239     0.16147813
X1          0.00526474              0.87         0.4083     0.00607025
X2         -0.03017242             -4.83         0.0009     0.00624867
X1*X2       0.00068770              9.06         0.0001     0.00007588
X1*X1      -0.00037952             -5.21         0.0006     0.00007286
X2*X2       0.00040726              6.67         0.0001     0.00006106
```

a Find the values of SSE and s^2.

b Find the prediction equation.

c Find R^2 and interpret its value.

d Do the data provide sufficient evidence to indicate that the model contributes information for the prediction of y? Test using $\alpha = .10$.

11.34 Nancy Kay Butts (1982) conducted a study to investigate the relationship between the physical characteristics of female cross-country runners and their completion time y for a 3000-meter race. Measurements were taken on 14 physical characteristics for each of 127 runners—age, height, weight, percentage fat, length of time on a treadmill, VO_2 max, maximal ventilation volume (VE max), and so on—along with the runner's completion time y for a 3000-meter race. Butts employed a stepwise regression analysis to develop a prediction equation. Variables entering the first-order stepwise regression model, in order of entry, were

$$x_1 = \text{minutes on treadmill}$$
$$x_2 = VO_2 \text{ max}$$
$$x_3 = \text{VE max}$$

The multiple correlation coefficient R values associated with these one-, two-, and three-variable models were 0.58, 0.66, and 0.67, respectively.

a How many degrees of freedom would be associated with SSE for this three-variable model?

b Does it appear that the three-variable stepwise regression model

$$E(y) = \beta_0 + \beta_1 x_1 + \beta_2 x_2 + \beta_3 x_3$$

provides a good fit to the data?

c Suppose that you want to determine whether the complete three-variable model contributes information for the prediction of completion time y. Calculate the value of the F statistic required for the test.

d Conduct the test in part (c) and state your conclusions. Use $\alpha = .05$.

11.35 Because dolphins (and other large marine mammals) are considered to be the top predators in the marine food chain, the heavy metal concentrations in striped dolphins were measured as part of a marine pollution study. The concentration of mercury, the heavy metal reported in this study, is expected to differ for males and females, since the mercury in a female is apparently transferred to her offspring during gestation and nursing. This study involved 28 males between the ages of 0.21 and 39.5 years, and 17 females between the ages of 0.80 and 34.5 years. For the data that follow,

$$x_1 = \text{age of the dolphin in years}$$

$$x_2 = \begin{cases} 0, & \text{if female} \\ 1, & \text{if male} \end{cases}$$

$$y = \text{mercury concentration} (\mu g/g) \text{ in the liver}$$

y	x_1	x_2	y	x_1	x_2
1.70	0.21	1	397.00	31.50	1
1.72	0.33	1	209.00	36.50	1
8.80	2.00	1	314.00	37.50	1
5.90	2.20	1	318.00	39.50	1
101.00	8.50	1	2.50	0.80	0
85.40	11.50	1	9.35	1.58	0
118.00	11.50	1	4.01	1.75	0
183.00	13.50	1	29.80	5.50	0
168.00	16.50	1	45.30	7.50	0
218.00	16.50	1	101.00	8.05	0
180.00	17.50	1	135.00	11.50	0
264.00	20.50	1	142.00	17.50	0
481.00	22.50	1	180.00	17.50	0
485.00	24.50	1	174.00	18.50	0
221.00	24.50	1	247.00	19.50	0
406.00	25.50	1	223.00	21.50	0
252.00	26.50	1	167.00	21.50	0
329.00	26.50	1	157.00	25.50	0
316.00	26.50	1	177.00	25.50	0
445.00	26.50	1	475.00	32.50	0
278.00	27.50	1	342.00	34.50	0
286.00	28.50	1			
315.00	29.50	1			
241.00	31.50	1			

```
MTB > REGRESS C1 4 C2-C5

THE REGRESSION EQUATION IS
Y = -17.7 + 10.7 X1 - 28.7 X2 + 12.7 X1*X2 - 0.371 X1SQ*X2

PREDICTOR      COEF       STDEV    T-RATIO        P
CONSTANT     -17.65       32.02      -0.55     0.585
X1           10.737       1.690       6.35     0.000
X2           -28.69       48.84      -0.59     0.560
X1*X2        12.722       4.440       2.87     0.007
X1SQ*X2     -0.3707      0.1053      -3.52     0.001

S = 71.29        R-SQ = 76.5%     R-SQ(ADJ) = 74.1%
```

```
ANALYSIS OF VARIANCE

SOURCE          DF        SS        MS        F        P
REGRESSION       4    661313    165328    32.53    0.000
ERROR           40    203285      5082
TOTAL           44    864598

SOURCE          DF    SEQ SS
X1               1    585782
X2               1     11185
X1*X2            1      1371
X1SQ*X2          1     62974

UNUSUAL OBSERVATIONS
OBS.      X1        Y       FIT    STDEV.FIT    RESIDUAL    ST.RESID
 13     22.5    481.0     293.8         18.1       187.2      2.71R
 14     24.5    485.0     305.9         17.2       179.1      2.59R
 28     39.5    318.0     302.0         41.8        16.0      0.28X
 44     32.5    475.0     331.3         32.9       143.7      2.27R

R DENOTES AN OBS. WITH A LARGE ST.RESID.
X DENOTES AN OBS. WHOSE X VALUE GIVES IT LARGE INFLUENCE.
```

Use the MINITAB printout to answer the following questions:

a What is the model that has been fitted?

b What is the fitted prediction equation?

c Find R^2 and interpret its value.

d Do the data provide sufficient evidence for us to conclude that the model contributes significant information for predicting y?

e What is the prediction equation for predicting the mercury concentration in a female dolphin as a function of her age?

f What is the prediction equation for predicting the mercury concentration in a male dolphin as a function of his age?

g Does the quadratic term in the prediction equation for males contribute significantly to the prediction of mercury concentration in a male dolphin?

h Are there any other important conclusions that you feel were not considered regarding the fitted prediction equation?

CASE STUDY

"Made in the U.S.A."—Another Look

The case study in Chapter 10 examined the effect of foreign competition in the automotive industry whereby the number of imported cars steadily increased during the 1970s and 1980s (*Automotive News*, 1992). The U.S. automobile industry has been besieged with complaints about product quality, worker layoffs, and high prices and has spent billions in advertising and research to produce an American-made car that will satisfy consumer demands. Have they been successful in stopping the flood of imported cars being purchased by American consumers? The data shown below represents the number of imported cars (y) sold in the United States (in millions) for the years 1969–1991. To simplify the analysis, we have coded the year using the coded variable $x =$ Year—1969.

Year	Year—1969 x	Number of Imported Cars y	Year	Year—1969 x	Number of Imported Cars y
1969	0	1.1	1981	12	2.3
1970	1	1.3	1982	13	2.2
1971	2	1.6	1983	14	2.4
1972	3	1.6	1984	15	2.4
1973	4	1.8	1985	16	2.8
1974	5	1.4	1986	17	3.2
1975	6	1.6	1987	18	3.1
1976	7	1.5	1988	19	3.1
1977	8	2.1	1989	20	2.8
1978	9	2.0	1990	21	2.5
1979	10	2.3	1991	22	2.1
1980	11	2.4			

By examining a scatterplot of these data, you will find that the number of imported cars does not appear to follow a linear relationship over time, but rather exhibits a curvilinear response. The question, then, is to decide whether a second-, third-, or higher-order model adequately describes the data. The results of fitting a linear, a quadratic, and a cubic regression analysis using the EXECUSTAT package for all $n = 23$ data points are given in Tables 11.10–11.12.

1 Plot the data and sketch what you would consider the best-fitting linear, quadratic, and cubic model.

2 Find the residuals using the fitted linear regression model. Does there appear to be any pattern in the residuals when plotted against x? What model do the residuals indicate would produce a better fit?

3 What is the increase in R^2 when fitting a quadratic rather than a linear model? Is the coefficient of the quadratic term significant? Is the fitted quadratic model significantly better than the fitted linear model? Plot the residuals from the fitted quadratic model. Does there seem to be any apparent pattern in the residuals when plotted against x?

4 What is the increase in R^2 when comparing the fitted cubic with the fitted quadratic model? Is the fitted cubic model significantly better than the fitted quadratic? Are there any patterns in a plot of the residuals versus x? What proportion of the variation in the response y is not accounted for by fitting a cubic model? Should any higher-order polynomial model be considered? Why, or why not?

TABLE 11.10
EXECUSTAT printout for a simple linear regression

Simple Regression Analysis for CASECH10

Linear model: IMPORTS = 1.32391 + 0.0756917*YEAR

Table of Estimates

	Estimate	Standard Error	t Value	P Value
Intercept	1.32391	0.130054	10.18	0.0000
Slope	0.0756917	0.0101247	7.48	0.0000

R-squared = 72.69%
Correlation coef. = 0.853
Standard error of estimation = 0.322087
Durbin-Watson statistic = 0.713625
Mean absolute error = 0.232205
Sample size (n) = 23

TABLE 11.11
EXECUSTAT printout for a second-order (quadratic) model

Polynomial Regression Analysis for CASECH10

Dependent variable: IMPORTS

Table of Estimates

	Estimates	Standard Error	t Value	P Value
Intercept	1.06739	0.171751	6.21	0.0000
YEAR	0.148984	0.0361672	4.12	0.0005
YEAR^2	-0.00333145	0.00158757	-2.10	0.0488

R-squared = 77.62%
Adjusted R-squared = 75.38%
Standard error of estimation = 0.298783
Durbin-Watson statistic = 0.797233
Mean absolute error = 0.224756
Sample size (n) = 23
Note: 0 incomplete cases have been excluded.

TABLE 11.12
EXECUSTAT printout for a third-order (cubic) model

Polynomial Regression Analysis for CASECH10

Dependent variable: IMPORTS

Table of Estimates

	Estimates	Standard Error	t Value	P Value
Intercept	1.38953	0.180325	7.71	0.0000
YEAR	-0.0490421	0.0725962	-0.68	0.5075
YEAR^2	0.0196786	0.00776906	2.53	0.0203
YEAR^3	-0.000697274	0.000231894	-3.01	0.0073

R-squared = 84.83%
Adjusted R-squared = 82.44%
Standard error of estimation = 0.252332
Durbin-Watson statistic = 1.05084
Mean absolute error = 0.199808
Sample size (n) = 23
Note: 0 incomplete cases have been excluded.

12

Analysis of Enumerative Data

Case Study

How do you rate your library? Is the atmosphere friendly, dull, or too quiet? Are the library staff helpful? Are the signposts clear and unambiguous? The modern consumer-led approach to marketing, in general, centers around the systematic study by organizations of their customers' wants and needs in order to improve their services or products. In the case study at the end of this chapter, we examine the results of a study to explore the attitudes of young adults toward the services provided by libraries.

General Objectives

Many types of surveys and experiments result in qualitative rather than quantitative response variables, so that the responses can be classified but not quantified. Data from these experiments consist of the count or number of observations falling in each of the response categories included in the experiment. In this chapter, we are concerned with methods of analysis for count (or enumerative) data.

Specific Topics

1 The multinomial experiment (12.1)
2 A test of specific cell probabilities (12.3)
3 Contingency tables (12.4)
4 A test for homogeneity of multinomial populations (12.5)
5 Other applications (12.6)
6 Assumptions (12.7)

12.1 A Description of the Experiment

Many experiments, particularly in the social sciences, result in enumerative (or count) data. For instance, the classification of people into five income brackets results in an enumeration or count corresponding to each of the five income classes. Or we might be interested in studying the reaction of a mouse to a particular stimulus in a psychological experiment. If the mouse were to react in one of three ways when the stimulus was applied and if a large number of mice were subjected to the stimulus, the experiment would yield three counts, indicating the number of mice falling in each of the reaction classes. Similarly, a traffic study might require a count and classification of the type of motor vehicles using a section of highway. An industrial process manufactures items that fall into one of three quality classes: acceptable, seconds, and rejects. A student of the arts might classify paintings in one of k categories according to style and period in order to study trends in style over time. We might wish to classify ideas in a philosophical study or style in the field of literature. The results of an advertising campaign would yield count data indicating a classification of consumer reaction. Indeed, many observations in the physical sciences are not amenable to measurement on a continuous scale and hence result in enumerative or classificatory data.

The illustrations in the preceding paragraph exhibit, to a reasonable degree of approximation, the following characteristics that define a **multinomial experiment**.

The Characteristics of a Multinomial Experiment

1. The experiment consists of n identical trials.
2. The outcome of each trial falls into one of k classes or cells.
3. The probability that the outcome of a single trial will fall in a particular cell—say, cell i— is $p_i (i = 1, 2, \dots, k)$ and remains the same from trial to trial. Note that $0 \leq p_i \leq 1$ for all i, and

$$p_1 + p_2 + p_3 + \cdots + p_k = 1$$

4. The trials are independent.
5. The experimenter is interested in $n_1, n_2, n_3, \dots, n_k$, where $n_i (i = 1, 2, \dots, k)$ is equal to the number of trials in which the outcome falls in cell i. Note that $n_1 + n_2 + n_3 + \cdots + n_k = n$.

A multinomial experiment is analogous to tossing n balls at k boxes, where each ball must fall into one of the boxes. The boxes are arranged so that the probability that a ball will fall into a box varies from box to box but remains the same for a particular box in repeated tosses. Finally, the balls are tossed in such a way that the trials are independent. At the conclusion of the experiment, we observe n_1 balls in the first box, n_2 in the second, ..., and n_k in the kth. The total number of balls is equal to

$$\sum_{i=1}^{k} n_i = n$$

Note the similarity between the binomial and multinomial experiments and, in particular, note that the binomial experiment represents the special case of the multinomial experiment when $k = 2$. The two cell probabilities p and q of the binomial experiment are replaced by the k cell probabilities, $p_1, p_2, \dots, p_k$ of the multinomial experiment. The objective of this chapter is to make inferences about the cell probabilities $p_1, p_2, \dots, p_k$. The inferences will be expressed in terms of a statistical test of hypothesis concerning their specific numerical values or their relationship to one another.

If we were to proceed as in Chapter 4, we would derive the probability of the observed sample $(n_1, n_2, \ldots, n_k)$ for use in calculating the probability of the type I and type II errors associated with a statistical test. Fortunately, we have been relieved of this chore by the British statistician Karl Pearson, who proposed a very useful test statistic for testing hypotheses concerning $p_1, p_2, \ldots, p_k$ and gave its approximate sampling distribution.

12.2 The Chi-Square Test

Suppose that $n = 100$ balls were tossed at the cells (boxes) and that we knew that p_1 was equal to .1. How many balls would be expected to fall into the first cell? Referring to Chapter 4 and utilizing knowledge of the binomial experiment, we would calculate

$$E(n_1) = np_1 = 100(.1) = 10$$

In like manner, the expected number falling into the remaining cells can be calculated using the formula

$$E(n_i) = np_i, \qquad i = 1, 2, \ldots, k$$

Now suppose that we hypothesize values for $p_1, p_2, \ldots, p_k$ and calculate the expected value for each cell. Certainly, if our hypothesis is true, the cell counts n_i should not deviate greatly from their expected values $np_i (i = 1, 2, \ldots, k)$. Hence, it would seem intuitively reasonable to use a test statistic involving the k deviations,

$$(n_i - np_i), \qquad i = 1, 2, \ldots, k$$

In 1900 Karl Pearson proposed the following test statistic, which is a function of the squares of the deviations of the observed counts from their expected values, weighted by the reciprocals of their expected values.

Chi-Square Test Statistic

$$X^2 = \sum_{i=1}^{k} \frac{[n_i - E(n_i)]^2}{E(n_i)} = \sum_{i=1}^{k} \frac{[n_i - np_i]^2}{np_i}$$

Although the mathematical proof is beyond the scope of this text, it can be shown that when n is large, X^2 has, approximately, a chi-square probability distribution in repeated sampling. Experience has shown that the cell counts n_i should not be too small in order that the chi-square distribution provide an adequate approximation to the distribution of X^2. As a rule of thumb, **we will require that all expected cell counts equal or exceed 5**, although Cochran has noted that this value can be as low as 1 for some situations.

Recall from Section 9.6 that the chi-square probability distribution is used for testing an hypothesis concerning a population variance σ^2 and that the shape of the chi-square distribution varies according to the number of degrees of freedom associated with s^2. We discussed the use of Table 5 in Appendix I, which presents the critical values of χ^2 corresponding to various right-hand tail areas of the distribution. Therefore, we must know which chi-square distribution to use in approximating the distribution of X^2. That is, we must determine the number of degrees of freedom associated with the appropriate chi-square distribution. Furthermore, we must determine whether the rejection region for the test is one-tailed or two-tailed.

The latter problem is easily solved. Since large deviations of the observed cell counts from those expected would tend to contradict the null hypothesis concerning the cell probabilities $p_1, p_2, \ldots, p_k$, we reject the null hypothesis when X^2 is large and employ a one-tailed statistical test using the upper-tail values of χ^2 to locate the rejection region.

The determination of the appropriate number of degrees of freedom for the test can be rather difficult and therefore will be specified for the practical applications described in the following sections. In addition, we will state the principle involved (which is fundamental to the mathematical proof of the approximation) so that you may understand why the number of degrees of freedom changes with various applications. The principle states that **the appropriate number of degrees of freedom (d.f.) equals the number of cells k less 1 d.f. for each independent linear restriction placed on the observed cell counts**. For example, one linear restriction is always present because the sum of the cell counts must equal n; that is,

$$n_1 + n_2 + n_3 + \cdots + n_k = n$$

Other restrictions will be introduced for some applications because of the necessity for estimating unknown parameters required in the calculation of the expected cell frequencies or because of the method in which the sample is collected. These restrictions will become apparent as we consider various practical examples.

12.3 A Test of Hypothesis Concerning Specified Cell Probabilities

The simplest hypothesis concerning the cell probabilities is one that specifies numerical values for each cell. For example, suppose that a psychologist performs the following experiment. One or more rats are attracted to the end of a ramp that divides, leading to three different doors. The objective of the experiment is to determine whether the rats have or acquire a preference for one of the three doors. We wish to test the hypothesis that the rats have no preference in the choice of a door, so that over the long run, we would expect the rats to choose each door approximately one-third of the time. Therefore,

$$H_0 : p_1 = p_2 = p_3 = \frac{1}{3}$$

versus

$$H_a : \text{ at least one } p_i \text{ is different from } \frac{1}{3}$$

where p_i is the probability that a rat will choose door i, $i = 1, 2$, or 3.

Suppose that the rat was sent down the ramp $n = 90$ times and that the three observed cell frequencies were $n_1 = 23$, $n_2 = 36$, and $n_3 = 31$. The expected cell frequencies are the same for each cell; namely, $E(n_i) = np_i = 90(1/3) = 30$.

The observed and expected cell frequencies are presented in Table 12.1. Noting the discrepancy between the observed and expected cell frequencies, we would wonder whether the data present sufficient evidence to warrant rejection of the hypothesis of no preference.

TABLE **12.1**
Observed and expected cell counts for the rat experiment

	Door		
	1	2	3
Observed cell frequency	$n_1 = 23$	$n_2 = 36$	$n_3 = 31$
Expected cell frequency	30	30	30

The chi-square test statistic for this example has $(k - 1) = 2$ d.f. since the only linear restriction on the cell frequencies is that

$$n_1 + n_2 + \cdots + n_k = n$$

or, for our example,

$$n_1 + n_2 + n_3 = 90$$

Therefore, if we choose $\alpha = .05$, we would reject the null hypothesis when $X^2 > 5.99147$ (see Table 5 in Appendix I).

Substituting into the formula for X^2, we obtain

$$X^2 = \sum_{i=1}^{k} \frac{[n_i - E(n_i)]^2}{E(n_i)} = \sum_{i=1}^{k} \frac{[n_i - np_i]^2}{np_i}$$

$$= \frac{(23 - 30)^2}{30} + \frac{(36 - 30)^2}{30} + \frac{(31 - 30)^2}{30} = 2.87$$

Since X^2 is less than the tabulated critical value of χ^2, 5.991, the null hypothesis is not rejected and we conclude that the data do not present sufficient evidence to indicate that the rats have a preference for a particular door.

Exercises

Basic Techniques

12.1 List the characteristics of a multinomial experiment.

12.2 Give the value of χ^2_α for

a $\alpha = .05$, d.f. $= 3$.
b $\alpha = .01$, d.f. $= 8$.
c $\alpha = .10$, d.f. $= 15$.
d $\alpha = .10$, d.f. $= 11$.

12.3 Give the rejection region for a chi-square test of specified cell probabilities if the experiment involves k cells, where

a $k = 7, \alpha = .10$.
b $k = 10, \alpha = .01$.
c $k = 14, \alpha = .05$.
d $k = 3, \alpha = .05$.

12.4 Suppose that a response can fall in one of $k = 5$ categories with probabilities $p_1, p_2, \ldots, p_5$, respectively, and that $n = 300$ responses produced the following category counts:

Category	1	2	3	4	5
Observed count	47	63	74	51	65

a If you were to test $H_0 : p_1 = p_2 = \cdots = p_5$ using the chi-square test, how many degrees of freedom would the test statistic have?

b If $\alpha = .05$, find the rejection region for the test.

c What is your alternative hypothesis?

d Conduct the test in part (a) using $\alpha = .05$. State your conclusions.

e Find the approximate observed significance level for the test and interpret its value.

12.5 Suppose that a response can fall in one of $k = 3$ categories with probabilities p_1, p_2, and p_3, respectively, and that $n = 300$ responses produced the following category counts:

Category	1	2	3
Observed count	130	98	72

a If you were to test $H_0 : p_1 = p_2 = p_3$ using the chi-square test, how many degrees of freedom would the test statistic have?

b If $\alpha = .05$, find the rejection region for the test.

c What is your alternative hypothesis?

d Conduct the test in part (a) using $\alpha = .05$. State your conclusions.

e Find the approximate observed significance level for the test and interpret its value.

Applications

12.6 A city expressway utilizing four lanes in each direction was studied to see whether drivers preferred to drive on the inside lanes. A total of 1000 automobiles were observed during the heavy early-morning traffic, and their respective lanes were recorded. The results were as follows:

Lane	1	2	3	4
Observed count	294	276	238	192

Do the data present sufficient evidence to indicate that some lanes are preferred over others? (Test the hypothesis that $p_1 = p_2 = p_3 = p_4 = 1/4$, using $\alpha = .05$.)

12.7 The Mendelian theory states that the number of peas of a certain type falling into the classifications round and yellow, wrinkled and yellow, round and green, and wrinkled and green should be in the ratio 9:3:3:1. Suppose that 100 such peas revealed 56, 19, 17, and 8 in the respective classes. Do these data disagree with the Mendelian theory? Use $\alpha = .05$.

12.8 Do you hate Mondays? Researchers from Germany provided another reason for you: They concluded that the risk of a heart attack for a working person may be as much as 50% greater than on any other day (Haney, 1992). The researchers kept track of heart attacks and coronary arrests over a period of 5 years among 330,000 people who lived near Augsburg, Germany. In an attempt to verify their claim, 200 working people who had recently had heart attacks were surveyed, and the day on which their heart attacks occurred was recorded. The results are shown below:

Sunday	Monday	Tuesday	Wednesday	Thursday	Friday	Saturday
24	36	27	26	32	26	29

Do the data present sufficient evidence to indicate that there is a difference in the incidence of heart attacks depending on the day of the week? Test using $\alpha = .05$.

12.9 Medical statistics show that deaths due to four major diseases—call them A, B, C, and D—account for 15, 21, 18, and 14%, respectively, of all nonaccidental deaths. A study of the causes of 308 nonaccidental deaths at a hospital gave the following counts of patients dying of diseases A, B, C, and D:

Disease	Number of Deaths
A	43
B	76
C	85
D	21
Other	83
Total	308

Do these data provide sufficient evidence to indicate that the proportions of people dying of diseases A, B, C, and D at this hospital differ from the proportions accumulated for the population at large?

12.10 Research has suggested a link between the prevalence of schizophrenia and birth during particular months of the year in which viral infections are prevalent. Suppose you are working on a similar problem and you suspect a linkage between a disease observed in later life and month of birth. You have records of 400 cases of the disease and you classify them according to month of birth. The data appear in the table. Do the data present sufficient evidence to indicate that the proportion of cases of the disease per month varies from month to month? Test with $\alpha = .10$.

Month	Jan	Feb	Mar	Apr	May	June	July	Aug	Sept	Oct	Nov	Dec
Number of births	38	31	42	46	28	31	24	29	33	36	27	35

12.11 Portable personal computers, sometimes called "laptops," represent a fast-growing segment of the PC market. According to Market Intelligence Research Company, the use of portable "laptop" computers can be classified in the following user segments ("Laptop's Three Musts," 1988):

Business–professional	69%
Government	21%
Education	7%
Home	3%

A sample of $n = 150$ laptop computer owners were surveyed this year, and the user segments were tabulated as follows:

Business–professional	102
Government	32
Education	12
Home	4

Do the data provide sufficient evidence to indicate that the figures given in 1988 by Market Intelligence Research Company are not accurate today? Calculate the approximate p-value for the test, and use it to make your decision.

12.4 Contingency Tables

A problem frequently encountered in the analysis of count data concerns the independence of two methods of classifying observed events. For example, suppose we wish to classify defects found on furniture produced in a manufacturing plant—first, according to the type of defect and, second, according to the production shift during which the piece of furniture was produced. If the proportions of the various types of defects are constant from shift to shift, then classification by defects is independent of the classification by production shift. On the other hand, if the proportions of the various defects vary from shift to shift, then the classification by defects is **contingent** upon the shift classification, and the classifications are dependent. In investigating whether one method of classification is contingent upon another, we display the data by using a cross-classification in an array called a **contingency table.**

A total of $n = 309$ furniture defects were recorded, and the defects were classified as being one of four types: A, B, C, or D. At the same time, each piece of furniture was identified according to the production shift in which it was manufactured. These counts are presented as a contingency table in Table 12.2. (NOTE: Numbers in parentheses are the expected cell frequencies.)

TABLE 12.2
Contingency table

	Type of Defect				
Shift	**A**	**B**	**C**	**D**	**Total**
1	15(22.51)	21(20.99)	45(38.94)	13(11.56)	94
2	26(22.99)	31(21.44)	34(39.77)	5(11.81)	96
3	33(28.50)	17(26.57)	49(49.29)	20(14.63)	119
Total	74	69	128	38	309

Let p_A be the unconditional probability that a defect will be of type A. Similarly, define p_B, p_C, and p_D as the probabilities of observing the three other types of defects. Then these probabilities, which we will call the **column probabilities** of Table 12.2, will satisfy the requirement

$$p_A + p_B + p_C + p_D = 1$$

In like manner, let p_i $(i = 1, 2, \text{ or } 3)$ equal the **row probability** that a defect will have occurred on shift i, where

$$p_1 + p_2 + p_3 = 1$$

Then, if the two classifications are independent of each other, a cell probability will equal the product of its respective row and column probabilities in accordance with the multiplicative law of probability.

For example, the probability that a particular defect will occur on shift 1 and be of type A is $(p_1)(p_A)$. Thus, we observe that the numerical values of the cell probabilities are unspecified in the problem under consideration. **The null hypothesis specifies only that each cell probability will equal the product of its respective row and column probabilities and therefore imply independence of the two classifications.** The alternative hypothesis is that this equality does not hold for at least one cell.

The analysis of the data obtained from a contingency table differs from the problem discussed in Section 12.3 because we must **estimate** the row and column probabilities so that we can estimate the expected cell frequencies.

If proper estimates of the cell probabilities are obtained, the estimated expected cell frequencies can be substituted for the $E(n_i)$ in X^2, and X^2 will continue to have a distribution in repeated sampling that is approximated by the chi-square probability distribution. The proof of this statement, as well as a discussion of the methods for obtaining the estimates, is beyond the scope of this text. Fortunately, the procedures for obtaining the estimates, known as the **method of maximum likelihood and the method of minimum chi-square**, yield estimates that are intuitively obvious for our relatively simple applications.

It can be shown that the maximum likelihood estimator of a column probability will equal the column total divided by $n = 309$. If we denote the total of column j as c_j, then

$$\hat{p}_A = \frac{c_1}{n} = \frac{74}{309} \qquad \hat{p}_C = \frac{c_3}{n} = \frac{128}{309}$$

$$\hat{p}_B = \frac{c_2}{n} = \frac{69}{309} \qquad \hat{p}_D = \frac{c_4}{n} = \frac{38}{309}$$

Similarly, the row probabilities p_1, p_2, and p_3 can be estimated using the row totals r_1, r_2, and r_3:

$$\hat{p}_1 = \frac{r_1}{n} = \frac{94}{309} \qquad \hat{p}_2 = \frac{r_2}{n} = \frac{96}{309} \qquad \hat{p}_3 = \frac{r_3}{n} = \frac{119}{309}$$

Denote the observed frequency of the cell in row i and column j of the contingency table by n_{ij}. Then the estimated expected value of n_{11} will be

$$\hat{E}(n_{11}) = n[\hat{p}_1\hat{p}_A] = n\left(\frac{r_1}{n}\right)\left(\frac{c_1}{n}\right) = \frac{r_1 c_1}{n}$$

where $(\hat{p}_1\hat{p}_A)$ is the estimated cell probability. Similarly, we can find the estimated expected value for any other cell, say $\hat{E}(n_{23})$:

$$\hat{E}(n_{23}) = n[\hat{p}_2\hat{p}_C] = n\left(\frac{r_2}{n}\right)\left(\frac{c_3}{n}\right) = \frac{r_2 c_3}{n}$$

Thus, we see that the estimated expected value of the observed cell frequency n_{ij} for a contingency table is equal to the product of its respective row and column totals divided by the total frequency; that is,

Estimated Expected Cell Frequency

$$\hat{E}(n_{ij}) = \frac{r_i c_j}{n}$$

where

$$r_i = \text{total for row } i$$
$$c_j = \text{total for column } j$$

The estimated expected cell frequencies for our example are shown in parentheses in Table 12.2.

We can now use the expected and observed cell frequencies shown in Table 12.2 to calculate the value of the test statistic

$$\begin{aligned} X^2 &= \sum_{j=1}^{4} \sum_{i=1}^{3} \frac{[n_{ij} - \hat{E}(n_{ij})]^2}{\hat{E}(n_{ij})} \\ &= \frac{(15 - 22.51)^2}{22.51} + \frac{(26 - 22.99)^2}{22.99} + \cdots + \frac{(20 - 14.63)^2}{14.63} \\ &= 19.18 \end{aligned}$$

The next step is to determine the appropriate number of degrees of freedom associated with the test statistic. We give the rule here, which we will later attempt to justify. **The degrees of freedom associated with a contingency table having r rows and c columns will always equal $(r - 1)(c - 1)$.** Thus, for our example, we will compare X^2 with the critical value of χ^2 with $(r - 1)(c - 1) = (3 - 1)(4 - 1) = 6$ d.f.

You will recall that the number of degrees of freedom associated with the X^2 statistic equals the number of cells (in this case, $k = rc$) less one degree of freedom for each independent linear restriction placed on the observed cell frequencies. The total number of cells for the data of Table 12.2 is $k = 12$. From this we subtract 1 d.f. because the sum of the observed cell frequencies must equal n; that is,

$$n_{11} + n_{12} + \cdots + n_{34} = 309$$

In addition, we used the cell frequencies to estimate three of the four column probabilities. Note that the estimate of the fourth column probability will be determined once we have estimated p_A, p_B, and p_C because

$$p_A + p_B + p_C + p_D = 1$$

Thus, we lose $(c - 1) = 3$ d.f. for estimating the column probabilities.

Finally, we used the cell frequencies to estimate $(r - 1) = 2$ row probabilities and, therefore, we lose $(r - 1) = 2$ additional degrees of freedom. The total number of degrees of freedom remaining will be

$$\text{d.f.} = 12 - 1 - 3 - 2 = 6$$

In general, we see that the total number of degrees of freedom associated with an $r \times c$ contingency table is

$$\begin{aligned} \text{d.f.} &= rc - 1 - (c - 1) - (r - 1) \\ &= rc - c - r + 1 = (r - 1)(c - 1) \end{aligned}$$

Therefore, if we use $\alpha = .05$, we will reject the null hypothesis that the two classifications are independent if $X^2 > 12.5916$. Since the value of the test statistic, $X^2 = 19.18$, exceeds the critical value of χ^2, we reject the null hypothesis. The data present sufficient evidence to indicate that the proportion of the various types of defects varies from shift to shift. A study of the production operations for the three shifts would likely reveal the cause or causes of the observed discrepancies.

EXAMPLE 12.1 A survey was conducted to evaluate the effectiveness of a new flu vaccine that had been administered in a small community. The vaccine was provided free of charge in a two-shot sequence over a period of 2 weeks. Some people received the two-shot sequence, some appeared only for the first shot, and others received neither.

A survey of 1000 local inhabitants the following spring provided the information shown in Table 12.3. Do the data present sufficient evidence to indicate that the vaccine was successful in reducing the number of flu cases in the community?

TABLE 12.3 Data tabulation for Example 12.1

	No Vaccine	One Shot	Two Shots	Total
Flu	24(14.4)	9(5.0)	13(26.6)	46
No flu	289(298.6)	100(104.0)	565(551.4)	954
Total	313	109	578	1000

Solution The question concerning the success of the vaccine in reducing flu cases can be restated in terms of whether the data provide sufficient evidence to indicate a dependence between the vaccine classification and the occurrence or nonoccurrence of flu. Therefore, we analyze the data as a contingency table.

The estimated expected cell frequencies may be calculated using the appropriate row and column totals,

$$\hat{E}(n_{ij}) = \frac{r_i c_j}{n}$$

Thus,

$$\hat{E}(n_{11}) = \frac{r_1 c_1}{n} = \frac{46(313)}{1000} = 14.4$$

$$\hat{E}(n_{12}) = \frac{r_1 c_2}{n} = \frac{46(109)}{1000} = 5.0$$

$$\vdots$$

$$\hat{E}(n_{23}) = \frac{r_2 c_3}{n} = \frac{954(578)}{1000} = 551.4$$

These values are shown in parentheses in Table 12.3.

The value of the test statistic X^2 can now be computed and compared with the critical value of χ^2 having $(r-1)(c-1) = (1)(2) = 2$ d.f. Then, for $\alpha = .05$, we will reject the null hypothesis when $X^2 > 5.99147$. Substituting into the formula for X^2, we obtain

$$\begin{aligned} X^2 &= \frac{(24-14.4)^2}{14.4} + \frac{(289-298.6)^2}{298.6} + \cdots + \frac{(565-551.4)^2}{551.4} \\ &= 17.35 \end{aligned}$$

Since the observed value of X^2 falls in the rejection region, we reject the null hypothesis of independence of the two methods of classification. Based on our sample evidence, we believe there is a dependence between the vaccine classification and the occurrence or nonoccurrence of the flu.

Has the vaccine been successful? A comparison of the percentage incidence of flu for each of the three categories would suggest that those receiving the two-shot sequence were less susceptible to the disease. Further analysis of the data could be obtained by deleting one of the three categories—the second column, for example—to compare the effect of the vaccine with that of no vaccine. This could be done by using a 2×2 contingency table or treating the two categories as two binomial populations and using the methods of Section 8.7. Or we might wish to analyze the data by comparing the results of the two-shot vaccine sequence with those of the combined no vaccine—one-shot group; that is, we would combine the first two columns of the 2×3 table into one. ▪

Most statistical packages include a program for analyzing data contained in an $r \times c$ contingency table. To illustrate, the SAS, EXECUSTAT, and MINITAB computer printouts for the analysis of the data in Example 12.1 are shown in Table 12.4. The computed value of X^2 and its observed significance level are boxed on the computer printouts.

At the top of the SAS printout, Table 12.4 (a), is the chi-square table showing the observed and expected values in their respective cells. The computed value of the test statistic, $X^2 = 17.313$, is shown directly below the table. The number of degrees of freedom for X^2, DF $= 2$, and the observed significance level, PROB $= 0.0002$, are shown to the right of X^2. The last four lines of the printout, PHI, CONTINGENCY COEFFICIENT, CRAMER'S V, and LIKELIHOOD RATIO CHISQUARE, are not pertinent to our analysis.

The MINITAB and EXECUSTAT printouts show the same 2×3 contingency table, except that the EXECUSTAT table displays the percentage of the total represented by each cell count rather than the expected cell count. The computed value, $X^2 = 17.313$, is shown below the table in both printouts along with its degrees of freedom.

The information shown on the printouts differs slightly from one printout to another, as do some of the computed numbers (due to rounding), but the basic information is the same. All three show the calculated value of X^2 and its degrees

TABLE 12.4 SAS, EXECUSTAT, and MINITAB computer printouts for the chi-square analysis of the data in Example 12.1

(a) SAS

```
                         TABLE OF FLU BY SHOTS
FLU          SHOTS
FREQUENCY |
EXPECTED  |   NONE   |    ONE   |    TWO   |  TOTAL
----------+----------+----------+----------+
YES       |     24   |      9   |     13   |     46
          |   14.4   |    5.0   |   26.6   |
----------+----------+----------+----------+
NO        |    289   |    100   |    565   |    954
          |  298.6   |  104.0   |  551.4   |
----------+----------+----------+----------+
TOTAL     |    313   |    109   |    578   |   1000

                        STATISTICS FOR 2-WAY TABLES

CHI-SQUARE                        17.313     DF=2      PROB =0.0002
PHI                                0.132
CONTINGENCY COEFFICIENT            0.130
CRAMER'S V                         0.132
LIKELIHOOD RATIO CHISQUARE        17.252     DF=2      PROB =0.0002
```

(b) EXECUSTAT

```
                                       Crosstabulation
              No vaccine   One shot     Two shots              Row
                                                               Total

Flu                   24          9            13                 46
                     2.4        0.9           1.3               4.60

No flu               289        100           565                954
                    28.9       10.0          56.5              95.40

      Column         313        109           578               1000
      Total        31.30      10.90         57.80             100.00


                    Summary Statistics for Crosstabulation

     Chi-square          D.F.           P Value
         17.31             2             0.0002
```

(c) MINITAB

```
MTB > CHISQUARE C1-C3

Expected counts are printed below observed counts

            ZERO       ONE       TWO     TOTAL
     1        24         9        13        46
           14.40      5.01     26.59

     2       289       100       565       954
          298.60    103.99    551.41

Total        313       109       578      1000

ChiSq =  6.404 +  3.169 +  6.944 +
         0.309 +  0.153 +  0.335 =  17.313
df = 2
```

of freedom. These two quantities, along with the table of the critical values of chi-square, Table 5 in Appendix I, are all that we need to test in order to detect dependence between the two qualitative variables represented in the contingency table. The value of the observed significance level, given in the SAS and the EXECUSTAT printouts, eliminates the need for the chi-square table and gives us a measure of the weight of evidence favoring rejection of the null hypothesis.

Exercises

Basic Techniques

12.12 Calculate the value and give the number of degrees of freedom for X^2 for the following contingency tables:

a

	Columns			
Rows	**1**	**2**	**3**	**4**
1	120	70	55	16
2	79	108	95	43
3	31	49	81	140

b

	Columns		
Rows	**1**	**2**	**3**
1	35	16	84
2	120	92	206

12.13 Suppose that a consumer survey summarizes the responses of $n = 307$ people in a contingency table that contains three rows and five columns. How many degrees of freedom will be associated with the chi-square test statistic?

12.14 A survey of 400 respondents produced the following cell counts in a 2×3 contingency table:

	Columns			
Rows	**1**	**2**	**3**	**Total**
1	37	34	93	164
2	66	57	113	236
Total	103	91	206	400

a If you wish to test the null hypothesis of "independence"—that the probability that a response falls in any one row is independent of the column it will fall in—and you plan to use a chi-square test, how many degrees of freedom will be associated with the χ^2 statistic?

b Find the value of the test statistic.

c Find the rejection region for $\alpha = .10$.

d Conduct the test and state your conclusions.

e Find the approximate observed significance level for the test and interpret its value.

Applications

12.15 A study conducted by J. E. Brush and colleagues (1985) suggests that the initial electrocardiogram (ECG) of a suspected heart attack victim can be used to predict in-hospital complications of an acute nature. The study included 469 patients with suspected myocardial infarction (heart attack). Each of these patients was categorized in the table below according to whether their initial ECG was positive or negative and whether the person suffered life-threatening complications of an acute myocardial infarction subsequently in the hospital. The tabled values give the number of persons in each category:

	Subsequent In-Hospital Life-Threatening Complications		
ECG	**No**	**Yes**	**Total**
Negative	166	1	167
Positive	260	42	302
Total	426	43	469

Do the data present sufficient evidence to indicate that the probability of having life-threatening acute myocardial infarctions is dependent on the ECG results? To answer this question,

a State the null hypothesis and the alternative hypothesis to be tested.

b Give the formula for the test statistic and substitute the numerical values into it.

c Give the rejection region for the test, using $\alpha = .05$.

d State your conclusions for each of the two possible outcomes of the test.

e What assumptions are necessary in order that this test be valid? Are they satisfied for these data?

f Complete the necessary calculations, conduct the test, and state your conclusions.

12.16 A study was conducted by Joseph Jacobson and Diane Wille to determine the effect of early child care on infant-mother attachment patterns (Schmittroth, 1991). In the study, 93 infants were classified as either "secure" or "anxious" using the Ainsworth strange situation paradigm. In addition, the infants were classified according to the average number of hours per week that they spent in child care. The data is presented in the table below:

Attachment Pattern	Low (0–3 hours)	Moderate (4–19 hours)	High (20–54 hours)
Secure	24	35	5
Anxious	11	10	8

f Do the data provide sufficient evidence to indicate that there is a difference in attachment pattern for the infants depending on the amount of time spent in child care? Test using $\alpha = .05$.

g What is the approximate p-value for the test in part (a)?

12.17 The data shown below are the results of a study by J. Farwell and J. T. Flannery (1984) to determine whether the parents, siblings, and offspring of children with central nervous system tumors have a higher incidence of cancer than the general public. Among a group of 643 children with this type of cancer, 73 had at least one relative with cancer. In contrast, of 360 healthy children (controls), 35 had at least one relative with cancer.

	At Least One Relative with Cancer	No Relatives with Cancer	Total
Children with central nervous system tumors	73	570	643
Controls	35	325	360
Total	108	895	1003

a Do the data present sufficient evidence to indicate a dependence between the occurrence of central nervous system tumors in children and the incidence of cancer in their parents, siblings, and children? Test using the chi-square test with $\alpha = .05$.

b Let p_1 represent the proportion of children with a central nervous system tumor who had at least one relative with cancer and let p_2 represent the corresponding proportion for the children in the control group. State the null and alternative hypotheses employed in the test in part (a).

c Suppose that you only wanted the detect p_2 larger than p_1. State the null and alternative hypotheses for this test.

d How would you conduct the test in part (c)?

12.18 Is there a difference in the spending patterns of high school seniors depending on their gender? A study to investigate this question focused attention on 196 employed high school seniors. Each student was asked to classify the amount of their earnings which they spent on car expenses during a given month. The results are shown below:

	None or Only a Little	Some	About Half	Most	All or Almost All
Male	73	12	6	4	3
Female	57	15	11	9	6

```
MTB  >  CHISQUARE  C1-C5


Expected counts are printed below observed counts


         NONE     SOME     HALF     MOST      ALL    Total
    1      73       12        6        4        3       98
        65.00    13.50     8.50     6.50     4.50


    2      57       15       11        9        6       98
        65.00    13.50     8.50     6.50     4.50


Total     130       27       17       13        9      196


ChiSq =  0.985 + 0.167 + 0.735 + 0.962 + 0.500 +
         0.985 + 0.167 + 0.735 + 0.962 + 0.500 = 6.696


df = 4
2 cells with expected counts less than 5.0
```

a Do the data provide sufficient evidence to indicate that there is a difference in spending patterns between males and females? Test using $\alpha = .05$; use the MINITAB printout.

b Find the approximate p-value of the test in part (a).

12.19 According to one theory, people with type A behavior—hard-driving, impatient, competitive people—are much more prone to heart problems that those classified as low-key, type B, individuals. In a study conducted by Robert Case and colleagues (1985), 516 heart attack patients were administered the Jenkins Activity Survey (JAS), a questionnaire that purports to provide an approximate measure of individual type A behavior. The scores ranged from a low of −21 (type B behavior) to a high of 23 (type A behavior). The higher the score, the greater the level of type A behavior in an individual. Each of the 516 patients were categorized according to whether the patient's JAS score was less than −5 (type B), −5 to 5 (neutral), or greater than 5 (type A). The number of patients in each of the three classes, along with the number who died during a 3-year follow-up period, are shown in the table:

	JAS Score		
	Less Than −5	**−5 to 5**	**Greater Than 5**
Total number of patients	180	171	165
Number dying in 3-year follow-up	21	17	11

The authors used the chi-square test to test for a dependence between the follow-up mortality rate and the level of type A behavior as measured by the JAS score.

a Compute the value of X^2.

b Using Table 5 in Appendix 1, would you conclude that the p-value for the test is larger than 0.10?

c Based on the data in the table, what do you think the authors concluded about the dependence between the mortality rate of heart attack patients and their JAS scores?

12.20 Has the rise over the last 20 years in the number of children involved in organized sports (Little League baseball, organized football, and soccer teams) produced a decrease in the numbers involved in self-directed physical games? To study this problem, D. A. Kleiber and G. C. Roberts (1983) interviewed 78 male and 77 female fourth and fifth grade students from public grammar schools in the Champaign–Urbana area of Illinois. Each student was questioned to determine whether they actively participated in one or more organized sports and the frequency of involvement in self-directed games. The following table shows the numbers of girls falling into three levels of participation in self-directed games for those participating in organized sports and for those who were nonparticipants:

Frequency of Participation in Self-Directed Physical Games	Participants	Nonparticipants
Less than once a week	4	16
More than once a week but less than daily	15	20
Every day	7	15
Total	26	51

a Do the data provide sufficient evidence to indicate that the frequencies in the three levels of involvement in self-directed games differ depending on whether the girls were or were not involved in organized sports? Test using $\alpha = .05$.

b Find the approximate p-value for the test.

c Check your answers with the results published in the paper by Kleiber and Roberts. They show $X^2 = 3.02$, d.f. $= 2$, p-value $= .22$. Do you agree?

12.5 Tables with Fixed Row or Column Totals: Tests of Homogeneity

In the previous section, we described the analysis of an $r \times c$ contingency table, using examples that for all practical purposes fit the multinomial experiment described in Section 12.1. While methods of collecting data in many surveys obviously satisfy the requirements of a multinomial experiment, other methods do not. For example, we may not wish to randomly sample the population described in Example 12.1 because we might find that, owing to chance, one category contains a very small number of people or is completely missing. To avoid such outcomes, we may decide beforehand to interview a specified number of people in each column category, thereby fixing the column totals in advance.

Suppose we decide to fix each column total at 300 people. With this sampling plan we would not have a multinomial experiment with $r \times c = 2 \times 3 = 6$ cells; instead we would have three binomial experiments. The first consists of $n_1 = 300$ people who received no vaccine, with the probability of an individual getting the flu given by $P(\text{flu}) = p_1$. The second consists of $n_2 = 300$ people who received one shot, with $P(\text{flu}) = p_2$. The third consists of $n_3 = 300$ people who received both shots, with $P(\text{flu}) = p_3$. By fixing the column totals at 300, we are assured that the unconditional probability that an individual in the study belongs to group 1, 2, or 3 is $300/900 = 1/3$.

Consider the joint multinomial cell probabilities p_{ij}, for $i = 1, 2$ and $j = 1, 2, 3$ when the vaccine has no effect in preventing the flu. The probability that an individual gets the flu is the same for the three groups and $p_1 = p_2 = p_3 = p$. Therefore, the unconditional probability of a person getting the flu is p, and the joint probability that a person in group j gets the flu is

$$p_{1j} = \left(\frac{1}{3}\right)p \qquad \text{for} \qquad j = 1, 2, 3$$

Similarly, the probability of a person in group j *not* getting the flu is

$$p_{2j} = \left(\frac{1}{3}\right)q \qquad \text{for} \qquad j = 1, 2, 3$$

If, on the other hand, the vaccine is effective, then p_1, p_2, and p_3 will not be equal to p. Consequently,

$$p_{1j} \neq \left(\frac{1}{3}\right)p \qquad \text{and} \qquad p_{2j} \neq \left(\frac{1}{3}\right)q$$

However, we see that this is the same as asking whether row and column classifications are independent. Thus, the test statistic is calculated as though you were testing for independence using Pearson's chi-square with $(r-1)(c-1)$ d.f.

As you can see, when the column totals are fixed, testing for independence comes down to a test of equality of three binomial p's. Tests of this sort are called **tests of homogeneity** of several binomial populations. If there were three or more row classifications with fixed column totals, the test would be one of **homogeneity** of several multinomial populations.

EXAMPLE **12.2** A survey of voter sentiment was conducted in four midcity political wards to compare the fraction of voters favoring candidate A. Random samples of 200 voters were polled in each of the four wards with results as shown in Table 12.5. Do the data present sufficient evidence to indicate that the fractions of voters favoring candidate A differ in the four wards?

TABLE **12.5**
Data tabulation for Example 12.2

	Ward				
	1	2	3	4	Total
Favor A	76(59)	53(59)	59(59)	48(59)	236
Do not favor A	124(141)	147(141)	141(141)	152(141)	564
Total	200	200	200	200	800

Solution You will observe that the test of hypothesis concerning the equivalence of the parameters of the four binomial populations corresponding to the four wards is identical to the test of hypothesis implying independence of the row and column classifications. If we denote the fraction of voters favoring A as p and hypothesize that p is the same for all four wards, we imply that the first- and second-row probabilities are equal to p and $(1 - p)$, respectively. The probability that a member of the sample of $n = 800$ voters falls in a particular ward will equal one-fourth, since this was fixed in advance. Then the cell probabilities for the table would be obtained as usual.

Notice that

$$\hat{p}_1 = .38 \qquad \hat{p}_3 = .30$$
$$\hat{p}_2 = .27 \qquad \hat{p}_4 = .24$$

while the pooled estimate of the unconditional probability of favoring A is

$$\hat{p} = \frac{236}{800} = .30$$

The estimated expected cell frequencies, calculated using the row and column totals, appear in parentheses in Table 12.5. We see that

$$X^2 = \sum_{j=1}^{4}\sum_{i=1}^{2} \frac{[n_{ij} - \hat{E}(n_{ij})]^2}{\hat{E}(n_{ij})}$$
$$= \frac{(76 - 59)^2}{59} + \frac{(124 - 141)^2}{141} + \cdots + \frac{(152 - 141)^2}{141}$$
$$= 10.72$$

The critical value of χ^2 for $\alpha = .05$ and $(r - 1)(c - 1) = (1)(3) = 3$ d.f. is 7.81473. Since X^2 exceeds this critical value, we reject the null hypothesis and conclude that the fraction of voters favoring candidate A is not the same for all four wards. ▪

Exercises

Basic Techniques

12.21 Random samples of 200 observations were selected from each of three populations, and each observation was classified according to whether it fell into one of three mutually exclusive categories. The cell counts are shown below:

	Category			
Population	**1**	**2**	**3**	**Total**
1	108	52	40	200
2	87	51	62	200
3	112	39	49	200

Do the data provide sufficient evidence to indicate that the proportions of observations in the three categories depend on the population from which they were drawn?

a Give the value of X^2 for the test.

b Give the rejection region for the test for $\alpha = .10$.

c State your conclusions.

d Find the approximate p-value for the test and interpret its value.

12.22 Suppose you wish to test the null hypothesis that three binomial parameters p_A, p_B, and p_C are equal against the alternative hypothesis that at least two of the parameters differ. Independent random samples of 100 observations were selected from each of the populations. The data are shown below:

	Population			
	A	**B**	**C**	**Total**
Number of successes	24	19	33	76
Number of failures	76	81	67	224
Total	100	100	100	300

a Find the value of X^2, the test statistic.

b Give the rejection region for the test for $\alpha = .05$.

c State your test conclusions.

d Find the observed significance level for the test and interpret its value.

Applications

12.23 Are baby-boomers more likely to increase their investing now that they are reaching middle age? A poll was conducted by Hal Riney & Partners (*Los Angeles Times*, June 11, 1990), in which 400 investors were classified according to their age group and their likely investment pattern over the next 5 years versus the last 5 years. The data are shown below. Notice that there were 200 investors included from each age group.

Age Group	More	Less	Same
35–54	90	18	92
55+	40	60	100

Source: Adapted from Miko and Weilant (1991), p. 258.

Do the data provide sufficient information to conclude that the investing patterns of the baby-boomer age group differs from that of the older age group? Test using $\alpha = .01$.

12.24 A study to determine the effectiveness of a drug (serum) for arthritis resulted in the comparison of two groups, each consisting of 200 arthritic patients. One group was inoculated with the serum; the other received a placebo (an inoculation that appears to contain serum but actually is nonactive). After a period of time, each person in the study was asked to state whether his or her arthritic condition had improved. The following results were observed.

	Treated	Untreated
Improved	117	74
Not improved	83	126

Do these data present sufficient evidence to indicate the serum was effective in improving the condition of arthritic patients?

e Test by means of the X^2 statistic. Use $\alpha = .05$.

f Test by use of the z test in Section 8.7.

12.25 A particular poultry disease is thought to be noncommunicable. To test this theory, 30,000 chickens were randomly partitioned into three groups of 10,000. One group had no contact with diseased chickens, one had moderate contact, and the third had heavy contact. After a 6-month period, the following data were collected on the number of diseased chickens in each group of 10,000. Do the data provide sufficient evidence to indicate a dependence between the amount of contact between diseased and nondiseased fowl and the incidence of the disease? Use $\alpha = .05$.

	No Contact	Moderate Contact	Heavy Contact
Number of diseased chickens	87	89	124
Number of nondiseased chickens	9,913	9,911	9,876
Total	10,000	10,000	10,000

12.26 A study of the purchase decisions for three stock portfolio managers A, B, and C was conducted to compare the rates of stock purchases that resulted in profits over a time period that was less than or equal to 1 year. One hundred randomly selected purchases obtained for each of the managers gave the following results:

	Manager		
	A	B	C
Purchases that resulted in a profit	63	71	55
Purchases that resulted in no profit	37	29	45
Total	100	100	100

Do the data provide evidence of differences among the rates of successful purchases for the three managers? Use $\alpha = .05$.

12.27 H. W. Menard (1976) has conducted research involving manganese nodules, a mineral-rich concoction found abundantly on the deep-sea floor. In one portion of his report, Menard provides data relating the magnetic age of the earth's crust to the "probability of finding manganese nodules." The data shown in the table give the number of samples of the earth's core and the percentage of those that contain manganese nodules for each of a set of magnetic-crust ages. Do the data provide sufficient evidence to indicate that the probability of finding manganese nodules in the deep-sea earth's crust is dependent on the magnetic-age classification? Test with $\alpha = .05$.

Age	Number of Samples	Percentage with Nodules
Miocene—recent	389	5.9
Oligocene	140	17.9
Eocene	214	16.4
Paleocene	84	21.4
Late Cretaceous	247	21.1
Early and Middle Cretaceous	1120	14.2
Jurassic	99	11.0

12.28 In this age of VCRs and videotaped movies, a new type of advertising has evolved whereby the advertisement is included on a commercially produced videocassette. Advertisers have found that, although many consumers fast-forward past the ads, the best chance of getting persons to watch lies in the 18- to 34-year-age group. Overall, however, most people find the ads annoying. The data below show the responses of two groups of consumers, 50 in the 18- to 34-year-age category and 50 in the 35- to 54-year-age category:

Age Group	Very Annoying	Somewhat Annoying	Not Annoying
18–34	18	15	17
35–54	21	19	10

Source: Adapted from Miko and Weilant (1991), p. 19.

Is there a significant difference in the proportions who find the ads very annoying or somewhat annoying for the two age categories? Test using $\alpha = .01$ and the MINITAB printout below:

```
MTB  >  CHISQUARE  C1-C3

Expected counts are printed below observed counts


            VERY  SOMEWHAT        NOT      Total
    1         18        15         17         50
           19.50     17.00      13.50

    2         21        19         10         50
           19.50     17.00      13.50

Total         39        34         27        100


ChiSq  =   0.115  +  0.235  +  0.907  +
           0.115  +  0.235  +  0.907  =  2.516

df  =  2
```

12.6 Other Applications

The applications of the chi-square test in analyzing enumerative data, which we have described, represent only a few of the interesting classification problems that can be approximated by one or more multinomial experiments and for which our method of analysis is appropriate. By and large, these applications are complicated to a greater or lesser degree because the numerical values of the cell probabilities are unspecified and hence require the estimation of one or more population parameters. Then, as in Section 12.4 and 12.5, we can estimate the expected cell frequencies and use X^2

as a test statistic. Although we omit the mechanics of the statistical tests, several additional applications of the chi-square test are worth mentioning.

Goodness-of-Fit Tests

Goodness-of-fit tests are used to determine whether observed cell counts are consistent with expected cell counts calculated under the hypothesis that a specified probability model is true. This type of problem was presented in Section 12.3.

As another example, suppose that we wish to test an hypothesis stating that a population has a normal probability distribution. The cells of a sample frequency histogram (e.g., Figure 2.1) would correspond to the k cells of the multinomial experiment and the observed cell frequencies would be the number of measurements falling in each cell of the histogram. Given the hypothesized normal probability distribution for the population, we could use the areas under the normal curve to calculate the theoretical cell probabilities and hence the expected cell frequencies. The difficulty arises when μ and σ are unspecified for the normal population, and these parameters must be estimated to obtain the estimated cell probabilities. This difficulty can, of course, be surmounted.

Time-Dependent Multinomials

A second and interesting application of our methodology is its use in the investigation of the rate of change of a multinomial (or binomial) population as a function of time. For example, we might study the decision-making ability of a human (or any animal) as he or she is subjected to an educational program and tested over time. If, for instance, he or she is tested at prescribed intervals of time and the test is of the yes or no type yielding a number of correct answers x that would follow a binomial probability distribution, we would be interested in the behavior of the probability of a correct response p as a function of time. If the number of correct responses was recorded for c time periods, the data would fall in a $2 \times c$ table similar to that in Example 12.2. We would then be interested in testing the hypothesis that p is equal to a constant (i.e., that no learning has occurred), and we would then proceed to more interesting hypotheses to determine whether the data present sufficient evidence to indicate a gradual (say, linear) change over time as opposed to an abrupt change at some point in time. The procedures we have described could be extended to decisions involving more than two alternatives.

Our learning example is common to business, to industry, and to many other fields, including the social sciences. For example, we might wish to study the rate of consumer acceptance of a new product for various types of advertising campaigns as a function of the length of time that the campaign has been in effect. Or we might wish to study the trend in the lot fraction defective in a manufacturing process as a function of time. Both of these examples, as well as many others, require a study of the behavior of a binomial (or multinomial) process as a function of time.

Multidimensional Contingency Tables

The construction of a two-way table to investigate dependency between two classifications can be extended to three or more classifications. For example, if we wish to test the mutual independence of three classifications, we would employ a three-dimensional "table" or rectangular parallelepiped. The reasoning and methodology associated with the analysis of both the two- and three-way tables are identical, although the analysis of the three-way table is a bit more complex.

One area of research, concerned with **log-linear models**, assumes that $\ln p_{ij}$ is a linear function of row and/or column parameters or a linear function of regressor variables.

The examples that we have just described are intended to suggest the relatively broad application of the chi-square analysis of enumerative data, a fact that should be borne in the mind by the experimenter concerned with this type of data. The statistical test employing X^2 as a test statistic requires care in the determination of the appropriate estimates and the number of degrees of freedom for X^2, which, for some of these problems, may be rather complex.

12.7 Assumptions

The following assumptions must be satisfied if X^2 is to have approximately a chi-square distribution and, consequently, if the tests described in this chapter are to be valid.

Assumptions

1 The cell counts $n_1, n_2, \ldots, n_k$ satisfy the conditions of a multinomial experiment (or a set of multinomial experiments created by restrictions on row or column totals).

2 The expected values of all cell counts should equal or exceed 5.

Assumption 1 must be satisfied. The chi-square goodness-of-fit tests, of which these tests are special cases, compare observed frequencies with expected frequencies and apply only to data generated by a multinomial experiment.

The larger the sample size n, the closer the chi-square distribution will approximate the distribution of X^2. We have stated in assumption 2 that n must be large enough so that all the expected cell frequencies will be equal to 5 or more. This is a safe figure. Actually, the expected cell frequencies can be smaller for some tests.

12.8 Summary

The preceding material concerned a test of hypothesis regarding the cell probabilities associated with one or more multinomial experiments. When the number of observations n is large, the test statistic X^2 can be shown to have, approximately, a chi-square probability distribution in repeated sampling, the number of degrees of freedom being dependent on the particular application. In general, we assume that n is large and that the minimum expected cell frequency is equal to or greater than five.

Several words of caution concerning the use of the X^2 statistic as a method of analyzing enumerative data are in order. The determination of the correct number of degrees of freedom associated with the X^2 statistic is very important in locating the rejection region. If the number is incorrectly specified, erroneous conclusions might result. Also, note that nonrejection of the null hypothesis does not imply that it should be accepted. We would have difficulty in stating a meaningful alternative hypothesis for many practical applications and, therefore, we would lack knowledge of the probability of making a type II error. For example, we hypothesize that the two classifications of a contingency table are independent. A specific alternative would have to specify some measure of dependence, which may or may not have practical significance to the experimenter. Finally, if parameters are missing and the expected cell frequencies must be estimated, the estimators of missing parameters should have certain properties in order that the test be valid.

MINITAB Commands

```
CHISQUARE test on table stored in C ...C
```

Supplementary Exercises

Starred (*) exercises are optional.

12.29 A manufacturer of floor polish conducted a consumer preference experiment to see whether a new floor polish A was superior to those produced by four of his competitors. A sample of 100 housekeepers viewed five patches of flooring that had received the five polishes, and each indicated the patch that he or she considered superior in appearance. The lighting, background, and so on were approximately the same for all five patches. The results of the survey are as follows:

Polish	A	B	C	D	E
Frequency	27	17	15	22	19

Do these data present sufficient evidence to indicate a preference for one or more of the polished patches of floor over the others? If one were to reject the hypothesis of "no preference" for this experiment, would this imply that polish A is superior to the others? Can you suggest a better method of conducting the experiment?

12.30 A survey was conducted to investigate interest of middle-aged adults in physical fitness programs in Rhode Island, Colorado, California, and Florida. The objective of the investigation was to determine whether adult participation in physical fitness programs varies from one region of the United States to another. A random sample of people were interviewed in each state and the following data were recorded:

	Rhode Island	Colorado	California	Florida
Participate	46	63	108	121
Do not participate	149	178	192	179

Do the data indicate a difference in adult participation in physical fitness programs from one state to another?

12.31 An analysis of accident data was made to determine the distribution of numbers of fatal accidents for automobiles of three sizes. The data for 346 accidents are as follows:

	Size of Auto		
	Small	Medium	Large
Fatal	67	26	16
Not fatal	128	63	46

Do the data indicate that the frequency of fatal accidents is dependent on the size of automobiles?

12.32 An experiment was conducted to investigate the effect of general hospital experience on the attitudes of physicians toward lower-class people. A random sample of 50 physicians who had just completed 4 weeks of service in a general hospital were categorized according to their concern for lower-class people before and after their general hospital experience. The data are shown below. Do the data provide sufficient evidence to indicate a change in "concern" due to the general hospital experience?

Concern Before Experience in a General Hospital	After Experience in a General Hospital		
	High	Low	Total
Low	27	5	32
High	9	9	18

12.33 In a study of alcohol advertising and adolescent drinking, R. L. Lieberman and M. A. Orlandi (1987) found that of $n = 1747$ young adolescents in the study, 63% recalled the specific brand of alcohol being advertised, and 89% of 1108 adolescents perceived the ads as depicting young adults drinking in social situations, partying, and having fun. The following were the results after $n_1 = 100$ adolescents and $n_2 = 100$ adults were shown these same advertisements, and were asked to identify the ages of the people in the advertisements:

Perceived Age	Adolescents	Adults
Young adults	78	60
Teens/kids	10	22
Mixed ages	7	15
Older adults	5	3

Do the data present sufficient evidence to indicate that there is a difference in the perceived age(s) of the people specified in alcohol advertisements between adolescents and adults? Test using $\alpha = .05$.

12.34 If you find yourself forgetting names and appointments, 2 tablespoons daily of the common health food staple lecithin may improve your memory, according to Florence Safford and Barry Baumel, who presented the results of their study to the Gerontological Society of America annual conference in 1988 (*Gainesville Sun*, Gainesville, Fla., Feb. 9, 1989). Their experiment involved 61 people age 50 to 80 years and excluded any Alzheimer's patients. Of the 41 subjects receiving lecithin, 37 reported a significant decrease in memory lapse. Among the 20 subjects given only a placebo, 12 reported more memory lapses. The data summary follows:

	Memory Lapses		
Groups	Decrease	No Decrease	Total
Lecithin	37	4	41
Placebo	8	12	20

Do the data present sufficient evidence to indicate that a decrease in memory lapses depends on whether a subject has or has not been on a daily regimen of lecithin? Use $\alpha = .05$.

12.35 In a study of changing evaluations of floodplain hazards, Robert Payne and John Pigram (1981) describe the attitudes of people at risk of flood hazards in the Hunter River Valley in Australia. In part of the study, each respondent was asked to classify the damage (major or minor) that the respondent expected would incur if a flood were to occur. The respondent was also asked whether his or her preparation for a flood would result in low, moderate, or high cost. The numbers of responses in the six cost-damage categories are shown below:

	Damage		
Cost	Major	Minor	Total
Low	43	16	59
Moderate	10	28	38
High	4	9	13
Total	57	53	110

a Do the proportions of people taking the three levels of preparation prior to a flood differ depending on whether the respondents perceive the prospective flood damages as major or minor? Test using $\alpha = .05$.

b Find the approximate observed significance level for the test and interpret its values.

12.36 A group of 306 people were interviewed to determine their opinion concerning a particular current U.S. foreign-policy issue. At the same time, their political affiliation was recorded. The data are as follows:

	Approve of Policy	Do Not Approve of Policy	No Opinion
Republicans	114	53	17
Democrats	87	27	8

Do the data present sufficient evidence to indicate a dependence between party affiliation and the opinion expressed for the sampled population?

12.37 A survey was conducted to determine student, faculty, and administration attitudes on a new university parking policy. The distribution of those favoring or opposing the policy is shown below:

	Student	Faculty	Administration
Favor	252	107	43
Oppose	139	81	40

Do the data provide sufficient evidence to indicate that attitudes regarding the parking policy are independent of student, faculty, or administration status?

12.38* The chi-square test used in Exercise 12.24 is equivalent to the two-tailed z test of Section 8.6, provided α is the same for the two tests. Show algebraically that the chi-square test statistic X^2 is the square of the test statistic z for the equivalent test.

12.39* It is often not clear whether all properties of a binomial experiment are actually met in a given application. A goodness-of-fit test is desirable for such cases. Suppose that an experiment consisting of four trials was repeated 100 times. The number of repetitions on which a given number of successes was obtained is recorded in the following table:

Possible results (number of successes)	0	1	2	3	4
Number of times obtained	11	17	42	21	9

Estimate p (assuming that the experiment was binomial), obtain estimates of the expected cell frequencies, and test for goodness of fit. To determine the appropriate number of degrees of freedom for X^2, note that p was estimated by a linear combination of the observed frequencies.

12.40 A problem that sometimes occurs during surgical operations is the occurrence of infections during blood transfusions. An experiment was conducted to determine whether the injection of antibodies reduced the probability of infection. An examination of the records of 138 patients produced the data shown in the accompanying table. Do the data provide sufficient evidence to indicate that injections of antibodies affect the likelihood of transfusional infections? Test by using $\alpha = .10$.

	Infection	No Infection
Antibody	4	78
No antibody	11	45

12.41 By tradition, U.S. labor unions have been content to leave the management of the company to the managers and corporate executives. But in Europe, worker participation in management decision making is an accepted idea and one that is continually spreading. To study the effect of worker participation in managerial decision making, 100 workers were interviewed in each of two separate West German manufacturing plants. One plant had active worker participation in managerial decision making; the other did not. Each selected worker was asked whether he or she generally approved of the managerial decisions made within the firm. The results of the interviews are shown in the table:

	Participative Decision Making	No Participative Decision Making
Generally approve of the firm's decisions	73	51
Do not approve of the firm's decisions	27	49

a Do the data provide sufficient evidence to indicate that approval or disapproval of management's decisions depends on whether workers participate in decision making? Test by using the X^2 test statistic. Use $\alpha = .05$.

b Do these data support the hypothesis that workers in a firm with participative decision making more generally approve of the firm's managerial decisions than those employed by firms without participative decision making? Test by using the z test presented in Section 8.6. This problem requires a one-tailed test. Why?

12.42 An occupant-traffic study was conducted to aid in the remodeling of an office building that contains three entrances. The choice of entrance was recorded for a sample of 200 persons entering the building. Do the data shown in the table indicate that there is a difference in preference for the three entrances? Find a 90% confidence interval for the proportion of persons favoring entrance 1.

	Entrance		
	1	2	3
Number entering	83	61	56

12.43 Most high school required reading lists have changed little over the last 25 years, despite conservative critics' allegations of watered-down curricula and a retreat from the classics. Arthur N. Applebee, the author of a survey by the Center for the Learning and Teaching of Literature at the State University of New York in Albany, (*Press Enterprise*, Riverside, Calif., May 21, 1989), indicated that only one of the ten most frequently assigned titles was written by a woman—*To Kill a Mockingbird* by Harper Lee—and none by minorities. This survey indicated that 69% of 322 public schools, 67% of 80 Catholic schools, and 47% of private schools included *To Kill a Mockingbird* on their required reading lists. Do these data provide sufficient evidence to indicate that the proportion of schools that include Harper Lee's book as required reading varies according to school classification?

	Public	Catholic	Private
Required	222	54	40
Not required	100	26	46

Source: Center for Learning and Teaching of Literature, State University of New York–Albany.

12.44 Refer to Exercise 12.43. A survey by the Center for Learning and Teaching of Literature at the State University of New York–Albany revealed that 84% of 322 public schools, 63% of Catholic schools, and 66% of private schools included *Romeo and Juliet* by William Shakespeare among the ten most frequently assigned titles. A summary of this information follows:

	Schools		
Required	**Public**	**Catholic**	**Private**
Yes	270	50	57
No	52	30	29
Total	322	80	86

Is this sufficient information to conclude that the proportion of schools requiring the reading of *Romeo and Juliet* varies significantly among the three categories of schools? Use $\alpha = .05$.

12.45 Refer to Exercise 12.43. The novel *Huckleberry Finn* by Mark Twain was also included among the ten most frequently assigned works in 70% of the 322 public schools, 76% of the 80 Catholic schools, and 56% of the 86 private schools in the survey. Do these data, summarized in the accompanying table, indicate that the proportion of schools with *Huckleberry Finn* on their required reading list varies significantly according to the classification of school? Use $\alpha = .05$.

	Schools		
	Public	**Catholic**	**Private**
Required	225	61	48

12.46 The following table shows the categorization of 204 men awaiting bypass heart surgery according to the relative degree of each man's coronary artery obstruction and according to his perceived level of discomfort due to angina pectoris (Jenkins et al., 1983). Do the data present sufficient evidence to indicate that the level of angina is dependent on the level of coronary artery obstruction? The authors report the p-value for a chi-square test to be $p = 0.01$.

a Compute the value of X^2 for the data.

b Find the p-value for the test and compare with the authors' value $p = 0.01$.

c What conclusions would you reach based on your analysis?

	Arteries Obstructed 75% or More			
Level of Angina	**0 or 1**	**2**	**3 to 6**	**Total**
None	3	21	20	44
Mild	2	12	9	23
Moderate	26	20	31	77
Moderate/severe	13	10	18	41
Severe	7	5	7	19
Total	51	68	85	204

12.47 Although white has long been the most popular car color, recent trends in fashion and home design have signaled the emergence of green as the new color of the 1990s. The growth in the popularity of green hues stems partially from an increased interest in the environment and increased feelings of uncertainty. According to an article in the *Press-Enterprise* ("White Cars Still Favored," 1993), "green symbolizes harmony and counteracts emotional stress." The article cites the top five colors and the percentage of the market share for four different classes of cars. These data are given below for the truck–van category:

Color	**White**	**Medium/Dark Red**	**Green**	**Red**	**Black**
Percentage	29.72	11.00	9.24	9.08	9.01

In an attempt to verify the accuracy of these figures, we take a random sample of 250 trucks and vans and record their color. Suppose that the number of vehicles falling in each of the five categories above were 82, 22, 27, 21, and 20, respectively.

a Is there any category that is missing in the above classification? How many cars and trucks fell in that category?

b Is there sufficient evidence to indicate that the percentages of trucks and vans differ from those given above? Find the approximate p-value for the test.

12.48 According to a Sports Participation Survey by the National Sporting Goods Association reported in *American Demographics* (May 1989, p. 41), the fitness activities in which people participate change as they get older. The following table gives the results of a survey of 1103 men and women in which individuals who were frequent participants in fitness activities were classified by sex and type of exercise.

Sex	Type of Activity					
	Walking	Cycling	Aerobics	Running	Calesthenics	Swimming
Male	60	85	28	113	79	179
Female	106	81	138	55	89	90

Do these data provide sufficient evidence to indicate that the type of fitness activity participation varies by sex? Use $\alpha = .05$.

12.49 In the academic world, students and their faculty advisors often collaborate on research papers, producing works in which publication credit can take several forms. In theory, the first authorship of a student's paper should be given to the student unless the input from the faculty advisor was substantial. In an attempt to see whether this is, in fact, the case, authorship credit was studied for several different levels of faculty input and two objectives (dissertation vs. nondegree research). The frequency of author assignment decisions for published dissertations is shown below as assigned by 60 faculty members and 161 students:

Faculty Respondents

Authorship Assignment	High Input	Medium Input	Low Input
Faculty first author, student mandatory second author	4	0	0
Student first author, faculty mandatory second author	15	12	3
Student first author, faculty courtesy second author	2	7	7
Student sole author	2	3	5

Student Respondents

Authorship Assignment	High Input	Medium Input	Low Input
Faculty first author, student mandatory second author	19	6	2
Student first author, faculty mandatory second author	19	41	27
Student first author, faculty courtesy second author	3	7	31
Student sole author	0	3	3

Source: Costa and Gatz (1992), p. 54.

a Is there sufficient evidence to indicate a dependence between the authorship assignment and the input of the faculty advisor as judged by faculty members? Test using $\alpha = .01$.

b Is there sufficient evidence to indicate a dependence between the authorship assignment and the input of the faculty advisor as judged by students? Test using $\alpha = .01$.

c If there is a dependence in the two classifications from parts (a) and (b), does it appear from looking at the data that students are more likely to assign a higher authorship to their faculty advisors than the advisors themselves?

d Have any of the assumptions necessary for the analysis used in parts (a) and (b) been violated? What affect might this have on the validity of your conclusions?

12.50 How would you rate yourself as a driver? According to a recent survey conducted by the Field Institute, most Californians think that they are good drivers but have little respect for others' driving ability. The data below shows the distribution of opinions, according to gender for two different questions, the first rating themselves as drivers and the second rating others as drivers. Although not stated in the source, we assume that there were 100 men and 100 women in the surveyed group.

Rating Self as a Driver

Sex	Excellent	Good	Fair
Male	43	48	9
Female	44	53	3

Rating Others as Drivers

Sex	Excellent	Good	Fair	Poor/Very Poor
Male	4	42	41	13
Female	3	48	35	14

Source: Adapted from Smith (1991).

a Is there sufficient evidence to indicate that there is a difference in the self-ratings between male and female drivers? Find the approximate p-value for the test.

b Is there sufficient evidence to indicate that there is a difference in the rating of other drivers between male and female drivers? Find the approximate p-value for the test.

c Have any of the assumptions necessary for the analysis used in parts (a) and (b) been violated? What affect might this have on the validity of your conclusions?

12.51 In Exercise 12.20, we examined the relationship between the involvement of fourth and fifth grade girls in self-directed physical games and their involvement in organized sports. Similar data for the 78 boys involved in the study are shown below:

Frequency of Participation in Self-Directed Physical Games	Participants	Nonparticipants
Less than once a week	19	12
More than once a week but less than daily	23	8
Every day	11	5
Total	53	25

Source: Kleiber and Roberts (1983).

a Do the data provide sufficient evidence to indicate that the frequencies in the three levels of involvement in self-directed games differ depending on whether the boys were or were not involved in organized sports? Test using $\alpha = .05$.

b Find the approximate p-value for the test.

c Check your answers with the results published in the paper by Kleiber and Roberts. They show $X^2 = 1.19$, d.f. $= 2$, $p = .55$. Do you agree?

CASE STUDY

Can a Marketing Approach Improve Library Services?

Carol Day and Del Loewenthal (1992) studied the responses of young adults in their evaluation of library services. Of the $n = 200$ young adults involved in the study, $n_1 = 152$ were students, and $n_2 = 48$ were nonstudents. The following table presents the number of favorable responses for each group to seven questions in which the atmosphere, staff, and the design of the library were examined.

Favorable Responses to Attitude Questions for Students and Nonstudents

Question	Question	% Student Favorable	$n_1 = 152$	Nonstudent Favorable	$n_2 = 48$	$P(\chi^2)$
3	Libraries are friendly	79.6	121	56.2	27	<.01
4	Libraries are dull	77	117	58.3	28	<.05
5	Library staff are helpful	91.4	139	87.5	42	N.S.
6	Library staff are less helpful to teenagers	60.5	92	45.8	22	<.01
7	Libraries are so quiet they feel uncomfortable	75.6	115	52.05	25	<.01
11	Libraries should be more brightly decorated	29	44	18.8	9	N.S.
13	Libraries are badly signposted	45.4	69	43.8	21	N.S.

The entry in the last column labeled $P(\chi^2)$ is the p-value for testing the hypothesis of no difference in the proportion of students and nonstudents answering each question favorably. Hence, each question gives rise to a 2×2 contingency table.

1 Perform a test of homogeneity for each question and verify the reported p-value of the test.

2 Questions 3, 4, and 7 are concerned with the atmosphere of the library; questions 5 and 6 are concerned with the library staff; and questions 11 and 13 are concerned with the library design. How would you summarize the results of your analyses regarding the seven questions concerning the image of the library?

3 With the information given, is it possible to do any further testing concerning the proportion favorable versus unfavorable responses for two or more questions simultaneously?

13

The Analysis of Variance

Case Study

Are offensive linemen more at risk for coronary heart disease (CHD) than football players who play other positions? A comparison of various CHD risk factors for football players in different playing positions provides the basis for the case study at the end of this chapter.

General Objective

In this chapter, we will demonstrate how certain factors affect the quantity of information contained in a sample. Methods for comparing two population means, based on two independent random samples, and on a paired-difference experiment, were presented in Chapter 9. In this chapter, we extend these analyses to the comparison of three or more means using a technique called the analysis of variance. We will explain the logic of an analysis of variance and give the analysis for two designs that are generalizations of the unpaired and paired experiments in Chapter 9.

Specific Topics

1. The analysis of variance (13.1, 13.2)
2. The completely randomized design (13.3)
3. The analysis of variance and estimation for the completely randomized design (13.4, 13.5)
4. The randomized block design (13.6)
5. The analysis of variance and estimation for the randomized block design (13.7, 13.8)
6. Tukey's method of paired comparisons (13.10)

13.1 The Motivation for an Analysis of Variance

Suppose that you want to compare the mean size of health insurance claims submitted by five groups of policyholders. Ten claims were randomly selected from among the claims for each group. The data are shown in Table 13.1. Do the data contained in the five samples provide sufficient evidence to indicate a difference in the mean levels of claims among the five health groups?

To answer this question, we might think of comparing the means in pairs by using repeated applications of the Student's t test of Section 9.2. If we were to detect a difference between any pair of means, then we would conclude that there is evidence of at least one difference among the means, and it would appear that we would have

TABLE 13.1
Insurance claims submitted by five health groups

Group 1	Group 2	Group 3	Group 4	Group 5
$ 763	$1335	$ 596	$3742	$1632
4365	1262	1448	1833	5078
2144	217	1183	375	3010
1998	4100	3200	2010	671
5412	2948	630	743	2145
957	3210	942	867	4063
1286	867	1285	1233	1232
311	3744	128	1072	1456
863	1635	844	3105	2735
1499	643	1683	1767	767
$\bar{x}_1 = 1959.8$	$\bar{x}_2 = 1996.1$	$\bar{x}_3 = 1193.9$	$\bar{x}_4 = 1674.7$	$\bar{x}_5 = 2278.9$

answered our question. The problem with this procedure is that there are $C_2^5 = 10$ different pairs of means that have to be tested. Even if all of the means are identical, we have a probability α of rejecting the null hypothesis that a particular pair of means are equal. When this test procedure is repeated ten times, the probability of incorrectly concluding that at least one pair of means differ is quite high. Because the risk of an erroneous decision may be quite large, we look for a single test of the null hypothesis, that the five group means $\mu_1, \mu_2, \ldots, \mu_5$ are equal, against the alternative hypothesis, that at least one pair of means differ.

The procedure for comparing more than two population means is known as an **analysis of variance**. The logic underlying an analysis of variance can be seen by examining the dot diagrams for two sets of *sample* data for two different cases, (a) and (b). In case (a), the first sample contains $n_1 = 3$ measurements, 1, 8, and 3. The second sample contains $n_2 = 2$ measurements, 2 and 10. The dot diagrams for these two samples appear as shown in Figure 13.1(a). For case (b), the first sample contains $n_1 = 3$ measurements, 4.5, 3, and 4.5. The second sample contains two measurements, 5.5 and 6.5. The dot diagrams for case (b) are shown in Figure 13.1(b). Examining the dot diagrams, you can see that for both cases, $\bar{x}_1 = 4$ and $\bar{x}_2 = 6$. Which case, do you think, suggests a difference between population means μ_1 and μ_2?

From a visual observation of the dot diagrams and your intuition, we think that you will agree with us. Case (b) suggests a difference between the population means; case (a) does not. To arrive at this conclusion, we compared the *variation* (difference) *between the sample means* with the *variation within the samples*. Although $\bar{x}_1 - \bar{x}_2 = -2$ for both cases, this difference is small in case (a) in comparison to the large amount of variation within the two samples. In contrast, the difference in sample means in case (b) is very large in comparison to the variation within the two samples.

An analysis of variance to compare more than two population means formalizes the visual comparison of the variation between means with the variation within samples. This comparison of two sources of variation will lead us to the analysis of variance F test in Section 9.7.

FIGURE 13.1 Dot diagrams for comparing sample means

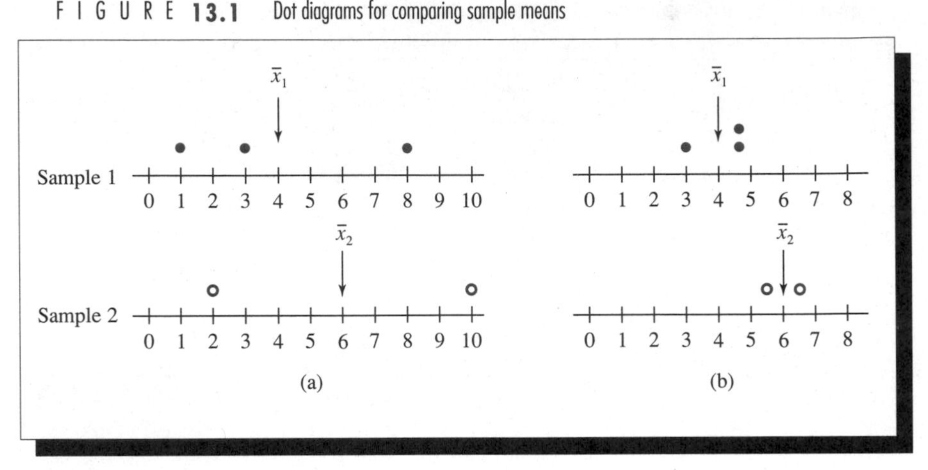

13.2 The Assumptions for an Analysis of Variance

The assumptions upon which the test and estimation procedures for an analysis of variance are based are similar to those required for the Student's t statistic of Chapter 9. Regardless of the sampling procedure employed to collect the data, we assume that the observations within each sample population are normally distributed with a common variance σ^2.

In this chapter, we will describe the analysis of variance for two different experimental designs. One is based on independent random sampling from the respective populations. The second is an extension of the matched-pairs design of Chapter 9 and involves the random assignment of treatments within matched sets of observations. The sampling procedure for each of these designs will be restated in their respective sections.

The assumptions for an analysis of variance are given in the following display.

Assumptions Underlying Analysis-of-Variance Test and Estimation Procedures

1. The observations within each population are normally distributed with a common variance σ^2.
2. Assumptions regarding the sampling procedure are specified for each design in the sections that follow.

13.3 The Completely Randomized Design: A One-Way Classification

The analysis of experimental data depends on the design of the experiment, which refers to the way the data were collected. A very useful and relatively simple design called the **completely randomized design** is one in which random samples are independently selected from each of k populations. This design results in observations that are classified only according to the population from which they came, hence, the designation as a one-way classification.

In some instances, this design is implemented by randomly selecting independent samples from the appropriate populations. For example, in assessing voter attitudes concerning various items on the ballot for the next state election, we may wish to select random samples of registered voters in each of k areas within the state. The k populations of registered voters exist in fact, and the design will specify how the registered voters in each of the k samples are to be selected. In this situation, we need a randomized device, such as a table of random numbers, to select those individuals who will be included in each of the samples from the k areas.

In other situations in which we wish to determine the effect of various treatments on a variable of interest, it may be that the populations exist only in concept, and our observations comprise samples from these conceptual populations. For example, we may be interested in evaluating several insecticides for use in agricultural pest control of the boll weevil in cotton. A completely randomized design in this situation would consist of the random assignment of the insecticides to experimental units consisting of a row or a fixed number of cotton plants. The sample observations from experimental units treated using insecticide number 1 constitute a sample from the conceptual population of all cotton plants treated with insecticide 1, and similarly for the other samples. If equal number of observations for each of the k treatments are to be taken, then a table of random numbers is used to first select the n experimental units to receive insecticide 1. The next n of the remaining units selected will receive insecticide 2, and so on. In both this example and the earlier example concerning voters, the design used was a completely randomized design resulting in a one-way classification of the data.

13.4 An Analysis of Variance for a Completely Randomized Design

The methodology, an analysis of variance, derives its name from the manner in which the quantities used to measure variation are acquired. Suppose we want to compare k population means $\mu_1, \mu_2, \ldots, \mu_k$, based on independent random samples of $n_1, n_2, \ldots, n_k$ observations selected from populations 1, 2, ..., k, respectively. Thus, for the health claims data in Table 13.1, we want to compare the means for $k = 5$ insurance groups based on random samples of $n_1 = n_2 = \cdots = n_5 = 10$ claims per group. Now let x_{ij} represent the jth measurement ($j = 1, \ldots, 10$) in the ith health group ($i = 1, \ldots, 5$). Then it can be shown (proof omitted) that the sum of squares of deviations of all $n = n_1 + n_2 + \cdots + n_5 = 50$ x values about their overall mean $\bar{x}$, often called the **total sum of squares** (or Total SS),

$$\text{Total SS} = \text{S}_{xx} = \sum_{i=1}^{k}\sum_{j=1}^{n_i}(x_{ij} - \bar{x})^2$$

can be partitioned into two components. The first component, called the **sum of squares for treatments (SST)** is used to calculate a measure of the variation between sample means. For our example, the **treatments** are the five insurance groups. The second component, called **sum of squares for error (SSE)**, is used to measure the variation within samples. Thus,

$$\text{Total SS} = \text{SST} + \text{SSE}$$

This partitioning of the Total SS into relevant components that measure sources of variation explains why the procedure is called an analysis of variance.

It is not difficult to calculate the sums of squares needed in an analysis of variance, but it is tedious. Since most statistical computer program packages contain programs to conduct analyses of variance for various experimental designs, we will discuss the procedure while explaining and interpreting the SAS printout for an analysis of variance of the health insurance claims data in Table 13.1. Corresponding MINITAB and EXECUSTAT printouts will also be presented so that you can compare the outputs for an analysis of variance for the same set of data. The computing formulas, which will enable you to perform the analysis of variance by using a calculator, are presented in optional Section 13.5.

EXAMPLE 13.1 Interpret the SAS analysis-of-variance printout and present corresponding MINITAB and EXECUSTAT printouts for the comparison of mean health insurance claims for the five groups (see Table 13.1).

Solution The SAS, MINITAB and EXECUSTAT computer printouts for an analysis of variance of the health insurance claims data are shown in Tables 13.2–13.4, respectively. Since the three printouts are similar, we will box and explain the meanings of the various quantities that appear in the SAS printout in Table 13.2 and we will box (where available) the corresponding quantities in the MINITAB and EXECUSTAT printouts in Tables 13.3 and 13.4.

1 The table in area ① in the SAS printout is called an **analysis-of-variance or ANOVA table**. This table contains five columns. The three sources of variation are listed under SOURCE in column 1:

MODEL: This source, sometimes identified as TREATMENTS, represents variation among the sample means. MINITAB calls it FACTOR, and EXECUSTAT identifies it as BETWEEN GROUPS.

TABLE 13.2 SAS computer printout for an analysis of variance of the health insurance data of Table 13.1

```
                         ANALYSIS-OF-VARIANCE PROCEDURE

DEPENDENT VARIABLE: COST
                                  ①
SOURCE            DF    SUM OF SQUARES      MEAN SQUARE   F VALUE      PR > F     R-SQUARE       C.V.

MODEL              4    6742554.48000000  1685638.62000000   0.98      0.4281     0.080122    72.0381

ERROR             45   77411264.40000000  1720250.32000000            ROOT MSE              COST MEAN
                                                                                ④
CORRECTED TOTAL   49   84153818.88000000                          1311.58313499         1820.68000000
                                   ②
SOURCE            DF           ANOVA SS  F VALUE    PR > F
                                                           ③
GROUP              4   6742554.48000000     0.98    0.4281
```

TABLE 13.3 MINITAB computer printout for an analysis of variance of the health insurance data of Table 13.1.

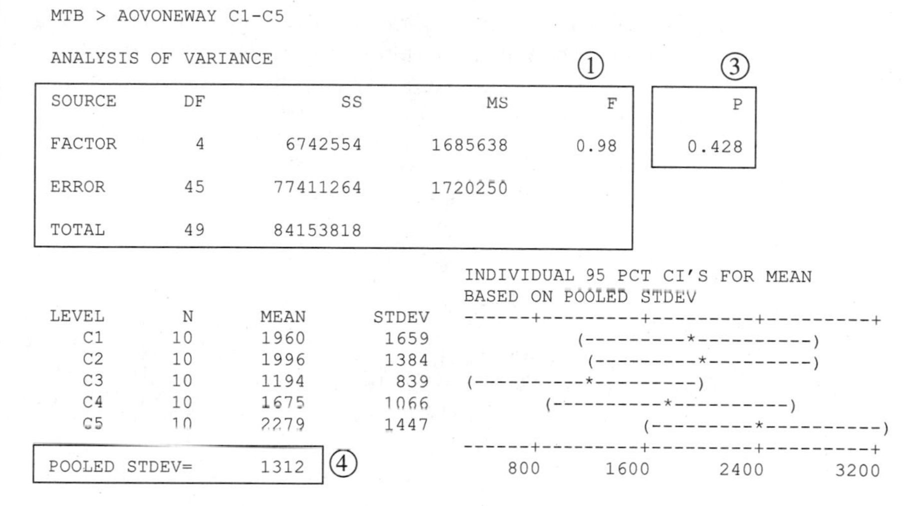

```
MTB > AOVONEWAY C1-C5

ANALYSIS OF VARIANCE                                        ①          ③
SOURCE      DF          SS            MS           F            P
FACTOR       4     6742554       1685638        0.98        0.428
ERROR       45    77411264       1720250
TOTAL       49    84153818

                                       INDIVIDUAL 95 PCT CI'S FOR MEAN
                                       BASED ON POOLED STDEV
LEVEL      N     MEAN     STDEV    ------+---------+---------+---------+
   C1     10     1960      1659               (----------*----------)
   C2     10     1996      1384                (----------*---------)
   C3     10     1194       839    (----------*---------)
   C4     10     1675      1066          (----------*----------)
   C5     10     2279      1447                    (----------*----------)
                                   ------+---------+---------+---------+
POOLED STDEV=     1312  ④               800      1600      2400      3200
```

TABLE 13.4 EXECUSTAT computer printout for an analysis of variance of the health insurance data of Table 13.1.

Analysis of Many Samples for IPS13-1.claims by IPS13-1.group

Class	Value	Sample Size	Mean	Standard Deviation
1	1	10	1959.8	1658.78
2	2	10	1996.1	1383.98
3	3	10	1193.9	838.92
4	4	10	1674.7	1066.33
5	5	10	2278.9	1446.88

① ③

Analysis of Variance

Source of Variation	Sum of Squares	D.F.	Mean Square	F-Ratio	P-Value
Between groups	6.74255e+006	4	1.68564e+006	0.98	0.4281
Within groups	7.74113e+007	45	1.72025e+006		
Total (corr.)	8.41538e+007	49			

95% Confidence Intervals (pooled)

Class	Value	Sample Size	Mean	+/- Interval
1	1	10	1.96e+003	835.4
2	2	10	2.00e+003	835.4
3	3	10	1.19e+003	835.4
4	4	10	1.67e+003	835.4
5	5	10	2.28e+003	835.4

ERROR: This source measures the variation within samples.

CORRECTED TOTAL: This source measures the variation of all x values about the overall mean of all $n = 50x$ values.

2 The sums of squares of deviations corresponding to the three sources of variation are shown in column 3 of area ① (column 2 in the EXECUSTAT printout). The sum of squares corresponding to MODEL, the sum of squares for treatments (or, for this example, insurance groups), SST, is

$$\text{SSE} = 6742554.48000000$$

The formula for computing SST is given in Section 13.5.

The sum of squares corresponding to ERROR is a measure of the variability of the x values within samples. Thus

$$\text{SSE} = 77411264.40000000$$

SSE is the pooled sum of squares of deviations of the observations about their respective sample means; that is,

$$\text{SSE} = \sum_{j=1}^{n_1}(x_{1j} - \bar{x}_1)^2 + \sum_{j=1}^{n_2}(x_{2j} - \bar{x}_2)^2 + \cdots + \sum_{j=1}^{n_5}(x_{5j} - \bar{x}_5)^2$$

This computing formula is given later in Section 13.5. Finally, the sum of squares of deviations corresponding to the CORRECTED TOTAL is what we have called Total SS or SS_{xx}, that is,

$$\text{Total SS} = \sum_{i=1}^{k}\sum_{j=1}^{n_1}(x_{ij} - \bar{x})^2 = 84153818.88$$

You can verify that

$$\text{SST} + \text{SSE} = \text{Total SS}$$

that is,

$$6742554.48 + 77411264.40 = 84153818.88$$

3 Each sum of squares of deviations, divided by the appropriate number of degrees of freedom, will provide an estimate of σ^2 when the null hypothesis is true, in other words, when

$$\mu_1 = \mu_2 = \cdots = \mu_5$$

These degrees of freedom are shown in column 2 (column 3 in EXECUSTAT).

a The degrees of freedom for the CORRECTED TOTAL (Total SS) will always be $(n - 1)$, or for this example, $50 - 1 = 49$.

b The degrees of freedom for MODEL (or insurance groups) will always equal one less than the number k of populations, in this case, $(k - 1) = 5 - 1 = 4$.

c The number of degrees of freedom for ERROR will always equal $n_1 + n_2 + \cdots + n_k - k = n - k$, or for this example, $50 - 5 = 45$. Note that the sum of the numbers of degrees of freedom for MODEL and ERROR will always equal the number of degrees of freedom for the CORRECTED TOTAL; that is, $4 + 45 = 49$.

4 Column 4 of the ANOVA table, headed MEAN SQUARE, gives the estimates of σ^2 based on the variation among the sample means (in the row corresponding to MODEL) and the variation within samples (in the row corresponding to ERROR) when the null hypothesis is true—that is, when $\mu_1 = \mu_2 = \mu_3 = \cdots = \mu_5$. These estimates are calculated by dividing a sum of squares by its corresponding degrees of freedom. Thus the **mean square for treatments** (MODEL), denoted as MST and shown in column 4, is

$$\text{MST} = \frac{\text{SST}}{k-1} = \frac{6742554.48}{4} = 1685638.62$$

Similarly, the **mean square for error**, denoted as MSE or s^2 and shown in column 4, is

$$\text{MSE} = s^2 = \frac{\text{SSE}}{n-k} = \frac{77411264.4}{45} = 1720250.32$$

This quantity s^2 is the pooled estimate of σ^2 based on the sum of squares of deviations of the x values about their respective sample means and is an extension (since it is based on $k = 5$ samples) of the pooled estimate of σ^2 given in Section 9.4.

5 The final step in testing $H_0 : \mu_1 = \mu_2 = \cdots = \mu_k$ is comparing the two estimates of σ^2: MST, which is based on the variation of the sample means about $\bar{x}$, and MSE $= s^2$, which is based on the variation of the x values about their respective sample means. We use the F statistic of Section 9.7. Thus, when H_0 is true, the sampling distribution of

$$F = \frac{\text{MST}}{\text{MSE}}$$

will be an F distribution with $v_1 = k - 1$ (for example, $k - 1 = 5 - 1 = 4$) numerator degrees of freedom and $v_2 = n - k$ (for our example, $n - k = 45$) denominator degrees of freedom. If H_0 is false—that is, if $\mu_1, \mu_2, \ldots, \mu_k$ are not all equal—the estimate of σ^2 based on MST will be overly large, and the calculated value of F will be larger than expected. Consequently, we reject H_0 for large values of F; that is, values of F larger than some critical value F_α (see Figure 13.2). The critical values of F corresponding to various values of v_1 and

FIGURE **13.2**
Rejection region for the analysis of variance F test

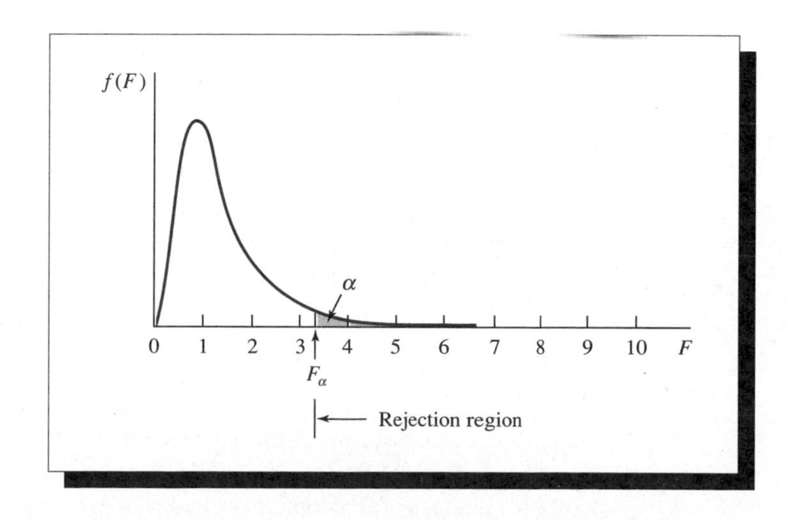

v_2 and for $\alpha = .10, .05, .025, .01$, and $.005$ are shown, respectively, in Table 6 of Appendix I. (The use of tables is explained in Section 9.7.)

For example, if we want to test

$$H_0 : \mu_1 = \mu_2 = \mu_3 = \mu_4 = \mu_5$$

for $\alpha = .05$, we consult Table 6 in Appendix I and look for the F value corresponding to $v_1 = 4$ and $v_2 = 45$. Table 6 does not give this F value, but it does give the F value for $v_1 = 4$ and $v_2 = 40$ as 2.61 and the F value for $v_1 = 4$ and $v_2 = 60$ as 2.53. Consequently, we will reject H_0 if the computed value of F is larger than 2.61 (or actually a number slightly smaller). The computed value of F,

$$F = \frac{\text{MST}}{\text{MSE}} = .98$$

is shown in column 4 of area (2). You can see that this computed value of F, .98, does not exceed the critical value and therefore does not fall in the rejection region. Consequently, there is not sufficient evidence to indicate that the mean claim size differs among the five insurance groups.

6 The observed significance level, the probability of observing a value of the F statistic as large or larger than .98, is shown in area (3) of the SAS printout. The p-value for the test is .4281. This large p-value is consistent with the results of the F test (step 5). For $\alpha = .05$, we will reject H_0 when the p-value is less than or equal to .05.

7 An analysis of variance can be conducted (as you will see in the following sections) to partition the Total SS into sums of squares corresponding to two or more sources of variation in addition to SSE. The SAS analysis of variance table in area (1) always combines these sources into a single source designated as MODEL. The MODEL source is partitioned and shown in area (2). For our example there is only one source in addition to SSE, namely, the sum of squares of deviations corresponding to treatments (insurance groups). Consequently, the sum of squares for MODEL is identical to the sum of squares for (insurance) GROUPS.

8 The standard deviation s, shown in area (4) is used to construct a confidence interval for a single mean or for the difference between a pair of means. Thus,

$$s = \sqrt{\text{MSE}} = \sqrt{1720250.32} = 1311.58313499$$

For example, if we want to find a $(1 - \alpha)$ 100% confidence interval for a population mean—say, that the mean size of a claim μ_4 for health group 4—we use the formula (given in Section 9.3)

$$\bar{x}_4 \pm t_{\alpha/2} \frac{s}{\sqrt{n_4}}$$

where $\bar{x}_4$ is the sample mean for insurance group 4, $S = 1311.58313499$, $n_4 = 10$, and $t_{\alpha/2}$ is based on $(n - k) = 50 - 5 = 45$ degrees of freedom (d.f.), the number of degrees of freedom associated with MSE $= s^2$. The t table, Table 4 in Appendix I does not give the t values for 45 d.f., but you can see that the value

$t_{.025}$ will be close to 2.0. Therefore, the 95% confidence interval for μ_4, the mean claim for health insurance group 4, is

$$\bar{x}_4 \pm t_{\alpha/2}\frac{s}{\sqrt{n_4}}$$

$$1674.7 \pm (2.0)\frac{1311.6}{\sqrt{10}} \qquad \text{or} \qquad \$845.2 \text{ to } \$2504.2$$

Means, standard deviations, and one-sample 95% confidence intervals are given on both the MINITAB and EXECUSTAT computer printouts.

If we want to estimate the difference in the size of the mean claims between health insurance groups 1 and 3 by using a $(1-\alpha)$ 100% confidence interval, we use the formula (Section 9.4)

$$(\bar{x}_1 - \bar{x}_3) \pm t_{\alpha/2}s\sqrt{\frac{1}{n_1}+\frac{1}{n_3}}$$

For a 95% confidence interval the value of $t_{.025}$ for d.f. $= 45$ will be approximately 2.0, $s = 1311.6$, and the values of $\bar{x}_1$ and $\bar{x}_3$ were shown in Table 13.1. Then the 95% confidence interval for $(\mu_1 - \mu_3)$ is

$$(\bar{x}_1 - \bar{x}_3) \pm t_{.025}s\sqrt{\frac{1}{n_1}+\frac{1}{n_3}}$$

$$(1959.8 - 1193.9) \pm (2.0)(1311.6)\sqrt{\frac{1}{10}+\frac{1}{10}}$$

$$765.9 \pm 1173.1$$

or from $-\$407.2$ to $\$1939.0$. Thus, we estimate the difference in mean claims for groups 1 and 3 to be in the interval $-\$407.2$ to $\$1939.0$. Because this interval includes 0 as a possible value, a t test of $H_0: \mu_1 = \mu_3$ would not lead to rejection of H_0. There is not sufficient evidence to indicate that μ_1 and μ_3 differ.

9. The quantity as R-SQUARE is related to a multiple regression analysis. The significance of R^2 was explained in Section 11.3.

Now that we have explained the SAS output, compare the SAS output with the MINITAB and the EXECUSTAT outputs in Tables 13.3 and 13.4. You will be able to locate the relevant sources of variation, degrees of freedom, sums of squares of deviations, and mean squares on the MINITAB and EXECUSTAT printouts, and you will see the quantities necessary to conduct the F test for comparing the $k = 5$ population means. ■

The typical ANOVA table for an analysis of variance for k independent random samples is shown below. The other two boxes summarize the analysis of variance F test and give the confidence intervals for treatment means.

ANOVA Table for k Independent Random Samples

Source	d.f.	SS	MS	F
Treatments	$k-1$	SST	MST = SST/$(k-1)$	MST/MSE
Error	$n-k$	SSE	MSE = SSE/$(n-k)$	
Total	$n-1$	Total SS		

F Test for Comparing k Population Means

1. *Null hypothesis*: $H_0 : \mu_1 = \mu_2 = \cdots = \mu_k$.
2. *Alternative hypothesis*: H_a : one or more pairs of population means differ.
3. *Test statistic*: $F =$ MST/MSE, where F is based on $v_1 = (k-1)$ and $v_2 = (n-k)$ d.f.
4. *Rejection region*: Reject H_0 if $F > F_\alpha$, where F_α lies in upper tail of the F distribution (with $v_1 = k-1$ and $v_2 = n-k$) and satisfies the expression $P(F > F_\alpha) = \alpha$.

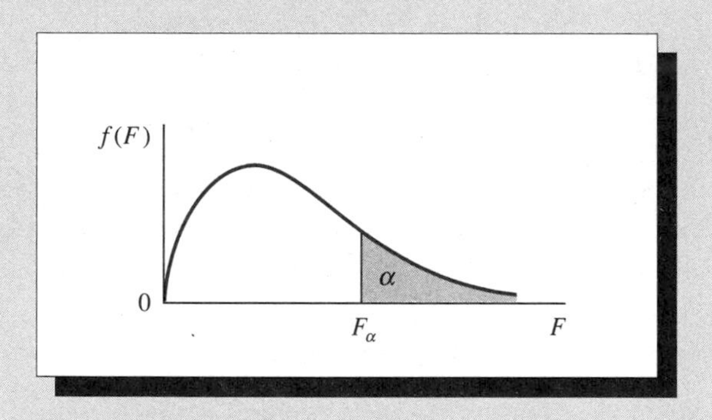

Assumptions

1. The samples have been randomly and independently selected from their respective populations.
2. The populations are normally distributed with means $\mu_1, \mu_2, \ldots, \mu_k$ and equal variances, $\sigma_1^2 = \sigma_2^2 = \cdots = \sigma_k^2 = \sigma^2$.

Independent Random Samples: $(1-\alpha)$ 100% Confidence Intervals for a Single Treatment Mean and the Difference between Two Treatment Means

Single Treatment Mean

$$\bar{x}_i \pm t_{\alpha/2}\frac{s}{\sqrt{n_i}}$$

Difference between Two Treatment Means

$$(\bar{x}_i - \bar{x}_j) \pm t_{\alpha/2}s\sqrt{\frac{1}{n_i}+\frac{1}{n_j}}$$

with

$$s = \sqrt{s^2} = \sqrt{\text{MSE}} = \sqrt{\frac{\text{SSE}}{n_1+n_2+\cdots+n_k-k}}$$

where $n = n_1 + n_2 + \cdots + n_k$ and $t_{\alpha/2}$ is based on $(n-k)$ d.f.

13.5 Computing Formulas: The Completely Randomized Design (Optional)

Suppose we want to compare k populations means $\mu_1, \mu_2, \ldots, \mu_k$, based on independent random samples, n_1 observations from population 1, n_2 from population 2, . . . , and n_k observations from population k. We will employ the following symbols for the quantities used to perform an analysis of variance of the data.

Notation

	Population			
Statistic	**1**	**2**	$\cdots$	k
Sample size	n_1	n_2	$\cdots$	n_k
Total of all observations in the sample	T_1	T_2	$\cdots$	T_k
Sample mean	$\bar{x}_1$	$\bar{x}_2$	$\cdots$	$\bar{x}_k$

$$T = \text{total of all } n = n_1 + n_2 + \cdots + n_k \text{ observations } = \sum_{i,j} x_{ij}$$

$$= T_1 + T_2 + \cdots + T_k$$

A summary of the computing formulas is given in the following display:

Summary of Computing Formulas: Analysis of Variance for an Independent Random Samples Design, k Treatments

$$\text{CM}^{\dagger} = \frac{(T)^2}{n}$$

$$\text{Total SS} = \sum_{ij} x_{ij}^2 - \text{CM}$$

$$= \text{sum of all squares of all } x \text{ values} - \text{CM}$$

where

$$n = n_1 + n_2 + \cdots + n_k$$

$$T = \text{total of all } n \text{ observations}$$

$$\text{SST} = \sum_{i=1}^{k} \frac{T_i^2}{n_i} - \text{CM} \qquad \text{MST} = \frac{\text{SST}}{k-1}$$

$$\text{SSE} = \text{Total SS} - \text{SST} \qquad \text{MSE} = \frac{\text{SSE}}{n-k}$$

Tips on Problem Solving

The following suggestions apply to all the analyses of variance in this chapter.

1. When calculating sums of squares, be certain to carry at least six significant figures before performing subtractions.
2. Remember, sums of squares can never be negative. If you obtain a negative sum of squares, you have made a mistake in arithmetic.
3. Always check your analysis-of-variance table to make certain that the degrees of freedom sum to the total degrees of freedom $n - 1$ and that the sum of squares sum to the Total SS.

† In the summary of formulas, the correction for the mean, abbreviated as CM, is used to convert a sum of squares into a sum of squared deviations about the mean.

Exercises

Basic Techniques

Starred (*) exercises are optional.

13.1 Suppose you wish to compare the means of six populations based on independent random samples, each of which contains ten observations. Insert, in an ANOVA table, the sources of variation and their respective degrees of freedom.

13.2 The values of Total SS and SSE for the experiment in Exercise 13.1 are Total SS $= 21.4$ and SSE $= 16.2$.

a Complete the ANOVA table for Exercise 13.1.

b How many degrees of freedom are associated with the F statistic for testing $H_0 : \mu_1 = \mu_2 = \cdots = \mu_6$?

c Give the rejection region for the test in part (b) for $\alpha = .05$.

d Do the data provide sufficient evidence to indicate differences among the population means?

13.3 The sample means corresponding to populations 1 and 2 in Exercise 13.1 are $\bar{x}_1 = 3.07$ and $\bar{x}_2 = 2.52$.

a Find a 95% confidence interval for μ_1.

b Find a 95% confidence interval for the difference $(\mu_1 - \mu_2)$.

13.4 Suppose you wish to compare the means of four populations based on independent random samples, each of which contains six observations. Insert, in an ANOVA table, the sources of variation and their respective degrees of freedom.

13.5 The values of Total SS and SST for the experiment in Exercise 13.4 are Total SS $= 473.2$ and SST $= 339.8$.

a Complete the ANOVA table for Exercise 13.4.

b How many degrees of freedom are associated with the F statistic for testing $H_0 : \mu_1 = \mu_2 = \mu_3 = \mu_4$?

c Give the rejection region for the test in part (b) for $\alpha = .10$.

d Do the data provide sufficient evidence to indicate differences among the population means?

13.6 The sample means corresponding to populations 1 and 2 in Exercise 13.4 are $\bar{x}_1 = 88.0$ and $\bar{x}_2 = 83.9$.

a Find a 90% confidence interval for μ_1.

b Find a 90% confidence interval for the difference $(\mu_1 - \mu_2)$.

13.7* The data shown below are observations collected using a completely randomized design:

Sample		
1	2	3
3	4	2
2	3	0
4	5	2
3	2	1
2	5	

a Calculate CM and Total SS.

b Calculate SST and MST.

c Calculate SSE and MSE.

d Construct an ANOVA table for the data.

e State the null and alternative hypotheses for an analysis-of-variance F test.

f Give the rejection region for the test using $\alpha = .05$.

g Conduct the test and state your conclusions.

13.8* Refer to Exercise 13.7. Do the data provide sufficient evidence to indicate a difference between μ_2 and μ_3? Test using the t test of Section 9.4 with $\alpha = .05$.

13.9* Refer to Exercise 13.7.

a Find a 90% confidence interval for μ_1.

b Find a 90% confidence interval for the difference $(\mu_1 - \mu_3)$.

Applications

13.10 A clinical psychologist wished to compare three methods for reducing hostility levels in university students. A certain psychological test (HLT) was used to measure the degree of hostility. High scores on this test were taken to indicate great hostility. Eleven students obtaining high and nearly equal scores were used in the experiment. Five were selected at random from among the 11 problem cases and treated by method A. Three were taken at random from the remaining six students and treated by method B. The other three students were treated by method C. All treatments continued throughout a semester. Each student was given the HLT test again at the end of the semester, with the following results:

Method	Scores on the HLT Tests				
A	73	83	76	68	80
B	54	74	71		
C	79	95	87		

```
MTB > ONEWAY C1 C2

ANALYSIS OF VARIANCE ON SCORES
SOURCE      DF          SS          MS          F          p
METHODS      2       641.9       320.9       5.15      0.037
ERROR        8       498.7        62.3
TOTAL       10      1140.5
                                        INDIVIDUAL 95 PCT CI'S FOR MEAN
                                        BASED ON POOLED STDEV
 LEVEL       N        MEAN       STDEV  ----+---------+---------+---------+--
     1       5      76.000       5.874              (-----*-----)
     2       3      66.333      10.786  (-------*--------)
     3       3      87.000       8.000                    (--------*-------)
                                        ----+---------+---------+---------+--
POOLED STDEV =       7.895                 60        72        84        96
```

Use the MINITAB analysis-of-variance computer printout given above to determine if the data provide sufficient evidence to indicate a difference in mean student response after treatment with one of the three methods.

a Perform an analysis of variance for this experiment.

b Do the data provide sufficient evidence to indicate a difference in mean student response to the three methods after treatment?

13.11 Refer to Exercise 13.10. Let μ_A and μ_B, respectively, denote the mean scores at the end of the semester for the populations of extremely hostile students who are treated throughout that semester by method A and method B.

a Find a 95% confidence interval for μ_A.

b Find a 95% confidence interval for μ_B.

c Find a 95% confidence interval for $\mu_A - \mu_B$.

d Is it correct to claim that the confidence intervals found in parts (a), (b), and (c) are jointly valid?

13.12 An experiment was conducted to compare the effectiveness of three training programs A, B, and C in training assemblers of a piece of electronic equipment. Fifteen employees were randomly assigned, five each, to the three programs. After completion of the courses, each person was required to assemble four pieces of the equipment, and the average length of time required to complete the assembly was recorded. Due to resignation from the company, only four employees completed program A, and only three completed B. The data are shown in the accompanying table. An SAS computer printout of the analysis of variance for the data is also shown below. Use the information in the printout to answer questions (a) through (d)

Training Program	Average Assembly Time (min)				
A	59	64	57	62	
B	52	58	54		
C	58	65	71	63	64

a Do the data provide sufficient evidence to indicate a difference in mean assembly time for people trained by the three programs? Give the p-value for the test and interpret its value.

b Find a 90% confidence interval for the difference in mean assembly time between persons trained by programs A and B.

c Find a 90% confidence interval for the mean assembly time for persons trained in program A.

d Do you think the data will satisfy (approximately) the assumption that they have been selected from normal populations? Why?

```
                        STATISTICAL ANALYSIS SYSTEM
                                  ANALYSIS-OF-VARIANCE PROCEDURE

DEPENDENT VARIABLE: X

SOURCE            DF    SUM OF SQUARES    MEAN SQUARE    F VALUE    PR > F    R-SQUARE          C.V.

MODEL              2      170.45000000    82.22500000       5.70    0.0251    0.559005        6.3802

ERROR              9      134.46666667    14.94074074                ROOT MSE                 Y MEAN

CORRECTED TOTAL   11      304.91666667                            3.86532544            60.58333333

SOURCE            DF          ANOVA SS        F VALUE     PR > F

TRTMENTS           2      170.45000000           5.70     0.0251
```

13.13 An ecological study was conducted to compare rate of growth of vegetation at four swampy undeveloped sites and to determine the cause of any differences that might be observed. Part of the study involved the measurement of leaf lengths of a particular plant species at a preselected date in May. Six plants were randomly selected at each of the four sites to be used in the comparison. The following data represent the mean leaf length per plant, in centimeters, for a random sample of ten leaves per plant:

Location	Mean Leaf Length (cm)					
1	5.7	6.3	6.1	6.0	5.8	6.2
2	6.2	5.3	5.7	6.0	5.2	5.5
3	5.4	5.0	6.0	5.6	4.9	5.2
4	3.7	3.2	3.9	4.0	3.5	3.6

The EXECUSTAT analysis of variance computer printout for these data is given below.

Analysis of Many Samples for Exercise 13.13

Class	Value	Sample Size	Mean	Standard Deviation
1	1	6	6.01667	0.231661
2	2	6	5.65	0.3937
3	3	6	5.35	0.408656
4	4	6	3.65	0.288097

Analysis of Variance

Source of Variation	Sum of Squares	D.F.	Mean Square	F-Ratio	P-Value
Between groups	19.74	3	6.58	57.38	0.0000
Within groups	2.29333	20	0.114667		
Total (corr.)	22.0333	23			

a You will recall that the test and estimation procedures for an analysis of variance require that the observations be selected from normally distributed (at least, roughly so) populations. Why might you feel reasonably confident that your data satisfy this assumption?

b Do the data provide sufficient evidence to indicate a difference in mean leaf length among the four locations?

c Suppose, prior to seeing our data, we decided to compare the mean leaf length of locations 1 and 4. Test the null hypothesis that $\mu_1 = \mu_4$ against the alternative that $\mu_1 \neq \mu_4$.

d Refer to part (c). Find a 90% confidence interval for $(\mu_1 - \mu_4)$.

e Rather than use an analysis-of-variance F test, it would seem simpler to examine one's data, select the two locations that have the smallest and largest sample mean lengths, and then compare these two means using a Student's t test. If there is evidence to indicate a difference in these means, there is clearly evidence of a difference among the four. (If you were to use this logic, there would be no need for the analysis-of-variance F test.) Explain why this procedure is invalid.

13.14 Water samples were taken at four different locations in a river to determine whether the quantity of dissolved oxygen, a measure of water pollution, varied from one location to another. Locations 1 and 2 were selected above an industrial plant, one near the shore and the other in midstream; location 3 was adjacent to the industrial water discharge for the plant; and location 4 was slightly downriver in midstream. Five water specimens were randomly selected at each location, but one specimen, corresponding to location 4, was lost in the laboratory. The data and a MINITAB analysis-of-variance computer printout follow (the greater the pollution, the lower the dissolved oxygen readings).

Location	Mean Dissolved Oxygen Content				
1	5.9	6.1	6.3	6.1	6.0
2	6.3	6.6	6.4	6.4	6.5
3	4.8	4.3	5.0	4.7	5.1
4	6.0	6.2	6.1	5.8	

```
ANALYSIS OF VARIANCE ON OXYGEN
SOURCE        DF          SS          MS          F          p
LOCATION       3      7.8361      2.6120      63.66      0.000
ERROR         15      0.6155      0.0410
TOTAL         18      8.4516
                                           INDIVIDUAL 95 PCT CI'S FOR MEAN
                                           BASED ON POOLED STDEV
LEVEL      N        MEAN       STDEV       ----+---------+---------+---------+--
    1      5      6.0800      0.1483                                (--*---)
    2      5      6.4400      0.1140                                     (--*---)
    3      5      4.7800      0.3114       (---*--)
    4      4      6.0250      0.1708                               (--*---)
                                           ----+---------+---------+---------+--
POOLED STDEV =    0.2026                     4.80      5.40      6.00      6.60
```

Use the MINITAB analysis-of-variance computer printout to answer the following questions.

a Do the data provide sufficient evidence to indicate a difference in mean dissolved oxygen content for the four locations?

b Compare the mean dissolved oxygen content in midstream above the plant with the mean content adjacent to the plant (location 2 versus location 3). Use a 95% confidence interval.

13.15 Larry R. Smeltzer (Louisiana State University) and Kittie W. Watson (Tulane University) conducted an experiment to investigate the effect of four different instructional approaches for improving learning. The treatments are listed below:

1 No instruction on listening

2 A 45-minute lecture on listening skills

3 A 30-minute video model on effective listening skills

4 Both of the educational exposures described in treatments 2 and 3

Ninety-nine subjects were randomly assigned to receive the treatments: 19 to the control group (treatment 1), 31 to treatment 2, 27 to treatment 3, and 22 to treatment 4. The educational messages presented in treatments 2, 3, and 4 emphasized the importance of asking questions and taking notes during discussions. After the treatments were applied, the students were scored on the basis of the numbers of questions asked. The analysis-of-variance table for the data is shown in the accompanying table:

Source	d.f.	SS	MS	F
Between groups	3	63.21	21.07	8.11[†]
Within groups	95	350.55	3.69	
Total	98	413.76		

[†] Significant at the .01 level.

Source: Smeltzer and Watson (1985).

a Why is "no instruction on listening" called the control group?

b Do the data present sufficient evidence to indicate differences in the mean number of questions asked among the four treatment groups? Test using $\alpha = .05$.

c The mean number of questions asked for each of the four treatments is shown in the accompanying table.

Treatment	Sample Size	Mean
1 Control group	19	1.36
2 Lecture group	31	1.87
3 Video role model	27	2.97
4 Lecture plus video role model	22	3.18

Compare the most intensive educational treatment, treatment 4, with the control, treatment 1. Do the data present sufficient evidence to indicate a difference in mean number of questions asked? Test using $\alpha = .05$.

13.6 The Randomized Block Design

The completely randomized design or one-way classification introduced in Sections 13.3 and 13.4 generalized the design involving two independent samples presented in Chapter 9. This design was deemed appropriate when the experimental material was homogeneous, and no sources of variation other than those due to the treatments and experimental error were expected to influence the response. If the experimental material is not homogeneous, we may be able to find groups of homogenous units, called **blocks**, within which the means associated with the treatments under investigation may be compared. This type of design, which is an extension of the paired-difference design of Chapter 9, is called a **randomized block design**. Its main purpose is to remove the block-to-block variability that might otherwise hide the effect of the treatments.

A randomized block design utilizes blocks of k-matched or homogeneous experimental units, with 1 unit within each block assigned to each treatment. The design is said to be randomized because the treatments are randomly assigned to the units within a block. If the randomized block design involves k treatments within each of b blocks, then the total number of observations in the experiment is $n = bk$.

For example, suppose the chief executive officer of a large construction corporation employs three experienced construction engineers to perform the time-consuming cost analyses, estimates, and bids for the work on large construction projects. It is important to know whether these three estimators tend to produce estimates at the same mean level or whether one or another tends to always submit a high (or low) bid on projects. One way to determine if differences in estimates exist would be to randomly select samples of projects estimated by each of the estimators and compare the means of theses estimates. However, project-to-project variability may mask any differences that actually might exist.

A much better experimental design for detecting differences in the average level of estimates for the three would be to conduct an experiment along the following lines. Each of the three estimators would be required to produce an analysis, and estimate, and a bid price for the *same* set of b projects. In this way, differences in bids for the same projects can be compared, thereby eliminating the project-to-project variation.

As another example, consider the problem of assessing the effects of three different package designs on the number or amount of sales. We might decide to use a completely randomized design and select 12 supermarkets and display each of the designs in four different markets. Unless the markets all had similar characteristics (size, sales volume, display areas, etc.), differences in sales for the three package designs might also reflect differences in the characteristics of the stores. One way to avoid this problem is to use, say, four stores and display each of the three designs in all four stores. In this way, by blocking on stores, the store-to-store variability has been eliminated by the choice of design.

Matching or blocking can take place in different ways. Comparisons of treatments are often made within blocks of time, within blocks of people, or within blocks of similar material or external environments.

13.7 An Analysis of Variance for a Randomized Block Design

An analysis of variance for a randomized block design partitions the total sum of squares of deviations for all x values about the overall mean $\bar{x}$ into **three** parts, the first measuring the variation among treatment means, the second measuring the variation among block means, and the third measuring the variation of the differences among the treatment observations *within* blocks (which measures experimental error). Thus,

$$\text{Total SS} = \text{SST} + \text{SSB} + \text{SSE}$$

where

$$\text{Total SS} = \sum_{ij}(x_{ij} - \bar{x})^2$$

$$\begin{aligned}
\text{SST} &= \text{sum of squares for treatments} \\
&= b(\text{sum of squares of deviations of the treatment means about } \bar{x}) \\
\text{SSB} &= \text{sum of squares for blocks} \\
&= k(\text{sum of squares of deviations of the block means about } \bar{x}) \\
\text{SSE} &= \text{sum of squares for error} = \text{Total SS} - \text{SST} - \text{SSB} \\
&= \text{unexplained variation}
\end{aligned}$$

To avoid distracting you with computational formulas and computations, we will explain how to perform an analysis of variance by explaining and interpreting an SAS analysis of variance printout for an example. We will present corresponding MINITAB and SPSS[x] printouts in case you have access to these program packages. The computing formulas and the calculator computations for the example are given in Section 13.8.

EXAMPLE 13.2 Refer to the comparison of the mean project bid price levels for the three construction project estimators described in Section 13.6. Each of the three estimators was required to analyze and determine a bid price for each of $b = 5$ projects. The data are shown in Table 13.5 and the SAS, MINITAB, and EXECUSTAT analysis-of-variance printouts are shown in Tables 13.6–13.8. Describe and interpret the SAS printout.

Solution 1 The SAS printout for an analysis of variance always presents the information for an ANOVA table in two stages. The first stage, area (1) in Table 13.6, shows only two sources of variation, MODEL and ERROR. The source MODEL *includes all sources of variation other than ERROR*.

TABLE 13.5 Bid price data (million $) for three estimators for each of five projects for Example 13.2

	Project					
Estimator	**1**	**2**	**3**	**4**	**5**	**Total**
1	3.52	4.71	3.89	5.21	4.14	21.47
2	3.39	4.79	3.82	4.93	3.96	20.89
3	3.64	4.92	4.19	5.10	4.20	22.05
Total	10.55	14.42	11.90	15.24	12.30	64.41

TABLE 13.6 SAS ANOVA printout for Example 13.2

```
                     ANALYSIS-OF-VARIANCE PROCEDURE

DEPENDENT VARIABLE: PRICE
                                        (1)
SOURCE            DF   SUM OF SQUARES   MEAN SQUARE   F VALUE    PR > F    R-SQUARE       C.V.

MODEL              6       5.02352000    0.83725333     99.32    0.0001    0.986753     2.1382

ERROR              8       0.06744000    0.0843000              ROOT MSE             PRICE TIME
                                                                           (4)
CORRECTED TOTAL   14       5.09096000                          0.09181503           4.294000000
                                        (2)
SOURCE            DF        ANOVA SS       F VALUE     PR> F   (3)

ESTIMATOR          2       0.13456000         7.98     0.0124

PROJECT            4       4.88896000       144.99     0.0001
```

TABLE 13.7 MINITAB ANOVA printout for Example 13.2

```
MTB > ANOVA C1 = C3 C2

Factor          Type Levels Values
ESTIMATR    fixed    3      1     2     3
PROJECT     fixed    5      1     2     3     4     5

Analysis of Variance for C1
                                                          (1)
Source        DF        SS          MS          F           P

ESTIMATR       2     0.13456     0.06728      7.98        0.012  (3)

PROJECT        4     4.88896     1.22224    144.99        0.000

Error          8     0.06744     0.00843

Total         14     5.09096
```

2 The source MODEL is broken down into its components, ESTIMATOR (treatments) and PROJECT (blocks), in the table area (2). This table gives the number of degrees of freedom for ESTIMATOR (treatments) and PROJECT (blocks). The number of degrees of freedom for treatments will always be one less than the number k of treatments, that is, $(k - 1)$. For our example, there were $k = 3$ estimators (treatments). Therefore, $(k - 1) = 3 - 1 = 2$. This number appears in the DF column in area (2) opposite ESTIMATOR. Similarly, if there are b blocks, the number of degrees of freedom for blocks is $(b - 1)$. For our example, there were $b = 5$ blocks. Therefore, the number of degrees of freedom for blocks is $(b - 1) = 5 - 1 = 4$. This number appears in the DF column in area (2) opposite PROJECT. The number of degrees of freedom for error will always equal

TABLE 13.8 EXECUSTAT ANOVA printout for Example 13.2

Two-way ANOVA for IPS13-2.PRICE

Source of Variation	Sum of Squares	D.F.	Mean Square	F-Ratio ①	P-Value ③
ESTIMATR	0.13456	2	0.06728	7.98102	0.0124
PROJECT	4.88896	4	1.22224	144.987	0.0000
Error	0.06744	8	0.00843		
Total (corr.)	5.09096	14			

Table of Means

ESTIMATR	Sample Size	Sample Mean	Standard Error	Estimated Effect
1	5	4.294	0.0410609	8.88178e-016
2	5	4.178	0.0410609	-0.116
3	5	4.41	0.0410609	0.116

PROJECT	Sample Size	Sample Mean	Standard Error	Estimated Effect
1	3	3.51667	0.0530094	-0.777333
2	3	4.80667	0.0530094	0.512667
3	3	3.96667	0.0530094	-0.327333
4	3	5.08	0.0530094	0.786
5	3	4.1	0.0530094	-0.194
Overall	15	4.294	0.0237065	

$(n - b - k + 1)$. This number, $(n - b - k + 1) = 15 - 5 - 3 + 1 = 8$, is shown in the DF column in area ① opposite ERROR. The number of degrees of freedom corresponding to Total SS is always equal to $(n - 1) = 15 - 1 = 14$. This number appears under DF in the row corresponding to CORRECTED TOTAL. Note that the sum of the *degrees of freedom corresponding to ESTIMATOR, PROJECT, and ERROR always equals the number of degrees of freedom corresponding to CORRECTED TOTAL*; that is, $(2 + 4 + 8) = 14$.

The traditional ANOVA format combines areas ① and ② into a single table by replacing the source MODEL in area ① with the sources in area ②. This traditional format is shown in the first box at the end of this section.

3 Column 3 of areas ① and ②, labeled SUM OF SQUARES and ANOVA SS, respectively, shows the sums of squares for the sources of variation. Thus, SSE, shown in area ① is

$$\text{SSE} = .06744000$$

The Total SS is also shown in area ① in the row corresponding to CORRECTED TOTAL. Thus,

$$\text{Total SS} = 5.09096000$$

The sums of squares for ESTIMATOR (treatments) and PROJECT (blocks) are shown in area (2). Thus,

$$\text{SST} = \text{SS(ESTIMATOR)} = .13456000$$
$$\text{SSB} = \text{SS(PROJECT)} = 4.88896000$$

4 Each mean square is obtained by dividing a sum of squares by its respective degrees of freedom. For example,

$$\text{MSE} = s^2 = \frac{\text{SSE}}{n - b - k + 1}$$

The mean squares for the sources of variation are shown in column 4 of area (1). Thus,

$$\text{MSE} = s^2 = .00843000$$
$$\text{MST} = \text{MS(ESTIMATOR)} = .06728000$$
$$\text{MSB} = \text{MS(PROJECT)} = 1.22224000$$

5 Under the null hypotheses that there are no differences among treatment means or that there are no differences among block means, the mean squares for treatments (MST), blocks (MSB), and error (MSE) provide independent estimates of the common population variance σ^2.

a To test H_0 : No differences among the k treatment means, we use $F =$ MST/MSE as the test statistic; we reject H_0 if $F > F_\alpha$, where F is based on the number of degrees of freedom associated with MST and MSE, namely, $v_1 = k - 1$ and $v_2 = n - b - k + 1$. The computed value of the F statistic, $F =$ MST/MSE, is shown in column 4 of area (2) as $F = 7.98$.[†] For $\alpha = .05$, the critical value of $F_{.05}$ for $v_1 = k - 1 = 2$ and $v_2 = n - b - k + 1 = 8$ is 4.46. Since the computed value of $F = 7.98$ exceeds this value, sufficient evidence indicates a difference among at least two of the treatment means. The observed significance level (p-value) for the test is shown in area (2) as .0124. Thus, the probability of observing an F value as large as or larger than $F = 7.98$, assuming H_0 true, is only .0214.

b To test H_0 : No differences among the b block means, we use $F =$ MSB/MSE as the test statistic; we reject H_0 if $F > F_\alpha$, where F is based on the number of degrees of freedom associated with MSB and MSE, namely, $v_1 = b - 1 = 4$ and $v_2 = n - b - k + 1 = 8$. The computed value of the F statistic, $F =$ MSB/MSE, is shown in column 4 of area (2) as $F = 144.99$[‡]. For $\alpha = .05$, the critical value of F for $v_1 = 4$ and $v_2 = 8$ d.f. is 3.84. Since the observed value of $F = 144.99$ greatly exceeds this critical value, ample evidence indicates differences among the block means. Since the sizes of the construction projects

[†] In the MINITAB and EXECUSTAT printouts, the calculated F values are in column 5 of area (1).

[‡] You cannot construct a confidence interval for a single mean unless the blocks have been randomly selected from among the population of all blocks. The procedure for constructing intervals for single means is beyond the scope of this text.

(in dollars) were known to vary over a wide range, we are not surprised to find differences in the mean values of the construction project bids.

6 The standard deviation, $s = \sqrt{\text{MSE}} = .09181503$, is shown in area (4). It can be used to construct a confidence interval for the difference between a pair of treatment means or between a pair of block means, and it can also be used to test for differences between pairs of means. The formulas and procedures are the same as those used for independent random samples. Since each treatment mean appears in each block, there are b observations per treatment. Therefore, a $(1-\alpha)$ 100% confidence interval for the difference between a pair of treatment means—say, i and j—is

$$(\bar{x}_i - \bar{x}_j) \pm t_{\alpha/2} s \sqrt{\frac{1}{b} + \frac{1}{b}}$$

$$(\bar{x}_i - \bar{x}_j) \pm t_{\alpha/2} s \sqrt{\frac{2}{b}}$$

where $t_{\alpha/2}$ is based on the number of degrees of freedom associated with s^2.

To illustrate, suppose we want to construct a 95% confidence interval for the difference between the mean estimates for treatments (estimators) 3 and 1. The means for these treatments, found in Table 13.8, are $\bar{x}_1 = 4.294$ and $\bar{x}_3 = 4.410$ and, from the printout in Table 13.6, $s = .09181503 \approx .0918$. Since s^2 is based on 8 d.f., $t_{\alpha/2} = t_{.025} = 2.306$, and the 95% confidence interval for $(\mu_3 - \mu_1)$ is

$$(\bar{x}_3 - \bar{x}_1) \pm t_{\alpha/2} s \sqrt{\frac{2}{b}}$$

$$(4.410 - 4.294) \pm (2.306)(.0918)\sqrt{\frac{2}{5}}$$

or $.116 \pm .134$. Therefore, we estimate the difference in mean level of estimates between estimators 3 and 1 to be from $-\$.018$ to $\$.250$ million. (Since this interval includes 0, there is not sufficient evidence to indicate a difference between μ_3 and μ_1). ■

The typical ANOVA table for k treatments in b blocks is shown below. The other two boxes summarize the analysis-of-variance F tests for comparing treatment and block means and give the confidence intervals for the differences between pairs of treatment and pairs of block means.

ANOVA Table for a Randomized Block Design, k Treatments and b Blocks

Source	d.f.	SS	MS	F
Treatments	$k-1$	SST	MST = SST/$(k-1)$	MST/MSE
Blocks	$b-1$	SSB	MSB = SSB/$(b-1)$	MSB/MSE
Error	$n-b-k+1$	SSE	MSE = SSE/$(n-b-k+1)$	
Total	$n-1$			

Tests for a Randomized Block Design

For comparing treatment means:

1. *Null hypothesis*: H_0 : the treatment means are equal.
2. *Alternative hypothesis*: H_a : at least two of the treatment means differ.
3. *Test Statistic*: $F =$ MST/MSE, where F is based on $v_1 = k-1$ and $v_2 = n-b-k+1$ d.f.
4. *Rejection region*: Reject if $F > F_\alpha$, where F_α lies in the upper tail of the F distribution (see the figure).

For comparing block means:

1. *Null hypothesis*: the block means are equal.
2. *Alternative hypothesis*: at least two of the block means differ.
3. *Test statistic*: $F =$ MSB/MSE, where F is based on $v_1 = b-1$ and $v_2 = n-b-k+1$ d.f.
4. *Rejection region*: Reject if $F > F_\alpha$, where F_α lies in the upper tail of the F distribution (see the figure).

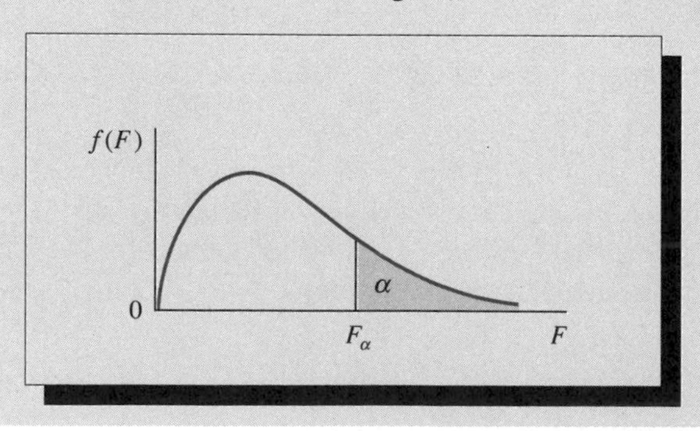

$(1-\alpha)$ 100% Confidence Intervals for the Difference between Pairs of Treatment and Block Means: A Randomized Design

Difference between Treatment Mean i and j:

$$(\bar{x}_i - \bar{x}_j) \pm t_{\alpha/2} s \sqrt{\frac{2}{b}}$$

where

$$b = \text{number of blocks}$$
$$s = \sqrt{\text{MSE}}$$

and $t_{\alpha/2}$ is based on $(n-b-k+1)$ d.f.

13.8 Computing Formulas: The Randomized Block Design (Optional)

We will use a notation given in the following box when conducting an analysis of variance for a randomized block design.

Notation

$$k = \text{number of treatments}$$
$$b = \text{number of blocks}$$
$$n = bk = \text{ total number of observations in the experiment}$$
$$T = \sum_{ij} x_{ij} = \text{ total of all observations in the experiment}$$
$$\bar{x} = \frac{T}{n} = \text{ mean of all observations in the experiment}$$
$$T_i = \text{total of all observations receiving treatment } i,\ i = 1, 2, \ldots, k$$
$$B_j = \text{total of all observations in block } j,\ j = 1, 2, \ldots, b$$

A summary of the computing formulas is given next.

Summary of Computing Formulas: Analysis of Variance for a Randomized Block Design, k Treatments in b Blocks

$$\text{CM} = \frac{(T)^2}{n}$$

where

$$n = bk$$

$$T = \text{sum of all } n \text{ observations}$$

$$\text{Total SS} = \sum_{ij} x_{ij}^2 - \text{CM}$$

$$= \text{sum of squares of all } x \text{ values} - \text{CM}$$

$$\text{SST} = \sum_{i=1}^{k} \frac{T_i^2}{b} - \text{CM} \qquad \text{MST} = \frac{\text{SST}}{k-1}$$

$$\text{SSB} = \sum_{j=1}^{b} \frac{B_j^2}{k} - \text{CM} \qquad \text{MSB} = \frac{\text{SSB}}{b-1}$$

Tips on Problem Solving

Be careful of this point: Unless the blocks have been randomly selected from a population of blocks, you cannot obtain a confidence interval for a single treatment mean. This limitation occurs because the sample treatment mean is biased by the positive and negative effects that the blocks have on the response.

Exercises

Basic Techniques

13.16 A randomized block design was conducted to compare the means of three treatments within six blocks. Construct an ANOVA table showing the sources of variation and their respective degrees of freedom.

13.17 Suppose that the analysis-of-variance calculations for Exercise 13.16 were SST $= 11.4$, SSB $= 17.1$, and Total SS $= 42.7$. Complete the ANOVA table, showing all sums of squares, mean squares, and pertinent F values.

13.18 Do the data of Exercise 13.16 provide sufficient evidence to indicate differences among the treatment means? Test using $\alpha = .05$.

13.19 Refer to Exercise 13.16. Find a 95% confidence interval for the difference between a pair of treatment means A and B if $\bar{x}_A = 21.9$ and $\bar{x}_B = 24.2$.

13.20 Do the data of Exercise 13.16 provide sufficient evidence to indicate that blocking increased the amount of information in the experiment about the treatment means? Justify your answer.

13.21 The data that follow are observations collected from an experiment that compared four treatments A, B, C, and D within each of three blocks, using a randomized block design.

	Treatment				
Block	**A**	**B**	**C**	**D**	**Total**
1	6	10	8	9	33
2	4	9	5	7	25
3	12	15	14	14	55
Total	22	34	27	30	113

Use the MINITAB printout below to answer the following questions.

```
MTB  > ANOVA C1=C2 C3;
SUBC > MEANS C3.

Factor      Type Levels Values
BLOCKS      fixed     3      1      2      3
TRTS        fixed     4      1      2      3      4

Analysis of Variance for Y

Source      DF         SS         MS        F        p
BLOCKS       2    120.667     60.333   135.75    0.000
TRTS         3     25.583      8.528    19.19    0.002
Error        6      2.667      0.444
Total       11    148.917

  MEANS

TRTS    N        Y
   1    3    7.333
   2    3   11.333
   3    3    9.000
   4    3   10.000
```

a Do the data present sufficient evidence to indicate differences among the treatment means? Test using $\alpha = .05$.

b Do the data present sufficient evidence to indicate differences among the block means? Test using $\alpha = .05$.

c Does it appear that the use of a randomized block design for this experiment was justified? Explain.

d Find a 90% confidence interval for the difference $(\mu_A - \mu_B)$.

13.22 The data shown below are observations collected from an experiment that compared three treatments A, B, and C within each of five blocks, using a randomized block design:

	Block					
Treatment	**1**	**2**	**3**	**4**	**5**	**Total**
A	2.1	2.6	1.9	3.2	2.7	12.5
B	3.4	3.8	3.6	4.1	3.9	18.8
C	3.0	3.6	3.2	3.9	3.9	17.6
Total	8.5	10.0	8.7	11.2	10.5	48.9

```
MTB  > ANOVA C1=C2 C3;
SUBC > MEANS C3.

Factor       Type Levels Values
BLOCKS      fixed      5      1      2      3      4      5
TRTS        fixed      3      1      2      3

Analysis of Variance for Y

Source       DF          SS          MS          F          p
BLOCKS        4      1.7960      0.4490      16.04      0.001
TRTS          2      4.4760      2.2380      79.93      0.000
Error         8      0.2240      0.0280
Total        14      6.4960


   MEANS

TRTS     N          Y
   1     5     2.5000
   2     5     3.7600
   3     5     3.5200
```

Use the preceding MINITAB printout to answer the following questions.

a Do the data present sufficient evidence to indicate differences among the treatment means? Test using $\alpha = 0.5$.

b Do the data present sufficient evidence to indicate differences among the block means? Test using $\alpha = .05$.

c Does it appear that the use of a randomized block design for this experiment was justified? Explain.

d Find a 99% confidence interval for the difference $(\mu_C - \mu_A)$.

13.23 The partially completed ANOVA table for a randomized block design is shown below:

Source	d.f.	SS	MS	F
Treatments	4	14.2		
Blocks		18.9		
Error	24			
Total	34	41.9		

a How many blocks were involved in the design?

b How many observations are in each treatment total?

c How many observations are in each block total?

d Fill in the blanks in the ANOVA table.

e Do the data present sufficient evidence to indicate differences among the treatment means? Test using $\alpha = .10$.

f Do the data present sufficient evidence to indicate differences among the block means? Test using $\alpha = .10$.

Applications

13.24 A study was conducted to compare automobile gasoline mileage for three brands of gasoline A, B, and C. Four automobiles, all of the same make and model, were employed in the experiment, and each gasoline brand was tested in each automobile. Using each brand within the same automobile has the effect of eliminating (blocking out) automobile-to-automobile variability. The data, in miles per gallon, are as follows:

Gasoline Brand	Automobile 1	2	3	4
A	15.7	17.0	17.3	16.1
B	17.2	18.1	17.9	17.7
C	16.1	17.5	16.8	17.8

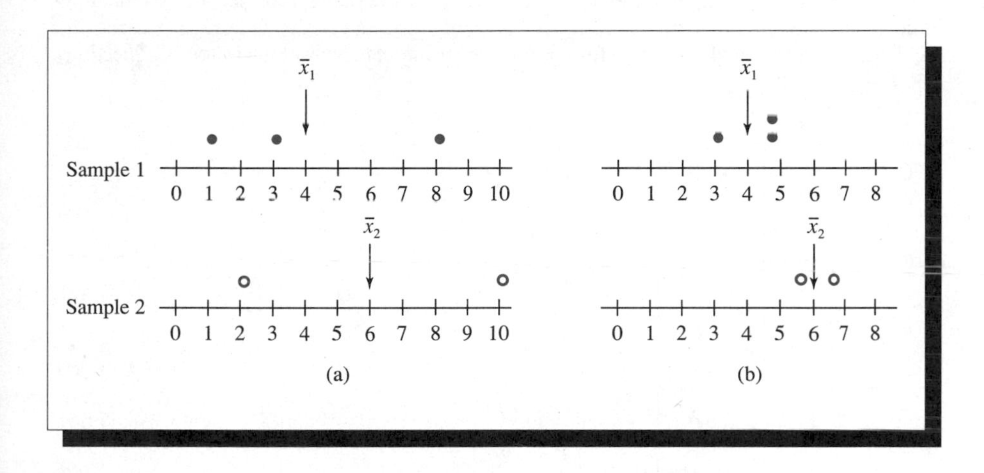

Two-way ANOVA for IPS13.24.MPG

Source of Variation	Sum of Squares	D.F.	Mean Square	F-Ratio	P-Value
TREATMENTS	2.895	2	1.4475	6.45725	0.0319
BLOCKS	2.52	3	0.84	3.74721	0.0792
Error	1.345	6	0.224167		
Total (corr.)	6.76	11			

Table of Means

TREATMENTS	Sample Size	Sample Mean	Standard Error	Estimated Effect
1	4	16.525	0.236731	-0.575
2	4	17.725	0.236731	0.625
3	4	17.05	0.236731	-0.05

BLOCKS	Sample Size	Sample Mean	Standard Error	Estimated Effect
1	3	16.3333	0.273354	-0.766667
2	3	17.5333	0.273354	0.433333
3	3	17.3333	0.273345	0.233333
4	3	17.2	0.273345	0.1
Overall	12	17.1	0.136677	

Use the preceding EXECUSTAT printout to answer the following questions:

a Do the data provide sufficient evidence to indicate a difference in mean mileage per gallon for the three gasolines?

b Is there evidence of a difference in mean mileage for the four automobiles?

c Suppose that *prior to looking at the data*, we had decided to compare the mean mileage per gallon for gasoline brands A and B. Find a 90% confidence interval for this difference.

13.25 An experiment was conducted to compare the effect of four different chemicals A, B, C, and D in producing water resistance in textiles. A strip of material, randomly selected from a bolt, was cut into four pieces, and the four pieces were randomly assigned to receive one of the four chemicals A, B, C, or D. This process was replicated three times, thus producing a randomized block design. The design, with moisture-resistance measurements, is as shown (low readings indicate low moisture penetration). An SAS computer printout of the analysis of variance for the data is also presented. Use the information in the printout to answer the following questions.

Blocks (bolt samples)

1	2	3
C 9.9	D 13.4	B 12.7
A 10.1	B 12.9	D 12.9
B 11.4	A 12.2	C 11.4
D 12.1	C 12.3	A 11.9

STATISTICAL ANALYSIS SYSTEM
ANALYSIS-OF-VARIANCE PROCEDURE

DEPENDENT VARIABLE: X

SOURCE	DF	SUM OF SQUARES	MEAN SQUARE	F VALUE	PR > F	R-SQUARE	C.V.
MODEL	5	12.37166667	2.47433333	27.75	0.0004	0.9548549	2.5023
ERROR	6	0.53500000	0.08916667		ROOT MSE		Y MEAN
CORRECTED TOTAL	11	12.90666667			0.29860788		11.93333333

SOURCE	DF	ANOVA SS	F VALUE	PR > F
BLOCKS	2	7.17166667	40.21	0.0003
TRTMENTS	3	5.20000000	19.44	0.0017

a Do the data provide sufficient evidence to indicate a difference in the mean moisture penetration for fabric treated with the four chemicals?

b Do the data provide evidence to indicate that blocking increased the amount of information in the experiment?

c Find a 95% confidence interval for the difference in mean moisture penetration for fabrics treated by chemicals A and D. Interpret the interval.

13.26 An experiment was conducted to compare the glare characteristics of four types of automobile rearview mirrors. Forty drivers were randomly selected to participate in the experiment. Each driver was exposed to the glare produced by a headlight located 30 feet behind the rear window of the experimental automobile. The driver then rated the glare produced by the rearview mirror on a scale of 1 (low) to 10 (high). Each of the four mirrors was tested by each driver;

the mirrors were assigned to a driver in random order. An analysis of variance of the data produced the following ANOVA table:

Source	d.f.	SS	MS	F
Mirrors		46.98		
Drivers			8.42	
Error				
Total		638.61		

a Fill in the blanks in the ANOVA table.

b Do the data present sufficient evidence to indicate differences in the mean glare ratings of the four rearview mirrors? Test using $\alpha = .10$.

c Do the data present sufficient evidence to indicate that the level of glare perceived by the drivers varied from driver to driver? Test using $\alpha = .10$.

13.27 An experiment was conducted to determine the effect of three methods of soil preparation on the first-year growth of slash pine seedlings. Four locations (state forest lands) were selected and each location was divided into three plots. Since it was felt that soil fertility within a location was more homogeneous than between locations, a randomized block design was employed using locations as blocks. The methods of soil preparation were A (no preparation), B (light fertilization), and C (burning). Each soil preparation was randomly applied to a plot within each location. On each plot the same number of seedlings were planted and the observation recorded was the average first-year growth of the seedlings on each plot.

Soil Preparation	Location 1	2	3	4
A	11	13	16	10
B	15	17	20	12
C	10	15	13	10

```
MTB  > ANOVA C1=C2 C3;
SUBC > MEANS C2 C3.

Factor          Type Levels Values
TREATMNTS      fixed      3      1      2      3
BLOCKS         fixed      4      1      2      3      4

Analysis of Variance for GROWTH

Source          DF         SS         MS         F         p
TREATMNTS        2     38.000     19.000     10.06     0.012
BLOCKS           3     61.667     20.556     10.88     0.008
Error            6     11.333      1.889
Total           11    111.000

    MEANS
TREATMNTS     N    GROWTH
        1     4    12.500
        2     4    16.000
        3     4    12.000
```

Use the preceding MINITAB printout to answer the following questions:

a Conduct an analysis of variance. Do the data provide evidence to indicate a difference in the mean growth for the three soil preparations?

b Is there evidence to indicate a difference in mean rates of growth for the four locations?

c Use a 90% confidence interval to estimate the difference in mean growth for methods A and B.

13.28 A study was conducted to compare the effect of three levels of digitalis on the level of calcium in the heart muscle of dogs. A description of the actual experimental procedure is omitted, but it is sufficient to note that the general level of calcium uptake varies from one animal to another so that comparison of digitalis levels (treatments) had to be blocked on heart muscles. That is, the tissue for a heart muscle was regarded as a block, and comparisons of the three treatments were made within a given muscle. The calcium uptakes for the three levels of digitalis A, B, and C were compared based on the heart muscle of four dogs. The results are as follows:

Dogs

1	2	3	4
A 1342	C 1698	B 1296	A 1150
B 1608	B 1387	A 1029	C 1579
C 1881	A 1140	C 1549	B 1319

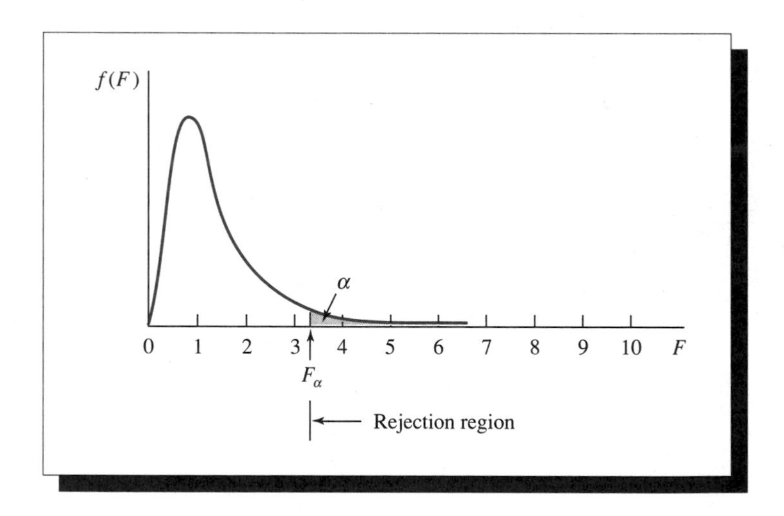

Two-way ANOVA for IPS13-28.CALCIUM

Source of Variation	Sum of Squares	D.F.	Mean Square	F-Ratio	P-Value
TREATMTS	524177	2	262089	258.237	0.0000
BLOCKS	173415	3	57805	56.9554	0.0001
Error	6089.5	6	1014.92		
Total (corr.)	703682	11			

Table of Means

TREATMTS	Sample Size	Sample Mean	Standard Error	Estimated Effect
1	4	1165.25	15.9289	-249.583
2	4	1402.5	15.9289	-12.3333
3	4	1676.75	15.9289	261.917

BLOCKS	Sample Size	Sample Mean	Standard Error	Estimated Effect
1	3	1610.33	18.3931	195.5
2	3	1408.33	18.3931	-6.5
3	3	1291.33	18.3931	-123.5
4	3	1349.33	18.3931	-65.5
Overall	12	1414.83	9.19654	

Use the preceding EXECUSTAT printout to answer the following questions:

a How many degrees of freedom are associated with SSE?

b Do the data present sufficient evidence to indicate a difference in the mean uptake of calcium for the three levels of digitalis?

c Do the data indicate a difference in the mean uptake in calcium for the four heart muscles?

d Give the standard deviation of the difference between the mean calcium uptake for two levels of digitalis.

e Find a 95% confidence interval for the difference in mean response between treatments A and B.

13.29 A building contractor employs three construction engineers A, B, and C to estimate and bid on jobs. To determine whether one tends to be a more conservative (or liberal) estimator than the others, the contractor selects four projected construction jobs, and has each estimator independently estimate the cost (dollars per square foot) of each job. The data are shown in the table:

Estimator (treatments)	Construction Job (blocks) 1	2	3	4	Total
A	35.10	34.50	29.25	31.60	130.45
B	37.45	34.60	33.10	34.40	139.55
C	36.30	35.10	32.45	32.90	136.75
Total	108.85	104.20	94.80	98.90	406.75

Two-way ANOVA for IPS13-29.COST

Source of Variation	Sum of Squares	D.F.	Mean Square	F-Ratio	P-Value
TREATMENTS	10.8617	2	5.43083	7.1958	0.0255
BLOCKS	37.6073	3	12.5358	16.6098	0.0026
Error	4.52833	6	0.754722		
Total (corr.)	52.9973	11			

Table of Means

TREATMENTS	Sample Size	Sample Mean	Standard Error	Estimated Effect
1	4	32.6125	0.434374	-1.28333
2	4	34.8875	0.434374	0.991667
3	4	34.1875	0.434374	0.291667
BLOCKS	**Sample Size**	**Sample Mean**	**Standard Error**	**Estimated Effect**
1	3	36.2833	0.501572	2.3875
2	3	34.7333	0.501572	0.8375
3	3	31.6	0.501572	-2.29583
4	3	32.9667	0.501572	-0.929167
Overall	12	33.8958	0.250786	

Use the preceding EXECUSTAT printout provided to answer the following questions:

a Do the data provide sufficient evidence to indicate a difference in the mean building costs estimated by the three estimators? Test by using $\alpha = .05$.

b Find a 90% confidence interval for the difference in the mean of the estimates produced by estimators A and B. Interpret the interval.

c Do the data support the contention that the mean estimate of the cost per square foot varies from job to job?

13.9 Some Cautionary Comments on Blocking

Two points need to be emphasized when using block designs. The first is that a randomized block design should not be used when treatments and blocks correspond to two **experimental** factors. In situations such as this, the effect of one factor usually depends on the level of the other, and the factors are said to **interact** rather than act independently. The randomized block design requires that the block and treatment factors **do not interact** in order to provide an unbiased estimate of the error variation. When there is interaction between the treatment and block factors, the analysis of variance could lead to erroneous conclusions regarding the relationship between the treatments and the response. Information on designs for multifactor experiments can be found in any standard text on the design of experiments.

The second point is that **blocking is not always beneficial**. A randomized block design produces a gain in information if the variation between blocks is greater than the variation within blocks. Then blocking removes this larger source of variation from SSE, and s^2 assumes a smaller value (as does the population variance σ^2). However, at the same time, you lose information because blocking reduces the number of degrees of freedom associated with SSE and σ^2. To see this, suppose we replicate each k treatments b times so that $n = bk$, first using a completely randomized design, and then using a randomized block design. The difference in the degrees of freedom for error are $(bk - k) - (b - 1)(k - 1) = b - 1$. Consequently, if blocking is to be beneficial, **the gain in information due to the elimination of block variation must outweigh the loss due to the reduction in the number of degrees of freedom for error**. However, unless blocking leaves only a small number of degrees of freedom for error, the loss of $(b - 1)$ d.f. for error causes only a small reduction in the information in the experiment. Consequently, if you have reason to suspect that the experimental units are not homogeneous, and can group the units into blocks, it usually pays to use the randomized block design.

13.10 Ranking Population Means

Many experiments are exploratory in nature. Thus, we have no preconceived notions about the results and have not decided (before conducting the experiment) to make specific treatment comparisons. Rather, we are searching for the treatment that possesses the largest treatment mean, possesses the smallest mean, or satisfies some other set of comparisons. When this situation occurs, we will want to rank the treatment means, determine which means differ, and identify sets of means for which no evidence of difference exists.

One way to achieve this goal is to order the sample means from the smallest to the largest and then to conduct t tests for adjacent means in the ordering. If two means differ by more than

$$t_{\alpha/2} s \sqrt{\frac{1}{n_1} + \frac{1}{n_2}}$$

you conclude that the pair of population means differ. The problem with this procedure is that the probability of making a type I error—that is, concluding that two means differ when, in fact, they are equal—is α for each test. If you compare a large number of pairs of means, the probability of detecting at least one difference in means, when, in fact, none exists, is quite large.

A simple way to avoid the high risk of proclaiming differences in multiple comparisons when they do not exist is to use the **studentized range**, the difference between smallest and the largest in a set of k sample means, as the yardstick for determining whether there is a difference in a pair of population means. This method, often called **Tukey's method for paired comparisons**, makes the probability of declaring that a difference exists between at least one pair in a set of k treatment means, when no difference exists, equal to α.

Tukey's method for making paired comparisons is based on the usual analysis-of-variance assumptions. **In addition, it assumes that the sample means are independent and based on samples of equal size.** The yardstick that determines whether a difference exists between a pair of treatment means is the quantity ω (Greek letter omega), which is presented next.

Yardstick for Making Paired Comparisons

$$\omega = q_\alpha(k, v)\frac{s}{\sqrt{n_t}}$$

where

k = number of treatments

s^2 = estimator of the common variance σ^2 (calculated in an analysis of variance)

v = number of degrees of freedom for s^2

n_t = common sample size, that is, the number of observations in each of the k treatment means

$q_\alpha(k, v)$ = tabulated value from Tables II(a) and II(b) in Appendix I, for $\alpha = .05$ and .01, respectively, and for various combinations of k and v

RULE: Two population means are judged to differ if the corresponding sample means differ by ω or more.

As noted in the box, the values of $q_\alpha(k, v)$ are listed in Tables II(a) and II(b) in Appendix I for $\alpha = .05$ and .01, respectively. A portion of Table II(a) in Appendix I is reproduced in Table 13.9. To illustrate the use of Tables I(a) and II(b), suppose you want to make pairwise comparisons of $k = 5$ means with $\alpha = .05$ for an analysis of variance, where s^2 possesses $v = 9$ d.f. The tabulated value for $k = 5$, $v = 9$, and $\alpha = .05$, shaded in Table 13.9, is $q_{.05}(5, 9) = 4.76$.

TABLE 13.9 A partial reproduction of Table II(a) in Appendix I; upper 5% points

v	2	3	4	5	6	7	8	9	10	11	12	...
1	17.97	26.98	32.82	37.08	40.41	43.12	45.40	47.36	49.07	50.59	51.96	
2	6.08	8.33	9.80	10.88	11.74	12.44	13.03	13.54	13.99	14.39	14.75	
3	4.50	5.91	6.82	7.50	8.04	8.48	8.85	9.18	9.46	9.72	9.95	
4	3.93	5.04	5.76	6.29	6.71	7.05	7.35	7.60	7.83	8.03	8.21	
5	3.64	4.60	5.22	5.67	6.03	6.33	6.58	6.80	6.99	7.17	7.32	
6	3.46	4.34	4.90	5.30	5.63	5.90	6.12	6.32	6.49	6.65	6.79	
7	3.34	4.16	4.68	5.06	5.36	5.61	5.82	6.00	6.16	6.30	6.43	
8	3.26	4.04	4.53	4.89	5.17	5.40	5.60	5.77	5.92	6.05	6.18	
9	3.20	3.95	4.41	4.76	5.02	5.24	5.43	5.59	5.74	5.87	5.98	
10	3.15	3.88	4.33	4.65	4.91	5.12	5.30	5.46	5.60	5.72	5.83	
11	3.11	3.82	4.26	4.57	4.82	5.03	5.20	5.35	5.49	5.61	5.71	
12	3.08	3.77	4.20	4.51	4.75	4.95	5.12	5.27	5.39	5.51	5.61	
⋮	⋮	⋮	⋮	⋮	⋮	⋮	⋮	⋮	⋮	⋮	⋮	

TABLE 13.10 Treatment means for Example 13.4

Estimator	2	1	3
Mean bid price	4.178	4.294	4.410

The following example will illustrate the use of Tables II(a) and II(b) in Appendix I in making paired comparisons.

EXAMPLE 13.3 In Example 13.2, we performed an analysis of variance on the bid price levels for three construction project estimators on the same five projects. The three estimator means are given in Table 13.10. Rank the means and determine which means, if any, differ from the others.

Solution For this example, there are $k = 3$ treatment means, each based on a sample of $b = n_t = 5$ observations. The standard deviation s, given by

$$s = \sqrt{\text{MSE}} = \sqrt{.06744} = .0918$$

is based on $v = 8$ d.f. (refer to area (4) on the SAS printout). Therefore, from Table II(a) in Appendix I, $q_{.05}(6, 12) = 4.04$. The yardstick for detecting a significant difference between a pair of treatment means is

$$\omega = q_{.05}(6, 12)\frac{s}{\sqrt{n}} = 4.04\frac{.0918}{\sqrt{5}} = .1659$$

The three treatment means are arranged in order from the smallest, 4.178 for estimator 2, to the largest, 4.410 for estimator 3. The next step is to check the difference between every pair of means. The difference between estimators 3 and 2 is $4.410 - 4.178 = .232$, which is greater than $\omega = .1659$ and therefore significant. The difference between estimators 3 and 1 is $4.410 - 4.294 = .116$, which is less than $\omega = .1659$ and therefore not significant. The third difference, between estimators 1 and 2 is .116 and is not significant. If there is no evidence of a significant difference between any pair of means, that fact is indicated by drawing a line under those two means, as in Figure 13.3.

The results here may seem confusing. However, it usually helps to think of ranking the means and interpreting nonsignificant differences as our inability to distinctly rank those means underlined by the same line. For this example, estimator

FIGURE 13.3

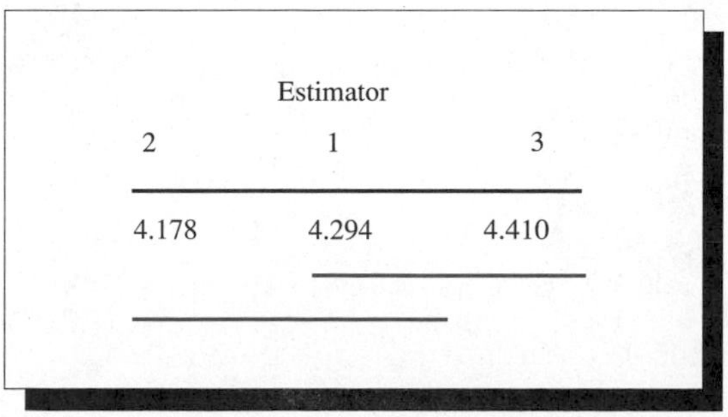

Estimator		
2	1	3
4.178	4.294	4.410

3 definitely ranked higher than estimator 2, but estimator 1 could not be ranked lower than estimator 3 nor higher than estimator 2. The probability that we make at least one error among the three comparisons is at most $\alpha = .05$. ▪

Exercises

Basic Techniques

13.30 Suppose that you wish to use Tukey's method of paired comparisons to rank a set of population means. In addition to the analysis-of-variance assumptions, what other property must the treatment means satisfy?

13.31 Consult Tables II(a) and II(b) in Appendix I and find the values of $q_\alpha(k, v)$:

a $\alpha = .05, k = 5, v = 7$ **b** $\alpha = .05, k = 3, v = 10$

c $\alpha = .01, k = 4, v = 8$ **d** $\alpha = .01, k = 7, v = 5$

13.32 If the sample size for each treatment is n_t and if s^2 is based on 12 d.f. find ω:

a $\alpha = .05, k = 4, n_t = 5$ **b** $\alpha = .01, k = 6, n_t = 8$.

13.33 An independent random sampling design was employed to compare the means of six treatments based on samples of four observations per treatment. The pooled estimator σ^2 is 9.12, and the sample means follow:

$$\bar{x}_1 = 101.6 \quad \bar{x}_2 = 98.4 \quad \bar{x}_3 = 112.3$$
$$\bar{x}_4 = 92.9 \quad \bar{x}_5 = 104.2 \quad \bar{x}_6 = 113.8$$

a Give the value of ω that you would use to make pairwise comparisons of the treatment means for $\alpha = .05$.

b Rank the treatment means using pairwise comparisons.

13.34 Refer to Exercise 13.28. Rank the mean calcium uptake for the three levels of digitalis with $\alpha = .05$ using Tukey's procedure.

13.35 Refer to Exercise 13.13. Rank the mean leaf growth for the four locations. Use $\alpha = .01$.

13.36 Refer to Exercise 13.24. Rank the three treatments using Tukey's procedure. Use $\alpha = .01$.

13.37 Refer to Exercise 13.25. Use Tukey's ranking procedure to rank the four means with respect to water resistance at level $\alpha = .05$.

13.38 Refer to Exercise 13.27. Use Tukey's ranking procedure to rank the three treatment means with respect to the average first-year growth. Use $\alpha = .05$

13.11 Assumptions

The assumptions about the probability distribution of the random response x that result in a valid analysis of variance are given in the following display.

Assumptions

1. For any treatment or block combination or combination of factor levels, x is normally distributed with variance equal to σ^2.
2. The observations have been selected independently so that any pair of x values, x_i and x_j, are independent in a probabilistic sense for all i and $j (i \neq j)$.

In a practical situation, you can never be certain that the assumptions are satisfied, but you will often have a fairly good idea whether the assumptions are reasonable for your data. To illustrate, the inferential methods of Chapters 7 through 11 are not seriously affected by moderate departures from the assumptions of normality, but you would want the probability of x to be at least mound-shaped. So if x is a discrete random variable that can assume only three values—say, $x = 10, 11, 12$—then it is *unreasonable* to assume that the probability distribution of x is approximately normal.

The assumption of a constant variance for x for the various experimental conditions should be approximately satisfied, although violation of this assumption is not too serious if the sample sizes for the various experimental conditions are equal. However, suppose the response is binomial—say, the proportion p of people who favor a particular type of investment. We know that the variance of a proportion is

$$\sigma_{\hat{p}}^2 = \frac{pq}{n}, \qquad \text{where } q = 1 - p$$

and therefore that the variance is dependent on the expected (or mean) value of $\hat{p}$, namely, p. Then the variance of $\hat{p}$ will change from one experimental setting to another, and the assumptions of the analysis of variance will have been violated.

A similar situation occurs when the response measurements are Poisson data (say, the number of industrial accidents per month in a manufacturing plant). If the response possesses a Poisson probability distribution, the variance of the response equals the mean. Consequently, Poisson response data also violate the analysis of variance assumptions.

Many kinds of data are not measurable and hence are unsuitable for an analysis of variance. For example, many responses cannot be measured but can be ranked. Product preference studies yield data of this type. You know you liked product A better than B, and B better than C, but you have difficulty assigning an exact value to the strength of your preferences.

What do you do when the assumptions of an analysis of variance are not satisfied? For example, suppose the variances of the responses for various experimental conditions are not equivalent. This situation can sometimes be remedied by transforming the response measurements. That is, instead of using the original response measurements, we might use their square roots, logarithms, or some other function of the response x. Transformations that tend to stabilize the variance of the response have been found to make the probability distributions of the transformed responses more nearly normal.

When nothing can be done to satisfy (even approximately) the assumptions of the analysis of variance, or if the data are rankings, you should use nonparametric testing and estimation procedures. These procedures, which rely on the comparative magnitudes of measurements (often ranks), are almost as powerful in detecting treatment differences as the tests presented in this chapter. When the assumptions are not satisfied, the nonparametric procedures may be more powerful. Nonparametric tests—what they are and how to use them—are presented in Chapter 14.

13.12 Summary

The completely randomized and the randomized block designs are procedures that specify the manner in which the treatments are assigned to the units to be used in an experiment. The **completely randomized design** is used to investigate the effect of one independent variable corresponding to treatments when the experimental units are relatively homogeneous. The **randomized block design**, which involves two independent variables corresponding to treatments and blocks, is used when the experimental units are not homogeneous but can be grouped into blocks such that the variability within blocks is less than the variability between blocks. The randomized block design is used to control the effect of the block variable and remove its effect from experimental error.

The **analysis of variance** is a technique that partitions the total sum of squares of deviations of the observations about their mean into portions associated with the independent variables in the experiment and a portion associated with error. The mean squares corresponding to the independent variables are tested against the mean square for error by using the F statistic. A significantly large value of F indicates that the mean square under test is larger than would be expected if the corresponding independent variable had no effect on the response of interest.

This chapter represents a brief introduction to the design of experiments and the widely used technique for analyzing designed experiments called the analysis of variance. Designs are available for experiments involving several design variables as well as several treatment factors. **Design variables** correspond to factors whose effect we wish to control and hence remove from experimental error; **treatment factors** are the experimental variables whose effects we wish to investigate. A properly designed experiment can be analyzed by using an analysis of variance. Experiments in which the levels of one or more variables are measured (as opposed to being set at preselected values) can be analyzed by using *multiple regression analysis*, which was the subject of Chapter 11.

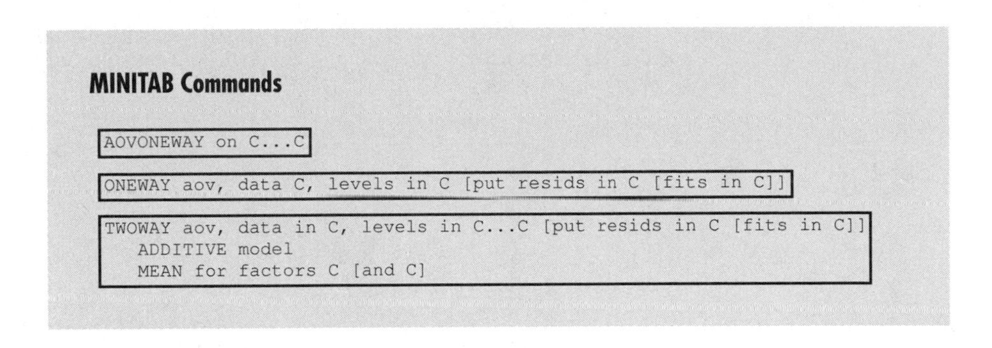

MINITAB Commands

```
AOVONEWAY on C...C

ONEWAY aov, data C, levels in C [put resids in C [fits in C]]

TWOWAY aov, data in C, levels in C...C [put resids in C [fits in C]]
   ADDITIVE model
   MEAN for factors C [and C]
```

Supplementary Exercises

13.39 State the assumptions underlying the analysis of variance of a completely randomized design.

13.40 A completely randomized design was conducted to compare the effect of five different stimuli on reaction time. Twenty-seven people were employed in the experiment, which was conducted using a completely randomized design. Regardless of the results of the analysis of variance, it was desired to compare stimuli A and D. The results of the experiment were as follows:

Stimulus	Reaction Time (sec)							Total	Mean
A	.8	.6	.6	.5				2.5	.625
B	.7	.8	.5	.5	.6	.9	.7	4.7	.671
C	1.2	1.0	.9	1.2	1.3	.8		6.4	1.067
D	1.0	.9	.9	1.1	.7			4.6	.920
E	.6	.4	.4	.7	.3			2.4	.48

```
MTB > ONEWAY C1 C2

ANALYSIS OF VARIANCE ON TIMES
SOURCE      DF        SS        MS         F        p
TREATMTS     4    1.2118    0.3030     11.67    0.000
ERROR       22    0.5711    0.0260
TOTAL       26    1.7830
                                    INDIVIDUAL 95 PCT CI'S FOR MEAN
                                    BASED ON POOLED STDEV
LEVEL       N      MEAN     STDEV   --------+---------+---------+---------
    1       4    0.6250    0.1258         (------*------)
    2       7    0.6714    0.1496             (----*----)
    3       6    1.0667    0.1966                            (-----*----)
    4       5    0.9200    0.1483                      (-----*-----)
    5       5    0.4800    0.1643   (-----*-----)
                                    -------+---------+---------+---------
POOLED STDEV = 0.1611                    0.50      0.75      1.00
```

Use the preceding MINITAB printout to answer the following questions:

a Conduct an analysis of variance and test for a difference in mean reaction time due to the five stimuli.

b Compare stimuli A and D to see if there is a difference in mean reaction time.

13.41 The experiment in Exercise 13.40 might have been more effectively conducted using a randomized block design with people as blocks, since we would expect mean reaction time to vary from one person to another. Hence, four people were used in a new experiment and each person was subjected to each of the five stimuli in a random order. The reaction times (in seconds) were as follows:

	Stimulus				
Subject	**A**	**B**	**C**	**D**	**E**
1	.7	.8	1.0	1.0	.5
2	.6	.6	1.1	1.0	.6
3	.9	1.0	1.2	1.1	.6
4	.6	.8	.9	1.0	.4

```
FACTOR        Type  Levels  Values
TREATMTS     fixed       5       1       2       3       4       5
BLOCKS       fixed       4       1       2       3       4

ANALYSIS OF VARIANCE FOR TIMES

SOURCE          DF           SS          MS          F          p
TREATMTS         4      0.78700     0.19675      27.78      0.000
BLOCKS           3      0.14000     0.04667       6.59      0.007
ERROR           12      0.08500     0.00708
TOTAL           19      1.01200


     MEANS

TREATMTS     N       TIMES
       1     4      0.7000
       2     4      0.8000
       3     4      1.0500
       4     4      1.0250
       5     4      0.5250
```

Use the preceding MINITAB printout and conduct an analysis of variance and test for differences in treatments (stimuli).

13.42 An experiment was conducted to examine the effect of age on heart rate when a person is subjected to a specific amount of exercise. Ten male subjects were randomly selected from four age groups: 10–19, 20–39, 40–59, and 60–69. Each subject walked on a treadmill at a fixed grade for a period of 12 minutes, and the increase in heart rate, the difference before and after exercise, was recorded (in beats per minute). These data are shown in the table:

	Age			
	10–19	**20–39**	**40–59**	**60–69**
	29	24	37	28
	33	27	25	29
	26	33	22	34
	27	31	33	36
	39	21	28	21
	35	28	26	20
	33	24	30	25
	29	34	34	24
	36	21	27	33
	22	32	33	32
Total	309	275	295	282

Use an appropriate computer program to answer the following questions:

a Do the data provide sufficient evidence to indicate a difference in mean increase in heart rate among the four age groups? Test by using $\alpha = .05$.

b Find a 90% confidence interval for the difference in mean increase in heart rate between age groups 10–19 and 60–69.

c Find a 90% confidence interval for the mean increase in heart rate for the age group 20–39.

d Approximately how many people would you need in each group if you wanted to be able to estimate a group mean correct to within two beats per minute with probability equal to .95?

13.43 A company wished to study the differences among four sales-training programs on the sales abilities of their sales personnel. Thirty-two people were randomly divided into four groups of equal size, and the groups were then subjected to the different sales-training programs. Because there were some dropouts during the training programs, due to illness, vacations, and so on, the number of trainees completing the programs varied from group to group. At the end of the training programs, each salesperson was randomly assigned a sales area from a group of sales areas that were judged to have equivalent sales potentials. The number of sales made by each of the four groups of salespeople during the first week after completing the training program are listed in the table:

	Training Program			
	1	2	3	4
	78	99	74	81
	84	86	87	63
	86	90	80	71
	92	93	83	65
	69	94	78	86
	73	85		79
		97		73
		91		70
Total	482	735	402	588

```
MTB > NAME C1 'SALES'
MTB > ONEWAY C1 C2

ANALYSIS OF VARIANCE ON SALES

SOURCE      DF        SS         MS         F         p
TREATMTS     3    1385.8      461.9      9.84     0.000
ERROR       23    1079.4       46.9
TOTAL       26    2465.2
                                     INDIVIDUAL 95 PCT CI'S FOR MEAN
                                     BASED ON POOLED STDEV
LEVEL        N      MEAN      STDEV  -----+---------+---------+---------+-
    1        6    80.333      8.595          (-------*--------)
    2        8    91.875      4.912                          (-----*-----)
    3        5    80.400      4.930          (--------*-------)
    4        8    73.500      7.964  (-----*-----)
                                     -----+---------+---------+---------+-
POOLED STDEV =     6.851                  72.0      80.0      88.0      96.0
```

Use the preceding MINITAB printout to answer the following questions:

a Do the data present sufficient evidence to indicate a difference in the mean achievement for the four training programs?

b Find a 90% confidence interval for the difference in the mean of sales that would be expected for persons subjected to training programs 1 and 4. Interpret the interval.

c Find a 90% confidence interval for the mean number of sales by persons subjected to training program 2.

13.44 The whitefly, which causes defoliation of shrubs and trees and a reduction in salable crop yields, has emerged as a pest in southern California. In a study to determine factors affecting the life cycle of the whitefly, an experiment was conducted in which whiteflies were placed on two different types of plants at three different temperatures. The observation of interest was the total number of eggs laid by caged females under one of the six possible treatment combinations. Each treatment combination was run using five cages.

	Cotton			Cucumber		
Temperature	**70°**	**77°**	**82°**	**70°**	**77°**	**82°**
	37	34	46	50	59	43
	21	54	32	53	53	62
	36	40	41	25	31	71
	43	42	36	37	69	49
	31	16	38	48	51	59

Use an appropriate computer program to answer the following questions:

a Do the data provide sufficient evidence to indicate that the mean number of eggs laid varies with differences in plant type and temperature?

b Plot the treatment means for cotton as a function of temperature. Plot the treatment means for cucumber as a function of temperature. Comment on the similarity or disparity of these two plots.

c Find the mean number of eggs laid on cotton and cucumber based on 15 observations each. Calculate a 95% confidence interval for the difference in the underlying population means. (HINT: The standard error is calculated as usual, but with $n_1 = 15$ and $n_2 = 15$.)

13.45 Four chemical plants, producing the same product and owned by the same company, discharge effluents into streams in the vicinity of their locations. To check on the extent of the pollution created by the effluents and to determine if this varies from plant to plant, the company collected random samples of liquid waste, five specimens for each of the four plants. The data are shown in the accompanying table. An SAS computer printout of the analysis of variance for the data is also shown. Use the information in the printout to answer the following questions.

Plant	Polluting Effluents (lb/gal of waste)				
A	1.65	1.72	1.50	1.37	1.60
B	1.70	1.85	1.46	2.05	1.80
C	1.40	1.75	1.38	1.65	1.55
D	2.10	1.95	1.65	1.88	2.00

a Do the data provide sufficient evidence to indicate a difference in the mean amount of effluents discharged by the four plants?

b If the maximum mean discharge of effluents is 1.5 lb/gal, do the data provide sufficient evidence to indicate that the limit is exceeded at plant A?

c Estimate the difference in the mean discharge of effluents between plants A and D, using a 95% confidence interval.

```
                      STATISTICAL ANALYSIS SYSTEM
                    ANALYSIS-OF-VARIANCE PROCEDURE

DEPENDENT VARIABLE: X

SOURCE            DF   SUM OF SQUARES   MEAN SQUARE    F VALUE    PR > F    R-SQUARE        C.V.

MODEL              3       0.46489500    0.15496500       5.20    0.0107    0.493679     10.1515

ERROR             16       0.47680000    0.02980000             ROOT MSE                  Y MEAN

CORRECTED TOTAL   19       0.94169500                         0.17262677              1.70050000

SOURCE            DF         ANOVA SS       F VALUE     PR > F

TRTMENTS           3       0.46489500          5.20     0.0107
```

13.46 In Exercise 9.36, we examined an advertisement for Albertson's, a supermarket chain in the western United States. The advertiser claims that Albertson's has been consistently lower than four other full service supermarkets. As part of a survey conducted by an "independent market basket price-checking company" (*Press-Enterprise*, Riverside, Calif., Feb. 11, 1993), the average weekly total based on the prices of approximately 95 items is given for five different supermarket chains recorded during four consecutive weeks in December 1992 and January, 1993.

Week	Albertson's	Ralphs	Vons	Alpha Beta	Lucky
1	254.26	256.03	267.92	260.71	258.84
2	240.62	255.65	251.55	251.80	242.14
3	231.90	255.12	245.89	246.77	246.80
4	234.13	261.18	254.12	249.45	248.99

a What type of design has been used in this experiment?

b Use an appropriate computer software program to conduct an analysis of variance for the data.

c Is there sufficient evidence to indicate that there is a difference in the average weekly totals for the five supermarkets? Use $\alpha = .05$.

d Use Tukey's method for paired comparisons to determine which of the means are significantly different from each other. Use $\alpha = .05$.

13.47 The yields of wheat in bushels per acre were compared for five different varieties A, B, C, D, and E at six different locations. Each variety was randomly assigned to a plot at each location. The results of the experiment are shown in the accompanying table. An SAS computer printout for the analysis of variance for the data is also shown below. Use the information in the printout to answer the following questions.

	Location					
Varieties	**1**	**2**	**3**	**4**	**5**	**6**
A	35.3	31.0	32.7	36.8	37.2	33.1
B	30.7	32.2	31.4	31.7	35.0	32.7
C	38.2	33.4	33.6	37.1	37.3	38.2
D	34.9	36.1	35.2	38.3	40.2	36.0
E	32.4	28.9	29.2	30.7	33.9	32.1

```
                         STATISTICAL ANALYSIS SYSTEM
                       ANALYSIS-OF-VARIANCE PROCEDURE

DEPENDENT VARIABLE: X

SOURCE              DF    SUM OF SQUARES    MEAN SQUARE     F VALUE     PR > F     R-SQUARE          C.V.

MODEL                9      210.81166667    23.42351852       12.22     0.0001     0.846152        4.0499

ERROR               20       38.33000000     1.91650000                ROOT MSE                    Y MEAN

CORRECTED TOTAL     29      249.14166667                             1.38437712               34.18333333

SOURCE              DF          ANOVA SS        F VALUE     PR >F

BLOCKS               5       68.14166667           7.11    0.0006

TRTMENTS             4      142.67000000          18.61    0.0001
```

a Do the data provide sufficient evidence to indicate a difference in the mean yield for the five varieties of wheat?

b Do the data provide sufficient evidence to indicate that blocking on "location" increased the amount of information in the experiment?

c Find a 95% confidence interval for the difference in mean yields for varieties C and E.

13.48 As Detroit automakers become more and more financially strapped, they are beginning to phase out the generous discounts that they had previously given to car rental companies. As a result, the cost of rental cars is going up, and the availability of cars is going down (Hirsch, 1992). A survey reported in the *Wall Street Journal* gave the reserved rates for midsized cars charged by the four largest rental-car companies in various cities on May 6, 1992.

Company	Boston	Chicago	San Francisco
Avis	$42.00	$45.99	$33.89
Hertz	38.99	49.99	35.99
National	40.90	48.90	35.90
Budget	32.49	44.99	31.50

a What type of design has been used in this experiment?

b Use an appropriate computer software program to conduct an analysis of variance for the data.

c Is there sufficient evidence to indicate that there is a difference in the average rates charged by the four rental car companies? What is the observed significance level of the test?

d Construct a 95% confidence interval for the difference in average rates between Avis and Hertz.

13.49 Refer to Exercise 13.48. Use Tukey's method for paired comparisons to rank the means for the experiment. What are the practical implications of your comparisons?

CASE STUDY

Coronary Heart Disease and College Football

In an article entitled "Coronary Heart Disease: Risk Profiles of College Football Players,"[†] Mindy Millard-Stafford and associates found, as expected, that linemen were somewhat taller and heavier than players of other positions. However, they also found that offensive linemen had the highest mean value for total cholesterol (TC) and the lowest mean value for high-density lipoprotein cholesterol (HDL-C) of all positions, independent of a player's race. More important, they indicated that although *average* blood pressures and blood lipids appear normal, individual football players—especially offensive linemen—may have an increased risk profile for coronary heart disease. A summary of their data concerning total cholesterol (TC) is given in the following table:

Position	Sample Size	Mean	Standard Deviation
Offensive linemen	17	206.1	34.2
Defensive linemen	15	181.2	42.2
Offensive backs	18	167.8	37.7
Defensive backs	19	196.0	36.2
Receivers	13	185.8	14.5
Linebackers	13	193.7	62.0

1 What type of experimental design was used in this experiment?

2 The ANOVA table that follows lists the degrees of freedom, sums of squares, and mean squares, derived from the information given in the table.

Source	d.f.	SS	MS
Treatments	5	15,292.4730	3058.4946
Error	89	140,046.8500	1573.5601

Complete the ANOVA table. What conclusions can you draw regarding the differences in mean TC levels among the various positions?

3 Use Tukey's procedure to determine which, if any, of the mean TC levels for the various positions differ from others.

4 Construct a confidence interval estimate for the difference in mean total cholesterol between offensive and defensive linemen and between offensive backs and receivers. Are your results surprising? Why, or why not?

[†]Millard-Stafford, Mindy, Linda B. Rosskopf, and Phillip B. Sparling, "Coronary Heart Disease: Risk Profiles of College Football Players." Reprinted by permission of *The Physician and Sportsmedicine*, 17(9), September 1989. Copyright McGraw-Hill, Inc.

14

Nonparametric Statistics

Case Study

How's your cholesterol level? Many of us have become more health conscious in the last few years as we read the nutritional labels on the food products we buy and choose foods that are low in fat and cholesterol and high in fiber. The case study at the end of this chapter involves a taste-testing experiment to compare three types of egg substitutes using nonparametric techniques.

General Objective

In Chapters 7–9 we presented statistical techniques for comparing two populations by comparing their respective population parameters (usually their respective population means). The techniques in Chapters 7–9 are applicable to data measured on a continuum, and the techniques in Chapter 9 are applicable to data having normal distributions. The purpose of this chapter is to present several statistical tests for comparing populations for the many types of data that do not satisfy the assumptions specified in Chapters 7–9.

Specific Topics

1. Parametric versus nonparametric tests (14.1)
2. The sign test (14.2)
3. The Mann-Whitney U test (14.4)
4. The Wilcoxon signed-rank test (14.5)
5. The Kruskal-Wallis H test (14.6)
6. The Friedman F_r test (14.7)
7. The rank correlation coefficient (14.8)

14.1 Introduction

Some experiments yield response measurements that defy quantification. That is, they generate response measurements that can be ordered (ranked), but the location of the response on a scale of measurement is arbitrary. For example, the data may only admit pairwise directional comparisons, that is, whether one observation is larger than another or vice versa. Or, we may be able to rank all observations in a data set but may not know the exact values of the measurements. To illustrate, suppose a judge is employed to evaluate and rank the sales abilities of four salespeople, the edibility and taste characteristics of five brands of cornflakes, or the relative appeal of five new automobile designs. Clearly, it is impossible to give an exact measure of sales competence, palatability of food, or design appeal, but it is usually possible to rank the salespeople, brands, and so on according to which one we think is best, second best, and so on. Thus, the response measurements here differ markedly from those presented in preceding chapters. Although experiments that produce this type of data occur in almost all fields of study, they are particularly evident in social science research and in studies of consumer preference. The data that they produce can be analyzed using **nonparametric statistical methods**.

Nonparametric statistical procedures are also useful in making inferences in situations where serious doubt exists about the assumptions that underlie standard methodology. For example, the t test for comparing a pair of means, Section 9.4, is based on the assumption that both populations are normally distributed with equal variances. The experimenter will never know whether these assumptions hold in a practical situation but will often be reasonably certain that departures from the assumptions will be small enough so that the properties of the statistical procedure will be undisturbed. That is, α and β will be approximately what the experimenter thinks they are. On the other hand, it is not uncommon for the experimenter to seriously question the assumptions and wonder whether he or she is using a valid statistical procedure. This difficulty may be circumvented by using a nonparametric statistical test and thereby avoiding reliance on a very uncertain set of assumptions.

Research has shown that nonparametric statistical tests are almost as capable of detecting differences among populations as the parametric methods of preceding chapters when normality and other assumptions are satisfied. They may be, and often are, more powerful in detecting population differences when the assumptions are not satisfied. For this reason, many statisticians advocate the use of nonparametric statistical procedures in preference to their parametric counterparts.

In this chapter, we will discuss only the most common situations in which nonparametric methods are used.

14.2 The Sign Test for Comparing Two Populations

Without emphasizing the point, we employed a nonparametric statistical test as an alternative procedure for determining whether evidence existed to indicate a difference in the mean wear for the two types of tires in the paired-difference experiment in Section 9.5. Each pair of responses was compared and x (the number of times A exceeded B) was used as the test statistic. This nonparametric test is known as the **sign test** because x is the number of positive (or negative) signs associated with the differences. **The implied null hypothesis is that the two population distributions are identical**, and the resulting technique is completely independent of the form of the distribution of differences. Thus, regardless of the distribution of differences, **the probability that A exceeds B for a given pair will be $p = .5$ when the null hypothesis is true** (i.e., when the distributions for A and B are identical). Then x will have a binomial probability distribution and a rejection region for x can be obtained using the binomial probability distribution of Chapter 4.

The Sign Test

1. *Null Hypothesis:* H_0 : The two population distributions are identical and $P(A$ exceeds B for a given pair$) = p = .5$.
2. *Alternative Hypothesis* H_a : The two population distributions are not identical and $p \neq .5$. Or H_a : The population of A measurements is shifted to the right of the population of B measurements and $p > .5$.[†]
3. *Test Statistic:* For n, the number of pairs in which no ties were reported, use x, the number of times that $(A - B)$ was positive, or equivalently, the number of times A exceeded (or was preferred to) B.
4. *Rejection Region:* For the two-tailed test $H_a : p \neq .5$ and a given value of α, reject H_0 if $x \leq x_L$ or $x \geq x_U$, where x_L and x_U are the lower- and upper-tailed values of a binomial distribution, with $p = .5$ satisfying $P(x \leq x_L) \leq \alpha/2$ and $P(x \geq x_U) \leq \alpha/2$. For the one-tailed test of $H_a : p > .5$ and a given value of α, reject H_0 if $x \geq x_U$, where x_U is the upper-tailed value of a binomial distribution, with $p = .5$ satisfying $P(x \geq x_U) \leq \alpha$.

The following example will help you understand how the sign is constructed and how it is used in a practical situation.

EXAMPLE 14.1 The number of defective electrical fuses proceeding from each of two production lines A and B was recorded daily for a period of ten days, with the following results:

	Production Lines	
Day	**A**	**B**
1	170	201
2	164	179
3	140	159
4	184	195
5	174	177
6	142	170
7	191	183
8	169	179
9	161	170
10	200	212

Assume that both production lines produced the same daily output. Compare the number x_A of defectives produced each day by production line A with the number x_B produced by the production line B and let x equal the number of days when

[†]For the one-tailed test when the alternative hypothesis is $H_a : p < .5$, simply interchange the letters A and B.

x_A exceeded x_B. Do the data present sufficient evidence to indicate that one production line tends to produce more defectives than the other or, equivalently, that $P(x_A > x_B) \neq 1/2$? State the null hypothesis to be tested and use x as a test statistic.

Solution Note that this is a paired-difference experiment. Let x be the number of times that the observation for production line A exceeds that for B in a given day. Under the null hypothesis that the two distributions of defectives are identical, the probability p that A exceeds B for a given pair is $p = .5$. Or, equivalently, we may wish to test the null hypothesis that the binomial parameter p equals .5 against the alternative hypothesis $p \neq .5$.

Very large or very small values of x are most contradictory to the null hypothesis. Therefore, the rejection region for this two-tailed test will be located by including the most extreme values of x that at the same time provide an α that is feasible for the test.

Suppose we want α to be about .05 or .10. We begin the selection of the rejection region by including $x = 0$ and $x = 10$ and calculating the α associated with this region using $p(x)$ (the probability distribution for the binomial random variable, Chapter 4). Thus, with $n = 10$, $p = .5$,

$$\alpha = p(0) + p(10) = C_0^{10}(.5)^{10} + C_{10}^{10}(.5)^{10} = .002$$

Since this value of α is too small, the rejection region is expanded by including the next pair of x values that are most contradictory to the null hypothesis, namely, $x = 1$ and $x = 9$. The value of α for this rejection region ($x = 0, 1, 9, 10$) is obtained from Table 1 in Appendix I.

$$\alpha = p(0) + p(1) + p(9) + p(10) = .022$$

This α is also small, so we again expand the region. This time we include $x = 0, 1, 2, 8, 9, 10$. You can verify that the corresponding value of α is .11. At this point, we have a choice. We can either choose α equal to .022 or .11, depending on the risk we are willing to take of making a type I error. For this example, we choose $\alpha = .11$ and, consequently, employ $x = 0, 1, 2, 8, 9, 10$ as the rejection region for the test (Figure 14.1).

From this data, we observe that $x = 1$, and therefore we reject the null hypothesis. Thus, we conclude that sufficient evidence exists to indicate that the population distributions of number of defects are not identical. A quick glance at the data suggests that the mean number of defects for production line B tends to exceed that for A. The probability of rejecting the null hypothesis when true is only $\alpha = .11$, and therefore we are reasonably confident of our conclusion.

FIGURE **14.1**
Rejection region for Example 14.1

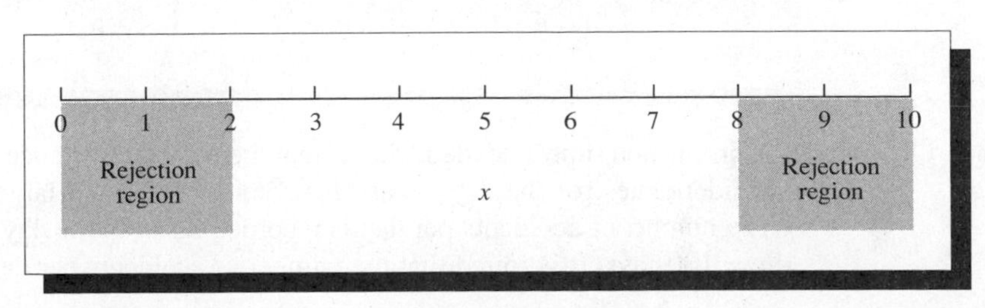

The experimenter in this example is using the test procedure as a rough tool for detecting faulty production lines. The rather large value of α is not likely to be disturbing because additional data can easily be collected if the experimenter is concerned about making a type I error in reaching a conclusion. ▪

One problem that may occur when you are conducting a sign test is that the observations associated with one or more pairs may be equal and therefore result in ties. When this situation occurs, delete the tied pairs and reduce *n*, the total number of pairs.

You will also encounter situations where n, the number of pairs, is large. Then the values of α associated with the sign test can be obtained by using the normal approximation to the binomial probability distribution discussed in Section 5.3. You can verify (by comparison of exact probabilities with their approximations) that these approximations will be quite adequate for n as small as 10. This is due to the symmetry of the binomial probability distribution for $p = .5$.

For $n \geq 25$, you can conduct the sign test by using the z statistic,

$$z = \frac{x - np}{\sqrt{npq}} = \frac{x - .5n}{.5\sqrt{n}}$$

as the test statistic. In using z, you are testing the null hypothesis $p = .5$ against the alternative $p \neq .5$ for a two-tailed test or against the alternative $p > .5$ (or $p < .5$) for a one-tailed test. The tests would utilize the familiar rejection regions of Chapter 8.

Sign Test for Large Samples: $n \geq 25$

1. *Null Hypothesis*: $H_0: p = .5$ (one treatment is not preferred to a second treatment, and vice versa).
2. *Alternative Hypothesis:* $H_a: p \neq .5$, for a two-tailed test. (NOTE: We use the two-tailed test as an example. Many analyses might require a one-tailed test.)
3. *Test Statistic:* $z = \dfrac{x - .5n}{.5\sqrt{n}}$
4. *Rejection Region:* Reject H_0 if $z \geq z_{\alpha/2}$ or if $z \leq -z_{\alpha/2}$, where $z_{\alpha/2}$ is the z value from Table 3, Appendix I, corresponding to an area of $\alpha/2$ in the upper tail of the normal distribution.

EXAMPLE 14.2 A production superintendent claims that there is no difference between the employee accident rates for the day versus the evening shifts in a large manufacturing plant. The number of accidents per day is recorded for both the day and evening shifts for $n = 100$ days. It is found that the number of accidents per day for the evening shift

x_E exceeded the corresponding number of accidents on the day shift x_D on 63 of the 100 days. Do these results provide sufficient evidence to indicate that more accidents tend to occur on one shift than on the other, or equivalently, that $P(x_E > x_D) \neq 1/2$?

Solution This study is a paired-difference experiment, with $n = 100$ pairs of observations corresponding to the 100 days. To test the null hypothesis that the two distributions of accidents are identical, we use the test statistic

$$z = \frac{x - .5n}{.5\sqrt{n}}$$

where x represents the number of days in which the number of accidents on the evening shift exceeded the number of accidents on the day shift. Then for $\alpha = .05$, we reject the null hypothesis if $z \geq 1.96$ or $z \leq 1.96$. Substituting into the formula for z, we have

$$z = \frac{x - .5n}{.5\sqrt{n}} = \frac{63 - (.5)(100)}{.5\sqrt{100}} = \frac{13}{5} = 2.60$$

Since the calculated value of z exceeds $z_{\alpha/2} = 1.96$, we reject the null hypothesis. The data provide sufficient evidence to indicate a difference between the accident rate distributions for the day versus evening shifts. ▪

Note that the data in Example 14.1 are the result of a paired-difference experiment. Suppose that the paired differences were normally distributed with variance σ_d^2. Would the sign test detect a shift in location of the two populations as effectively as the Student's t test? Intuitively, we would suspect the answer to be no, and this is correct because the Student's t test utilizes comparatively more information. In addition to giving the sign of the difference, the t test uses the magnitudes of the observations to obtain a more accurate value of the sample means and variances. Thus, one might say that the sign test is not as "efficient" as the Student's t test, but this statement is meaningful only if the populations conform to the assumption stated above; that is, the differences in paired observations are normally distributed with variance σ_d^2. The sign test might be more efficient when these assumptions are not satisfied.

Exercises

Basic Techniques

14.1 Suppose you wish to use the sign test to test $H_a: p > .5$ for a paired-difference experiment with $n = 25$ pairs.

a State the practical situation that would dictate the alternative hypothesis given above.

b Use Table 1 in Appendix I to find values of $\alpha (\alpha < .15)$ available for the test.

14.2 Repeat the instructions of Exercise 14.1 for $H_a: p \neq .5$.

14.3 Repeat the instruction of Exercises 14.1 and 14.2 for $n = 10$, 15, and 20.

14.4 A paired-difference experiment was conducted to compare two populations. The data are shown below. Use a sign test to determine whether the population distributions are different.

	Pairs						
Population	**1**	**2**	**3**	**4**	**5**	**6**	**7**
1	8.9	8.1	9.3	7.7	10.4	8.3	7.4
2	8.8	7.4	9.0	7.8	9.9	8.1	6.9

a State the null and alternative hypotheses for the test.

b Determine an appropriate rejection region with $\alpha \approx .01$.

c Calculate the observed value of the test statistic.

d Do the data present sufficient evidence to indicate that populations 1 and 2 are different?

Applications

14.5 In Exercise 9.43, we compared property evaluations of two tax assessors A and B. Their assessments for eight properties are shown in the table:

	Assessor	
Property	**A**	**B**
1	76.3	75.1
2	88.4	86.8
3	80.2	77.3
4	94.7	90.6
5	68.7	69.1
6	82.8	81.0
7	76.1	75.3
8	79.0	79.1

a Use the sign test to determine whether the data present sufficient evidence to indicate that one of the assessors tends to be consistently more conservative than the other; that is, $P(x_A > x_B) \neq 1/2$. Test by using a value of α near .05. Find the p-value for the test and interpret its value.

b Exercise 9.43 uses the t statistic to test the null hypothesis that there is no difference in the mean level of property assessments between assessors A and B. Check the answer (in the answer section) for Exercise 9.43 and compare it with your answer to part (a). Do the test results agree? Explain why the answers are (or are not) consistent.

14.6 Two gourmets rated 22 meals on a scale of 1 to 10. The data are shown in the table. Do the data provide sufficient evidence to indicate that one of the gourmets tends to give higher ratings than the other? Test by using the sign test with a value of α near .05.

Meal	A	B	Meal	A	B
1	6	8	12	8	5
2	4	5	13	4	2
3	7	4	14	3	3
4	8	7	15	6	8
5	2	3	16	9	10
6	7	4	17	9	8
7	9	9	18	4	6
8	7	8	19	4	3
9	2	5	20	5	4
10	4	3	21	3	2
11	6	9	22	5	3

a Use the binomial tables in Appendix I to find the exact rejection region for the test.

b Use the large-sample z statistic. (NOTE: Although the large-sample approximation is suggested for $n \geq 25$, it will work fairly well for values of n as small as 15.)

c Compare the results of parts (a) and (b).

14.7 A report in the August *American Journal of Public Health* (*Science News*, 1989, p. 126)—the first to follow blood lead levels in law-abiding handgun hobbyists using indoor firing ranges—documents a significant risk of lead poisoning. Lead exposure measurements were made on 17 members of a law enforcement trainee class before, during, and after a 3-month period of firearm instruction at a state-owned indoor firing range. No trainee had elevated blood lead levels before the training, but 15 of the 17 ended their training with blood lead levels deemed "elevated" by the Occupational Safety and Health Administration (OSHA). If the use of an indoor firing range has no positive effect on blood lead levels, then p, the probability that a person's blood lead level increases, is less than or equal to .5. If, however, use of the indoor firing range causes an increase in a person's blood lead levels, then $p > .5$. Use the sign test to determine whether using an indoor firing range has the effect of increasing a person's blood lead level with $\alpha = .05$. (HINT: The normal approximation to binomial probabilities will be fairly accurate for $n = 17$.)

14.8 Clinical data concerning the effectiveness of two drugs in treating a particular disease were collected from ten hospitals. The numbers of patients treated with the drugs varied from one hospital to another. The data, in percentage recovery, are shown below:

	Drug A			Drug B		
Hospital	Number in Group	Number Recovered	Percentage Recovered	Number in Group	Number Recovered	Percentage Recovered
1	84	63	75.0	96	82	85.4
2	63	44	69.8	83	69	83.1
3	56	48	85.7	91	73	80.2
4	77	57	74.0	47	35	74.5
5	29	20	69.0	60	42	70.0
6	48	40	83.3	27	22	81.5
7	61	42	68.9	69	52	75.4
8	45	35	77.8	72	57	79.2
9	79	57	72.2	89	76	85.4
10	62	48	77.4	46	37	80.4

Do the data present sufficient evidence to indicate a higher recovery rate for one of the two drugs?

a Test using the sign test. Choose your rejection region so that α is near .10.

b Why might it be inappropriate to use a Student's t test in analyzing the data?

14.3 A Comparison of Statistical Tests

We have introduced two different statistical tests to test an hypothesis based on the same set of data. Which test, if either, is better? One way to answer this question would be to hold the sample size and α constant for both procedures and compare β, the probability of a type II error. Actually, statisticians prefer a comparison of the **power of a test** where

$$\text{power} = 1 - \beta$$

Since β is the probability of failing to reject the null hypothesis when it is false, the power of the test is the probability of rejecting the null hypothesis when it is false and some specified alternative is true. It is the probability that the test will do what it was designed to do, that is, detect a departure from the null hypothesis when a departure exists.

Probably the most common method of comparing two test procedures is in terms of the relative efficiency of a pair of tests. **Relative efficiency** is the ratio of the sample sizes for the two test procedures required to achieve the same α and β for a given alternative to the null hypothesis.

In some situations, you may not be too concerned whether you are using the most powerful test. For example, you might choose to use the sign test over a more powerful competitor because of its ease of application. Thus, you might view tests as microscopes that are utilized to detect departures from an hypothesized theory. One need not know the exact power of a microscope to use it in a biological investigation, and the same applies to statistical tests. If the test procedure detects a departure from the null hypothesis, we are delighted. If not, we can reanalyze the data by using a more powerful test, or we can increase the power of the microscope (test) by increasing the sample size.

14.4 The Mann-Whitney *U* Test: Independent Random Samples

The t test for comparing two population means, Section 9.4, is a test to detect differences in the location of two normal population frequency distributions. The Mann-Whitney U test is a nonparametric alternative to this test. It is used when we have doubts that the assumptions of normality and/or equal variances required for the Student's t test are satisfied. The Mann-Whitney U test (and all of the tests that follow in this chapter) is based on an analysis of the ranks of the sample observations.

Assume that you have independent random samples of sizes n_1 and n_2 from two populations—say, 1 and 2. The first step in finding the Mann-Whitney U statistic is to rank all $(n_1 + n_2)$ observations in order of magnitude, assigning a 1 to the smallest observation, a 2 to the second smallest, and so on. **Ties in the observations can be handled by averaging the ranks that would have been assigned to the tied observations and assigning this average to each.** Then calculate the sums of the ranks, T_1 and T_2, for the two samples.

For example, two samples of four observations each ($n_1 = n_2 = 4$) are shown in Table 14.1(a). The ranks of the $n_1 + n_2 = 8$ observations are shown in parentheses. Notice that the similar samples each contain some of the larger and some of the smaller observations. There is little evidence to indicate a difference in the two population distributions, and the rank sums, $T_1 = T_2 = 18$, are equal. In contrast, Table 14.1(b) shows different samples, in which all of the small observations are in one sample and all of the large observations are in the other. The rank sums $T_1 = 10$ and $T_2 = 26$ differ by the maximum amount and provide the greatest amount of evidence to indicate a difference in location for the two population distributions.

The formula for the Mann-Whitney U statistic can be given in terms of T_1 or T_2; one value of U will be larger than the other, but the sum of the two U values will always equal n_1n_2. Since it is easier to construct a table of probabilities of U for only one tail of the U distribution, we will agree always to use the smaller value of U as a test statistic. The formulas for the two values of U, which we will denote as U_1 and U_2, are given in the display.

TABLE 14.1
Hypothetical rankings of eight observations

	(a) Similar Samples		(b) Different Samples	
	1	2	1	2
	28 (6)	25 (4)	18 (2)	27 (5)
	16 (1)	18 (2)	21 (3)	35 (8)
	35 (8)	27 (5)	16 (1)	28 (6)
	21 (3)	30 (7)	25 (4)	30 (7)
Rank sum	$T_1 = 18$	$T_2 = 18$	$T_1 = 10$	$T_2 = 26$

Formulas for the Mann-Whitney U Statistic

$$U_1 = n_1 n_2 + \frac{n_1(n_1 + 1)}{2} - T_1$$

$$U_2 = n_1 n_2 + \frac{n_2(n_2 + 1)}{2} - T_2$$

where

n_1 = number of observations in sample 1

n_2 = number of observations in sample 2

T_1 and T_2 are the rank sums of samples 1 and 2, respectively. (NOTE: It can be shown that $U_1 + U_2 = n_1n_2$ and $T_1 + T_2 = \frac{n(n+1)}{2}$, where $n = n_1 + n_2$.)

As you can see from the formulas for U_1 and U_2, U_1 will be small when T_1 is large, a situation that likely will occur when the population 1 distribution of measurements is shifted to the right of the population 2 distribution of measurements. Consequently, to conduct a one-tailed test to detect a shift in distribution 1 to the right of distribution 2, you will reject the null hypothesis of "no difference in the population distributions" if U_1 is less than some specified value U_0; that is, you will reject H_0 for small values of U_1. Similarly, to conduct a one-tailed test to detect a shift of distribution 2 to the right of distribution 1, you would reject H_0 if U_2 is less than some specified value, say U_0. Consequently, the rejection region for the Mann-Whitney U test would appear as shown in Figure 14.2.

FIGURE **14.2**
Rejection region for a Mann-Whitney U test

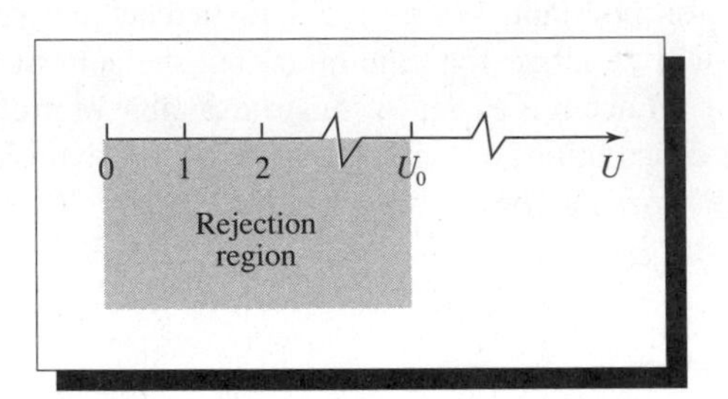

Table 7 in Appendix I gives the probability that an observed value of U will be less than some specified value, say, U_0. This is the value of α for a one-tailed test. To conduct a two-tailed test—that is, to detect a shift in the population distributions for the measurements in either direction—we use U, the smaller of U_1 or U_2, as the test statistic and reject H_0 for $U < U_0$ (see Figure 14.2). The value of α for the two-tailed test will be double the tabulated value given in Table 7 in Appendix I.

To see how to locate the rejection region for the Mann-Whitney U test, suppose that $n_1 = 4$ and $n_2 = 5$. Then you would consult the table corresponding to $n_2 = 5$ in Table 7 of Appendix I. The first few lines of Table 7 for $n_2 = 5$ are shown in Table 14.2. Note that the table is constructed assuming that $n_1 \le n_2$. Therefore, we identify the population with the smaller sample size as population 1.

TABLE **14.2**
An abbreviated version of Table 7 in Appendix I; $P(U \le U_0)$ for $n_2 = 5$

	n_1				
U_0	1	2	3	4	5
0	.1667	.0476	.0179	.0079	.0040
1	.3333	.0952	.0357	.0159	.0079
2	.5000	.1905	.0714	.0317	.0159
3		.2857	.1250	.0556	.0278
4		.4286	.1964	.0952	.0476
5		.5714	.2857	.1429	.0754
⋮			⋮	⋮	⋮

Across the top of Table 14.2 you see values of n_1. Values of U_0 are shown down the left side of the table. The entries give the probability that U will assume a small value, namely, the probability that $U \leq U_0$. Since, for this example, $n_1 = 4$, we move across the top of the table to $n_1 = 4$. Move to the second row of the table corresponding to $U_0 = 2$ for $n_1 = 4$. The probability that U will be less than or equal to 2 is found as .0317. Similarly, moving across the row for $U_0 = 3$, you see that the probability that U is less than or equal to 3 is .0556. (This value is shaded in Table 14.2.) So, if you want to conduct a one-tailed Mann-Whitney U test with $n_1 = 4$ and $n_2 = 5$ and would like α to be near .05, you would reject the null hypothesis of equality of population relative frequency distributions when $U \leq 3$. The probability of a type I error for the test would be $\alpha = .0556$. If you use this same rejection region for a two-tailed test, that is, $U \leq 3$, α would be double the tabulated value, or $\alpha = 2(.0556) = .1112$.

Table 7 in Appendix I can also be used to find the observed significance level for a test. For example, if $n_1 = 5$, $n_2 = 5$, and $U = 4$, then from Table 14.2, the p-value for a one-tailed test would be

$$P\{U \leq 4\} = .0476$$

If the test were two-tailed, the p-value would be

$$2(.0476) \quad \text{or} \quad .0952$$

When applying the test to a set of data, you may find that some of the observations are of equal value. **Recall that ties in the observations can be handled by averaging the ranks that would have been assigned to the tied observations and assigning this average to each.** Thus, if three observations were tied and were due to receive ranks 3, 4, and 5, we would assign the rank of 4 to all three. The next observation in the sequence would receive the rank of 6, and ranks 3 and 5 would not appear. Similarly, if two observations were tied for ranks 3 and 4, each would receive a rank of 3.5, and ranks 3 and 4 would not appear.

The Mann-Whitney U Test

1. *Null Hypothesis:* H_0 : The population relative frequency distributions 1 and 2 are identical.
2. *Alternative Hypothesis:* H_a : The two population relative frequency distributions are shifted with respect to their relative locations (a two-tailed test). Or H_a : The population relative frequency distribution 1 is shifted to the right of the relative frequency distribution for population 2 (a one-tailed test).[†]

[†]For the sake of convenience, we will describe the one-tailed test as one designed to detect a shift in distribution 1 measurements to the right of distribution 2 measurements. To detect a shift in distribution 2 to the right of distribution 1, just interchange the numbers 1 and 2 in the discussion.

3 *Test Statistic:* For a two-tailed test, use U, the smaller of

$$U_1 = n_1 n_2 + \frac{n_1(n_1+1)}{2} - T_1$$

and

$$U_2 = n_1 n_2 + \frac{n_2(n_2+1)}{2} - T_2$$

where T_1 and T_2 are the rank sums for samples 1 and 2, respectively. For a one-tailed test, use U_1.

4 *Rejection Region:*

a For the two-tailed test and a given value of α, reject H_0 if $U \le U_0$, where $P(U \le U_0) = \alpha/2$. [NOTE: U_0 is the value such that $P(U \le U_0)$ is equal to half of α.]

b For a one-tailed test and a given value of α, reject H_0 if $U_1 \le U_0$, where $P(U_1 \le U_0) = \alpha$.

Ties are assigned a rank equal to the average of the ranks that the observations would have received if they had been slightly different in value.

EXAMPLE **14.3** The bacteria counts per unit volume are shown for two types of cultures A and B. Four observations were made for each culture.

Culture A	Culture B
27(3)	32(7)
31(6)	29(5)
26(2)	35(8)
25(1)	28(4)

Do the data present sufficient evidence to indicate a difference in the population distributions of bacteria counts? Test using a value of α near .05.

Solution We wish to test the null hypothesis

H_0 : The population distributions of bacteria counts are identical

against the alternative hypothesis

H_a : The population distributions differ in location

This alternative hypothesis implies a two-tailed test. Since Table 7, Appendix I, gives values of $P(U \le U_0)$ for specified sample sizes and values of U_0, we must double the tabulated value to find α. Suppose that we desire a value of α near .05. Checking Table 7, Appendix I, for $n_1 = n_2 = 4$, we find $P(U \le 1) = .0286$. Using $U \le 1$ as the rejection region, α will equal $2(.0286) = .0572$ or, rounding to three decimal places, $\alpha = .057$. The ranks for the $n_1 + n_2 = 8$ bacteria counts are shown in parentheses next to each data value, and the rank sums are $T_1 = 12$ and $T_2 = 24$. Then

$$\begin{aligned} U_1 &= n_1 n_2 + \frac{n_1(n_1+1)}{2} - T_1 \\ &= (4)(4) + \frac{4(4+1)}{2} - 12 \\ &= 14 \end{aligned}$$

and

$$\begin{aligned} U_2 &= n_1 n_2 + \frac{n_2(n_2+1)}{2} - T_2 \\ &= (4)(4) + \frac{4(4+1)}{2} - 24 \\ &= 2 \end{aligned}$$

The smaller of U_1 and U_2 is the test statistic, $U = 2$, which does not fall in the rejection region. Hence, there is not sufficient evidence to show a shift in locations of the population distributions of bacteria counts for cultures 1 and 2. ▪

The Mann-Whitney test can be implemented by using the MANN-WHITNEY command in the MINITAB package. As input, the program uses the values from the first and second samples, which are stored in two separate columns. The MINITAB output for the Mann-Whitney test using the data in Example 14.3 is given in Table 14.3. Data summaries appear in lines 1 and 2; point and interval estimates for the difference in population medians are given in lines 3 and 4. The value $W = 12.0$ in line 5 is the sum of the ranks for the values stored in column 1—and in general, for the column number given first in the program command. The value of W can be used to calculate U_1 and U_2 if desired. The significance level (.1124) of the test is given in line 6. Since the significance level of .1124 is greater than $\alpha = .05$, consistent with our earlier results, we cannot conclude that the underlying population distributions are different.

TABLE **14.3**
MINITAB output for the data of Example 14.3

```
MTB > MANNWHITNEY C1 C2

Mann-Whitney Confidence Interval and Test

C1           N =    4      Median =        26.500
C2           N =    4      Median =        30.500
Point estimate for ETA1-ETA2 is          -3.500
97.0 pct c.i. for ETA1-ETA2 is (-9.998,2.998)
W = 12.0
Test of ETA1 = ETA2  vs.  ETA1 n.e. ETA2 is significant at 0.1124

Cannot reject at alpha = 0.05
```

EXAMPLE 14.4 An experiment was conducted to compare the strength of two types of kraft papers, one a standard kraft paper of a specified weight and the other the same standard kraft paper treated with a chemical substance. Ten pieces of each type of paper, randomly selected from production, produced the strength measurements shown in the accompanying table. Test the hypothesis of "no difference in the distributions of strengths for the two types of paper" against the alternative hypothesis that the treated paper tends to be of greater strength (i.e., its distribution of strength measurements is shifted to the right of the corresponding distribution for the untreated paper).

Solution The ranks are shown in parentheses alongside the $n_1 + n_2 = 10 + 10 = 20$ strength measurements, and the rank sums T_1 and T_2 are shown below the columns. Since we wish to detect a shift in the distribution 1 measurements to the right of the distribution 2 measurements, we will conduct a one-tailed test and reject the null hypothesis of "no difference in population strength distributions" when T_2 is excessively large, or equivalently when U_2 is small.

	Standard 1	Treated 2
	1.21 (2)	1.49 (15)
	1.43 (12)	1.37 (7.5)
	1.35 (6)	1.67 (20)
	1.51 (17)	1.50 (16)
	1.39 (9)	1.31 (5)
	1.17 (1)	1.29 (3.5)
	1.48 (14)	1.52 (18)
	1.42 (11)	1.37 (7.5)
	1.29 (3.5)	1.44 (13)
	1.40 (10)	1.53 (19)
Rank sum	$T_1 = 85.8$	$T_2 = 124.5$

Suppose we choose a value of α near .05. Then we can find U_0 by consulting the portion of Table 7, Appendix I, corresponding to $n_2 = 10$. The probability, $P(U \leq U_0)$, nearest .05 is .0526 and corresponds to $U_0 = 28$. Hence, we will reject if $U_2 \leq 28$.

Calculating U_2, we have

$$
\begin{aligned}
U_2 &= n_1 n_2 + \frac{n_2(n_2+1)}{2} - T_2 \\
&= (10)(10) + \frac{(10)(11)}{2} - 124.5 \\
&= 30.5
\end{aligned}
$$

As you can see, U_2 is not less than $U_0 = 28$. Therefore, we cannot reject the null hypothesis. At the $\alpha = .05$ level of significance, there is not sufficient evidence to indicate that the treated kraft paper is stronger than the standard. ▪

A simplified large-sample test ($n_1 \geq 10$ and $n_2 \geq 10$) can be obtained by using the familiar z statistic of Chapter 8. When the population distributions are identical, it can be shown that the U statistic has expected value and variance,

$$E(U) = \frac{n_1 n_2}{2} \quad \text{and} \quad \sigma_U^2 = \frac{n_1 n_2 (n_1 + n_2 + 1)}{12}$$

The distribution of

$$z = \frac{U - E(U)}{\sigma_U}$$

tends to normality with mean 0 and variance equal to 1 as n_1 and n_2 become large. **This approximation would be adequate when n_1 and n_2 are both greater than or equal to 10.** Thus, for a two-tailed test with $\alpha = .05$, we would reject the null hypothesis if $|z| \geq 1.96$.

Observe that by using the z statistic, you will reach the same conclusions as when using the exact U test for Example 14.4. Thus,

$$z = \frac{30.5 - [(10)(10)/2]}{\sqrt{[(10)(10)(10 + 10 + 1)]/12}} = \frac{30.5 - 50}{\sqrt{2100/12}} = -1.47$$

For a one-tailed test with $\alpha = .05$ located in the lower tail of the z distribution, we will reject the null hypothesis if $z < -1.645$. Since $z = -1.47$ does not fall in the rejection region, we reach the same conclusion as the exact U test of Example 14.4.

The Mann-Whitney U Test for Large Samples: $n_1 \geq 10$ and $n_2 \geq 10$

1. *Null Hypothesis:* H_0 : The population relative frequency distributions are identical.
2. *Alternative Hypothesis:* H_a : The two population relative frequency distributions differ in location (a two-tailed test). Or H_a : The population 1 relative frequency distribution is shifted to the right (or left) of the relative frequency distribution for population 2 (a one-tailed test).
3. *Test Statistic:* $z = \dfrac{U - (n_1 n_2/2)}{\sqrt{n_1 n_2 (n_1 + n_2 + 1)/12}}$
   Let $U = U_1$.
4. *Rejection Region:* Reject H_0 if $z > z_{\alpha/2}$ or $z < -z_{\alpha/2}$ for a two-tailed test. For a one-tailed test, place all of α in one tail of the z distribution. To detect a shift in distribution 1 to the right of distribution 2, let $U = U_1$ and reject H_0 when $z < -z_\alpha$. To detect a shift in the opposite direction, let $U = U_1$ and reject H_0 when $z > z_\alpha$. Tabulated values of z are given in Table 3, Appendix I.

Exercises

Basic Techniques

14.9 Suppose you wish to detect a shift in distribution 1 to the right of distribution 2 based on sample sizes $n_1 = 6, n_2 = 8$.

a Should you use U_1 or U_2 for your test statistic?

b Give the rejection region for the test if you wish α to be close to but less than .10.

c Give the value of α for the test.

14.10 Suppose that the alternative hypothesis for Exercise 14.9 is that distribution 1 is shifted either to the left or to the right of distribution 2.

a Should you use U_1 or U_2 for your statistic?

b Give the rejection region for the test if you wish α to be close to but less than .10.

c Give the value of α for the test.

14.11 Observations from two random and independent samples, drawn from populations 1 and 2, are reproduced below:

Sample 1	Sample 2
1	4
3	7
2	6
3	8
5	6

Use the Mann-Whitney U test to determine whether population 1 is shifted to the left of population 2.

a State the null and alternative hypotheses to be tested.

b Find the rejection region for $\alpha \approx .05$.

c Rank the combined sample from smallest to largest. Calculate T_1, T_2, U_1, and U_2.

d What is the observed value of the test statistic?

e Do the data provide sufficient evidence to indicate that population 1 is shifted to the left of population 2?

14.12 Independent random samples of size $n_1 = 20$ and $n_2 = 25$ are drawn from nonnormal populations 1 and 2. The combined sample is ranked and $T_1 = 252$, $T_2 = 783$. Use the large-sample approximation to the Mann-Whitney U test to determine whether there is a difference in the two population distributions. Calculate the observed significance level, or p-value, for the test.

14.13 Suppose you wish to detect a shift in direction 1 to the right of distribution 2 based on sample sizes $n_1 = 12$ and $n_2 = 14$. If $T_1 = 193$, what do you conclude? Use $\alpha = .05$.

Applications

14.14 In Exercise 9.86 we presented data on the wing stroke frequencies of two species of Euglossine bees (Casey, May, and Morgan, 1985). The wing frequencies for a sample of $n_1 = 4$ *Euglossa mandibularis* Friese and $n_2 = 6$ *Euglossa imperialis* Cockerell are shown in the table:

Wing Stroke Frequencies

Euglossa mandibularis Friese	*Euglossa imperialis* Cockerell
235	180
225	169
190	180
188	185
	178
	182

Do the data present sufficient evidence to indicate that the distributions of wing stroke frequencies differ in location for the two species?

a Test using the Mann-Whitney U test with α less than but close to .10.

b Find the approximate p-value for the test.

14.15 The observations below represent the dissolved oxygen content in water. The higher the dissolved oxygen, the greater the ability of a river, lake, or stream to support aquatic life. In this experiment, a pollution-control inspector suspected that a river community was releasing amounts of semitreated sewage into a river. To check this theory, five randomly selected specimens of river water were selected at a location above the town, and another five below. The dissolved oxygen readings, in parts per million, are as follows:

Above town	4.8	5.2	5.0	4.9	5.1
Below town	5.0	4.7	4.9	4.8	4.9

a Use a one-tailed Mann-Whitney U test with $\alpha = .05$ to answer the question.

b Use a Student's t test (with $\alpha = .05$) to analyze the data. Compare the conclusions reached in parts (a) and (b).

14.16 In an investigation of visual scanning behavior of deaf children, measurements of eye-movement rate were taken on nine deaf and nine hearing children. From the data given, does it appear that the distributions of eye-movement rates for deaf children (A) and hearing children (B) differ?

	Deaf Children A	Hearing Children B
	(15)2.75	.89(1)
	(11)2.14	1.43(7)
	(18)3.23	1.06(4)
	(10)2.07	1.01(3)
	(14)2.49	.94(2)
	(12)2.18	1.79(8)
	(17)3.16	1.12(5.5)
	(16)2.93	2.01(9)
	(13)2.20	1.12(5.5)
Rank sum	126	45

14.17 The following table depicts the life in months of service before failure of the color television picture tube for eight television sets manufactured by firm A and ten sets manufactured by firm B.

Firm	Life of Picture Tube (months)									
A	32	25	40	31	35	29	37	39		
B	41	39	36	47	45	34	48	44	43	33

Use the Mann-Whitney U test to analyze the data and test to see if the life in months of service before failure of the picture tube differs for the picture tubes produced by the two manufacturers. (Use $\alpha = .10$.)

14.18 A comparison of the weights of turtles caught in two different lakes was conducted to compare the effects of the two lake environments on turtle growth. All the turtles were of the same age and were tagged before being released in the lakes. The weight measurements for $n_1 = 10$ tagged turtles caught in lake 1 and $n_2 = 8$ caught in lake 2 are as follows:

Lake	Weight (oz)									
1	14.1	15.2	13.9	14.5	14.7	13.8	14.0	16.1	12.7	15.3
2	12.2	13.0	14.1	13.6	12.4	11.9	12.5	13.8		

Do the data provide sufficient evidence to indicate a difference in the distribution of weights for the tagged turtles exposed to the two lake environments? Use a Mann-Whitney U test with $\alpha = .05$ to answer the question.

14.19 Cancer treatment by means of chemicals—chemotherapy—utilizes chemicals that kill both cancerous and normal cells. In some instances, the toxicity of the cancer drug—that is, its effect on normal cells—can be reduced by the simultaneous injection of a second drug. A study was conducted to determine whether a particular drug injection was beneficial in reducing the harmful effects of a chemotherapy treatment on the survival time for rats. Two randomly selected groups of 12 rats were used in an experiment in which both groups, call them A and B, received the toxic drug in a dosage large enough to cause death, but in addition, group B received the antitoxin, which was to reduce the toxic effect of the chemotherapy on normal cells. The test was terminated at the end of 20 days, or 480 hours. The lengths of survival time for the two groups of rats, to the nearest 4 hours, are shown in the table. Do the data provide sufficient evidence to indicate that rats receiving the antitoxin tend to survive longer after chemotherapy than those not receiving the toxin? Use the Mann-Whitney U test with a value of α near .05.

Chemotherapy Only A	Chemotherapy Plus Drug B
84	140
128	184
168	368
92	96
184	480
92	188
76	480
104	244
72	440
180	380
144	480
120	196

14.5 The Wilcoxon Signed-Rank Test for a Paired Experiment

A signed-rank test proposed by F. Wilcoxon can be used to analyze the paired-difference experiment of Section 9.5 by considering the paired differences of the two treatments 1 and 2. Under the null hypothesis of "no differences in the distributions for 1 and 2," you would expect (on the average) half of the differences in pairs to be negative and half to be positive; that is, the expected number of negative differences between pairs would be $n/2$ (where n is the number of pairs). Furthermore, it would follow that positive and negative differences of equal absolute magnitude should occur with equal probability. If one were to order the differences according to their absolute values and rank them from smallest to largest, the expected rank sums for the negative and positive differences would be equal. Sizable differences in the sums of the ranks assigned to the positive and negative differences would provide evidence to indicate a shift in location between the distributions of responses for the two treatments 1 and 2.

Calculation of the Test Statistic for the Wilcoxon Signed-Rank Test

1. Calculate the differences $(x_A - x_B)$ for each of the n pairs. Differences equal to 0 are eliminated and the number of pairs, n, is reduced accordingly.
2. Rank the **absolute values** of the differences, assigning 1 to the smallest, 2 to the second smallest, and so on. Tied observations are assigned the average of the ranks that would have been assigned with no ties.
3. Calculate the **rank sum** for the **negative** differences and label this value T^-. Similarly, calculate T^+, the **rank sum** for the **positive** differences.

For a **two-tailed test**, we use the **smaller of these two quantities T as a test statistic** to test the null hypothesis that the two population relative frequency histograms are identical. The smaller the value of T, the greater will be the weight of evidence favoring rejection of the null hypothesis. **Therefore, we will reject the null hypothesis if T is less than or equal to some value, say T_0.**

To detect the **one-sided alternative**, that **distribution 1 is shifted to the right of distribution 2, use the rank sum T^-** of the negative differences and reject the null hypothesis for small values of T^-—say, $T^- \leq T_0$. If we wish to detect a **shift of distribution 2 to the right of distribution 1, we use the rank sum T^+** of the positive differences as a test statistic and reject for small values of T^+—say, $T^+ \leq T_0$.

The probability that T is less than or equal to some value T_0 has been calculated for a combination of sample sizes and values of T_0. These probabilities, given in Table 8 in Appendix I, can be used to find the rejection region for the T test.

An abbreviated version of Table 8, Appendix I, is shown in Table 14.4. Across the top of the table you see the number of differences (the number of pairs) n. Values of α for a one-tailed test appear in the first column of the table. The second column gives values of α for a two-tailed test. Table entries are the critical values of T. You will recall that the critical value of a test statistic is the value that locates the boundary of the rejection region.

For example, suppose you have $n = 7$ pairs and you are conducting a two-tailed test of the null hypothesis that the two population relative frequency distributions are identical. Checking the $n = 7$ column of Table 14.4 and using the second row (corresponding to $\alpha = .05$ for a two-tailed test), you see the entry 2 (shaded in Table 14.4). This value is T_0, the critical value of T. As noted earlier, the smaller the value of T, the greater will be the evidence to reject the null hypothesis. Therefore,

TABLE 14.4 An abbreviated version of Table 8 in Appendix I; critical values of T

One-Sided	Two-Sided	$n = 5$	$n = 6$	$n = 7$	$n = 8$	$n = 9$	$n = 10$
$\alpha = .05$	$\alpha = .10$	1	2	4	6	8	11
$\alpha = .025$	$\alpha = .05$		1	2	4	6	8
$\alpha = .01$	$\alpha = .02$			0	2	3	5
$\alpha = .005$	$\alpha = .01$				0	2	3

One-Sided	Two-Sided	$n = 11$	$n = 12$	$n = 13$	$n = 14$	$n = 15$	$n = 16$
$\alpha = .05$	$\alpha = .10$	14	17	21	26	30	36
$\alpha = .025$	$\alpha = .05$	11	14	17	21	25	30
$\alpha = .01$	$\alpha = .02$	7	10	13	16	20	24
$\alpha = .005$	$\alpha = .01$	5	7	10	13	16	19

One-Sided	Two-Sided	$n = 17$
$\alpha = .05$	$\alpha = .10$	41
$\alpha = .025$	$\alpha = .05$	35
$\alpha = .01$	$\alpha = .02$	28
$\alpha = .005$	$\alpha = .01$	23

you will reject the null hypothesis for all values of T less than or equal to 2. The rejection region for the Wilcoxon signed-rank test for a paired experiment is always of the form: reject H_0 if $T \leq T_0$, where T_0 is the critical value of T. The rejection region is shown symbolically in Figure 14.3.

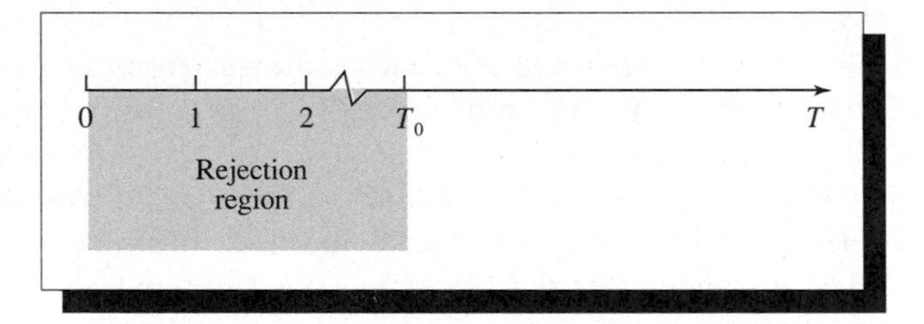

FIGURE 14.3 Rejection region for the Wilcoxon signed-rank test for a paired experiment (reject H_0 if $T \leq T_0$)

Wilcoxon Signed-Rank Test for a Paired Experiment

1. *Null Hypothesis:* H_0 : The two population relative frequency distributions are identical.
2. *Alternative Hypothesis:* H_a : The two population relative frequency distributions differ in location (a two-tailed test). Or H_a : The population 1 relative frequency distribution is shifted to the right of the relative frequency distribution for population 2 (a one-tailed test).

3 *Test Statistic:*

a For a two-tailed test, use T, the smaller of the rank sum for positive and the rank sum for negative differences.

b For a one-tailed test (to detect the alternative hypothesis described above), use the rank sum T^- of the negative differences.

4 *Rejection Region:*

a For a two-tailed test, reject H_0 if $T \leq T_0$, where T_0 is the critical value given in Table 8 in Appendix I.

b For a one-tailed test (to detect the alternative hypothesis described above), use the rank sum T^- of the negative differences. Reject H_0 if $T^- \leq T_0$.[†]
(NOTE: It can be shown that $T^+ + T^- = \frac{n(n+1)}{2}$.)

EXAMPLE 14.5 Test an hypothesis of no difference in population distributions of cake density for the paired-difference experiment in Exercise 9.104.

Solution The original data and differences in density for the six pairs of cakes are as follows:

Density (oz/in.^{3})

x_1	.135	.102	.098	.141	.131	.144
x_2	.129	.120	.112	.152	.135	.163
Difference $(x_1 - x_2)$	.006	−.018	−.014	−.011	−.004	−.019
Rank	2	5	4	3	1	6

As with other nonparametric tests, the null hypothesis to be tested is that the two population frequency distributions of cake densities are identical. The alternative hypothesis, which implies a two-tailed test, is that the distributions are different.

Because the amount of data is small, we will conduct our test using $\alpha = .10$. From Table 8, Appendix I, the critical value of T for a two-tailed test, $\alpha = .10$, is $T_0 = 2$. Hence, we will reject H_0 if $T \leq 2$.

The differences $x_1 - x_2$ are calculated and ranked according to their absolute value in the table. The sum of the positive ranks is $T^+ = 2$, and the rank sum of the negative ranks is $T^- = 19$. The test statistic is the smaller of these two rank sums, or $T = 2$. Since $T = 2$ falls in the rejection region, we reject H_0 and conclude that the two population frequency distributions of cake densities differ. ▪

Although Table 8 in Appendix I is applicable for values of n (number of data pairs) as large as $n = 50$, it is worth noting that T^+, **like the Mann-Whitney U, will**

[†]To detect a shift of distribution 2 to the right of the distribution 1, use the rank sum T^+ of the positive differences as the test statistic and reject H_0 if $T^+ \leq T_0$.

be approximately normally distributed when the null hypothesis is true and n is large (say, 25 or more). This enables us to construct a large-sample z test, where

$$E(T) = \frac{n(n+1)}{4}$$

$$\sigma_T^2 = \frac{n(n+1)(2n+1)}{24}$$

Then the z statistic

$$z = \frac{T^+ - E(T^+)}{\sigma_{T^+}} = \frac{T^+ - \dfrac{n(n+1)}{4}}{\sqrt{\dfrac{n(n+1)(2n+1)}{24}}}$$

can be used as a test statistic. Thus, for a two-tailed test and $\alpha = .05$, we would reject the hypothesis of "identical population distributions" when $|z| \geq 1.96$.

A Large-Sample Wilcoxon Signed-Rank Test for a Paired Experiment: $n \geq 25$

1. *Null Hypothesis:* H_0 : The population relative frequency distributions 1 and 2 are identical.
2. *Alternative Hypothesis:* H_a : The two population relative frequency distributions differ in location (a two-tailed test). Or H_a : The population 1 relative frequency distribution is shifted to the right (or left) of the relative frequency distribution for population 2 (a one-tailed test).
3. *Test Statistic:* $z = \dfrac{T^+ - [n(n+1)/4]}{\sqrt{[n(n+1)(2n+1)]/24}}$
4. *Rejection Region:* Reject H_0 if $z > z_{\alpha/2}$ or $z < -z_{\alpha/2}$ for a two-tailed test. For a one-tailed test, place all of α in one tail of the z distribution. To detect a shift in distribution 1 to the right of distribution 2, reject H_0 when $z > z_\alpha$. To detect a shift in the opposite direction, reject H_0 if $z < -z_\alpha$. Tabulated values of z are given in Table 3 in Appendix 1.

The MINITAB package includes a command WTEST for implementing the Wilcoxon signed-rank test for a paired experiment. The initial input and output for this program using the data in Example 14.5 is given in Table 14.5. The differences of the paired values are set in column 3. The command WTEST 0 C3 specifies that the Wilcoxon signed-rank procedure will be used to test for a zero median using the values in column 3. Notice that none of the six differences were equal to 0, so all six were used in calculating the value of the test statistic given as 2.0. Since the p-value of .093 exceeds $\alpha = .05$, there is not sufficient evidence to conclude that the underlying population distributions differ significantly.

TABLE **14.5**
MINITAB output for the data of Example 14.5

```
MTB > PRINT C3
C3
  0.0060000    -0.0180000    -0.0140000    -0.0110000    -0.0040000    -0.0190000
MTB > WTEST 0 C3

TEST OF MEDIAN = 0 VERSUS MEDIAN N.E. 0

                  N FOR      WILCOXON                  ESTIMATED
          N        TEST     STATISTIC      P-VALUE        MEDIAN
C3        6           6           2.0        0.093      -0.01100
```

Exercises

Basic Techniques

14.20 Suppose you wish to detect a difference in location between two population distributions based on a paired-difference experiment consisting of $n = 30$ pairs.

a Give the null and alternative hypotheses for the Wilcoxon signed-rank test.

b Give the test statistic.

c Give the rejection region for the test for $\alpha = .05$.

d If $T^+ = 249$, what are your conclusions? (NOTE: $T^+ + T^- = n(n+1)/2$.)

14.21 Refer to Exercise 14.20. Suppose you wish to detect only a shift in distribution 1 to the right of distribution 2.

a Give the null and alternative hypotheses for the Wilcoxon signed-rank test.

b Give the test statistic.

c Give the rejection region for the test for $\alpha = .05$.

d If $T^+ = 249$, what are your conclusions? (NOTE: $T^+ + T^- = n(n+1)/2$.)

14.22 Refer to Exercise 14.20. Conduct the test using the large-sample z test. Compare your results with the nonparametric test results in Exercise 14.20(d).

14.23 Refer to Exercise 14.21. Conduct the test using the large-sample z test. Compare your results with the nonparametric test results in Exercise 14.21(d).

14.24 Refer to Exercise 14.4. The data reproduced below are from a paired-difference experiment with $n = 7$ pairs of observations.

	Pairs						
Population	**1**	**2**	**3**	**4**	**5**	**6**	**7**
1	8.9	8.1	9.3	7.7	10.4	8.3	7.4
2	8.8	7.4	9.0	7.8	9.9	8.1	6.9

a Use Wilcoxon's signed-rank test to determine whether there is a significant difference between the two populations.

b Compare the results of part (a) with the result obtained in Exercise 14.4. Are the results the same? Explain.

Applications

14.25 In Exercise 14.5 we used the sign test to determine whether the data provided sufficient evidence to indicate a shift in the distributions of property assessments for assessors A and B.

a Use the Wilcoxon signed-rank test for a paired experiment to test the null hypothesis that there is no difference in the distributions of property assessments between assessors A and B. Test by using a value of α near .05.

b Compare the conclusion of the test in part (a) with the conclusions derived from the t test in Exercise 9.43 and the sign test in Exercise 14.5. Explain why these test conclusions are (or are not) consistent.

14.26 The number of machine breakdowns per month was recorded for 9 months on two identical machines used to make wire rope. The data are shown in the table:

	Machines	
Month	**A**	**B**
1	3	7
2	14	12
3	7	9
4	10	15
5	9	12
6	6	6
7	13	12
8	6	5
9	7	13

a Do the data provide sufficient evidence to indicate a difference in the monthly breakdown rates for the two machines? Test by using a value of α near .05.

b Can you think of a reason why the breakdown rates for the two machines might vary from month to month?

14.27 Refer to the comparison of gourmet meal ratings in Exercise 14.6, and use the Wilcoxon signed-rank test to determine whether the data provide sufficient evidence to indicate a difference in the ratings of the two gourmets. Test by using a value of α near .05. Compare the results of this test with the results of the sign test in Exercise 14.6. Are the test conclusions consistent?

14.28 Two methods for controlling traffic, A and B, were employed at each of $n = 12$ intersections for a period of 1 week, and the numbers of accidents occurring during this time period were recorded. The order of use (which method would be employed for the first week) was selected in a random manner.

Intersection	Method A	Method B	Intersection	Method A	Method B
1	5	4	7	2	3
2	6	4	8	4	1
3	8	9	9	7	9
4	3	2	10	5	2
5	6	3	11	6	5
6	1	0	12	1	1

Do the data provide sufficient evidence to indicate a shift in the distributions of accident rates for traffic control methods A and B?

a Analyze using a sign test.

b Analyze using the Wilcoxon signed-rank test for a paired experiment.

14.29 One explanation for the bright coloration of male birds versus the dull coloration of females and chicks is the predator-deflection theory, which suggests that males have developed a bright color to attract predators to themselves and away from their nesting females and young chicks. To test this theory, G. S. Butcher (1984) placed a stuffed and mounted sharp-shinned hawk (*Accipiter striatus*) a distance of 12 meters from each of 12 northern orioles' nests. For each nest, he recorded the total number of hits and swoops on the hawk by both the male and female oriole. He also noted which of the pair made the first attack on the hawk and which was more aggressive. The data are shown in the accompanying table.

Presentation Number†	Total Number of Hits and Swoops		More Aggressive Sex (M or F)‡	Sex Seen First
	M	F		
1	4	18	F	F
2	11	47	F	F
3	16	33	F	F
4	14	0	M	F
5	27	2	M	F
6	71	59	M	M
7	76	5	M	M
8	61	0	M	M
9	0	25	F	F
10	6	21	F	M
11	35	50	F	F
12	3	59	F	F

Do the data present sufficient evidence to indicate a difference in the distributions of numbers of hits and swoops for male versus female members of the nest?

a Test using Wilcoxon's signed-rank test with $\alpha = .05$. State your conclusions.

b Give the approximate p-value for the test and interpret it.

14.30 Refer to Exercise 14.29. Do the data present sufficient evidence to indicate that the male member of a nest is more aggressive than the female? Explain.

14.31 Eight subjects were asked to perform a simple puzzle-assembly task under normal conditions and under stressful conditions. During the stressful time, a mild shock was delivered to subjects 3 minutes after the start of the experiment and every 30 seconds thereafter until the task was completed. Blood pressure readings were taken under both conditions. The following data represent the highest readings during the experiment:

Subject	Normal	Stressful
1	126	130
2	117	118
3	115	125
4	118	120
5	118	121
6	128	125
7	125	130
8	120	120

Do the data present sufficient evidence to indicate higher blood pressure readings under stressful conditions? Analyze the data using the Wilcoxon signed-rank test for a paired experiment.

†For each of the 12 trials, the mounted *Accipiter* was placed within 12 meters of the nest for a 10-minute period.

‡M = male and F = female of the pair who fed offspring in the nest closest to the mounted *Accipiter*.

14.6 The Kruskal-Wallis H Test for Completely Randomized Designs

Just as the Mann-Whitney U test is the nonparametric alternative to the Student's t test for a comparison of population means, the Kruskal-Wallis H test is the nonparametric alternative to the analysis-of-variance F test for a completely randomized design. It is used to detect differences in location among more than two population distributions based on independent random sampling.

The procedure for conducting the Kruskal-Wallis H test is similar to that used for the Mann-Whitney U test. Suppose we are comparing k populations based on independent random samples n_1 from population 1, n_2 from population 2, ..., and n_k from population k where

$$n_1 + n_2 + \cdots + n_k = n$$

The first step is to rank all n observations from the smallest (rank 1) to the largest (rank n). **Tied observations are assigned a rank equal to the average of the ranks they would have received if they had been nearly equal but not tied.** We then calculate the rank sums $T_1, T_2, \ldots, T_k$ for the k samples and calculate the test statistic

$$H = \frac{12}{n(n+1)} \sum_{i=1}^{k} \frac{T_i^2}{n_i} - 3(n+1)$$

The greater the differences in location among the k population distributions, the larger will be the value of the H statistic. Thus, we reject the null hypothesis that the k population distributions are identical for large values of H.

How large is large? It can be shown (proof omitted) that when the sample sizes are moderate to large, say each sample size is equal to 5 or larger, and when H_0 is true, the H statistic will have approximately a chi-square distribution with $(k - 1)$ degrees of freedom (d.f.). Therefore, for a given value of α, we reject H_0 when the H statistic exceeds χ^2_α (Figure 14.4).

FIGURE **14.4** Approximate distribution of the H statistic when H_0 is true

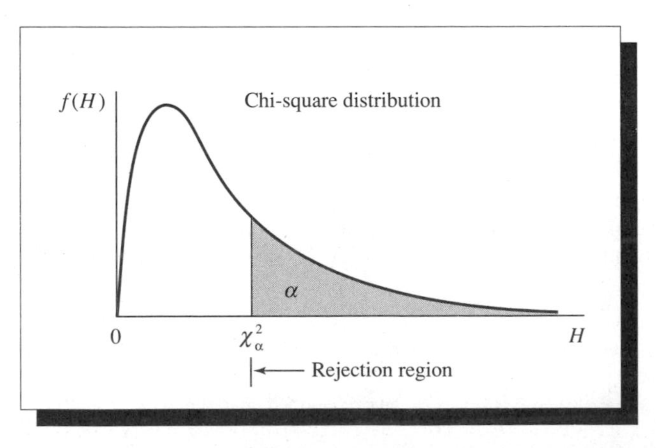

EXAMPLE **14.6** The following data were collected using a completely randomized design. They represent the achievement test scores for four different groups of students, each

group taught by a different teaching technique. The objective of the experiment is to test the hypothesis of "no difference in the population distributions of achievement test scores" against the alternative that they differ in location—that is, at least one of the distributions is shifted above the others. Conduct the test using the Kruskal-Wallis H test with $\alpha = .05$.

Techniques (ranks in parentheses)

	1	2	3	4
	65(3)	75(9)	59(1)	94(23)
	87(19)	69(5.5)	78(11)	89(21)
	73(8)	83(17.5)	67(4)	80(14)
	79(12.5)	81(15.5)	62(2)	88(20)
	81(15.5)	72(7)	83(17.5)	
	69(5.5)	79(12.5)	76(10)	
		90(22)		
Rank sum	$T_1 = 63.5$	$T_2 = 89$	$T_3 = 45.5$	$T_4 = 78$

Solution The first step is to rank the $n = 23$ observations from the smallest (rank 1) to the largest (rank 23). These ranks are shown in parentheses in the table. Note how the ties are handled. For example, two observations at 69 were tied for rank 5. Therefore, they were assigned the average 5.5 of the two ranks (5 and 6) that they would have occupied if they had been slightly different. The rank sums T_1, T_2, T_3, and T_4 for the four samples are shown in the bottom row of the table. Substituting rank sums and sample sizes into the formula for the H statistic, we obtain

$$\begin{aligned} H &= \frac{12}{n(n+1)} \sum_{i=1}^{k} \frac{T_i^2}{n_i} - 3(n+1) \\ &= \frac{12}{23(24)} \left[\frac{(63.5)^2}{6} + \frac{(89)^2}{7} + \frac{(45.5)^2}{6} + \frac{(78)^2}{4} \right] - 3(24) \\ &= 79.775102 - 72 \\ &= 7.775102 \end{aligned}$$

The rejection region for the H statistic for $\alpha = .05$ includes values of $H \geq \chi^2_{.05}$ where $\chi^2_{.05}$ is based on $k - 1 = 4 - 1 = 3$ d.f. The value of χ^2, given in Table 5 in Appendix I, is $\chi^2_{.05} = 7.81473$. The observed value of the H statistic, $H = 7.775102$, does not fall in the rejection region for the test. Therefore, there is insufficient evidence to indicate differences in the distributions of achievement test scores for the four teaching techniques. ■

EXAMPLE **14.7** Compare the results of the analysis-of-variance F test and the Kruskal-Wallis H test for testing for differences in the distributions of achievement test scores for the four teaching techniques.

Solution Examining Example 13.4, you can see that the F test detected differences among the population means (for $\alpha = .05$). The H test did not detect a shift in population distributions. Although these conclusions seem to be far apart, the test results are very close. The p-value corresponding to $F = 3.77$, $v_1 = 3$, and $v_2 = 10$, is slightly less than .05 in contrast to the p-value for $H = 7.78$, $v = 3$, which is slightly larger than .05. A person viewing the p-values for the two tests would see little difference in the results of the F and H tests. However, if we adhere to our choice of $\alpha = .05$, we could not reject H_0 using the H test. ■

The Kruskal-Wallis H Test for Comparing More than Two Populations: Completely Randomized Design (Independent Random Samples)

1 *Null Hypothesis:* The k population distributions are identical.

2 *Alternative Hypothesis:* At least two of the k population distributions differ in location.

3 *Test Statistic:* $H = \frac{12}{n(n+1)} \sum_{i=1}^{k} \frac{T_i^2}{n_i} - 3(n+1)$

where n_i = sample size for population i

T_i = rank sum for population i

n = total number of observations

$= n_1 + n_2 + \cdots + n_k$

4 *Rejection Region for a Given α:* $H > \chi^2_\alpha$

Assumptions

1 All sample sizes are greater than or equal to 5.

2 Ties assume the average of the ranks that they would have occupied if they had not been tied.

Table 14.6 contains the output generated using the MINITAB command KRUSKAL-WALLIS for the data in Example 14.6. The data were stored in C1, and the treatment subscripts in C2.

TABLE **14.6**
MINITAB printout for the data in Example 14.6

```
MTB > KRUSKAL-WALLIS C1 C2

LEVEL     NOBS     MEDIAN   AVE. RANK   Z VALUE
    1        6      76.00        10.6     -0.60
    2        7      79.00        12.7      0.33
    3        6      71.50         7.6     -1.86
    4        4      88.50        19.5      2.43
OVERALL     23                   12.0

H = 7.78   d.f. = 3  p = 0.051
H = 7.79   d.f. = 3  p = 0.051 (adj. for ties)

* NOTE   * One or more small samples
```

Exercises

Basic Techniques

14.32 Three treatments were compared using a completely randomized design. The data are shown below:

Treatment		
1	2	3
26	27	25
29	31	24
23	30	27
24	28	22
28	29	24
26	32	20
	30	21
	33	

Do the data provide sufficient evidence to indicate a difference in location for at least two of the population distributions? Test using the Kruskal-Wallis H statistic with $\alpha = .05$.

14.33 Four treatments were compared using a completely randomized design. The data are shown below:

Treatment			
1	2	3	4
124	147	141	117
167	121	144	128
135	136	139	102
160	114	162	119
159	129	155	128
144	117	150	123
133	109		

Do the data provide sufficient evidence to indicate a difference in location for at least two of the population distributions? Test using the Kruskal-Wallis H statistic with $\alpha = .05$.

Applications

14.34 Exercise 13.13 presents data on the rate of growth of vegetation at four swampy underdeveloped sites. Six plants were randomly selected at each of the four sites to be used in the comparison. The following data represent the mean leaf length per plant, in centimeters, for a random sample of ten leaves per plant.

Location	Mean Leaf Length (cm)					
1	5.7	6.3	6.1	6.0	5.8	6.2
2	6.2	5.3	5.7	6.0	5.2	5.5
3	5.4	5.0	6.0	5.6	4.9	5.2
4	3.7	3.2	3.9	4.0	3.5	3.6

a Do the data present sufficient evidence to indicate differences in location for at least two of the distributions of mean leaf length corresponding to the four locations? Test using the Kruskal-Wallis H test with $\alpha = .05$.

b Find the approximate p-value for the test.

c We analyzed this same set of data in Exercise 13.13 using an analysis of variance. Find the p-value for the F test used to compare the four location means in Exercise 13.13.

d Compare the p-values in parts (b) and (c) and explain the implications of the comparison.

14.35 Exercise 13.42 presents data on the heart rate for samples of ten men randomly selected from each of four age groups. Each subject walked a treadmill at a fixed grade for a period of 12 minutes, and the increase in heart rate (the difference before and after exercise) was recorded in beats per minute. The data are shown in the table.

	Age			
	10–19	20–39	40–59	60–69
	29	24	37	28
	33	27	25	29
	26	33	22	34
	27	31	33	36
	39	21	28	21
	35	28	26	20
	33	24	30	25
	29	34	34	24
	36	21	27	33
	22	32	33	32
Total	309	275	295	282

a Do the data present sufficient evidence to indicate differences in location for at least two of the four age groups? Test using the Kruskal-Wallis H test with $\alpha = .10$.

b Find the approximate p-value for the test in part (a).

c Since the F test in Exercise 13.42 and the H test in part (a) are both tests to detect differences in location of the four heart-rate populations, how do the test results compare? Compare the p-values for the two tests and explain the implications of the comparison.

14.36 A sampling of the acidity of rain for 10 randomly selected rainfalls was recorded at three different locations in the United States, the Northeast, the Middle Atlantic region, and the Southeast. The pH readings for these 30 rainfalls are shown below (NOTE: pH readings range from 0 to 14; 0 is acid, 14 is alkaline. Pure water falling through clean air has a pH reading of 5.7.)

Northeast	Middle Atlantic	Southeast
4.45	4.60	4.55
4.02	4.27	4.31
4.13	4.31	4.84
3.51	3.88	4.67
4.42	4.49	4.28
3.89	4.22	4.95
4.18	4.54	4.72
3.95	4.76	4.63
4.07	4.36	4.36
4.29	4.21	4.47

a Do the data present sufficient evidence to indicate differences in the levels of acidity in rainfalls for the three different locations? Test using the Kruskal-Wallis H test with $\alpha = .05$.

b Find the approximate p-value for the test in part (a), and interpret it.

14.7 The Friedman F_r Test for Randomized Block Designs

The Friedman F_r test, proposed by Nobel prize-winning economist Milton Friedman, is a nonparametric test for comparing the distributions of measurements for k treatments laid out in b blocks using a randomized block design. The procedure for conducting the test is very similar to that used for the Kruskal-Wallis H test. The first step in the procedure is to rank the k treatment observations within each block. Ties are treated in the usual way; that is, they receive an average of the ranks occupied by the tied observations. The rank sums $T_1, T_2, \ldots, T_k$ are then obtained and the test statistic

$$F_r = \frac{12}{bk(k+1)} \sum_{i=1}^{k} T_i^2 - 3b(k+1)$$

is calculated.

The value of the F_r statistic will be at a minimum when the rank sums are equal—that is, $T_1 = T_2 = \cdots = T_k$—and will increase in value as the differences among the rank sums increase. When either the number k of treatments or the number b of blocks is larger than 5, the sampling distribution of F_r can be approximated by a chi-square distribution with $(k-1)$ d.f. Therefore, like the Kruskal-Wallis H test, the rejection region for the F_r test consists of values of F_r for which

$$F_r > \chi^2_\alpha$$

EXAMPLE **14.8** Suppose that you wished to compare the reaction times of people subjected to one of six different stimuli. A reaction time measurement is obtained by subjecting a person to a stimulus and then measuring the length of time until the person presents some specified reaction. The objective of the experiment is to determine whether differences exist in the magnitudes of the reaction times for the stimuli employed in the experiment.

To eliminate the person-to-person variation in reaction time, four persons (subjects) were employed in the experiment and each person's reaction time (in seconds) was measured for each of the six stimuli. The data are reproduced below (ranks of the observations within each block are shown in parentheses). Use the Friedman F_r test to determine whether the data present sufficient evidence to indicate differences in the locations of the distributions of reaction times for the six stimuli. Test using $\alpha = .05$.

Subject	A	B	C	D	E	F
1	.6(2.5)	.9(6)	.8(5)	.7(4)	.5(1)	.6(2.5)
2	.7(3.5)	1.1(6)	.7(3.5)	.8(5)	.5(1.5)	.5(1.5)
3	.9(3)	1.3(6)	1.0(4.5)	1.0(4.5)	.7(1)	.8(2)
4	.5(2)	.7(5)	.8(6)	.6(3.5)	.4(1)	.6(3.5)
Rank sum	$T_1 = 11$	$T_2 = 23$	$T_3 = 19$	$T_4 = 17$	$T_5 = 4.5$	$T_6 = 9.5$

Solution We wish to test

H_0 : The distributions of reaction times for the six stimuli are identical.

against the alternative hypothesis

H_a : At least two of the distributions of reaction times for the six stimuli differ in location.

The table shows the ranks (in parentheses) of the observations within each block and the rank sums for each of the six stimuli (the treatments). The value of the F_r statistic for these data is

$$
\begin{aligned}
F_r &= \frac{12}{bk(k+1)} \sum_{i=1}^{k} T_i^2 - 3b(k+1) \\
&= \frac{12}{(4)(6)(7)} [(11)^2 + (23)^2 + (19)^2 + \cdots + (9.5)^2] - 3(4)(7) \\
&= 100.75 - 84 = 16.75
\end{aligned}
$$

Since the number $k = 6$ of treatments exceeds 5, the sampling distribution of F_r can be approximated by a chi-square distribution with $k - 1 = 6 - 1 = 5$ d.f. Therefore, for $\alpha = .05$, we reject H_0 if

$$F_r > \chi^2_{.05} \qquad \text{where} \qquad \chi^2_{.05} = 11.0705$$

This rejection region is shown in Figure 14.5.

Since the observed value $F_r = 16.75$ exceeds $\chi^2_{.05} = 11.0705$, it falls in the rejection region. We therefore reject H_0 and conclude that at least two of the distributions in reaction times differ in location. ■

EXAMPLE 14.9 Find the approximate p-value for the test in Example 14.8.

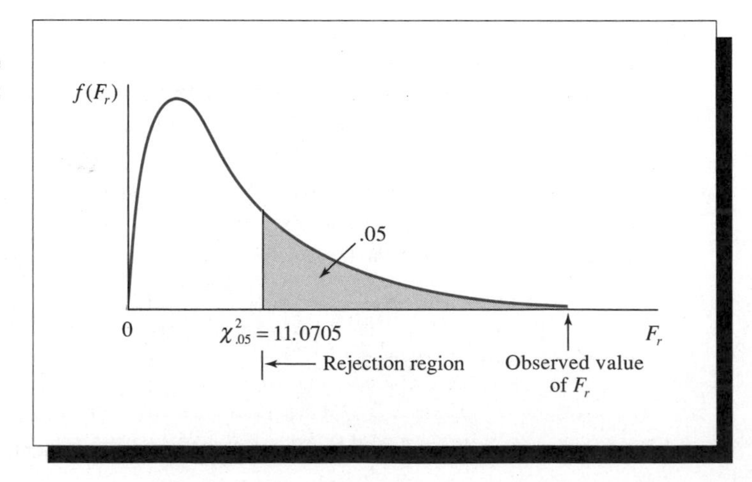

FIGURE 14.5 Rejection region for Example 14.8

Solution Consulting Table 5 in Appendix I with 5 d.f., we see that the observed value of $F_r = 16.75$ exceeds the table value $\chi^2_{.005} = 16.7496$. Hence, the p-value is very close to, but slightly less than, .005. ■

The Friedman F_r Test for a Randomized Block Design

1. *Null Hypothesis:* The k population distributions are identical.
2. *Alternative Hypothesis:* At least two of the k population distributions differ in location.
3. *Test Statistic:* $F_r = \dfrac{12}{bk(k+1)} \sum_{i=1}^{k} T_i^2 - 3b(k+1)$

   where b = number of blocks

   k = number of treatments

   T_i = rank sum for treatment i, $i = 1, 2, \ldots, k$
4. *Rejection Region:* $F_r > \chi^2_\alpha$ where χ^2_α is based on $(k-1)$ d.f.

   Assumption: Either the number k of treatments or the number b of blocks is larger than 5.

Table 14.7 contains the output generated using the MINITAB command FRIEDMAN for the data in Example 14.8. The data were stored in C1, the treatment subscripts in C2, and the block subscripts in C3.

TABLE **14.7**
MINITAB printout for the data in Example 14.8

```
MTB > FRIEDMAN C1 C2 C3

Friedman test of C1 by C2 blocked by C3

S = 16.75  d.f. = 5  p = 0.005
S = 17.37  d.f. = 5  p = 0.004 (adjusted for ties)

                     Est.    Sum of
   C2         N    Median     RANKS
    1         4    0.6500      11.0
    2         4    1.0000      23.0
    3         4    0.8000      19.0
    4         4    0.7500      17.0
    5         4    0.5000       4.5
    6         4    0.6000       9.5

Grand median   =   0.7167
```

Exercises

Basic Techniques

14.37 A randomized block design is employed to compare three treatments in six blocks. The data are shown below:

	Treatment		
Block	**1**	**2**	**3**
1	3.2	3.1	2.4
2	2.8	3.0	1.7
3	4.5	5.0	3.9
4	2.5	2.7	2.6
5	3.7	4.1	3.5
6	2.4	2.4	2.0

a Use the Friedman F_r test to detect differences in location among the three treatment distributions. Test using $\alpha = .05$.

b Find the approximate p-value for the test.

c Perform an analysis of variance and give the ANOVA table for the analysis.

d Give the value of the F statistic for testing the equality of the three treatment means.

e Give the approximate p-value for the F statistic in part (d).

f Compare the p-values for the tests in parts (a) and (d) and explain the practical implications of the comparison.

14.38 A randomized block design is employed to compare four treatments in eight blocks. The data are shown below:

	Treatment			
Block	**1**	**2**	**3**	**4**
1	89	81	84	85
2	93	86	86	88
3	91	85	87	86
4	85	79	80	82
5	90	84	85	85
6	86	78	83	84
7	87	80	83	82
8	93	86	88	90

a Use the Friedman F_r test to detect differences in location among the four treatment distributions. Test using $\alpha = .05$.

b Find the approximate p-value for the test.

c Perform an analysis of variance and give the ANOVA table for the analysis.

d Give the value of the F statistic for testing the equality of the four treatment means.

e Give the approximate p-value for the F statistic in part (d).

f Compare the p-values for the tests in parts (a) and (d) and explain the practical implications of the comparison.

Applications

14.39 In a comparison of the prices of items at five supermarkets, six items were randomly selected for the comparison, and the price of each was recorded for each of the five supermarkets. The objective of the study was to see whether the data indicated differences in the level of prices among the five supermarkets. The data are shown below:

	Supermarket				
Item	**Kash n' Karry**	**Publix**	**Winn-Dixie**	**Albertson's**	**Food 4 Less**
Celery	.33	.34	.69	.59	.58
Colgate toothpaste	1.28	1.49	1.44	1.37	1.28
Campbell's beef soup	1.05	1.19	1.23	1.19	1.10
Crushed pineapple	.83	.95	.95	.87	.84
Mueller's spaghetti	.68	.79	.83	.69	.69
Heinz ketchup	1.41	1.69	1.79	1.65	1.49

a Does the distribution of the prices of items differ in location from one supermarket to another? Test using the Friedman F_r test with $\alpha = .05$.

b Find the approximate p-value for the test and interpret it.

14.40 An experiment was conducted to compare the effects of three toxic chemicals A, B, and C on the skin of rats. One-inch squares of skin were treated with the chemicals and then scored from 0 to 10 depending on the degree of irritation. Three adjacent 1-inch squares were marked on the backs of eight rats, and each of the three chemicals was applied to each rat. Thus, the experiment was blocked on rats to eliminate the variation in skin sensitivity from rat to rat. The data are as follows:

Rats							
1	**2**	**3**	**4**	**5**	**6**	**7**	**8**
B 5	A 9	A 6	C 6	B 8	C 5	C 5	B 7
A 6	C 4	B 9	B 8	C 8	A 5	B 7	A 6
C 3	B 9	C 3	A 5	A 7	B 7	A 6	C 7

a Do the data provide sufficient evidence to indicate a difference in the toxic effect of the three chemicals? Test using the Friedman F_r test with $\alpha = .05$.

b Find the approximate p-value for the test and interpret it.

14.8 Rank Correlation Coefficient

In the preceding sections, we have used ranks to indicate the relative magnitude of observations in nonparametric tests for comparison of treatments. We will now employ the same technique in testing for a relation between two ranked variables. Two common rank-correlation coefficients are the **Spearman r_s** and the **Kendall τ**. We will present the Spearman r_s because its computation is identical to that for the sample correlation coefficient r of Chapter 10.

Suppose eight elementary science teachers have been ranked by a judge according to their teaching ability and all have taken a "national teachers' examination." The data are as follows:

Teacher	Judge's Rank	Examination Score
1	7	44
2	4	72
3	2	69
4	6	70
5	1	93
6	3	82
7	8	67
8	5	80

Do the data suggest an agreement between the judge's ranking and the examination score? That is, is there a correlation between ranks and test scores?

The two variables of interest are rank and test score. The former is already in rank form, and the test scores can be ranked similarly, as shown below. **The ranks for tied observations are obtained by averaging the ranks that the tied observations would have had if no ties had been observed.**

Teacher	Judge's Rank x_i	Test Rank (y_i)
1	7	1
2	4	5
3	2	3
4	6	4
5	1	8
6	3	7
7	8	2
8	5	6

The Spearman rank correlation coefficient r_s is calculated by using the ranks as the paired measurements on the two variables x and y in the formula for r, Chapter 10.

Spearman's Rank Correlation Coefficient

$$r_s = \frac{S_{xy}}{\sqrt{S_{xx}S_{yy}}}$$

where x_i and y_i represent the ranks of the ith pair of observations and

$$S_{xy} = \sum_{i=1}^{n}(x_i - \bar{x})(y_i - \bar{y}) = \sum_{i=1}^{n} x_i y_i - \frac{\left(\sum_{i=1}^{n} x_i\right)\left(\sum_{i=1}^{n} y_i\right)}{n}$$

$$S_{xx} = \sum_{i=1}^{n}(x_i - \bar{x})^2 = \sum_{i=1}^{n} x_i^2 - \frac{\left(\sum_{i=1}^{n} x_i\right)^2}{n}$$

$$S_{yy} = \sum_{i=1}^{n} (y_i - \bar{y})^2 = \sum_{i=1}^{n} y_i^2 - \frac{\left(\sum_{i=1}^{n} y_i\right)^2}{n}$$

When there are no ties in either the x observations or the y observations, the above expression for r_s algebraically reduces to the simpler expression

$$r_s = 1 - \frac{6 \sum_{i=1}^{n} d_i^2}{n(n^2 - 1)}, \qquad \text{where } d_i = x_i - y_i$$

If the number of ties is small in comparison with the number of data pairs, little error will result in using this shortcut formula.

EXAMPLE 14.10 Calculate r_s for the teacher–judge test-score data.

Solution The differences and squares of differences between the two rankings are as follows:

Teacher	x_i	y_i	d_i	d_i^2
1	7	1	6	36
2	4	5	−1	1
3	2	3	−1	1
4	6	4	2	4
5	1	8	−7	49
6	3	7	−4	16
7	8	2	6	36
8	5	6	−1	1
Total				144

Substituting into the formula for r_s,

$$r_s = 1 - \frac{6 \sum_{i=1}^{n} d_i^2}{n(n^2 - 1)} = 1 - \frac{6(144)}{8(64 - 1)}$$
$$= -.714$$

The Spearman rank correlation coefficient can be employed as a test statistic to test an hypothesis of "no association" between two populations. We assume that the n pairs of observations (x_i, y_i) have been randomly selected and, therefore, "no association between the populations" implies a random assignment of the n ranks within each sample. Each random assignment (for the two samples) represents a simple event associated with the experiment, and a value of r_s can be calculated for each. Thus, it is possible to calculate the probability that r_s assumes a large absolute value due solely to chance and thereby suggests an association between populations when none exists.

The rejection region for a two-tailed test is shown in Figure 14.6. If the alternative hypothesis is that the correlation between x and y is negative, you would reject H_0 for negative values of r_s that are close to -1 (in the lower tail of Figure 14.6). Similarly, if the alternative hypothesis is that the correlation between x and y is positive, you would reject H_0 for large positive values of r_s (in the upper tail of Figure 14.6).

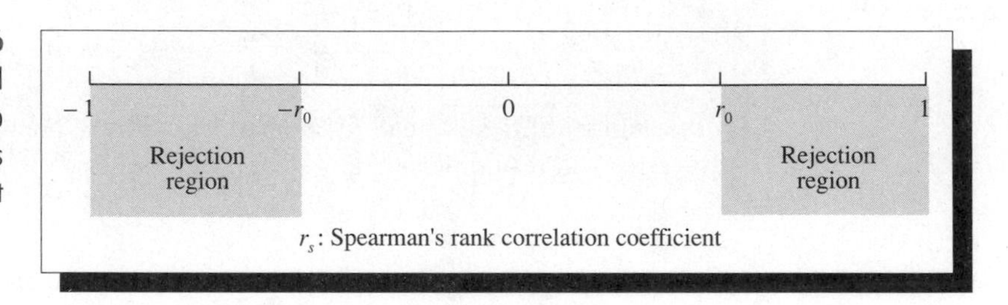

FIGURE 14.6 Rejection region for a two-tailed test of the null hypothesis of "no association," using Spearman's rank correlation test

The critical values of r_s are given in Table 9 in Appendix I. An abbreviated version of Table 9 is shown here in Table 14.8.

TABLE 14.8 An abbreviated version of Table 9 in Appendix I; for Spearman's rank correlation test

n	$\alpha = .05$	$\alpha = 0.025$	$\alpha = .01$	$\alpha = .005$
5	0.900	—	—	—
6	0.829	0.886	0.943	—
7	0.714	0.786	0.893	—
8	0.643	0.738	0.833	0.881
9	0.600	0.683	0.783	0.833
10	0.564	0.648	0.745	0.794
11	0.523	0.623	0.736	0.818
12	0.497	0.591	0.703	0.780
13	0.475	0.566	0.673	0.745
14	0.457	0.545		
15	0.441	0.525		
16	0.425			
17	0.412			
18	0.399	⋮	⋮	⋮
19	0.388			
20	0.377			
⋮	⋮			

Across the top of Table 14.8 (and Table 9 in Appendix I) are recorded values of α that you might wish to use for a one-tailed test of the null hypothesis of "no association" between x and y. The number of rank pairs n appears at the left side of the table. The table entries give the critical value r_0 for a one-tailed test. Thus, $P(r_s \geq r_0) = \alpha$.

For example, suppose you have $n = 8$ rank pairs and the alternative hypothesis is that the correlation between the ranks is positive. Then you would want to reject the null hypothesis of "no association" only for large positive values of r_s and would use a one-tailed test. Referring to Table 14.8 and using the row corresponding to $n = 8$ and the column for $\alpha = .05$, you read $r_0 = 0.643$. Therefore, you would reject H_0 for all values of r_s greater than or equal to 0.643.

The test is conducted in exactly the same manner if you wish to test only the alternative hypothesis that the ranks are negatively correlated. The only difference is that you would reject the null hypothesis if $r_s \le -0.643$. That is, you just use the negative of the tabulated value of r_0 to get the lower-tail critical value.

To conduct a two-tailed test, you reject the null hypothesis if $r_s \ge r_0$ or $r_s \le -r_0$. The value of α for the test will be double the value shown at the top of the table. For example, if $n = 8$ and you choose the 0.025 column, you will reject H_0 if $r_s \ge .738$ or $r_s \le -.738$. The α value for the test will be $2(.025) = .05$.

Spearman's Rank Correlation Test

1 *Null Hypothesis:* H_0 : There is no association between the rank pairs.
2 *Alternative Hypothesis:* H_a : There is an association between the rank pairs (a two-tailed test). Or H_a : The correlation between the rank pairs is positive or negative (a one-tailed test).
3 *Test Statistic:* $r_s = \dfrac{S_{xy}}{\sqrt{S_{xx}S_{yy}}}$

   where x_i and y_i represent the ranks of the ith pair of observations.
4 *Rejection Region:* For a two-tailed test, reject H_0 if $r_s \ge r_0$ or $r_s \le -r_0$, where r_0 is given in Table 9 in Appendix I. Double the tabulated probability to obtain the value of α for the two-tailed test. For a one-tailed test, reject H_0 if $r_s \ge r_0$ (for an upper-tailed test) or $r_s \le -r_0$ (for a lower-tailed test). The α value for a one tailed test is the value shown in Table 9, Appendix I.

EXAMPLE **14.11** Test the hypothesis of "no association" between the populations for Example 14.10.

Solution The critical value of r_s for a one-tailed test with $\alpha = .05$ and $n = 8$ is .643. Let us assume that a correlation between judge's rank and the teachers' test scores could not possibly be positive. (Low rank means good teaching and should be associated with a high test score if the judge and test measure teaching ability.) The alternative hypothesis is that the population rank correlation coefficient ρ_s is less than 0, and we are concerned with a one-tailed statistical test. Thus, α for the test is the tabulated value for .05, and we will reject the null hypothesis if $r_s \le -.643$.

The calculated value of the test statistic, $r_s = -.714$, is less than the critical value for $\alpha = .05$. Hence, the null hypothesis would be rejected at the $\alpha = .05$ level of significance. It appears that some agreement does exist between the judge's rankings and the test scores. However, it should be noted that this agreement could exist when *neither* provides an adequate yardstick for measuring teaching ability. For example, the association could exist if both the judge and those who constructed the teacher's examination had a completely erroneous, but similar, concept of the characteristics of good teaching. ▪

Exercises

Basic Techniques

14.41 Give the rejection region for a test to detect positive rank correlation if the number of pairs of ranks is 16 and

a $\alpha = .05$ **b** $\alpha = .01$

14.42 Give the rejection region for a test to detect negative rank correlation if the number of pairs of ranks is 12 and

a $\alpha = .05$ **b** $\alpha = .01$

14.43 Give the rejection region for a test to detect rank correlation if the number of pairs of ranks is 25 and

a $\alpha = .05$ **b** $\alpha = .01$

14.44 The following paired observations were obtained on two variables x and y.

x	1.2	.8	2.1	3.5	2.7	1.5
y	1.0	1.3	.1	−.8	−.2	.6

a Calculate the Spearman's rank correlation coefficient r_s.

b Do the data present sufficient evidence to indicate a correlation between x and y? Test using $\alpha = .05$.

Applications

14.45 A political scientist wished to examine the relationship of the voter image of a conservative political candidate and the distance between the residences of the voter and the candidate. Each of 12 voters rated the candidate on a scale of 1 to 20. The data are as follows:

Voter	Rating	Distance	Voter	Rating	Distance
1	12	75	7	9	120
2	7	165	8	18	60
3	5	300	9	3	230
4	19	15	10	8	200
5	17	180	11	15	130
6	12	240	12	4	130

a Calculate the Spearman rank correlation coefficient r_s.

b Do these data provide sufficient evidence to indicate a negative correlation between rating and distance?

14.46 Is the number of years of competitive running experience related to a runner's distance running performance? The data on nine runners, obtained from the study by Scott Powers and colleagues (1983), are shown below:

Runner	Years of Competitive Running	10-Kilometer Finish Time (min)
1	9	33.15
2	13	33.33
3	5	33.50
4	7	33.55
5	12	33.73
6	6	33.86
7	4	33.90
8	5	34.15
9	3	34.90

a Calculate the rank correlation coefficient between years of competitive running x and a runner's finish time y in the 10-kilometer race.

b Do the data provide sufficient evidence to indicate a rank correlation between y and x? Test using $\alpha = .10$.

14.47 The data shown in the accompanying table give measures of bending stiffness and twisting stiffness as measured by engineering tests for 12 tennis racquets.

Racquet	Bending Stiffness	Twisting Stiffness
1	419	227
2	407	231
3	363	200
4	360	211
5	257	182
6	622	304
7	424	384
8	359	194
9	346	158
10	556	225
11	474	305
12	441	235

a Calculate the rank correlation coefficient r_s between bending stiffness and twisting stiffness.

b If a racquet has bending stiffness, is it also likely to have twisting stiffness? Use the rank correlation coefficient to determine whether there is a significant positive relationship between bending stiffness and twisting stiffness. Use $\alpha = .05$.

14.48 A school principal suspected that a teacher's attitude toward a first grader depended on his original judgment of the child's ability. The principal also suspected that much of that judgment was based on the first grader's IQ score, which was usually known to the teacher. After three weeks of teaching, a teacher was asked to rank the nine children in his class from 1 (highest) to 9 (lowest) as to his opinion of their ability. Calculate r_s for the following teacher–IQ ranks:

Teacher	1	2	3	4	5	6	7	8	9
IQ	3	1	2	4	5	7	9	6	8

14.49 Refer to Exercise 14.48. Do the data provide sufficient evidence to indicate a positive correlation between the teacher's ranks and the ranks of the IQs? Use $\alpha = .05$.

14.50 Two art critics each ranked ten paintings by contemporary (but anonymous) artists in accordance with their appeal to the respective critics. The ratings are shown in the table. Do the critics seem to agree on their ratings of contemporary art? That is, do the data provide sufficient evidence to indicate a positive correlation between critics A and B? Test by using a value of α near .05.

Paintings	Critic A	Critic B
1	6	5
2	4	6
3	9	10
4	1	2
5	2	3
6	7	8
7	3	1
8	8	7
9	5	4
10	10	9

14.51 An experiment was conducted to study the relationship between the ratings of a tobacco leaf grader and the moisture content of the corresponding tobacco leaves. Twelve leaves were rated by the grader on a scale of 1 to 10, and corresponding readings of moisture content were made. The data are as follows:

Leaf	Grader's Rating	Moisture Content
1	9	.22
2	6	.16
3	7	.17
4	7	.14
5	5	.12
6	8	.19
7	2	.10
8	6	.12
9	1	.05
10	10	.20
11	9	.16
12	3	.09

Calculate r_s. Do the data provide sufficient evidence to indicate an association between the grader's ratings and the moisture content of the leaves?

14.52 The table below gives the average seller's asking price, the average buyer's bid, and the average closing bid for each of ten types of used computer equipment ("Used Computer Prices," p. 96). The data are reproduced below:

Machine	Average Seller's Asking Price	Average Buyer's Bid	Average Closing Bid
20MB PC XT	$400	$200	$300
20MB PC AT	700	400	575
IBM XT 089	450	200	325
IBM AT 339	700	350	600
20MB IBM PS/2 30	950	500	725
20MB IBM PS/2 50	1050	700	875
60MB IBM PS/2 70	2000	1600	1725
20MB Compaq SLT	1200	700	875
Toshiba 1600	1000	700	900
Toshiba 1200HB	1150	800	975

a Calculate the rank correlation between the average seller's asking price and the average closing bid. Is this correlation significantly positive? Test using $\alpha = .01$.

b Calculate the rank correlation between the average buyer's bid and the average closing bid. Is this correlation significantly positive? Test using $\alpha = .01$.

c Do you think that it was necessary to use the nonparametric correlation analysis rather than the parametric test presented in Chapter 10 in this instance? Explain.

14.53 A social skills training program was implemented with seven mildly handicapped students in a study to determine whether the program caused improvements in pre/post measures and behavior ratings. For one such test (Exercise 1.58), the pre- and posttest scores for the seven students are given below:

Subject	SSRS-T (standard score) Pretest	Posttest
Earl	101	113
Ned	89	89
Jasper	112	121
Charlie	105	99
Tom	90	104
Susie	91	94
Lori	89	99

a Use a nonparametric test to determine whether there is a significant positive relationship between the pre- and posttest scores. Use $\alpha = .05$.

b Do these results agree with the results of the parametric test in Exercise 10.35?

14.9 Summary

The nonparametric statistical tests presented in the preceding pages represent only a few of the many nonparametric statistical methods of inference available.

We have indicated that nonparametric statistical procedures are particularly useful when the experimental observations can be ordered but cannot be measured on a quantitative scale. Parametric statistical procedures usually cannot be applied to this type of data; hence, all inferential procedures must be based on nonparametric methods.

A second application of nonparametric statistical methods is in testing hypotheses associated with populations of quantitative data when uncertainty exists concerning the satisfaction of assumptions about the form of the population distributions. Just how useful are nonparametric methods for this situation?

In this chapter, we presented a number of useful nonparametric methods along with illustrations of their applications. The Mann-Whitney U test can be used to compare the locations of two population frequency distributions when the observations can be ranked according to their relative magnitudes and when the samples have been randomly and independently selected from the two populations. The Kruskal-Wallis H test provides similar methodology for comparing the locations of three or more population frequency distributions. The simplest nonparametric test—the sign test—provides a rapid procedure for comparing the locations of two population distributions when the observations have been independently selected in matched pairs. If the differences between pairs can be ranked according to their relative magnitudes, you can use the Wilcoxon signed-rank test for comparing the two populations. This latter test utilizes more sample information than the sign test and consequently is more likely to detect a difference in location if a difference exists. The Friedman F_r test enables us to extend this comparison to more than two population distributions when the data have been collected in matched sets, that is, according to a randomized block design. Finally, we presented a nonparametric method—Spearman's rank correlation test—for testing the correlation between the ranks of two variables when the observations associated with each variable can be ranked according to their relative magnitudes.

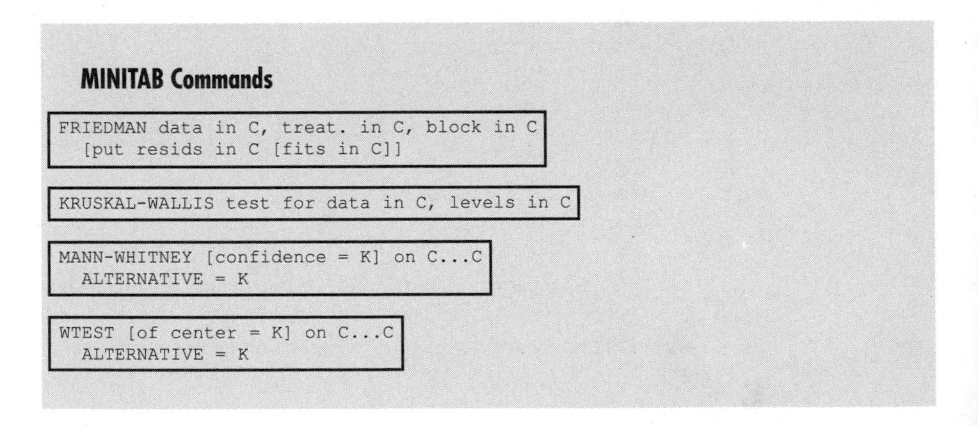

MINITAB Commands

```
FRIEDMAN data in C, treat. in C, block in C
  [put resids in C [fits in C]]
```

```
KRUSKAL-WALLIS test for data in C, levels in C
```

```
MANN-WHITNEY [confidence = K] on C...C
  ALTERNATIVE = K
```

```
WTEST [of center = K] on C...C
  ALTERNATIVE = K
```

Supplementary Exercises

14.54 A psychological experiment was conducted to compare the lengths of response time (in seconds) for two different stimuli. To remove natural person-to-person variability in the responses, both stimuli were applied to each of nine subjects, thus permitting an analysis of the difference between stimuli *within* each person.

Subject	Stimulus 1	Stimulus 2
1	9.4	10.3
2	7.8	8.9
3	5.6	4.1
4	12.1	14.7
5	6.9	8.7
6	4.2	7.1
7	8.8	11.3
8	7.7	5.2
9	6.4	7.8

a Use the sign test to determine whether sufficient evidence exists to indicate a difference in mean response for the two stimuli. Use a rejection region for which $\alpha \leq .05$.

b Test the hypothesis of no difference in mean response using Student's t test.

14.55 Refer to Exercise 14.54. Test the hypothesis that no difference exists in the distributions of responses for the two stimuli, using the Wilcoxon signed-rank test. Use a rejection region for which α is as near as possible to the α achieved in Exercise 14.54(a).

14.56 To compare two junior high schools A and B in academic effectiveness, an experiment was designed requiring the use of ten sets of identical twins, each twin having just completed the sixth grade. In each case, the twins in the same set had obtained their schooling in the same classrooms at each grade level. One child was selected at random from each pair of twins and assigned to school A. The remaining children were sent to school B. Near the end of the ninth grade, a certain achievement test was given to each child in the experiment. The results are shown in the following table:

	Twin Pair									
School	1	2	3	4	5	6	7	8	9	10
A	67	80	65	70	86	50	63	81	86	60
B	39	75	69	55	74	52	56	72	89	47

a Test (using the sign test) the hypothesis that the two schools are the same in academic effectiveness, as measured by scores on the achievement test, against the alternative that the schools are not equally effective. Use a level of significance as near as possible to $\alpha = .05$.

b Suppose it was known that junior high school A had a superior faculty and better learning facilities. Test the hypothesis of equal academic effectiveness against the alternative that school A is superior. Use a level of significance as near as possible to $\alpha = .05$.

14.57 Refer to Exercise 14.56. What answers are obtained if Wilcoxon's signed-rank test is used in analyzing the data? Compare with the answers to Exercise 14.56.

14.58 The coded values for a measure of brightness in paper (light reflectivity), prepared by two different processes, are given below for samples of nine observations drawn randomly from each of the two processes:

Process	Brightness								
A	6.1	9.2	8.7	8.9	7.6	7.1	9.5	8.3	9.0
B	9.1	8.2	8.6	6.9	7.5	7.9	8.3	7.8	8.9

Do the data present sufficient evidence ($\alpha = .10$) to indicate a difference in the populations of brightness measurements for the two processes?

a Use the Mann-Whitney U test.

b Use the Student's t test.

14.59 If (as in the case of measurements produced by two well-calibrated measuring instruments) the means of two populations are equal, it is possible to use the Mann-Whitney U statistic for testing hypotheses concerning the population variances as follows:

a Rank the combined sample.

b Number the ranked observations "from the outside in"; that is, number the smallest observation, 1; the largest, 2; the next-to-smallest, 3; the next-to-largest, 4; and so on. This final sequence of numbers induces an ordering on the symbols A (population A items) and B (population B items). If $\sigma_A^2 > \sigma_B^2$, one would expect to find a preponderance of A's near the first of the sequences, and thus a relatively small "sum of ranks" for the A observations.

c Given the following measurements produced by well-calibrated precision instruments A and B, test at near the $\alpha = .05$ level to determine whether the more expensive instrument B is more precise than A. (Note that this would imply a one-tailed test.) Use the Mann-Whitney U test.

Instrument A	Instrument B
1060.21	1060.24
1060.34	1060.28
1060.27	1060.32
1060.36	1060.30
1060.40	

d Test, using the F statistic of Section 9.7.

14.60 An experiment was conducted to compare the tenderness of meat cuts subjected to two different meat tenderizers A and B. To reduce the effect of extraneous variables, the data were paired by the specific meat cut, by applying the tenderizers to two cuts taken from the same steer, by cooking paired cuts together, and by using a single judge for each pair. After cooking, each cut was rated by a judge on a scale of 1 to 10, with 10 corresponding to the most tender condition of the cooked meat. The data, shown for a single judge, are given below. Do the data provide sufficient evidence to indicate that one of the two tenderizers tends to receive higher ratings than the other? Use a value of α near .05. [NOTE: $(1/2)^8 = .003906$.] Would a Student's t test be appropriate for analyzing these data? Explain.

	Tenderizer	
Cut	A	B
Shoulder roast	5	7
Chuck roast	6	5
Rib steak	8	9
Brisket	4	5
Club steak	9	9
Round steak	3	5
Rump roast	7	6
Sirloin steak	8	8
Sirloin tip steak	8	9
T-bone steak	9	10

14.61 A large corporation selects college graduates for employment, using both interviews and a psychological achievement test. Interviews conducted at the home office of the company were far more expensive than the tests that could be conducted on campus. Consequently, the personnel office was interested in determining whether the test scores were correlated with interview ratings and whether tests could be substituted for interviews. The idea was not to eliminate interviews but to reduce their number. To determine whether correlation was present, ten prospects were ranked during interviews and tested. The paired scores are as follows:

Subject	Interview Rank	Test Score
1	8	74
2	5	81
3	10	66
4	3	83
5	6	66
6	1	94
7	4	96
8	7	70
9	9	61
10	2	86

Calculate the Spearman rank correlation coefficient r_s. Rank 1 is assigned to the candidate judged to be the best.

14.62 Refer to Exercise 14.61. Do the data present sufficient evidence to indicate that the correlation between interview rankings and test scores is less than 0? If this evidence does exist, can we say that tests could be used to reduce the number of interviews?

14.63 A comparison of reaction times for two different stimuli in a psychological word-association experiment produced the following results when applied to a random sample of 16 people:

Stimulus	Reaction Time (sec)							
1	1	3	2	1	2	1	3	2
2	4	2	3	3	1	2	3	3

Do the data present sufficient evidence to indicate a difference in mean reaction time for the two stimuli? Use the Mann-Whitney U statistic and test using $\alpha = .05$. (NOTE: This test was conducted using Student's t in the exercises for Chapter 9. Compare your results.)

14.64 The following table gives the scores of a group of 15 students in mathematics and art:

Student	Math	Art	Student	Math	Art
1	22	53	9	62	55
2	37	68	10	65	74
3	36	42	11	66	68
4	38	49	12	56	64
5	42	51	13	66	67
6	58	65	14	67	73
7	58	51	15	62	65
8	60	71			

Use Wilcoxon's signed-rank test to determine if the median scores for these students differ significantly for the two subjects.

14.65 Refer to Exercise 14.64. Compute Spearman's rank correlation coefficient for these data and test $H_0 : \rho_s = 0$ at the 10% level of significance.

14.66 Exercise 13.47 presents an analysis of variance of the yields of five different varieties of wheat, observed on one plot each at each of six different locations. The data from this randomized block design are shown below:

Varieties	Location 1	2	3	4	5	6
A	35.3	31.0	32.7	36.8	37.2	33.1
B	30.7	32.2	31.4	31.7	35.0	32.7
C	38.2	33.4	33.6	37.1	37.3	38.2
D	34.9	36.1	35.2	38.3	40.2	36.0
E	32.4	28.9	29.2	30.7	33.9	32.1

a Use the Friedman F_r test to determine whether the data provide sufficient evidence to indicate a difference in the levels of yield for the five different varieties of wheat. Test using $\alpha = .05$.

b Exercise 13.47 presents a computer printout of the analysis of variance for comparing the mean yield for the five varieties of wheat. How do the results of the analysis-of-variance F test compare with the Friedman F_r test in part (a)? Explain.

14.67 In Exercise 13.43, we compared the number of sales per trainee after completing one of four different sales training programs. Six trainees completed training program 1, eight completed 2, and so on. The number of sales per trainee is shown in the following table:

	Training Program 1	2	3	4
	78	99	74	81
	84	86	87	63
	86	90	80	71
	92	93	83	65
	69	94	78	86
	73	85		79
		97		73
		91		70
Total	482	735	402	588

a Do the data present sufficient evidence to indicate that the location of the distribution of number of sales per trainee differs from one training program to another? Test using the Kruskal-Wallis H test with $\alpha = .05$.

b How do the test results in part (a) compare with the results of the analysis-of-variance F test in Exercise 13.43?

14.68 In Exercise 13.45, we performed an analysis of variance to compare the mean level of effluents in water at four different industrial plants. Five samples of the liquid waste (lb/gal) were taken at the output of each of the four industrial plants. The data are shown below:

Plant	Polluting Effluents (lb/gal of waste)				
A	1.65	1.72	1.50	1.37	1.60
B	1.70	1.85	1.46	2.05	1.80
C	1.40	1.75	1.38	1.65	1.55
D	2.10	1.95	1.65	1.88	2.00

a Do the data present sufficient evidence to indicate a difference in the level of pollutants for the four different industrial plants? Test using the Kruskal-Wallis H test with $\alpha = .05$.

b Find the approximate p-value for the test and interpret its value.

c Compare the test results in part (a) with the analysis-of-variance test in Exercise 13.45. Do the results agree? Explain.

14.69 In comparing the 1989 and 1991 median prices of a single-family home in 20 metropolitan statistical areas as defined by the U.S. Office of Budget and Management (Wright, 1992), 14 areas showed an increase in median price, and 6 showed a decrease. Do these data contain sufficient evidence to conclude that the median price of a single-family dwelling has increased significantly from 1989 to 1991? Use $\alpha = .05$. What is the observed significance level of this test?

14.70 Scientists have shown that a newly developed vaccine can shield rhesus monkeys from infection by a virus closely related to the AIDS-causing human immunodeficiency virus (HIV). In their work, Ronald C. Resrosiers and his colleagues of the New England Regional Primate Research Center gave each of the $n = 6$ rhesus monkeys five inoculations with the simian immunodeficiency virus (SIV) vaccine. One week after the last vaccination, each monkey received an injection of live SIV. Two of the six vaccinated monkeys showed no evidence of SIV infection for as long as a year and a half after the SIV injection (*Science News*, 1989, p. 116). Scientists were able to isolate the SIV virus from the other four vaccinated monkeys, although these animals showed no sign of the disease. Does this information contain sufficient evidence to indicate that the vaccine is effective in protecting monkeys from SIV? Use $\alpha = .10$.

14.71 In the traditional American Western, you could immediately recognize the bad guys by their black hats. Mark Frank and Thomas Gilovich (1988) ran an experiment to determine whether the color of a person's clothing has a significant effect on others' perception of his or her behavior. As part of this experiment, they had 25 subjects rate the uniforms of all National Football League (NFL) teams on several 7-point scales. Five of the 28 teams have uniforms that were considered "black"—namely, the Pittsburgh Steelers, the New Orleans Saints, the Los Angeles Raiders, the Cincinnati Bengals, and Chicago Bears. The resulting "malevolence indexes" are given in the following table:

TABLE **14.9**
Malevolence ratings of the uniforms of professional football teams

Football Team	Rating	Football Team	Rating
LA Raiders	5.10	Minnesota	3.90
Pittsburgh	5.00	Atlanta	3.87
Cincinnati	4.97	San Francisco	3.83
New Orleans	4.83	Indianapolis	3.83
Chicago	4.68	Seattle	3.82
Kansas City	4.58	Denver	3.80
Washington	4.40	Tampa Bay	3.77
St. Louis	4.27	New England	3.60
NY Jets	4.12	Buffalo	3.53
LA Rams	4.10	Detroit	3.38
Cleveland	4.05	NY Giants	3.27
San Diego	4.05	Dallas	3.15
Green Bay	4.00	Houston	2.88
Philadelphia	3.97	Miami	2.80

NOTE : Teams in boldface letters are those with black uniforms. The malevolence ratings represent the average rating of three semantic differential scales: good/bad, timid/aggressive, and nice/mean.

Use the large-sample Mann-Whitney U test to determine whether there is sufficient information to conclude that "black" uniforms receive higher malevolence scores. Use $\alpha = .05$.

CASE STUDY

How's Your Cholesterol Level?

As consumers become more and more interested in eating healthy foods, many "light," "fat-free," and "cholesterol-free" products are appearing in the marketplace. One such product is the frozen egg substitute, a cholesterol-free product that can be used in cooking and baking in many of the same ways that regular eggs can—although not all. Some consumers even use egg substitutes for Caesar salad dressings and other recipes calling for raw eggs because these products are pasteurized and worries about bacterial contamination are eliminated.

Unfortunately, the products currently on the market exhibit strong differences in both flavor and texture when tasted in their primary preparation as scrambled eggs. Five panelists, all experts in nutrition and food preparation, were asked to rate each of three egg substitutes on the basis of taste, appearance, texture, and whether they would buy the product (Sakekel, 1993). The judges tasted the three egg substitutes and rated them on a scale of 0 to 20. The results, shown in the table, indicate that the highest rating, by 23 points, when to ConAgra's Healthy Choice Egg Product, which the tasters unanimously agreed most closely resembled eggs as they come from the hen. The second-place product, Morningstar Farms' Scramblers struck several tasters as having an "oddly sweet flavor" . . . "similar to carrots." Finally, none of the tasters indicated that they would be willing to buy Fleishmann's Egg Beaters, which was described by the testers as "watery," "slippery," and "unpleasant." Oddly enough, these results are contrary to a similar taste test 4 years ago, in which Egg Beaters were considered better than other competing egg substitutes.

Tasters	Healthy Choice	Scramblers	Egg Beaters
Dan Bowe	16	9	7
John Carroll	16	7	8
Donna Katzl	14	8	4
Rick O'Connell	15	16	9
Roland Passot	13	11	2
Totals	74	51	30

1 What type of design has been used in this taste-testing experiment?

2 Do you think that the data satisfy the assumptions required for a parametric analysis of variance? Explain.

3 Use the appropriate nonparametric technique to determine whether there is a significant difference between the average scores for the three brands of egg substitutes. Use $\alpha = .05$.

I

Tables

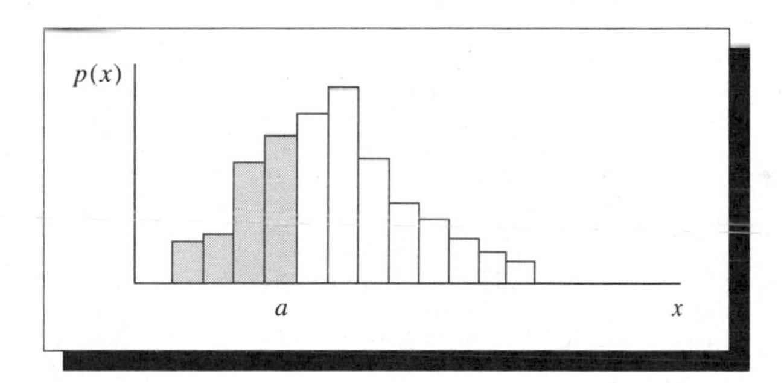

TABLE 1 Cumulative Binomial Probabilities

Tabulated values are $P(x \leq a) = \sum_{x=0}^{a} p(x)$. (Computations are rounded at the third decimal place.)

$n = 2$

	p													
a	**0.01**	**0.05**	**0.10**	**0.20**	**0.30**	**0.40**	**0.50**	**0.60**	**0.70**	**0.80**	**0.90**	**0.95**	**0.99**	**a**
0	.980	.902	.810	.640	.490	.360	.250	.160	.090	.040	.010	.002	.000	0
1	1.000	.998	.990	.960	.910	.840	.750	.640	.510	.360	.190	.098	.020	1
2	1.000	1.000	1.000	1.000	1.000	1.000	1.000	1.000	1.000	1.000	1.000	1.000	1.000	2

$n = 3$

	p													
a	**0.01**	**0.05**	**0.10**	**0.20**	**0.30**	**0.40**	**0.50**	**0.60**	**0.70**	**0.80**	**0.90**	**0.95**	**0.99**	**a**
0	.970	.857	.729	.512	.343	.216	.125	.064	.027	.008	.001	.000	.000	0
1	1.000	.993	.972	.896	.784	.648	.500	.352	.216	.104	.028	.007	.000	1
2	1.000	1.000	.999	.992	.973	.936	.875	.784	.657	.488	.271	.143	.030	2
3	1.000	1.000	1.000	1.000	1.000	1.000	1.000	1.000	1.000	1.000	1.000	1.000	1.000	3

TABLE **1** (Continued)

$n = 4$

a	p = 0.01	0.05	0.10	0.20	0.30	0.40	0.50	0.60	0.70	0.80	0.90	0.95	0.99	a
0	.961	.815	.656	.410	.240	.130	.062	.026	.008	.002	.000	.000	.000	0
1	.999	.986	.948	.819	.652	.475	.312	.179	.084	.027	.004	.000	.000	1
2	1.000	1.000	.996	.973	.916	.821	.688	.525	.348	.181	.052	.014	.001	2
3	1.000	1.000	1.000	.998	.992	.974	.938	.870	.760	.590	.344	.185	.039	3
4	1.000	1.000	1.000	1.000	1.000	1.000	1.000	1.000	1.000	1.000	1.000	1.000	1.000	4

$n = 5$

a	p = 0.01	0.05	0.10	0.20	0.30	0.40	0.50	0.60	0.70	0.80	0.90	0.95	0.99	a
0	.951	.774	.590	.328	.168	.078	.031	.010	.002	.000	.000	.000	.000	0
1	.999	.977	.919	.737	.528	.337	.188	.087	.031	.007	.000	.000	.000	1
2	1.000	.999	.991	.942	.837	.683	.500	.317	.163	.058	.009	.001	.000	2
3	1.000	1.000	1.000	.993	.969	.913	.812	.663	.472	.263	.081	.023	.001	3
4	1.000	1.000	1.000	1.000	.998	.990	.969	.922	.832	.672	.410	.226	.049	4
5	1.000	1.000	1.000	1.000	1.000	1.000	1.000	1.000	1.000	1.000	1.000	1.000	1.000	5

$n = 6$

a	p = 0.01	0.05	0.10	0.20	0.30	0.40	0.50	0.60	0.70	0.80	0.90	0.95	0.99	a
0	.941	.735	.531	.262	.118	.047	.016	.004	.001	.000	.000	.000	.000	0
1	.999	[illegible]67	.886	.655	.420	.233	.109	.041	.011	.002	.000	.000	.000	1
2	1.000	.[illegible]98	.984	.901	.744	.544	.344	.179	.070	.017	.001	.000	.000	2
3	1.000	1.000	.999	.983	.930	.821	.656	.456	.256	.099	.016	.002	.000	3
4	1.000	1.000	1.000	.998	.989	.959	.891	.767	.580	.345	.114	.033	.001	4
5	1.000	1.000	1.000	1.000	.999	.996	.984	.953	.882	.738	.469	.265	.059	5
6	1.000	1.000	1.000	1.000	1.000	1.000	1.000	1.000	1.000	1.000	1.000	1.000	1.000	6

TABLE 1 (Continued)

$n = 7$

	p													
a	**0.01**	**0.05**	**0.10**	**0.20**	**0.30**	**0.40**	**0.50**	**0.60**	**0.70**	**0.80**	**0.90**	**0.95**	**0.99**	*a*
0	.932	.698	.478	.210	.082	.028	.008	.002	.000	.000	.000	.000	.000	0
1	.998	.956	.850	.577	.329	.159	.062	.019	.004	.000	.000	.000	.000	1
2	1.000	.996	.974	.852	.647	.420	.227	.096	.029	.005	.000	.000	.000	2
3	1.000	1.000	.997	.967	.874	.710	.500	.290	.126	.033	.003	.000	.000	3
4	1.000	1.000	1.000	.995	.971	.904	.773	.580	.353	.148	.026	.004	.000	4
5	1.000	1.000	1.000	1.000	.996	.981	.938	.841	.671	.423	.150	.044	.002	5
6	1.000	1.000	1.000	1.000	1.000	.998	.992	.972	.918	.790	.522	.302	.068	6
7	1.000	1.000	1.000	1.000	1.000	1.000	1.000	1.000	1.000	1.000	1.000	1.000	1.000	7

$n = 8$

	p													
a	**0.01**	**0.05**	**0.10**	**0.20**	**0.30**	**0.40**	**0.50**	**0.60**	**0.70**	**0.80**	**0.90**	**0.95**	**0.99**	*a*
0	.923	.663	.430	.168	.058	.017	.004	.001	.000	.000	.000	.000	.000	0
1	.997	.943	.813	.503	.255	.106	.035	.009	.001	.000	.000	.000	.000	1
2	1.000	.994	.962	.797	.552	.315	.145	.050	.011	.001	.000	.000	.000	2
3	1.000	1.000	.995	.944	.806	.594	.363	.174	.058	.010	.000	.000	.000	3
4	1.000	1.000	1.000	.990	.942	.826	.637	.406	.194	.056	.005	.000	.000	4
5	1.000	1.000	1.000	.999	.989	.950	.855	.685	.448	.203	.038	.006	.000	5
6	1.000	1.000	1.000	1.000	.999	.991	.965	.894	.745	.497	.187	.057	.003	6
7	1.000	1.000	1.000	1.000	1.000	.999	.996	.983	.942	.832	.570	.337	.077	7
8	1.000	1.000	1.000	1.000	1.000	1.000	1.000	1.000	1.000	1.000	1.000	1.000	1.000	8

$n = 9$

	p													
a	**0.01**	**0.05**	**0.10**	**0.20**	**0.30**	**0.40**	**0.50**	**0.60**	**0.70**	**0.80**	**0.90**	**0.95**	**0.99**	*a*
0	.914	.630	.387	.134	.040	.010	.002	.000	.000	.000	.000	.000	.000	0
1	.997	.929	.775	.436	.196	.071	.020	.004	.000	.000	.000	.000	.000	1
2	1.000	.992	.947	.738	.463	.232	.090	.025	.004	.000	.000	.000	.000	2
3	1.000	.999	.992	.914	.730	.483	.254	.099	.025	.003	.000	.000	.000	3
4	1.000	1.000	.999	.980	.901	.733	.500	.267	.099	.020	.001	.000	.000	4
5	1.000	1.000	1.000	.997	.975	.901	.746	.517	.270	.086	.008	.001	.000	5
6	1.000	1.000	1.000	1.000	.996	.975	.910	.768	.537	.262	.053	.008	.000	6
7	1.000	1.000	1.000	1.000	1.000	.996	.980	.929	.804	.564	.225	.071	.003	7
8	1.000	1.000	1.000	1.000	1.000	1.000	.998	.990	.960	.866	.613	.370	.086	8
9	1.000	1.000	1.000	1.000	1.000	1.000	1.000	1.000	1.000	1.000	1.000	1.000	1.000	9

TABLE 1 (Continued)

$n = 10$

	p													
a	0.01	0.05	0.10	0.20	0.30	0.40	0.50	0.60	0.70	0.80	0.90	0.95	0.99	a
0	.904	.599	.349	.107	.028	.006	.001	.000	.000	.000	.000	.000	.000	0
1	.996	.914	.736	.376	.149	.046	.011	.002	.000	.000	.000	.000	.000	1
2	1.000	.988	.930	.678	.383	.167	.055	.012	.002	.000	.000	.000	.000	2
3	1.000	.999	.987	.879	.650	.382	.172	.055	.011	.001	.000	.000	.000	3
4	1.000	1.000	.998	.967	.850	.633	.377	.166	.047	.006	.000	.000	.000	4
5	1.000	1.000	1.000	.994	.953	.834	.623	.367	.150	.033	.002	.000	.000	5
6	1.000	1.000	1.000	.999	.989	.945	.828	.618	.350	.121	.013	.001	.000	6
7	1.000	1.000	1.000	1.000	.998	.988	.945	.833	.617	.322	.070	.012	.000	7
8	1.000	1.000	1.000	1.000	1.000	.998	.989	.954	.851	.624	.264	.086	.004	8
9	1.000	1.000	1.000	1.000	1.000	1.000	.999	.994	.972	.893	.651	.401	.096	9
10	1.000	1.000	1.000	1.000	1.000	1.000	1.000	1.000	1.000	1.000	1.000	1.000	1.000	10

$n = 11$

	p													
a	0.01	0.05	0.10	0.20	0.30	0.40	0.50	0.60	0.70	0.80	0.90	0.95	0.99	a
0	.895	.569	.314	.086	.020	.004	.000	.000	.000	.000	.000	.000	.000	0
1	.995	.898	.697	.322	.113	.030	.006	.001	.000	.000	.000	.000	.000	1
2	1.000	.985	.910	.617	.313	.119	.033	.006	.001	.000	.000	.000	.000	2
3	1.000	.998	.981	.839	.570	.296	.113	.029	.004	.000	.000	.000	.000	3
4	1.000	1.000	.997	.950	.790	.533	.274	.099	.022	.002	.000	.000	.000	4
5	1.000	1.000	1.000	.988	.922	.754	.500	.246	.078	.012	.000	.000	.000	5
6	1.000	1.000	1.000	.998	.978	.901	.726	.467	.210	.050	.003	.000	.000	6
7	1.000	1.000	1.000	1.000	.996	.971	.887	.704	.430	.161	.019	.002	.000	7
8	1.000	1.000	1.000	1.000	.999	.994	.967	.881	.687	.383	.090	.015	.000	8
9	1.000	1.000	1.000	1.000	1.000	.999	.994	.970	.887	.678	.303	.102	.005	9
10	1.000	1.000	1.000	1.000	1.000	1.000	1.000	.996	.980	.914	.686	.431	.105	10
11	1.000	1.000	1.000	1.000	1.000	1.000	1.000	1.000	1.000	1.000	1.000	1.000	1.000	11

TABLE 1 (Continued)

$n = 12$

a	0.01	0.05	0.10	0.20	0.30	0.40	0.50	0.60	0.70	0.80	0.90	0.95	0.99	a
0	.886	.540	.282	.069	.014	.002	.000	.000	.000	.000	.000	.000	.000	0
1	.994	.882	.659	.275	.085	.020	.003	.000	.000	.000	.000	.000	.000	1
2	1.000	.980	.889	.558	.253	.083	.019	.003	.000	.000	.000	.000	.000	2
3	1.000	.998	.974	.795	.493	.225	.073	.015	.002	.000	.000	.000	.000	3
4	1.000	1.000	.996	.927	.724	.438	.194	.057	.009	.001	.000	.000	.000	4
5	1.000	1.000	.999	.981	.882	.665	.387	.158	.039	.004	.000	.000	.000	5
6	1.000	1.000	1.000	.996	.961	.842	.613	.335	.118	.019	.001	.000	.000	6
7	1.000	1.000	1.000	.999	.991	.943	.806	.562	.276	.073	.004	.000	.000	7
8	1.000	1.000	1.000	1.000	.998	.985	.927	.775	.507	.205	.026	.002	.000	8
9	1.000	1.000	1.000	1.000	1.000	.997	.981	.917	.747	.442	.111	.020	.000	9
10	1.000	1.000	1.000	1.000	1.000	1.000	.997	.980	.915	.725	.341	.118	.006	10
11	1.000	1.000	1.000	1.000	1.000	1.000	1.000	.998	.986	.931	.718	.460	.114	11
12	1.000	1.000	1.000	1.000	1.000	1.000	1.000	1.000	1.000	1.000	1.000	1.000	1.000	12

$n = 15$

a	0.01	0.05	0.10	0.20	0.30	0.40	0.50	0.60	0.70	0.80	0.90	0.95	0.99	a
0	.860	.463	.206	.035	.005	.000	.000	.000	.000	.000	.000	.000	.000	0
1	.990	.829	.549	.167	.035	.005	.000	.000	.000	.000	.000	.000	.000	1
2	1.000	.964	.816	.398	.127	.027	.004	.000	.000	.000	.000	.000	.000	2
3	1.000	.995	.944	.648	.297	.091	.018	.002	.000	.000	.000	.000	.000	3
4	1.000	.999	.987	.836	.515	.217	.059	.009	.001	.000	.000	.000	.000	4
5	1.000	1.000	.998	.939	.722	.403	.151	.034	.004	.000	.000	.000	.000	5
6	1.000	1.000	1.000	.982	.869	.610	.304	.095	.015	.001	.000	.000	.000	6
7	1.000	1.000	1.000	.996	.950	.787	.500	.213	.050	.004	.000	.000	.000	7
8	1.000	1.000	1.000	.999	.985	.905	.696	.390	.131	.018	.000	.000	.000	8
9	1.000	1.000	1.000	1.000	.996	.966	.849	.597	.278	.061	.002	.000	.000	9
10	1.000	1.000	1.000	1.000	.999	.991	.941	.783	.485	.164	.013	.001	.000	10
11	1.000	1.000	1.000	1.000	1.000	.998	.982	.909	.703	.352	.056	.005	.000	11
12	1.000	1.000	1.000	1.000	1.000	1.000	.996	.973	.873	.602	.184	.036	.000	12
13	1.000	1.000	1.000	1.000	1.000	1.000	1.000	.995	.965	.833	.451	.171	.010	13
14	1.000	1.000	1.000	1.000	1.000	1.000	1.000	1.000	.995	.965	.794	.537	.140	14
15	1.000	1.000	1.000	1.000	1.000	1.000	1.000	1.000	1.000	1.000	1.000	1.000	1.000	15

TABLE 1 (Continued)

$n = 20$

	p													
a	0.01	0.05	0.10	0.20	0.30	0.40	0.50	0.60	0.70	0.80	0.90	0.95	0.99	a
0	.818	.358	.122	.012	.001	.000	.000	.000	.000	.000	.000	.000	.000	0
1	.983	.736	.392	.069	.008	.001	.000	.000	.000	.000	.000	.000	.000	1
2	.999	.925	.677	.206	.035	.004	.000	.000	.000	.000	.000	.000	.000	2
3	1.000	.984	.867	.411	.107	.016	.001	.000	.000	.000	.000	.000	.000	3
4	1.000	.997	.957	.630	.238	.051	.006	.000	.000	.000	.000	.000	.000	4
5	1.000	1.000	.989	.804	.416	.126	.021	.002	.000	.000	.000	.000	.000	5
6	1.000	1.000	.998	.913	.608	.250	.058	.006	.000	.000	.000	.000	.000	6
7	1.000	1.000	1.000	.968	.772	.416	.132	.021	.001	.000	.000	.000	.000	7
8	1.000	1.000	1.000	.990	.887	.596	.252	.057	.005	.000	.000	.000	.000	8
9	1.000	1.000	1.000	.997	.952	.755	.412	.128	.017	.001	.000	.000	.000	9
10	1.000	1.000	1.000	.999	.983	.872	.588	.245	.048	.003	.000	.000	.000	10
11	1.000	1.000	1.000	1.000	.995	.943	.748	.404	.113	.010	.000	.000	.000	11
12	1.000	1.000	1.000	1.000	.999	.979	.868	.584	.228	.032	.000	.000	.000	12
13	1.000	1.000	1.000	1.000	1.000	.994	.942	.750	.392	.087	.002	.000	.000	13
14	1.000	1.000	1.000	1.000	1.000	.998	.979	.874	.584	.196	.011	.000	.000	14
15	1.000	1.000	1.000	1.000	1.000	1.000	.994	.949	.762	.370	.043	.003	.000	15
16	1.000	1.000	1.000	1.000	1.000	1.000	.999	.984	.893	.589	.133	.016	.000	16
17	1.000	1.000	1.000	1.000	1.000	1.000	1.000	.996	.965	.794	.323	.075	.001	17
18	1.000	1.000	1.000	1.000	1.000	1.000	1.000	.999	.992	.931	.608	.264	.017	18
19	1.000	1.000	1.000	1.000	1.000	1.000	1.000	1.000	.999	.988	.878	.642	.182	19
20	1.000	1.000	1.000	1.000	1.000	1.000	1.000	1.000	1.000	1.000	1.000	1.000	1.000	20

TABLE 1 (Continued)

$n = 25$

a	0.01	0.05	0.10	0.20	0.30	0.40	0.50	0.60	0.70	0.80	0.90	0.95	0.99	a
0	.778	.277	.072	.004	.000	.000	.000	.000	.000	.000	.000	.000	.000	0
1	.974	.642	.271	.027	.002	.000	.000	.000	.000	.000	.000	.000	.000	1
2	.998	.873	.537	.098	.009	.000	.000	.000	.000	.000	.000	.000	.000	2
3	1.000	.966	.764	.234	.033	.002	.000	.000	.000	.000	.000	.000	.000	3
4	1.000	.993	.902	.421	.090	.009	.000	.000	.000	.000	.000	.000	.000	4
5	1.000	.999	.967	.617	.193	.029	.002	.000	.000	.000	.000	.000	.000	5
6	1.000	1.000	.991	.780	.341	.074	.007	.000	.000	.000	.000	.000	.000	6
7	1.000	1.000	.998	.891	.512	.154	.022	.001	.000	.000	.000	.000	.000	7
8	1.000	1.000	1.000	.953	.677	.274	.054	.004	.000	.000	.000	.000	.000	8
9	1.000	1.000	1.000	.983	.811	.425	.115	.013	.000	.000	.000	.000	.000	9
10	1.000	1.000	1.000	.994	.902	.586	.212	.034	.002	.000	.000	.000	.000	10
11	1.000	1.000	1.000	.998	.956	.732	.345	.078	.006	.000	.000	.000	.000	11
12	1.000	1.000	1.000	1.000	.983	.846	.500	.154	.017	.000	.000	.000	.000	12
13	1.000	1.000	1.000	1.000	.994	.922	.655	.268	.044	.002	.000	.000	.000	13
14	1.000	1.000	1.000	1.000	.998	.966	.788	.414	.098	.006	.000	.000	.000	14
15	1.000	1.000	1.000	1.000	1.000	.987	.885	.575	.189	.017	.000	.000	.000	15
16	1.000	1.000	1.000	1.000	1.000	.996	.946	.726	.323	.047	.000	.000	.000	16
17	1.000	1.000	1.000	1.000	1.000	.999	.978	.846	.488	.109	.002	.000	.000	17
18	1.000	1.000	1.000	1.000	1.000	1.000	.993	.926	.659	.220	.009	.000	.000	18
19	1.000	1.000	1.000	1.000	1.000	1.000	.998	.971	.807	.383	.033	.001	.000	19
20	1.000	1.000	1.000	1.000	1.000	1.000	1.000	.991	.910	.579	.098	.007	.000	20
21	1.000	1.000	1.000	1.000	1.000	1.000	1.000	.998	.967	.766	.236	.034	.000	21
22	1.000	1.000	1.000	1.000	1.000	1.000	1.000	1.000	.991	.902	.463	.127	.002	22
23	1.000	1.000	1.000	1.000	1.000	1.000	1.000	1.000	.998	.973	.729	.358	.026	23
24	1.000	1.000	1.000	1.000	1.000	1.000	1.000	1.000	1.000	.996	.928	.723	.222	24
25	1.000	1.000	1.000	1.000	1.000	1.000	1.000	1.000	1.000	1.000	1.000	1.000	1.000	25

TABLE 2(a) Cumulative Poisson Probabilities

Tabulated values are $P(x \le a) = \sum_{x=0}^{a} p(x)$. (Computations are rounded at the third decimal place.)

	μ										
a	0.1	0.2	0.3	0.4	0.5	0.6	0.7	0.8	0.9	1.0	1.5
0	.905	.819	.741	.670	.607	.549	.497	.449	.407	.368	.223
1	.995	.982	.963	.938	.910	.878	.844	.809	.772	.736	.558
2	1.000	.999	.996	.992	.986	.977	.966	.953	.937	.920	.809
3		1.000	1.000	.999	.998	.997	.994	.991	.987	.981	.934
4				1.000	1.000	1.000	.999	.999	.998	.996	.981
5							1.000	1.000	1.000	.999	.996
6										1.000	.999
7											1.000

TABLE 2(a) (Continued)

	μ										
a	2.0	2.5	3.0	3.5	4.0	4.5	5.0	5.5	6.0	6.5	7.0
0	.135	.082	.055	.033	.018	.011	.007	.004	.003	.002	.001
1	.406	.287	.199	.136	.092	.061	.040	.027	.017	.011	.007
2	.677	.544	.423	.321	.238	.174	.125	.088	.062	.043	.030
3	.857	.758	.647	.537	.433	.342	.265	.202	.151	.112	.082
4	.947	.891	.815	.725	.629	.532	.440	.358	.285	.224	.173
5	.983	.958	.916	.858	.785	.703	.616	.529	.446	.369	.301
6	.995	.986	.966	.935	.889	.831	.762	.686	.606	.563	.450
7	.999	.996	.988	.973	.949	.913	.867	.809	.744	.673	.599
8	1.000	.999	.996	.990	.979	.960	.932	.894	.847	.792	.729
9		1.000	.999	.997	.992	.983	.968	.946	.916	.877	.830
10			1.000	.999	.997	.993	.986	.975	.957	.933	.901
11				1.000	.999	.998	.995	.989	.980	.966	.947
12					1.000	.999	.998	.996	.991	.984	.973
13						1.000	.999	.998	.996	.993	.987
14							1.000	.999	.999	.997	.994
15								1.000	.999	.999	.998
16									1.000	1.000	.999
17											1.000

TABLE 2(a) (Continued)

	μ								
a	7.5	8.0	8.5	9.0	9.5	10.0	12.0	15.0	20.0
0	.001	.000	.000	.000	.000	.000	.000	.000	.000
1	.005	.003	.002	.001	.001	.000	.000	.000	.000
2	.020	.014	.009	.006	.004	.003	.001	.000	.000
3	.059	.042	.030	.021	.015	.010	.002	.000	.000
4	.132	.100	.074	.055	.040	.029	.008	.001	.000
5	.241	.191	.150	.116	.089	.067	.020	.003	.000
6	.378	.313	.256	.207	.165	.130	.046	.008	.000
7	.525	.453	.386	.324	.269	.220	.090	.018	.001
8	.662	.593	.523	.456	.392	.333	.155	.037	.002
9	.776	.717	.653	.587	.522	.458	.242	.070	.005
10	.862	.816	.763	.706	.645	.583	.347	.118	.011
11	.921	.888	.849	.803	.752	.697	.462	.185	.021
12	.957	.936	.909	.876	.836	.792	.576	.268	.039
13	.978	.966	.949	.926	.898	.864	.682	.363	.066
14	.990	.983	.973	.959	.940	.917	.772	.466	.105
15	.995	.992	.986	.978	.967	.951	.844	.568	.157
16	.998	.996	.993	.989	.982	.973	.899	.664	.221
17	.999	.998	.997	.995	.991	.986	.937	.749	.297
18	1.000	.999	.999	.998	.996	.993	.963	.819	.381
19		1.000	.999	.999	.998	.997	.979	.875	.470
20			1.000	1.000	.999	.998	.988	.917	.559
21					1.000	.999	.994	.947	.644
22						1.000	.997	.967	.721
23							.999	.981	.787
24							.999	.989	.843
25							1.000	.994	.888
26								.997	.922
27								.998	.948
28								.999	.966
29								1.000	.978
30									.987
31									.992
32									.995
33									.997
34									.999
35									.999
36									1.000

TABLE 2(b)
Values of $e^{-\mu}$

μ	$e^{-\mu}$	μ	$e^{-\mu}$	μ	$e^{-\mu}$
0.00	1.000000	1.40	.246597	2.80	.060810
0.05	.951229	1.45	.234570	2.85	.057844
0.10	.904837	1.50	.223130	2.90	.055023
0.15	.860708	1.55	.212248	2.95	.052340
0.20	.818731	1.60	.201897	3.00	.049787
0.25	.778801	1.65	.192050	3.05	.047359
0.30	.740818	1.70	.182684	3.10	.045049
0.35	.704688	1.75	.173774	3.15	.042852
0.40	.670320	1.80	.165299	3.20	.040762
0.45	.637628	1.85	.157237	3.25	.038774
0.50	.606531	1.90	.149569	3.30	.036883
0.55	.576950	1.95	.142274	3.35	.035084
0.60	.548812	2.00	.135335	3.40	.033373
0.65	.522046	2.05	.128735	3.45	.031746
0.70	.496585	2.10	.122456	3.50	.030197
0.75	.472367	2.15	.116484	3.55	.028725
0.80	.449329	2.20	.110803	3.60	.027324
0.85	.427415	2.25	.105399	3.65	.025991
0.90	.406570	2.30	.100259	3.70	.024724
0.95	.386741	2.35	.095369	3.75	.023518
1.00	.367879	2.40	.090718	3.80	.022371
1.05	.349938	2.45	.086294	3.85	.021280
1.10	.332871	2.50	.082085	3.90	.020242
1.15	.316637	2.55	.078082	3.95	.019255
1.20	.301194	2.60	.074274	4.00	.018316
1.25	.286505	2.65	.070651	4.05	.017422
1.30	.272532	2.70	.067206	4.10	.016573
1.35	.259240	2.75	.063928	4.15	.015764

TABLE 2(b) (Continued)

μ	$e^{-\mu}$	μ	$e^{-\mu}$	μ	$e^{-\mu}$
4.20	.014996	6.15	.002133	8.10	.000304
4.25	.014264	6.20	.002029	8.15	.000289
4.30	.013569	6.25	.001930	8.20	.000275
4.35	.012907	6.30	.001836	8.25	.000261
4.40	.012277	6.35	.001747	8.30	.000249
4.45	.011679	6.40	.001661	8.35	.000236
4.50	.011109	6.45	.001581	8.40	.000225
4.55	.010567	6.50	.001503	8.45	.000214
4.60	.010052	6.55	.001430	8.50	.000204
4.65	.009562	6.60	.001360	8.55	.000194
4.70	.009095	6.65	.001294	8.60	.000184
4.75	.008652	6.70	.001231	8.65	.000175
4.80	.008230	6.75	.001171	8.70	.000167
4.85	.007828	6.80	.001114	8.75	.000158
4.90	.007447	6.85	.001059	8.80	.000151
4.95	.007083	6.90	.001008	8.85	.000143
5.00	.006738	6.95	.000959	8.90	.000136
5.05	.006409	7.00	.000912	8.95	.000130
5.10	.006097	7.05	.000867	9.00	.000123
5.15	.005799	7.10	.000825	9.05	.000117
5.20	.005517	7.15	.000785	9.10	.000112
5.25	.005248	7.20	.000747	9.15	.000106
5.30	.004992	7.25	.000710	9.20	.000101
5.35	.004748	7.30	.000676	9.25	.000096
5.40	.004517	7.35	.000643	9.30	.000091
5.45	.004296	7.40	.000611	9.35	.000087
5.50	.004087	7.45	.000581	9.40	.000083
5.55	.003887	7.50	.000553	9.45	.000079
5.60	.003698	7.55	.000526	9.50	.000075
5.65	.003518	7.60	.000501	9.55	.000071
5.70	.003346	7.65	.000476	9.60	.000068
5.75	.003183	7.70	.000453	9.65	.000064
5.80	.003028	7.75	.000431	9.70	.000061
5.85	.002880	7.80	.000410	9.75	.000058
5.90	.002739	7.85	.000390	9.80	.000056
5.95	.002606	7.90	.000371	9.85	.000053
6.00	.002479	7.95	.000353	9.90	.000050
6.05	.002358	8.00	.000336	9.95	.000048
6.10	.002243	8.05	.000319	10.00	.000045

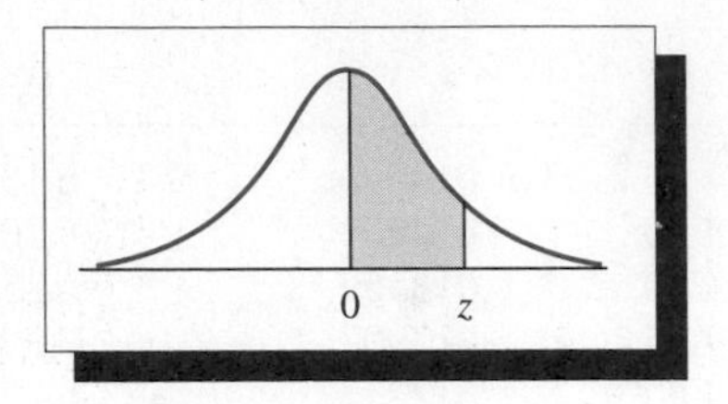

TABLE 3
Normal Curve Areas

z	.00	.01	.02	.03	.04	.05	.06	.07	.08	.09
0.0	.0000	.0040	.0080	.0120	.0160	.0199	.0239	.0279	.0319	.0359
0.1	.0398	.0438	.0478	.0517	.0557	.0596	.0636	.0675	.0714	.0753
0.2	.0793	.0832	.0871	.0910	.0948	.0987	.1026	.1064	.1103	.1141
0.3	.1179	.1217	.1255	.1293	.1331	.1368	.1406	.1443	.1480	.1517
0.4	.1554	.1591	.1628	.1664	.1700	.1736	.1772	.1808	.1844	.1879
0.5	.1915	.1950	.1985	.2019	.2054	.2088	.2123	.2157	.2190	.2224
0.6	.2257	.2291	.2324	.2357	.2389	.2422	.2454	.2486	.2517	.2549
0.7	.2580	.2611	.2642	.2673	.2704	.2734	.2764	.2794	.2823	.2852
0.8	.2881	.2910	.2939	.2967	.2995	.3023	.3051	.3078	.3106	.3133
0.9	.3159	.3186	.3212	.3238	.3264	.3289	.3315	.3340	.3365	.3389
1.0	.3413	.3438	.3461	.3485	.3508	.3531	.3554	.3577	.3599	.3621
1.1	.3643	.3665	.3686	.3708	.3729	.3749	.3770	.3790	.3810	.3830
1.2	.3849	.3869	.3888	.3907	.3925	.3944	.3962	.3980	.3997	.4015
1.3	.4032	.4049	.4066	.4082	.4099	.4115	.4131	.4147	.4162	.4177
1.4	.4192	.4207	.4222	.4236	.4251	.4265	.4279	.4292	.4306	.4319
1.5	.4332	.4345	.4357	.4370	.4382	.4394	.4406	.4418	.4429	.4441
1.6	.4452	.4463	.4474	.4484	.4495	.4505	.4515	.4525	.4535	.4545
1.7	.4554	.4564	.4573	.4582	.4591	.4599	.4608	.4616	.4625	.4633
1.8	.4641	.4649	.4656	.4664	.4671	.4678	.4686	.4693	.4699	.4706
1.9	.4713	.4719	.4726	.4732	.4738	.4744	.4750	.4756	.4761	.4767
2.0	.4772	.4778	.4783	.4788	.4793	.4798	.4803	.4808	.4812	.4817
2.1	.4821	.4826	.4830	.4834	.4838	.4842	.4846	.4850	.4854	.4857
2.2	.4861	.4864	.4868	.4871	.4875	.4878	.4881	.4884	.4887	.4890
2.3	.4893	.4896	.4898	.4901	.4904	.4906	.4909	.4911	.4913	.4916
2.4	.4918	.4920	.4922	.4925	.4927	.4929	.4931	.4932	.4934	.4936
2.5	.4938	.4940	.4941	.4943	.4945	.4946	.4948	.4949	.4951	.4952
2.6	.4953	.4955	.4956	.4957	.4959	.4960	.4961	.4962	.4963	.4964
2.7	.4965	.4966	.4967	.4968	.4969	.4970	.4971	.4972	.4973	.4974
2.8	.4974	.4975	.4976	.4977	.4977	.4978	.4979	.4979	.4980	.4981
2.9	.4981	.4982	.4982	.4983	.4984	.4984	.4985	.4985	.4986	.4986
3.0	.4987	.4987	.4987	.4988	.4988	.4989	.4989	.4989	.4990	.4990

Source: This table is abridged from Table 1 of *Statistical Tables and Formulas,* by A. Hald (New York: Wiley, 1952). Reproduced by permission of A. Hald and the publisher, John Wiley & Sons, Inc.

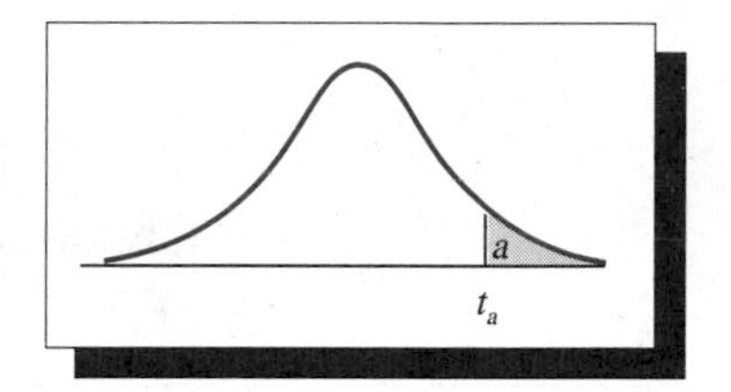

TABLE 4
Critical Values of t

d.f.	$t_{.100}$	$t_{.050}$	$t_{.025}$	$t_{.010}$	$t_{.005}$	d.f.
1	3.078	6.314	12.706	31.821	63.657	1
2	1.886	2.920	4.303	6.965	9.925	2
3	1.638	2.353	3.182	4.541	5.841	3
4	1.533	2.132	2.776	3.747	4.604	4
5	1.476	2.015	2.571	3.365	4.032	5
6	1.440	1.943	2.447	3.143	3.707	6
7	1.415	1.895	2.365	2.998	3.499	7
8	1.397	1.860	2.306	2.896	3.355	8
9	1.383	1.833	2.262	2.821	3.250	9
10	1.372	1.812	2.228	2.764	3.169	10
11	1.363	1.796	2.201	2.718	3.106	11
12	1.356	1.782	2.179	2.681	3.055	12
13	1.350	1.771	2.160	2.650	3.012	13
14	1.345	1.761	2.145	2.624	2.977	14
15	1.341	1.753	2.131	2.602	2.947	15
16	1.337	1.746	2.120	2.583	2.921	16
17	1.333	1.740	2.110	2.567	2.898	17
18	1.330	1.734	2.101	2.552	2.878	18
19	1.328	1.729	2.093	2.539	2.861	19
20	1.325	1.725	2.086	2.528	2.845	20
21	1.323	1.721	2.080	2.518	2.831	21
22	1.321	1.717	2.074	2.508	2.819	22
23	1.319	1.714	2.069	2.500	2.807	23
24	1.318	1.711	2.064	2.492	2.797	24
25	1.316	1.708	2.060	2.485	2.787	25
26	1.315	1.706	2.056	2.479	2.779	26
27	1.314	1.703	2.052	2.473	2.771	27
28	1.313	1.701	2.048	2.467	2.763	28
29	1.311	1.699	2.045	2.462	2.756	29
inf.	1.282	1.645	1.960	2.326	2.576	inf.

Source: From "Table of Percentage Points of the t-Distribution," *Biometrika* 32 (1941) 300. Reproduced by permission of the *Biometrika* Trustees.

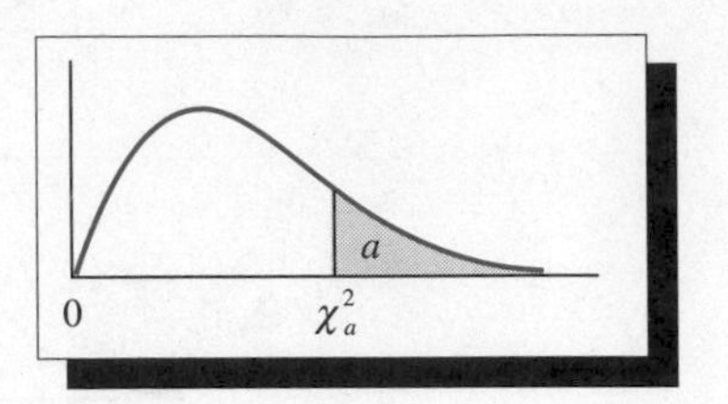

TABLE 5
Critical Values of Chi-Square

d.f.	$\chi^2_{0.995}$	$\chi^2_{0.990}$	$\chi^2_{0.975}$	$\chi^2_{0.950}$	$\chi^2_{0.900}$
1	0.0000393	0.0001571	0.0009821	0.0039321	0.0157908
2	0.0100251	0.0201007	0.0506356	0.102587	0.210720
3	0.0717212	0.114832	0.215795	0.351846	0.584375
4	0.206990	0.297110	0.484419	0.710721	1.063623
5	0.411740	0.554300	0.831211	1.145476	1.61031
6	0.675727	0.872085	1.237347	0.63539	2.20413
7	0.989265	1.239043	1.68987	2.16735	2.83311
8	1.344419	1.646482	2.17973	2.73264	3.48954
9	1.734926	2.087912	2.70039	3.32511	4.16816
10	2.15585	2.55821	3.24697	3.94030	4.86518
11	2.60321	3.05347	3.81575	4.57481	5.57779
12	3.07382	3.57056	4.40379	5.22603	6.30380
13	3.56503	4.10691	5.00874	5.89186	7.04150
14	4.07468	4.66043	5.62872	6.57063	7.78953
15	4.60094	5.22935	6.26214	7.26094	8.54675
16	5.14224	5.81221	6.90766	7.96164	9.31223
17	5.69724	6.40776	7.56418	8.67176	10.0852
18	6.26481	7.01491	8.23075	9.39046	10.8649
19	6.84398	7.63273	8.90655	10.1170	11.6509
20	7.43386	8.26040	9.59083	10.8508	12.4426
21	8.03366	8.89720	10.28293	11.5913	13.2396
22	8.64272	9.54249	10.9823	12.3380	14.0415
23	9.26042	10.19567	11.6885	13.0905	14.8479
24	9.88623	10.8564	12.4011	13.8484	15.6587
25	10.5197	11.5240	13.1197	14.6114	16.4734
26	11.1603	12.1981	13.8439	15.3791	17.2919
27	11.8076	12.8786	14.5733	16.1513	18.1138
28	12.4613	13.5648	15.3079	16.9279	18.9392
29	13.1211	14.2565	16.0471	17.7083	19.7677
30	13.7867	14.9535	16.7908	18.4926	20.5992
40	20.7065	22.1643	24.4331	26.5093	29.0505
50	27.9907	29.7067	32.3574	34.7642	37.6886
60	35.5346	37.4848	40.4817	43.1879	46.4589
70	43.2752	45.4418	48.7576	51.7393	55.3290
80	51.1720	53.5400	57.1532	60.3915	64.2778
90	59.1963	61.7541	65.6466	69.1260	73.2912
100	67.3276	70.0648	74.2219	77.9295	82.3581

Source: From "Tables of the Percentage Points of the χ^2-Distribution," *Biometrika Tables for Statisticians* 1, 3d ed. (1966). Reproduced by permission of the *Biometrika* Trustees.

TABLE 5
(Continued)

$\chi^2_{1.000}$	$\chi^2_{0.050}$	$\chi^2_{0.025}$	$\chi^2_{0.010}$	$\chi^2_{0.005}$	d.f.
2.70554	3.84146	5.02389	6.63490	7.87944	1
4.60517	5.99147	7.37776	9.21034	10.5966	2
6.25139	7.81473	9.34840	11.3449	12.8381	3
7.77944	9.48773	11.1433	13.2767	14.8602	4
9.23635	11.0705	12.8325	15.0863	16.7496	5
10.6446	12.5916	14.4494	16.8119	18.5476	6
12.0170	14.0671	16.0128	18.4753	20.2777	7
13.3616	15.5073	17.5346	20.0902	21.9550	8
14.6837	16.9190	19.0228	21.6660	23.5893	9
15.9871	18.3070	20.4831	23.2093	25.1882	10
17.2750	19.6751	21.9200	24.7250	26.7569	11
18.5494	21.0261	23.3367	26.2170	28.2995	12
19.8119	22.3621	24.7356	27.6883	29.8194	13
21.0642	23.6848	26.1190	29.1413	31.3193	14
22.3072	24.9958	27.4884	30.5779	32.8013	15
23.5418	26.2962	28.8485	31.9999	34.2672	16
24.7690	27.8571	30.1910	33.4087	35.7185	17
25.9894	28.8693	31.5264	34.8053	37.1564	18
27.2036	30.1435	32.8523	36.1908	38.5822	19
28.4120	31.4104	34.1696	37.5662	39.9968	20
29.6151	32.6705	35.4789	38.9321	41.4010	21
30.8133	33.9244	36.7807	40.2894	42.7956	22
32.0069	35.1725	38.0757	41.6384	44.1813	23
33.1963	36.4151	39.3641	42.9798	45.5585	24
34.3816	37.6525	40.6465	44.3141	46.9278	25
35.5631	38.8852	41.9232	45.6417	48.2899	26
36.7412	40.1133	43.1944	46.9630	49.6449	27
37.9159	41.3372	44.4607	48.2782	50.9933	28
39.0875	42.5569	45.7222	49.5879	52.3356	29
40.2560	43.7729	46.9792	50.8922	53.6720	30
51.8050	55.7585	59.3417	63.6907	66.7659	40
63.1671	67.5048	71.4202	76.1539	79.4900	50
74.3970	79.0819	83.2976	88.3794	91.9517	60
85.5271	90.5312	95.0231	100.425	104.215	70
96.5782	101.879	106.629	112.329	116.321	80
107.565	113.145	118.136	124.116	128.299	90
118.498	124.342	129.561	135.807	140.169	100

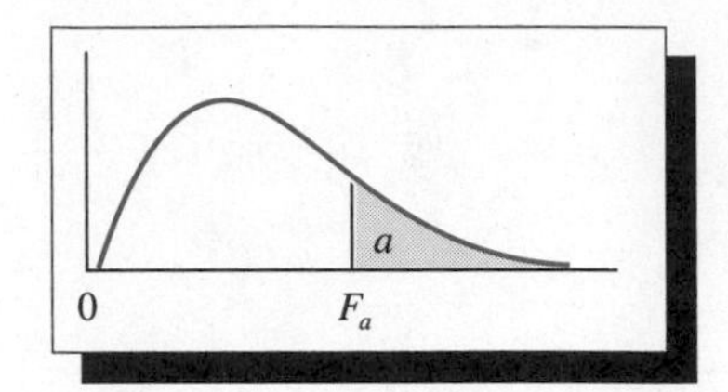

TABLE 6 Percentage Points of the F Distribution

ν_2	a	ν_1 = 1	2	3	4	5	6	7	8	9
1	.100	39.86	49.50	53.59	55.83	57.24	58.20	58.91	59.44	59.86
	.050	161.4	199.5	215.7	224.6	230.2	234.0	236.8	238.9	240.5
	.025	647.8	799.5	864.2	899.6	921.8	937.1	948.2	956.7	963.3
	.010	4052	4999.5	5403	5625	5764	5859	5928	5982	6022
	.005	16211	20000	21615	22500	23056	23437	23715	23925	24091
2	.100	8.53	9.00	9.16	9.24	9.29	9.33	9.35	9.37	9.38
	.050	18.51	19.00	19.16	19.25	19.30	19.33	19.35	19.37	19.38
	.025	38.51	39.00	39.17	39.25	39.30	39.33	39.36	39.37	39.39
	.010	98.50	99.00	99.17	99.25	99.30	99.33	99.36	99.37	99.39
	.005	198.5	199.0	199.2	199.2	199.3	199.3	199.4	199.4	199.4
3	.100	5.54	5.46	5.39	5.34	5.31	5.28	5.27	5.25	5.24
	.050	10.13	9.55	9.28	9.12	9.01	8.94	8.89	8.85	8.81
	.025	17.44	16.04	15.44	15.10	14.88	14.73	14.62	14.54	14.47
	.010	34.12	30.82	29.46	28.71	28.24	27.91	27.64	27.49	27.35
	.005	55.55	49.80	47.47	46.19	45.39	44.84	44.43	44.13	43.88
4	.100	4.54	4.32	4.19	4.11	4.05	4.01	3.98	3.95	3.94
	.050	7.71	6.94	6.59	6.39	6.26	6.16	6.09	6.04	6.00
	.025	12.22	10.65	9.98	9.60	9.36	9.20	9.07	8.98	8.90
	.010	21.20	18.00	16.69	15.98	15.52	15.21	14.98	14.80	14.66
	.005	31.33	26.28	24.26	23.15	22.46	21.97	21.62	21.35	21.14
5	.100	4.06	3.78	3.62	3.52	3.45	3.40	3.37	3.34	3.32
	.050	6.61	5.79	5.41	5.19	5.05	4.95	4.88	4.82	4.77
	.025	10.01	8.43	7.76	7.39	7.15	6.98	6.85	6.76	6.68
	.010	16.26	13.27	12.06	11.39	10.97	10.67	10.46	10.29	10.16
	.005	22.78	18.31	16.53	15.56	14.94	14.51	14.20	13.96	13.77
6	.100	3.78	3.46	3.29	3.18	3.11	3.05	3.01	2.98	2.96
	.050	5.99	5.14	4.76	4.53	4.39	4.28	4.21	4.15	4.10
	.025	8.81	7.26	6.60	6.23	5.99	5.82	5.70	5.60	5.52
	.010	13.75	10.92	9.78	9.15	8.75	8.47	8.26	8.10	7.98
	.005	18.63	14.54	12.92	12.03	11.46	11.07	10.79	10.57	10.39
7	.100	3.59	3.26	3.07	2.96	2.88	2.83	2.78	2.75	2.72
	.050	5.59	4.74	4.35	4.12	3.97	3.87	3.79	3.73	3.68
	.025	8.07	6.54	5.89	5.52	5.29	5.12	4.99	4.90	4.82
	.010	12.25	9.55	8.45	7.85	7.46	7.19	6.99	6.84	6.72
	.005	16.24	12.40	10.88	10.05	9.52	9.16	8.89	8.68	8.51
8	.100	3.46	3.11	2.92	2.81	2.73	2.67	2.62	2.59	2.56
	.050	5.32	4.46	4.07	3.84	3.69	3.58	3.50	3.44	3.39
	.025	7.57	6.06	5.42	5.05	4.82	4.65	4.53	4.43	4.36
	.010	11.26	8.65	7.59	7.01	6.63	6.37	6.18	6.03	5.91
	.005	14.69	11.04	9.60	8.81	8.30	7.95	7.69	7.50	7.34
9	.100	3.36	3.01	2.81	2.69	2.61	2.55	2.51	2.47	2.44
	.050	5.12	4.26	3.86	3.63	3.48	3.37	3.29	3.23	3.18
	.025	7.21	5.71	5.08	4.72	4.48	4.32	4.20	4.10	4.03
	.010	10.56	8.02	6.99	6.42	6.06	5.80	5.61	5.47	5.35
	.005	13.61	10.11	8.72	7.96	7.47	7.13	6.88	6.69	6.54

Source: A portion of "Tables of percentage points of the inverted beta (E) distribution" *Biometrika,* vol. 33 (1943) by M. Merrington and C. M. Thompson and from Table 18 of *Biometrika Tables for Statisticians*, vol. 1, Cambridge University Press, 1954, edited by E. S. Pearson and H. O. Hartley. Reproduced with permission of the authors, editors, and *Biometrika* trustees.

TABLE 6 (Continued)

ν_1											
10	**12**	**15**	**20**	**24**	**30**	**40**	**60**	**120**	∞	a	ν_2
60.19	60.71	60.22	61.74	62.00	62.26	62.53	62.79	63.06	63.33	.100	1
241.9	243.9	245.9	248.0	249.1	250.1	251.2	252.2	253.3	254.3	.050	
968.6	976.7	984.9	993.1	997.2	1001	1006	1010	1014	1018	.025	
6056	6106	6157	6209	6235	6261	6287	6313	6339	6366	.010	
24224	24426	24630	24836	24940	25044	25148	25253	25359	25465	.005	
9.39	9.41	9.42	9.44	9.45	9.46	9.47	9.47	9.48	9.49	.100	2
19.40	19.41	19.43	19.45	19.45	19.46	19.47	19.48	19.49	19.50	.050	
39.40	39.41	39.43	39.45	39.46	39.46	39.47	39.48	39.49	39.50	.025	
99.40	99.42	99.43	99.45	99.46	99.47	99.47	99.48	99.49	99.50	.010	
199.4	199.4	199.4	199.4	199.5	199.5	199.5	199.5	199.5	199.5	.005	
5.23	5.22	5.20	5.18	5.18	5.17	5.16	5.15	5.14	5.13	.100	3
8.79	8.74	8.70	8.66	8.64	8.62	8.59	8.57	8.55	8.53	.050	
14.42	14.34	14.25	14.17	14.12	14.08	14.04	13.99	13.95	13.90	.025	
27.23	27.05	26.87	26.69	26.60	26.50	26.41	26.32	26.22	26.13	.010	
43.69	43.39	43.08	42.78	42.62	42.47	42.31	42.15	41.99	41.83	.005	
3.92	3.90	3.87	3.84	3.83	3.82	3.80	3.79	3.78	3.76	.100	4
5.96	5.91	5.86	5.80	5.77	5.75	5.72	5.69	5.66	5.63	.050	
8.84	8.75	8.66	8.56	8.51	8.46	8.41	8.36	8.31	8.26	.025	
14.55	14.37	14.20	14.02	13.93	13.84	13.75	13.65	13.56	13.46	.010	
20.97	20.70	20.44	20.17	20.03	19.89	19.75	19.61	19.47	19.32	.005	
3.30	3.27	3.24	3.21	3.19	3.17	3.16	3.14	3.12	3.10	.100	5
4.74	4.68	4.62	4.56	4.53	4.50	4.46	4.43	4.40	4.36	.050	
6.62	6.52	6.43	6.33	6.28	6.23	6.18	6.12	6.07	6.02	.025	
10.05	9.89	9.72	9.55	9.47	9.38	9.29	9.20	9.11	9.02	.010	
13.62	13.38	13.15	12.90	12.78	12.66	12.53	12.40	12.27	12.14	.005	
2.94	2.90	2.87	2.84	2.82	2.80	2.78	2.76	2.74	2.72	.100	6
4.06	4.00	3.94	3.87	3.84	3.81	3.77	3.74	3.70	3.67	.050	
5.46	5.37	5.27	5.17	5.12	5.07	5.01	4.96	4.90	4.85	.025	
7.87	7.72	7.56	7.40	7.31	7.23	7.14	7.06	6.97	6.88	.010	
10.25	10.03	9.81	9.59	9.47	9.36	9.24	9.12	9.00	8.88	.005	
2.70	2.67	2.63	2.59	2.58	2.56	2.54	2.51	2.49	2.47	.100	7
3.64	3.57	3.51	3.44	3.41	3.38	3.34	3.30	3.27	3.23	.050	
4.76	4.67	4.57	4.47	4.42	4.36	4.31	4.25	4.20	4.14	.025	
6.62	6.47	6.31	6.16	6.07	5.99	5.91	5.82	5.74	5.65	.010	
8.38	8.18	7.97	7.75	7.65	7.53	7.42	7.31	7.19	7.08	.005	
2.54	2.50	2.46	2.42	2.40	2.38	2.36	2.34	2.32	2.29	.100	8
3.35	3.28	3.22	3.15	3.12	3.08	3.04	3.01	2.97	2.93	.050	
4.30	4.20	4.10	4.00	3.95	3.89	3.84	3.78	3.73	3.67	.025	
5.81	5.67	5.52	5.36	5.28	5.20	5.12	5.03	4.95	4.86	.010	
7.21	7.01	6.81	6.61	6.50	6.40	6.29	6.18	6.06	5.95	.005	
2.42	2.38	2.34	2.30	2.28	2.25	2.23	2.21	2.18	2.16	.100	9
3.14	3.07	3.01	2.94	2.90	2.86	2.83	2.79	2.75	2.71	.050	
3.96	3.87	3.77	3.67	3.61	3.56	3.51	3.45	3.39	3.33	.025	
5.26	5.11	4.96	4.81	4.73	4.65	4.57	4.48	4.40	4.31	.010	
6.42	6.23	6.03	5.83	5.73	5.62	5.52	5.41	5.30	5.19	.005	

TABLE 6 (Continued)

ν_2	α	ν_1 = 1	2	3	4	5	6	7	8	9
10	.100	3.29	2.92	2.73	2.61	2.52	2.46	2.41	2.38	2.35
	.050	4.96	4.10	3.71	3.48	3.33	3.22	3.14	3.07	3.02
	.025	6.94	5.46	4.83	4.47	4.24	4.07	3.95	3.85	3.78
	.010	10.04	7.56	6.55	5.99	5.64	5.39	5.20	5.06	4.94
	.005	12.83	9.43	8.08	7.34	6.87	6.54	6.30	6.12	5.97
11	.100	3.23	2.86	2.66	2.54	2.45	2.39	2.34	2.30	2.27
	.050	4.84	3.98	3.59	3.36	3.20	3.09	3.01	2.95	2.90
	.025	6.72	5.26	4.63	4.28	4.04	3.88	3.76	3.66	3.59
	.010	9.65	7.21	6.22	5.67	5.32	5.07	4.89	4.74	4.63
	.005	12.23	8.91	7.60	6.88	6.42	6.10	5.86	5.68	5.54
12	.100	3.18	2.81	2.61	2.48	2.39	2.33	2.28	2.24	2.21
	.050	4.75	3.89	3.49	3.26	3.11	3.00	2.91	2.85	2.80
	.025	6.55	5.10	4.47	4.12	3.89	3.73	3.61	3.51	3.44
	.010	9.33	6.93	5.95	5.41	5.06	4.82	4.64	4.50	4.39
	.005	11.75	8.51	7.23	6.52	6.07	5.76	5.52	5.35	5.20
13	.100	3.14	2.76	2.56	2.43	2.35	2.28	2.23	2.20	2.16
	.050	4.67	3.81	3.41	3.18	3.03	2.92	2.83	2.77	2.71
	.025	6.41	4.97	4.35	4.00	3.77	3.60	3.48	3.39	3.31
	.010	9.07	6.70	5.74	5.21	4.86	4.62	4.44	4.30	4.19
	.005	11.37	8.19	6.93	6.23	5.79	5.48	5.25	5.08	4.94
14	.100	3.10	2.73	2.52	2.39	2.31	2.24	2.19	2.15	2.12
	.050	4.60	3.74	3.34	3.11	2.96	2.85	2.76	2.70	2.65
	.025	6.30	4.86	4.24	3.89	3.66	3.50	3.38	3.29	3.21
	.010	8.86	6.51	5.56	5.04	4.69	4.46	4.28	4.14	4.03
	.005	11.06	7.92	6.68	6.00	5.56	5.26	5.03	4.86	4.72
15	.100	3.07	2.70	2.49	2.36	2.27	2.21	2.16	2.12	2.09
	.050	4.54	3.68	3.29	3.06	2.90	2.79	2.71	2.64	2.59
	.025	6.20	4.77	4.15	3.80	3.58	3.41	3.29	3.20	3.12
	.010	8.68	6.36	5.42	4.89	4.56	4.32	4.14	4.00	3.89
	.005	10.80	7.70	6.48	5.80	5.37	5.07	4.85	4.67	4.54
16	.100	3.05	2.67	2.46	2.33	2.24	2.18	2.13	2.09	2.06
	.050	4.49	3.63	3.24	3.01	2.85	2.74	2.66	2.59	2.54
	.025	6.12	4.69	4.08	3.73	3.50	3.34	3.22	3.12	3.05
	.010	8.53	6.23	5.29	4.77	4.44	4.20	4.03	3.89	3.78
	.005	10.58	7.51	6.30	5.64	5.21	4.91	4.69	4.52	4.38
17	.100	3.03	2.64	2.44	2.31	2.22	2.15	2.10	2.06	2.03
	.050	4.45	3.59	3.20	2.96	2.81	2.70	2.61	2.55	2.49
	.025	6.04	4.62	4.01	3.66	3.44	3.28	3.16	3.06	2.98
	.010	8.40	6.11	5.18	4.67	4.34	4.10	3.93	3.79	3.68
	.005	10.38	7.35	6.16	5.50	5.07	4.78	4.56	4.39	4.25
18	.100	3.01	2.62	2.42	2.29	2.20	2.13	2.08	2.04	2.00
	.050	4.41	3.55	3.16	2.93	2.77	2.66	2.58	2.51	2.46
	.025	5.98	4.56	3.95	3.61	3.38	3.22	3.10	3.01	2.93
	.010	8.29	6.01	5.09	4.58	4.25	4.01	3.84	3.71	3.60
	.005	10.22	7.21	6.03	5.37	4.96	4.66	4.44	4.28	4.14
19	.100	2.99	2.61	2.40	2.27	2.18	2.11	2.06	2.02	1.98
	.050	4.38	3.52	3.13	2.90	2.74	2.63	2.54	2.48	2.42
	.025	5.92	4.51	3.90	3.56	3.33	3.17	3.05	2.96	2.88
	.010	8.18	5.93	5.01	4.50	4.17	3.94	3.77	3.63	3.52
	.005	10.07	7.09	5.92	5.27	4.85	4.56	4.34	4.18	4.04
20	.100	2.97	2.59	2.38	2.25	2.16	2.09	2.04	2.00	1.96
	.050	4.35	3.49	3.10	2.87	2.71	2.60	2.51	2.45	2.39
	.025	5.87	4.46	3.86	3.51	3.29	3.13	3.01	2.91	2.84
	.010	8.10	5.85	4.94	4.43	4.10	3.87	3.70	3.56	3.46
	.005	9.94	6.99	5.82	5.17	4.76	4.47	4.26	4.09	3.96

TABLE 6 (Continued)

ν_1											
10	**12**	**15**	**20**	**24**	**30**	**40**	**60**	**120**	∞	a	ν_2
2.32	2.28	2.24	2.20	2.18	2.16	2.13	2.11	2.08	2.06	.100	10
2.98	2.91	2.85	2.77	2.74	2.70	2.66	2.62	2.58	2.54	.050	
3.72	3.62	3.52	3.42	3.37	3.31	3.26	3.20	3.14	3.08	.025	
4.85	4.71	4.56	4.41	4.33	4.25	4.17	4.08	4.00	3.91	.010	
5.85	5.66	5.47	5.27	5.17	5.07	4.97	4.86	4.75	4.64	.005	
2.25	2.21	2.17	2.12	2.10	2.08	2.05	2.03	2.00	1.97	.100	11
2.85	2.79	2.72	2.65	2.61	2.57	2.53	2.49	2.45	2.40	.050	
3.53	3.43	3.33	3.23	3.17	3.12	3.06	3.00	2.94	2.88	.025	
4.54	4.40	4.25	4.10	4.02	3.94	3.86	3.78	3.69	3.60	.010	
5.42	5.24	5.05	4.86	4.76	4.65	4.55	4.44	4.34	4.23	.005	
2.19	2.15	2.10	2.06	2.04	2.01	1.99	1.96	1.93	1.90	.100	12
2.75	2.69	2.62	2.54	2.51	2.47	2.43	2.38	2.34	2.30	.050	
3.37	3.28	3.18	3.07	3.02	2.96	2.91	2.85	2.79	2.72	.025	
4.30	4.16	4.01	3.86	3.78	3.70	3.62	3.54	3.45	3.36	.010	
5.09	4.91	4.72	4.53	4.43	4.33	4.23	4.12	4.01	3.90	.005	
2.14	2.10	2.05	2.01	1.98	1.96	1.93	1.90	1.88	1.85	.100	13
2.67	2.60	2.53	2.46	2.42	2.38	2.34	2.30	2.25	2.21	.050	
3.25	3.15	3.05	2.95	2.89	2.84	2.78	2.72	2.66	2.60	.025	
4.10	3.96	3.82	3.66	3.59	3.51	3.43	3.34	3.25	3.17	.010	
4.82	4.64	4.46	4.27	4.17	4.07	3.97	3.87	3.76	3.65	.005	
2.10	2.05	2.01	1.96	1.94	1.91	1.89	1.86	1.83	1.80	.100	14
2.60	2.53	2.46	2.39	2.35	2.31	2.27	2.22	2.18	2.13	.050	
3.15	3.05	2.95	2.84	2.79	2.73	2.67	2.61	2.55	2.49	.025	
3.94	3.80	3.66	3.51	3.43	3.35	3.27	3.18	3.09	3.00	.010	
4.60	4.43	4.25	4.06	3.96	3.86	3.76	3.66	3.55	3.44	.005	
2.06	2.02	1.97	1.92	1.90	1.87	1.85	1.82	1.79	1.76	.100	15
2.54	2.48	2.40	2.33	2.29	2.25	2.20	2.16	2.11	2.07	.050	
3.06	2.96	2.86	2.76	2.70	2.64	2.59	2.52	2.46	2.40	.025	
3.80	3.67	3.52	3.37	3.29	3.21	3.13	3.05	2.96	2.87	.010	
4.42	4.25	4.07	3.88	3.79	3.69	3.58	3.48	3.37	3.26	.005	
2.03	1.99	1.94	1.89	1.87	1.84	1.81	1.78	1.75	1.72	.100	16
2.49	2.42	2.35	2.28	2.24	2.19	2.15	2.11	2.06	2.01	.050	
2.99	2.89	2.79	2.68	2.63	2.57	2.51	2.45	2.38	2.32	.025	
3.69	3.55	3.41	3.26	3.18	3.10	3.02	2.93	2.84	2.75	.010	
4.27	4.10	3.92	3.73	3.64	3.54	3.44	3.33	3.22	3.11	.005	
2.00	1.96	1.91	1.86	1.84	1.81	1.78	1.75	1.72	1.69	.100	17
2.45	2.38	2.31	2.23	2.19	2.15	2.10	2.06	2.01	1.96	.050	
2.92	2.82	2.72	2.62	2.56	2.50	2.44	2.38	2.32	2.25	.025	
3.59	3.46	3.31	3.16	3.08	3.00	2.92	2.83	2.75	2.65	.010	
4.14	3.97	3.79	3.61	3.51	3.41	3.31	3.21	3.10	2.98	.005	
1.98	1.93	1.89	1.84	1.81	1.78	1.75	1.72	1.69	1.66	.100	18
2.41	2.34	2.27	2.19	2.15	2.11	2.06	2.02	1.97	1.92	.050	
2.87	2.77	2.67	2.56	2.50	2.44	2.38	2.32	2.26	2.19	.025	
3.51	3.37	3.23	3.08	3.00	2.92	2.84	2.75	2.66	2.57	.010	
4.03	3.86	3.68	3.50	3.40	3.30	3.20	3.10	2.99	2.87	.005	
1.96	1.91	1.86	1.81	1.79	1.76	1.73	1.70	1.67	1.63	.100	19
2.38	2.31	2.23	2.16	2.11	2.07	2.03	1.98	1.93	1.88	.050	
2.82	2.72	2.62	2.51	2.45	2.39	2.33	2.27	2.20	2.13	.025	
3.43	3.30	3.15	3.00	2.92	2.84	2.76	2.67	2.58	2.49	.010	
3.93	3.76	3.59	3.40	3.31	3.21	3.11	3.00	2.89	2.78	.005	
1.94	1.89	1.84	1.79	1.77	1.74	1.71	1.68	1.64	1.61	.100	20
2.35	2.28	2.20	2.12	2.08	2.04	1.99	1.95	1.90	1.84	.050	
2.77	2.68	2.57	2.46	2.41	2.35	2.29	2.22	2.16	2.09	.025	
3.37	3.23	3.09	2.94	2.86	2.78	2.69	2.61	2.52	2.42	.010	
3.85	3.68	3.50	3.32	3.22	3.12	3.02	2.92	2.81	2.69	.005	

TABLE 6 (Continued)

ν_2	α	ν_1 = 1	2	3	4	5	6	7	8	9
21	.100	2.96	2.57	2.36	2.23	2.14	2.08	2.02	1.98	1.95
	.050	4.32	3.47	3.07	2.84	2.68	2.57	2.49	2.42	2.37
	.025	5.83	4.42	3.82	3.48	3.25	3.09	2.97	2.87	2.80
	.010	8.02	5.78	4.87	4.37	4.04	3.81	3.64	3.51	3.40
	.005	9.83	6.89	5.73	5.09	4.68	4.39	4.18	4.01	3.88
22	.100	2.95	2.56	2.35	2.22	2.13	2.06	2.01	1.97	1.93
	.050	4.30	3.44	3.05	2.82	2.66	2.55	2.46	2.40	2.34
	.025	5.79	4.38	3.78	3.44	3.22	3.05	2.93	2.84	2.76
	.010	7.95	5.72	4.82	4.31	3.99	3.76	3.59	3.45	3.35
	.005	9.73	6.81	5.65	5.02	4.61	4.32	4.11	3.94	3.81
23	.100	2.94	2.55	2.34	2.21	2.11	2.05	1.99	1.95	1.92
	.050	4.28	3.42	3.03	2.80	2.64	2.53	2.44	2.37	2.32
	.025	5.75	4.35	3.75	3.41	3.18	3.02	2.90	2.81	2.73
	.010	7.88	5.66	4.76	4.26	3.94	3.71	3.54	3.41	3.30
	.005	9.63	6.73	5.58	4.95	4.54	4.26	4.05	3.88	3.75
24	.100	2.93	2.54	2.33	2.19	2.10	2.04	1.98	1.94	1.91
	.050	4.26	3.40	3.01	2.78	2.62	2.51	2.42	2.36	2.30
	.025	5.72	4.32	3.72	3.38	3.15	2.99	2.87	2.78	2.70
	.010	7.82	5.61	4.72	4.22	3.90	3.67	3.50	3.36	3.26
	.005	9.55	6.66	5.52	4.89	4.49	4.20	3.99	3.83	3.69
25	.100	2.92	2.53	2.32	2.18	2.09	2.02	1.97	1.93	1.89
	.050	4.24	3.39	2.99	2.76	2.60	2.49	2.40	2.34	2.28
	.025	5.69	4.29	3.69	3.35	3.13	2.97	2.85	2.75	2.68
	.010	7.77	5.57	4.68	4.18	3.85	3.63	3.46	3.32	3.22
	.005	9.48	6.60	5.46	4.84	4.43	4.15	3.94	3.78	3.64
26	.100	2.91	2.52	2.31	2.17	2.08	2.01	1.96	1.92	1.88
	.050	4.23	3.37	2.98	2.74	2.59	2.47	2.39	2.32	2.27
	.025	5.66	4.27	3.67	3.33	3.10	2.94	2.82	2.73	2.65
	.010	7.72	5.53	4.64	4.14	3.82	3.59	3.42	3.29	3.18
	.005	9.41	6.54	5.41	4.79	4.38	4.10	3.89	3.73	3.60
27	.100	2.90	2.51	2.30	2.17	2.07	2.00	1.95	1.91	1.87
	.050	4.21	3.35	2.96	2.73	2.57	2.46	2.37	2.31	2.25
	.025	5.63	4.24	3.65	3.31	3.08	2.92	2.80	2.71	2.63
	.010	7.68	5.49	4.60	4.11	3.78	3.56	3.39	3.26	3.15
	.005	9.34	6.49	5.36	4.74	4.34	4.06	3.85	3.69	3.56
28	.100	2.89	2.50	2.29	2.16	2.06	2.00	1.94	1.90	1.87
	.050	4.20	3.34	2.95	2.71	2.56	2.45	2.36	2.29	2.24
	.025	5.61	4.22	3.63	3.29	3.06	2.90	2.78	2.69	2.61
	.010	7.64	5.45	4.57	4.07	3.75	3.53	3.36	3.23	3.12
	.005	9.28	6.44	5.32	4.70	4.30	4.02	3.81	3.65	3.52
29	.100	2.89	2.50	2.28	2.15	2.06	1.99	1.93	1.89	1.86
	.050	4.18	3.33	2.93	2.70	2.55	2.43	2.35	2.28	2.22
	.025	5.59	4.20	3.61	3.27	3.04	2.88	2.76	2.67	2.59
	.010	7.60	5.42	4.54	4.04	3.73	3.50	3.33	3.20	3.09
	.005	9.23	6.40	5.28	4.66	4.26	3.98	3.77	3.61	3.48
30	.100	2.88	2.49	2.28	2.14	2.05	1.98	1.93	1.88	1.85
	.050	4.17	3.32	2.92	2.69	2.53	2.42	2.33	2.27	2.21
	.025	5.57	4.18	3.59	3.25	3.03	2.87	2.75	2.65	2.57
	.010	7.56	5.39	4.51	4.02	3.70	3.47	3.30	3.17	3.07
	.005	9.18	6.35	5.24	4.62	4.23	3.95	3.74	3.58	3.45

TABLE 6 (Continued)

ν_1											
10	**12**	**15**	**20**	**24**	**30**	**40**	**60**	**120**	∞	α	ν_2
1.92	1.87	1.83	1.78	1.75	1.72	1.69	1.66	1.62	1.59	.100	21
2.32	2.25	2.18	2.10	2.05	2.01	1.96	1.92	1.87	1.81	.050	
2.73	2.64	2.53	2.42	2.37	2.31	2.25	2.18	2.11	2.04	.025	
3.31	3.17	3.03	2.88	2.80	2.72	2.64	2.55	2.46	2.36	.010	
3.77	3.60	3.43	3.24	3.15	3.05	2.95	2.84	2.73	2.61	.005	
1.90	1.86	1.81	1.76	1.73	1.70	1.67	1.64	1.60	1.57	.100	22
2.30	2.23	2.15	2.07	2.03	1.98	1.94	1.89	1.84	1.78	.050	
2.70	2.60	2.50	2.39	2.33	2.27	2.21	2.14	2.08	2.00	.025	
3.26	3.12	2.98	2.83	2.75	2.67	2.58	2.50	2.40	2.31	.010	
3.70	3.54	3.36	3.18	3.08	2.98	2.88	2.77	2.66	2.55	.005	
1.89	1.84	1.80	1.74	1.72	1.69	1.66	1.62	1.59	1.55	.100	23
2.27	2.20	2.13	2.05	2.01	1.96	1.91	1.86	1.81	1.76	.050	
2.67	2.57	2.47	2.36	2.30	2.24	2.18	2.11	2.04	1.97	.025	
3.21	3.07	2.93	2.78	2.70	2.62	2.54	2.45	2.35	2.26	.010	
3.64	3.47	3.30	3.12	3.02	2.92	2.82	2.71	2.60	2.48	.005	
1.88	1.83	1.78	1.73	1.70	1.67	1.64	1.61	1.57	1.53	.100	24
2.25	2.18	2.11	2.03	1.98	1.94	1.89	1.84	1.79	1.73	.050	
2.64	2.54	2.44	2.33	2.27	2.21	2.15	2.08	2.01	1.94	.025	
3.17	3.03	2.89	2.74	2.66	2.58	2.49	2.40	2.31	2.21	.010	
3.59	3.42	3.25	3.06	2.97	2.87	2.77	2.66	2.55	2.43	.005	
1.87	1.82	1.77	1.72	1.69	1.66	1.63	1.59	1.56	1.52	.100	25
2.24	2.16	2.09	2.01	1.96	1.92	1.87	1.82	1.77	1.71	.050	
2.61	2.51	2.41	2.30	2.24	2.18	2.12	2.05	1.98	1.91	.025	
3.13	2.99	2.85	2.70	2.62	2.54	2.45	2.36	2.27	2.17	.010	
3.54	3.37	3.20	3.01	2.92	2.82	2.72	2.61	2.50	2.38	.005	
1.86	1.81	1.76	1.71	1.68	1.65	1.61	1.58	1.54	1.50	.100	26
2.22	2.15	2.07	1.99	1.95	1.90	1.85	1.80	1.75	1.69	.050	
2.59	2.49	2.39	2.28	2.22	2.16	2.09	2.03	1.95	1.88	.025	
3.09	2.96	2.81	2.66	2.58	2.50	2.42	2.33	2.23	2.13	.010	
3.49	3.33	3.15	2.97	2.87	2.77	2.67	2.56	2.45	2.33	.005	
1.85	1.80	1.75	1.70	1.67	1.64	1.60	1.57	1.53	1.49	.100	27
2.20	2.13	2.06	1.97	1.93	1.88	1.84	1.79	1.73	1.67	.050	
2.57	2.47	2.36	2.25	2.19	2.13	2.07	2.00	1.93	1.85	.025	
3.06	2.93	2.78	2.63	2.55	2.47	2.38	2.29	2.20	2.10	.010	
3.45	3.28	3.11	2.93	2.83	2.73	2.63	2.52	2.41	2.29	.005	
1.84	1.79	1.74	1.69	1.66	1.63	1.59	1.56	1.52	1.48	.100	28
2.19	2.12	2.04	1.96	1.91	1.87	1.82	1.77	1.71	1.65	.050	
2.55	2.45	2.34	2.23	2.17	2.11	2.05	1.98	1.91	1.83	.025	
3.03	2.90	2.75	2.60	2.52	2.44	2.35	2.26	2.17	2.06	.010	
3.41	3.25	3.07	2.89	2.79	2.69	2.59	2.48	2.37	2.25	.005	
1.83	1.78	1.73	1.68	1.65	1.62	1.58	1.55	1.51	1.47	.100	29
2.18	2.10	2.03	1.94	1.90	1.85	1.81	1.75	1.70	1.64	.050	
2.53	2.43	2.32	2.21	2.15	2.09	2.03	1.96	1.89	1.81	.025	
3.00	2.87	2.73	2.57	2.49	2.41	2.33	2.23	2.14	2.03	.010	
3.38	3.21	3.04	2.86	2.76	2.66	2.56	2.45	2.33	2.21	.005	
1.82	1.77	1.72	1.67	1.64	1.61	1.57	1.54	1.50	1.46	.100	30
2.16	2.09	2.01	1.93	1.89	1.84	1.79	1.74	1.68	1.62	.050	
2.51	2.41	2.31	2.20	2.14	2.07	2.01	1.94	1.87	1.79	.025	
2.98	2.84	2.70	2.55	2.47	2.39	2.30	2.21	2.11	2.01	.010	
3.34	3.18	3.01	2.82	2.73	2.63	2.52	2.42	2.30	2.18	.005	

TABLE 6 (Continued)

ν_2	α	ν_1 1	2	3	4	5	6	7	8	9
40	.100	2.84	2.44	2.23	2.09	2.00	1.93	1.87	1.83	1.79
	.050	4.08	3.23	2.84	2.61	2.45	2.34	2.25	2.18	2.12
	.025	5.42	4.05	3.46	3.13	2.90	2.74	2.62	2.53	2.45
	.010	7.31	5.18	4.31	3.83	3.51	3.29	3.12	2.99	2.89
	.005	8.83	6.07	4.98	4.37	3.99	3.71	3.51	3.35	3.22
60	.100	2.79	2.39	2.18	2.04	1.95	1.87	1.82	1.77	1.74
	.050	4.00	3.15	2.76	2.53	2.37	2.25	2.17	2.10	2.04
	.025	5.29	3.93	3.34	3.01	2.79	2.63	2.51	2.41	2.33
	.010	7.08	4.98	4.13	3.65	3.34	3.12	2.95	2.82	2.72
	.005	8.49	5.79	4.73	4.14	3.76	3.49	3.29	3.13	3.01
120	.100	2.75	2.35	2.13	1.99	1.90	1.82	1.77	1.72	1.68
	.050	3.92	3.07	2.68	2.45	2.29	2.17	2.09	2.02	1.96
	.025	5.15	3.80	3.23	2.89	2.67	2.52	2.39	2.30	2.22
	.100	6.85	4.79	3.95	3.48	3.17	2.96	2.79	2.66	2.56
	.005	8.18	5.54	4.50	3.92	3.55	3.28	3.09	2.93	2.81
∞	.100	2.71	2.30	2.08	1.94	1.85	1.77	1.72	1.67	1.63
	.050	3.84	3.00	2.60	2.37	2.21	2.10	2.01	1.94	1.63
	.025	5.02	3.69	3.12	2.79	2.57	2.41	2.29	2.19	2.11
	.010	6.63	4.61	3.78	3.32	3.02	2.80	2.64	2.51	2.41
	.005	7.88	5.30	4.28	3.72	3.35	3.09	2.90	2.74	2.62

TABLE 7
Distribution Function of U, $P(U \leq U_0)$; U_0 is the Argument; $n_1 \leq n_2$; $3 \leq n_2 \leq 10$

$n_2 = 3$

U_0	n_1 1	2	3
0	.25	.10	.05
1	.50	.20	.10
2		.40	.20
3		.60	.35
4			.50

$n_2 = 4$

U_0	n_1 1	2	3	4
0	.2000	.0667	.0286	.0143
1	.4000	.1333	.0571	.0286
2	.6000	.2667	.1143	.0571
3		.4000	.2000	.1000
4		.6000	.3143	.1714
5			.4286	.2429
6			.5714	.3429
7				.4429
8				.5571

Note: Computed by M. Pagano, Department of Statistics, University of Florida.

TABLE 6 (Continued)

				ν_1							
10	**12**	**15**	**20**	**24**	**30**	**40**	**60**	**120**	**∞**	**a**	**ν_2**
1.76	1.71	1.66	1.61	1.57	1.54	1.51	1.47	1.42	1.38	.100	40
2.08	2.00	1.92	1.84	1.79	1.74	1.69	1.64	1.58	1.51	.050	
2.39	2.29	2.18	2.07	2.01	1.94	1.88	1.80	1.72	1.64	.025	
2.80	2.66	2.52	2.37	2.29	2.20	2.11	2.02	1.92	1.80	.010	
3.12	2.95	2.78	2.60	2.50	2.40	2.30	2.18	2.06	1.93	.005	
1.71	1.66	1.60	1.54	1.51	1.48	1.44	1.40	1.35	1.29	.100	60
1.99	1.92	1.84	1.75	1.70	1.65	1.59	1.53	1.47	1.39	.050	
2.27	2.17	2.06	1.94	1.88	1.82	1.74	1.67	1.58	1.48	.025	
2.63	2.50	2.35	2.20	2.12	2.03	1.94	1.84	1.73	1.60	.010	
2.90	2.74	2.57	2.39	2.29	2.19	2.08	1.96	1.83	1.69	.005	
1.65	1.60	1.55	1.48	1.45	1.41	1.37	1.32	1.26	1.19	.100	120
1.91	1.83	1.75	1.66	1.61	1.55	1.50	1.43	1.35	1.25	.050	
2.16	2.05	1.94	1.82	1.76	1.69	1.61	1.53	1.43	1.31	.025	
2.47	2.34	2.19	2.03	1.95	1.86	1.76	1.66	1.53	1.38	.010	
2.71	2.54	2.37	2.19	2.09	1.98	1.87	1.75	1.61	1.43	.005	
1.60	1.55	1.49	1.42	1.38	1.34	1.30	1.24	1.17	1.00	.100	∞
1.83	1.75	1.67	1.57	1.52	1.46	1.39	1.32	1.22	1.00	.050	
2.05	1.94	1.83	1.71	1.64	1.57	1.48	1.39	1.27	1.00	.025	
2.32	2.18	2.04	1.88	1.79	1.70	1.59	1.47	1.32	1.00	.010	
2.52	2.36	2.19	2.00	1.90	1.79	1.67	1.53	1.36	1.00	.005	

TABLE 7 (Continued)

$n_2 = 5$

			n_1		
U_0	**1**	**2**	**3**	**4**	**5**
0	.1667	.0476	.0179	.0079	.0040
1	.3333	.0952	.0357	.0159	.0079
2	.5000	.1905	.0714	.0317	.0159
3		.2857	.1250	.0556	.0278
4		.4286	.1964	.0952	.0476
5		.5714	.2857	.1429	.0754
6			.3929	.2063	.1111
7			.5000	.2778	.1548
8				.3651	.2103
9				.4524	.2738
10				.5476	.3452
11					.4206
12					.5000

TABLE 7 (Continued)

$n_2 = 6$

	n_1					
U_0	1	2	3	4	5	6
0	.1429	.0357	.0119	.0048	.0022	.0011
1	.2857	.0714	.0238	.0095	.0043	.0022
2	.4286	.1429	.0476	.0190	.0087	.0043
3	.5714	.2143	.0833	.0333	.0152	.0076
4		.3214	.1310	.0571	.0260	.0130
5		.4286	.1905	.0857	.0411	.0206
6		.5714	.2738	.1286	.0628	.0325
7			.3571	.1762	.0887	.0465
8			.4524	.2381	.1234	.0660
9			.5476	.3048	.1645	.0898
10				.3810	.2143	.1201
11				.4571	.2684	.1548
12				.5429	.3312	.1970
13					.3961	.2424
14					.4654	.2944
15					.5346	.3496
16						.4091
17						.4686
18						.5314

$n_2 = 7$

	n_1						
U_0	1	2	3	4	5	6	7
0	.1250	.0278	.0083	.0030	.0013	.0006	.0003
1	.2500	.0556	.0167	.0061	.0025	.0012	.0006
2	.3750	.1111	.0333	.0121	.0051	.0023	.0012
3	.5000	.1667	.0583	.0212	.0088	.0041	.0020
4		.2500	.0917	.0364	.0152	.0070	.0035
5		.3333	.1333	.0545	.0240	.0111	.0055
6		.4444	.1917	.0818	.0366	.0175	.0087
7		.5556	.2583	.1152	.0530	.0256	.0131
8			.3333	.1576	.0745	.0367	.0189
9			.4167	.2061	.1010	.0507	.0265
10			.5000	.2636	.1338	.0688	.0364
11				.3242	.1717	.0903	.0487
12				.3939	.2159	.1171	.0641
13				.4636	.2652	.1474	.0825
14				.5364	.3194	.1830	.1043
15					.3775	.2226	.1297
16					.4381	.2669	.1588
17					.5000	.3141	.1914
18						.3654	.2279
19						.4178	.2675
20						.4726	.3100
21						.5274	.3552
22							.4024
23							.4508
24							.5000

TABLE 7 (Continued)

$n_2 = 8$

	n_1							
U_0	1	2	3	4	5	6	7	8
0	.1111	.0222	.0061	.0020	.0008	.0003	.0002	.0001
1	.2222	.0444	.0121	.0040	.0016	.0007	.0003	.0002
2	.3333	.0889	.0242	.0081	.0031	.0013	.0006	.0003
3	.4444	.1333	.0424	.0141	.0054	.0023	.0011	.0005
4	.5556	.2000	.0667	.0242	.0093	.0040	.0019	.0009
5		.2667	.0970	.0364	.0148	.0063	.0030	.0015
6		.3556	.1394	.0545	.0225	.0100	.0047	.0023
7		.4444	.1879	.0768	.0326	.0147	.0070	.0035
8		.5556	.2485	.1071	.0466	.0213	.0103	.0052
9			.3152	.1414	.0637	.0296	.0145	.0074
10			.3879	.1838	.0855	.0406	.0200	.0103
11			.4606	.2303	.1111	.0539	.0270	.0141
12			.5394	.2848	.1422	.0709	.0361	.0190
13				.3414	.1772	.0906	.0469	.0249
14				.4040	.2176	.1142	.0603	.0325
15				.4667	.2618	.1412	.0760	.0415
16				.5333	.3108	.1725	.0946	.0524
17					.3621	.2068	.1159	.0652
18					.4165	.2454	.1405	.0803
19					.4716	.2864	.1678	.0974
20					.5284	.3310	.1984	.1172
21						.3773	.2317	.1393
22						.4259	.2679	.1641
23						.4749	.3063	.1911
24						.5251	.3472	.2209
25							.3894	.2527
26							.4333	.2869
27							.4775	.3227
28							.5225	.3605
29								.3992
30								.4392
31								.4796
32								.5204

TABLE 7 (Continued)

$n_2 = 9$

U_0	n_1 = 1	2	3	4	5	6	7	8	9
0	.1000	.0182	.0045	.0014	.0005	.0002	.0001	.0000	.0000
1	.2000	.0364	.0091	.0028	.0010	.0004	.0002	.0001	.0000
2	.3000	.0727	.0182	.0056	.0020	.0008	.0003	.0002	.0001
3	.4000	.1091	.0318	.0098	.0035	.0014	.0006	.0003	.0001
4	.5000	.1636	.0500	.0168	.0060	.0024	.0010	.0005	.0002
5		.2182	.0727	.0252	.0095	.0038	.0017	.0008	.0004
6		.2909	.1045	.0378	.0145	.0060	.0026	.0012	.0006
7		.3636	.1409	.0531	.0210	.0088	.0039	.0019	.0009
8		.4545	.1864	.0741	.0300	.0128	.0058	.0028	.0014
9		.5455	.2409	.0993	.0415	.0180	.0082	.0039	.0020
10			.3000	.1301	.0559	.0248	.0115	.0056	.0028
11			.3636	.1650	.0734	.0332	.0156	.0076	.0039
12			.4318	.2070	.0949	.0440	.0209	.0103	.0053
13			.5000	.2517	.1199	.0567	.0274	.0137	.0071
14				.3021	.1489	.0723	.0356	.0180	.0094
15				.3552	.1818	.0905	.0454	.0232	.0122
16				.4126	.2188	.1119	.0571	.0296	.0157
17				.4699	.2592	.1361	.0708	.0372	.0200
18				.5301	.3032	.1638	.0869	.0464	.0252
19					.3497	.1942	.1052	.0570	.0313
20					.3986	.2280	.1261	.0694	.0385
21					.4491	.2643	.1496	.0836	.0470
22					.5000	.3035	.1755	.0998	.0567
23						.3445	.2039	.1179	.0680
24						.3878	.2349	.1383	.0807
25						.4320	.2680	.1606	.0951
26						.4773	.3032	.1852	.1112
27						.5227	.3403	.2117	.1290
28							.3788	.2404	.1487
29							.4185	.2707	.1701
30							.4591	.3029	.1933
31							.5000	.3365	.2181
32								.3715	.2447
33								.4074	.2729
34								.4442	.3024
35								.4813	.3332
36								.5187	.3652
37									.3981
38									.4317
39									.4657
40									.5000

TABLE 7 (Continued)

$n_2 = 10$

	n_1									
U_0	1	2	3	4	5	6	7	8	9	10
0	.0909	.0152	.0035	.0010	.0003	.0001	.0001	.0000	.0000	.0000
1	.1818	.0303	.0070	.0020	.0007	.0002	.0001	.0000	.0000	.0000
2	.2727	.0606	.0140	.0040	.0013	.0005	.0002	.0001	.0000	.0000
3	.3636	.0909	.0245	.0070	.0023	.0009	.0004	.0002	.0001	.0000
4	.4545	.1364	.0385	.0120	.0040	.0015	.0006	.0003	.0001	.0001
5	.5455	.1818	.0559	.0180	.0063	.0024	.0010	.0004	.0002	.0001
6		.2424	.0804	.0270	.0097	.0037	.0015	.0007	.0003	.0002
7		.3030	.1084	.0380	.0140	.0055	.0023	.0010	.0005	.0002
8		.3788	.1434	.0529	.0200	.0080	.0034	.0015	.0007	.0004
9		.4545	.1853	.0709	.0276	.0112	.0048	.0022	.0011	.0005
10		.5455	.2343	.0939	.0376	.0156	.0068	.0031	.0015	.0008
11			.2867	.1199	.0496	.0210	.0093	.0043	.0021	.0010
12			.3462	.1518	.0646	.0280	.0125	.0058	.0028	.0014
13			.4056	.1868	.0823	.0363	.0165	.0078	.0038	.0019
14			.4685	.2268	.1032	.0467	.0215	.0103	.0051	.0026
15			.5315	.2697	.1272	.0589	.0277	.0133	.0066	.0034
16				.3177	.1548	.0736	.0351	.0171	.0086	.0045
17				.3666	.1855	.0903	.0439	.0217	.0110	.0057
18				.4196	.2198	.1099	.0544	.0273	.0140	.0073
19				.4725	.2567	.1317	.0665	.0338	.0175	.0093
20				.5275	.2970	.1566	.0806	.0416	.0217	.0116
21					.3393	.1838	.0966	.0506	.0267	.0144
22					.3839	.2139	.1148	.0610	.0326	.0177
23					.4296	.2461	.1349	.0729	.0394	.0216
24					.4765	.2811	.1574	.0864	.0474	.0262
25					.5235	.3177	.1819	.1015	.0564	.0315
26						.3564	.2087	.1185	.0667	.0376
27						.3962	.2374	.1371	.0782	.0446
28						.4374	.2681	.1577	.0912	.0526
29						.4789	.3004	.1800	.1055	.0615
30						.5211	.3345	.2041	.1214	.0716
31							.3698	.2299	.1388	.0827
32							.4063	.2574	.1577	.0952
33							.4434	.2863	.1781	.1088
34							.4811	.3167	.2001	.1237
35							.5189	.3482	.2235	.1399
36								.3809	.2483	.1575
37								.4143	.2745	.1763
38								.4484	.3019	.1965
39								.4827	.3304	.2179
40								.5173	.3598	.2406
41									.3901	.2644
42									.4211	.2894
43									.4524	.3153
44									.4841	.3421
45									.5159	.3697
46										.3980
47										.4267
48										.4559
49										.4853
50										.5147

TABLE 8 Critical Values of T in the Wilcoxon Signed-Rank Test, $n = 5(1)50$

One-Sided	Two-Sided	$n = 5$	$n = 6$	$n = 7$	$n = 8$	$n = 9$	$n = 10$
$\alpha = .05$	$\alpha = .10$	1	2	4	6	8	11
$\alpha = .025$	$\alpha = .05$		1	2	4	6	8
$\alpha = .01$	$\alpha = .02$			0	2	3	5
$\alpha = .005$	$\alpha = .01$				0	2	3

One-Sided	Two-Sided	$n = 11$	$n = 12$	$n = 13$	$n = 14$	$n = 15$	$n = 16$
$\alpha = .05$	$\alpha = .10$	14	17	21	26	30	36
$\alpha = .025$	$\alpha = .05$	11	14	17	21	25	30
$\alpha = .01$	$\alpha = .02$	7	10	13	16	20	24
$\alpha = .005$	$\alpha = .01$	5	7	10	13	16	19

One-Sided	Two-Sided	$n = 17$	$n = 18$	$n = 19$	$n = 20$	$n = 21$	$n = 22$
$\alpha = .05$	$\alpha = .10$	41	47	54	60	68	75
$\alpha = .025$	$\alpha = .05$	35	40	46	52	59	66
$\alpha = .01$	$\alpha = .02$	28	33	38	43	49	56
$\alpha = .005$	$\alpha = .01$	23	28	32	37	43	49

One-Sided	Two-Sided	$n = 23$	$n = 24$	$n = 25$	$n = 26$	$n = 27$	$n = 28$
$\alpha = .05$	$\alpha = .10$	83	92	101	110	120	130
$\alpha = .025$	$\alpha = .05$	73	81	90	98	107	117
$\alpha = .01$	$\alpha = .02$	62	69	77	85	93	102
$\alpha = .005$	$\alpha = .01$	55	68	68	76	84	92

One-Sided	Two-Sided	$n = 29$	$n = 30$	$n = 31$	$n = 32$	$n = 33$	$n = 34$
$\alpha = .05$	$\alpha = .10$	141	152	163	175	188	201
$\alpha = .025$	$\alpha = .05$	127	137	148	159	171	183
$\alpha = .01$	$\alpha = .02$	111	120	130	141	151	162
$\alpha = .005$	$\alpha = .01$	100	109	118	128	138	149

One-Sided	Two-Sided	$n = 35$	$n = 36$	$n = 37$	$n = 38$	$n = 39$
$\alpha = .05$	$\alpha = .10$	214	228	242	256	271
$\alpha = .025$	$\alpha = .05$	195	208	222	235	250
$\alpha = .01$	$\alpha = .02$	174	186	198	211	224
$\alpha = .005$	$\alpha = .01$	160	171	183	195	208

One-Sided	Two-Sided	$n = 40$	$n = 41$	$n = 42$	$n = 43$	$n = 44$	$n = 45$
$\alpha = .05$	$\alpha = .10$	287	303	319	336	353	371
$\alpha = .025$	$\alpha = .05$	264	279	295	311	327	344
$\alpha = .01$	$\alpha = .02$	238	252	267	281	297	313
$\alpha = .005$	$\alpha = .01$	221	234	248	262	277	292

One-Sided	Two-Sided	$n = 46$	$n = 47$	$n = 48$	$n = 49$	$n = 50$
$\alpha = .05$	$\alpha = .10$	389	408	427	446	466
$\alpha = .025$	$\alpha = .05$	361	379	397	415	434
$\alpha = .01$	$\alpha = .02$	329	345	362	380	398
$\alpha = .005$	$\alpha = .01$	307	323	339	356	373

Source: From "Some Rapid Approximate Statistical Procedures" (1964) 28, by F. Wilcoxon and R. A. Wilcox. Reproduced with the kind permission of Lederle Laboratories, a division of American Cyanamid Company.

TABLE 9
Critical Values of Spearman's Rank Correlation Coefficient for a One-Tailed Test

n	$\alpha = .05$	$\alpha = .025$	$\alpha = .01$	$\alpha = .005$
5	0.900	—	—	—
6	0.829	0.886	0.943	—
7	0.714	0.786	0.893	—
8	0.643	0.738	0.833	0.881
9	0.600	0.683	0.783	0.833
10	0.564	0.648	0.745	0.794
11	0.523	0.623	0.736	0.818
12	0.497	0.591	0.703	0.780
13	0.475	0.566	0.673	0.745
14	0.457	0.545	0.646	0.716
15	0.441	0.525	0.623	0.689
16	0.425	0.507	0.601	0.666
17	0.412	0.490	0.582	0.645
18	0.399	0.476	0.564	0.625
19	0.388	0.462	0.549	0.608
20	0.377	0.450	0.534	0.591
21	0.368	0.438	.0521	0.576
22	0.359	0.428	0.508	0.562
23	0.351	0.418	0.496	0.549
24	0.343	0.409	.0485	0.537
25	0.336	0.400	0.475	0.526
26	0.329	0.392	0.465	0.515
27	0.323	0.385	0.456	0.505
28	0.317	0.377	0.448	0.496
29	0.311	0.370	0.440	0.487
30	0.305	0.364	0.432	0.478

Source: From "Distribution of Sums of Squares of Rank Differences for Small Samples," by E. G. Olds, *Annals of Mathematical Statistics* 9 (1938). Reproduced with the permission of the editor, *Annals of Mathematical Statistics*.

TABLE 10 Random Numbers

Line	1	2	3	4	5	6	7	8	9	10	11	12	13	14
							Column							
1	10480	15011	01536	02011	81647	91646	69179	14194	62590	36207	20969	99570	91291	90700
2	22368	46573	25595	85393	30995	89198	27982	53402	93965	34095	52666	19174	39615	99505
3	24130	48360	22527	97265	76393	64809	15179	24830	49340	32081	30680	19655	63348	58629
4	42167	93093	06243	61680	07856	16376	39440	53537	71341	57004	00849	74917	97758	16379
5	37570	39975	81837	16656	06121	91782	60468	81305	49684	60672	14110	06927	01263	54613
6	77921	06907	11008	42751	27756	53498	18602	70659	90655	15053	21916	81825	44394	42880
7	99562	72905	56420	69994	98872	31016	71194	18738	44013	48840	63213	21069	10634	12952
8	96301	91977	05463	07972	18876	20922	94595	56869	69014	60045	18425	84903	42508	32307
9	89579	14342	63661	10281	17453	18103	57740	84378	25331	12566	58678	44947	05585	56941
10	84575	36857	53342	53988	53060	59533	38867	62300	08158	17983	16439	11458	18593	64952
11	28918	69578	88231	33276	70997	79936	56865	05859	90106	31595	01547	85590	91610	78188
12	63553	40961	48235	03427	49626	69445	18663	72695	52180	20847	12234	90511	33703	90322
13	09429	93969	52636	92737	88974	33488	36320	17617	30015	08272	84115	27156	30613	74952
14	10365	61129	87529	85689	48237	52267	67689	93394	01511	26358	85104	20285	29975	89868
15	07119	97336	71048	08178	77233	13916	47564	81056	97735	85977	29372	74461	28551	90707
16	51085	12765	51821	51259	77452	16308	60756	92144	49442	53900	70960	63990	75601	40719
17	02368	21382	52404	60268	89368	19885	55322	44819	01188	65255	64835	44919	05944	55157
18	01011	54092	33362	94904	31273	04146	18594	29852	71585	85030	51132	01915	92747	64951
19	52162	53916	46369	58586	23216	14513	83149	98736	23495	64350	94738	17752	35156	35749
20	07056	97628	33787	09998	42698	06691	76988	13602	51851	46104	88916	19509	25625	58104
21	48663	91245	85828	14346	09172	30168	90229	04734	59193	22178	30421	61666	99904	32812
22	54164	58492	22421	74103	47070	25306	76468	26384	58151	06646	21524	15227	96909	44592
23	32639	32363	05597	24200	13363	38005	94342	28728	35806	06912	17012	64161	18296	22851
24	29334	27001	87637	87308	58731	00256	45834	15398	46557	41135	10367	07684	36188	18510
25	02488	33062	28834	07351	19731	92420	60952	61280	50001	67658	32586	86679	50720	94953
26	81525	72295	04839	96423	24878	82651	66566	14778	76797	14780	13300	87074	79666	95725
27	29676	20591	68086	26432	46901	20849	89768	81536	86645	12659	92259	57102	80428	25280
28	00742	57392	39064	66432	84673	40027	32832	61362	98947	96067	64760	64585	96096	98253
29	05366	04213	25669	26422	44407	44048	37937	63904	45766	66134	75470	66520	34693	90449
30	91921	26418	64117	94305	26766	25940	39972	22209	71500	64568	91402	42416	07844	69618
31	00582	04711	87917	77341	42206	35126	74087	99547	81817	42607	43808	76655	62028	76630
32	00725	69884	62797	56170	86324	88072	76222	36086	84637	93161	76038	65855	77919	88006
33	69011	65795	95876	55293	18988	27354	26575	08625	40801	59920	29841	80150	12777	48501
34	25976	57948	29888	88604	67917	48708	18912	82271	65424	69774	33611	54262	85963	03547
35	09763	83473	73577	12908	30883	18317	28290	35797	05998	41688	34952	37888	38917	88050
36	91567	42595	27958	30134	04024	86385	29880	99730	55536	84855	29080	09250	79656	73211
37	17955	56349	90999	49127	20044	59931	06115	20542	18059	02008	73708	83517	36103	42791
38	46503	18584	18845	49618	02304	51038	20655	58727	28168	15475	56942	53389	20562	87338
39	92157	89634	94824	78171	84610	82834	09922	25417	44137	48413	25555	21246	35509	20468
40	14577	62765	35605	81263	39667	47358	56873	56307	61607	49518	89656	20103	77490	18062
41	98427	07523	33362	64270	01638	92477	66969	98420	04880	45585	46565	04102	46880	45709
42	34914	63976	88720	82765	34476	17032	87589	40836	32427	70002	70663	88863	77775	69348
43	70060	28277	39475	46473	23219	53416	94970	25832	69975	94884	19661	72828	00102	66794
44	53976	54914	06990	67245	68350	82948	11398	42878	80287	88267	47363	46634	06541	97809
45	76072	29515	40980	07391	58745	25774	22987	80059	39911	96189	41151	14222	60697	59583
46	90725	52210	83974	29992	65831	38857	50490	83765	55657	14361	31720	57375	56228	41546
47	64364	67412	33339	31926	14883	24413	59744	92351	97473	89286	35931	04110	23726	51900
48	08962	00358	31662	25388	61642	34072	81249	35648	56891	69352	48373	45578	78547	81788
49	95012	68379	93526	70765	10592	04542	76463	54328	02349	17247	28865	14777	62730	92277
50	15664	10493	20492	38391	91132	21999	59516	81652	27195	48223	46751	22923	32261	85653

Source: Abridged from *Handbook of Tables for Probability and Statistics,* 2d ed. Edited by William H. Beyer (Cleveland: The Chemical Rubber Company, 1968). Reproduced by permission of CRC Press, Inc.

TABLE 10 (Continued)

Line	Column 1	2	3	4	5	6	7	8	9	10	11	12	13	14
51	16408	81899	04153	53381	79401	21438	83035	92350	36693	31238	59649	91754	72772	02338
52	18629	81953	05520	91962	04739	13092	97662	24822	94730	06496	35090	04822	86774	98289
53	73115	35101	47498	87637	99016	71060	88824	71013	18735	20286	23153	72924	35165	43040
54	57491	16703	23167	49323	45021	33132	12544	41035	80780	45393	44812	12515	98931	91202
55	30405	83946	23792	14422	15059	45799	22716	19792	09983	74353	68668	30429	70735	25499
56	16631	35006	85900	98275	32388	52390	16815	69298	82732	38480	73817	32523	41961	44437
57	96773	20206	42559	78985	05300	22164	24369	54224	35033	19687	11052	91491	60383	19746
58	38935	64202	14349	82674	66523	44133	00697	35552	35970	19124	63318	29686	03387	59846
59	31624	76384	17403	53363	44167	64486	64758	75366	76554	31601	12614	33072	60332	92325
60	78919	19474	23632	27889	47914	02584	37680	20801	72152	39339	34806	08930	85001	87820
61	03931	33309	57047	74211	63445	17361	62825	39908	05607	91284	68833	25570	38818	46920
62	74426	33278	43972	10119	89917	15665	52872	73823	73144	88662	88970	74492	51805	99378
63	09066	00903	20795	95452	92648	45454	09552	88815	16553	51125	79375	97596	16296	66092
64	42238	12426	87025	14267	20979	04508	64535	31355	86064	29472	47689	05974	52468	16834
65	16153	08002	26504	41744	81959	65642	74240	56302	00033	67107	77510	70625	28725	34191
66	21457	40742	29820	96783	29400	21840	15035	34537	33310	06116	95240	15957	16572	06004
67	21581	57802	02050	89728	17937	37621	47075	12080	97403	48626	68995	43805	33386	21597
68	55612	78095	83197	33732	05810	24813	86902	60397	16489	03264	88525	42786	05269	92532
69	44657	66999	99324	51281	84463	60563	79312	93454	68876	25471	93911	25650	12682	73572
70	91340	84979	46949	81973	37949	61023	43997	15263	80644	43942	89203	71795	99533	50501
71	91227	21199	31935	27022	84067	05462	35216	14486	29891	68607	41867	14951	91696	85065
72	50001	38140	66321	19924	72163	09538	12151	06878	91903	18749	34405	56087	82790	70925
73	65390	05224	72958	28609	81406	39147	25549	48542	42627	45233	57202	94617	23772	07896
74	27504	96131	83944	41575	10573	08619	64482	73923	36152	05184	94142	25299	84387	34925
75	37169	94851	39117	89632	00959	16487	65536	49071	39782	17095	02330	74301	00275	48280
76	11508	70225	51111	38351	19444	66499	71945	05422	13442	78675	84081	66938	93654	59894
77	37449	30362	06694	54690	04052	53115	62757	95348	78662	11163	81651	50245	34971	52924
78	46515	70331	85922	38329	57015	15765	97161	17869	45349	61796	66345	81073	49106	79860
79	30986	81223	42416	58353	21532	30502	32305	86482	05174	07901	54339	58861	74818	46942
80	63798	64995	46583	09785	44160	78128	83991	42865	92520	83531	80377	35909	81250	54238
81	82486	84846	99254	67632	43218	50076	21361	64816	51202	88124	41870	52689	51275	83556
82	21885	32906	92431	09060	64297	51674	64126	62570	26123	05155	59194	52799	28225	85762
83	60336	98782	07408	53458	13564	59089	26445	29789	85205	41001	12535	12133	14645	23541
84	43937	46891	24010	25560	86355	33941	25786	54990	71899	15475	95434	98227	21824	19585
85	97656	63175	89303	16275	07100	92063	21942	18611	47348	20203	18534	03862	78095	50136
86	03299	01221	05418	38982	55758	92237	26759	86367	21216	98442	08303	56613	91511	75928
87	79626	06486	03574	17668	07785	76020	79924	25651	83325	88428	85076	72811	22717	50585
88	85636	68335	47539	03129	65651	11977	02510	26113	99447	68645	34327	15152	55230	93448
89	18039	14367	61337	06177	12143	46609	32989	74014	64708	00533	35398	58408	13261	47908
90	08362	15656	60627	36478	65648	16764	53412	09013	07832	41574	17639	82163	60859	75567
91	79556	29068	04142	16268	15387	12856	66227	38358	22478	73373	88732	09443	82558	05250
92	92608	82674	27072	32534	17075	27698	98204	63863	11951	34648	88022	56148	34925	57031
93	23982	25835	40055	67006	12293	02753	14827	23235	35071	99704	37543	11601	35503	85171
94	09915	96306	05908	97901	28395	14186	00821	80703	70426	75647	76310	88717	37890	40129
95	59037	33300	26695	62247	69927	76123	50842	43834	86654	70959	79725	93872	28117	19233
96	42488	78077	69882	61657	34136	79180	97526	43092	04098	73571	80799	76536	71255	64239
97	46764	86273	63003	93017	31204	36692	40202	35275	57306	55543	53203	18098	47625	88684
98	03237	45430	55417	63282	90816	17349	88298	90183	36600	78406	06216	95787	42579	90730
99	86591	81482	52667	61582	14972	90053	89534	76036	49199	43716	97548	04379	46370	28672
100	38534	01715	94964	87288	65680	43772	39560	12918	86737	62738	19636	51132	25739	56947

TABLE 11(a) Percentage Points of the Studentized Range, $q(t, \nu)$; Upper 5% Points

ν	t = 2	3	4	5	6	7	8	9	10	11
1	17.97	26.98	32.82	37.08	40.41	43.12	45.40	47.36	49.07	50.59
2	6.08	8.33	9.80	10.88	11.74	12.44	13.03	13.54	13.99	14.39
3	4.50	5.91	6.82	7.50	8.04	8.48	8.85	9.18	9.46	9.72
4	3.93	5.04	5.76	6.29	6.71	7.05	7.35	7.60	7.83	8.03
5	3.64	4.60	5.22	5.67	6.03	6.33	6.58	6.80	6.99	7.17
6	3.46	4.34	4.90	5.30	5.63	5.90	6.12	6.32	6.49	6.65
7	3.34	4.16	4.68	5.06	5.36	5.61	5.82	6.00	6.16	6.30
8	3.26	4.04	4.53	4.89	5.17	5.40	5.60	5.77	5.92	6.05
9	3.20	3.95	4.41	4.76	5.02	5.24	5.43	5.59	5.74	5.87
10	3.15	3.88	4.33	4.65	4.91	5.12	5.30	5.46	5.60	5.72
11	3.11	3.82	4.26	4.57	4.82	5.03	5.20	5.35	5.49	5.61
12	3.08	3.77	4.20	4.51	4.75	4.95	5.12	5.27	5.39	5.51
13	3.06	3.73	4.15	4.45	4.69	4.88	5.05	5.19	5.32	5.43
14	3.03	3.70	4.11	4.41	4.64	4.83	4.99	5.13	5.25	5.36
15	3.01	3.67	4.08	4.37	4.60	4.78	4.94	5.08	5.20	5.31
16	3.00	3.65	4.05	4.33	4.56	4.74	4.90	5.03	5.15	5.26
17	2.98	3.63	4.02	4.30	4.52	4.70	4.86	4.99	5.11	5.21
18	2.97	3.61	4.00	4.28	4.49	4.67	4.82	4.96	5.07	5.17
19	2.96	3.59	3.98	4.25	4.47	4.65	4.79	4.92	5.04	5.14
20	2.95	3.58	3.96	4.23	4.45	4.62	4.77	4.90	5.01	5.11
24	2.92	3.53	3.90	4.17	4.37	4.54	4.68	4.81	4.92	5.01
30	2.89	3.49	3.85	4.10	4.30	4.46	4.60	4.72	4.82	4.92
40	2.86	3.44	3.79	4.04	4.23	4.39	4.52	4.63	4.73	4.82
60	2.83	3.40	3.74	3.98	4.16	4.31	4.44	4.55	4.65	4.73
120	2.80	3.36	3.68	3.92	4.10	4.24	4.36	4.47	4.56	4.64
∞	2.77	3.31	3.63	3.86	4.03	4.17	4.29	4.39	4.47	4.55

TABLE 11(a)
(Continued)

				t					
12	**13**	**14**	**15**	**16**	**17**	**18**	**19**	**20**	ν
51.96	53.20	54.33	55.36	56.32	57.22	58.04	58.83	59.56	1
14.75	15.08	15.38	15.65	15.91	16.14	16.37	16.57	16.77	2
9.95	10.15	10.35	10.52	10.69	10.84	10.98	11.11	11.24	3
8.21	8.37	8.52	8.66	8.79	8.91	9.03	9.13	9.23	4
7.32	7.47	7.60	7.72	7.83	7.93	8.03	8.12	8.21	5
6.79	6.92	7.03	7.14	7.24	7.34	7.43	7.51	7.59	6
6.43	6.55	6.66	6.76	6.85	6.94	7.02	7.10	7.17	7
6.18	6.29	6.39	6.48	6.57	6.65	6.73	6.80	6.87	8
5.98	6.09	6.19	6.28	6.36	6.44	6.51	6.58	6.64	9
5.83	5.93	6.03	6.11	6.19	6.27	6.34	6.40	6.47	10
5.71	5.81	5.90	5.98	6.06	6.13	6.20	6.27	6.33	11
5.61	5.71	5.80	5.88	5.95	6.02	6.09	6.15	6.21	12
5.53	5.63	5.71	5.79	5.86	5.93	5.99	6.05	6.11	13
5.46	5.55	5.64	5.71	5.79	5.85	5.91	5.97	6.03	14
5.40	5.49	5.57	5.65	5.72	5.78	5.85	5.90	5.96	15
5.35	5.44	5.52	5.59	5.66	5.73	5.79	5.84	5.90	16
5.31	5.39	5.47	5.54	5.61	5.67	5.73	5.79	5.84	17
5.27	5.35	5.43	5.50	5.57	5.63	5.69	5.74	5.79	18
5.23	5.31	5.39	5.46	5.53	5.59	5.65	5.70	5.75	19
5.20	5.28	5.36	5.43	5.49	5.55	5.61	5.66	5.71	20
5.10	5.18	5.25	5.32	5.38	5.44	5.49	5.55	5.59	24
5.00	5.08	5.15	5.21	5.27	5.33	5.38	5.43	5.47	30
4.90	4.98	5.04	5.11	5.16	5.22	5.27	5.31	5.36	40
4.81	4.88	4.94	5.00	5.06	5.11	5.15	5.20	5.24	60
4.71	4.78	4.84	4.90	4.95	5.00	5.04	5.09	5.13	120
4.62	4.68	4.74	4.80	4.85	4.89	4.93	4.97	5.01	∞

Source: From *Biometrika Tables for Statisticians,* Vol. 1, 3rd ed., edited by E. S. Pearson and H. O. Hartley (Cambridge University Press, 1966). Reproduced by permission of the Biometrika Trustees.

TABLE 11(b) Percentage Points of the Studentized Range, $q(t, \nu)$; Upper 1% Points

	t									
ν	2	3	4	5	6	7	8	9	10	11
1	90.03	135.0	164.3	185.6	202.2	215.8	227.2	237.0	245.6	253.2
2	14.04	19.02	22.29	24.72	26.63	28.20	29.53	30.68	31.69	32.59
3	8.26	10.62	12.17	13.33	14.24	15.00	15.64	16.20	16.69	17.13
4	6.51	8.12	9.17	9.96	10.58	11.10	11.55	11.93	12.27	12.57
5	5.70	6.98	7.80	8.42	8.91	9.32	9.67	9.97	10.24	10.48
6	5.24	6.33	7.03	7.56	7.97	8.32	8.61	8.87	9.10	9.30
7	4.95	5.92	6.54	7.01	7.37	7.68	7.94	8.17	8.37	8.55
8	4.75	5.64	6.20	6.62	6.96	7.24	7.47	7.68	7.86	8.03
9	4.60	5.43	5.96	6.35	6.66	6.91	7.13	7.33	7.49	7.65
10	4.48	5.27	5.77	6.14	6.43	6.67	6.87	7.05	7.21	7.36
11	4.39	5.15	5.62	5.97	6.25	6.48	6.67	6.84	6.99	7.13
12	4.32	5.05	5.50	5.84	6.10	6.32	6.51	6.67	6.81	6.94
13	4.26	4.96	5.40	5.73	5.98	6.19	6.37	6.53	6.67	6.79
14	4.21	4.89	5.32	5.63	5.88	6.08	6.26	6.41	6.54	6.66
15	4.17	4.84	5.25	5.56	5.80	5.99	6.16	6.31	6.44	6.55
16	4.13	4.79	5.19	5.49	5.72	5.92	6.08	6.22	6.35	6.46
17	4.10	4.74	5.14	5.43	5.66	5.85	6.01	6.15	6.27	6.38
18	4.07	4.70	5.09	5.38	5.60	5.79	5.94	6.08	6.20	6.31
19	4.05	4.67	5.05	5.33	5.55	5.73	5.89	6.02	6.14	6.25
20	4.02	4.64	5.02	5.29	5.51	5.69	5.84	5.97	6.09	6.19
24	3.96	4.55	4.91	5.17	5.37	5.54	5.69	5.81	5.92	6.02
30	3.89	4.45	4.80	5.05	5.24	5.40	5.54	5.65	5.76	5.85
40	3.82	4.37	4.70	4.93	5.11	5.26	5.39	5.50	5.60	5.69
60	3.76	4.28	4.59	4.82	4.99	5.13	5.25	5.36	5.45	5.53
120	3.70	4.20	4.50	4.71	4.87	5.01	5.12	5.21	5.30	5.37
∞	3.64	4.12	4.40	4.60	4.76	4.88	4.99	5.08	5.16	5.23

TABLE 11(b) (Continued)

				t					
12	**13**	**14**	**15**	**16**	**17**	**18**	**19**	**20**	ν
260.0	266.2	271.8	277.0	281.8	286.3	290.0	294.3	298.0	1
33.40	34.13	34.81	35.43	36.00	36.53	37.03	37.50	37.95	2
17.53	17.89	18.22	18.52	18.81	19.07	19.32	19.55	19.77	3
12.84	13.09	13.32	13.53	13.73	13.91	14.08	14.24	14.40	4
10.70	10.89	11.08	11.24	11.40	11.55	11.68	11.81	11.93	5
9.48	9.65	9.81	9.95	10.08	10.21	10.32	10.43	10.54	6
8.71	8.86	9.00	9.12	9.24	9.35	9.46	9.55	9.65	7
8.18	8.31	8.44	8.55	8.66	8.76	8.85	8.94	9.03	8
7.78	7.91	8.03	8.13	8.23	8.33	8.41	8.49	8.57	9
7.49	7.60	7.71	7.81	7.91	7.99	8.08	8.15	8.23	10
7.25	7.36	7.46	7.56	7.65	7.73	7.81	7.88	7.95	11
7.06	7.17	7.26	7.36	7.44	7.52	7.59	7.66	7.73	12
6.90	7.01	7.10	7.19	7.27	7.35	7.42	7.48	7.55	13
6.77	6.87	6.96	7.05	7.13	7.20	7.27	7.33	7.39	14
6.66	6.76	6.84	6.93	7.00	7.07	7.14	7.20	7.26	15
6.56	6.66	6.74	6.82	6.90	6.97	7.03	7.09	7.15	16
6.48	6.57	6.66	6.73	6.81	6.87	6.94	7.00	7.05	17
6.41	6.50	6.58	6.65	6.72	6.79	6.85	6.91	6.97	18
6.34	6.43	6.51	6.58	6.65	6.72	6.78	6.84	6.89	19
6.28	6.37	6.45	6.52	6.59	6.65	6.71	6.77	6.82	20
6.11	6.19	6.26	6.33	6.39	6.45	6.51	6.56	6.61	24
5.93	6.01	6.08	6.14	6.20	6.26	6.31	6.36	6.41	30
5.76	5.83	5.90	5.96	6.02	6.07	6.12	6.16	6.21	40
5.60	5.67	5.73	5.78	5.84	5.89	5.93	5.97	6.01	60
5.44	5.50	5.56	5.61	5.66	5.71	5.75	5.79	5.83	120
5.29	5.35	5.40	5.45	5.49	5.54	5.57	5.61	5.65	∞

Source: From *Biometrika Tables for Statisticians,* Vol. 1, 3rd ed., edited by E. S. Pearson and H. O. Hartley (Cambridge University Press, 1966). Reproduced by permission of the Biometrika Trustees.

II

Primary Data Sources

"Americans Round Retreating from Healthy Eating Habits." *New York Times,* 14 March 1993, p. Y13.

"Anatomy of a Taking." *American Demographics,* February 1993, p. 54.

"Angioplasty Much Riskier for Women." Riverside, Calif.: *Press-Enterprise,* 9 March 1993, p. A5.

"Athletic Grad Rate Same as Others." Riverside, Calif.: *Press-Enterprise,* 3 July, 1992.

Automotive News: 1992 Market Data Book, 27 May 1992, p. 17.

Baell, Wendy K., and E. H. Wertheim. "Predictors of Outcome in the Treatment of Bulimia Nervosa." *British Journal of Clinical Psychology* 31 (1992): 330–332.

Beckham, Susan J., W. A. Grana, P. Buckley, J. E. Breasile, and P. L. Claypool. "A Comparison of Anterior Compartment Pressures in Competitive Runners and Cyclists." *American Journal of Sports Medicine* 21(1) (1993):36.

Blair, R., and R. Miser. "Biotin Bioavailability from Protein Supplements and Cereal Grains for Growing Broiler Chickens." *International Journal of Vitamin and Nutrition Research* 59 (1989):55–58.

Bovee, Tim, Associated Press. "College Campuses Graying: 1 in 4 Students 30 or Older." Riverside, Calif.: *Press-Enterprise,* 1 December 1991, p. A19.

Brush, J. E., et al. "Use of the Initial Electrocardiogram to Predict In-Hospital Complications of Acute Myocardial Infarction." *New England Journal of Medicine,* 2 May 1985.

Butcher, G. S. "The Predator-Deflection Hypothesis for Sexual Colour Dimorphism: A Test on the Northern Oriole." *Animal Behaviour* 32 (1984).

Butts, Nancy K. "Physiological Profiles of High School Female Cross Country Runners." *Research Quarterly for Exercise and Sport* 53(1) (1982).

Calabrese, Raymond L., and Edgar J. Raymond. "Alienation: Its impact on Adolescents from Stable Environments." *Journal of Psychology* 123(4) (1989):397–404.

"Call It in the Air." Riverside, Calif.: *Press-Enterprise,* 19 October 1992.

Case, Robert B., et al. "Type A Behavior and Survival After Acute Myocardial Infarction." *New England Journal of Medicine* 312(12) (1985).

Casey, T. M., and M. L. May, and K. R. Morgan. "Flight Energetics of Euglossine Bees in Relation to Morphology and Wing Stroke Frequency." *Journal of Experimental Biology* 116 (1985).

Chang, Alice F., T. L. Rosenthal, E. S. Bryant, R. H. Rosenthal, R. M. Heidlage, and B. K. Fritzler. "Comparing High School and College Students' Leisure Interests and Stress Ratings." *Behavioral Research Therapy* 31(2) (1993):179–184.

"The Changing American Household." *American Demographics Desk Reference,* July 1992, p. 2.

Costa, M. Martin, M. Gatz. "Determination of Authorship Credit in Published Dissertations." *Psychological Science* 3(6) (1992):54.

Crichton, Michael. *Congo.* New York: Knopf, 1980.

Curhan, R. C., W. J. Salmon, and R. D. Buzzell. "Sales and Profitability of Health and Beauty Aids and General Merchandise in Supermarkets." *Journal of Retailing* 59(1) (1983).

Day, Carole, and Del Lowenthal. "The Use of Open Group Discussions in Marketing Library Services to Young Adults." *British Journal of Educational Psychology* 62 (1992):324–340.

Dickey, Charles. "A Strategy for Big Bucks." *Field and Stream,* October 1990.

Dubin, Murray. "U.S. Households Shrink, Age." Riverside, Calif.: *Press-Enterprise,* 23 April 1992, p. A16.

Dworkin, J. B., C. J. Hobson, E. Frieling, and D. M. Oakes. "How German Workers View Their Jobs." *Columbia Journal of World Business* (Summer 1983).

Ellis, A. J. "Geothermal Systems." *American Scientist,* September/October 1975.

Eskey, Kenneth. "Sexes Near Parity in Medical School Entry," Riverside, Calif.: *Press-Enterprise,* 4 November 1992, p. A13.

Farwell, J., and J. T. Flannery. "Cancer in Relatives of Children with Central-Nervous-System Neoplasms." *New England Journal of Medicine* 311(12) (1984).

"Filling a Hole on Sesame Street." *American Demographics,* August 1992, p. 22.

Flora, Stephen, T. R. Schieferecke, and H. G. Bremenkamp III. "Effects of Aversive Noise on Human Self-Control for Positive Reinforcement." *Psychological Record* 42 (1992):505–517.

Frank, Mark, and T. Gilovich. "The Dark Side of Self- and Social Perception: Black Uniforms and Aggression in Professional Sports." *Journal of Personality and Social Psychology* 54(1) (1988):74–85. Copyright 1988 by the American Psychological Association. Reprinted with permission.

Gellene, Denise. "Burger Safety on Back Burner in State." *Los Angeles Times,* 29 January 1993, p. D2.

Good, Karen. "Doing Whatever It Takes for a Job." *National College Magazine,* March 1993, p. 19.

Good, W. S. "Productivity in the Retail Grocery Trade." *Journal of Retailing* 60(3) (1984).

Grannis, Chandler B. "Title Production for '91 Rebounds." *International News Magazine of Book Publishing* 239(47) (1992):24.

Greenhouse, Joel B., and Samuel W. Greenhouse. "An Aspirin a Day. . . ?" *Chance: New Directions for Statistics and Computing* 1(4) (1988):24–31.

Hackl, J. *Journal of Quality Technology,* April 1991.

Haertel, Ursala, U. Keil, et al. *Science News* 135 (June 1989):389.

Haney, Daniel Q. "Studies Caution on Milk, Vitamin D." Riverside, Calif.: *Press-Enterprise,* 30 April 1992, p. A9.

Haney, Daniel Q. "Mondays May Be Hazardous." Riverside, Calif.: *Press-Enterprise,* 17 November 1992, p. A16.

Hill, Loren. *Bassmaster,* September/October 1980.

Hirsch, James S. "Car Rental Companies Driving Up Prices." Riverside, Calif.: *Press-Enterprise,* 9 November 1992, p. D1.

Hugick, Larry, and Leslie McAneny. "A Gloomy America Sees a Nation in Decline, No Easy Solutions." *Gallup Monthly Poll,* September 1992, p. 2.

"If Fido Takes a Piece of the Postman, Feds May Take a Chunk Out of You." *Wall Street Journal,* 10 July 1981.

"In the Eye of the Political Storm." *Time,* 7 September 1992, p. 18.

Jeffries, W. B., H. K. Voris, and C. M. Yang. "Diversity and Distribution of the Pedunculate Barnacle *Octolasmis* Gray, 1825 Epizoic on the Scyllarid Lobster, *Thenus orientalis* (Lund, 1793)." *Crustaceana* 46(3) (1984).

Jenkins, C. David, et al. "Correlates of Angina Pectoris Among Men Awaiting Coronary Bypass Surgery." *Psychomatic Medicine* 45(2) (1983).

Karalekas, P. C., Jr., C. R. Ryan, and F. B. Taylor. "Control of Lead, Copper, and Iron Pipe Corrosion in Boston." *American Water Works Journal,* February 1983.

Kelly, Dennis. "More Colleges See Rise in New Applicants." *USA Today,* 18 August 1992, p. D1.

Kleiber, D. A., and G. C. Roberts. "The Relationship Between Game and Sport Involvement in Later Childhood: A Preliminary Investigation." *Research Quarterly for Exercise and Sport* 54(2) (1983).

Knight-Ridder Newspapers. "Hot News: 98.6 Not Normal." Riverside, Calif.: *Press-Enterprise,* 23 September 1992.

Knudson, Ronald J., W. T. Kaltenborn, and B. Burrows. *American Review of Respiratory Diseases* 140 (1989):645–651.

Kraul, Chris. "Bayer Hopes New 'Select' Line Will Ease Its Headache." *Los Angeles Times,* 27 March 1993, p. D1.

"Laptop's Three Musts for Success in Sales." *Sales and Marketing Management,* February 1988, p. 91.

Larson, Jan. "Destiny Is Destiny." *American Demographics,* February 1993, p. 38.

Lieberman, L. R., and M. A. Orlandi. "Alcohol Advertising and Adolescent Drinking." *Alcohol Health and Research World* 12(1) (1987):30.

Lindhe, Jan D. "Clinical Assessment of Antiplaque Agents." *Compendium of Continuing Education in Dentistry,* Suppl. 5 (1984).

Macellari, Carlos E. "Revision of Serpulids of the Genus *Rotularia* (*Annelida*) at Seymour Island (Antarctic Peninsula) and Their Value in Stratigraphy." *Journal of Paleontology* 58(4) (1984).

Malvy, J., et al. "Retinol, β-Carotene and α-Tocopherol Status in a French Population of Healthy Children." *International Journal of Vitamin and Nutrition Research* 59 (1989):29.

Manning, C. A., J. L. Hall, and P. E. Gold. "Glucose Effects on Memory and Other Neuropsychological Tests in Elderly Humans." *Psychological Science* 1(5) (1990):307.

Marino, G. Wayne. "Selected Mechanical Factors Associated with Acceleration in Ice Skating." *Research Quarterly for Exercise and Sport* 54(3) (1983).

Menard, W. W. "Time, Chance and the Origin of Manganese Nodules." *American Scientist,* September/October 1976.

Miko, Chris John, and Edward Weilant (eds.). *Opinions '90 Cumulation.* Detroit and London: Gale Research, 1991.

Millard-Stafford, Mindy, Linda B. Rosskopf, and Phillip B. Sparling. "Coronary Heart Disease: Risk Profiles of College Football Players." *Physician and Sportsmedicine* 17(9) (1989).

Miller, W. H. "News from Chemistry." *News Journal of the College of Chemistry* (University of California-Berkeley) 1(1) (1993):5.

Neergaard, Lauran. "Survey Shows Smoking Decline Reversal." Riverside, Calif.: *Press-Enterprise,* 2 April 1993, p. A9.

Niklas, Karl J., and T. G. Owens. "Physiological and Morphological Modifications of *Plantago Major (Plantaginaceae)* in Response to Light Conditions." *American Journal of Botany* 76(3) (1989):370–382.

Ohlsson, Stellan. "The Learning Curve for Writing Books: Evidence from Professor Asimov." *Psychological Science* 3(6) (1992):380–382.

Otten, Alan L. "People Patterns." *Wall Street Journal,* 8 December 1988.

"Overdrive Boosts 486SX-25s to Peak Performance Levels." *PC World,* November 1992, p. 192.

Payne, Robert J., and J. J. Pigram. "Changing Evaluations of Flood Plain Hazard: The Hunter Valley, Australia." *Environment and Behavior* 13(4) (1981):461–480.

Pettit, Michael J., and Thomas L. Beitinger. "Oxygen Acquisition of the Reedfish, *Erpetoichthys calabaricus.*" *Journal of Experimental Biology* 114 (1985).

Physician and Sportsmedicine 17(9) (1989):55.

Pillay, Y. G. "International Comparisons: Selected Mental Health Data," Part 1. *Psychological Reports* 71(3) (1992):723–726.

Pittman, Karen E., and Donald A. Levin. "Effects of Parental Identities and Environment on Components of Crossing Success in *Phlox drummondii.*" *American Journal of Botany* 76(3) (1989):409–418.

Polakovic, Gary. "Revised Car-Pool Rules Give Business a Break." Riverside, Calif.: *Press-Enterprise,* 13 February 1993, p. A3.

"Poll Shows Candidates in Statistical Dead Heat." Riverside, Calif.: *Press-Enterprise,* 9 July 1992, p. A3.

"Polls Say Clinton Keeps Lead Despite G.O.P. Fire." *New York Times,* 1 November 1992.
Powers, Scott K., et al. "Ventilatory Threshold, Running Economy and Distance Running Performance of Trained Athletes." *Research Quarterly for Exercise and Sport* 54(3) (1983).
Powers, Scott K., and M. B. Walker. "Physiological and Anatomical Characteristics of Outstanding Female Junior Tennis Players." *Research Quarterly for Exercise and Sport* 53(2) (1983).
Prewett, Peter N., and D. B. Fowler. "Predictive Validity of the Slosson Intelligence Test with the WISC-R and WRAT-R Level 1." *Psychology in the Schools* 29 (January 1992):17.
"Pump Prices and Tax Collectors—IV." *Time,* 8 February 1993.
"Quarter of Americans Lack Health Insurance." Riverside, Calif.: *Press-Enterprise,* 25 June 1992, p. A13.
Robichaux, Mark. "Boom in Fancy Coffee Pits: Big Marketers, Little Firms." *Wall Street Journal,* 6 November 1989, p. B1.
Robinson, John P. *American Demographics,* April 1989, p. 68.
Rosenbaum, David E. "Older More Alienated Politically Than Young, Poll Shows." Riverside, Calif.: *Press-Enterprise,* 8 July 1992. p. A5.
Ross, Sonya, Associated Press. "More Graduate from High School." Riverside, Calif.: *Press-Enterprise,* 8 July 1992, p. A3.
Rothenberg, Randall. "Market Researchers Finding Deaf Ears." Riverside, Calif.: *Press-Enterprise,* 15 April 1991.
Rowland, Mary. "Anticipating a Family Leave Law." *New York Times,* 31 January 1993, p. F16.
Sakekel, Karola. "Egg Substitutes Range in Quality." *San Francisco Chronicle,* 10 February 1993, p. 8.
"SAT Scores by Intended Field of Study." Riverside, Calif.: *Press-Enterprise,* 8 April 1993.
Sanders, Larry. "EPA Report Spurs Confusion." *USA Today,* 1989.
Schmale, Arthur H., et al. "Well-Being of Cancer Survivors." *Psychomatic Medicine* 45(2) (1983).
Schmittroth, Linda (ed.). *Statistical Record of Women Worldwide.* Detroit and London: Gale Research, 1991, pp. 8, 9, 335.
Schreiner, T., and R. Hamel. "Upstate, Downstate." *American Demographics,* September 1989, p. 60.
Science News 136 (August 1989).
"Seeing the World Through Tinted Lenses." *Washington Post,* 16 March 1993, p. 5.
Seligman, Daniel. "Keeping Up." *Fortune,* 27 July 1981.
Seligman, Daniel. "The Road to Monte Carlo." *Fortune,* 15 April 1985.
Shapiro, Eben. "Soft Drink Companies Still Looking for the Next Real Thing." Riverside, Calif.: *Press-Enterprise,* 11 May 1992, p. C1.
Smeltzer, K. R., and K. W. Watson. "A Test of Instructional Strategies for Listening Improvement in a Simulated Business Setting." *Journal of Business Communication* 22(2) (1985).
Smith, Dan. "Motorists Have Little Respect for Others' Skills." Riverside, Calif.: *Press-Enterprise,* 15 May 1991.
Sosin, Michael R. "Homeless and Vulnerable Meal Program Users: A Comparison Study." *Social Problems* 39(2) (1992):170.
"Teachers Earn Average $34,213." Riverside, Calif.: *Press-Enterprise,* 28 August 1992.
"The Three VCR-Family." *American Demographics,* July 1992, p. 10.
Torrey, Gregory K., S. F. Vasa, J. W. Maag, and J. J. Kramer. "Social Skills Interventions Across School Settings: Case Study Reviews of Students with Mild Disabilities." *Psychology in the Schools* 29 (July 1992):248.
"Upward Mobility." *American Demographics,* April 1989, p. 13.
U.S. Department of Commerce, Bureau of the Census. *Statistical Abstract of the United States, 1992,* 112th ed. Washington, D.C.: Government Printing Office, 1993, p. 94.

U.S. Department of Commerce, Bureau of the Census. In *World Almanac and Book of Facts,* 1993. New York, Pharos Books, 1993.

U.S. Department of Labor, Bureau of Labor Statistics. *Youth Policy* 14 (7/8) (1992):50.

"Used Computer Prices." *PC Source,* June 1992, p. 96.

Ward, P. I. "*Gammarus pulex* Control Their Moult Timing to Secure Mates." *Animal Behaviour* 32 (1984).

Wessel, David. "American Way of Buying Survey." *Wall Street Journal,* 6 November 1989.

"What Americans Are Saying About Taxes." *Public Opinion,* March/April 1989, p. 21.

"When Parents Stop Reading, Kids Turn to TV, According to Survey." Riverside, Calif.: *Press-Enterprise,* 17 November 1992, p. A5.

"White Cars Still Favored, but Green Fast Approaching." Riverside, Calif.: *Press-Enterprise,* 19 April 1993.

"Who's Tops in Service?" *PC World,* November 1992, p. 198.

Wilder, Robert P., D. Brennan, and D. E. Schotte. "A Standard Measure for Exercise Prescription for Aqua Running." *American Journal of Sports Medicine* 21(1) (1993):45.

Wright, John W. (ed.). *The Universal Almanac, 1993.* Kansas City, Mo., and New York: Andrews and McNeel, 1992.

"You Aren't Paranoid If You Think Someone Eyes Your Every Move." *Wall Street Journal,* 19 March 1985.

Answers to Selected Exercises

Chapter 1

1.1 **a** quantitative **b** quantitative **c** qualitative **d** qualitative

1.3 **c** yes; no **1.5** yes

1.9 **a** 8–10 classes **c** 43/50 **d** 33/50

1.11 **b** .30 **c** .70 **d** .30

1.13 **a** number of attempts—discrete; total number of rushing yards—continuous; average number of yards per carry—continuous

1.15 **b** .12 **c** .44 **1.17** **b** 25/60 **1.19** **b** yes

1.23 **b** as x increases, y increases

1.31 **a** skewed **b** symmetric **c** symmetric **d** symmetric
e skewed **f** skewed

1.33 **a** continuous **b** continuous **c** discrete **d** discrete
e discrete

1.35 **a** year—quantitative; age of children—quantitative; whether or not mother is in the paid work force—qualitative **c** comparative side-by-side bar chart
d stretch the vertical scale

1.37

```
 7|89
 8|017
 9|01244566688
10|179
11|2
```

1.41 **a** added arrow; stretched vertical scale **b** shrink vertical axis; remove arrow

1.43 **a** GDP per capita, expenditures per capita, beds/1000 population, length of stay
b positive but not strong **c** little or no relationship

1.45 **a** no **b** symmetric; no

1.49 **b** slightly skewed left **c** yes

Chapter 2

2.1 **b** $\bar{x} = 2$; $m = 1$; mode $= 1$ **c** skewed

2.3 **a** 5.8 **b** 5.5 **c** 5 and 6

2.5 **a** skewed slightly right **c** $\bar{x} = 1.08$; $m = 1$; mode $= 1$

2.9 **a** skewed to the right **b** mean greater than the median

2.11 **a** 2.125 **b** $s^2 = 1.2679$; $s = 1.126$

2.13 **a** 5 **b** 3.875 **c** $s^2 = 2.4107$; $s = 1.55$

2.15 **a** $s \approx .15$ **b** $\bar{x} = .76$; $s = .165$

2.17 **a** 7617 to 12,417 **b** 6417 to 13,617

2.19 **a** .68 **b** .95 **c** ≈ 0

2.21 **a** 1.4; 1.4 **b** 1.4; 1.4

2.23 **a** 2.04; 2.806

b–c

k	$\bar{x} \pm ks$	**Actual**	**Tchebysheff's Theorem**	**Empirical Rule**
1	(−.77, 4.85)	.84	≥ 0	$\approx$ .68
2	(−3.57, 7.65)	.92	$\geq 3/4$	$\approx$ .95
3	(−6.38, 10.46)	1.00	$\geq 8/9$	$\approx$ 1.00

2.25 **a** $\bar{x} = 4.75$; $s = 2.454$ **b** −1.94; 1.32; no

2.27 **a** $m = 1175$; $Q_L = 1039.25$; $Q_U = 1364.75$

2.29 \$20 is the 10th percentile; \$55 is the 90th percentile

2.31 84th; 87th

2.33 inner fences: 17.25 and 31.25; outer fences: 12 and 36.5; 12 is a mild outlier

2.35 **a** upper hinge = 1358; lower hinge = 1040
b inner fences: 563 and 1835; outer fences: 86 and 2312
c 2325 and 3168 are extreme outliers

2.37 **a** inner fences: −49 and 959; outer fences: −427 and 1337; 5200 is an extreme outlier
b yes

2.39 **a** $s \approx 5789$ **b** .815

2.41 **a** 32.1 **b** $s \approx 8.025$ **c** 7.671

2.43 $m = 6.35$; $Q_L = 2.325$; $Q_U = 12.825$

2.45 at least 3/4 of the measurements in the interval 0 to 140; at least 8/9 of the measurements in the interval 0 to 193

2.47 **a** 7.75 **b** 59.2; 10.369 vs. 7.75

2.49 (0, 32)

2.51

k	$\bar{x} \pm ks$	**Approximate Fraction in Interval**
1	(415, 425)	$\approx$ .68
2	(410, 430)	$\approx$ .95
3	(405, 435)	$\approx$ 1.00

2.53 **a** .025 **b** .84

2.55 **a** at least 3/4 have between 145 and 205 teachers **b** .16

2.59 **b** $Q_L = 75.5$; $Q_U = 129.7$ **c** San Francisco, Boston, Los Angeles, NY/NJ, San Diego

2.61 **a** skewed **b** inner fences: −2.9 to 10.7; outer fences: −8.0 to 15.8; Los Angeles and San Francisco are mild outliers; New York is an extreme outlier

Chapter 3

3.1 $P(D) = 1/6$; $P(E) = 1/2$; $P(F) = 0$

3.3 $P(E_1) = .45$; $P(E_2) = .15$; $P(E_i) = .05$ for $i = 3, 4, \ldots, 10$

3.5 **b** $P(E_i) = 1/38$ **c** $P(A) = 1/19$ **d** 9/19

3.7 **a** 36 **b** 1/6; 1/18

3.9 **a** choose 2 of 4 cans without replacement
b (N_1N_2), (N_1W_1), (N_1W_2), (N_2W_1), (N_2W_2), (W_1W_2) where N = no water, W = water
c 1/6

3.11 **a** rank A, B, C
b (ABC), (ACB), (BAC), (BCA), (CAB), (CBA)
c 1/3, 1/3

3.13 **a** 2/27 **b** 20/27

3.15 **a** 16
b $(EEEE)$, $(EEEN)$, $(EENE)$, $(ENEE)$, $(NEEE)$, $(EENN)$, $(ENEN)$, $(NEEN)$, $(ENNE)$, $(NENE)$, $(NNEE)$, $(ENNN)$, $(NENN)$, $(NNEN)$, $(NNNE)$, $(NNNN)$
d $P(A) = 11/16$; $P(B) = 3/8$; $P(C) = 1/4$

3.17 **a** 3/5 **b** 1/5 **c** 1/5 **d** 1/5
e 1/5 **f** 1 **g** 3/5 **h** 1/2 **i** 1/4 **j** 1
k 4/5 **l** 0

3.19 **a** 1/4 **b** 1/2

3.21 **a** no **b** no

3.23 **a** no; no **b** no; yes

3.25 **a** .0004 **b** .9996 **c** .0004

3.29 **a** .413 **b** .400 **c** .091

3.31 **a** .0005 **b** .0002 **c** .3685 **d** .0207

3.33 .0214 **3.35** .255

3.37 **a** 406 **b** .0025 **c** .0690 **d** .0027

3.39 **a** $P(A) = .9918$; $P(B) = .0082$ **b** $P(A) = .9836$; $P(B) = .0164$

3.41 **a** discrete **b** continuous **c** continuous **d** continuous
e discrete

3.43 **a** $p(4) = .05$

3.45 **a** (B_1B_2), (B_1W_1), (B_1W_2), (B_2W_1), (B_2W_2), (W_1W_2); $P(E_i) = 1/6$, $i = 1, 2, \ldots, 6$ **c** $p(0) = 1/6$; $p(1) = 2/3$; $p(2) = 1/6$

3.47 **a** (S), (FS), (FFS), $(FFFS)$ **c** $p(1) = p(2) = p(3) = p(4) = 1/4$

3.49 **a** $p(0) = 3/10$; $p(1) = 6/10$; $p(2) = 1/10$

3.51 **a** .1; .09; .081 **b** $p(x) = (.9)^{x-1}(.1)$

3.53 **a** $p(x) = .27(.73)^{x-1}$ **3.55** **a** 3.45; 2.0475; 1.4309
c (.59, 6.31); .95 **d** yes

3.57 $E(\text{gain}) = -\$0.26$ **3.59** **a** 2.98 **b** 3.94 **c** 5.16

3.61 **a** 7.9 **b** 2.1749 **c** .96 **3.63** \$2050

3.65 **a** 2.5 **b** 7.5 **c** 1.25; 1.118

3.69 **a** (C_1C_2), (C_1A_1), (C_1A_2), (C_2C_1), (C_2A_1), (C_2A_2), (A_1C_1), (A_1C_2), (A_1A_2), (A_2C_1), (A_2C_2), (A_2A_1)
b (C_1C_2), (C_1A_1), (C_1A_2), (C_2C_1), (C_2A_1), (C_2A_2)
c (C_1A_1), (C_1A_2), (C_2A_1), (C_2A_2), (A_1C_1), (A_1C_2), (A_2C_1), (A_2C_2)
d (A_1A_2), (A_2A_1)

3.71 **b** $(HHHT)$, $(HHTH)$, $(HTHH)$, $(THHH)$ **c** 1/16; 1/4

3.73 **a** S contains 21 simple events **b** A contains 6 simple events **c** 2/7

3.75 $p(0) = .0256$; $p(1) = .1536$; $p(2) = .3456$; $p(3) = .3456$; $p(4) = .1296$; .4752

3.77 2.4; .48 **3.79** 1/6 **3.81** 0; 1/3

3.83 .3913 **3.85** .999999 **3.87** 1; .4783; .00781

3.89 .6 **3.91** .2; .1 **3.93** **a** .128 **b** .488

3.95 **a** 1/8 **b** 1/64 **c** not necessarily; they could have studied together, etc.

3.97 **a** 5/6 **b** 25/36 **c** 11/36 **3.99** 1023.7; 40.9%

3S.1 80 **3S.3** 45 **3S.5** 8

3S.7 $5.720645(10^{12})$ **3S.9** 2,598,960 **3S.11** 1/42

3S.13 .6087; .3913 **3S.15** .2432 **3S.17** .6585

3S.19 .3130 **3S.21** **a** .5182 **b** .1136 **c** .7091 **d** .3906

Chapter 4

4.1 not binomial; dependent trials; p varies from trial to trial

4.3 **a** .2965 **b** .8145 **c** .1172 **d** .3670

4.5 **a** .3670; .0547; .0004 **b** .7903; .9922; 1.0000
c .4233; .9375; .9996 **d** .5767; .0625; .0004 **e** 1.4; 3.5; 5.6
f 1.0583; 1.3229; 1.0583

4.7 **a** .967 **b** .005 **c** .000 **d** 1.000

4.9 **a** .748 **b** .610 **c** .367 **d** .966 **e** .656

4.11 **a** 1; .99 **b** 90; 3 **c** 30; 4.58 **d** 70; 4.58 **e** 50; 5

4.13 **a** .9568 **b** .957 **c** .9569 **d** $\mu = 2$; $\sigma = 1.342$
e .7455; .9569; .9977 **f** yes

4.15 binomial, assuming random sampling; $n = 1162$, $p = P(\text{adequate security})$

4.17 not binomial; experiment does not result in one of two outcomes

4.19 their performance would seem to be below the 80% rate of success

4.21 600; 420; 20.4939; $P(x > 700) \leq .04$

4.23 **a** .9606 **b** .9994

4.25 **a** $\mu - 3\sigma \approx 256$ **b** $\mu + 3\sigma \approx 320$

4.27 $\mu = 360$; $\sigma = 10.04$; $\mu \pm 3\sigma$ or 329.88 to 390.12

4.29 **a** .135335 **b** .270671 **c** .593994 **d** .036089

4.31 **a** .677 **b** .6767

4.33 **a** .0067 **b** .1755 **c** .560

4.35 **a** .271 **b** .594 **c** .406

4.37 no

4.39 **a** .6 **b** .5143 **c** .0714

4.41 **a** $p(0) = 165/455; p(1) = 220/455; p(2) = 66/455; p(3) = 4/455$
c $\mu = .8; \sigma^2 = .50286$
d .99 fall between −.618 and 2.218; .99 fall between −1.327 and 2.927; yes

4.43 $p(0) = 1/5; p(1) = 3/5; p(2) = 1/5$ **4.45** .0096; .9996

4.47 .0035; a very rare event has occurred, or the sample may not have been randomly chosen

4.51 **a** $p(0) = .125; p(1) = .375; p(2) = .375; p(3) = .125$
c 1.5; .866 **d** .75; 1.00

4.53 **a** .9606 **b** .9994

4.55 **a** .234 **b** .136 **c** claim is not unlikely

4.57 **a** .228 **b** no indication that people are more likely to choose middle numbers

4.59 .6384; $\mu = 3$ **4.61** **a** 30,000 **b** 122.474 **c** highly unlikely if $p = .5$ **d** yes

4.63 .017; the percentage is probably less than 80%

4.65 **a** .3 **b** 7.5; 2.291 **c** 1.09 **d** no

4.67 **a** .0038 **b** .9560

4.69 **a** binomial; $n = 1000$, $p = .00019$ **b** .1731
c $\mu = .19; \sigma = .4358$ **d** yes

4.71 **a** hypergeometric, or approximately binomial
b Poisson
c approximately .85; .72; .61

4.73 **a** yes **b** $1/8192 = .00012$

Chapter 5

5.1 **a** .4452 **b** .4664 **c** .3159 **d** .3159

5.3 **a** .6753 **b** .2401 **c** .2694

5.5 **a** −1.96 **b** .36 **5.7** **a** 1.65 **b** −1.645

5.9 **a** .0401 **b** .1841 **c** .2358 **5.11** $\mu = 9.2$

5.13 $\mu = 8; \sigma = 2$ **5.15** **a** .1056; .5000 **b** .8944

5.17 .0505 **5.19** **a** .3446 **b** .0047 **c** 144.775°

5.21 **a** .0104 **b** .2206 **5.23** **a** .0174 **b** .1112

5.25 $\mu = 1940.119$

5.27 **a** .1469 **b** .1362 **c** $39,333

5.29 .0175 **5.31** .3745 **5.33** .2676

5.35 **a** .546 **b** .5468 **5.37** .3531

5.39 .1251 **5.41** .0901; no **5.43** .9441

5.45 **a** ≈ 0 **b** yes; .026 or 2.6%

5.47 **a** .3849 **b** .3159 **c** .4279 **d** .1628

5.49 **a** .7734 **b** .9115

5.51 .8612 **5.53** **a** .1056 **b** .8944 **c** .1056

5.55 .16 **5.57** **a** .0778 **b** .0274

5.59 3.44% **5.61** 85.36 **5.63** z-score = −1.26; no

5.65 **a** no **b** .0179 **5.67** **a** .0951 **b** .0681

5.69 **a** 141 **b** .0401 **5.71** .9474 **5.73** .1725; .4052

5.75 z-score = 1.461; no

5.77 **a** ≈ 0 **b** .6026 **c** survey not reliable

Chapter 6

6.1 **a** $\mu = 10; \sigma_{\bar{x}} = .6$
b $\mu = 5; \sigma_{\bar{x}} = .2$
c $\mu = 120; \sigma_{\bar{x}} = .3536$

6.3 **c** .5468

6.5 **a** $\mu_{\bar{x}} = 3.5; \sigma_{\bar{x}} = .541$
b $\mu_{\bar{x}} = 3.5; \sigma_{\bar{x}} = .442$
c $\mu_{\bar{x}} = 3.5; \sigma_{\bar{x}} = .342$

6.7

n	1	2	4	9	16	25	100
$\sigma_{\bar{x}}$	1	.707	.500	.333	.250	.200	.100

6.9 **a** 106; 2.4 **b** .0475 **c** .9050

6.13 **b** ≈ 0 **c** no

6.15 **a** normal **b** .0571 **c** $\mu = 21.948$

6.17 **a** approximately normal **b** $\mu_x = 255; \sigma_x = 13.693$

6.21 **a** yes **b** .0681 **c** .5 **d** .8638

6.23 **a** yes **b** .0668 **c** .9544

6.25 **a** approximately normal; $\mu_{\hat{p}} = .4; \sigma_{\hat{p}} = .0152$ **b** .8098 **c** .9372

6.27 **a** approximately normal; $\mu_{\hat{p}} = p; \sigma_{\hat{p}} \approx .0146$ **b** .0062

6.29 **a** $LCL = 150.13; UCL = 161.67$

6.31 **a** $LCL = 0; UCL = .090$

6.33 **a** $LCL = 8598.7; UCL = 12{,}905.3$

6.35 $LCL = .078; UCL = .316$

6.37 $LCL = .0155; UCL = .0357$

6.41 **a** .0114 **b** ≈ 0 **6.45** **a** 21 **b** 792 **c** 161,700

6.47 1/3921225 **6.51** .1922

6.53 **a** no **b** not necessarily **c** applications at private colleges are not increasing as fast as those at colleges in general

6.59 limit total number of passengers to $n = 11$

6.61 **a** 1312; 11.628 **b** .003 **6.63** .6736

6.65 **a** $LCL = 19.6023; UCL = 22.8043$

6.67 **b** $\mu_{\bar{x}} = 110.390; \sigma_{\bar{x}} = 6.288$

Chapter 7

7.3 **a** (12.496, 13.704) **b** (2.651, 2.809)

7.5 **a** (32.550, 35.450) **b** (1047.543, 1050.457) **c** (65.973, 66.627)

7.7 **a** 3.92 **b** 2.772 **c** 1.96

7.9 **a** 3.29 **b** 5.16 **7.11** 4.2 ± .339

7.13 **b** (1.756, 1.924) **c** (486.610, 619.390) **d** no

7.15 **a** $\bar{x} = 86.11$ with margin of error = 3.61
b (86.41, 94.53) **c** inference not valid for general population

7.17 (3.496, 3.904); random sampling

7.19 **a** (−2.525, −1.875) **b** (−2.710, −1.690)

7.21 (−1.032, −0.368) **7.23** (6.748, 9.052)

7.25 **a** $\bar{x}_1 - \bar{x}_2 = 2206$ with margin of error = 902.08 **b** yes

7.27 **a** (−2.65, −0.81) **b** (−4.62, −3.20) **c** yes; in both (a) and (b)

7.29 (.846, .908)

7.31 **a** (.507, .553) **b** (.474, .546)
c $\hat{p} = .46$ with margin of error = .028 **d** yes

7.33 $\hat{p} = 2/3$ with margin of error = .022 **7.35** **a** no **b** nothing; no

7.37 (.66, .70) **7.39** **a** (.668, .672) **b** (.708, .712) **7.41** .055

7.43 **a** (.086, .178) **b** random samples

7.45 (.001, .079), yes; (.031, .109), yes; yes

7.47 **a** .11 with margin of error = .043
b −.12 with margin of error = .035
c there are significant differences in the proportions for the two years

7.49 (.061, .259) **7.51** 505 **7.53** 1086

7.55 **b** 9604 **7.57** 97 **7.59** 70

7.63 (28.298, 29.902) **7.65** (3.867, 4.533)

7.67 **a** .48 ± .044 **b** (.443, .517)

7.69 (−.0157, .2907) **7.71** 9.7 ± 2.274

7.73 no; .031; .044; .052 **7.75** 97

7.77 (.753, .847) **7.79** (20.447, 23.553)

7.81 2401; yes **7.83** 9604 **7.85** 193

7.87 (33.41, 34.59) **7.89** (.059, .141)

7.91 **a** $\bar{x} \pm 1.96\, s/\sqrt{n}$ **b** (1.922, 1.984)

7.93 **a** (5.543, 5.963) **b** 530

7.95 (−5.785, −0.215) **7.97** (6.735, 9.905)

Chapter 8

8.1 **a** H_0: $\mu = 2.3$; H_a: $\mu > 2.3$ **c** $z = 1.645$ **d** yes; $z = 2.04$

8.3 H_0: $\mu = 2.9$; H_a: $\mu \neq 2.9$; two-tailed

8.5 **a** H_0: $\mu = 60$; H_a: $\mu < 60$ **b** one-tailed
c yes; $z = -1.992$

8.7 **a** H_0: $\mu = 80$; H_a: $\mu \neq 80$ **c** $z = -3.75$; reject H_0

8.9 no; $z = -.992$

8.11 **a** H_0: $\mu_1 - \mu_2 = 0$; H_a: $\mu_1 - \mu_2 > 0$
b one-tailed **c** $z > 1.28$ **d** reject H_0; $z = 2.087$

8.13 $\alpha = .20$ **8.15** yes; $z = -2.26$

8.17 **a** yes; $z = 4.79$ **b** yes

8.19 **a** H_0: $\mu_1 - \mu_2 = 0$; H_a: $\mu_1 - \mu_2 \neq 0$
b two-tailed **c** no; $z = -.954$

8.21 **a** H_0: $p = .4$; H_a: $p \neq .4$ **b** two-tailed **c** reject H_0; $z = -1.680$

8.23 no; $z = -1.23$

8.25 **a** reject H_0; $z = 3.07$; figure is not accurate
b no; sample is probably not representative of all college students in the U.S.

8.27 **a** do not reject H_0; $z = 1.77$ **b** (.17, .43)

8.29 **a** no, $z = .52$ **b** no; $z = -1.10$ (with $\hat{p} = .1875$) **c** no

8.31 **a** H_0: $p_1 - p_2 = 0$; H_a: $p_1 - p_2 < 0$ **b** one-tailed **c** do not reject H_0; $z = -.84$

8.33 yes; $z = -2.40$

8.35 **a** reject H_0; $z = 5.03$

b reject H_0; $z = -6.71$; ibuprofen has significantly increased its market share
c yes

8.37 fat: significant decrease; $z = 2.004$; salt: significant decrease; $z = 3.501$; no

8.39 yes; $z = 2.99$ **8.41** **a** .0268 **b** reject H_0

8.43 **a** .0718 **b** do not reject H_0 **8.45** .0658

8.47 p-value $< .001$; reject H_0 **8.49** .0104; reject H_0

8.51 **a** H_0: $p = 2/3$ **b** H_a: $p > 2/3$ **c** yes; $z = 4.6$
d p-value $< .001$

8.55 yes; $z = -25.298$; reject H_0

8.57 **a** H_0: $\mu = 7.5$; H_a: $\mu < 7.5$ **b** one-tailed **d** $z = -5.477$; reject H_0

8.59 **a** no; $z = .283$ **b** no; $z = -.80$

8.61 yes; $z = 2.858$

8.63 **a** H_0: $\mu = 35$; H_a: $\mu < 35$ **b** $z < -2.33$ **c** $z = -6.19$; mean is less than 35

8.65 309 **8.67** yes; $z = 4.00$

8.69 **a** yes; $z = 4.333$ **b** (7.12, 18.88)

8.71

μ	β
873	.6141
875	.7794
877	.8906

8.73 **a** p-value $< .002$ **b** reject H_0 **8.75** no; $z = -2.779$

Chapter 9

9.1 **a** 2.015 **b** 2.306 **c** 1.330 **d** ≈ 1.960

9.3 **a** $t = -1.438$; do not reject H_0
b $.10 < p$-value $< .20$
c (45.976, 48.224)

9.5 **a** (16.412, 19.824) **b** (.026, .042) **c** (26.964, 43.218)

9.7 no; $t = -1.195$; do not reject H_0 **9.9** yes; $t = -3.044$

9.11 (3.652, 3.912) **9.13** **a** reject H_0; $t = -3.40$ **b** (343.47, 386.53)

9.15 **a** reject H_0; $t = -4.31$ **b** (23.23, 29.97)

9.17 **a** 3.775 **b** 21.2258

9.19 **a** H_0: $\mu_1 - \mu_2 = 0$; H_a: $\mu_1 - \mu_2 \neq 0$
b $|t| > 1.703$ **c** $t = 2.795$
d p-value $< .01$ **e** reject H_0

9.21 **a** H_0: $\mu_1 - \mu_2 = 0$; H_a: $\mu_1 - \mu_2 > 0$
b yes; $t = 2.806$ **c** $.005 < p$-value $< .01$

9.23 **a** yes; $t = -2.17$ **b** p-value $= .042$

9.25 **a** yes; $t = -5.54$ **b** no; $t = 1.56$

9.27 assumption of equal variances is violated

9.29 $t = 2.372$; reject H_0; $.02 < p$-value $< .05$ **9.31** 62

9.33 **a** no; $t = 1.177$ **b** p-value $> .20$ **c** $(-.082, .202)$

9.35 **a** no significant difference; $t = 1.984$ with approximately 14 d.f.; $(-6.66, 170.32)$
b significant difference; $t = 2.307$ with approximately 14 d.f.; (7.567, 207.733)
c reject H_0; $t = 4.38$ **d** $(-1.6, 8.2)$; yes

9.37 **b** yes; $t = 9.150$ **c** p-value $< .01$ **d** (80.472, 133.328)

9.41 **a** yes; $t = -4.326$; reject H_0 **b** $(-2.594, -.566)$

9.43 **a** yes; $t = 2.82$; reject H_0 **b** 1.487 **d** yes

9.45 no; $\chi^2 = 34.24$

9.47 **a** .699 **b** (.291, 3.390) **c** $\chi^2 = 5.24$; do not reject H_0
d p-value $> .20$

9.49 **a** no; $\chi^2 = 12.8618$ **b** p-value $> .10$

9.51 (1.408, 31.264) **9.53** (.00476, .00980)

9.55 yes; $\chi^2 = 57.281$; reject H_0 for $\alpha = .05$

9.57 **a** no, $F = 1.774$ **b** p-value $> .20$

9.59 **a** no, $F = 2.316$ **b** $.05 < p$-value $< .10$

9.61 **a** yes, $F = 2.486$; reject H_0 **b** p-value $< .005$

9.63 **a** $\sigma_1^2 = \sigma_2^2$ **b** no, $F = 1.64$

9.65 **a** no, $F = 2.904$ **b** (.050, .254)

9.71 (9.860, 12.740) **9.73** yes, $t = 5.985$; reject H_0; (28.375, 33.625)

9.75 yes, $F = 3.268$ **9.77** $n = 72$

9.79 **a** do not reject H_0; $t = -.39$; $(-10.6, 7.5)$
b reject H_0; $t = -4.11$
c do not reject H_0; $t = -.85$; $(-8.31, 4.03)$
d reject H_0; $t = -3.59$

9.81 $(60.87, 98.07)$ **9.83** $(-10.246, -2.354)$

9.85 **b** yes, $F = 19.52$ **c** p-value $< .01$
d assumption of equal variances is violated

9.87 **a** no **b** yes, $t = 3.237$ **c** yes, $t = 59.342$; p-value $< .01$

9.89 **a** no, $F = 1.331$ **b** $t = -4.537$, p-value $< .01$; reject H_0

9.91 $(33.465, 34.107)$

9.93 **a** H_0: $\mu_1 - \mu_2 = 0$; H_a: $\mu_1 - \mu_2 > 0$
b $t > 2.552$ **c** no, $t = 1.69$
d paired difference test **e** no, the analysis was incorrect

9.95 no, $t = -1.712$ **9.97** $(-1.522, .022)$ **9.99** $(-1.842, .092)$; yes

9.101 $.259 \pm .047$ unpaired; $.259 \pm .050$ paired

9.103 **a** no, $t = 2.571$ **b** $(.000, .02)$

9.105 **a** two-tailed; H_a: $\sigma_1^2 \neq \sigma_2^2$ **b** lower-tailed; H_a: $\sigma_1^2 < \sigma_2^2$
c upper-tailed; H_a: $\sigma_1^2 > \sigma_2^2$

9.107 yes, $F = 2.407$ **9.109** $(.185, 2.465)$ **9.111** yes, $t = 2.425$

9.113 yes, $t = -2.945$ **9.115** yes, $t = 3.038$; reject H_0

9.117 $(24.582, 73.243)$

9.119 **a** no, $t = 1.120$ **b** no, $t = -1.519$

9.121 **a** $(80.207, 372.186)$ **b** yes

Chapter 10

10.1 y intercept $= 1$, slope $= 2$ **10.3** $y = 3 - x$

10.7 **a** $\hat{y} = 6.00 - .557x$ **b** yes **c** 4.05

10.9 **a** $\hat{y} = -156{,}162 + 78.8485x$ **10.11** $(.657, 1.743)$

10.13 $(-.653, -.461)$

10.15 **a** H_0: $\beta_1 = 0$; H_a: $\beta_1 > 0$ **b** reject H_0; $t = 70.29$ with p-value $= 0.000$

10.17 yes; $t = 3.79$; no **10.19** **a** $\hat{y} = 21.4 - .0063x$ **b** 122
c yes; $t = -9.00$

10.21 $(4.667, 5.605)$

10.23 **a** $(325.928, 400.939)$ **b** $(456.365, 672.58)$

10.25 **a** $(24.094, 30.718)$ **b** $(104676, 107360)$ **d** $(96602.48, 103375.52)$

10.27 **a** 6 **b** yes; $t = -14.49$
c $\hat{y} = 3393.93 - 1.319x$ **d** $\hat{y} = 2305.755$; no

10.31 **a** positive **b** $r = .9487$; $r^2 = .9$

10.33 **b** $r = .982$ **c** 96.5% **10.35** **a** positive **b** yes; $t = 2.615$

10.37 **a** yes; $t = 3.158$ **b** p-value $< .01$

10.39 **a** $\hat{y} = 46 - .317x$ **c** 19.033 **d** $(-.419, -.215)$
e (27.67, 32.67) **f** (21.88, 38.46)

10.41 **a** $r^2 = .2720$ **b** yes; $t = 4.234$

10.43 $\hat{y} = 0.067 + .517x$; significant positive linear relationship

10.45 **a** $r = .9803$ **b** $r^2 = .9610$
c $\hat{y} = 21.867 + 14.967x$ **d** yes

Chapter 11

11.1 **b** parallel lines

11.3 **b** the lines are not parallel **c** allows for interaction

11.5 x_1, x_2, and x_3 **11.7** (.536, 2.044)

11.9 **a** $\hat{y} = 1.21 + 7.60x - .94x^2$

11.11 **a** $\beta_2 = 0$ **b** yes, $t = -2.848$

11.13 **a** negative sign; $\hat{\beta}_2 = -0.8198$
b SSE = 1.05985748; $s^2 = .35328583$ **c** 3
d yes, $F = 332.53$ **e** p-value = .0003 **f** yes, $t = -4.49$
g p-value = .0206 **h** $\hat{y} = -44.192 + 16.334x - 0.820x^2$
i $R^2 = .9955$ **j** 27.34

11.15 **a** SSE = 2,238,508; $s^2 = 559{,}627$
b $\hat{y} = -383 + 2.418x_1 + 69.1x_2$ **c** .597 **d** 39.5%
e no, $F = 2.96$ **f** yes **g** the fit is not good

11.17 **a** $E(y) = \beta_0 + \beta_1x_1 + \beta_2x_2 + \beta_3x_3 + \beta_4x_4$
b $F = 6.345$; yes **c** p-value $< .01$ **d** x_1 and x_3
e $F = 6.43$ **g** relatively good fit; eliminate nonsignificant variables and refit

11.19 **a** $E(y) = \beta_0 + \beta_1x_1 + \beta_2x_2 + \beta_3x_3 + \beta_4x_4$
b $F = 5.389$; yes **c** $.01 < p\text{-value} < .025$
d x_2, x_3, and x_4 **e** $F = 6.12$
g relatively good fit; eliminate nonsignificant variables and refit

11.21 **a** x_2 **b** $\hat{y} = 12.6 + 3.9x_2^2$ or $\hat{y} = 13.14 - 1.2x_2 + 3.9x_2^2$

11.23 **a** $E(y) = \beta_0 + \beta_1x_1 + \beta_2x_2 + \beta_3x_3 + \beta_4x_4 + \beta_5x_5 + \beta_6x_6 + \beta_7x_7 + \beta_8x_8 + \beta_9x_9$
b 125 **c** x_1, x_4, x_5, and x_6 **e** yes, $F = 5.428$

11.25 **a** SSE = 152.17748; $s^2 = 8.4543$ **b** 18
c yes, $F = 31.85$ **d** p-value $< .005$
e $S_{yy} = 1498.625$; $R^2 = .898$ **f** $\hat{y} = 40.184$

11.29 When $\nu_1 = 1$

11.31 **b** 9 **d** $R^2 = .984$
f $\hat{y} = 4.10 + 1.04x_1 + 3.53x_2 + 4.76x_3 - 0.43x_1x_2 - 0.08x_1x_3$
g $t = -2.613$; reject H_0 **h** $(-0.802, -0.058)$
i do not reject H_0; $t = -.486$

11.33 **a** SSE = .05368; $s^2 = .00596$
b $\hat{y} = .4381 + .0053x_1 - .0302x_2 + .0007x_1x_2 - .00038x_1^2 + .0004x_2^2$
c $R^2 = .9888$ **d** yes, $F = 159.23$

11.35 **a** $y = \beta_0 + \beta_1 x_1 + \beta_2 x_2 + \beta_3 x_1 x_2 + \beta_4 x_1^2 x_2 + \epsilon$
b $\hat{y} = -17.65 + 10.737x_1 - 28.69x_2 + 12.722x_1x_2 - .3707x_1^2x_2$
c .765 **d** yes, $F = 32.53$ **e** $\hat{y} = -17.65 + 10.737x_1$
f $\hat{y} = -46.34 + 23.459x_1 - .3707x_1^2$ **g** yes, $t = -3.52$

Chapter 12

12.3 **a** 10.6446 **b** 21.6660 **c** 22.3621 **d** 5.99147

12.5 **a** 2 **b** $X^2 > 5.99147$
c H_a: $p_i \neq p_j$ for some pair i, j ($i \neq j$)
d $X^2 = 16.88$; reject H_0 **e** p-value $< .005$

12.7 no, $X^2 = .658$ ($\chi^2_{.05} = 7.81$)

12.9 yes, $X^2 = 31.77$ ($\chi^2_{.05} = 9.49$)

12.11 no, $X^2 = .2995$, p-value $> .10$ **12.13** 8

12.15 **c** $X^2 > 3.84$ **f** $X^2 = 22.87$; reject H_0

12.17 **a** no, $X^2 = 0.639$ **b** H_0: $p_1 = p_2$; H_a: $p_1 \neq p_2$
c H_0: $p_1 = p_2$; H_a: $p_1 < p_2$
d z test for the difference between two population proportions

12.19 **a** $X^2 = 2.56$ **b** yes **c** insufficient evidence to indicate a dependence

12.21 **a** $X^2 = 10.597$ **b** $X^2 > 7.779$ **c** reject H_0
d $.025 < p$-value $< .05$

12.23 yes, $X^2 = 42.179$ ($\chi^2_{.01} = 9.21$)

12.25 yes, $X^2 = 8.75$ ($\chi^2_{.05} = 5.99$)

12.27 yes, $X^2 = 38.12$ ($\chi^2_{.05} = 12.59$)

12.29 no, $X^2 = 4.4$ ($\chi^2_{.10} = 7.78$)

12.31 no, $X^2 = 1.89$ ($\chi^2_{.10} = 4.61$)

12.33 yes, $X^2 = 10.26$ ($\chi^2_{.05} = 7.81$)

12.35 **a** yes, $X^2 = 22.690$ ($\chi^2_{.05} = 5.99$) **b** p-value $< .005$

12.37 yes, $X^2 = 6.18$ ($\chi^2_{.05} = 5.99$)

12.39 $\hat{p} = .5$; 6.25, 25, 37.5, 25, 6.25; reject H_0: $X^2 = 8.56$ ($\chi^2_{.05} = 7.81$)

12.41 **a** yes, $X^2 = 10.27$ ($\chi^2_{.05} = 3.84$)
b yes, $z = 3.205$ ($z_{.05} = 1.645$)

12.43 yes, $X^2 = 15.28$; p-value $< .005$

12.45 yes, $X^2 = 8.91$ ($\chi^2_{.05} = 5.99$)

12.47 **a** yes, 78 **b** no, $X^2 = 3.01$; p-value $> .10$

12.49 **a** yes, $X^2 = 19.043$ ($\chi^2_{.01} = 16.8119$)
b yes, $X^2 = 60.139$ ($\chi^2_{.01} = 16.8119$)
d some expected cell counts are small; yes

12.51 **a** no, $X^2 = 1.19$ **b** p-value $> .10$ **c** yes

Chapter 13

13.1

Source	d.f.
Treatments	5
Error	54
Total	59

13.3 **a** (2.731, 3.409) **b** (.07, 1.03)

13.5 **a**

Source	d.f.	SS	MS	F
Treatments	3	339.8	113.267	16.98
Error	20	133.4	6.67	
Total	23			

b $\nu_1 = 3$ and $\nu_2 = 20$ **c** $F > 2.38$ **d** yes, $F = 16.98$

13.7 **a** CM = 103.142857; Total SS = 26.8571
b SST = 14.5071; MST = 7.2536
c SSE = 12.3500; MSE = 1.1227
f $F > 3.98$ **g** $F = 6.46$; reject H_0

13.9 **a** (1.95, 3.65) **b** (.27, 2.83)

13.11 **a** (67.86, 84.14) **b** (55.82, 76.84)
c (−3.629, 22.963) **d** no, they are not independent

13.13 **a** each observation is the mean length of 10 leaves
b yes, $F = 57.38$ **c** reject H_0; $t = 12.105$ **d** (2.030, 2.704)

13.15 **b** yes, $F = 8.11$ **c** yes, $t = 3.025$

13.17

Source	d.f.	SS	MS	F
Treatments	2	11.4	5.70	4.01
Blocks	5	17.1	3.42	2.41
Error	10	14.2	1.42	
Total	17	42.7		

13.19 (−3.833, −.767)

13.21 **a** yes, $F = 19.19$ **b** yes, $F = 135.75$ **c** yes
d (−5.057, −2.943)

13.23 **a** 7 **b** 7 **c** 5 **e** yes, $F = 9.68$ **f** yes, $F = 8.59$

13.25 **a** yes, $F = 19.44$ **b** yes, $F = 40.21$ **c** (−1.997, −.803)

13.27 **a** yes, $F = 10.06$ **b** yes, $F = 10.88$ **c** (1.612, 5.388)

13.29 **a** yes, $F = 7.20$ **b** (−3.469, −1.081) **c** yes, $F = 16.61$

13.31 **a** 5.06 **b** 3.88 **c** 6.20 **d** 9.32

13.33 **a** $\omega = 6.78$

b

$\bar{x}_4$	$\bar{x}_2$	$\bar{x}_1$	$\bar{x}_5$	$\bar{x}_3$	$\bar{x}_6$
92.9	98.4	101.6	104.2	112.3	113.8

13.35 $\omega = .694$; $\bar{x}_4$ is significantly different from the other three means

13.37 $\omega = .845$;

$\bar{x}_C$	$\bar{x}_A$	$\bar{x}_B$	$\bar{x}_D$
11.2	11.4	12.33	12.8

13.41 treatment means differ; $F = 27.78$

13.43 **a** yes, $F = 9.84$ **b** (.491, 13.175) **c** (87.723, 96.027)

13.45 **a** yes, $F = 5.20$ **b** no, $t = .88$ **c** $(-.579, -.117)$

13.47 **a** yes, $F = 18.61$ **b** yes, $F = 7.11$ **c** $(-6.767, -3.433)$

13.49 no significant differences at $\alpha = .05$; $\omega = 5.654$

Chapter 14

14.1 **b** $\alpha = .002, .007, .022, .054, .115$

14.3 One-tailed: $n = 10$, $\alpha = .001, .011, .055$; $n = 15$, $\alpha = .004, .018, .059$; $n = 20$, $\alpha = .001, .006, .021, .058, .132$ Two-tailed: $n = 10$, $\alpha = .002, .022, .110$; $n = 15$, $\alpha = .008, .036, .118$; $n = 20$, $\alpha = .002, .012, .042, .116$

14.5 **a** $H_0: p = 1/2$; $H_a: p \neq 1/2$; rejection region: $\{0, 1, 7, 8\}$; $x = 6$; do not reject H_0 at $\alpha = .07$; p-value $= .290$.

14.7 $z = 3.15$; reject H_0 **14.9** **a** U_1 **b** $U \leq 13$ **c** $\alpha = .0906$

14.11 **a** H_0: populations 1 and 2 are identical; H_a: population 1 is shifted to the left of population 2
b $U \leq 4$ with $\alpha = .0476$
c $T_1 = 16$, $T_2 = 39$, $U_1 = 24$, $U_2 = 1$
d $U = 1$ **e** yes

14.13 do not reject H_0; $z = -1.59$

14.15 **a** $U_1 = 6$; do not reject H_0; rejection region: $\{0, 1, \ldots, 4\}$ (one-tailed) for $\alpha = .0476$
b do not reject H_0: $\mu_A - \mu_B = 0$; $t = 1.606$ ($t_{.05} = 1.86$)

14.17 $U_2 = 12.5$; reject H_0; rejection region: $\{0, 1, \ldots, 21\}$ for $\alpha = .1012$

14.19 yes, $z = 3.49$

14.21 **a** H_0: populations 1 and 2 are identical; H_a: population 1 is shifted to the right of population 2
b T^- **c** $T \leq 152$ **d** $T^- = 216$; do not reject H_0

14.23 $z = -.34$; do not reject H_0

14.25 **a** $T = 3$; rejection region: $\{T \leq 4\}$; reject H_0; consistent with results of Exercise 9.43

14.27 no, $T = 97$; rejection region: $\{T \leq 52\}$; consistent with results of Exercise 14.6

14.29 **a** do not reject H_0; $T = 4$; rejection region: $\{T \leq 14\}$ for $\alpha = .05$
b p-value $> .10$

14.31 yes, $T = 3.5$; rejection region: $\{T \leq 4\}$ for $\alpha = .05$

14.33 yes, $H = 13.90$ ($\chi^2_{.05} = 7.81$)

14.35 **a** no, $H = 2.63$ ($\chi^2_{.10} = 6.25$)
b p-value $> .10$ **c** p-value $> .10$

14.37 **a** reject H_0; $F_r = 7.58$ ($\chi^2_{.05} = 5.99$)
b $.01 < p\text{-value} < .025$ **d** $F = 10.83$ **e** $p\text{-value} < .005$

14.39 **a** yes, $F_r = 19.57$ ($\chi^2_{.05} = 9.49$) **b** $p\text{-value} < .005$

14.41 **a** $r_s \geq .425$ **b** $r_s \geq .601$

14.43 **a** $|r_s| \geq .400$ **b** $|r_s| \geq .526$

14.45 **a** $-.593$ **b** yes **14.47** **a** $r_s = .811$ **b** yes

14.49 yes **14.51** yes, $r_s = .9118$

14.53 **a** no, $r_s = .645$; rejection region: $r_s \geq .714$ **b** no

14.55 do not reject H_0; $T = 10.5$; rejection region: $\{T \leq 6\}$ with $\alpha = .05$

14.57 **a** reject H_0; $T = 6$; rejection region: $\{T \leq 8\}$ with $\alpha = .05$
b reject H_0; $T = 6$; rejection region: $\{T \leq 11\}$ with $\alpha = .05$

14.59 **a** $U_B = 3$; reject H_0; rejection region: $\{U \leq 3\}$ with $\alpha = .0556$
b $F = 4.91$; do not reject H_0; rejection region: $F \geq 9.12$

14.61 $-.845$

14.63 $U = 17.5$; no; rejection region: $\{U \leq 13\}$ with $\alpha = .0498$

14.65 .67684; reject H_0; rejection region: $|r_s| \geq .441$ with $\alpha = .10$

14.67 **a** yes, $H = 14.52$ ($\chi^2_{.05} = 7.81$) **b** results are identical

14.69 yes, $x = 6$; rejection region: $x \leq 6$ with $\alpha = .058$; $p\text{-value} = .058$

14.71 $z = -3.45$; reject H_0; black uniforms receive higher malevolence scores

INDEX

Table of Normal Curve Areas

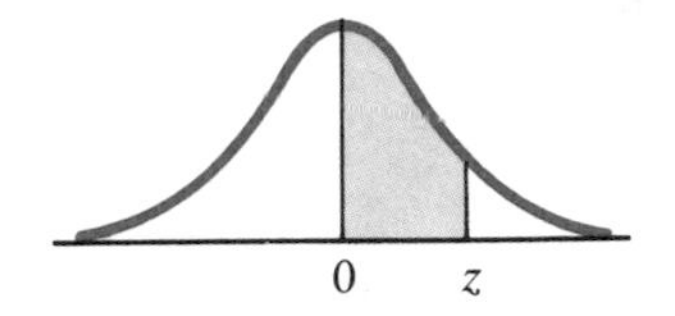

z	.00	.01	.02	.03	.04	.05	.06	.07	.08	.09
0.0	.0000	.0040	.0080	.0120	.0160	.0199	.0239	.0279	.0319	.0359
0.1	.0398	.0438	.0478	.0517	.0557	.0596	.0636	.0675	.0714	.0753
0.2	.0793	.0832	.0871	.0910	.0948	.0987	.1026	.1064	.1103	.1141
0.3	.1179	.1217	.1255	.1293	.1331	.1368	.1406	.1443	.1480	.1517
0.4	.1554	.1591	.1628	.1664	.1700	.1736	.1772	.1808	.1844	.1879
0.5	.1915	.1950	.1985	.2019	.2054	.2088	.2123	.2157	.2190	.2224
0.6	.2257	.2291	.2324	.2357	.2389	.2422	.2454	.2486	.2517	.2549
0.7	.2580	.2611	.2642	.2673	.2704	.2734	.2764	.2794	.2823	.2852
0.8	.2881	.2910	.2939	.2967	.2995	.3023	.3051	.3078	.3106	.3133
0.9	.3159	.3186	.3212	.3238	.3264	.3289	.3315	.3340	.3365	.3389
1.0	.3413	.3438	.3461	.3485	.3508	.3531	.3554	.3577	.3599	.3621
1.1	.3643	.3665	.3686	.3708	.3729	.3749	.3770	.3790	.3810	.3830
1.2	.3849	.3869	.3888	.3907	.3925	.3944	.3962	.3980	.3997	.4015
1.3	.4032	.4049	.4066	.4082	.4099	.4115	.4131	.4147	.4162	.4177
1.4	.4192	.4207	.4222	.4236	.4251	.4265	.4279	.4292	.4306	.4319
1.5	.4332	.4345	.4357	.4370	.4382	.4394	.4406	.4418	.4429	.4441
1.6	.4452	.4463	.4474	.4484	.4495	.4505	.4515	.4525	.4535	.4545
1.7	.4554	.4564	.4573	.4582	.4591	.4599	.4608	.4616	.4625	.4633
1.8	.4641	.4649	.4656	.4664	.4671	.4678	.4686	.4693	.4699	.4706
1.9	.4713	.4719	.4726	.4732	.4738	.4744	.4750	.4756	.4761	.4767
2.0	.4772	.4778	.4783	.4788	.4793	.4798	.4803	.4808	.4812	.4817
2.1	.4821	.4826	.4830	.4834	.4838	.4842	.4846	.4850	.4854	.4857
2.2	.4861	.4864	.4868	.4871	.4875	.4878	.4881	.4884	.4887	.4890
2.3	.4893	.4896	.4898	.4901	.4904	.4906	.4909	.4911	.4913	.4916
2.4	.4918	.4920	.4922	.4925	.4927	.4929	.4931	.4932	.4934	.4936
2.5	.4938	.4940	.4941	.4943	.4945	.4946	.4948	.4949	.4951	.4952
2.6	.4953	.4955	.4956	.4957	.4959	.4960	.4961	.4962	.4963	.4964
2.7	.4965	.4966	.4967	.4968	.4969	.4970	.4971	.4972	.4973	.4974
2.8	.4974	.4975	.4976	.4977	.4977	.4978	.4979	.4979	.4980	.4981
2.9	.4981	.4982	.4982	.4983	.4984	.4984	.4985	.4985	.4986	.4986
3.0	.4987	.4987	.4987	.4988	.4988	.4989	.4989	.4989	.4990	.4990

Source: This table is abridged from Table 1 of *Statistical Tables and Formulas*, by A. Hald (New York: Wiley, 1952). Reproduced by permission of A. Hald and the publisher, John Wiley & Sons, Inc.

Table of Critical Values of t

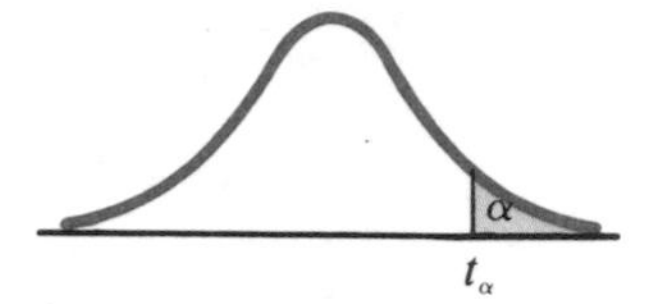

d.f.	$t_{.100}$	$t_{.050}$	$t_{.025}$	$t_{.010}$	$t_{.005}$	d.f.
1	3.078	6.314	12.706	31.821	63.657	1
2	1.886	2.920	4.303	6.965	9.925	2
3	1.638	2.353	3.182	4.541	5.841	3
4	1.533	2.132	2.776	3.747	4.604	4
5	1.476	2.015	2.571	3.365	4.032	5
6	1.440	1.943	2.447	3.143	3.707	6
7	1.415	1.895	2.365	2.998	3.499	7
8	1.397	1.860	2.306	2.896	3.355	8
9	1.383	1.833	2.262	2.821	3.250	9
10	1.372	1.812	2.228	2.764	3.169	10
11	1.363	1.796	2.201	2.718	3.106	11
12	1.356	1.782	2.179	2.681	3.055	12
13	1.350	1.771	2.160	2.650	3.012	13
14	1.345	1.761	2.145	2.624	2.977	14
15	1.341	1.753	2.131	2.602	2.947	15
16	1.337	1.746	2.120	2.583	2.921	16
17	1.333	1.740	2.110	2.567	2.898	17
18	1.330	1.734	2.101	2.552	2.878	18
19	1.328	1.729	2.093	2.539	2.861	19
20	1.325	1.725	2.086	2.528	2.845	20
21	1.323	1.721	2.080	2.518	2.831	21
22	1.321	1.717	2.074	2.508	2.819	22
23	1.319	1.714	2.069	2.500	2.807	23
24	1.318	1.711	2.064	2.492	2.797	24
25	1.316	1.708	2.060	2.485	2.787	25
26	1.315	1.706	2.056	2.479	2.779	26
27	1.314	1.703	2.052	2.473	2.771	27
28	1.313	1.701	2.048	2.467	2.763	28
29	1.311	1.699	2.045	2.462	2.756	29
inf.	1.282	1.645	1.960	2.326	2.576	inf.

Source: From "Table of Percentage Points of the t-Distribution," *Biometrika* 32 (1941) 300. Reproduced by permission of the *Biometrika* Trustees.